Pearson Australia
(a division of Pearson Australia Group Pty Ltd)
459–471 Church Street, Level 1, Building B, Richmond, Victoria 3121
PO Box 23360, Melbourne, Victoria 8012
www.pearson.com.au

First published 2018 by Pearson Australia

Lead Publisher: Misal Belvedere
Project Manager: Michelle Thomas
Production Manager: Casey McGrath
Lead Development Editor: Amy Sparkes
Development Editor: Haeyean Lee
Lead Editor: Sally Woollett
Editors: Jeanette Birtles, Sally McInnes, Sally Woollett
Designers: Anne Donald and iEnergizer Aptara Limited
Rights & Permissions Editor: Samantha Russell-Tulip
Senior Publishing Services Analyst: Rob Curulli
Proofreader: Mahtab Dubash
Indexer: Bruce Gillespie
Illustrator: DiacriTech
Printed in Singapore by Markono/04 (2025)

ISBN 978 1 4886 1927 4

Pearson Australia Group Pty Ltd ABN 40 004 245 943

A catalogue record for this book is available from the National Library of Australia

The Chemistry Education Association (CEA) was formed in 1977 by a group of teachers from secondary and tertiary institutions. It aims to promote the teaching of chemistry, particularly in secondary schools. The CEA has established a tradition of providing up-to-date text, electronic material and support resources for both students and teachers, and professional development opportunities for teachers.

PEARSON

CHEMISTRY

NEW SOUTH WALES

Writing and development team

We are grateful to the following people for their time and expertise in contributing to the Pearson Chemistry 11 New South Wales project.

Drew Chan
President of the CEA and Teacher
Coordinating Author

Chris Commons
Educator
Author and Reviewer

Richard Hecker
Science Writer
Author

Kathryn Hillier
Teacher and Lecturer
Author

Bob Hogendoorn
Science Consultant and Assessor
Author

Louise Lennard
Teacher
Author and Reviewer

Mick Moylan
Chemistry Outreach Fellow
Author

Pat O'Shea
Teacher
Author

Maria Porter
Teacher
Author

Patrick Sanders
Teacher
Author

Jim Sturgiss
Science Consultant
Author

Paul Waldron
Teacher
Author

Reuben Bolt
Director of the Nura Gili Indigenous Programs Unit, UNSW
Reviewer

Erin Bruns
Teacher
Contributing Author

Donna Chapman
Laboratory Technician
Safety Consultant

Warrick Clarke
Science Communicator
Contributing Author

Penny Commons
Lecturer
Contributing Author

Lanna Derry
Teacher
Contributing Author

Jane Dove
Teacher
Reviewer

Vicky Ellis
Teacher
Contributing Author

Emma Finlayson
Teacher
Reviewer

Elizabeth Freer
Teacher
Contributing Author

Simon Gooding
Teacher
Contributing Author

Elissa Huddart
Teacher
Contributing Author and Skills and Assessment Author

Aaron Jaraba
Teacher
Answer Checker

Raphael Johns
Laboratory Technician
Safety Consultant

Katrina Liston
Scientist
Answer Checker

Gary Molloy
Teacher
Reviewer

Geoff Quinton
President of ASTA and Teacher
Contributing Author

Bob Ross
Teacher
Contributing Author

Robert Sanders
Education Consultant
Contributing Author

Sophie Selby-Pham
Scientist
Answer Checker

Trish Weekes
Science Literacy Consultant

Gregory White
Scientist
Answer Checker

Maria Woodbury
Teacher
Reviewer

Contents

Working scientifically

Module 1 Properties and structure of matter

Module 2 Introduction to quantitative chemistry

Module 3 Reactive chemistry

Module 4 Drivers of reactions

How to use this book

Pearson Chemistry 11 New South Wales

Pearson Chemistry 11 New South Wales has been written to fully align with the new Stage 6 Syllabus for New South Wales Chemistry. The book covers Modules 1 to 4 in an easy-to-use resource. Explore how to use this book below.

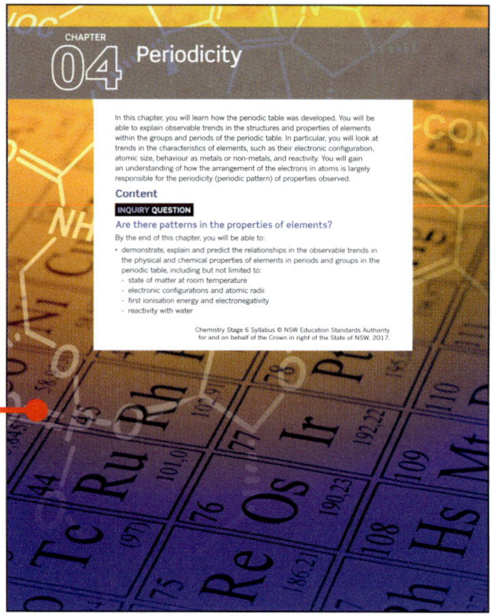

Chapter opener

The chapter opening page links the Syllabus to the chapter content. Key content addressed in the chapter is clearly listed.

Section

Each chapter is clearly divided into manageable sections of work. Best-practice literacy and instructional design are combined with high-quality, relevant photos and illustrations to help students better understand the ideas or concepts being developed.

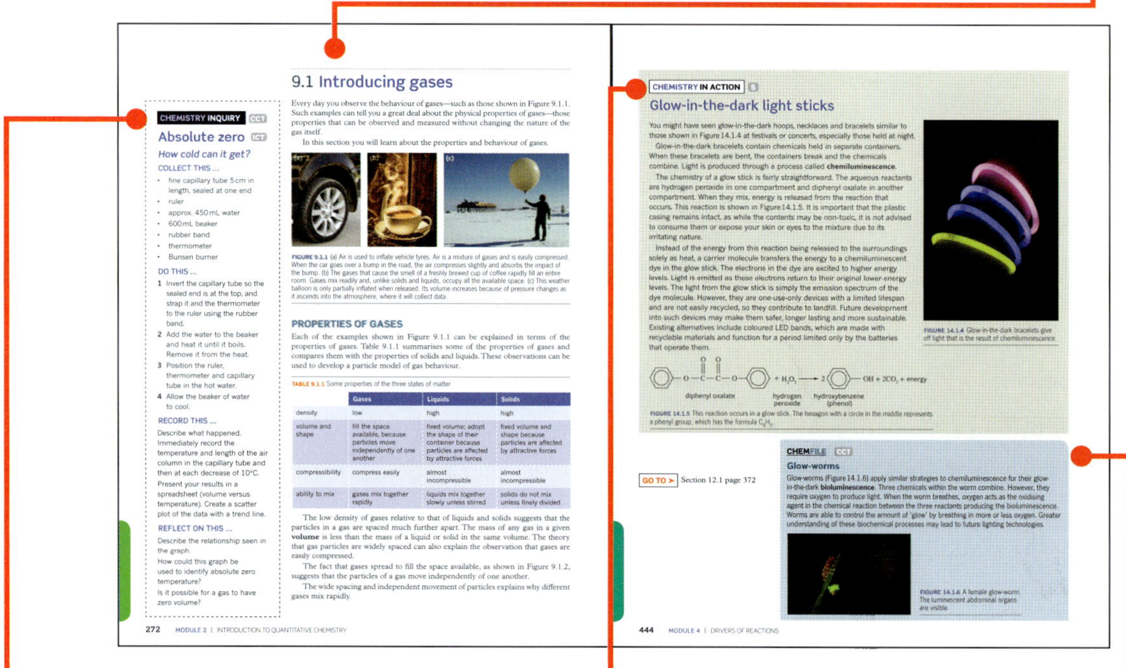

Chemistry Inquiry

Chemistry Inquiry features are inquiry-based activities that pre-empt the theory and allow students to engage with the concepts through a simple activity that sets them up to 'discover' the science before they learn about it. They encourage students to think about what happens in the world and how science can provide explanations.

Chemistry in Action

Chemistry in Action boxes place chemistry in an applied situation or a relevant context. They refer to the nature and practice of chemistry, its applications and associated issues, and the historical development of its concepts and ideas.

ChemFile

ChemFiles include a range of interesting and real-world examples to engage students.

Highlight box

Highlight boxes focus students' attention on important information, such as key definitions, formulae and summary points.

Worked examples

Worked examples are set out in steps that show thinking and working. This format greatly enhances student understanding by clearly linking underlying logic to the relevant calculations.

Each Worked example is followed by a Try yourself activity. This mirror problem allows students to immediately test their understanding. Fully worked solutions to all Worked example: Try yourself activities are available on *Pearson Chemistry 11 New South Wales* Reader+.

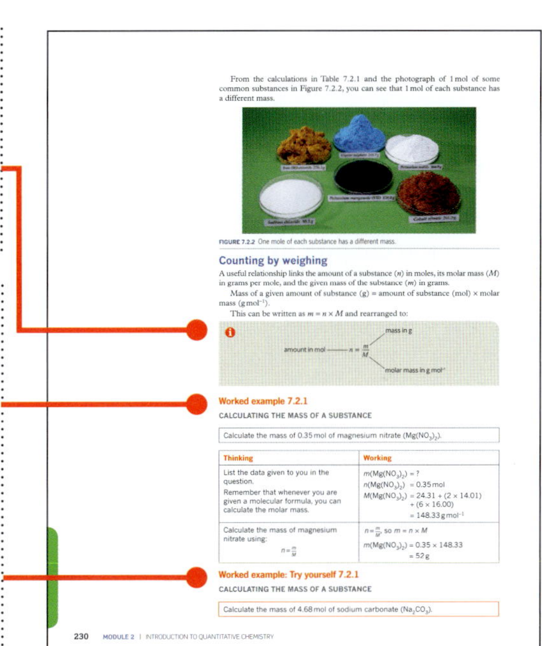

From the calculations in Table 7.2.1 and the photograph of 1 mol of some common substances in Figure 7.2.2, you can see that 1 mol of each substance has a different mass.

FIGURE 7.2.2 One mole of each substance has a different mass.

Counting by weighing

A useful relationship links the amount of a substance (n) in moles, its molar mass (M) in grams per mole, and the given mass of the substance (m) in grams.

Mass of a given amount of substance (g) = amount of substance (mol) × molar mass ($g\ mol^{-1}$).

This can be written as $m = n \times M$ and rearranged to:

amount in mol — $n = \frac{m}{M}$ — mass in g; molar mass in $g\ mol^{-1}$

Worked example 7.2.1

CALCULATING THE MASS OF A SUBSTANCE

Calculate the mass of 0.35 mol of magnesium nitrate ($Mg(NO_3)_2$).

Thinking	Working
List the data given to you in the question. Remember that whenever you are given a molecular formula, you can calculate the molar mass.	$m(Mg(NO_3)_2) = ?$ $n(Mg(NO_3)_2) = 0.35\ mol$ $M(Mg(NO_3)_2) = 24.31 + (2 \times 14.01) + (6 \times 16.00)$ $= 148.33\ g\ mol^{-1}$
Calculate the mass of magnesium nitrate using: $n = \frac{m}{M}$	$n = \frac{m}{M}$, so $m = n \times M$ $m(Mg(NO_3)_2) = 0.35 \times 148.33$ $= 52\ g$

Worked example: Try yourself 7.2.1

CALCULATING THE MASS OF A SUBSTANCE

Calculate the mass of 4.68 mol of sodium carbonate (Na_2CO_3).

230 MODULE 2 | INTRODUCTION TO QUANTITATIVE CHEMISTRY

Additional content

Additional content features include material that goes beyond the core content of the Syllabus. They are intended for students who wish to expand their depth of understanding in a particular area.

Section summary

Each section has a section summary to help students consolidate the key points and concepts of the section.

+ ADDITIONAL

Triads and octaves

In 1829, the German chemist Johann Wolfgang Döbereiner noticed that many of the known elements could be arranged in groups of three based on their chemical properties. He called these groups 'triads'. Within each of these triads, the properties of one element were intermediate between those of the other two. The intermediate element's relative atomic mass was almost exactly the average of the others.

One of Dobereiner's triads was lithium, sodium and potassium. Sodium is more reactive than lithium, but less reactive than potassium. Sodium's atomic mass is 23, which is the average of lithium's (atomic mass 7) and potassium's (atomic mass 39) atomic masses.

However, Dobereiner's theory was limited—not all elements could be included in triads. However, his work was quite remarkable, given he had fewer than 50 elements to work with at the time.

Decades later, English chemist John Alexander Newlands noticed a pattern in the atomic mass of elements. Newlands' law of octaves was published in 1865 and predicted properties of new elements such as germanium. His patterns worked well for the lighter elements, but did not fit for the heavier elements or allow for the discovery of new elements.

Four years later Mendeleev, working independently, published his periodic law, which, with a few modifications, was similar to Newlands' law of octaves.

SKILLBUILDER

Transforming decimal notation into scientific notation

Scientists use scientific notation to handle very large and very small numbers. For example, instead of writing 0.000000035, scientists would write 3.5×10^{-8}.

A number in scientific notation (also called standard form or power of 10 notation) is written as:

$a \times 10^n$

where

a is a number equal to or greater than 1 and less than 10, that is, $1 \leq a < 10$

n is an integer (a positive or negative whole number)

n is the power that 10 is raised to and is called the index value.

To transform a very large or very small number into scientific notation:

1 Write the original number as a decimal number greater than or equal to 1 but less than 10.

2 Multiply the decimal number by the appropriate power of 10.

The index value is determined by counting the number of places the decimal point needs to be moved to form the original number again.

- If the decimal point has to be moved n places to the right, n will be a positive number. For example:
  $51 = 5.1 \times 10^1$
- If the decimal point has to be moved n places to the left, n will be a negative number. For example:
  $0.51 = 5.1 \times 10^{-1}$

You will notice from these examples that when large numbers are written in scientific notation, the 10 has a positive index value. Very small numbers are written by multiplying by 10 with a negative index.

116 MODULE 1 | PROPERTIES AND STRUCTURE OF MATTER

1.1 Review

SUMMARY

- Before you begin your research, it is important to conduct a literature review. By using data from primary and/or secondary sources, you will better understand the context of your investigation and create an informed inquiry question.
- The purpose is a statement describing in detail what will be investigated; for example: 'The purpose of the experiment is to investigate the relationship between the concentration, mass and volume of a solution.'
- A hypothesis is a testable statement that is based on previous knowledge and evidence or observations; it attempts to answer the research question, for example: 'If increasing the concentration of a reactant increases the rate of reaction, and the concentration of this reactant is increased, then the rate of reaction will increase.'
- After a question has been formulated, it should be evaluated. The question may need further refinement before it is suitable as a basis for an achievable and worthwhile investigation. During planning, it is important to check whether the investigation can be completed using the time and resources available.
- There are three main types of variable.
  - The independent variable is determined by the researcher. This is the variable that is selected and changed.
  - The dependent variable may change in response to a change in the independent variable, and is the variable that will be measured or observed.
  - Controlled variables are the variables that must be kept constant during the investigation.
- Only one variable should be tested at a time. Otherwise, it is not possible to say whether the changes in the dependent variable are the result of changes in the independent variable.

KEY QUESTIONS

1 Scientists make observations from which a hypothesis is stated, and this is then experimentally tested. Define what a 'hypothesis' is.

2 Which of the following is an inquiry question?
A How are chemicals in solutions measured?
B A compound consists of two or more elements.
C Decreasing the volume of a container of gas will increase the pressure.
D The mass of the reactants equalled the mass of the products.

3 For each of the following hypotheses, select the dependent variable.
a If filtering water decreases electrical conductivity, and water is filtered through a domestic water purifier, then its electrical conductivity will decrease.
b If waterways near industrial sites are contaminated with lead, and the concentration of lead in waterways near industrial sites is tested and compared with the concentration of lead in waterways away from industrial sites, then the concentration of lead will be higher in the waterways closer to industrial sites.
c If increasing the salt concentration increases the electrical conductivity of water, and the electrical conductivity of water from Sydney Harbour is tested, then the electrical conductivity of the water will be greater where more ocean water is mixed in.
d If the pH of sparkling mineral water is higher than that of non-sparkling mineral water, and the pH of commercially available sparkling and non-sparkling mineral water is tested, then the pH will be lower in the commercially available non-sparkling mineral water.

4 In an experiment, a student uses the following descriptions for flame tests of ionic compounds: yellow, lilac, red and green.
Is the variable 'colour' a qualitative observation or a quantitative measurement?

5 Which of the following is likely to give the most accurate and quantitative measure of the pH of water?
A pH paper (e.g. litmus paper)
B universal indicator and a colour chart
C a calibrated pH meter at a particular temperature
D a conductivity meter

6 Select the best of the following hypotheses. Give reasons for your choice.
A If the pressure of a gas is affected by changes in volume and temperature, and the volume or temperature of a gas is changed, then the pressure of the gas will change.
B Concentration of solutions can be expressed using different units.
C If filtering water decreases its electrical conductivity, and water is filtered through a domestic water purifier, then its electrical conductivity will decrease.

10 CHAPTER 1 | WORKING SCIENTIFICALLY

SkillBuilder

Skillbuilders outline methods or techniques. They are instructive and self-contained. They step students through the skill to support science application.

Section review questions

Each section finishes with key questions to test students' understanding of and ability to recall the key concepts of the section.

How to use this book

Chapter review

Each chapter finishes with a list of key terms covered in the chapter and a set of questions to test students' ability to apply the knowledge gained from the chapter.

Module review

Each module finishes with a comprehensive set of questions, including multiple-choice, short-answer and extended-response questions. These assist students in drawing together their knowledge and understanding, and applying it to these types of questions.

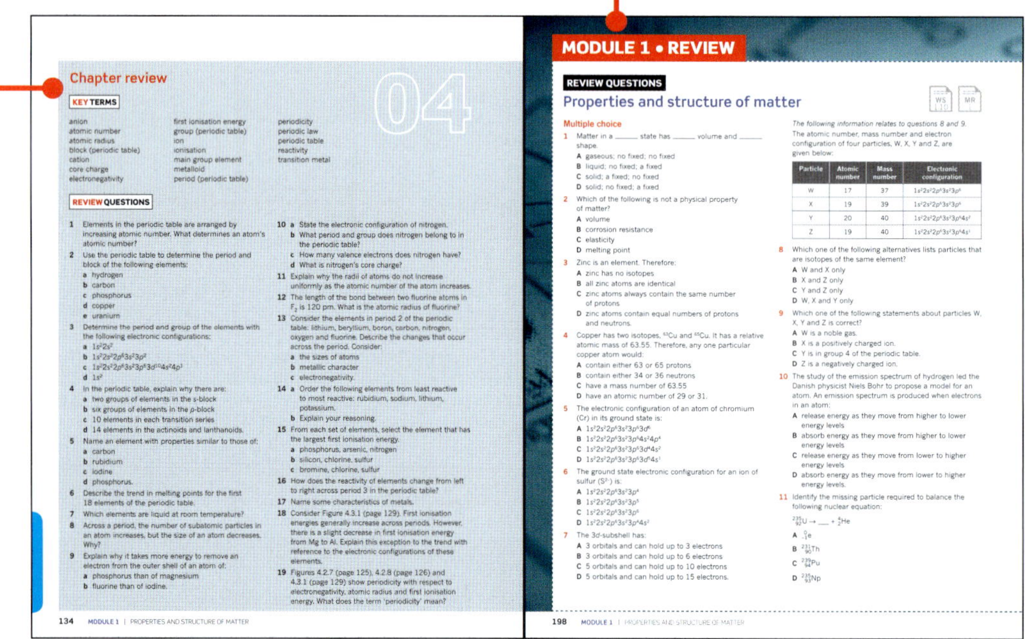

Icons

The New South Wales Stage 6 Syllabus 'Learning across the curriculum' and 'General capabilities' content are addressed throughout the series and are identified using the following icons.

'Go to' icons are used to make important links to relevant content within the Student Book.

This icon indicates when it is the best time to engage with a worksheet (WS), a practical activity (PA), a depth study (DS) or module review (MR) questions in the *Pearson Chemistry 11 Skills and Assessment Book*.

This icon indicates the best time to engage with a practical activity on *Pearson Chemistry 11 New South Wales* Reader+.

Glossary

Key terms are shown in **bold** in sections and listed at the end of each chapter. A comprehensive glossary at the end of the book includes and defines all the key terms.

Answers

Numerical answers and short-response answers are included at the back of the book. Comprehensive answers and fully worked solutions for all section review questions, Worked examples: Try yourself activities, chapter review questions and module review questions are provided via *Pearson Chemistry 11 New South Wales* Reader+.

Pearson Chemistry 11 New South Wales

Student Book

Pearson Chemistry 11 New South Wales has been written to fully align with the new Stage 6 Syllabus for New South Wales. The Student Book includes the very latest developments in and applications of chemistry and incorporates best-practice literacy and instructional design to ensure the content and concepts are fully accessible to all students.

Skills and Assessment Book

The *Skills and Assessment Book* gives students the edge in preparing for all forms of assessment. Key features include a toolkit, key knowledge summaries, worksheets, practical activities, suggested depth studies and module review questions. It provides guidance, assessment practice and opportunities for developing key skills.

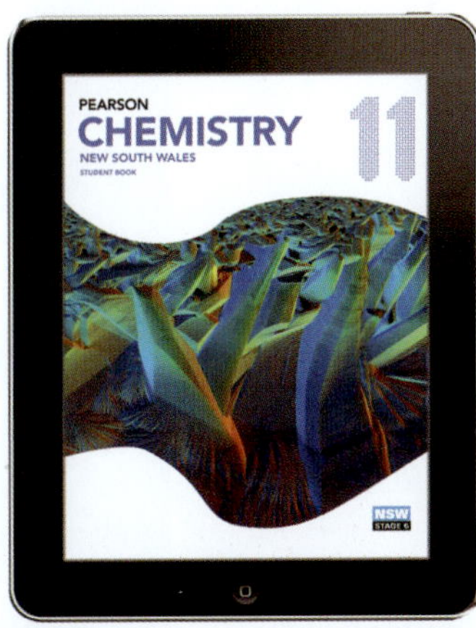

Reader+ the next generation eBook

Pearson Reader+ lets you use your Student Book online or offline on any device. Pearson Reader+ retains the look and integrity of the printed book. Practical activities, interactives and videos are available on Pearson Reader+, along with fully worked solutions for the Student Book questions.

Teacher Support

The Teacher Support available includes syllabus grids and a scope and sequence plan to support teachers with programming. It also includes fully worked solutions and answers to all Student Book and *Skills and Assessment Book* questions, including worksheets, practical activities, depth studies and module review questions. Teacher notes, safety notes, risk assessments and a laboratory technician checklist and recipes are available for all practical activities. Depth studies are supported with suggested marking schemes and exemplar answers.

PEARSON
CHEMISTRY
NEW SOUTH WALES
11

Pearson Digital
Access your digital resources at **pearsonplaces.com.au**
Browse and buy at **pearson.com.au**

CHAPTER

01 Working scientifically

This chapter covers the skills needed to successfully plan and conduct primary-and secondary-sourced investigations.

1.1 Questioning and predicting explains how to develop, propose and evaluate inquiry questions and hypotheses. When creating a hypothesis, it is necessary to consider the relevant variables.

1.2 Planning investigations is a guide to planning your investigation. You will learn to identify risks, to make sure all ethical concerns have been addressed, to choose appropriate materials and technology to carry out your investigation, and to check that your choice of variables allows you to collect the data you need.

1.3 Conducting investigations is a guide to conducting investigations. It describes methods for accurately collecting and recording data so as to reduce errors. Appropriate procedures that need to be carried out when disposing of waste are also described.

1.4 Processing data and information is a guide to processing your data. From an array of visual representations, you will learn how best to represent your information and how to identify trends and patterns in your data.

1.5 Analysing data and information explains how to analyse your results. It explains error and uncertainty and how to construct mathematical models to better understand the scientific principles of your research.

1.6 Problem solving is a guide to solving problems. Utilising critical thinking, you will demonstrate an understanding of the scientific principles underlying the solution to your inquiry question.

1.7 Communicating explains how to communicate an investigation clearly and accurately using appropriate scientific language, nomenclature and scientific notation.

Outcomes

By the end of this chapter, you will be able to:

- develop and evaluate questions and hypotheses for scientific investigation (CH11-1)
- design and evaluate investigations in order to obtain primary and secondary data and information (CH11-2)
- conduct investigations to collect valid and reliable primary and secondary data and information (CH11-3)
- select and process appropriate qualitative and quantitative data and information using a range of appropriate media (CH11-4)
- analyse and evaluate primary and secondary data and information (CH11-5)
- solve scientific problems using primary and secondary data, critical thinking skills and scientific processes (CH11-6)
- communicate scientific understanding using suitable language and terminology for a specific audience or purpose (CH11-7)

Content

In this chapter, you will learn how to design, plan and conduct investigations, including how to write a hypothesis and identify variables. You will also assess the validity, reliability and accuracy of results and research.

Finally, you will learn how to discuss your investigation and draw evidence-based conclusions in relation to your hypothesis and research question. By the end of this chapter, you will be able to:

- develop and evaluate inquiry questions and hypotheses to identify a concept that can be investigated scientifically, involving primary and secondary data (ACSCH001, ACSCH061, ACSCH096) L
- modify questions and hypotheses to reflect new evidence CCT
- assess risks, consider ethical issues and select appropriate materials and technologies when designing and planning an investigation (ACSCH031, ACSCH097) EU PSC
- justify and evaluate the use of variables and experimental controls to ensure that a valid procedure is developed that allows for the reliable collection of data (ACSCH002)
- evaluate and modify an investigation in response to new evidence CCT
- employ and evaluate safe work practices and manage risks (ACSCH031) PSC WE
- use appropriate technologies to ensure and evaluate accuracy ICT N
- select and extract information from a wide range of reliable secondary sources and acknowledge them using an accepted referencing style L
- select qualitative and quantitative data and information and represent them using a range of formats, digital technologies and appropriate media (ACSCH004, ACSCH007, ACSCH064, ACSCH101) L N
- apply quantitative processes where appropriate N
- evaluate and improve the quality of data CCT N
- derive trends, patterns and relationships in data and information
- assess error, uncertainty and limitations in data (ACSCH004, ACSCH005, ACSCH033, ACSCH099) CCT
- assess the relevance, accuracy, validity and reliability of primary and secondary data and suggest improvements to investigations (ACSCH005) CCT N
- use modelling (including mathematical examples) to explain phenomena, make predictions and solve problems using evidence from primary and secondary sources (ACSCH006, ACSCH010) CCT
- use scientific evidence and critical thinking skills to solve problems CCT
- select and use suitable forms of digital, visual, written and/or oral forms of communication L N
- select and apply appropriate scientific notations, nomenclature and scientific language to communicate in a variety of contexts (ACSCH008, ACSCH036, ACSCH067, ACSCH102) L N
- construct evidence-based arguments and engage in peer feedback to evaluate an argument or conclusion (ACSCH034, ACSCH036) CC DD

1.1 Questioning and predicting

TYPES OF INVESTIGATION

Many different types of investigations can be conducted in chemistry. You are probably most familiar with practical investigations or experiments (primary research). Other types of investigation involve researching and evaluating primary and/or secondary sources of information to answer an inquiry question. Tables 1.1.1 and 1.1.2 contain examples of the different types of investigation.

TABLE 1.1.1 Examples of primary research

Type of primary research	Example tasks
conducting experiments in a laboratory	• planning a valid experiment • conducting a risk assessment • working safely • recording observations and results • analysing and evaluating data and information
conducting field work	• conducting a risk assessment • working safely • recording observations and results • analysing and evaluating data and information
conducting surveys	• writing questions • collecting data • analysing data and information
designing a model	• identifying a problem to be modelled • summarising key findings • identifying advantages and limitations of the model

TABLE 1.1.2 Examples of secondary research

Type of secondary research	Example tasks
researching published data from primary and secondary sources	• finding published information in scientific magazines and journals, books, databases, media texts, labels of commercially available products • analysing and evaluating data and information

All of these investigations have things in common: an inquiry question or idea, a hypothesis and a purpose (aim).

QUESTIONS, PREDICTION AND PURPOSE

The inquiry question, hypothesis and purpose are linked, and they can be refined as the planning of the investigation continues.

Inquiry questions: defining an investigation

An inquiry question defines what is being investigated. For example, 'What is the relationship between the surface area of solid reactants and the reaction rate?'

It is important that you can interpret what an inquiry question is asking. To do this, you need to:

- identify a 'guiding' word, such as *who*, *what*, *where*, *why*
- link the guiding word to appropriate command verbs such as *identify*, *describe*, *compare*, *contrast*, *distinguish*, *analyse*, *evaluate* and *create*.

Table 1.1.3 contains examples of inquiry questions that could be investigated.

TABLE 1.1.3 Examples of guiding words and inquiry questions

Guiding word(s)	Example inquiry questions	What are you being asked to do? What are the command verbs?
what	• What difference can nanomaterials make to society and the environment? • What are electrons, protons and neutrons made of?	Identify and describe specific examples, evidence, reasons and analogies from a variety of possibilities. Evaluate possible applications. *identify, describe, evaluate*
where	• Where would an element with an atomic number of 130 be placed in the modern periodic table, what properties would it have and how likely is it to be discovered?	Identify and describe the location, giving reasons. *identify, describe*
how	• How are atoms 'seen'? • How can lead be transformed into gold? • How do different crude oil extraction methods compare in terms of their ease of extraction and environmental impacts?	Identify and describe in detail a process or mechanism. Give examples using evidence and reasons. Evaluate. *identify, describe, compare, contrast, evaluate*
why	• Why are the 10 most abundant elements in the universe not the same as the 10 most abundant elements on Earth? • Why do transition metals have multiple oxidation states? • Why does the composition of crude oil vary between different oil wells?	Identify the elements etc. Describe in detail the causes, reasons, mechanisms and evidence. Compare and contrast. Analyse the data. *identify, describe, compare, contrast, analyse*
would	• Would there be life if elements did not form compounds?	Evaluate giving reasons for and against (using evidence, analogies and comparisons). *evaluate, assess*
is/are	• Are there more elements to be discovered? • Is it an advantage or a disadvantage for elements to be unreactive? • Is it worth sending people to the Moon to mine for lanthanoids and actinoids?	Evaluate giving reasons and evidence. *evaluate, assess*
on what basis	• On what basis are alternative forms of the periodic table constructed?	Evaluate, giving reasons and evidence. *evaluate, assess*
can	• Can we live without lanthanoids and actinoids?	Evaluate and assess. Is it possible? Give reasons and suggest possible alternatives. *evaluate, assess, justify, create*
do/does	• Does surfactant biodegradability affect performance? • Do lanthanoids and actinoids rust or corrode? • Do we need crude oil?	Evaluate, giving reasons and evidence for and against. *evaluate, assess*
should	• Should cars be made from shape memory metals?	Evaluate the pros and cons, implications and limitations. Make a judgement. *evaluate, assess, create*
might	• What might we do if crude oil supplies run out? • What might life be like without glass? What might have to be used instead?	Evaluate, giving reasons for and against (using evidence, analogies and comparisons). *evaluate, assess, compare, contrast, create*

Formulating an inquiry question

Compile a list of possible topic ideas. Do not reject ideas that initially might seem impossible. It is sometimes easier to start by modifying an existing inquiry question, particularly if you are conducting research for the first time (Figure 1.1.1). You could choose one of the inquiry questions in Table 1.1.3 on page 5 and then create a mind map of how the question could be modified.

Before formulating a question, it is good practice to conduct a literature review of the topic to be investigated. You should become familiar with the relevant scientific concepts and key terms. During this review, write down questions or correlations as they occur to you.

Use your ideas to generate questions that are answerable.

Your question will lead to a hypothesis when:

- it relates to measurable variables
- you can make a prediction based on knowledge and experience.

More information about research sources to consult before and during an investigation is on page 9.

FIGURE 1.1.1 The many aspects of a practical investigation may appear overwhelming to begin with. Taking a step-by-step approach will make the process easier and help to keep the investigation focused.

Evaluating your question

After a question has been chosen, evaluate the question before progressing. The question may need further refinement or even further investigation before it is a suitable basis for an achievable and worthwhile investigation. During planning, consider whether you can complete your investigation in the time available or with the resources on hand. For example, could you construct a complex device with the facilities available in your school laboratory?

To evaluate your question, consider:

- *relevance*: Is the question related to the appropriate area of study?
- *clarity and measurability*: Can the question be framed as a clear hypothesis? If the question cannot be stated as a specific hypothesis, then completing the research will be difficult.
- *time frame*: Can the question be answered within a reasonable period of time? Is the question too broad?
- *knowledge and skills*: Do your knowledge and laboratory skills allow you to explore the question? Keep the question simple and achievable.
- *practicality*: Are resources, such as laboratory equipment and materials, likely to be readily available? Keep things simple. Avoid investigations that require sophisticated or rare equipment. Pipettes and burettes, timing devices and top-loading balances may be more readily available than more sophisticated equipment.
- *safety and ethics*: Consider the safety and ethical issues associated with the question you will be investigating. If there are issues, can these be addressed?
- *advice*: Seek advice from your teacher about your question. Their input and experience may prove very useful. They may consider aspects of the question that you have not thought about.

Hypothesis—a scientific prediction

A hypothesis is a testable prediction based on previous knowledge, evidence or observations that attempts to answer the inquiry question. It often takes the form of a proposed cause-and-effect relationship between two or more variables; in other words, 'If x is true and this is tested, then y will occur.'

Here are some examples of a hypothesis.

- If the rate of reaction increases when temperature is increased, and the temperature of a hydrochloric acid solution is increased (for a constant particle size of marble chips and a constant concentration and volume of hydrochloric acid), then the rate of reaction will increase.
- If increasing the surface area increases the rate of reaction, and the surface area of marble chips is increased (for a constant concentration and volume of hydrochloric acid and a constant temperature), then the rate of reaction will increase.

Hypotheses can be written in a variety of ways, such as 'x happens because of y' or 'when x happens, y will happen'. However they are written, hypotheses must always be testable and clearly state the independent and dependent variables.

- If decreasing the concentration of a solution decreases the rate of reaction, and the concentration of a hydrochloric acid solution is decreased (for a constant particle size of marble chips and volume of hydrochloric acid at a particular temperature), then the rate of reaction will decrease.
- If adding a catalyst increases the rate of reaction, and a catalyst is added to a constant concentration and volume of hydrogen peroxide at a constant temperature, then the rate of reaction will increase.

Formulating a hypothesis

After an inquiry question is finalised, a hypothesis is formulated. A hypothesis needs to include a proposed relationship between two variables. It should predict that a relationship exists or does not exist.

First, identify the two variables in your question. Second, name the independent and dependent variables involved.

For example, 'If x is true and I do this (to the independent variable), then y will happen (to the dependent variable).'

A good hypothesis should:

- be a statement
- be based on information contained in the research question (purpose)
- be worded so that it can be tested in the experiment
- include an independent and a dependent variable (page 8)
- include variables that are measurable.

The hypothesis should also be falsifiable. This means that a negative outcome is possible and the hypothesis can be rejected. For example, the hypothesis that all apples are round cannot be proved beyond doubt, but it can be disproved (rejected); in other words, it is falsifiable. Only one conical apple is needed to reject this hypothesis. Unfalsifiable hypotheses cannot be proved or disproved by science. These include hypotheses with ethical or moral aspects, or other subjective judgements. Scientific investigations cannot prove that a hypothesis is correct, they can only find information to support or reject a hypothesis.

Modifying a hypothesis

As you collect new evidence from secondary sources, you may need to adjust your inquiry question or hypothesis. Imagine you have a hypothesis that states, 'If increasing the temperature increases the rate of reaction, and the temperature of a hydrochloric acid solution is increased (for a constant particle size of calcium carbonate), the rate of reaction will increase.' You continue your research but realise that you didn't take into account the concentration and volume of hydrochloric acid, so you must modify your investigation.

Purpose—the aim of an investigation

A purpose is a statement describing in detail what will be investigated. It is also known as the aim of your investigation. For example, 'The purpose of the experiment is to investigate the relationship between the surface area of calcium carbonate and its rate of reaction with hydrochloric acid.'

The purpose includes the key steps required to test the hypothesis. Each purpose should directly relate to the variables in the hypothesis, and describe how each will be measured. The purpose does not need to include the details of the procedure. Here is another example.

- *Hypothesis*: If diluting a coloured solution reduces the intensity of the colour, and water is added to a coloured solution, then the intensity of the colour will be reduced.
- *Purpose*: To qualitatively investigate the relationship between intensity of colour and dilution.

In the first stage of the experiment, a standard solution of potassium permanganate will be prepared.

In the second stage of the experiment, a series of dilutions will be performed and the intensity of colour will be observed.

Extension: If the school has a colorimeter, the concentration of the original and diluted solutions can be measured.

VARIABLES: KNOWING WHAT TO MEASURE

A good scientific hypothesis can be tested (and supported or refuted) through investigation. In a testable hypothesis, it should be possible to measure both what is changed or carried out and what happens. The factors that are monitored during an experiment or investigation are called **variables**. An experiment or investigation includes measurements and/or observations and determines the relationship between variables.

There are three main types of variable.

- The **independent variable** is the variable that is determined by the researcher (the variable that is selected and changed).
- The **dependent variable** is the variable that may change in response to a change in the independent variable. This is the variable that is measured or observed.
- **Controlled variables** are the variables that must be kept constant during the investigation.

Only one variable at a time should be tested; otherwise, it cannot be stated that the changes in the dependent variable are the result of changes in the independent variable. Completing a table like Table 1.1.4 will help to determine the variables for your question(s).

TABLE 1.1.4 Determining the variables of an inquiry question

Inquiry question	What is the relationship between the concentration of a liquid reactant and reaction rate?
Independent variable	concentration of hydrochloric acid solution
Dependent variable	mass change during the reaction
Controlled variables	starting temperature of hydrochloric acid, volume of hydrochloric acid, surface area of solid reactants, equipment (including beakers, thermometers and weighing balance)
Potential hypothesis	If decreasing the concentration of a solution decreases the rate of reaction, and the concentration of the hydrochloric acid solution is decreased (for a constant particle size of marble chips, volume of hydrochloric acid and temperature), then the rate of reaction will decrease.

Qualitative and quantitative variables

Variables can be qualitative or quantitative, with further subsets within each category.

- **Qualitative** variables can be observed but not measured. They can be sorted into groups or into categories such as brightness, type of construction material and type of device.
  - Nominal variables are categorical variables in which the order is not important (for example: colours of a flame (Figure 1.1.2), states of matter, batteries and types of cell).
  - Ordinal variables are categorical variables in which order is important and groups have an obvious ranking or level (for example: the activity series of metals, and trends in the periodic table).
- **Quantitative** variables can be measured. Mass, volume, temperature, pH and time are all examples of quantitative variables.
  - Discrete variables consist of integers, not fractions; for example: number of protons in an atom, number of atoms of each element in a compound and number of isotopes of a particular element.
  - Continuous variables allow for any numerical value within a given range, for example: measurement of temperature, volume, mass, pH and conductivity.

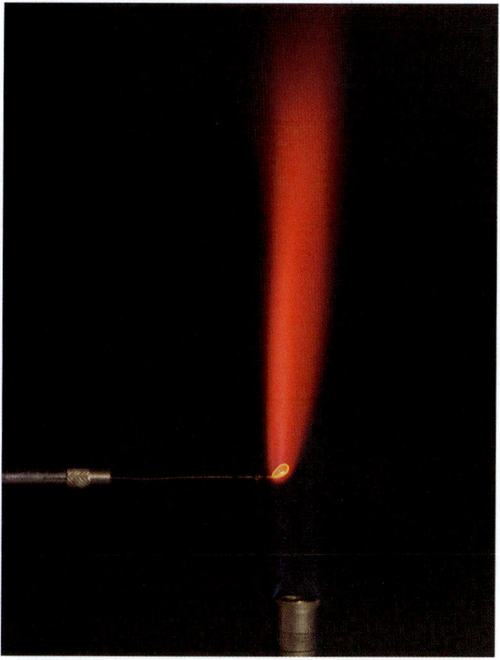

FIGURE 1.1.2 When recording qualitative data, describe in detail how each variable will be defined. For example, if recording colour during flame tests, take pictures to clearly define what each assigned term represents.

SOURCING INFORMATION

Finding reliable information is important both when choosing your topic and during your investigation. Some of the steps involved in sourcing information are:

- identifying key terms
- locating information
- evaluating the credibility of sources
- evaluating experimental data or evidence.

Sources can be:

- **primary sources**—original sources of data and evidence; for example: articles containing research findings that have been published in peer-reviewed scientific journals, or research presented at a scientific conference
- **secondary sources**—analyses and interpretations of primary sources; for example: textbooks, magazine articles and newspaper articles.

Sources that may contain useful information include:

- newspaper articles and opinion pieces
- journal articles (from peer-reviewed journals)
- magazine articles
- government reports
- global databases, statistics and surveys
- laboratory work
- computer simulations and modelling
- interviews with relevant professionals.

Some reputable science journals and magazines are:

- *Cosmos*
- *Double Helix*
- *ECOS*
- *Nature*
- *New Scientist*
- *Popular Science*
- *Scientific American.*

1.1 Review

SUMMARY

- Before you begin your research, it is important to conduct a literature review. By using data from primary and/or secondary sources, you will better understand the context of your investigation and create an informed inquiry question.
- The purpose is a statement describing in detail what will be investigated; for example: 'The purpose of the experiment is to investigate the relationship between the concentration, mass and volume of a solution.'
- A hypothesis is a testable statement that is based on previous knowledge and evidence or observations; it attempts to answer the research question, for example: 'If increasing the concentration of a reactant increases the rate of reaction, and the concentration of this reactant is increased, then the rate of reaction will increase.'
- After a question has been formulated, it should be evaluated. The question may need further refinement before it is suitable as a basis for an achievable and worthwhile investigation. During planning, it is important to check whether the investigation can be completed using the time and resources available.
- There are three main types of variable.
  - The independent variable is determined by the researcher. This is the variable that is selected and changed.
  - The dependent variable may change in response to a change in the independent variable, and is the variable that will be measured or observed.
  - Controlled variables are the variables that must be kept constant during the investigation.
- Only one variable should be tested at a time. Otherwise, it is not possible to say whether the changes in the dependent variable are the result of changes in the independent variable.

KEY QUESTIONS

1 Scientists make observations from which a hypothesis is stated, and this is then experimentally tested. Define what a 'hypothesis' is.

2 Which of the following is an inquiry question?

A How are chemicals in solutions measured?

B A compound consists of two or more elements.

C Decreasing the volume of a container of gas will increase the pressure.

D The mass of the reactants equalled the mass of the products.

3 For each of the following hypotheses, select the dependent variable.

a If filtering water decreases electrical conductivity, and water is filtered through a domestic water purifier, then its electrical conductivity will decrease.

b If waterways near industrial sites are contaminated with lead, and the concentration of lead in waterways near industrial sites is tested and compared with the concentration of lead in waterways away from industrial sites, then the concentration of lead will be higher in the waterways closer to industrial sites.

c If increasing the salt concentration increases the electrical conductivity of water, and the electrical conductivity of water from Sydney Harbour is tested, then the electrical conductivity of the water will be greater where more ocean water is mixed in.

d If the pH of sparkling mineral water is higher than that of non-sparkling mineral water, and the pH of commercially available sparkling and non-sparkling mineral water is tested, then the pH will be lower in the commercially available non-sparkling mineral water.

4 In an experiment, a student uses the following descriptions for flame tests of ionic compounds: yellow, lilac, red and green.
Is the variable 'colour' a qualitative observation or a quantitative measurement?

5 Which of the following is likely to give the most accurate and quantitative measure of the pH of water?

A pH paper (e.g. litmus paper)

B universal indicator and a colour chart

C a calibrated pH meter at a particular temperature

D a conductivity meter

6 Select the best of the following hypotheses. Give reasons for your choice.

A If the pressure of a gas is affected by changes in volume and temperature, and the volume or temperature of a gas is changed, then the pressure of the gas will change.

B Concentration of solutions can be expressed using different units.

C If filtering water decreases its electrical conductivity, and water is filtered through a domestic water purifier, then its electrical conductivity will decrease.

1.2 Planning investigations

After you have formulated your hypothesis, defined the purpose of your investigation and determined your variables, you will need to plan and design your investigation. Taking the time to carefully plan and design a practical investigation before beginning will help you to maintain focus throughout. Preparation is essential. This section is a guide to some of the key steps that should be taken when planning and designing a practical investigation.

WRITING A METHODOLOGY

The methodology of an investigation is a step-by-step procedure. When detailing a methodology, make sure that it has the following elements so that it is a valid, reliable, **precise** and **accurate** investigation.

Methodology elements

Validity

Validity refers to whether an experiment or investigation is in fact testing the set hypothesis and purposes. Is the investigation obtaining data that is relevant to the question?

A valid investigation is designed so that only one variable is being changed at a time. The other variables remain constant so that meaningful conclusions can be drawn about the effect of each variable in turn.

To ensure validity, carefully determine:

- the independent variable (the variable that will be changed) and how it will change
- the dependent variable (the variable that will be measured)
- the controlled variables (variables that must remain constant) and how they will be maintained.

Reliability

Reliability refers to the idea that an experiment can be repeated many times, and the average of the results from all the repeated experiments will be consistent. This can be enhanced by:

- defining the control
- ensuring there is sufficient replication of the experiment.

The control is an identical experiment carried out at the same time, except that in the control experiment the independent variable is not changed. The two types are:

- negative control: The effect or change is expected in the experimental group but not in the control group.
- positive control: The effect or change is expected in the control group but not in the experimental group.

The expectations are based on previous experiments or observations. When the controls do not behave as expected, the data obtained from an experiment or observation is not reliable.

It is important to determine how many times an experiment needs to be replicated (Figure 1.2.1). Many scientific investigations lack sufficient repetition to ensure that the results can be considered reliable and repeatable. To ensure that your results are reliable:

- *Take several readings*: Repeat each reading at least three times, record each measurement and then average the three measurements. This allows random errors to be identified. If a reading differs too much from the others (known as an outlier), discard it before averaging.
- *Take care when sampling*: If there might be differences in the characteristics or construction of a sample, include multiple samples of each type in the same experiment. The greater the sample size, the more reliable the data.

FIGURE 1.2.1 Replication increases the reliability of your investigation. Reliable results mean that anyone repeating the investigation will obtain similar data.

- *Repeat the experiment*: If possible, repeat the experiment on a different day. Don't change anything. If the results are not the same, think about what could have happened. For example, was the equipment faulty, and were all the controlled variables correctly identified? Repeat the experiment a third time to confirm which run was correct. More repeats is better. Three is a good number; but if time and resources allow, aim for five.

Accuracy and precision

In science and statistics, the terms 'accuracy' and 'precision' have very specific meanings and they are quite different.

- **Accuracy** is the ability to obtain the correct measurement. To obtain accurate results, you must minimise systematic errors.
- **Precision** is the ability to consistently obtain the same measurement. To obtain precise results, you must minimise random errors.

To understand more clearly the difference between accuracy and precision, imagine that you are shooting arrows at an archery target (Figure 1.2.2). Accuracy is being able to hit the bullseye, whereas precision is being able to hit the same spot every time you shoot. If you hit the bullseye every time you shoot, you are both accurate and precise (Figure 1.2.2a). If you hit the same area of the target every time, but not the bullseye, you are precise but not accurate (Figure 1.2.2b). If you hit the area around the bullseye each time, but don't always hit the bullseye, you are accurate but not precise (Figure 1.2.2c). If you hit a different part of the target every time you shoot, you are neither accurate nor precise (Figure 1.2.2d).

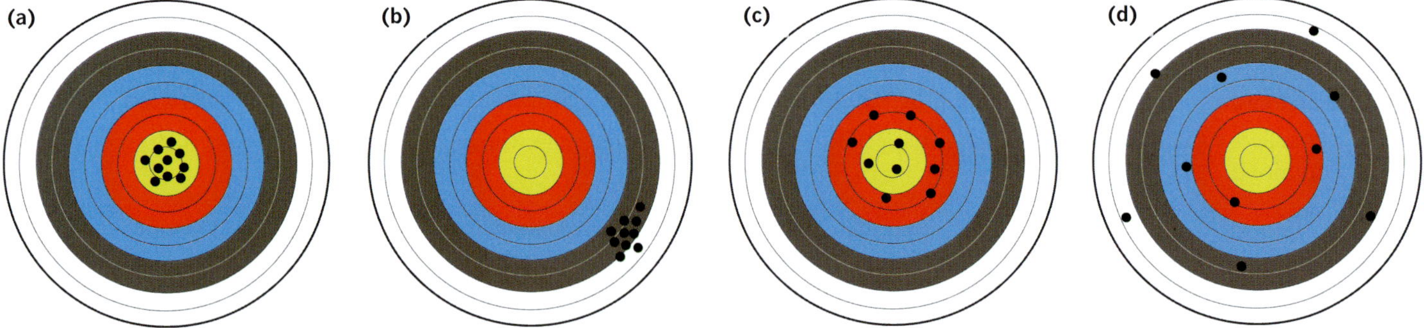

FIGURE 1.2.2 Examples of accuracy and precision: (a) both accurate and precise, (b) precise but not accurate, (c) accurate but not precise, (d) neither accurate nor precise.

Scientific data

All scientists strive to measure and report accurate and precise results. However, very precise measurements can be unwieldly. Imagine entering a calculation with five values that were all measured to 20 decimal places! Scientists therefore restrict some measurements to a certain number of significant figures or decimal places.

For example, the periodic table at the end of this book lists the atomic weight of elements to four significant figures. Zinc (Zn) is listed as having an atomic weight of 65.38. In a different periodic table, the atomic weight of zinc is listed as 65.4 (this has been rounded to three significant figures). Neither measurement is incorrect, but 65.38 is the more precise measurement.

It is important that you are aware that some scientific data can vary depending on the source. Always check that the data you are using has come from a reliable source.

Recording numerical data

Are the instruments to be used in your experiments sensitive enough? What units of measurement will be used? Build some testing into your investigation to confirm the accuracy and reliability of the equipment and your ability to interpret the information obtained.

To ensure the accuracy of the investigation, consider:

- the units in which the independent and dependent variables will be measured
- the instruments that will be used to measure the variables.

Select and use appropriate equipment, materials and procedures. For example, select equipment that measures to smaller degrees of precision to reduce uncertainty, and repeat the measurements to confirm them.

Describe the materials and procedure in appropriate detail. This should ensure that every measurement can be repeated and the same result obtained within reasonable margins of experimental error or uncertainty (less than 5% is reasonable). Percentage error (also known as percentage uncertainty) is a way to quantify the accuracy of a measurement. This will be discussed in Section 1.4.

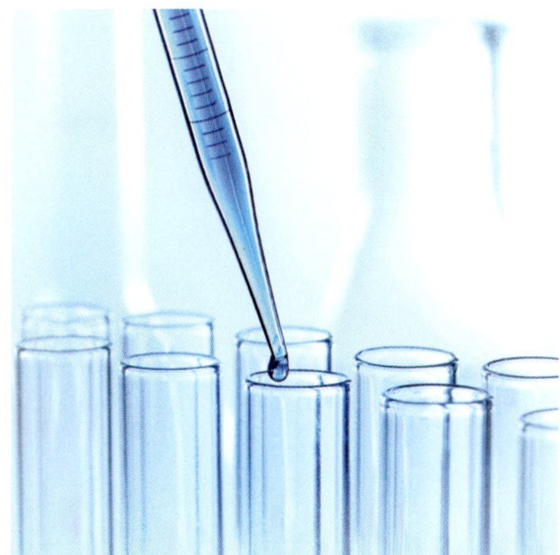

FIGURE 1.2.3 A 5 mL graduated pipette can measure volumes to an accuracy of one-hundredth of a millilitre, or 5.00±0.01 mL. The pipette has major divisions of 1 mL and minor divisions of 0.1 mL. You can estimate to 0.01 mL and record volumes to two decimal places, for example: 3.80 mL or 4.52 mL.

When using measuring instruments, the number of significant figures (or digits) and decimal places you use is determined by the precision of your measurements.

This depends on the scale, accuracy and precision of the instrument and the technique you are using (Figure 1.2.3). For example, a beaker is used to measure volumes approximately and has limited accuracy, generally ±5%. A graduated pipette, which is a more specialised piece of glassware, is more accurate, with accuracies of ±0.1% or ±0.2%. Your pipette may be accurate, but if your technique when using the pipette is variable, the overall accuracy and precision will be limited.

When you record raw data and report processed data, use only the number of significant figures possible for your equipment or observation (see Section 1.3). Using either a greater or smaller number of significant figures can be misleading. Table 1.2.1 shows measurements of five samples weighed on an electronic balance accurate to two decimal places. The data was entered into a spreadsheet to calculate the mean, which was displayed with four decimal places. You would record the mean as 20.83 g, not 20.8260 g, because two decimal places is the precision limit of the instrument. Recording 20.8260 g would be an example of false precision.

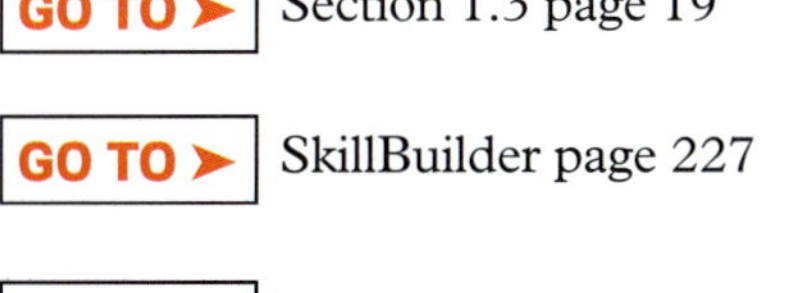

GO TO ➤ Section 1.3 page 19

GO TO ➤ SkillBuilder page 227

GO TO ➤ Appendix 2 page 528

TABLE 1.2.1 An example of false precision in data analysis

Sample	1	2	3	4	5	
Mass (g)	20.13	20.62	21.22	20.99	21.17	mean = 20.8260

Data analysis

Data analysis is part of the procedure. It is important to consider how the data will be presented and analysed. A wide range of analysis tools are available. For example, tables can be used to arrange data so that patterns can be seen. Graphs can show relationships and enable comparisons. Preparing an empty table with headings for the data to be obtained will help in the planning of the investigation.

The nature of the data being collected, such as whether the variables are qualitative or quantitative, influences the type of method or tool needed to analyse the data. The purpose and the hypothesis will also influence the choice of analysis tool.

Sourcing appropriate materials and technology

Part of designing an investigation is deciding on the materials, technology and instrumentation needed to carry out the research. It is important to find the right balance between items that are easily accessible and those that will give accurate results. As you move onto conducting your investigation, it will be important to take note of the precision of your chosen instrumentation and how this affects the accuracy and validity of your results. This will be discussed in greater detail in Section 1.3.

GO TO ➤ Section 1.3 page 19

Modifying a procedure

The procedure may need modifying as the investigation is carried out. The following actions will help to determine any issues in the methodology and how to modify them.

- Record everything.
- Be prepared to make changes to the approach.
- Note any difficulties encountered and the ways they were overcome. What were the failures and successes? Every test carried out can contribute to the understanding of the investigation as a whole, no matter how much of a disaster it may first appear.
- Do not panic. Go over the theory again, and talk to the teacher and other students. A different perspective can lead to a solution.

If the expected data is not obtained, don't worry. As long as it is critically and objectively evaluated, the limitations of the investigation are identified, and further investigations proposed, the work is worthwhile.

ETHICAL AND SAFETY GUIDELINES

Ethical considerations

When deciding on an investigation, identify all possible ethical issues and consider their relevance and ways to address them. Some investigations require an ethics approval; consult with your teacher.

The following questions relate to some ethical issues that might arise.

- How might this research affect the wider society?
- Who will the benefits/applications of this research be available to?
- Will one individual or group of individuals benefit at the expense of another?
- Does this research prevent anyone from obtaining their basic needs?
- How might it impact on future ethical issues? For example, even if your investigation is ethical, could it clear a path to other applications that are unethical?

Risk assessments

When planning an investigation, it is important for the safety of the experimenter and of others that potential risks are considered (Figure 1.2.4).

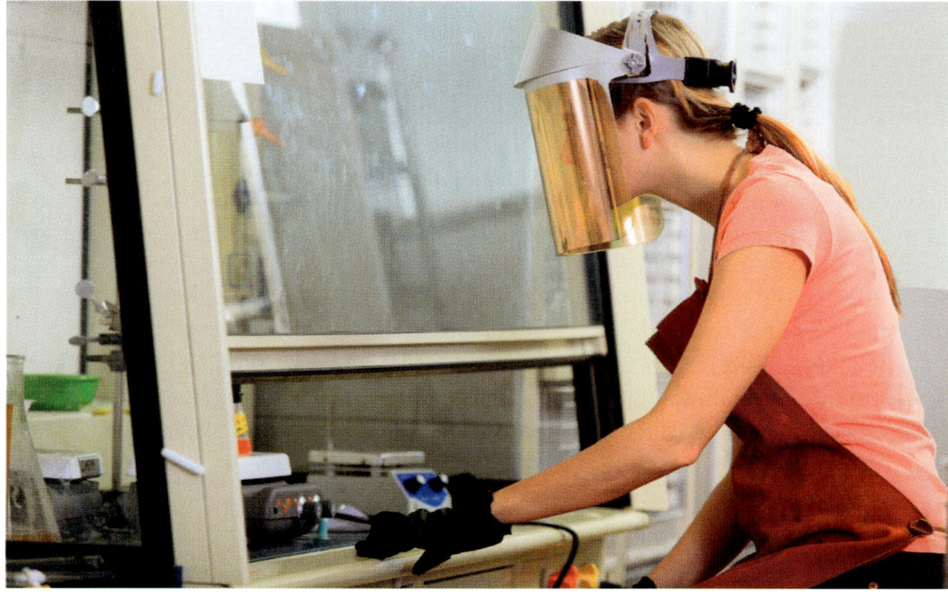

FIGURE 1.2.4 When planning an investigation, it is essential to identify, assess and control hazards.

Everything we do has some risk involved. Risk assessments are performed to identify, assess and control hazards. A risk assessment should be performed for any experimental situation, whether in the laboratory or outside in the field. Always identify and control the risks to keep everyone as safe as possible.

To identify risks, think about:

- the activity that will be carried out
- the equipment or chemicals that will be used.

The following hierarchy of risk controls is organised from most effective to least effective:

1. Elimination: Eliminate dangerous equipment, procedures or substances.
2. Substitution: Find equipment, procedures or substances that will achieve the same result, but have less risk.
3. Isolation: Ensure there is a barrier between the person and the hazard. Examples include physical barriers such as guards in machines, and fume hoods for work with volatile substances.
4. Administrative controls: Provide guidelines, special procedures and warning signs, and explain safe behaviours to participants.
5. Personal protective equipment: Wear safety glasses, lab coats, gloves and respirators, for example, where appropriate, and provide these to other participants.

Figure 1.2.5 is a flow chart showing how to consider and assess the risks involved in a research investigation.

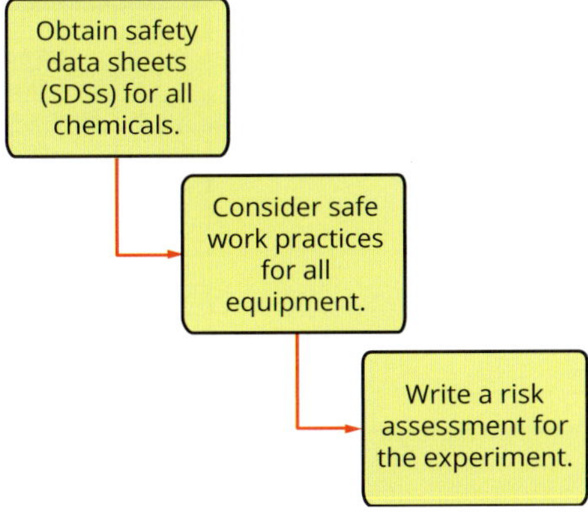

FIGURE 1.2.5 These steps must be taken when identifying risks.

Science outdoors

Sometimes investigations and experiments will be conducted outdoors. Working outdoors has its own set of potential risks, and it is important to consider ways of eliminating or reducing these risks. Table 1.2.2 contains examples of risks associated with field work in a national park.

TABLE 1.2.2 Risks associated with field work in a national park

Risk	Control measure
sunburn	Wear sunscreen, a hat and sunglasses.
exposure	Wear clothing to protect against heat or cold.
falls	Minimise the use of computer and equipment cables, and cover them up with matting. Be aware of tree roots, rocks etc.
drowning	Be cautious near deep water when taking water samples.

First aid

Minimising the risk of injury reduces the chance of requiring first aid assistance. However, it is still important to have someone with first aid training present during practical investigations. Always tell the teacher or laboratory technician if an injury or accident happens.

Personal protective equipment

Everyone who works in a laboratory wears items that help keep them safe. This is called **personal protective equipment (PPE)** and includes:

- safety glasses
- shoes with covered tops
- disposable gloves for handling chemicals
- a disposable apron or a lab coat if there is risk of damage to clothing or skin
- ear protection if there is a risk to hearing.

Chemical hazard codes

The chemicals at school or at the hardware shop have a warning symbol on the label. These are **chemical hazard codes** (HAZCHEM codes) or GHS (Globally Harmonized System of Classification and Labelling of Chemicals) pictograms. Some common codes and their meanings are shown in Figures 1.2.6 and 1.2.8. Trucks that carry chemicals display hazard symbols, as shown in Figure 1.2.7.

HAZCHEM INTERPRETATION

NUMBER			
1		Water Jets	
2		Water Fog	
3		Foam	
4		Dry Agent	
FIRST LETTER			
P	V	Full Protective Clothing*	DILUTE
R		Full Protective Clothing*	DILUTE
S	V	Breathing Apparatus	DILUTE
S	V	Breathing Apparatus for Fire Only	DILUTE
T		Breathing Apparatus	DILUTE
T		Breathing Apparatus for Fire Only	DILUTE
W	V	Full Protective Clothing*	CONTAIN
X		Full Protective Clothing*	CONTAIN
Y	V	Breathing Apparatus	CONTAIN
Y	V	Breathing Apparatus for Fire Only	CONTAIN
Z		Breathing Apparatus	CONTAIN
Z		Breathing Apparatus for Fire Only	CONTAIN
SECOND LETTER			
E		Consider Evacuation	

Note V: Danger of violent reaction or explosion
* Full Protective Clothing includes Breathing Apparatus

Class	Label
1	EXPLOSIVE (Gunpowder, flares)
2.1	FLAMMABLE GAS (LP gas, acetylene)
2.2	NON-FLAMMABLE NON-TOXIC GAS (Carbon dioxide)
2.3	TOXIC GAS (Chlorine gas)
3	FLAMMABLE LIQUID (Petrol, kerosene)
4.1	FLAMMABLE SOLID (Firelighters, matches)
4.2	SPONTANEOUSLY COMBUSTIBLE (Carbon, white phosphorus)
4.3	DANGEROUS WHEN WET (Calcium carbide)
5.1	OXIDIZING AGENT (Calcium hypochlorite)
5.2	ORGANIC PEROXIDE
6	TOXIC (Arsenic)
7	RADIOACTIVE MATERIAL (Uranium)
8	CORROSIVE (Hydrochloric acid)
9	MISCELLANEOUS DANGEROUS GOODS (Dry ice, asbestos)
	MIXED CLASS LABEL (For road transport)

FIGURE 1.2.6 HAZCHEM signs have very specific meanings.

FIGURE 1.2.7 Trucks transporting hazardous substances such as flammable liquids have hazard symbols attached.

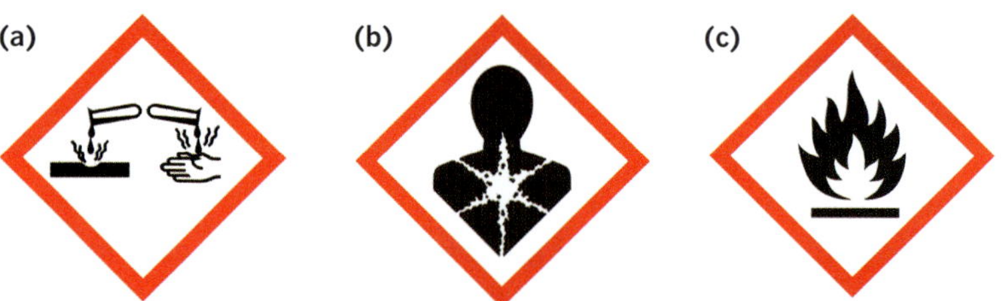

FIGURE 1.2.8 On 1 January 2017, the GHS (Globally Harmonized System of Classification and Labelling of Chemicals) pictograms were introduced in Australia. These are used for labelling chemical containers and in safety data sheets. Some of the pictograms you may see denote chemicals that (a) are corrosive, (b) pose a health hazard or (c) are flammable.

Safety data sheets

Each chemical substance has an accompanying document called a **safety data sheet (SDS)** (Figure 1.2.9), previously known as a material safety data sheet (MSDS). An SDS contains important safety and first aid information about each chemical you commonly use in the laboratory. For example, if the products of a reaction are toxic to the environment, you must pour your waste into a special container and not down the sink.

The SDS helps employers, workers and other health and safety representatives safely manage the risks of hazardous substance exposure.

1. IDENTIFICATION OF THE MATERIAL AND SUPPLIER

Product Name: **HYDROCHLORIC ACID - 20% OR GREATER**

Recommended use of the chemical and restrictions on use: Precursor for generation of chlorine dioxide gas used in water treatment.

Please ensure you refer to the limitations of this Safety Data Sheet as set out in the "Other Information" section at the end of this Data Sheet.

2. HAZARDS IDENTIFICATION

Classified as Dangerous Goods by the criteria of the Australian Dangerous Goods Code (ADG Code) for Transport by Road and Rail; DANGEROUS GOODS.

This material is hazardous according to Safe Work Australia; HAZARDOUS SUBSTANCE.

Classification of the substance or mixture:
Corrosive to Metals - Category 1
Skin Corrosion - Sub-category 1B
Eye Damage - Category 1
Specific target organ toxicity (single exposure) - Category 3

SIGNAL WORD: DANGER

Hazard Statement(s):
H290 May be corrosive to metals.
H314 Causes severe skin burns and eye damage.
H335 May cause respiratory irritation.

Precautionary Statement(s):

Prevention:
P234 Keep only in original container.
P260 Do not breathe mist / vapours / spray.
P264 Wash hands thoroughly after handling.
P271 Use only outdoors or in a well-ventilated area.
P280 Wear protective gloves / protective clothing / eye protection / face protection.

FIGURE 1.2.9 Part of a safety data sheet (SDS) for concentrated hydrochloric acid. The SDS alerts the reader to any potential hazards associated with the use of a substance, and lists appropriate measures to reduce risk of harm.

1.2 Review

SUMMARY

- The methodology of your investigation is the step-by-step procedure. When detailing the methodology, ensure it works as a valid, reliable and accurate investigation.
- Determine how many times an experiment needs to be replicated. Many scientific investigations lack sufficient repetition to ensure that the results can be considered reliable and repeatable.
- Risk assessments must be carried out before conducting an investigation so that everyone involved is kept as safe as possible. If elements of an investigation are high risk, then the experimental design needs to be re-evaluated.
- It is important to choose appropriate equipment for an experiment. This includes personal protective equipment that will help keep you safe and instrumentation that will give you accurate results.

KEY QUESTIONS

1 A journal article reported the materials and procedure used in an experiment. The experiment was repeated three times, and all values were reported in the results section of the article.

Repeating an experiment and reporting results supports which of the following?

A validity
B reliability
C credibility
D systematic errors

2 **a** Explain what is meant by the term 'controlled experiment'.
b Using an example, distinguish between independent and dependent variables.

3 You are conducting an experiment to determine the pH of various soft drinks. Identify:
a the independent variable
b the dependent variable
c at least one controlled variable.

4 You are conducting an experiment to determine the pH of a solution. Discuss the accuracy of your results if you are:
a using litmus paper or universal indicator
b recording the pH using a calibrated pH meter.

5 Give the correct term to describe an experiment with each of the following conditions.
a The experiment addresses the hypothesis and purposes.
b The experiment is repeated and consistent results are obtained.
c Appropriate equipment is chosen to reduce systematic errors.

1.3 Conducting investigations

After a practical investigation has been planned and designed, the investigation can begin and the results can be recorded. Certain key steps and skills will help you maintain high standards and minimise potential errors (Figure 1.3.1). This section will focus on the best methods for conducting a practical investigation, and for systematically generating, recording and processing data.

FIGURE 1.3.1 Read the bottom of the meniscus at eye level to avoid parallax error (an error due to viewing the level of the solution from an angle). This student is showing how a piece of white card (or a tile) improves the contrast between the solution and the scale.

COLLECTING AND RECORDING DATA

A scientific investigation must be objective and systematic. Being familiar with the methodology and protocols before beginning will help you to achieve this.

When working, keep asking questions. Is the work biased in any way? If changes are made, how will they affect the study? Will the investigation still be valid for the purpose and hypothesis?

During the investigation, record in your logbook:

- all quantitative and qualitative data collected
- methods used to collect the data
- any incident, feature or unexpected event that may have affected the quality or validity of the data.

The data recorded in the logbook is the **raw data**. Usually this data needs to be processed in some manner before it can be presented. If an error occurs in the processing of the data, or if you decide to present the data in an alternative format, the recorded raw data will always be available for you to refer back to.

Safe work practices

Remember to employ safe work practices at all times while conducting your experiment. See Section 1.2 about conducting risk assessments. Keep in mind safe procedures to follow when disposing of waste. These procedures will depend on the types of waste produced throughout your experiment. Your teacher will usually be able to show how best to approach waste disposal. Education or government websites are another useful source of information.

GO TO ➤ Section 1.2 page 11

IDENTIFYING ERRORS

Most practical investigations have errors associated with them. Errors can occur for a variety of reasons. Being aware of potential errors helps you to avoid or minimise them. For an investigation to be valid, it is important to identify and record any errors.

The three types of errors are:

- mistakes
- systematic errors
- random errors.

Types of errors

Mistakes

A **mistake** is an avoidable error. Mistakes include:

- misreading the numbers on a scale
- not labelling a sample adequately (Figure 1.3.2)
- spilling a portion of a sample.

A measurement that involves a mistake must be rejected. It must not be included in any calculations or averaged with other measurements of the same quantity.

FIGURE 1.3.2 Adequate labelling of samples is important for avoiding mistakes.

Systematic errors

A **systematic error** is consistent and will occur again if the investigation is repeated in the same way.

Systematic errors are usually a result of instruments that are not calibrated correctly or methods that are flawed. An example of a systematic error would be if a ruler mark indicating 5 cm was actually only 4.9 cm from the 0 cm mark due to a manufacturing error or shrinkage of the wood. Another example would be a researcher repeatedly using a piece of equipment incorrectly.

Random errors

Random errors occur in an unpredictable manner and are generally small. A random error would occur if an instrument was affected by a power cut or low battery power.

Techniques for reducing error

Care in designing the procedure, and in selecting and using equipment, will help to reduce errors.

Appropriate equipment

Use equipment that is best suited to the data needed for testing the hypothesis. Determining the units and the scale of the data being collected will help in selecting the correct equipment. Using the right unit and scale will ensure that measurements are more accurate and precise (with smaller systematic errors).

GO TO ➤ SkillBuilder page 227

GO TO ➤ Appendix 2 page 528

Significant figures are the numbers that convey meaning and precision. The number of significant figures used depends on the scale of the instrument. It is important to record data to the number of significant figures available from the equipment or observation. Using either a greater or smaller number of significant figures can be misleading.

Review the following examples of significant figures.

- 15 has two significant figures.
- 3.5 has two significant figures.
- 3.50 has three significant figures.
- 0.037 has two significant figures.
- 1401 has four significant figures.

The formula used to calculate heat energy (q) is $q = mc\Delta T$.

If $m = 1000$ g, $c = 4.18\,\text{J g}^{-1}\text{K}^{-1}$ and $\Delta T = 4.8\,\text{K}$, $q = 20\,064\,\text{J}$.

Only write the answer to the least number of significant figures in the data (that is, two significant figures):

$q = 20\,\text{kJ}$ or $2.0 \times 10^4\,\text{J}$

Although digital scales can measure to many more than two figures, and calculators can give 12 figures, use common sense and follow the significant figure rules.

Calibrated equipment

Some equipment, such as a pH meter, needs to be calibrated before use. Before carrying out an investigation, make sure that instruments or measuring devices are properly calibrated and functioning correctly. Record the precision of the glassware that you intend to use. If you are preparing a solution of known concentration, you might have access to a pipette, which has less measurement uncertainty than a beaker (Figure 1.3.3).

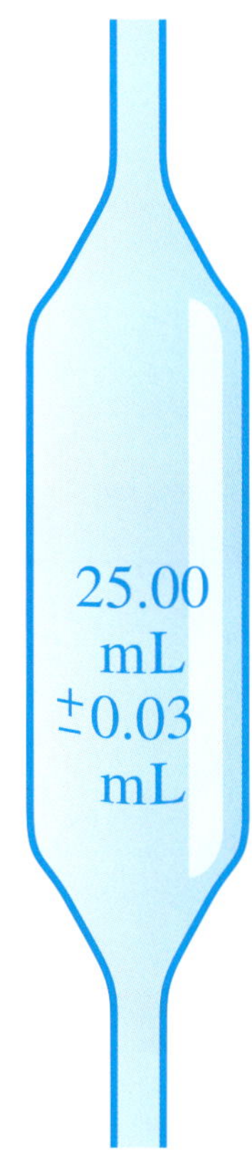

FIGURE 1.3.3 Record the uncertainty for glassware and instruments. This pipette can dispense a volume (aliquot) of 25.00 ± 0.03 mL. When it is used correctly, the volume dispensed will be between 24.97 mL and 25.03 mL.

Correctly using equipment

Use equipment properly. Ensure you have been trained to use the equipment and that you have practised using it before beginning the investigation. Improper use of equipment can result in inaccurate, imprecise data with large errors, and the validity of data can be compromised.

Incorrect reading of measurements is a common misuse of equipment. Make sure that you can use all of the equipment you need, and record the instructions in detail so that you can refer back to them if the data doesn't appear to be correct.

Repeating an investigation

As discussed in Section 1.2, reliability is ensured by repeating your experiment. If you have doubts about reliability, review your procedure before repeating the investigation to ensure all variables are being tested under the same conditions.

Referencing secondary-sourced information

As you conduct your investigation, make note of any secondary-sourced information that you use. Include it in your written report. This is discussed further in Section 1.7.

Categorising the information and evidence you find while you are working will make it easier to locate information later and to write up your final investigation. Categories you might use include:

- research methods
- key findings
- evidence
- research relevance
- use
- **stakeholders** and impacts
- future concerns.

Record information from resources in a clear way so you can retrieve it and use it.

1.3 Review

SUMMARY

- During an investigation, the following must be recorded in a logbook:
  - all quantitative and qualitative data collected
  - the procedures used to collect the data
  - any incident, feature or unexpected event that may have affected the quality or validity of the data.
- A systematic error is an error that is consistent and will occur again if the investigation is repeated in the same way. Systematic errors are usually the result of instruments that are not calibrated correctly or procedures that are flawed.
- Random errors occur in an unpredictable manner and are generally small. A random error could be the result of a researcher reading the same result correctly one time and incorrectly another time.
- The number of significant figures used when recording data depends on the scale of the instrument used. It is important to record data to the number of significant figures available from the equipment or observation.

KEY QUESTIONS

1. Both sets of the data shown contain errors. Identify which set is more likely to contain a systematic error and which is more likely to contain a mistake.
   Data set A: 11.4, 10.9, 11.8, 10.6, 1.5, 11.1
   Data set B: 25, 27, 22, 26, 23, 25, 27
2. What type of error is associated with:
   a inaccurate measurements
   b imprecise measurements?
3. A student carried out a titration (a type of experiment) during a research investigation. The volumes dispensed from a piece of specialised glassware, called a burette, are called titres. Concordant (acceptably consistent) titres have a maximum range of 0.10 mL. The following titres were recorded: 18.34 mL, 17.34 mL, 17.38 mL, 17.84 mL and 17.44 mL. Which three are concordant titres?
4. Identify whether each of these errors is a mistake, a systematic error or a random error.
   a A pipette that should dispense volumes of 25.00 $\pm$ 0.03 mL actually dispenses aliquots of 25.87 $\pm$ 0.03 mL.
   b A student misreads the value of the burette for the second titration.
   c A sample of sodium carbonate powder is weighed three times with these results: 1.5791 g, 1.5792 g and 1.5790 g.
5. A scientist carries out a set of experiments, analyses the results and publishes them in a scientific journal. Other scientists in different laboratories repeat the experiment, but do not get the same results as the original scientist. Suggest several reasons for this.

1.4 Processing data and information

Once you have conducted your investigation and collected data, you will need to find the best way of collating it. This section is a guide to the different forms of representation that will help you to better understand your data.

RECORDING, ORGANISING AND PRESENTING QUANTITATIVE DATA

The raw data that has been obtained needs to be presented in a way that is clear, concise and accurate. Data can be presented in a number of ways, including as tables, graphs, flow charts and diagrams.

Raw data is unlikely to be used directly to validate a hypothesis. However, raw data is essential to the investigation, and plans for collecting the raw data should be made carefully. Consider the formulae or graphs that will help you to analyse the data at the end of the investigation. This will help to determine the type of raw data that needs to be collected in order to validate the hypothesis.

Presenting raw and processed data in tables

Tables can be used to organise raw data and processed data or to summarise results. The simplest form of a table is a two-column format. In a two-column table, the first column contains the independent variable (the one being changed) and the second column contains the dependent variable (the one that may change in response to a change in the independent variable).

Tables should have:

- a descriptive title
- column headings (including the units of measurement)
- numbers aligned at the decimal point
- the independent variable in the left column
- the dependent variable in the right column.

Look at Table 1.4.1, which has been used to organise raw and processed data about the water in Sydney Harbour.

TABLE 1.4.1 Analysis of Sydney Harbour water and how the data could be recorded

Location	Date and time	Temperature (°C) (±0.1)	pH (±0.1)	Conductivity (ms m^{-1}) (±1)	Turbidity (NTU) (±2)	Dissolved oxygen (mg L^{-1}) (±0.1)	Biological oxygen demand (mg L^{-1}) (±0.1)
A	26 Feb, 11 am	10.4	6.5	72	5	11.8	0.9
B	26 Feb, 8 am	9.2	6.6	73	10	11.4	0.7
C	27 Feb, 8 am	9.5	6.4	77	10	10.9	0.9
D	27 Feb, 9 am	9.9	6.5	75	10	11.3	1.0

Mean and uncertainty

A table of processed data usually presents the average value of replicates—the **mean**. However, the mean on its own does not provide a complete picture of the results. To report processed data more usefully, the **uncertainty** should be presented as well. In other words, the mean must be accompanied by a description of the range of data obtained.

Uncertainty is calculated as:

uncertainty = ±(maximum variance from the mean)

Worked example 1.4.1

CALCULATING UNCERTAINTY

The temperature of the water in the Murrumbidgee River at Carrathool was measured in March 2017. The temperatures (in °C) recorded were:

21, 22, 20, 20, 21, 20, 19

Find the uncertainty for these values to the nearest degree.

Thinking	Working
Calculate the average temperature.	average = $\frac{21+22+20+20+21+20+19}{7} = 20°C$
Calculate the maximum variance from the mean.	22°C is 2°C above the average temperature, so the uncertainty is 2.
Write the average temperature and include the uncertainty.	The average temperature is 20 ± 2°C.

Worked example: Try yourself 1.4.1

CALCULATING UNCERTAINTY

The temperature of the water in the Murrumbidgee River at Wagga Wagga was measured in February 2017. The temperatures (in °C) recorded were:

23, 22, 24, 21, 26, 24, 23

Find the uncertainty for these values to the nearest degree.

Percentage error

Percentage error (also known as percentage uncertainty) is a way to quantify the accuracy of a measurement. It is the data uncertainty divided by the measurement. Suppose you have measured a distance of 230.0 cm using a ruler, and have calculated that the uncertainty for this measurement is ±0.5 cm.

$$\text{percentage error} = \left(\frac{0.5}{230.0}\right) \times 100 = \pm 0.2\%$$

Mode and median

The mean and the uncertainty are statistical measures that help describe data accurately. Other statistical measures that can be used, depending on the data obtained, are:

- **mode**—the value that appears most often in a data set. The mode is useful for describing qualitative or discrete quantitative data (e.g. the mode of the values 0.01, 0.01, 0.02, 0.02, 0.02, 0.03, 0.04 is 0.02).
- **median**—the middle value of an ordered list of values (e.g. the median of the values 5, 5, 8, 8, 9, 10, 20 is 8). The median is used when the data range is spread; for example: due to the presence of unusual results, making the mean unreliable.

Graphs

In general, tables provide more detailed data than graphs, but graphs make it easier to observe trends and patterns in data.

Graphs are often used when two variables are being considered and one variable is dependent on the other. A graph shows the relationship between the variables.

There are several types of graphs that can be used, including line graphs, bar graphs and pie charts. The best one to use will depend on the type of data. This is discussed further in Section 1.5.

When making a graph (Figure 1.4.1):

- keep the graph simple and uncluttered
- use a descriptive title
- represent the independent variable on the *x*-axis and the dependent variable on the *y*-axis

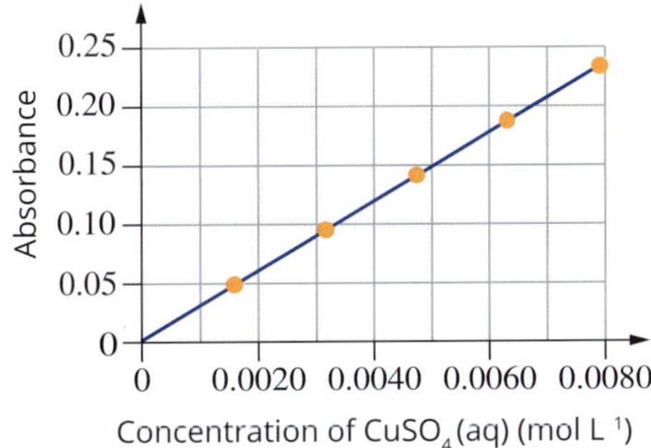

FIGURE 1.4.1 A graph is a good way to show trends and patterns in data. Well-constructed graphs have a title and labelled axes, including units.

- make axes proportionate to the data
- clearly label each axis with both the variable and the unit in which it is measured (if applicable).

Line graphs

Line graphs are a good way of representing continuous quantitative data. In a line graph, the values are plotted as a series of points on the graph. There are two ways of joining these points.

- A line can be ruled from each point to the next (Figure 1.4.2a). This shows the overall **trend**, or relationship; it is not meant to be used to predict the value of the points between the plotted data.
- The points can be joined with a single smooth straight (Figure 1.4.3) or curved line (Figure 1.4.2b). This creates a **trend line**, also known as a line of best fit. The line of best fit does not have to pass through every point, but should go close to as many points as possible. It is used when there is an obvious trend in the data.

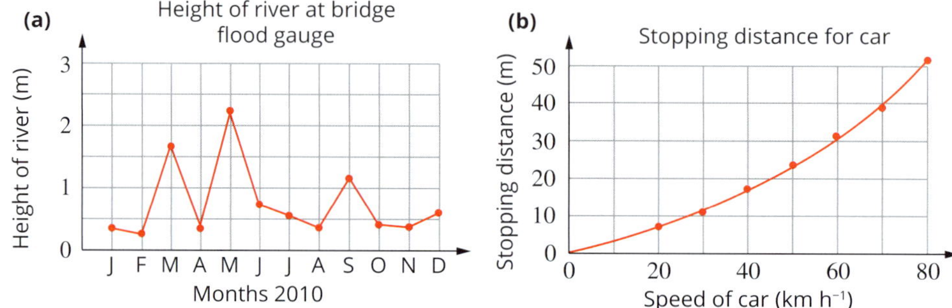

FIGURE 1.4.2 (a) The data in the graph is joined from point to point. (b) The data in the graph is joined with a line of best fit, which shows the general trend.

Outliers

Sometimes when the data is collected, one piece of data may not fit with the trend and is clearly an error. This is called an **outlier**. An outlier is often caused by a mistake made in measuring or recording data, or by a random error in the measuring equipment. If there is an outlier, include it on the graph, but ignore it when adding a line of best fit (Figure 1.4.3).

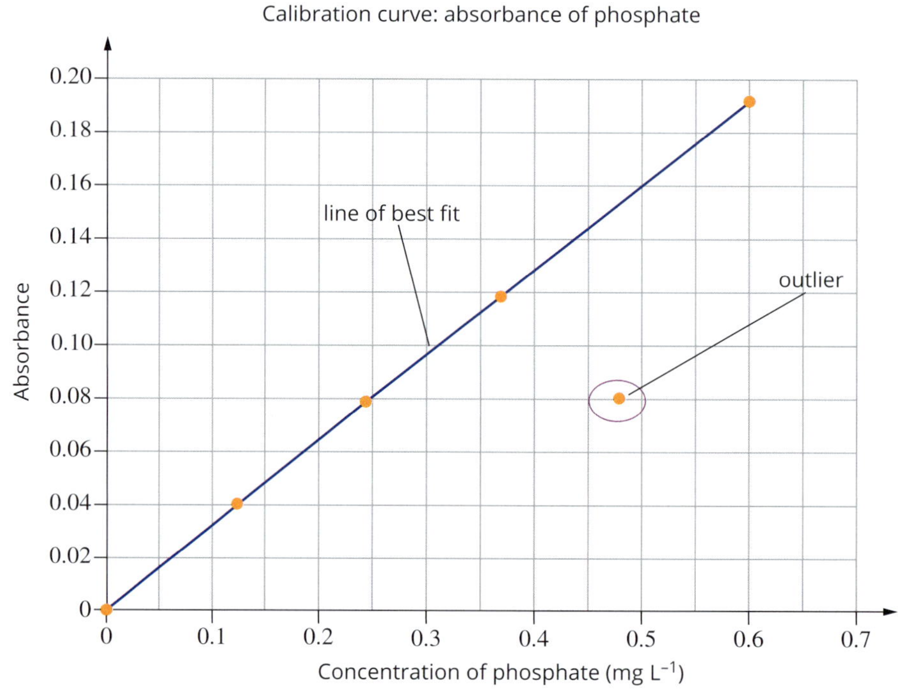

FIGURE 1.4.3 Outliers should be ignored when constructing a line of best fit.

1.4 Review

SUMMARY

- A wide range of analysis tools can be used to organise and present data. Tables can show data in more detail, and graphs can show relationships and comparisons.
- The simplest form of a table is a two-column format in which the first column contains the independent variable (the one being changed) and the second column contains the dependent variable (the one that may change in response to a change in the independent variable).
- For a range of measurements of a particular value, the mean must be accompanied by the uncertainty, to indicate precision.
- When making a graph:
  - keep the graph simple and uncluttered
  - use a descriptive title
  - represent the independent variable on the *x*-axis and the dependent variable on the *y*-axis
  - make axes proportionate to the data
  - clearly label each axis with both the variable and the unit in which it is measured.

KEY QUESTIONS

1 For the data set 21, 28, 19, 19, 25 and 24, determine the:
   a mean
   b mode
   c median.

2 How can the general trend of a graph be represented after the points are plotted?

3 **a** Use the following values to plot a calibration curve to determine the concentration of phosphate in samples of water. The absorbance values were obtained during an experiment at 450 nm.

Standard phosphate concentration ($mg L^{-1}$)	Absorbance
0.00	0.000
0.12	0.038
0.24	0.079
0.36	0.159
0.48	0.154
0.60	0.191

   b From the graph you have drawn, select the data point that is an outlier.
   c Define the term 'outlier'.

4 State the mean and uncertainty for the data set 20, 28, 19, 19, 25 and 24.

1.5 Analysing data and information

Now that the investigation has been conducted and the data collected, it is time to analyse and discuss your findings. Analysing your results allows you to better understand the physical processes behind them.

FIGURE 1.5.1 Analyse both your raw and processed data before drawing conclusions from your investigation.

WRITING A DISCUSSION

The discussion is the part of the investigation in which the evaluation and explanation of the procedure and results takes place. It is the interpretation of what the results mean.

The key sections of the discussion are:

- analysing and evaluating the data
- evaluating the investigative procedure
- explaining the link between the investigation findings and the relevant chemical concepts.

Consider the message to be conveyed to the audience when writing the discussion. Statements need to be clear and concise. At the conclusion of the discussion, the audience must have a clear idea of the context, results and implications of the investigation.

ANALYSING AND EVALUATING THE DATA

In the discussion, the findings of the investigation need to be analysed and interpreted.

- State whether a pattern, trend or relationship was observed between the independent and dependent variables. Describe what kind of pattern it was and specify under what conditions it was observed.
- Were there discrepancies, deviations or anomalies in the data? If so, these should be acknowledged and explained.
- Identify any limitations in the data. Perhaps a larger sample or greater variation in the independent variable would lead to a stronger conclusion.

You can use the trends, patterns and mathematical models observed in your results to discuss whether the data supported or refuted the hypothesis. Ask questions such as:

- was the hypothesis supported?
- has the hypothesis been fully investigated? (If not, give an explanation of why this is so and suggest what could be done to either improve or complement the investigation.)
- do the results contradict the hypothesis? If so, why? (The explanation must be plausible and must be based on the results and previous evidence.)

Trends in line graphs

Graphs show the relationship between variables.

- Variables that change linearly (or in direct proportion to each other produce a straight, sloping trend line (Figure 1.5.2).
- Variables that change exponentially (or in indirect proportion to each other) produce a curved trend line (Figure 1.5.3).
- When there is an inverse relationship, one variable increases as the other variable decreases (Figure 1.5.4).
- When there is no relationship between two variables, one variable will not change when the other changes (Figure 1.5.5).

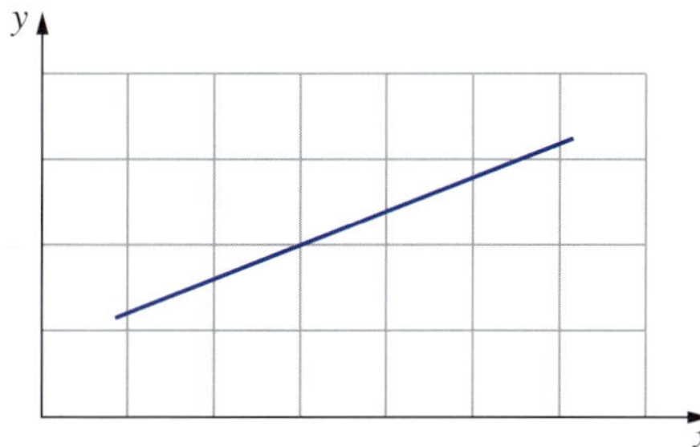

FIGURE 1.5.2 This graph indicates variables that change in direct response to each other.

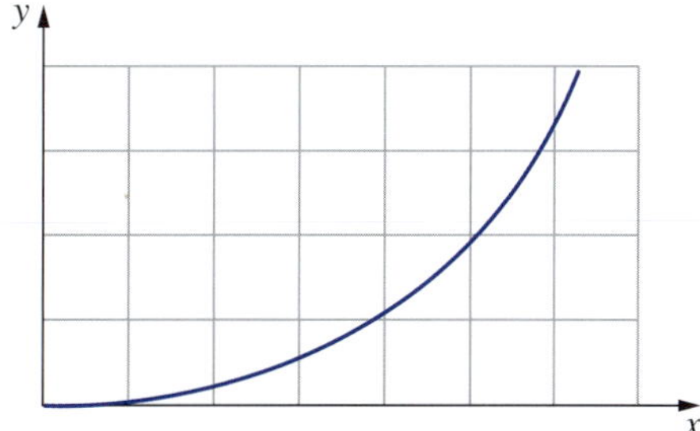

FIGURE 1.5.3 This graph indicates variables that change in response to each other in a non-linear way.

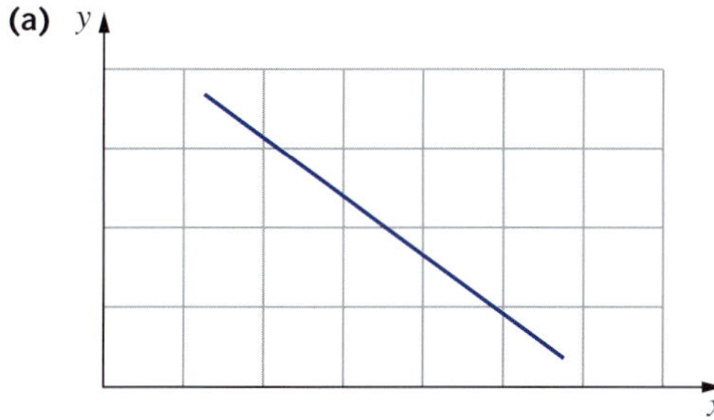

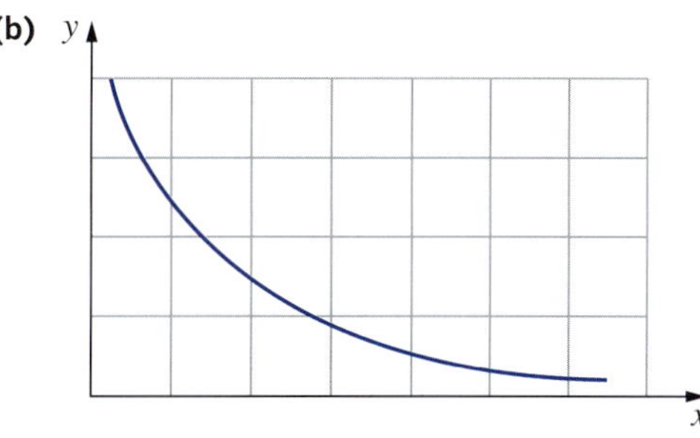

FIGURE 1.5.4 This graph indicates variables with an inverse relationship: one variable decreases in response to the other variable increasing. The relationship may be (a) direct (linear) or (b) indirect (non-linear).

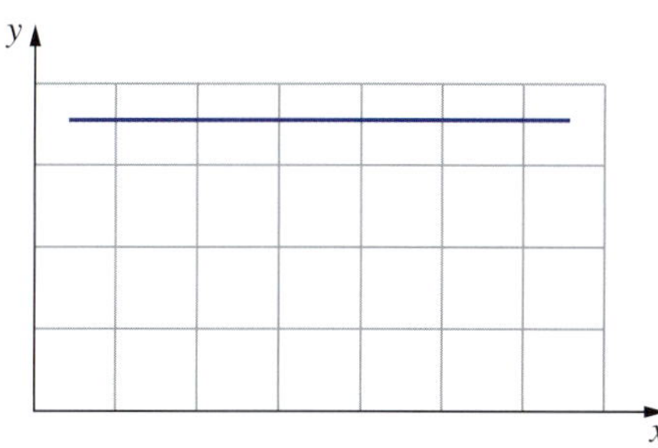

FIGURE 1.5.5 When two variables show no relationship, there is no trend in the data.

Remember that results may be unexpected. This does not make the investigation a failure. However, the findings must be related to the hypothesis, purpose and inquiry question.

Table 1.5.1 on page 28 shows the types of graph used for presenting various kinds of data.

TABLE 1.5.1 Types of graph that can be used in an investigation discussion

Type of graph	When to use	Example
scatter graph with a line of best fit	showing quantitative data where one variable is dependent on another variable	calibration curve in colorimetry
line graph	presenting continuous quantitative data	concentration of dissolved oxygen in a creek at a particular location over a period of time
bar graph	displaying data in an investigation with a qualitative independent variable	turbidity of water at various locations (Figure 1.5.6)
pie diagram	summarising qualitative data; displaying proportions	relative proportion of various pesticides in a sample of water

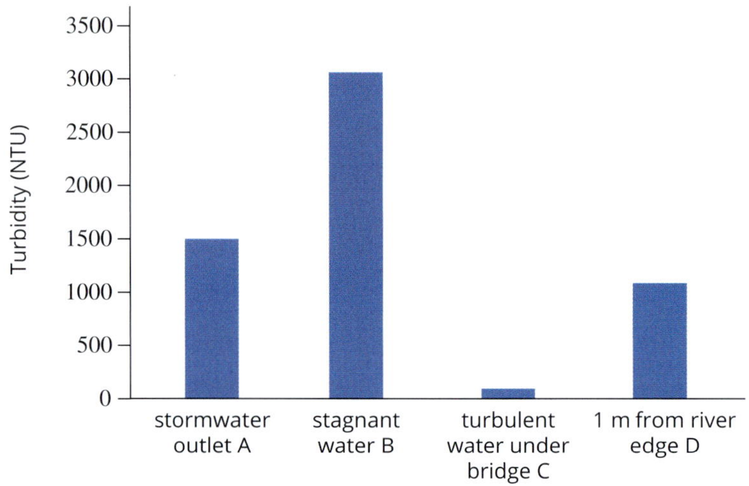

FIGURE 1.5.6 A bar graph can be used to compare data. This graph shows the turbidity (a measure of water quality) of water samples taken from various locations along the Murrumbidgee River.

EVALUATING THE PROCEDURE

It is important to discuss the limitations of the investigation procedure. Evaluate the procedure and identify any issues that could have affected the validity, accuracy, precision or reliability of the data. Sources of errors and uncertainty must also be included in the discussion.

If any limitations or problems in the methodology are identified, recommend improvements to the investigation; for example, suggest how a different measurement technique could be more accurate.

Addressing bias

Bias is a form of systematic error resulting from the researcher's personal preferences or motivations. There are many types of bias, including:

- poor definitions of either concepts or variables (for example, not defining pH)
- incorrect assumptions (for example, measuring the pH of a solution without also measuring its temperature)
- errors in the investigation design and procedure (for example, taking more water samples from one site compared with all other sites).

Bias may occur in any part of the investigation. Some biases cannot be eliminated, but they should be addressed in the discussion.

Evaluating accuracy and precision

In the discussion, evaluate the degree of accuracy and precision of the measurements of each variable in the hypothesis. Comment on the uncertainties obtained.

When relevant, compare the chosen measurement technique with any other techniques that might have been selected. Evaluate the advantages and disadvantages of the selected methodology and its effect on the results.

FIGURE 1.5.7 Honest evaluation and reflection are important when analysing the procedure.

Indicating reliability

When discussing the results, indicate the range of the data obtained from replicates. Explain how the sample size was selected. Larger samples are usually more reliable, but time and resources might have been scarce. Discuss whether the results of the investigation have been limited by the sample size.

The control group is important for the reliability of the investigation. A control group helps determine whether a variable that should have been controlled has been overlooked, and it may help explain any unexpected results.

Explaining errors

Discuss any sources of systematic or random error and suggest ways of improving the investigation (Figure 1.5.7).

CRITICALLY EVALUATING RESOURCES

Not all sources are **credible**. It is essential to critically evaluate information content and its origin. Here are some examples of questions you should ask when evaluating a source.

- Who created this information? What are the qualifications, **expertise**, **reputation** and **affiliations** of the authors?
- Why was it written?
- Where was the information published?
- When was the information published?
- How often is the information referred to by other researchers?
- Are the conclusions supported by data or evidence?
- What is implied?
- What is omitted?
- Are any opinions or beliefs being presented?
- Is the writing objectively and accurately describing a scientific concept or **phenomenon**?
- How might other people understand or interpret this message differently?

  When evaluating the validity or bias of a website, consider its domain extension:
  - '.gov' refers to government.
  - '.edu' refers to educational groups such as universities.
  - '.org' refers to non-profit organisations.
  - '.com' refers to commercial/business entities.

1.5 Review

SUMMARY

- After completing an investigation, the data must be analysed and interpreted. A discussion of the results is required.
  - State whether a pattern or trend was observed in the data, suggesting a relationship between the independent and dependent variables. Describe what kind of pattern it was and specify the conditions under which it was observed.
  - If possible, create a mathematical model to describe the data.
  - Were there discrepancies, deviations or anomalies in the data? If so, these should be acknowledged and explained.
  - Identify any limitations in the data collected. Perhaps a larger sample or further variations in the independent variable would lead to a stronger conclusion.
- It is important to discuss the limitations of the investigation methodology. Evaluate the methodology and identify any issues that could have affected the validity, accuracy, precision or reliability of the data. Sources of errors and uncertainty must also be stated in the discussion.
- When discussing the results, indicate the range of the data obtained from replicates. Explain how the sample size was selected. Larger samples are usually more reliable, but time and resources may have been scarce. Discuss whether the results of the investigation have been limited by the sample size.

KEY QUESTIONS

1. What relationship between the variables is indicated by a sloping linear graph?
2. What relationship exists if one variable decreases as the other increases?
3. What relationship exists if both variables increase or both decrease at the same rate?
4. What might cause a sample size to be limited in an investigation?
5. Consider this investigation hypothesis: If precipitation of calcium carbonate (limescale) on the heating element of a kettle decreases its efficiency, and the precipitation in various kettles is quantified, then the kettles with less limescale will be more efficient (heat more quickly) than kettles with more limescale.

   Improve this response to the hypothesis: 'When 10% of the heating element was covered by limescale, the kettle worked at 90% efficiency, and when 20% of the heating element was covered by limescale, the kettle worked at 76% efficiency.'

1.6 Problem solving

After the results have been analysed, critical thinking can be used to relate them to chemistry concepts to formulate conclusions. In this section, you will learn how this leads to a better understanding of the scientific principles underlying the problem you set out to solve.

DISCUSSING RELEVANT CHEMICAL CONCEPTS

To make the investigation more meaningful, it should be explained within the right context—the related chemical ideas, concepts, theories and models. Within this context, you can explain the basis for the hypothesis. For example, if you were studying the impact of dissolved carbon dioxide on the pH of sparkling mineral water, you could include the information given in Table 1.6.1 in your discussion.

TABLE 1.6.1 Examples of chemical concepts in a scientific investigation

Key idea	Example
definitions of key terms	'pH', 'dissolved carbon dioxide' and 'sparkling mineral water'
function of added carbon dioxide	to create 'sparkling' water
relationship between variables	dissolved carbon dioxide and pH of water; temperature was controlled in the experiment
chemical principle; include relevant equations	dissolving carbon dioxide involves formation of carbonic acid (H_2CO_3)
sources of error	reduce random error by repeating measurements and calculating averages

Relating findings to a chemical concept

Using a theoretical context, you can compare your results with existing relevant research and knowledge. After identifying the major findings of the investigation, ask questions such as:

- how does the data fit with the literature?
- does the data contradict the literature?
- do the findings fill a gap in the literature?
- do the findings lead to further questions?
- can the findings be extended to another situation?

Be sure to discuss the broader implications of the findings. Implications are the bigger picture. Outlining them for the audience is an important part of the investigation. Ask questions such as:

- do the findings contribute to or impact on the existing literature and knowledge of the topic?
- are there any practical applications for the findings?

DRAWING EVIDENCE-BASED CONCLUSIONS

A conclusion is usually a paragraph that links the collected evidence to the hypothesis and provides a justified response to the research question.

Indicate whether your hypothesis was supported or refuted and the evidence on which this is based (that is, the results). Do not provide irrelevant information. Only refer to the specifics of the hypothesis and the research question, and do not make generalisations.

Read the examples of weak and strong conclusions in Table 1.6.2 for the hypothesis and research question shown.

TABLE 1.6.2 Strong and weak conclusions to a hypothesis and research question

	Strong conclusion	Weak conclusion
Research question: Does temperature affect the pH of water?	Analysis of the results for the effect of an increase in temperature of water from 5°C to 40°C showed an inverse relationship, in which the pH of water decreased from 7.4 to 6.7. These results support the current knowledge that an increase in water temperature results in a decrease in its pH.	The results show that temperature does affect the pH of water.
Hypothesis: If pH decreases as water temperature increases, and the temperature of the water in various ponds is measured, then pond water with a higher temperature will have a lower pH.	An increase in water temperature from 5°C to 40°C resulted in a decrease in the pH of the water from 7.4 to 6.8.	The pH of water decreased as temperature increased.

INTERPRETING INFORMATION FROM SCIENTIFIC SOURCES

GO TO ➤ Section 1.5 page 26

Sometimes you may be required to investigate claims and conclusions made by other sources, such as scientific and media texts. As discussed in the previous section, some sources are more credible than others. Once you have analysed the validity of the primary or secondary source, you will be able to follow the steps already described to evaluate the conclusions drawn.

1.6 Review

SUMMARY

- A meaningful investigation is explained within the right context, which means together with related chemical ideas, concepts, theories and models. The basis for the hypothesis is explained within this context.
- The discussion should indicate whether the hypothesis has been supported or refuted and on what evidence this is based (that is, the results). Only the specifics of the hypothesis and the research question should be referred to.

KEY QUESTIONS

1. Which of the following would not form part of a strong conclusion to a report?
   A The concluding paragraphs are relevant and provide evidence.
   B The concluding paragraphs are written in emotive language.
   C The concluding paragraphs include reference to limitations of the research.
   D The concluding paragraphs include suggestions for further avenues of research.
2. Explain whether generalisations or implications should be included in an investigation.
3. A procedure was repeated 30 times. How should the following statement be rewritten?
   'Many repeats of the procedure were conducted.'
4. Which of the following would form part of a strong conclusion to a practical report?
   A a statement specifically referring to whether the hypothesis was supported or refuted
   B reference to the results of the investigation
   C reference to relevant chemical theory
   D a statement that the hypothesis has been proved to be correct.

1.7 Communicating

The way to approach communication of the results will depend on the audience. For example, investigation results intended for a general audience may adopt the style of a news article or blog post that doesn't use too much scientific language.

You will need to present your research using appropriate scientific language and notation. There are many different presentation formats, such as posters, oral presentations and reports. This section discusses the main characteristics of effective science communication and report writing, including objectivity, clarity, conciseness and coherence.

STRUCTURING A REPORT

Scientific reports should have a clear, logical structure.

Introduction

In the first paragraph, introduce your research topic and define the key terms.

Body paragraphs

Subsequent paragraphs each cover one main idea.

- Summarise the content of the paragraph in the first sentence of each paragraph.
- Use evidence to support your statements.
- Avoid very long or very short paragraphs.

Conclusion

The final section summarises the main findings. You should:

- relate it to the title of the investigation
- include mention of the limitations of the investigation
- discuss implications and applications, and potential future research.

Analysing relevant information

Scientific research should always be objective and neutral. Any premise presented must be supported with facts and evidence to allow the audience to decide for themselves. In your report, identify the evidence supporting or contradicting each point you make. Explain connections between ideas, concepts, theories and models, and make sure your discussion is relevant to the question under investigation. Figure 1.7.1 shows the questions to consider when writing an investigation report.

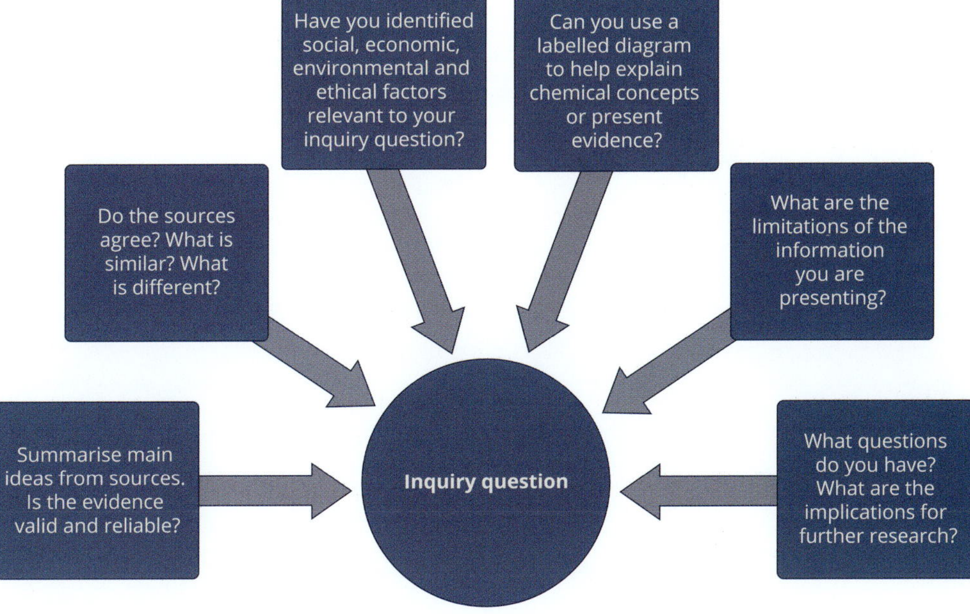

FIGURE 1.7.1 Ask yourself these questions when preparing to write a scientific report.

SKILLBUILDER 

Structuring body paragraphs

The body paragraphs of a report or essay need to be structured such that each idea is presented in a clear way. Good paragraphs create a report that has a logical flow.

One way to ensure each paragraph is structured well is to use the acronym TEEL.

Topic sentence

This establishes the key idea or argument that will be put forward in the paragraph. It supports the main proposition of the overall report.

Elaborate on the idea

Add further detail to the initial topic sentence.

Evidence

Provide evidence to support the idea or argument presented in the topic sentence.

Link back to the topic sentence

Summarise the argument in the paragraph and indicate how it links to the overall proposition set out by the report as a whole.

After you have analysed your sources, make notes on your report outline, indicating where you will use evidence and the source of that evidence. Try to introduce only one idea per sentence and one theme per paragraph.

For example, for a report on 'Experimental research into biodegradability of plastics', the third paragraph might contain notes like this:

- Selke et al. (2015), who reported no significant degradation
- Chiellini et al. (2007) reported significant degradation.

A report should include an analysis and synthesis of your sources. The information from different sources needs to be explicitly connected and it should be clear where sources agree and disagree. In this example, the final sentences could be:

Selke et al. (2015) reported that tests of plastic polymers treated with biodegradation additives resulted in no significant biodegradation after three years. This finding contrasts with that of Chiellini et al. (2007), who reported significant biodegradability of additive-treated polymers.

The different results can be explained by differences in the studies. The 2007 study tested degradation in natural river water, whereas the 2015 study tested degradation under ultraviolet light, aerobic soil burial and anaerobic aqueous conditions (Chiellini et al. 2007; Selke et al. 2015). As well as using different additives and different experimental techniques, Selke et al. (2015) used additive rates of 1–5% and tested both polyethylene terephthalate (PET) and polyethylene, whereas Chiellini et al. (2007) used additive rates of 10–15% and tested only polyethylene.

Both studies were conducted under laboratory conditions, so they may not reflect what happens in the natural environment.

WRITING FOR SCIENCE

Scientific reports are usually written in an objective and unbiased style. This is in contrast with persuasive writing, which often uses the subjective techniques of **rhetoric** or **persuasion**. Table 1.7.1 contrasts persuasive and scientific writing styles.

TABLE 1.7.1 Examples of persuasive writing and scientific writing styles

Persuasive writing	Scientific writing
Biased and subjective language: *The results were extremely bad/atrocious/ wonderful.* *This is terrible because ...* *This produced a disgusting odour.* *Health crisis!*	Unbiased and objective language: *The results showed ...* *The implications of these results are ...* *The results imply ...* *This produced a pungent odour.* *Health issue*
Exaggeration: *The object weighed a colossal amount, like an elephant.*	Non-emotive language: *The object weighed 256 kg.*
Everyday or colloquial language: *The bacteria passed away.* *The results don't ...* *The researchers had a sneaking suspicion ...*	Formal language: *The bacteria died.* *The results do not ...* *The researchers predicted/hypothesised/ theorised ...*

Consistent reporting narrative

Scientific writing can be written either in first-person or in third-person narrative. Your teacher may advise you on which to select. In either case, ensure that you keep the narrative point of view consistent. Read the examples of first-person and third-person narrative in Table 1.7.2.

TABLE 1.7.2 Examples of first-person and third-person narrative

First person	Third person
I put 50 g marble chips in a conical flask and then added 10 mL of 2 mol L^{-1} hydrochloric acid.	*First, 50 g of marble chips was weighed into the conical flask, and then 10 mL of 2 mol L^{-1} hydrochloric acid was added.*
After I observed the reaction, I found that ...	*After the reaction was completed, the results showed ...*
My colleagues and I found ...	*It was found ...*

Qualified writing

Be careful of words that are absolute, such as *always*, *never*, *shall*, *will* and *proved*. Sometimes it may be more accurate and appropriate to use qualifying words, such as *may*, *might*, *possible*, *probably*, *likely*, *suggests*, *indicates*, *appears*, *tends*, *can* and *could*.

Concise writing

It is important to write concisely, particularly if you want to engage and maintain the interest of your audience. Use shorter sentences that are less verbose (that is, they don't contain too many words). Table 1.7.3 shows some examples of verbose and concise writing.

TABLE 1.7.3 Examples of verbose and concise writing

Verbose writing	Concise writing
Due to the fact that ...	*Because ...*
Smith undertook an investigation into ...	*Smith investigated ...*
It is possible that the cause could be ...	*The cause may be ...*
A total of five experiments ...	*Five experiments ...*
end result	*result*
In the event that ...	*If ...*
shorter in length	*shorter*

Visual support

Identify concepts that can be explained using visual models, and information that can be presented in graphs or diagrams. This will not only reduce the word count of your work, but also make it more accessible for your audience.

For example, flow charts convey the steps in a process or method. The flow chart in Figure 1.7.2 shows how polymers used in the production of a consumer item can be decomposed. A limitation of flow charts is that the details of the process are omitted. Of course, simplification can be one of the benefits of visual models.

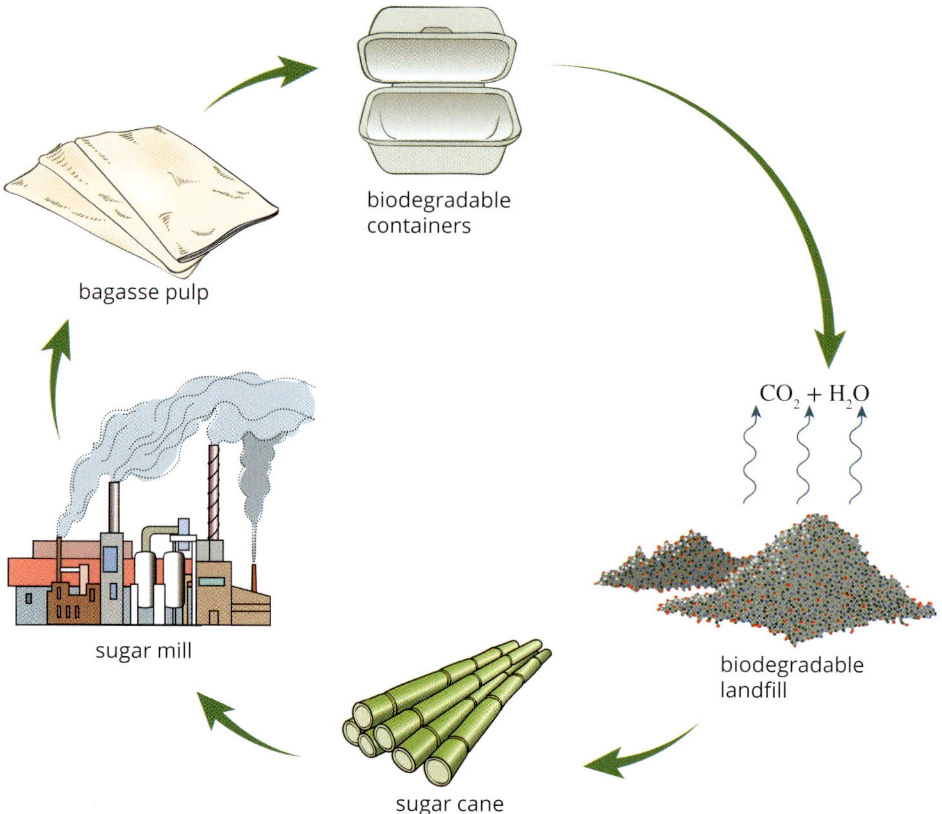

FIGURE 1.7.2 This flow chart shows how polymers used in the production of a consumer item can be decomposed. A limitation of this diagram is that it does not indicate the timeline or details involved in this process.

EDITING A REPORT

Editing is an important part of the report-writing process. After editing your report, save new drafts with a different file name, and always back up your files in another location.

Pretend you are reading your report for the first time when editing. After you have completed a draft, it is a good idea to put it aside and return to it with 'fresh eyes' later. This will help you find areas that need further work and give you the opportunity to improve them. Look for content that is:

- ambiguous or unclear
- repetitive
- awkwardly phrased
- too lengthy
- not relevant to your research question
- poorly structured
- lacking evidence
- lacking a reference (if it is another researcher's work).

Use a spellchecker tool to help you identify typographical errors. Make sure it is set to Australian English. Be wary of words that are commonly confused, for example:

- where/were
- their/they're/there
- affect/effect
- which/that.

REFERENCES AND ACKNOWLEDGEMENTS

All the quotations, documents, publications and ideas used in the investigation need to be acknowledged in a References list and an Acknowledgements section to ensure authors are credited for their work. References and acknowledgements also give credibility to the study and allow the audience to locate information sources should they wish to investigate further. The standard referencing style used is the American Psychological Association (APA) academic referencing style.

When referencing a book, include (in this order) the:

- author's surname and initials
- date of publication
- title
- place of publication
- publisher's name.

For example:

Rickard, G., et al. (2016). *Pearson Science 9 Student Book* (2nd ed.). Melbourne: Pearson Education.

When referencing a website, include (in this order):

- author's surname and initials, or name of organisation, or title
- year the website was written or last revised
- title of webpage
- website address.

For example:

National Geographic. (2015). *Killer fungus that's devastating bats may have met its match.* Retrieved from http://news.nationalgeographic.com/2015/05/150527-bats-white-nose-syndrome-treatment-conservation-animalsscience

In-text citations

Each time you write about the findings of other people or organisations, you need to provide an in-text citation and full details of the source in the References list. In the APA style, in-text citations include the first author's last name and the date in brackets (author, date). List the full details in your list of references.

The following examples show the use of in-text citation.

It was reported that in the testing of five pro-oxidant additives added to commonly manufactured polymers, none resulted in significant biodegradation after three years (Selke et al., 2015).

or

Selke et al. (2015) reported that in the testing of five pro-oxidant additives added to commonly manufactured polymers, none resulted in significant biodegradation after three years.

The reference details of this example would be:

Selke, S., Auras, R., Nguyen, T.A., Aguirre, E.C., Cheruvathur, R., & Liu, Y. (2015). Evaluation of biodegradation-promoting additives for plastics. *Environmental Science & Technology, 49*(6), 3769–3777.

USING APPROPRIATE CHEMICAL TERMINOLOGY

In every area of chemistry, you should attempt to quantify the phenomena you study. In practical demonstrations and investigations, you generally make measurements and process those measurements to come to some conclusions. Scientists have a number of conventional ways of interpreting and analysing data from their investigations. There are also conventional ways of writing numerical measurements and their units.

The use of appropriate chemical **nomenclature**, scientific notations and units is important. Table 1.7.4 on page 38 lists a few examples of common terms used in chemistry.

TABLE 1.7.4 Some common terms used in chemistry

Chemical term	Definition	Examples
element	An element is a substance whose atoms have the same atomic number. Atoms of different elements have different atomic numbers.	sodium, chlorine, tin
compound	A compound is a substance consisting of two or more elements combined in fixed proportions. A chemical formula can be written for a pure compound.	H_2O, NaCl
substance	element, compound	H_2, diamond, H_2O
particle	atom, ion, molecule, proton, neutron, electron	Ne, Na^+, H_2O, H^+, n^2, e^-
atom	building block of matter; smallest unit of an element	Na, He, C, Sn
molecule	two or more atoms covalently bonded together	water (H_2O), butane (C_4H_{10}), glucose ($C_6H_{12}O_6$)
ion	positively charged or negatively charged atom or group of atoms, resulting from the loss or gain of one or more electrons	Na^+, Cl^-, NO_3^-
cation	positively charged ion	Na^+, Mg^{2+}, Al^{3+}, NH_4^+
anion	negatively charged ion	Cl^-, O^{2-}, PO_4^{3-}
covalent network	covalently bonded network in which the bonds extend in three dimensions	diamond, silicon dioxide

Table 1.7.5 shows chemical nomenclature commonly used in chemistry.

TABLE 1.7.5 Common chemical nomenclature

Application	Convention	Examples
naming elements	Do not capitalise the first letter, unless at the start of a sentence.	carbon, hydrogen
element symbols	Capitalise the first letter; the subsequent letter (if present) is lower case.	N, Na, Ne, Ni
naming transition metals that can form ions of different charges (oxidation states)	Write the name of the transition metal, immediately followed by a Roman numeral in brackets representing the number of charges on the ion it forms.	Cu^+ (copper(I) ion) Cu^{2+} (copper(II) ion)
naming ionic compounds	Name the cation before the anion.	sodium chloride tin(IV) chloride
writing ionic formulae	Write the cation before the anion.	NaCl $SnCl_4$
using brackets in chemical formulae	Use brackets to indicate atoms that need to be considered together.	$Al_2(CO_3)_3$
using brackets in condensed structural (semi-structural) formulae	Use of brackets can indicate which groups of atoms are attached to a central atom (such as a carbon atom). Brackets can indicate which group of atoms is repeated; for example, repeating CH_2 groups.	$CH_3CH(CH_3)CH_3$ $CH_3(CH_2)_5CH_3$

Examples of the uses and limitations of some commonly used chemical conventions and representations are shown in Table 1.7.6.

TABLE 1.7.6 Some uses and limitations of common chemical conventions and representations

Representation	Use	Limitations	Example
empirical formula	This is determined by experiment and shows the simplest ratio of atoms of each element in a compound.	It does not necessarily show the actual number of atoms in a molecule.	C_2H_5, H_2O, NaCl
molecular formula	It shows the number of atoms of each element in a molecule of a compound.	It does not show the arrangement of atoms in a molecule.	C_4H_{10}, H_2O
structural formula	It shows the relative location of atoms within a molecule in two dimensions.	It does not show the arrangement of atoms in three dimensions.	tetracycline structural formula
condensed structural (semi-structural) formula	It enables a structural formula to be drawn.	It does not show the arrangement of atoms in three dimensions.	$CH_3CH_2CH_2CH_3$ or $CH_3(CH_2)_2CH_3$

Physical quantities can be represented by symbols (most of which are shown in italics). Most physical quantities have units, and each of these also has a symbol. Table 1.7.7 presents some quantities and units used in chemistry.

TABLE 1.7.7 Some quantities and units used in chemistry

Name of quantity	Symbol for quantity	Unit name and symbol
mass	m	gram (g)
volume	V	litre (L)
amount of substance	n	mole (mol)
molar mass	M	grams per mole ($g\,mol^{-1}$)
relative molecular or formula mass	M_r	no units; relative to one atom of ^{12}C exactly
relative atomic mass	A_r	no units, relative to one atom of ^{12}C exactly
relative isotopic mass	I_r	no units, relative to one atom of ^{12}C exactly
density	d	grams per millilitre or kilograms per litre ($g\,mL^{-1}$ or $kg\,L^{-1}$)
molarity	c	moles per litre ($mol\,L^{-1}$)

Correct use of unit symbols

The correct use of unit symbols removes ambiguity, as symbols are recognised internationally. Unit symbols should not be followed by a full stop unless they are at the end of a sentence.

Uppercase letters are not used for the names of any physical quantities or units. Uppercase letters are only used for the symbols of units named after people. For example, the name for the unit of energy is joule and the symbol is J.

The joule was named after James Joule, who is best known for his studies on energy conversions. The exception to this rule is 'L' for litre. This is done because a lowercase 'l' looks like the numeral '1'.

The product of a number of units is shown by separating the symbol for each unit with a centered dot or a space (Table 1.7.8). Most teachers prefer a space, but a dot is also correct. The division or ratio of two or more units can be shown in fraction form, using a slash, or using negative indices. Most teachers prefer negative indices. Prefixes should not be separated by a space.

TABLE 1.7.8 Examples of the use of symbols for derived units

Preferred notation	Incorrect notation
J g^{-1} K^{-1}	$Jg^{-1}K^{-1}$
g mol^{-1}	$gmol^{-1}$
kPa ('kilo' is a prefix)	k Pa

Most unit names can take the plural form by adding an 's' when used with numbers greater than 1. Never do this with the unit symbols. It is acceptable to say 'two kilojoules' but it is wrong to write 2 kJs (the correct way to write this is 2 kJ).

Numbers and symbols should not be mixed with words for units and numbers. For example, thirty grams and 30 g are correct, but 30 grams and thirty g are incorrect.

Scientific notation

GO TO ➤ SkillBuilder page 80

To overcome confusion or ambiguity, measurements are often written using scientific notation. Quantities are written as a number between 1 and 10 and then multiplied by an appropriate power of 10. 'Scientific notation', 'standard notation' and 'standard form' all have the same meaning.

Examples of some values written in scientific notation are:

6.022×10^{23} particles

$25.25\,\text{mL} = 2.525 \times 10^{-2}\,\text{L}$

$0.00302\,\text{mol} = 3.02 \times 10^{-3}\,\text{mol}$

You should be routinely using scientific notation to express numbers. This also involves learning to use your calculator intelligently. Scientific and graphics calculators (Figure 1.7.3) can be put into a mode whereby all numbers are displayed in scientific notation. It is useful when doing calculations to use this mode, rather than frequently attempting to convert to scientific notation by counting digits on the calculator display.

FIGURE 1.7.3 A scientific calculator.

An important reason for using scientific notation is that it removes ambiguity about the precision of measurements. For example, a measurement recorded as 240 g could be a measurement to the nearest 10 g (that is, it is somewhere between 235 g and 255 g), or it could be to the nearest gram (between 239.5 g and 240.5 g).

PREFIXES AND CONVERSION FACTORS

You should be familiar with the prefixes and conversion factors in Table 1.7.9. Conversion factors should be used carefully. A common question when converting between units is whether to multiply or divide with the conversion factor. To convert small units to large units, you would divide by the conversion factor. Conversely, to convert large units to small units, you would multiply by the conversion factor.

It is important to give a symbol the correct case (upper or lower case). There is a big difference between 1 mL and 1 ML.

TABLE 1.7.9 Prefixes and conversion factors

Conversion factor	Index form	Prefix	Symbol
1 000 000 000 000	10^{12}	tera	T
1 000 000 000	10^{9}	giga	G
1 000 000	10^{6}	mega	M
1000	10^{3}	kilo	k
0.01	10^{-2}	centi	c
0.001	10^{-3}	milli	m
0.000 001	10^{-6}	micro	μ
0.000 000 001	10^{-9}	nano	n
0.000 000 000 001	10^{-12}	pico	p

1.7 Review

SUMMARY

- A scientific report must include an introduction, body paragraphs and a conclusion.
- The conclusion should include a summary of the main findings, a concluding statement related to the issue being investigated, limitations of the research, implications and applications of the research, and potential future research.
- Scientific writing uses unbiased, objective, accurate, formal language. Scientific writing should be concise and qualified.
- Visual support can assist in conveying scientific concepts and processes efficiently.
- Reports should be edited before finalising.
- Scientific notation needs to be used when communicating results.

KEY QUESTIONS

1 Which of the following statements is written in scientific style?

A The results were fantastic ...
B The data in Table 2 indicates ...
C The researchers felt ...
D The smell was awful ...

2 Which of the following statements is written in first-person narrative?

A The researchers reported ...
B Samples were analysed using ...
C The experiment was repeated three times ...
D I reported ...

3 The variables molar mass, specific heat capacity and molarity each have different units. Write the units for each of the following in correct scientific notation.

a molar mass; grams per mole
b specific heat capacity; joules per gram per kelvin
c molarity; mole per litre

4 Describe how to convert grams into kilograms.

5 Discuss why you might need to convert between different units; for example litres to millilitres.

Chapter review

01

KEY TERMS

accurate
affiliation
bias
chemical hazard code
controlled variable
credible
dependent variable
expertise
independent variable
mean
median
mistake
mode
nomenclature
outlier
percentage error
personal protective equipment (PPE)
persuasion
phenomenon
precise
primary source
qualitative
quantitative
random error
raw data
reliability
reputation
rhetoric
safety data sheet (SDS)
secondary source
significant figures
stakeholder
systematic error
trend
trend line
uncertainty
validity
variable

REVIEW QUESTIONS

1 Consider the following research question:
'Is the concentration of lead in water sampled from Sydney Harbour within acceptable limits?'
From the following options, identify the independent, dependent and controlled variables:
   a concentration of lead
   b analytical technique, temperature of water sample, type of sampling container
   c source and location of water.

2 Match the following command verbs with their definitions:
describe, analyse, apply, create, identify, reflect on, investigate
   a Think deeply about.
   b Produce or make new.
   c Identify connections and relationships; interpret to reach a conclusion.
   d Observe, study, examine, inquire systematically in order to establish facts or derive conclusions.
   e Use knowledge and understanding in a new situation.
   f Recognise or indicate what or who.
   g Give a detailed account.

3 Consider the following hypothesis:
'The phosphate concentration of laundry wastewater will be greater than that of drinking water.'
Name the independent, dependent and controlled variables for an experiment with this hypothesis.

4 Which of these graph types would be best to use for each of the following data types?
pie diagram, scatter graph (with line of best fit), bar graph, line graph
   a levels of a pesticide detected in drinking water at various locations
   b temperature of water sampled at the same time of day over a period of one month
   c calibration curve showing absorbance of standard solutions of phosphate measured using UV–visible spectroscopy
   d proportion of specific contaminants detected in water

5 What are the meanings of these chemical codes?
   a

   b

   c

6 Explain the terms 'accuracy' and 'validity.'

7 The values for some mass measurements (in g) obtained in an experiment were 7.00, 6.50, 6.08, 7.20, 6.50 and 6.50.
What is the uncertainty for the average of these values?

8 Which of the statistical measurements of mean, mode and median is most affected by an outlier?

9 Identify whether the following are mistakes, systematic errors or random errors.
   a A student spills some solution during a titration.
   b Reported measurements are greater than and less than the true value.
   c A weighing balance has not been calibrated.

10 What relationship between variables is indicated by a curved trend line?

11 A scientist designed and completed an experiment to test the following hypothesis:
'If the electrical conductivity of water increases with increasing temperature, and the temperature of water in various containers is quantified, then the water with the highest temperature will have the highest conductivity of electricity.'
The discussion section of the scientist's report included comments on the reliability, validity, accuracy and precision of the investigation.
Determine which of these sentences comment on reliability, validity, accuracy or precision.
   a Three water samples from the same source were examined at each temperature. Each water sample was analysed and the measurements were recorded.
   b The temperature and the electrical conductivity of the water samples were recorded using data-logging equipment. The temperature of some of the water samples was measured using a glass thermometer.
   c The data logging equipment was calibrated for electrical conductivity against a known standard. The equipment was calibrated before measurements were taken.
   d The temperature probe (data-logger) measured temperature to the nearest 0.1°C. The glass thermometer measured temperature to the nearest 0.5°C.

12 What factors can you look at to ensure you discuss the limitations of a method?

13 Explain the meaning of the term 'trend' in a scientific investigation, and describe the types of trend that might exist.

14 What is meant by the 'limitations' of an investigation procedure?

15 What is 'bias' in an investigation?

16 What is the purpose of referencing and acknowledging documents, ideas and quotations in an investigation?

17 Which of these lists consists only of secondary sources of information?
   A interactive periodic table, article published in a science magazine, science documentary, practical report written by a Year 11 student
   B article published in a peer-reviewed science journal, article published in a science magazine, science documentary
   C interactive periodic table, article published in a science magazine, science documentary, this Year 11 textbook
   D science article published in a newspaper, article published in a science magazine, science documentary, practical report written by a Year 11 student.

18 What is the correct way to cite in text the following source in APA style?
Selke, S., Auras, R., Nguyen, T.A., Aguirre, E.C., Cheruvathur, R., & Liu, Y. (2015). Evaluation of biodegradation-promoting additives for plastics. *Environmental Science & Technology, 49*(6), 3796–3777.
   A However, Selke et al. (2015) did not find any significant difference in biodegradability.
   B However, Selke et al. did not find any significant difference in biodegradability[1].
   C However, Selke et al. did not find any significant difference in biodegradability (Selke, S., Auras, R., Nguyen, T.A., Aguirre, E.C., Cheruvathur, R., & Liu, Y. (2015). Evaluation of biodegradation-promoting additives for plastics. *Environmental Science & Technology, 49*(6), 3769–3777).
   D However, Selke et al. (2015) did not find any significant difference in biodegradability (Evaluation of biodegradation-promoting additives for plastics. *Environmental Science & Technology*).

19 Convert 30.00 mL so that it is expressed using the unit of litres.

20 Please refer to the experiment described in Question **11**.
   a Write a possible purpose for the scientist's experiment.
   b What would be the independent, dependent and controlled variables in this investigation?
   c What kind of data would be collected? Would it be qualitative or quantitative?
   d Explain the difference between raw data and processed data, using this as an example. What would you expect the graph of the results to look like if the scientist's hypothesis was correct?

APPROX

MODULE 1 Properties and structure of matter

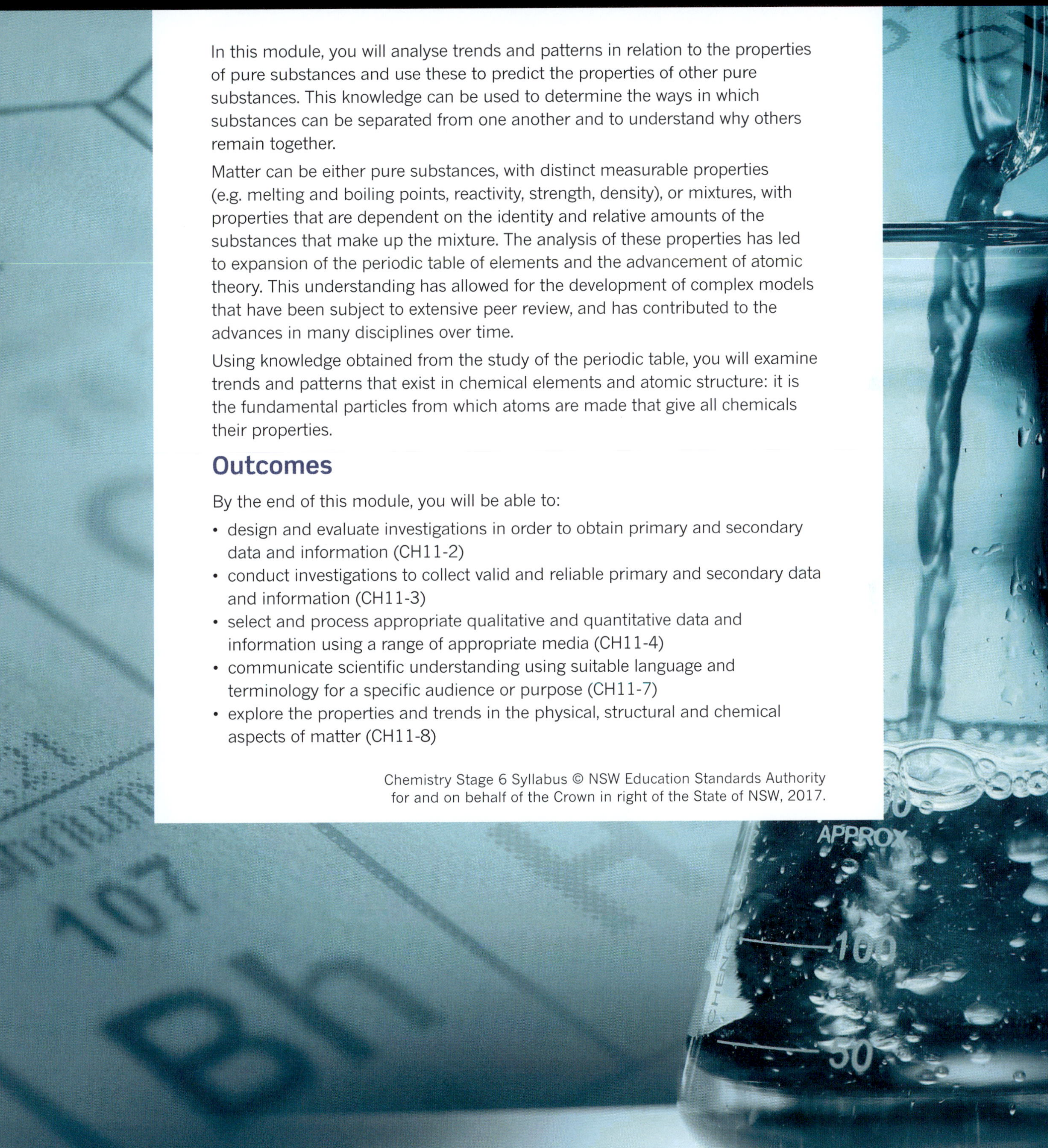

In this module, you will analyse trends and patterns in relation to the properties of pure substances and use these to predict the properties of other pure substances. This knowledge can be used to determine the ways in which substances can be separated from one another and to understand why others remain together.

Matter can be either pure substances, with distinct measurable properties (e.g. melting and boiling points, reactivity, strength, density), or mixtures, with properties that are dependent on the identity and relative amounts of the substances that make up the mixture. The analysis of these properties has led to expansion of the periodic table of elements and the advancement of atomic theory. This understanding has allowed for the development of complex models that have been subject to extensive peer review, and has contributed to the advances in many disciplines over time.

Using knowledge obtained from the study of the periodic table, you will examine trends and patterns that exist in chemical elements and atomic structure: it is the fundamental particles from which atoms are made that give all chemicals their properties.

Outcomes

By the end of this module, you will be able to:

- design and evaluate investigations in order to obtain primary and secondary data and information (CH11-2)
- conduct investigations to collect valid and reliable primary and secondary data and information (CH11-3)
- select and process appropriate qualitative and quantitative data and information using a range of appropriate media (CH11-4)
- communicate scientific understanding using suitable language and terminology for a specific audience or purpose (CH11-7)
- explore the properties and trends in the physical, structural and chemical aspects of matter (CH11-8)

CHAPTER 02

Properties of matter

In this chapter, you will develop an understanding of the concept of matter as being composed of elements, compounds and mixtures. You will learn that matter has various states—solid, liquid or gas—and how energy affects matter and can lead to transitions between these states. The distinction between compounds combined chemically and mixtures combined physically will be clarified by showing how mixtures can be separated, and how separation makes it possible to analyse the components of matter. The chapter concludes with an overview of elements as the building blocks of chemistry, and of how they can be grouped by physical and chemical characteristics into the periodic table.

Content

INQUIRY QUESTION

How do the properties of substances help us to classify and separate them?

By the end of this chapter, you will be able to:

- explore homogeneous mixtures and heterogeneous mixtures through practical investigations:
  - using separation techniques based on physical properties (ACSCH026)
  - calculating percentage composition by weight of component elements and/or compounds (ACSCH007)
- investigate the nomenclature of inorganic substances using International Union of Pure and Applied Chemistry (IUPAC) naming conventions
- classify the elements based on their properties and position in the periodic table through their:
  - physical properties
  - chemical properties ICT

2.1 Types of matter

CHEMISTRY INQUIRY CCT

Comparing the states and properties of matter

COLLECT THIS ...

- candle
- matches
- balloon
- 4 tsp baking power
- empty soft drink can
- 100 mL vegetable oil in a glass
- 100 mL water in a separate glass
- food dye
- 1 large crystal of rock salt
- pliers

DO THIS ...

1 Draw up the table shown below and predict what will happen after completing the listed tasks.
2 Light the candle with a match.
3 Place the baking powder inside the balloon. Add a few drops of water and quickly tie the balloon closed.
4 Pour the water onto the vegetable oil, and add a few drops of food dye.
5 Use the pliers to try to bend the crystal of rock salt and the soft drink can.

RECORD THIS ...

Copy the table and record your predictions.

Description of experiment	Prediction	Observations
lighting a candle		
adding baking powder and water to a balloon and sealing it		
pouring water onto vegetable oil, then adding drops of food dye		
using pliers to bend rock salt using pliers to bend a soft drink can		

Carry out the tasks.

Describe what happened by recording your observations.

Present a table of your results.

REFLECT ON THIS ...

Explain the meaning of your observations.

What could you do next time to improve your experiment?

The development of advanced materials such as nanoparticles is the result of centuries of scientific discovery and research. Over time, scientists have gained a deep understanding of the structure of atoms, which are the basic building blocks of **matter**. Matter is anything that has mass and occupies space. As scientists' understanding has increased, their ability to control matter on the atomic scale has grown. An unaided human eye can only decipher objects that are larger than 0.04 mm in width, and matter at the atomic level is impossible to see. Therefore, much of what scientists know about atoms has come from theoretical **models** and indirect observations.

A scientific model is a description that scientists use to represent the important features of what they are trying to describe. They are able to test the consistency of their observations against various predictions of the model.

ATOMIC THEORY

In 1802, the English scientist John Dalton (Figure 2.1.1) presented the first **atomic theory of matter**. Dalton proposed that all matter is made up of tiny spherical particles that are indivisible and indestructible.

Dalton accurately described **elements** as materials containing just one type of atom and **compounds** as materials containing different types of atoms in fixed ratios. Elements cannot be broken down into simpler substances.

Subsequent experiments have led to the rejection of some of Dalton's ideas, but aspects of his theory are still accepted. Scientists now know that atoms are not indivisible or indestructible. Atoms are made up of even smaller **subatomic particles**.

FIGURE 2.1.1 John Dalton (1766–1844) proposed that matter is composed of atoms.

CHEMFILE ICT

Viewing atoms

Dalton's atomic theory of matter assumed that atoms are spherical. Atoms cannot be seen with conventional microscopes; therefore, in Dalton's time there was no way to confirm the shape of atoms. It wasn't until 1981 that a microscope capable of viewing atoms was developed by IBM researchers Gerd Binnig and Heinrich Rohrer. This type of microscope is known as a scanning tunnelling microscope (STM). Using STMs, scientists were able to confirm that atoms are indeed spherical.

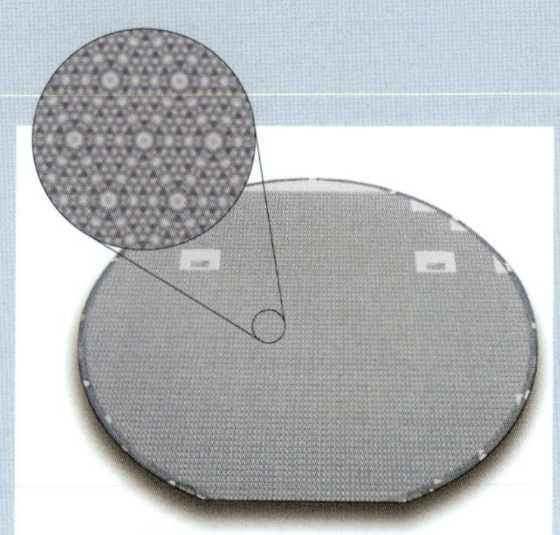

FIGURE 2.1.2 This image of the silicon atoms in a silicon wafer was taken by a scanning tunnelling microscope.

STMs use an extremely sharp metal tip to detect atoms. The tip scans, line by line, across the surface of a crystal. As the tip moves, it measures tiny height differences in the crystal's surface due to the individual atoms. This is similar to the way people with little or no vision use their fingers to sense braille on a page. The data from the tip is then sent to a computer, which constructs an image of the atoms. An STM image of silicon atoms on the surface of a silicon wafer is shown in Figure 2.1.2.

ELEMENTS

As Dalton predicted, elements are made of just one type of atom. Most non-metallic elements form **discrete** (individual) **molecules**, each with a definite number of atoms. For example, sulfur forms a discrete molecule of eight sulfur atoms. Some non-metals form **covalent networks** or **giant molecules**. Carbon is an example of such a non-metallic element. Diamond and graphite are both examples of covalent networks formed of sheets of carbon. Graphene is a giant molecule formed of sheets of carbon. (You will learn more about covalent networks of carbon in Chapter 5.) A comparison of the numbers of atoms in two different molecules is shown in Figure 2.1.3.

The metallic elements form a different type of network, which you will look at in detail in Chapter 5.

Several non-metallic elements are **monatomic**, which means they exist as individual atoms. On Earth, the monatomic elements are helium, neon, argon, krypton, xenon and radon; they are also known as the **noble gases** because they are generally chemically inert (unreactive).

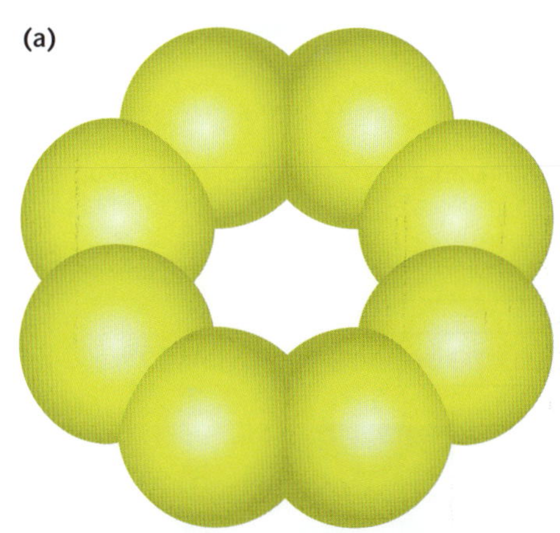

FIGURE 2.1.3 Most non-metal elements, such as (a) sulfur, form discrete molecules. (b) Other non-metals, such as carbon, form bonded networks or giant molecules, such as the example here showing the molecular structure of graphene.

Monatomic elements are those made up of only one atom. The prefix mon- (or mono-) is frequently used in science. It means only one, or single.

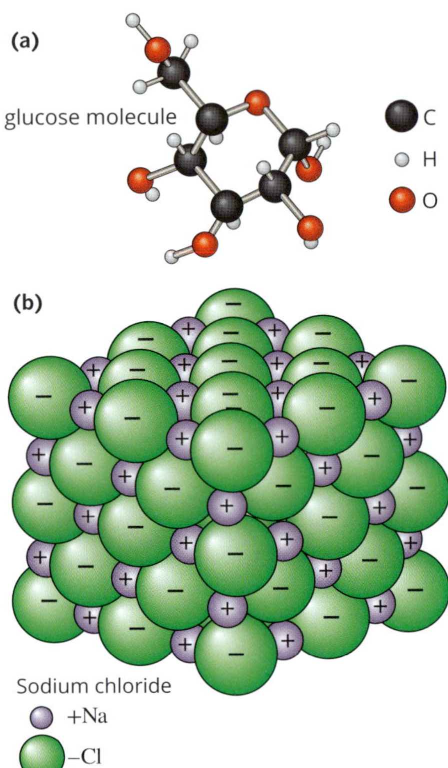

FIGURE 2.1.4 Compounds can be made up of (a) molecules, e.g. glucose, which contains carbon, oxygen and hydrogen atoms, or (b) networks, e.g. sodium chloride (table salt).

GO TO ➤ Section 2.3 page 59

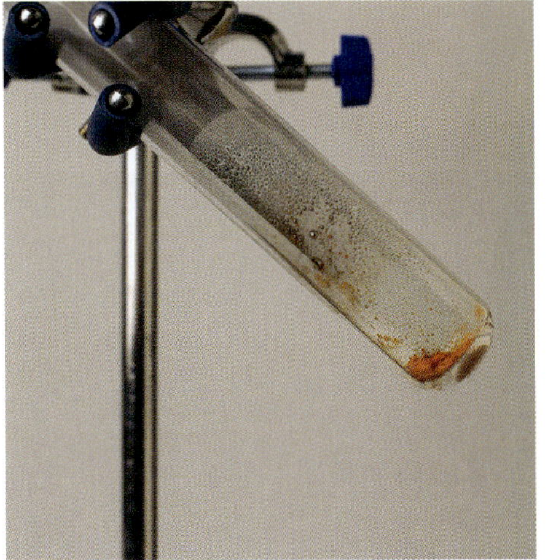

FIGURE 2.1.5 The red powder in this test-tube is mercuric oxide. If you look closely at the test-tube, you will see beads of liquid mercury forming from the decomposition of the compound.

COMPOUNDS

Dalton also correctly predicted that different types of atom could combine to form new substances. These substances are known as compounds. The atoms in compounds can also form **molecules**, or a network of atoms, as shown in Figure 2.1.4.

To determine whether a pure substance is an element or a compound, you must determine whether the substance can be broken down into simpler substances. For example, when heated, mercuric oxide (HgO) decomposes to liquid mercury (Hg) and oxygen gas (O_2). If it were not a compound, the mercuric oxide would not break down. You can see the compound mercuric oxide and the element mercury in Figure 2.1.5. As oxygen is a colourless gas, you cannot see it. Oxygen can be detected by placing a glowing splint at the top of the test-tube. When oxygen is present, the splint reignites.

Compounds contain different types of atom chemically bonded together in definite proportions.

HOMOGENEOUS AND HETEROGENEOUS MIXTURES

Matter can be classified as a pure substance (element or compound) or a **mixture**. Mixtures are combinations of either two or more elements, two or more compounds or two or more elements and compounds. A mixture has no definite composition; the proportions of each substance can vary.

The important difference between a compound and a mixture is that a compound is made up of two or more elements chemically bonded together, whereas a mixture has two or more pure substances physically combined together. It is more difficult for chemists to separate a compound into its elements, because a process such as heating is required. Chemists can separate a mixture more easily, such as by filtering. You will explore how mixtures are separated in Section 2.3.

The physical properties of compounds do not vary, but those of mixtures do, meaning the properties of mixtures can be changed. The freezing points of the pure compounds methanol and water are –98°C and 0°C, respectively. A 20% methanol, mixture freezes at –11°C; a 50% methanol mixture freezes at –40°C; yet an 80% methanol mixture freezes at –102°C—lower even than the freezing point of pure methanol (Figure 2.1.6).

FIGURE 2.1.6 Two beakers of different methanol–water mixtures cooled to –20°C. (a) This mixture freezes, as shown by the crystals at the bottom of the beaker. (b) This mixture does not freeze.

Mixtures can be classified as homogeneous or heterogeneous.

- A **homogeneous** mixture is uniform in composition. Any sample of the mixture will be just like any other sample. For example, a teaspoon of salt stirred into a litre of water forms a homogeneous mixture because the salt becomes evenly distributed throughout the water.

- A **heterogeneous** mixture has a composition that varies within the mixture. For example, if you place a few pebbles in a jar of sand and give the jar a few shakes, the heavier pebbles will tend to fall to the bottom of the jar. A sample of the mixture taken from the top of the jar will be different from a sample taken from the bottom.

The most well known homogeneous mixture is a solution (Figure 2.1.7). A **solution** has one or more solutes that are **dissolved** in a solvent. A **solvent** is a substance in which a solute dissolves, and a **solute** is a substance that dissolves in a solvent. The solvent is the substance present in the greater amount. Solvents can be gases, liquids or solids, but in all cases the solute is distributed evenly through the solvent. A solution can be dilute or concentrated, depending on the amount of solute in the solvent. Many familiar mixtures have a liquid solvent, which is usually water.

In a homogeneous solution, each part of the solution is the same as any other part.

Water as a solvent

Water is an excellent solvent. This is one of its most important properties. Almost all biological processes and many industrial processes occur in water. These systems are known as **aqueous** environments.

When substances are dissolved in water, the particles are free to move throughout the solution. When two aqueous reactants are combined in a reaction vessel, the dissolved reactant particles mix freely. This increases the chances of the reactants coming into contact. Because of the increased movement of the reactant particles, interactions between reactants are generally much more effective than if the same reactants were mixed as solids.

In Figure 2.1.8, you can see how the deep purple colour of potassium permanganate spreads through the water as the solid dissolves and the particles move and mix. Eventually, the liquid will appear completely purple.

FIGURE 2.1.8 When solid potassium permanganate is added to water, it dissolves. The particles disperse into the solution and move around freely.

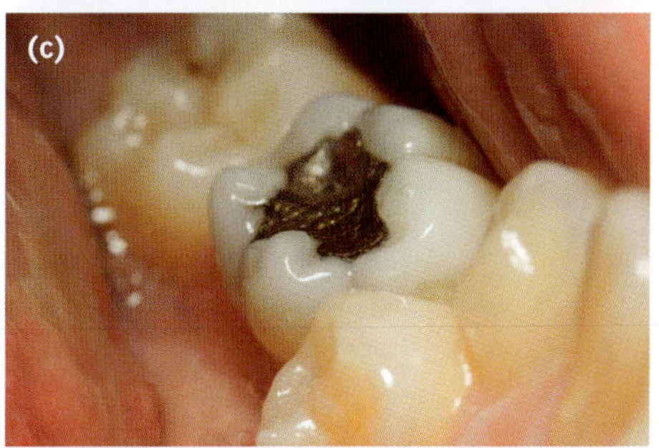

FIGURE 2.1.7 Solutions come in many forms. (a) Oxyacetylene welding uses a gas-in-gas solution. (b) Hydrogen fuel can be held as a gas-in-solid solution. (c) A dental filling is a liquid-in-solid solution.

Aqueous solutions

An aqueous solution is formed when a solid, liquid or gas is dissolved in water. The solute of an aqueous solution may vary, but the solvent will always be water. You can have an aqueous solution of salt (saline) or an aqueous solution of sugar. Table 2.1.1 lists some common aqueous solutions.

TABLE 2.1.1 Some everyday aqueous solutions

Solution	Solute(s)	Solvent
saline	sodium chloride	water
soft drink	carbon dioxide, sugar, flavouring agent, colouring agent	water
coffee	coffee, sugar	water

Suspensions and colloids

When a substance is mixed with a solvent, it does not always dissolve to form a solution. It could form a suspension or a colloid.

A suspension is a heterogeneous mixture, with visible distinguishable parts, that forms when a substance does not dissolve significantly. Some particles will settle out over time, and in many cases you can separate them from the solvent by using filter paper. Chalk in water and red blood cells in plasma are suspensions. Figure 2.1.9 shows blood before and after it is separated into its component parts.

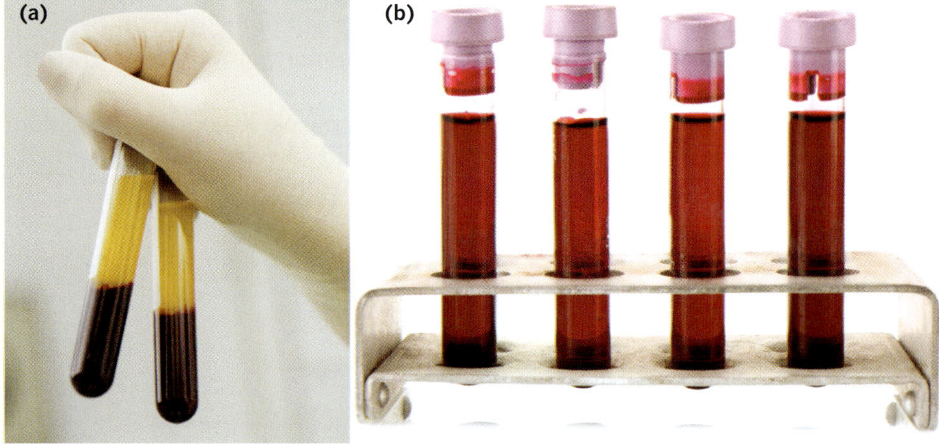

FIGURE 2.1.9 (a) At first glance, blood looks like a homogeneous mixture. (b) Centrifuging blood separates it into clear layers, demonstrating that blood is a heterogeneous suspension. Here you can see the plasma on the top and the red blood cells on the bottom.

A colloid is a mixture of particles that consists of smaller clusters of ions or molecules. These are evenly dispersed throughout the solvent and do not settle on standing. For example, milk is a colloid of a fat in water (Figure 2.1.10). Milk solids and fat are finely dispersed throughout the more aqueous components of the milk. Other common colloids include mayonnaise, paint and ink.

FIGURE 2.1.10 Milk is a colloid. Small particles are dispersed throughout the liquid.

2.1 Review

SUMMARY

- All substances are made up of atoms.
- The atoms in elements can exist as individual atoms (monatomic), discrete molecules, giant molecules or networks.
- The atoms in compounds can exist as molecules or large networks of atoms.
- Mixtures are combinations of different elements or compounds combined physically, not chemically.
- Mixtures may be classified as homogeneous (evenly mixed) or heterogeneous (unevenly mixed).
- Solutions are a type of homogenous mixture. If the main component (solvent) is water, the mixture is an aqueous solution.

KEY QUESTIONS

1 Explain the difference between an element and a compound.

2 Classify each of the following as elements or compounds.
 a copper
 b sulfur
 c water
 d carbon dioxide
 e diamond
 f sodium chloride
 g gold
 h silicon carbide.

3 Consider a soft drink. Identify each of the following as solute, solvent or solution.
 a water
 b soft drink
 c flavouring
 d sugar

4 What is the one thing that all aqueous solutions have in common?

5 **a** You are preparing a chocolate cake and spoon out a sample from the middle of the mixing bowl. The batter tastes as you expect. Another sample from the left of the bowl is sweeter, and a third sample from the right of the bowl has a lump of cocoa. How would you classify this mixture?
 b You continue making your cake and mix the batter thoroughly. What would you expect if you sample the batter now? How would you describe this mixture?

6 Water is made from hydrogen and oxygen. If hydrogen gas was mixed with oxygen gas, would water form?

7 Analysis of a material shows it is 15% iron, 40% oxygen, 25% carbon, 15% hydrogen and 5% silicon. Four samples are taken and all four give the same results. Explain whether it is a compound or a mixture.

2.2 Physical properties and changes of state

Of all the properties of matter, the physical state is the easiest to observe. The three most common physical states of matter are solid, liquid and gas.

Within any state, physical changes can take place. A physical change may alter the physical properties of a substance, but not the chemical nature of the substance. For example, crushing an aluminum can change the shape of the material, but the nature of the material and its physical and chemical properties do not change. Physical changes are also reversible. A metal rod can be beaten back into shape, and warm air can cool again, without changing the nature of the material.

Mixing is another type of physical change. While it may be difficult to 'un-mix' the components, each component retains its chemical properties, and no new materials form during the mixing.

FIGURE 2.2.1 Heat makes metals expand. Engineers need to add expansion joints into their designs to prevent railway tracks and bridges from buckling in hot conditions.

Adding heat to a substance leads to a physical change. The substance usually expands in volume to take up more space. Figure 2.2.1 shows the effect of heat on railway tracks, which can make them expand and buckle. The same mass of a gas put into a container with a greater volume will have a lower density. Removing heat from a substance usually leads to the material reducing its volume (contracting) and so increasing its density.

If enough heat is applied, a substance may undergo a change of state, also known as a phase transition. A solid melts to a liquid and a liquid vaporises to a gas (Figure 2.2.2). This is a physical change, and is reversible. If heat is removed, a gas condenses to a liquid, and a liquid freezes to a solid. Depending on the material and conditions, it is possible to go directly from solid to gas in a reversible process known as **sublimation**. For example, a solid such as carbon dioxide (dry ice) **sublimes** to a gas, and the gas deposits to a solid.

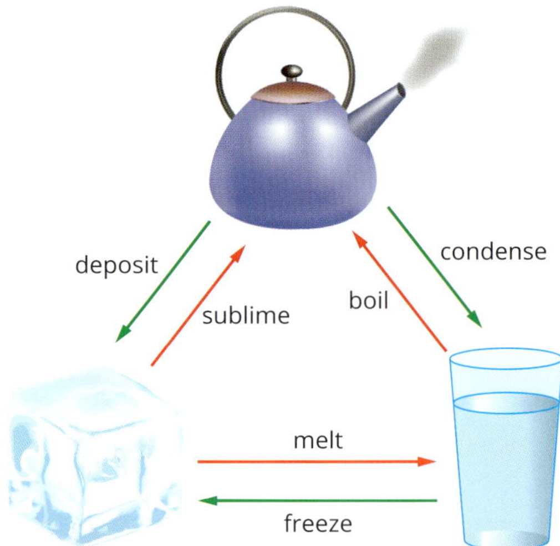

FIGURE 2.2.2 Heating and cooling leads to a change of state. On heating, a solid melts into a liquid and a liquid boils to form a gas. On cooling, a gas condenses to a liquid and a liquid freezes into a solid.

Within a given state of matter, physical changes happen gradually. A thermometer works on this principle. The liquid within, usually an alcohol with an added dye, expands along a marked tube to a known volume at a given temperature, and the marks indicate the temperature. A phase transition is different, with some physical properties changing drastically. Liquefied petroleum gas (LPG), a commonly available fuel, will expand around 200 times in volume when it becomes a gas.

SOLIDS, LIQUIDS AND GASES

The particle model is a simple but effective model for understanding the differences between solids, liquids and gases. According to this model, matter can be considered as made up of hard units ('particles') that cannot be made smaller by either compression or further division. The particles are in constant motion and are attracted to one another, some strongly and some weakly. The particles are not necessarily atoms. The three states of matter are explained by the balance between the particles' motion and the attraction between the particles. Table 2.2.1 summarises how the particle interactions give rise to the properties of the different states of matter.

TABLE 2.2.1 Properties of solids, liquids and gases

Property	Solids	Liquids	Gases
particle packing	Particles are close together.	Particles are loosely packed.	Particles are far apart.
density	high	fairly high	low
compressible	No—there is no space to push particles closer together.	No—there is no space to push particles closer together.	Yes—there is space to push particles closer together.
shape	Fixed—particles are held together.	Not fixed—particles can slide over each other.	Not fixed—particles move in all directions.
mixing and diffusion	No—particles remain in a fixed position.	Yes—particles can change position.	Yes—particles move in all directions.
pressure on container	no	some	Yes—particles impact container.
diagram of state			
description	Solids are incompressible and hold their shape.	Liquids are incompressible and flow to adopt the shape of the container.	Gases are compressible and take the shape of the container.

In solids, the attraction between the particles is strong and this keeps the particles tightly packed together. The particles vibrate in a fixed position, and vibrate more vigorously as the temperature increases, but they do not shift their position. This model explains the properties of solids—they have a fixed shape, are difficult to break apart, and cannot be compressed.

In liquids, the attraction between the particles is weaker. The particles remain packed closely, but may shift their positions and move past one another. This model explains the properties of liquids—they are able to flow, have no fixed shape, are easy to break apart, but cannot be compressed.

In gases, the attraction between particles is weak and the particles are not packed closely. This model explains the properties of gases—they have no fixed shape, occupy the volume of any container, and can be compressed.

Both liquids and gases have no fixed shape, and are called **fluids**.

+ ADDITIONAL

Physical transitions and energy

Changes of state require the addition or removal of energy.

For a gram of liquid water, 4.2 J of heat energy has to be absorbed to increase the temperature of the water by 1°C. It takes 336 J of energy to heat 1 g of water from 20°C to 100°C. At 100°C, water will then boil to steam, but the change of state needs 2200 J. While water is transitioning from a boiling liquid to a gas (steam), the temperature of the water will not rise—the added energy is all used for the physical transformation.

Energy is required to transform solids to liquids. For water, 334 J of energy transforms 1 g of ice to 1 g of water. During the transition, the added heat does not increase the temperature of the ice.

The heat required to change a solid into a liquid or liquid into vapour without changing the temperature of the material is known as **latent heat**. For the solid–liquid and liquid–gas transitions, it is known as the **latent heat of fusion** and the **latent heat of vaporisation**, respectively. You will examine this topic in more detail in Chapter 15.

Latent heat is one reason why steam burns are so dangerous. When steam condenses back to water on your skin, all of the energy used to transform water to steam is released. A scald from water at 99°C is painful and will burn. A scald from steam at 101°C is much worse because the energy of the latent heat of vaporisation is also released.

Temperature affects how much the particles in a substance move, both in vibrations and in particle motion. An increase in temperature means an increase in the movement of the particles. An effect of increasing temperature is that the distance between particles increases.

For solids, the greater movement of the particles means that the particles remain in contact, but the distance between the particles increases. This explains why a solid material expands as it is heated and contracts as it cools.

For liquids, the greater movement of the particles means they flow past each other more effectively. This explains why a liquid expands, generally to a greater extent than for solids, and why the liquid becomes runnier (less viscous).

For gases, the greater movement of the particles means the particles themselves move faster, and the particles impact the sides of their container with more force and more frequently. This models the increase in pressure on the container walls; if the container is flexible, like a balloon, the container expands. You will explore the nature and properties of gases in more detail in Chapter 9.

CHANGES OF STATE

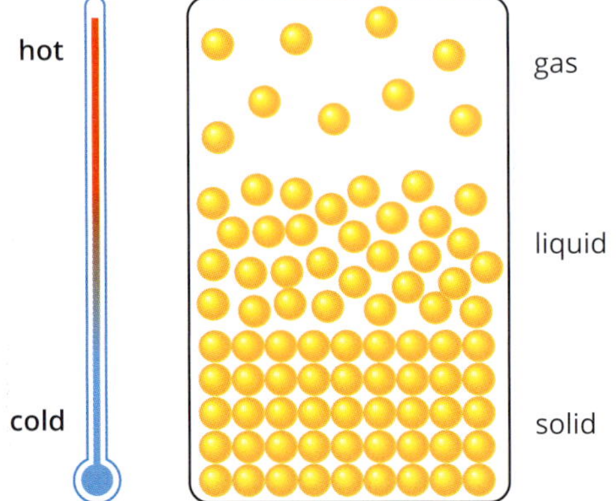

FIGURE 2.2.3 On heating, a solid becomes a liquid and a liquid becomes a gas. According to the particle model, adding heat enables particles to move out of a fixed position (when a substance melts) and are able to overcome attractive forces (when a substance boils).

The particle model explains changes of state—the vibrations and motion of the particles oppose the force of attraction between particles.

When heated, a solid expands because the particles move more, causing them to move apart (Figure 2.2.3). Eventually, the movements become so energetic that the forces of attraction can no longer hold the particles in position. At this point, the solid melts to a liquid. When a liquid is cooled, the motion of the particles can no longer overcome the forces of attraction between them, and the material freezes back to a solid.

When a liquid is heated, some particles on the surface of the liquid move so energetically that they have enough energy to overcome the forces of attraction (Figure 2.2.3). The liquid particles escape and form a gas. When a gas is cooled, the motion of the particles can no longer overcome the forces of attraction between particles. The gas particles stick together and the material condenses back to a liquid.

Sublimation occurs when the state of a substance changes directly from a solid to a gas without becoming a liquid. This occurs when heating a substance in the solid phase causes all forces of attraction between the particles to be overcome. When pressure is low enough, such as in a vacuum, some common materials that normally melt to a liquid instead sublime into a gas.

The particle model explains some properties of the states of matter, but not others. The model does not include a measure of the forces between the particles, so it cannot be used to predict the temperature at which a material may change state.

+ ADDITIONAL

Glass, ferrofluids and liquid crystals

The particle model explains the difference between solids and liquids in terms of whether the particles are fixed in place or are able to move past one another. Heating can overcome interparticle forces, and for certain sensitive materials the interparticle effects can be reduced so that the materials can change between a liquid and a solid without a change in temperature.

Solids such as crystals and metals have particles that are ordered and arranged in a fixed pattern. Glass has particles that are fixed in place like a solid, but there is no long-range order (Figure 2.2.4a). Heating glass will transform it from being brittle to rubbery before it eventually melts.

Magnetic materials are usually solids. However, a ferrofluid is a liquid mixture that responds to magnetic fields (Figure 2.2.4b). When placed close to a magnet, the ferrofluid behaves like a solid, and when placed away from a magnet, the ferrofluid behaves like a liquid. Ferrofluids are used in computer hard disks and loudspeakers.

Liquid crystals (Figure 2.2.4c) are usually liquids, but in the presence of an electrical field they become ordered like a solid.

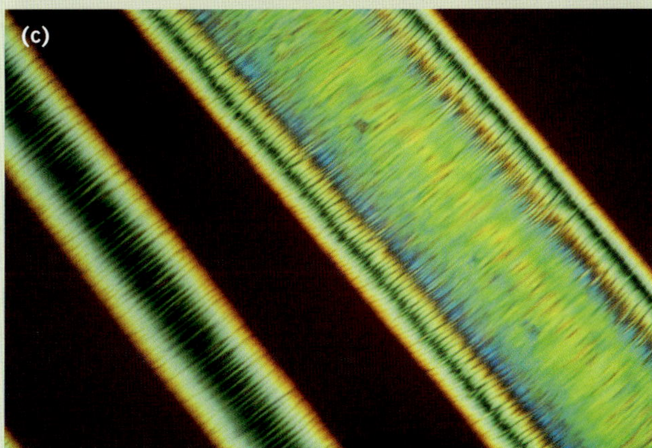

FIGURE 2.2.4 According to the particle model, in solids the particles remain fixed in place, and in liquids the particles move past one another. (a) Glass has particles without long-range ordering, like a liquid, but which are strongly attracted to one another, as in a solid. (b) Ferrofluids and (c) liquid crystals are materials in which the interparticle forces can be changed. These substances can be transformed between solids and liquids without heating.

2.2 Review

SUMMARY

- Matter exists in three main states: solid, liquid and gas.
- The particle model can explain some of the differences between the states.
- The particle model considers matter to be made up of small, hard, indivisible spheres that are always moving, with a force of attraction acting between the particles.
- Temperature affects the physical properties of a material and can lead to physical changes.
- A substance melts or boils when it is heated. An increase in temperature causes particles to move more rapidly, overcoming the forces of attraction between particles.

KEY QUESTIONS

1. The particle model assumes that all forms of matter can be considered to be made up of particles. What are the four properties of the particles?
2. In which states are the particles able to move positions?
3. In which state are particles held tightly by bonds and arranged in a regular pattern?
4. When 1 kg of water is boiled, what mass of steam is made?
5. Solids and liquids are fairly incompressible. How does the particle model explain this observation? How does the particle model explain why gases are compressible?
6. Use the particle model to explain why a puddle of water on the footpath disappears eventually on a hot, windy day.

2.3 Separating mixtures

It is often useful to isolate one of the **components** (chemicals) from a mixture. There are many methods for separating the components of a mixture, based on their physical properties. Separation based on magnetic properties is used in recycling plants to separate iron and steel from other non-magnetic metals, glass and plastic. Separation based on sedimentation is used in panning for gold from river mud, or in a centrifuge to separate blood cells from whole blood. In a chemical laboratory, two commonly used methods for physical separation are filtration and evaporation. Filtration is usually used to separate a solid from a fluid mixture, and evaporation typically separates a fluid from a fluid mixture.

CHEMISTRY IN ACTION WE

Analytical chemistry and separation science

Analytical chemistry is the science of obtaining information about the composition and structure of matter.

There are two branches of analytical chemistry: qualitative analysis and quantitative analysis. Qualitative analysis is the determination of those elements and compounds that are present in a sample of an unknown material. Quantitative analysis is the determination of the amount of each element or compound present.

Analytical chemists (Figure 2.3.1) are skilled in chemistry, instrumentation, computers and statistics, and they solve problems in almost all areas of chemistry and for all kinds of industries. Their measurements and recommendations are used to assure the safety and quality of food, medicines and water, and to ensure that environmental and pollution regulations are met. They also work in the field of forensic science and provide quality control for industrial processes.

FIGURE 2.3.1 Analytical chemists work in the laboratory or out in the field collecting samples, and in areas such as medicine, industry and environmental science.

FILTRATION

A filter is a barrier that traps the solid components of a mixture, while allowing the fluid component to pass through. The fluid that passes through the filter is called the **filtrate**. Filtration does not completely separate the solid from the mixture. Some of the fluid from the original mixture will remain trapped in the solids, and the smallest particles of the solid may pass through the filter into the filtrate. A finer filter can be used to retain the smaller particles, but the filtration process takes longer.

Filtration techniques are frequently used in the chemical laboratory. Within the kidneys, blood is filtered to remove excess wastes.

FIGURE 2.3.3 Many foods are made by adding a reagent to a liquid to precipitate a solid. A small amount of acid and coagulant added to milk forms liquid whey and solid curds, and the curds form the basis for cheese. Tofu, from soybean pulp, is made in a similar process.

For a homogeneous fluid mixture, it is often possible to add a reagent or change the temperature in order to make one of the mixture's components change state to a solid (Figure 2.3.3). This then allows the mixture to be separated by filtration.

More sophisticated filtration techniques may use a thin, charged filter, a membrane, or pressure to drive a mixture through a filter. An example that utilises a few of these techniques is the process of reverse osmosis, which allows fresh, drinkable water to be separated from sea water.

CHEMISTRY IN ACTION S

Wastewater treatment

You take the plug out of a basin or press the flush button on the toilet. What then happens to the water and its contents?

After passing through pipes, the heterogeneous wastewater mixture reaches a treatment plant, where it undergoes two or sometimes three treatments.

The primary treatment starts with a filtration process to remove solids: the water passes a mesh screen to collect larger pieces of materials, such as paper or plastics. The water then flows into a tank. Oils and grease float to the top, and grit and other sludgy solids settle to the bottom. The oils and sludges are removed, completing the primary treatment.

FIGURE 2.3.2 During secondary wastewater treatment, bacteria help break down solids and clean the water. Cleaned water overflows around the rim of the tank, from where it is channelled off for more treatment.

During secondary treatment, the water remains in a tank where bacteria grow (Figure 2.3.2). The bacteria break down dissolved organic material and any remaining solid wastes. The water then flows into clarifiers, where any suspended materials not caught in the primary treatment settle out. Tertiary treatment is a disinfection step involving chlorine or ultraviolet light.

In Sydney, more than 80% of the treated water is released into the ocean for further disinfection by sunlight. The other 20% of treated waste water ('recycled water') is used for irrigating farms, watering gardens, or for toilets.

EVAPORATION

A small fraction of any liquid is also present as a vapour. A liquid that easily becomes a vapour is said to be **volatile**. A liquid's boiling point is a measure of a liquid's volatility—a more volatile liquid has a lower boiling point. In an open container, the vapour can move away from the liquid's surface, to be replaced by more vapour from the liquid. Eventually, all the liquid will become vapour, and the liquid will have evaporated. Evaporation occurs more quickly for volatile liquids.

Evaporation can be accelerated by various methods, such as by warming the mixture or increasing the surface area over which evaporation can occur.

If a homogeneous mixture is a solution in which one component is not volatile, evaporation can be used to separate the liquid from the solid, such as occurs when salts are left behind after the water in sea water has evaporated. As the liquid evaporates, the mixture becomes increasingly concentrated, and eventually the solid may **precipitate** (form and separate) from the solution.

Some solids precipitate as crystals in a process known as **crystallisation**. Smaller crystals form when the solvent evaporates quickly, and larger crystals form when the solvent evaporates slowly. After precipitation, the solids can be separated from the liquid (Figure 2.3.4). The solids can be separated by filtration or, if the solids have settled out of solution, by carefully pouring off the liquid in a process known as **decanting**. Crystallisation is used in industry to purify substances such as pharmaceuticals and in manufacturing advanced materials.

If the homogeneous mixture is of two liquids, and one is substantially more volatile than the other, the evaporative process can lead to separation in a process called **distillation**. For a mixture of liquids, the vapours produced are rarely of only one pure component, but rather are a mixture. The composition of the vapour is richer in the more volatile component. The vapours can be collected and condensed to a liquid mixture richer in the volatile component, and the distillation process can then be repeated to enrich the more volatile component still further. Eventually, the composition of the liquid and of the vapour is the same, so the components cannot be further separated by distillation.

FIGURE 2.3.4 When water is evaporated from a solution of copper(II) sulfate, crystals of copper(II) sulfate precipitate out. The crystals can then be filtered to separate them from the remaining solution. The filtrate remains blue, which means that not all of the copper(II) sulfate has precipitated.

+ ADDITIONAL

Combining separation techniques

A combination of filtration and evaporation is frequently used to purify compounds. A chemist wanting to separate a physical mixture of salt and sand would see that the solids cannot be distilled and that filtration would not work because there is no liquid present. By adding water to the mixture, the salt would readily dissolve but the sand would not. Filtration could then be used to separate the sand from the salt water. Evaporation would then separate the salt from the saltwater solution. The two components of the original mixture are now separate.

Thinking about the problem further, the chemist might see that the choice of water as the solvent could be improved, as water is not particularly volatile and the evaporation step would be slow. The chemist could select methanol as the solvent. Methanol is cheap and readily available. Methanol does not dissolve salt as well as water does, so more methanol would be needed to dissolve the salt; however, methanol is substantially more volatile than water, so evaporation would be faster.

2.3 Review

SUMMARY

- The components of a mixture can be separated based on their different physical properties.
- Filtration typically uses a physical barrier to separate solids from liquids.
- Evaporation allows a liquid component to be separated from a solution.
- The components in a mixture may not always be completely separated.

KEY QUESTIONS

1. Which of the following is not an example of a separation method?
   A boiling an egg
   B straining tea
   C using a surgical (face) mask
   D fishing with a net
2. Crude oil is a mixture of several liquids with different boiling points. How can they be separated?
3. To remove iron filings from a mixture with sand, is evaporation or using a magnet the best solution?
4. Outline a method to recover a tablespoon of magnesium sulfate that is dissolved into a pot of water.
5. By accident, Arif drops 10 g of sodium phosphate into a litre of water. He works hard to evaporate most of the water. When the solution cools, crystals of sodium phosphate precipitate out. After filtering the crystals from the solution, he dries them. When he weighs the crystals, he finds they only weigh 8 g. Where is the missing 2 g of material? (Arif is a tidy student and has not dropped any on the floor.)

2.4 Calculating percentage composition

PERCENTAGE COMPOSITION OF COMPOUNDS

Compounds are substances that contain two or more different elements chemically bonded together. One way of describing the composition of elements in a chemical compound is using its percentage composition by mass, which gives the mass of each element that would be found in a 100 g sample of the compound.

Being able to determine percentages by mass is important in chemistry. For example, the iron ore mined in Western Australia is iron(III) oxide, and it has the formula Fe_2O_3. A company that is producing iron from this iron ore will want to know the mass of iron that can be extracted from a given quantity of iron ore. The pie chart in Figure 2.4.1 shows that the percentage of iron in iron(III) oxide is 70%. This means that the company can extract a maximum of 70 g iron per 100 g of iron ore.

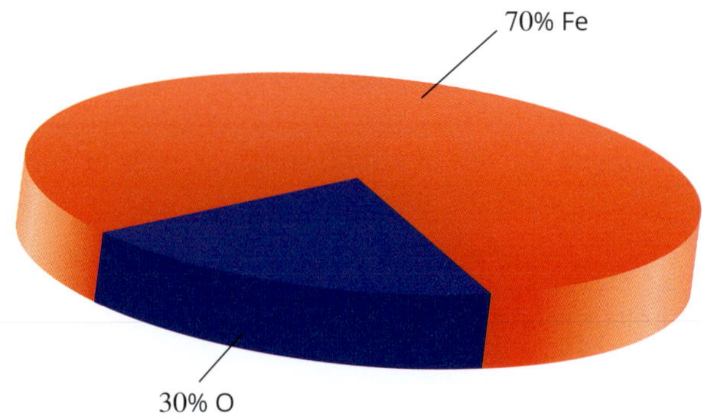

FIGURE 2.4.1 This pie chart shows the percentage composition by mass of iron(III) oxide (Fe_2O_3).

Consider a 56 g sample of calcium oxide, of which 40 g is calcium. The percentage composition of calcium by mass is $\frac{40}{56} \times 100 = 71\%$. Because there is only calcium and oxygen in calcium oxide, the remainder of the mass of calcium oxide comes from oxygen. Oxygen must contribute $56 - 40 = 16$ g, so the percentage by mass of oxygen is $\frac{16}{56} \times 100 = 29\%$.

A useful check can be done by recalling that the total percentage composition by mass for all elements in a compound must add up to 100%.

To calculate the percentage composition of an element in a compound:

$$\text{percentage composition by mass of element} = \frac{\text{mass of element}}{\text{mass of compound}} \times 100$$

SKILLBUILDER N

Calculating percentages

Scientists use percentages to express a ratio or fraction of a quantity. To express one quantity as a percentage of another, use the second quantity to represent 100%.

For example, expressing 6 as a percentage of 24 is like saying '6 is to 24 as *x* is to 100':

$$\frac{6}{24} = \frac{x}{100}$$
$$x = \frac{6}{24} \times 100$$
$$= 25\%$$

To calculate a percentage of a quantity, the percentage is expressed as a fraction or a decimal, then multiplied by the quantity.

For example, to calculate 40% of 20:

$$x = \frac{40}{100} \times 20$$
$$= 0.4 \times 20$$
$$= 8$$

Worked example 2.4.1

CALCULATING THE MASS OF ELEMENTS IN A COMPOUND, BASED ON PERCENTAGE COMPOSITION BY MASS

A 152 g sample of iron(II) sulfate was analysed and found to contain 37% iron and 21% sulfur, and the remainder was oxygen. Determine the mass of each element in the compound.

Thinking	Working
Calculate the mass of each element in the compound using the formula: $\text{mass of element} = \frac{\text{element \%} \times \text{mass of compound}}{100}$	Mass of iron: $\text{mass of iron} = \frac{\text{iron \%} \times \text{mass of iron(II) sulfate}}{100} = \frac{37 \times 152}{100} = 56\text{ g}$ Mass of sulfur: $\text{mass of sulfur} = \frac{\text{sulfur \%} \times \text{mass of iron(II) sulfate}}{100} = \frac{21 \times 152}{100} = 32\text{ g}$
The mass of oxygen can be determined by subtracting the mass of iron and sulfur from the mass of iron(II) sulfate.	Mass of oxygen: $\text{mass of oxygen} = \text{mass of iron(II) sulfate} - \text{mass of iron} - \text{mass of sulfur} = 152 - 56 - 32 = 64\text{ g}$
State the masses of each element in the sample.	The sample is made up of 56 g of iron, 32 g of sulfur and 64 g of oxygen.

Worked example: Try yourself 2.4.1

CALCULATING THE MASS OF ELEMENTS IN A COMPOUND, BASED ON PERCENTAGE COMPOSITION BY MASS

An 85 g sample of sodium nitrate was analysed and found to contain 16% nitrogen and 56% oxygen, and the remainder was sodium. Determine the mass of each element in the compound.

FIGURE 2.4.2 These earrings are made of rose gold, which is more durable than yellow or white gold.

PERCENTAGE COMPOSITION OF MIXTURES

One way of describing the composition of a mixture is by describing its percentage composition by mass. For example, gold is often mixed with other metals to change its properties, such as hardness, scratch resistance or colour. Rose gold (Figure 2.4.2) is a mixture containing 75% gold, 22% copper and 3% silver by mass.

If you know the percentage composition by mass of each component in a mixture, and the identity of each component, you will also be able to work out the percentage composition by mass of every element in the mixture.

Worked example 2.4.2

CALCULATING THE PERCENTAGE COMPOSITION BY MASS OF EACH ELEMENT IN A MIXTURE

A 50 g mixture contains 40% iron(II) oxide (70% Fe, 30% O) and 60% calcium oxide (71% Ca, 29% O). Determine the percentage composition by mass of each element in this mixture.

Thinking	Working
Calculate the mass of each component in the mixture using the formula: $\text{mass of component} = \frac{\text{component \%} \times \text{mass of mixture}}{100}$	Mass of iron(II) oxide: $\text{mass of iron(II) oxide} = \frac{\text{iron(II) oxide \%} \times \text{mass of mixture}}{100}$ $= \frac{40 \times 50}{100}$ $= 20\text{ g}$ Mass of calcium oxide: $\text{mass of calcium oxide} = \frac{\text{calcium oxide \%} \times \text{mass of mixture}}{100}$ $= \frac{60 \times 50}{100}$ $= 30\text{ g}$
Calculate the mass of iron in iron(II) oxide using the formula: $\text{mass of element} = \frac{\text{element \%} \times \text{mass of compound}}{100}$	Mass of iron: $\text{mass of iron} = \frac{\text{iron \%} \times \text{mass of iron(II) oxide}}{100}$ $= \frac{70 \times 20}{100}$ $= 14\text{ g}$
Calculate the mass of calcium in calcium oxide using the formula: $\text{mass of element} = \frac{\text{element \%} \times \text{mass of compound}}{100}$	Mass of calcium: $\text{mass of calcium} = \frac{\text{calcium \%} \times \text{mass of calcium oxide}}{100}$ $= \frac{71 \times 30}{100}$ $= 21\text{ g}$
Calculate the percentage by mass of iron in the mixture.	$\text{percentage by mass of iron} = \frac{\text{mass of iron}}{\text{mass of mixture}} \times 100$ $= \frac{14}{50} \times 100$ $= 28\%$
Calculate the percentage by mass of calcium in the mixture.	$\text{percentage by mass of calcium} = \frac{\text{mass of calcium}}{\text{mass of mixture}} \times 100$ $= \frac{21}{50} \times 100$ $= 42\%$
The remaining balance must be oxygen.	$\text{oxygen \%} = 100 - \text{iron \%} - \text{calcium \%}$ $= 100 - 28 - 42$ $= 30\%$
State the percentage composition by mass of each element in the sample.	The sample is made up of 28% iron, 42% calcium and 30% of oxygen.

Worked example: Try yourself 2.4.2

CALCULATING THE PERCENTAGE COMPOSITION BY MASS OF EACH ELEMENT IN A MIXTURE

A 75 g glucose solution contains 5% glucose (40% C, 7% H, 53% O) and 95% water (11% H, 89% O). Determine the percentage composition by mass of each element in this mixture.

N

Understanding pie charts, frequency graphs and histograms

It is essential in science to collect data and arrange it in an orderly way. Tables are often used to organise data, which can then be displayed in a graph.

Pie charts

A pie chart is a circle that is divided into sectors (Figure 2.4.3). Each sector represents one type of item in a data set and is shown as a percentage or fraction of the total data set.

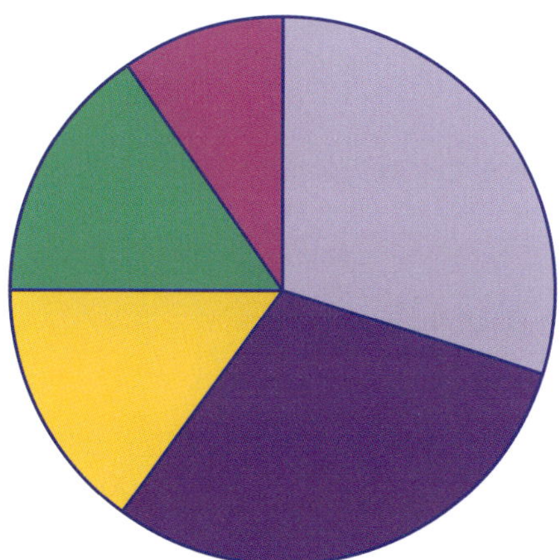

FIGURE 2.4.3 Each sector of a pie chart represents one type of item in a data set.

Column graphs and histograms

Column graphs and histograms are another way of representing data visually.

If the data are discrete (i.e. can be counted), each column in a column graph will represent one category, for example, 'apples' or 'strawberries'. Often these columns have a gap between them.

If the data are continuous, such as the heights of the students in a class, each column will represent a range of possible heights, for example, 140–160 cm, and there will be no gaps between the columns. These are called histograms (Figure 2.4.4).

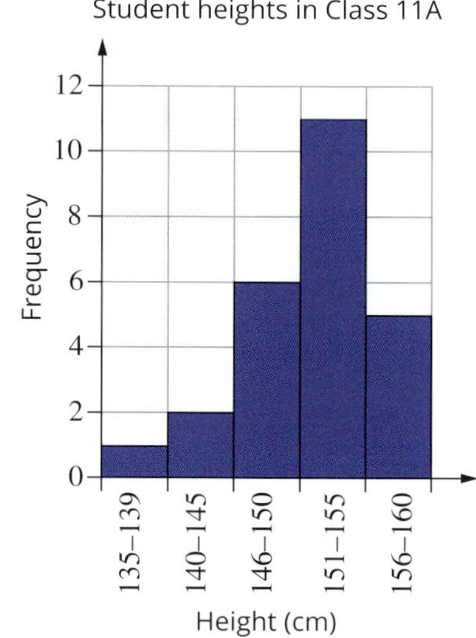

FIGURE 2.4.4 A histogram has no gaps between the columns. The data are continuous.

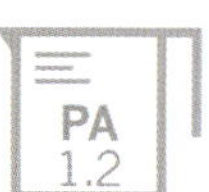

2.4 Review

SUMMARY

- The percentage by mass of an element in a compound can be calculated as follows:

$$\text{percentage composition by mass of element} = \frac{\text{mass of element}}{\text{mass of compound}} \times 100\%$$

KEY QUESTIONS

1. An 88 g sample of carbon dioxide is found to contain 24 g of carbon and 64 g of oxygen. What is the percentage by mass of all elements? Give answers to the nearest whole number.
2. After analysis, a 106 g sample of ammonium chloride is shown to contain 28 g of nitrogen and 71 g of chlorine, and the remainder is hydrogen. What is the percentage by mass of all elements? Give answers to the nearest whole number.
3. A 57 g sample of sodium carbonate reportedly contains 25 g of sodium, 18 g of carbon and 20 g of oxygen. Do you accept this analysis? Explain why or why not.
4. A 10 g sample of carbon dioxide is mixed with 3 g of oxygen. What is the percentage by mass of each component?
5. Pink gold contains 75% gold, 20% copper and 5% silver by mass. Determine the mass of each metal in a pink gold necklace that weighs 20 g.
6. **a** Carbon monoxide is 43% carbon by mass; nitrous oxide is 30% nitrogen by mass. For both gases, the remainder is oxygen. In a mixture containing 5 g of each gas, what is the mass of oxygen present?

   b A 5 g sample of carbon monoxide is mixed with 4 g of nitrous oxide and 3 g of sulfur dioxide. Sulfur dioxide is 50% sulfur by mass, the balance is oxygen. What is the percentage by mass of oxygen and of sulfur present in the mixture? Use the mass compositions given in part **a**.
7. In a carbon monoxide and nitrous oxide mixture, an analyst determines that the mixture is 10% nitrogen by mass. What is the percentage of carbon by mass? Use the mass compositions given in part **a** of Question **6**. (Hint: Consider a 100 g sample to work out the percentages.)

2.5 Elements and the periodic table

ELEMENT NAMES AND ELEMENT SYMBOLS

Scientists have identified 118 different elements. Only about 98 of these occur in nature (the exact number is debatable). The other elements have only been observed in the laboratory.

Each element has a unique name and **chemical symbol**. Table 2.5.1 shows the chemical symbols of some of the most common elements.

TABLE 2.5.1 Chemical symbols and names of some of the most common elements

Element	Symbol	Element	Symbol
aluminium	Al	mercury	Hg
argon	Ar	nitrogen	N
carbon	C	oxygen	O
chlorine	Cl	potassium	K
copper	Cu	silver	Ag
hydrogen	H	sodium	Na
iron	Fe	uranium	U

The chemical symbol is usually made up of one or two letters. The first letter is always capitalised and subsequent letters are always in lower case. In many cases, the chemical symbol corresponds to the name of the element. For example, nitrogen has the chemical symbol N, chlorine has the chemical symbol Cl and uranium has the chemical symbol U.

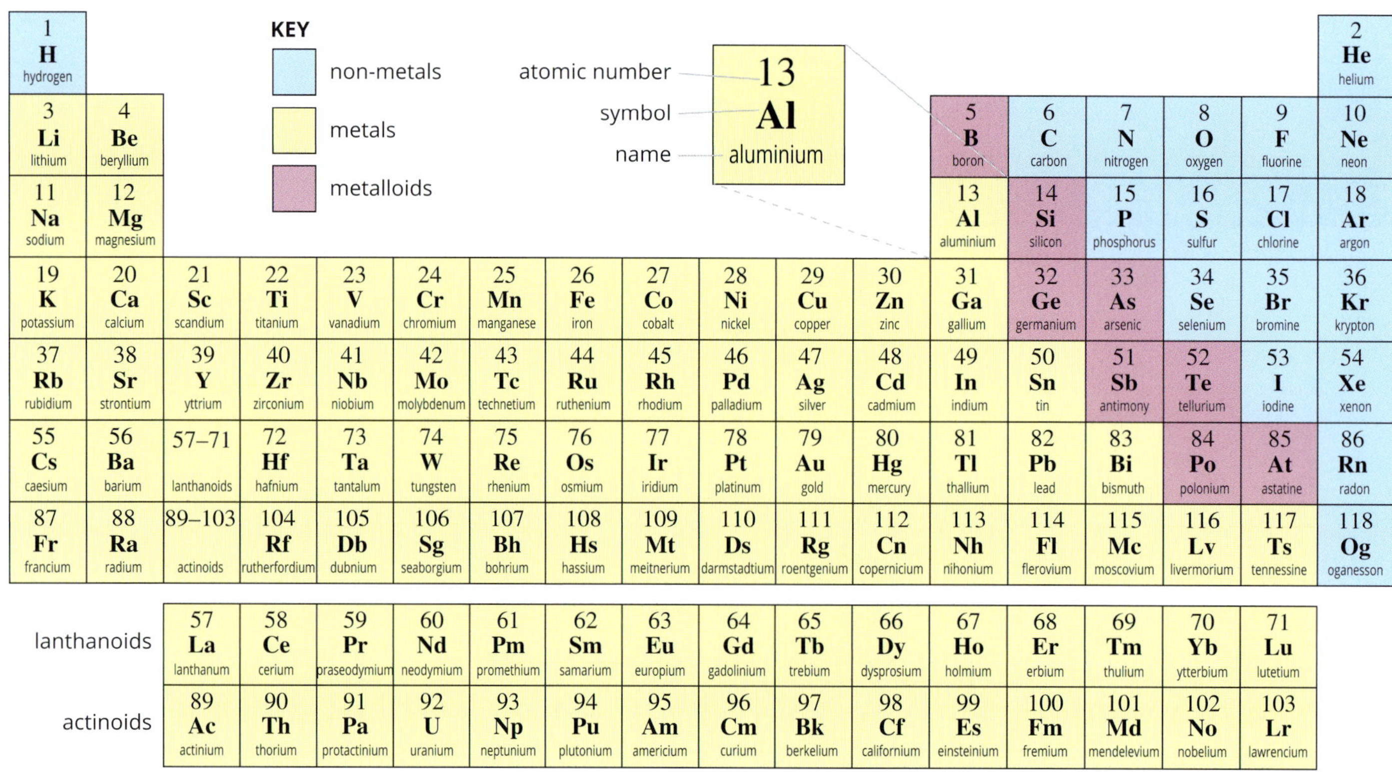

FIGURE 2.5.1 A periodic table groups chemical elements according to their chemical properties.

CHEMFILE IU

Element names

Ancient elements

Twelve elements were known in ancient times. These included gold, silver, mercury and sulfur. The symbols for most of these elements are derived from Latin names. People at this time did not have the modern definition of an element, so materials like soda (sodium oxide) and calomel (mercury chloride) were thought to be elemental.

Searching for new elements

Later in the 18th century and in the 19th century, scientists could extract gaseous elements from the air and isolate solid elements from the ground. Elements often took names from the Latin words for where they were found. Such elements included silicon (found in sand, known in Latin as *silex*, so given the symbol Si) and calcium (found in limestone, known in Latin as *calx*, so given the symbol Ca).

As analytical techniques improved and more elements were found, naming became more imaginative, with names being derived from Greek roots, detected characteristics, places or mythology: the inert gas argon (from the Greek *άργόν*, inactive), rubidium (from the Latin *rubidius*, deep red), germanium (from the Latin *germania*, meaning Germany), and vanadium (after the Scandinavian goddess Vanadis).

By 1829, 55 elements were known. The first modern periodic table, constructed in 1869, included 64 elements, and by 1914 a total of 72 elements had been discovered.

Synthetic elements and atomic science

The first synthetic element, technetium, was made in 1936. New elements were discovered after 1940; of these, many were discovered during thermonuclear bomb tests. This took the total known number of elements to 118. The synthetic elements have large, unstable nuclei and not much is known about their chemistry. Many, such as einsteinium or americium, were named after a scientist or place. The Cold War made naming of these elements contentious, but since 1999 the names of all new elements have been decided by a panel of the International Union of Pure and Applied Chemistry (IUPAC).

Some chemical symbols are not obviously related to the name of the element. For example, sodium has the chemical symbol Na, potassium has the chemical symbol K and iron has the chemical symbol Fe. This is because the chemical symbols have been derived from the Latin or Greek names of the elements. In Latin, sodium is *natrium*, potassium is *kalium* and iron is *ferrum*.

The atomic symbols are usually displayed in a **periodic table** like the one shown in Figure 2.5.1. The periodic table groups elements that have similar chemical properties. You will learn more about the periodic table in Chapter 4.

METALS, NON-METALS AND METALLOIDS

The periodic table (Figure 2.5.1) highlights three basic classes of elements—metals, non-metals and metalloids. Lithium is an example of a metal; oxygen, a non-metal; and boron, a metalloid.

Metals

Most elements are **metals**. Metals are usually solids at room temperature. Some can be found as pure elements in the Earth's crust, like gold, but most are combined with other elements and need to be extracted and processed to achieve their pure elemental form.

Typically, metals are:

- lustrous (shiny and reflect light)
- malleable (can be beaten into new shapes without breaking)
- ductile (can be drawn into a wire)
- silver-coloured
- dense (heavy for their volume)
- of high melting point (solid at room temperature)
- of high boiling point
- of high tensile strength (can be stretched without breaking)
- good conductors of electricity
- good conductors of heat.

Some metals vary to some degree from this, such as liquid mercury, gold and copper (which are not silver in colour), low-density aluminium and brittle bismuth. However, the listed properties make metals very useful. Table 2.5.2 gives some examples of metals often used in their pure form.

TABLE 2.5.2 Fitting the metal to the task

Metal	Example	Use	Properties that make the metal suited to this use
copper (Cu)		electrical wiring	excellent conductor and very ductile
gold (Au)		jewellery	soft and malleable, easily polished
aluminium (Al)		aeroplanes	light for its strength, does not corrode
zinc (Zn)		coating for steel ('galvanising')	coated onto steel to protect it from rusting

GO TO ➤ Section 5.1 page 138

The bonding between metallic atoms, which gives rise to these properties, will be examined more closely in Chapter 5. Many of the materials in everyday use that are called metals are not pure metallic elements but **alloys**. An alloy is a physical mixture of a metal with a small amount of some other element, and it may have physical properties that are preferred over those of the pure metal.

Non-metals

The next largest group of elements is the **non-metals**. Many are found in the atmosphere as gases; a few (such as sulfur) are found in their solid form. Non-metals have a much greater variety of physical properties than metals, as shown in Figure 2.5.2.

FIGURE 2.5.2 Non-metals have very different physical appearances, for example, yellow sulfur, brown bromine in a flask, red phosphorus, dull grey iodine and solid black carbon.

Typically, non-metals are:

- not malleable
- not ductile
- dull in colour, not shiny
- not dense (compared with metals)
- lower in melting and boiling points (compared with metals)
- poor conductors of electricity
- poor conductors of heat.

The bonding between non-metallic atoms that gives rise to these properties will be examined more closely in Chapter 5.

+ ADDITIONAL WE

Alloys

An alloy is a physical mixture, and the properties of the mixture can be quite different from the properties of the components. The properties of an alloy can also vary depending on how the alloy is prepared. Iron-based alloys are the most well known, and the alloy of iron and carbon is known as steel. Table 2.5.3 and Figure 2.5.3 shows how a small amount of added carbon has important effects.

TABLE 2.5.3 Iron alloys and their uses

Alloy	Percentage composition by mass	Properties of this alloy	Example uses
wrought iron	Nearly 100% Fe, trace C	very malleable	handmade (blacksmith) products
mild steel	99.5% Fe, 0.5% C	cheap to manufacture	car parts, buildings
hardened steel	99% Fe, 1% C	hard, holds an edge	springs, blades, tools
cast iron	97% Fe, 3% C	low melting temperature	moulded forms like cookware and tubes

FIGURE 2.5.3 (a) Blacksmiths and (b) engineers both work with iron alloys. The difference between these materials is just a small amount of carbon.

Alloys such as brass (70% copper, 30% zinc), bronze (95% copper, 5% tin) and electrum (50–90% gold, 10–50% silver) have been known since ancient times (Figure 2.5.4).

FIGURE 2.5.4 One of the oldest known coins, the *trite*, was made from electrum and struck around 600 BCE in the kingdom of Lydia. At the time, this coin could buy 11 sheep or 10 goats.

Modern alloys include solder (65% tin, 35% lead), duralumin (95% aluminium, 4% copper, 1% magnesium and manganese) and 'silver' coins (75% copper, 25% nickel). Sophisticated alloys are used in supermagnets, high-performance vehicles and computer components, and for biomedical purposes.

With many dozens of metals to select from for the mixtures, it is no surprise that novel uses for new alloys is a rich field of research.

Metalloids

The third and smallest group is the **metalloids**. There is no generally accepted definition of a metalloid, and no single characteristic that is associated with defines a metalloid. For example, tin and selenium can be classified as either metals or metalloids.

Metalloids have some metallic and some non-metallic properties. The metalloid silicon and the non-metal carbon can form covalent networks, while antimony conducts nearly as well as a metal. Metalloids tend to be shiny like metals, but brittle like non-metals. They can conduct heat and electricity like metals, but usually not as well. Their ability to moderately conduct electricity is known as semiconductivity.

DISTINGUISHING BETWEEN PHYSICAL AND CHEMICAL PROPERTIES

FIGURE 2.5.6 A ribbon of magnesium burning in air. When heated in air, it burns with a bright, white flame.

As you have already seen, elements have different physical attributes. Some are solids at room temperature, some are liquids. Some elements are hard and shiny, some are odorous.

Physical properties describe features that can be observed or measured. Common physical properties of elements include appearance, colour, density, brittleness, electrical conductivity, magnetic effects and crystal shape. These can be observed or measured without changing the element.

The temperature at which an element melts or boils is also a physical property. When measuring the melting point of sulfur, the element transforms from solid sulfur to liquid sulfur, but it remains sulfur; as the liquid sulfur cools, it transforms reversibly back to solid sulfur.

Elements react chemically in different ways. Some react with water, some do not react and some don't. Some are acidic and some alkaline. Some burn easily. These and other characteristics make up the elements' chemical properties.

Chemical properties relate to how easily an element undergoes chemical change. A chemical change results in substances changing from one substance to another. In a flame, the metal magnesium suddenly glows bright white as it reacts with oxygen in the air (Figure 2.5.6). Therefore, a chemical property of magnesium is that it burns easily. This is a chemical change: when the reaction is complete, the atoms of magnesium and oxygen have rearranged and combined to form a new material, magnesium oxide.

CHEMFILE ICT

Semiconductors

The core of any gadget containing a computer or using radio waves depends on semiconductors. Practically all semiconductor chips are created from the metalloid silicon—so much so that the name Silicon Valley is used to describe a US hub of computer technology.

Silicon atoms can form a network of crystalline silicon. A silicon crystal looks like a metal, but unlike a metal, pure silicon conducts electricity poorly— it almost qualifies as an electrical insulator. In a process called doping, tiny amounts of other elements—usually germanium, phosphorus, arsenic, boron or gallium—are added, and the material transforms from an insulator into a semiconductor.

A single semiconductor by itself is not that interesting, but a sandwich of three layers of semiconductors forms a device called a transistor. A transistor controls the flow of electricity. A careful arrangement of many transistors on a wafer-thin slice of silicon forms a silicon chip. A single chip may hold tens of millions of transistors. Modern computer devices depend totally on silicon chips (Figure 2.5.5) manufactured from ultra-high-purity silicon crystals.

FIGURE 2.5.5 Many electronics depend on silicon chips made from silicon.

The properties of metals, non-metals and metalloids discussed so far are physical properties. It is much harder to create a general list of chemical properties. However, the location of an element in the periodic table gives you clues as to how the element will react chemically.

PROPERTIES AND POSITION ON THE PERIODIC TABLE

The most usual form of the periodic table arranges elements into horizontal rows (periods) and vertical columns (groups), as shown in Figure 2.5.7. Periods are numbered from 1 to 7 and groups from 1 to 18. Each element has its own place in the periodic table.

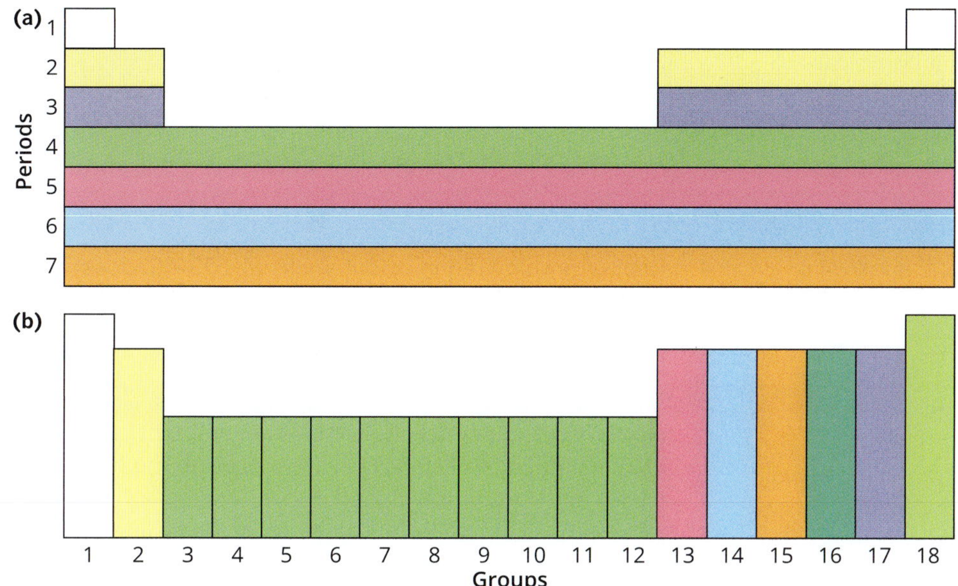

FIGURE 2.5.7 The periodic table is usually arranged in (a) periods 1–7 and (b) groups 1–18.

Carbon (element symbol C) is in period 2 and group 14. Oxygen (element symbol O) is also in period 2 but in group 16. Calcium (element symbol Ca) is in period 4 and group 2.

Elements in the same group have, in general, similar physical and chemical properties, but each element has its own characteristics. For example, elements of group 11—copper (Cu), silver (Ag), gold (Au)—are dense, malleable, corrosion-resistant metals that can be found in the ground as pure elements. Elements of group 17—fluorine (F), chlorine (Cl), bromine (Br), iodine (I)—are highly reactive and toxic to life in surprisingly small quantities. Elements of group 18—helium (He), neon (Ne), argon (Ar), krypton (Kr)—are colourless, odourless and generally unreactive gases, the noble gases.

FIGURE 2.5.8 The group 1 metal lithium reacts in water to produce a gas and heat.

Knowing the properties of one element in a group allows you to make predictions about other elements in the group. For example, when the group 1 metal lithium is added to water, it begins to react with the water, producing a gas and a lot of heat (Figure 2.5.8). Likewise, sodium is in group 1 and reacts similarly but more vigorously when added to water. Therefore, you could predict that if potassium was added to water, it would react similarly to lithium and sodium but even more vigorously.

Using groups to make predictions on properties can however, be difficult. Group 15 contains the non-metals nitrogen and phosphorus, metalloids arsenic and antimony, and metallic bismuth. These elements each have very different physical and chemical properties. It might be expected that all elements of group 17 will be reactive non-metals, but at room temperature fluorine and chlorine are gases, bromine is a liquid and iodine is a solid.

Why groups have similar chemical reactivity is due to the arrangement of electrons within the atom, which will be investigated in Chapters 3 and 4.

2.5 Review

SUMMARY

- Every element has a chemical symbol that is made up of one or two letters. The first letter is capitalised and the second letter is in lower case.
- Elements are organised in the periodic table.
- Elements can be classified as metals, non-metals or metalloids.
- Metallic elements have many common physical properties.
- Non-metallic elements have a wide range of diverse physical properties.
- In the periodic table, elements are arranged into groups with similar physical and chemical properties.

KEY QUESTIONS

1. Identify the common name of each of the following elements from its non-English name.
   a ferrum
   b kalium
   c wolfram
   d plumbum.
2. If an element can be hammered flat, would it be said to be conductive, ductile, malleable or calorific?
3. Of the elements oxygen (O, element 8), silicon (Si, element 14), rhodium (Rh, element 45) and calcium (Ca, element 20), which is least likely to conduct heat and electricity?
4. Copper is shiny red in colour, melts at 1084°C, does not react with water, and can be mixed with zinc to form a brass alloy. Which of these is a chemical property?
5. Aluminium is a good thermal conductor and a good electrical conductor, it has a high melting point and is ductile. Which of these properties makes aluminium a useful material for making cookware?
6. An electrical conductor is a poor electrical insulator. Arrange the elements silicon (Si, element 14), sulfur (S, element 16) and scandium (Sc, element 21) in order of their ability to insulate, from best to worst.

Chapter review

02

KEY TERMS

alloy
aqueous
atom
atomic theory of matter
chemical symbol
component
compound
covalent network structure
crystallisation
decanting
discrete molecule
dissolved
distillation
element
filtrate
fluid
giant molecule
latent heat
latent heat of fusion
latent heat of vaporisation
matter
metal
metalloid
mixture
model
molecule
monatomic
noble gas
non-metal
periodic table
precipitate
solute
solution
solvent
subatomic particle
sublimation
sublime
volatile

REVIEW QUESTIONS

1 What did Dalton's theory of elements predict for compounds?

2 Which of Dalton's predictions about the nature of atoms was later proven to be incorrect?

3 What is the fundamental difference between an atom and a molecule?

4 What is the difference between a discrete molecule and a network?

5 Classify the following elements according to whether they are monatomic, made up of molecules, or form a large network of atoms bonded together: sulfur, copper, carbon, tin, helium, neon, gold, oxygen, krypton, nitrogen.

6 What is the fundamental difference between a compound and a mixture?

7 Fresh milk is composed of milk fats and water. Some of the milk fats float on the top of the milk as cream. What type of mixture is milk?

8 Dalton referred to hard particles in his atomic theory, and particle theory also refers to hard particles. Are these particles the same? Consider solid sulfur in your answer.

9 Is the air inside a blown-up balloon more or less dense than air outside the balloon?

10 A solid metal can be heated to create a metal vapour. Is this a physical or a chemical change? Refer to the arrangement of particles in the metal in your answer.

11 **a** During a change of state, what happens to the measured temperature?

b During a change of state do the attractive forces between particles stay the same or change?

12 Toothpaste is a physical mixture of a powder in a solvent. It cannot be easily classified either as a solid or a liquid. List the properties of toothpaste that could classify it as a solid and as a liquid.

13 Could ink be separated into its components using filtration?

14 Blood can be approximated as a mixture of red blood cells and a liquid. Could blood cells be separated by filtration? Is there a better option?

15 Two cups of water are on the ground on a warm day. One cup is knocked over and spills the water to form a puddle. Will the water in the cup or in the puddle evaporate more quickly?

16 When a homogeneous solution made of two volatile liquids evaporates, will the vapour be composed of a greater or lesser proportion of the liquid with the lower boiling point?

17 An elemental analysis of sodium hypochlorite (NaOCl) shows it is 31% sodium and 21% oxygen by mass. What percentage by mass is chlorine?

18 If the measured mass of chlorine in Question 17 is 3.0 g, what was the original mass of the sample?

19 A stainless steel object of mass 150 g was analysed and found to contain 111 g iron, 27 g chromium and the remainder nickel. What is the percentage composition by mass of the components in this stainless steel?

20 Baking powder is a raising agent commonly used when making baked goods. One formulation of baking powder contains 30% sodium hydrogen carbonate, 12% calcium phosphate, 26% sodium aluminium sulfate and the remainder cornstarch. A recipe for scones requires 20 g of baking powder. What is the mass of calcium phosphate in the 20 g of baking powder?

21 One type of methylated spirits contains 10% methanol and 90% ethanol by mass. The table shows the percentage composition by mass of elements in each compound.

Alcohol	%C	%H	%O
methanol	38	13	49
ethanol	52	13	35

What is the percentage composition by mass of carbon in methylated spirits?

22 State the symbols for the following elements:
- **a** oxygen
- **b** carbon
- **c** helium
- **d** nickel
- **e** potassium
- **f** gold.

23 Give the names of the elements corresponding to the following symbols:
- **a** N
- **b** Ca
- **c** Cl
- **d** Ag
- **e** Hg.

24 **a** An element is shiny but a poor conductor of heat. Is the element most likely to be a metal, non-metal or metalloid?
- **b** There are four unknown elements. One is black and gritty, one shatters easily, one is an insulator and one conducts electricity. From these observations, which one is most likely to be a metal?

25 Why can a blade be made of iron but not of sulfur?

26 Which one or more of the following properties are expected of the group 18 noble gases?
shiny and lustrous, poor conductor of electricity, soft and malleable, good conductor of heat

27 Classify the following statements as reflecting a physical property or a chemical property.
- **a** Diamonds can cut glass.
- **b** Sugar dissolves in water.
- **c** Wood burns.
- **d** Aluminium has a low density.
- **e** Lithium reacts with water.

28 **a** In the periodic table, lithium is in group 1, period 2, sodium is in group 1, period 3, and magnesium is in group 2, period 3. Which pair of elements are likely to have the most similar chemical properties? What could be predicted about these elements' physical properties?
- **b** In the periodic table in group 14, lead is in period 6, germanium in period 4 and carbon in period 2. Would you expect the physical properties of these elements to be similar?
- **c** Considering period 4 of the periodic table, from potassium (K) to krypton (Kr), comment on the trends in heat conductivity.

29 Some metallic elements are found in nature as metals, some as compounds. Aluminium, element 13, comprises about 8% by mass of Earth's crust, but was only isolated by Hans Oersted as late as 1825. Conduct some research to find out why such an abundant metal should have been isolated so late in the discovery of common elements.

30 In the atmosphere, the three main elemental gases are nitrogen (78%), oxygen (21%) and argon (1%). Using your knowledge of separations and of the states of matter, describe how this homogeneous mixture could be separated into its components.

31 Reflect on the Inquiry activity on page 48. What property was tested when bending the aluminium can with pliers? Would this property indicate that aluminium is a metal, metalloid or non-metal?

Atomic structure and atomic mass

In this chapter, you will learn that atoms are the fundamental building blocks of matter. You will develop a detailed understanding of the structure of atoms, which is the foundation for all chemistry. You will also learn that different types of atoms make up the various elements, and that different atoms of the same element are called isotopes. Some isotopes, known as radioisotopes, are unstable and release radiation when they decay. To determine the relative masses of isotopes and atoms, chemists use an instrument called a mass spectrometer.

By the end of this chapter, you will also have a greater understanding of the various models used to describe the arrangement of electrons in atoms.

Content

INQUIRY QUESTION

Why are atoms of elements different from one another?

By the end of this chapter, you will be able to:

- investigate the basic structure of stable and unstable isotopes by examining:
  - their position in the periodic table
  - the distribution of electrons, protons and neutrons in the atom
  - representation of the symbol, atomic number and mass number (nucleon number) ICT
- model the atom's discrete energy levels, including electronic configuration and *spdf* notation (ACSCH017, ACSCH018, ACSCH020, ACSCH022) ICT
- calculate the relative atomic mass from isotopic composition (ACSCH024) ICT N
- investigate energy levels in atoms and ions through:
  - collecting primary data from a flame test using different ionic solutions of metals (ACSCH019) ICT
  - examining spectral evidence for the Bohr model and introducing the Schrödinger model
- investigate the properties of unstable isotopes using natural and human-made radioisotopes as examples, including but not limited to:
  - types of radiation
  - types of balanced nuclear reactions

3.1 Inside atoms

CHEMISTRY INQUIRY 

In what way do atoms of the various elements differ from one another?

COLLECT THIS ...

- marbles and ball bearings of various sizes
- toy car plastic race track with a 90° curve
- stopwatch
- top-loading weighing balance
- camera

DO THIS ...

1 Set up the race track as shown, with a raised starting position and a 90° curve at the base of the track.

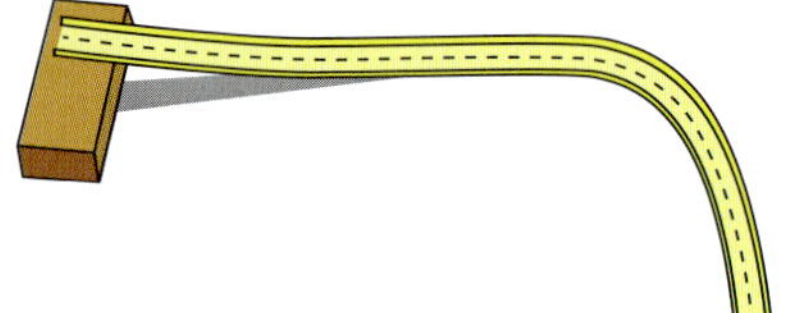

2 Predict what will happen when you roll balls of various masses down the ramp and around the curved track.

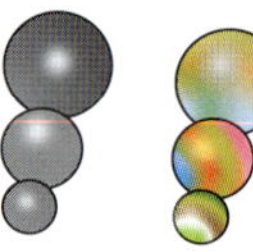

3 Draw up a table in your workbook for recording the results of your experiment.

4 Roll the balls down the track one at a time. Record the time each ball takes to reach the end of the track and the path each ball follows when it exits the track.

5 If possible, take a video of your experiment.

RECORD THIS ...

Present a table of your results.

Describe what happened when you rolled the different balls down the track.

REFLECT ON THIS ...

What makes the balls different from one another?

Identify any patterns in your results and try to explain them.

What could you do next time to improve your experiment?

ATOMIC MODELS

Until the mid-19th century, scientists believed that **atoms** were hard spheres that couldn't be broken down into smaller parts. By the end of the 19th century, there was increasing evidence to suggest that atoms are made up of smaller particles. This led to a series of atomic **models** that attempted to explain the structure of atoms.

The current model of the atom is based on the work of many scientists. All atoms are made up of a small, positively charged **nucleus** surrounded by a much larger cloud of negatively charged **electrons**, as shown in Figure 3.1.1. The nucleus is made up of two types of **subatomic particles—protons** and **neutrons**. The protons are positively charged and the neutrons have no charge.

FIGURE 3.1.1 A simplified model of the atom. The central glow represents the nucleus, where the protons and neutrons are housed. The nucleus is surrounded by a cloud of orbiting electrons.

Electrons

Electrons are negatively charged particles. You can imagine them forming a cloud of negative charge around the nucleus. This cloud gives the atom its size and volume.

An electron has a mass that is approximately 1800 times smaller than that of a proton or neutron. Therefore, electrons contribute very little to the total mass of an atom. However, the space occupied by the cloud of electrons is 10 000–100 000 times larger than the nucleus.

Negative particles attract positive particles. This is called **electrostatic attraction**. Electrons are bound to the nucleus by electrostatic attraction to the protons within the nucleus. The charge of an electron is equal but opposite to the charge of a proton. Electrons are said to have a charge of −1 and protons to have a charge of +1.

In some circumstances, electrons can be easily removed from an atom. For example, when you rub a rubber balloon on a woollen jumper or dry hair, electrons are transferred to the balloon, and the balloon develops a negative charge.

The negative charge is observed as an electrostatic force that can stick the balloon to a wall or even move an aluminium can. You will look more closely at the transfer of electrons when looking at ionic compounds in Chapter 5.

GO TO ➤ Section 5.2 page 147

The electricity that powers lights and appliances is the result of electrons moving in a current through wires. Sparks and lightning are also caused by electrons moving through air.

+ ADDITIONAL

Discovering the electron

In the late 1890s, English physicist Joseph John Thomson conducted a series of experiments in which an electric current was passed through different gases in sealed tubes at very low pressures. When a high voltage was applied, the tubes glowed with a coloured light and a coating on the glass tube wall opposite the negative electrode became fluorescent. This fluorescence was caused by invisible rays that became known as 'cathode rays' because they came from the negative cathode. They could be deflected by a magnet. This is shown in Figure 3.1.2.

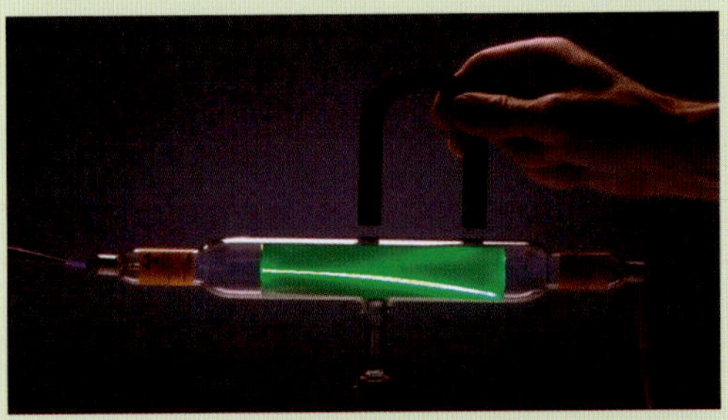

FIGURE 3.1.2 This glass tube contains a gas at low pressure with a high voltage applied across it. The resulting cathode ray can be seen being deflected by a magnet.

When an electric field was applied at right angles near the cathode, the position of the fluorescence moved away from the negative pole of the electric field, indicating the charge was negative. This can be seen in Figure 3.1.3.

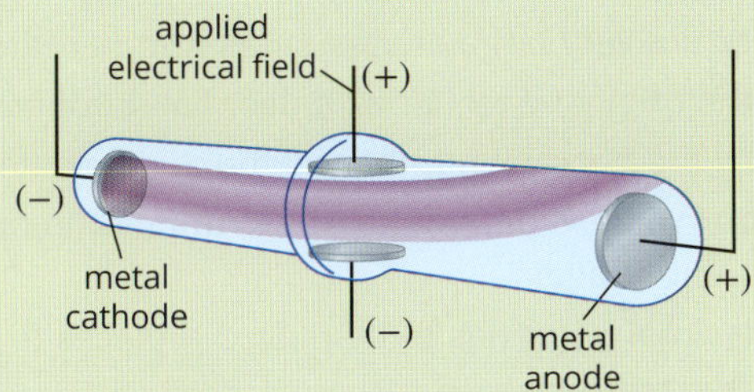

FIGURE 3.1.3 Cathode rays are deflected by an applied electric field.

From his experiments, Thomson deduced that:

- the rays were a stream of particles
- the particles came from the negative electrode (cathode) and were negatively charged
- the particles must be found in all matter and therefore were subatomic particles
- atoms must also contain positive subatomic particles, since atoms have no overall charge.

Thomson's research laid the foundation for other scientists to build the modern model of the atom. For his discovery of the electron, Thomson was awarded the 1906 Nobel Prize in Physics.

The nucleus of an atom

The nucleus of an atom is approximately 10 000–100 000 times smaller than the size of the atom. To put this in perspective, if an atom were the size of the Sydney Cricket Ground (Figure 3.1.4), then the nucleus would be about the size of a pea. Nonetheless, the nucleus contributes about 99.97% of the atom's mass. This means that atomic nuclei are extremely dense.

FIGURE 3.1.4 If an atom were the size of the Sydney Cricket Ground, then the nucleus would be the size of a pea.

The subatomic particles in the nucleus, the protons and neutrons, are referred to collectively as **nucleons**. Protons are positively charged particles with a mass of approximately 1.673×10^{-27} kg. Neutrons and protons are almost identical in mass.

Table 3.1.1 summarises the properties of protons, neutrons and electrons.

TABLE 3.1.1 Properties of the subatomic particles

Particle	Symbol	Charge	Size relative to a proton	Mass (kg)
proton	p	+1	1	1.673×10^{-27}
neutron	n	0	1	1.675×10^{-27}
electron	e	−1	$\frac{1}{1800}$	9.109×10^{-31}

Transforming decimal notation into scientific notation

Scientists use scientific notation to handle very large and very small numbers. For example, instead of writing 0.000 000 035, scientists would write 3.5×10^{-8}.

A number in scientific notation (also called standard form or power of 10 notation) is written as:

$a \times 10^n$

where

a is a number equal to or greater than 1 and less than 10, that is, $1 \leq a < 10$

n is an integer (a positive or negative whole number)

n is the power that 10 is raised to and is called the index value.

To transform a very large or very small number into scientific notation:

1 Write the original number as a decimal number greater than or equal to 1 but less than 10.

2 Multiply the decimal number by the appropriate power of 10.

The index value is determined by counting the number of places the decimal point needs to be moved to form the original number again.

- If the decimal point has to be moved n places to the right, n will be a positive number. For example:

  $51 = 5.1 \times 10^1$

- If the decimal point has to be moved n places to the left, n will be a negative number. For example:

  $0.51 = 5.1 \times 10^{-1}$

You will notice from these examples that when large numbers are written in scientific notation, the 10 has a positive index value. Very small numbers are written by multiplying by 10 with a negative index.

3.1 Review

SUMMARY

- All atoms are composed of a small, positively charged nucleus surrounded by a negatively charged cloud of electrons.
- The mass of an atom is mostly determined by the mass of the nucleus, while the size of an atom is determined by the cloud of electrons.
- The nucleus is made up of two types of subatomic particles—protons and neutrons. These particles are collectively referred to as nucleons.
- Protons have a positive charge, electrons have a negative charge and neutrons have no charge.
- Protons and neutrons are similar in size and mass, while electrons are approximately 1800 times smaller.
- The charges on protons and electrons are equal but opposite.

KEY QUESTIONS

1 How many times larger is the atom compared with its nucleus?

2 What subatomic particles make up the most mass of an atom and where are they found?

3 How are electrons held within the cloud surrounding the nucleus?

4 State the charge of:

a an electron

b a proton

c a neutron.

3.2 Classifying atoms

DIFFERENT TYPES OF ATOM

Each **element** is made up of one type of atom. The type of atom that makes up each element is determined by the number of protons in the nucleus. The number of protons in an atom's nucleus is known as the **atomic number** and is represented by the symbol Z.

All atoms that belong to the same element have the same number of protons and therefore have the same atomic number, Z. For example, all hydrogen atoms have $Z = 1$, all carbon atoms have $Z = 6$ and all gold atoms have $Z = 79$ (Figure 3.2.1).

The number of protons and neutrons in the nucleus of an atom is known as the **mass number** and is represented by the symbol A. The mass number represents the total mass of the nucleus.

As all atoms are electrically neutral, the number of protons in an atom is equal to the number of electrons in that atom. The atomic number therefore indicates both the number of protons and the number of electrons. For example, carbon atoms, with $Z = 6$, have six protons and six electrons.

FIGURE 3.2.1 Gold has an atomic number of 79. This example, the Welcome Stranger nugget, weighing over 78 kg, was found in 1869 in Victoria.

REPRESENTING ATOMIC STRUCTURE

The number of protons, neutrons and electrons forms the basic structure of an atom. The standard way of representing an atom of a particular element is with its atomic and mass numbers, as shown in Figure 3.2.2.

mass number → $^{A}_{Z}X$ ← symbol of element
atomic number →

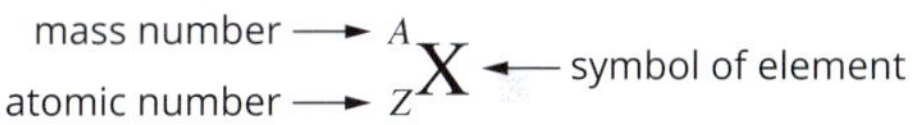

FIGURE 3.2.2 The standard way of representing an atom of a particular element is to show its atomic number and mass number.

For an aluminium atom, this would be written as shown in Figure 3.2.3.

$$^{27}_{13}Al$$

FIGURE 3.2.3 This representation of an aluminium atom indicates the number of protons, neutrons and electrons in the atom.

From this representation, you can determine that:

- The number of protons is 13 because the number of protons is equal to the atomic number (Z).
- The number of neutrons is 14 because the number of neutrons plus the number of protons is equal to the mass number. Therefore, you can subtract the atomic number from the mass number to determine the number of neutrons ($A - Z$).
- The number of electrons is 13 because atoms have no overall charge. Therefore, the number of electrons must equal the number of protons.

Worked example 3.2.1

CALCULATING THE NUMBER OF SUBATOMIC PARTICLES

Calculate the number of protons, neutrons and electrons for the atom with this atomic symbol: $^{40}_{18}Ar$

Thinking	Working
The atomic number is equal to the number of protons.	number of protons = Z = 18
Calculate the number of neutrons. number of neutrons = mass number − atomic number	number of neutrons = $A - Z$ = 40 − 18 = 22
Calculate the number of electrons. The number of electrons is equal to the atomic number because the total negative charge is equal to the total positive charge.	number of electrons = Z = 18

Worked example: Try yourself 3.2.1

CALCULATING THE NUMBER OF SUBATOMIC PARTICLES

Calculate the number of protons, neutrons and electrons for the atom with this atomic symbol: $^{235}_{92}U$

ISOTOPES

All atoms that belong to the same element have the same number of protons in the nucleus and therefore the same atomic number, Z. However, not all atoms that belong to the same element have the same mass number, A. For example, hydrogen atoms can have mass numbers of 1, 2 or 3. In other words, hydrogen atoms may contain just a single proton, a proton and a neutron, or a proton and two neutrons, as shown in Figure 3.2.4. Atoms that have the same number of protons (atomic number) but different numbers of neutrons (and therefore different mass numbers) are known as **isotopes**. Isotopes have identical chemical properties but different physical properties, for example different masses and densities.

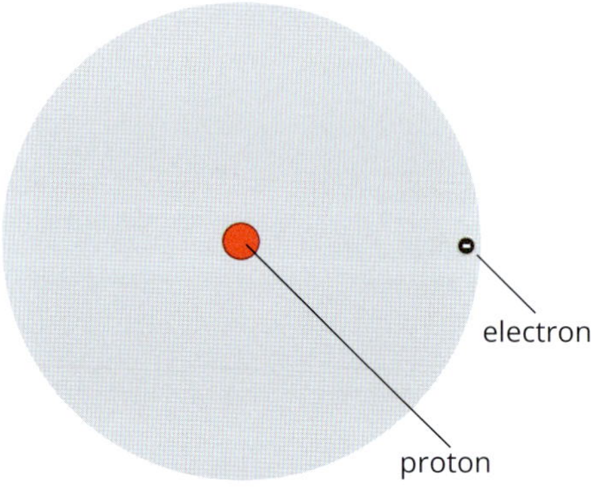

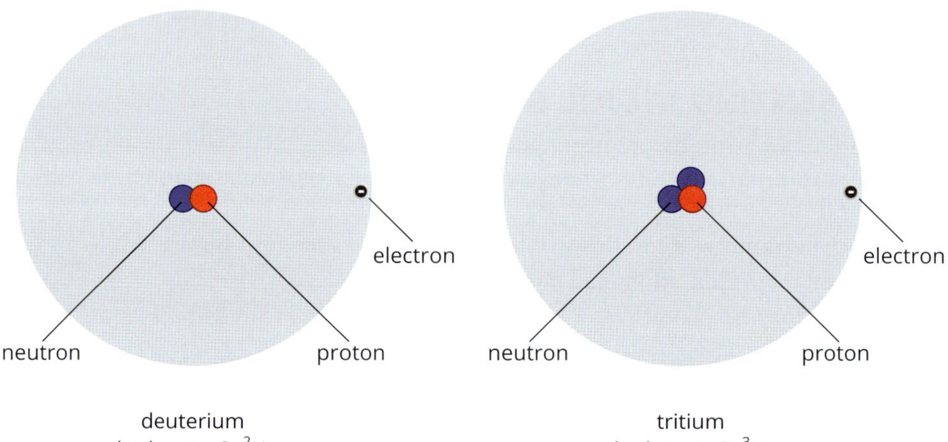

FIGURE 3.2.4 The three isotopes of hydrogen are given special names. A hydrogen atom with just one proton in its nucleus is known as hydrogen or protium. A hydrogen atom with one proton and one neutron is known as deuterium. A hydrogen atom with one proton and two neutrons is known as tritium.

Carbon has three naturally occurring isotopes. These three isotopes are known as carbon-12, carbon-13 and carbon-14. Carbon-12 atoms have a mass number of 12, carbon-13 atoms have a mass number of 13 and carbon-14 atoms have a mass number of 14. In the 1950s and 1960s, nuclear weapons testing caused a spike in carbon-14 in the atmosphere. This has been declining in the last 50 years. These three carbon isotopes can be represented as shown in Figure 3.2.5.

$^{12}_{6}C$	$^{13}_{6}C$	$^{14}_{6}C$
carbon-12	carbon-13	carbon-14

FIGURE 3.2.5 A scientific representation of the three isotopes of carbon: carbon-12, carbon-13 and carbon-14.

RADIOISOTOPES

Some isotopes are **radioactive**, so they are referred to as **radioisotopes**.

Figure 3.2.6 shows a plot of the number of neutrons versus the number of protons in various nuclei. It shows that there is a region called the **band of stability** where the nuclei are stable. Nuclei that lie outside of this region are unstable so they are radioactive.

Radioisotopes will decay (break down) to form stable nuclei. During the decay process, radiation is emitted in the form of **alpha particles**, **beta particles** or **gamma radiation**.

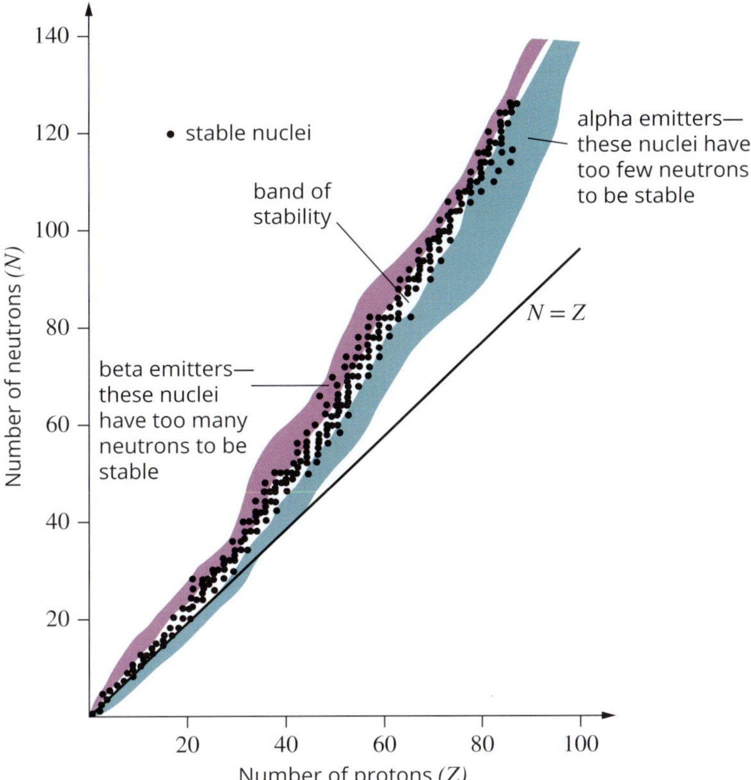

FIGURE 3.2.6 A plot of number of neutrons versus number of protons in various nuclei. The region of stable nuclei is called the band of stability (white). Nuclei that lie outside this region (in the purple and blue regions) are unstable, so they are radioactive.

Alpha particles are made up of two protons and two neutrons, which is the same as a helium nucleus. They are emitted by nuclei that have too few neutrons to be stable. Alpha particles are a lower energy form of radiation, travelling no further than a few centimetres from their source. They can be stopped by a sheet of paper or an equivalent material (Figure 3.2.7a).

Beta particles are high-energy electrons formed from the decay of nuclei with too many neutrons. They have more penetrating power than alpha particles, but can be blocked by, for example, an aluminium plate several centimetres thick (Figure 3.2.7b). In the process of producing a beta particle, a neutron is converted to a proton, so the atomic number of the element increases by one.

Gamma radiation is a type of high-energy **electromagnetic radiation**. It has great penetrating power and can only be stopped by something as dense as a lead plate several centimetres thick (Figure 3.2.7c).

Natural and artificial radioisotopes and balanced nuclear reactions

The decay of radioisotopes to produce different types of radiation can be represented by equations of **balanced nuclear reactions**.

For example, the decay of naturally occurring uranium-238 produces thorium-234, an alpha particle and gamma radiation. This decay is represented by the following equation:

$$^{238}_{92}\text{U} \rightarrow \, ^{234}_{90}\text{Th} + \, ^{4}_{2}\text{He} + \gamma$$

The Greek letter γ is used to represent gamma radiation. Notice that when the decay produces alpha particles, a new element is also formed.

(a)

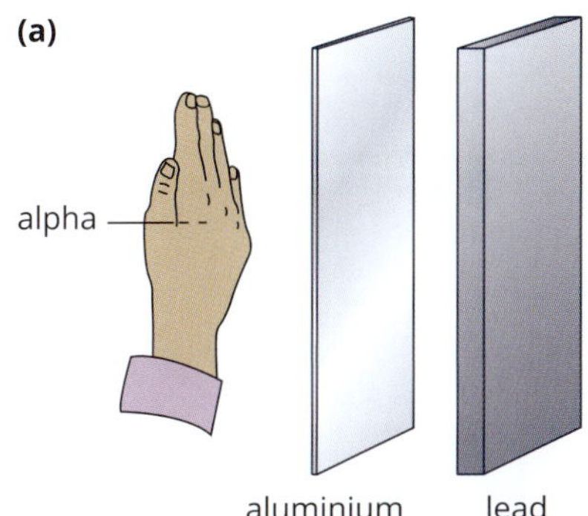

(b)

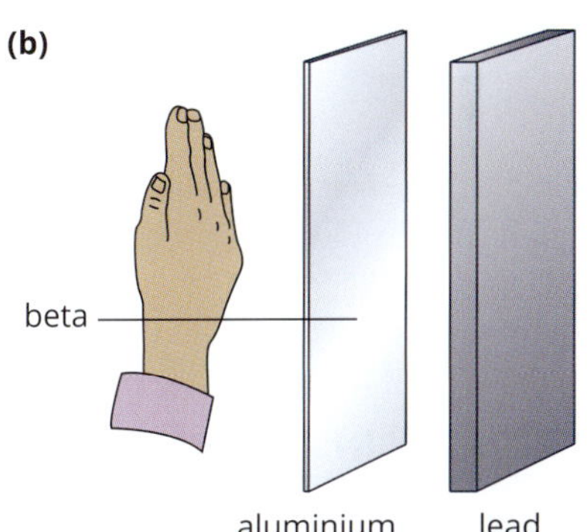

(c)

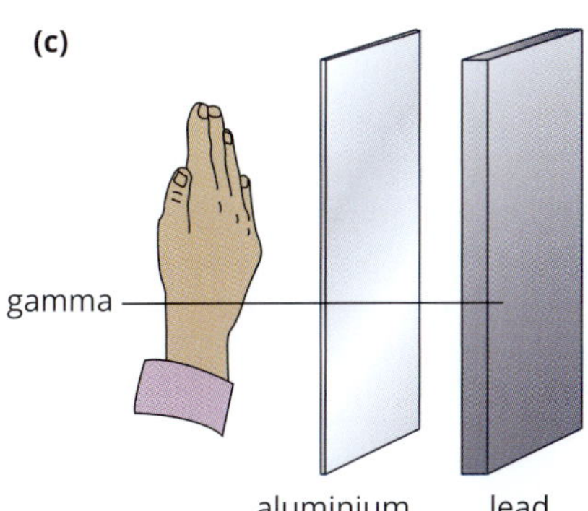

FIGURE 3.2.7 (a) Alpha particles, (b) beta particles and (c) gamma particles have different penetrating powers.

When writing an equation for a balanced nuclear reaction, check that the atomic numbers add up to the same value on both sides of the equation. You will also need to do the same check for mass numbers.

Carbon-14, mentioned earlier in this section, is a radioisotope that decays to produce beta particles (electrons). Its decay can be represented by the following equation:

$$^{14}_{6}C \rightarrow {}^{14}_{7}N + {}^{0}_{-1}e$$

where

${}^{0}_{-1}e$ represents a beta particle (an electron).

The decay of carbon-14 can be used to determine the age of a once-living organism, such as a fossil, in a technique known as carbon dating. Living organisms take up carbon-14 from the environment, so the ratio of carbon-12 to carbon-14 in the organism remains fairly constant over their lifetime. After the organism dies, it can no longer take up carbon-14, so the amount of carbon-14 in the organism decreases as the isotope decays. By looking at the ratio of carbon-12 to carbon-14, the age of the fossilised object can be determined.

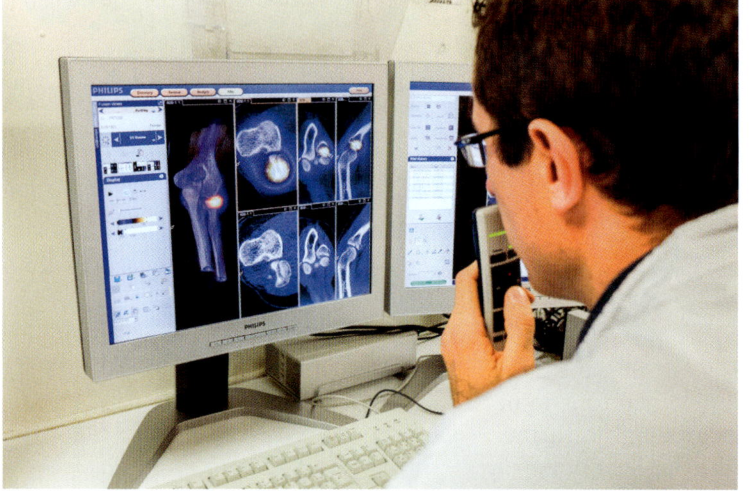

FIGURE 3.2.9 A doctor reviews the results of a bone scintigraphy scan.

Technetium, which was discovered in 1936, is an artificial radioisotope that is routinely used in medicine. The medical radioisotope technetium-99m (the 'm' representing a nuclear isomer) is produced in nuclear reactors from the decay of uranium-238 to molybdenum-99. The molybdenum-99 is isolated and, through further decay, technetium-99m is formed, as shown in the following equation:

$$^{99}_{42}Mo \rightarrow {}^{99m}_{43}Tc + {}^{0}_{-1}e$$

Compounds containing technetium-99m can be injected into various body organs or into a cancerous tumour in a patient. For example, a type of bone scintigraphy scan involves injecting the technetium-99m compound into a vein. When the technetium-99m decays, it releases gamma radiation, which is then detected by nuclear imaging equipment (Figure 3.2.9).

CHEMISTRY IN ACTION 

Carbon-14 and the illegal ivory trade

Isotopes are important in combatting ivory poaching in Africa. By performing carbon dating on the elephant tusks and ivory (Figure 3.2.8), scientists can determine the age of the tusks.

If the authorities know the age of the ivory, then legal action can be taken against the poachers and sellers of the product. Ivory products crafted prior to 1989 are allowed to be traded. Since 1989, the trade of ivory has been made illegal worldwide.

FIGURE 3.2.8 These tusks are from African elephants killed for their ivory.

3.2 Review

SUMMARY

- The number of subatomic particles in an atom can be determined from an element's atomic number (Z) and mass number (A).
- The atomic number indicates how many protons or electrons an atom has.
- The mass number indicates how many protons and neutrons are in the nucleus of the atom.
- Isotopes are atoms with the same atomic number but different mass numbers, i.e. they have the same number of protons but different numbers of neutrons.
- Isotopes have the same chemical properties but different physical properties such as mass, density and radioactivity.
- Unstable isotopes that decay and release radiation are called radioisotopes.
- The three types of radiation released are alpha particles, beta particles and gamma radiation.
- The decay of a radioisotope can be represented by equations of balanced nuclear reactions.

KEY QUESTIONS

1. What term is given to the number of protons and neutrons in the nucleus of an atom?
2. Calculate the numbers of protons, neutrons and electrons in the atom ${}^{31}_{15}P$.
3. What is the correct name of the element that has an atom with seven protons and eight neutrons?
4. Explain the similarities and differences between isotopes of the same element.
5. How many more neutrons does an atom of carbon-14 have than an atom of the carbon-12 isotope?
6. Describe the difference between an alpha particle, a beta particle and gamma radiation.
7. Use the periodic table to balance the equations for the following nuclear reactions:
   - **a** ${}^{234}_{90}Th \rightarrow {}^{234}_{91}Pa + \_\_\_$
   - **b** ${}^{234}_{91}Pa \rightarrow \_\_\_ + {}^{4}_{2}He$
   - **c** $\_\_\_ \rightarrow {}^{216}_{84}Po + {}^{4}_{2}He$
   - **d** ${}^{216}_{84}Po \rightarrow \_\_\_ + {}^{0}_{-1}e$

3.3 Masses of particles

Atoms are particles so small that they cannot be counted individually or even in groups of thousands or millions. The mass of one atom is incredibly small. For example, one atom of carbon has a mass of approximately 2×10^{-23} g.

Figure 3.3.1a shows a model of a glucose **molecule**. Glucose is a type of sugar used as an energy source by almost all living organisms on Earth. One glucose molecule contains 6 carbon atoms, 12 hydrogen atoms and 6 oxygen atoms bonded together. One glucose molecule has an actual mass of 3×10^{-22} g. This means that the teaspoon of glucose crystals shown in Figure 3.3.1b contains approximately 1.4×10^{22} glucose molecules.

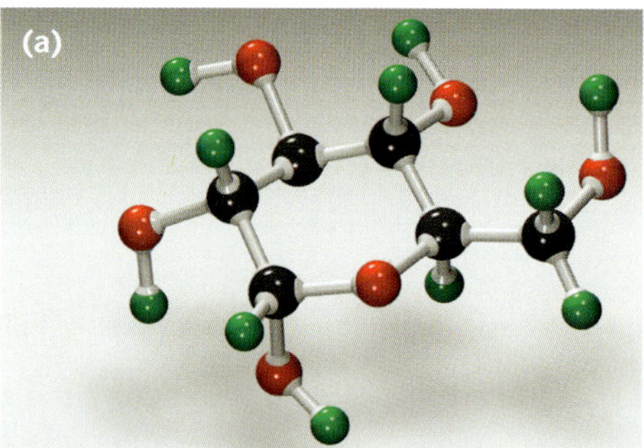

FIGURE 3.3.1 (a) A model of a single glucose molecule, which contains 6 carbon atoms (black), 12 hydrogen atoms (green) and 6 oxygen atoms (red) bonded together. (b) A teaspoon of glucose crystals contains an incredibly large number of extremely small glucose molecules.

Such small masses are not easily measured and can be inconvenient to use in calculations. This section will introduce you to the ways scientists determine and use the masses of various particles.

RELATIVE MASSES

From earlier in the chapter, you will remember that isotopes are atoms of the same element that have different numbers of neutrons in their nuclei. So isotopes have the same atomic number but different mass number.

The masses used most frequently in chemistry are relative masses, rather than actual masses. The standard to which all masses are compared is the mass of an atom of the common isotope of carbon, **carbon-12** or ^{12}C, which is given a mass of exactly 12.

Carbon-12 is the universal reference standard for relative atomic mass.

Carbon-12 was selected as the standard in 1961 (Figure 3.3.2). Before then, oxygen was used as the standard. Chemists and physicists could not agree on a way of assigning a standard mass to oxygen. Chemists assigned a mass of exactly 16 to the average mass of oxygen atoms. Physicists assigned a mass of exactly 16 to the oxygen-16 isotope. This resulted in two different tables of slightly different atomic masses.

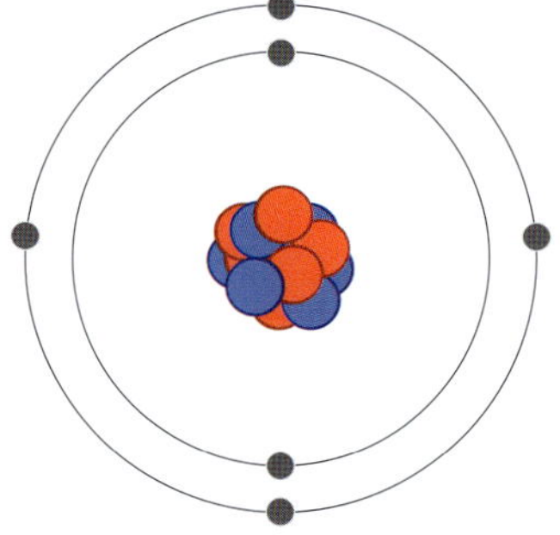

FIGURE 3.3.2 A Bohr-style diagram of carbon-12.

Making carbon-12 the standard was a compromise both agreed to. Assigning carbon-12 a mass of exactly 12 created a new scale that was adopted universally.

RELATIVE ISOTOPIC MASSES

The individual isotopes of each element have a **relative isotopic mass**, symbol I_r. Table 3.3.1 shows the approximate mass of various isotopes relative to the carbon-12 isotope, taken as 12 units exactly.

TABLE 3.3.1 Approximate mass of various isotopes relative to the ^{12}C isotope, which is taken as exactly 12 units

Isotope	Diagram of nucleus	Number of protons in nucleus	Number of neutrons in nucleus	Total number of protons and neutrons	Approximate mass of atom relative to carbon-12 isotope
hydrogen $^{1}_{1}H$		1	0	1	$\frac{1}{12}$ of 12 = 1
helium $^{4}_{2}He$		2	2	4	$\frac{4}{12}$ of 12 = 4
lithium $^{7}_{3}Li$		3	4	7	$\frac{7}{12}$ of 12 = 7
carbon $^{12}_{6}C$		6	6	12	12

As shown in Figure 3.3.3, there are two different isotopes of the element chlorine. They are:

- $^{35}_{17}Cl$, which contains 17 protons and 18 neutrons
- $^{37}_{17}Cl$, which contains 17 protons and 20 neutrons.

The isotopes have different masses because of the different numbers of neutrons. Remember that the mass of an atom is mainly determined by the number of protons and neutrons in its nucleus, because the mass of the electrons is relatively small.

The relative isotopic masses of the two chlorine isotopes are experimentally determined to be 34.969 (^{35}Cl) and 36.966 (^{37}Cl). Since the masses of a proton and a neutron are similar and close to 1 on the ^{12}C = 12 scale, the relative isotopic mass of an isotope is almost, but not exactly, equal to the mass number of that isotope.

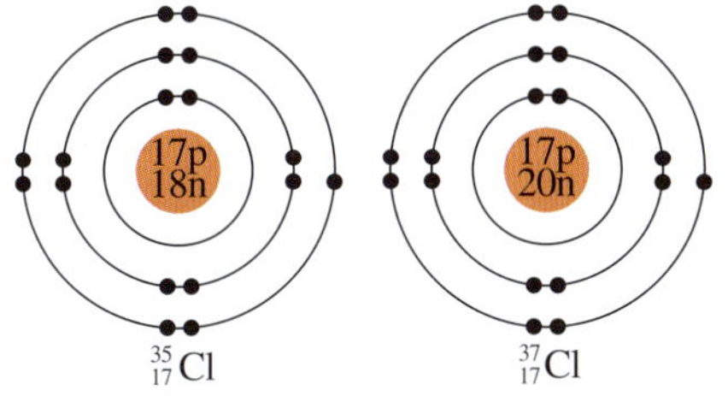

FIGURE 3.3.3 The two isotopes of the element chlorine. Different numbers of neutrons in the nuclei of these atoms give the atoms different masses.

 The relative isotopic mass of an isotope is the mass of an atom of that isotope relative to the mass of an atom of carbon-12 (^{12}C), taken as 12 units exactly.

Relative isotopic abundance

Naturally occurring chlorine is made up of the two isotopes shown in Figure 3.3.3: 75.80% of the lighter isotope and 24.20% of the heavier isotope. This composition is virtually the same, no matter the source of the chlorine. The percentage abundance of an isotope in the natural environment is called its **relative isotopic abundance**.

Most elements, like chlorine, are a mixture of two or more isotopes. Details of the isotopes of some common elements are shown in Table 3.3.2 on page 88.

TABLE 3.3.2 Isotopic composition of some common elements

Element	Isotopes	Relative isotopic mass	Relative isotopic abundance (%)
hydrogen	^{1}H	1.008	99.986
	^{2}H	2.014	0.014
	^{3}H	3.016	0.0001
oxygen	^{16}O	15.995	99.76
	^{17}O	16.999	0.04
	^{18}O	17.999	0.20
silver	^{107}Ag	106.9	51.8
	^{109}Ag	108.9	48.2

The mass spectrometer

Relative isotopic masses of elements and their isotopic abundances are determined by using an instrument called a **mass spectrometer**, which was invented by Francis Aston in 1919.

A mass spectrometer, as seen in Figure 3.3.4, separates the individual isotopes in a sample of an element and determines the mass of each isotope, relative to the carbon-12 isotope, and the relative abundances of the isotopes.

FIGURE 3.3.4 A mass spectrometer is used to determine the relative masses and abundances of different isotopes.

CHEMISTRY IN ACTION 

Mass spectrometry reveals history

Scientists can use mass spectrometry to determine the abundance of one isotope relative to that of another in tissues taken from a dead organism. This can help them determine important features about that organism's life.

For example, archaeologists can analyse the fossil jawbone shown in Figure 3.3.5 to calculate how long ago the short-necked giraffe lived, to study its diet, or to determine the habitat of the animal. Isotope analysis of a number of related fossils can help track the evolution of this species.

FIGURE 3.3.5 An ancient fossil jawbone from an extinct short-necked giraffe. Isotope analysis using a mass spectrometer provides archaeologists with information about the organism's life.

+ ADDITIONAL ICT

Operation of a mass spectrometer

A mass spectrometer, like the one in Figure 3.3.4, is a laboratory instrument used to detect isotopes of an element and to determine both the relative mass and abundance of each isotope.

The construction of a mass spectrometer is shown in Figure 3.3.6.

Several different stages are involved in the operation of a mass spectrometer.

1 A sample of the element is vaporised and the gas is injected into the ionisation chamber.
2 In the ionisation chamber, the sample is **ionised** (positive ions are formed) when a beam of bombarding electrons dislodges electrons from the sample atoms.
3 The positive ions enter an electric field, which accelerates them to high speeds.
4 The fast-moving positive ions then enter a magnetic field perpendicular to their path. This causes the ions to be deflected, or move in a curved path with a radius that depends upon the mass-to-charge ratio (*m/z* or *m/e*) of the ions.
5 Ions of a certain mass and charge reach a detector, which measures the current produced by a particular ion.
6 The accelerating voltage and strength of the magnetic field are varied to allow different ions to reach the detector, which measures the mass of the ions and their relative abundance. This data is recorded as a mass spectrum.

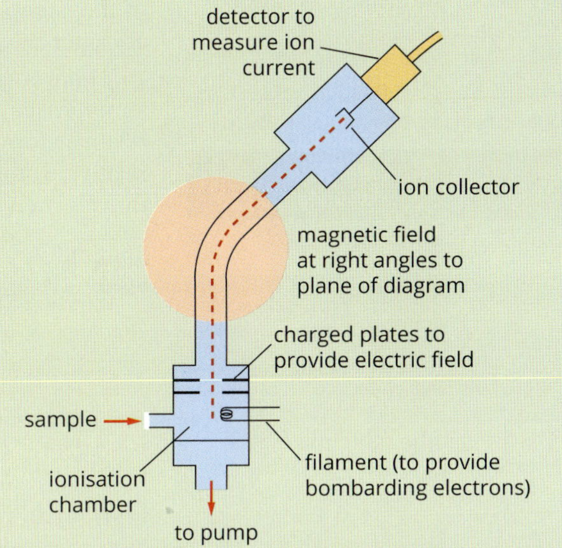

FIGURE 3.3.6 The construction of a mass spectrometer. The sample is ionised and the ions pass through an electric field and a magnetic field before reaching the detector.

Mass spectra

The output from a mass spectrometer is called a **mass spectrum**. Figure 3.3.7 shows the mass spectrum of chlorine. In this mass spectrum:

- The number of peaks indicates the number of isotopes, in this case, two.
- The horizontal axis indicates the relative mass of each isotope present in an element according to the isotope's mass-to-charge ratio (*m/z* or *m/e*). (The charge on most ions reaching the detector is +1, so the mass of an isotope can be read directly from the horizontal axis.) The two isotopes of chlorine have relative masses of 35 and 37.
- The vertical axis indicates the abundance of each isotope in the sample: 75% ^{35}Cl and 25% ^{37}Cl.

In summary, in the mass spectrum of an element:

- the number of peaks indicates the number of isotopes
- the position of each peak on the horizontal axis indicates the relative isotopic mass
- the relative heights of the peaks correspond to the relative abundances of the isotopes.

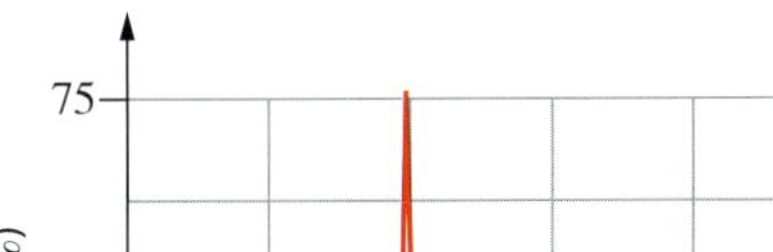
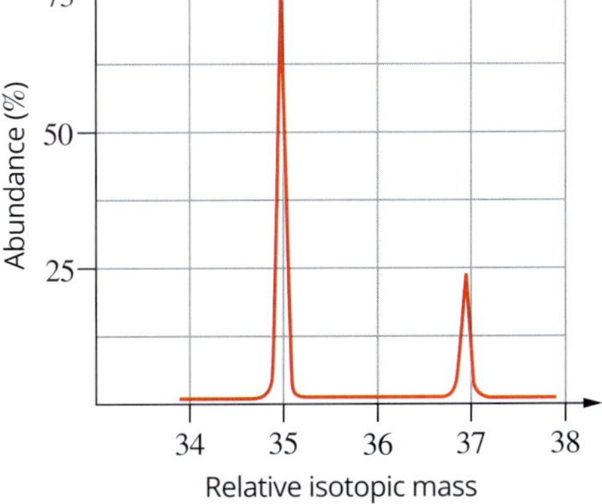

FIGURE 3.3.7 The mass spectrum of chlorine. The two peaks indicate there are two isotopes. The most abundant isotope has a relative isotopic mass of approximately 35 and an abundance of approximately 75%. The other isotope has a relative isotopic mass of approximately 37 and an abundance of approximately 25%.

RELATIVE ATOMIC MASS

> The relative atomic mass of an element is the weighted average of the relative masses of the isotopes of the element on the ^{12}C scale.

Most elements consist of a mixture of isotopes. For the purpose of the calculations you will be doing later in this chapter, it is convenient to know the average relative mass of an atom in this mixture. This average is called the **relative atomic mass** of an element, and given the symbol A_r. To calculate the average of the relative masses of the isotopes that exist in a naturally occurring mixture of an element, you must consider the relative abundances of each isotope.

Data obtained from the mass spectrum of chlorine, shown in Figure 3.3.7, is summarised in Table 3.3.3. The data about the two isotopes is used to calculate the relative atomic mass of chlorine. A weighted average mass is calculated by using the relative isotopic masses and abundances to find the total mass of 100 atoms. This mass is then divided by 100 to find the average mass of one atom.

TABLE 3.3.3 Isotopic composition of chlorine

Isotope	I_r	Relative abundance (%)
^{35}Cl	34.969	75.80
^{37}Cl	36.966	24.20

> The relative atomic mass (A_r) of an element with two isotopes can be calculated using the formula:
>
> $$A_r = \frac{(I_r \times \% \text{ abundance}) + (I_r \times \% \text{ abundance})}{100}$$

GO TO ➤ Periodic table page 553

The **periodic table** at the end of the book provides relative atomic masses, calculated by taking into account the relative abundances of all the natural isotopes of each element.

Worked example 3.3.1

CALCULATING RELATIVE ATOMIC MASS FROM ISOTOPIC MASSES AND PERCENTAGE ABUNDANCES

Chlorine has two isotopes. Their relative isotopic masses and percentage abundances are provided in Figure 3.3.7 and Table 3.3.3. Calculate the relative atomic mass of chlorine correct to 2 decimal places.

Thinking	Working
Determine the relative isotopic masses (I_r) and abundances of each isotope.	Two peaks on the spectrum indicate two isotopes: First isotope: I_r 34.969; abundance 75.80% Second isotope: I_r 36.966; abundance 24.20%
Substitute the relative isotopic masses and abundances into the formula for calculating relative atomic mass (A_r): $A_r = \frac{(I_r \times \% \text{ abundance}) + (I_r \times \% \text{ abundance})}{100}$	$A_r = \frac{(34.969 \times 75.80) + (36.966 \times 24.20)}{100}$
Calculate the relative atomic mass.	$A_r = \frac{2650.65 + 894.58}{100}$ $= 35.452$
Express the answer to two decimal places.	$A_r(Cl) = 35.45$

Worked example: Try yourself 3.3.1

CALCULATING RELATIVE ATOMIC MASS FROM ISOTOPIC MASSES AND PERCENTAGE ABUNDANCES

Boron has two isotopes. Their relative isotopic masses and percentage abundances are provided. Calculate the relative atomic mass of boron correct to 2 decimal places.

Isotope	I_r	Relative abundance (%)
^{10}B	10.013	19.91
^{11}B	11.009	80.09

Worked example 3.3.2

CALCULATING PERCENTAGE ABUNDANCE OF EACH ISOTOPE FROM THE MASS SPECTRUM

Use the simplified mass spectrum of magnesium to calculate the percentage abundance of each of its three isotopes.

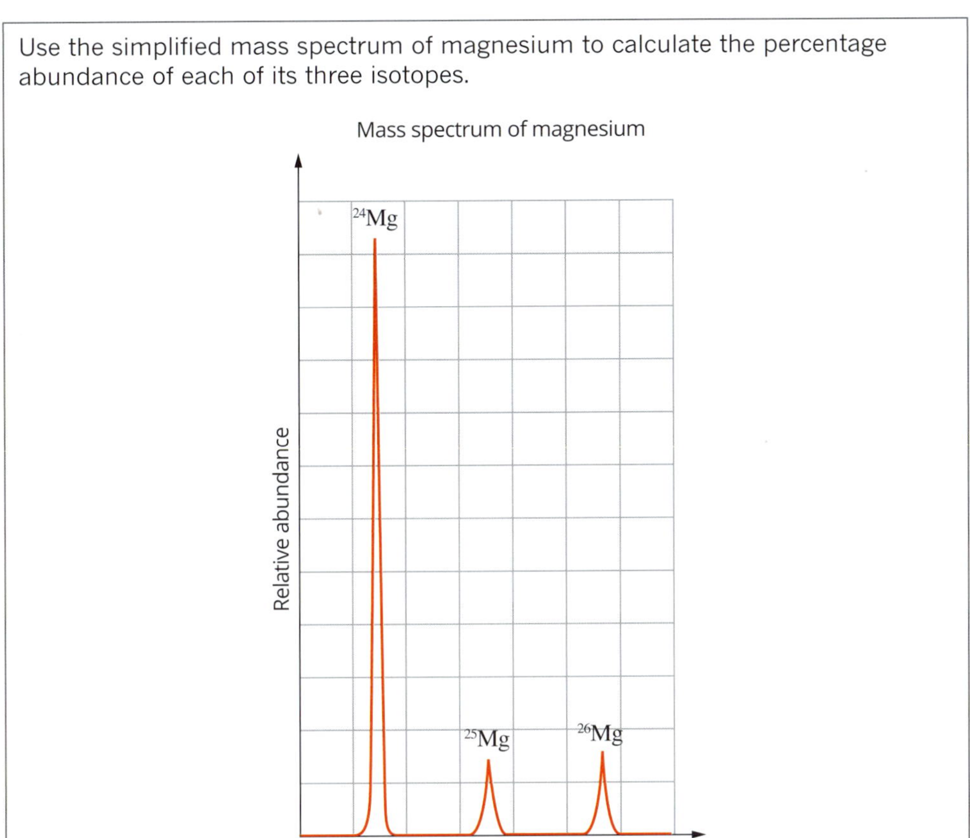

Thinking	Working
Measure the peak height for each isotope using a ruler.	From the spectrum, the height of each peak is: $^{24}Mg = 7.9$ cm $^{25}Mg = 1.0$ cm $^{26}Mg = 1.1$ cm
Calculate the total peak height for the three isotopes by adding the individual peak heights.	Total peak height = 7.9 + 1.0 + 1.1 = 10.0 cm
Substitute the peak height for each isotope into the formula: $\% \text{ abundance} = \frac{\text{peak height}}{\text{total peak height}} \times 100\%$	$\% \text{ abundance } ^{24}Mg = \frac{7.9}{10.0} \times 100 = 79\%$ $\% \text{ abundance } ^{25}Mg = \frac{1.0}{10.0} \times 100 = 10\%$ $\% \text{ abundance } ^{26}Mg = \frac{1.01}{10.0} \times 100 = 11\%$

Worked example: Try yourself 3.3.2

CALCULATING PERCENTAGE ABUNDANCE OF EACH ISOTOPE FROM THE MASS SPECTRUM

The graph shows a simplified mass spectrum of lead. From this mass spectrum, calculate the percentage abundance of each of its three isotopes.

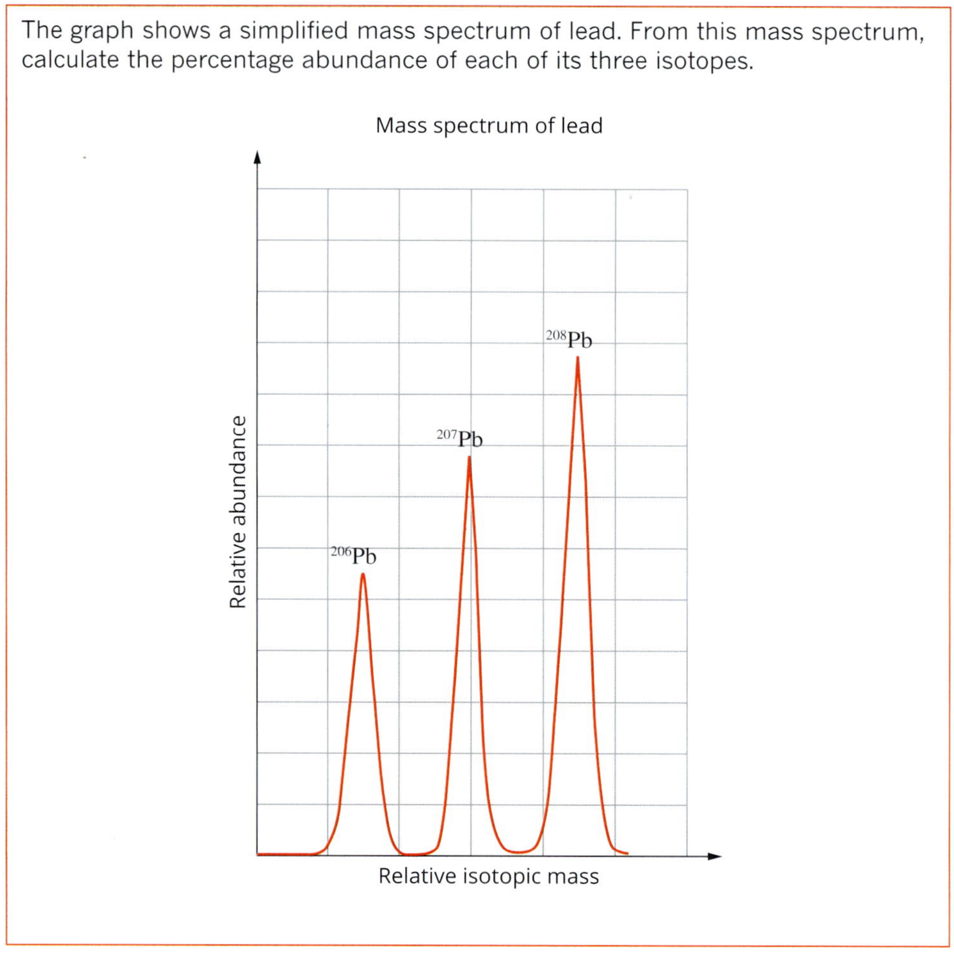

Worked example 3.3.3

CALCULATING PERCENTAGE ABUNDANCES FROM RELATIVE ATOMIC MASS AND RELATIVE ISOTOPIC MASSES

The relative atomic mass of rubidium is 85.47. The relative isotopic masses of its two isotopes are 84.95 and 86.94. Calculate the relative abundances of the isotopes in naturally occurring rubidium.

Thinking	Working
State the relative abundances of the isotopes in terms of *x*, where *x* is the abundance of the lighter isotope. Abundance of lighter isotope = *x*. The abundance of the heavier isotope must equal 100 – *x*.	Abundance of 84.95 isotope = *x*. Abundance of 86.94 isotope = 100 – *x*.
Substitute the relative isotopic masses, relative abundances and relative atomic mass into the formula: $A_r = \frac{(I_r \times \% \text{ abundance}) + (I_r \times \% \text{ abundance})}{100}$	$85.47 = \frac{84.95x + (86.94 \times (100 - x))}{100}$
Expand the top line of the equation.	$85.47 = \frac{84.95x + 8694 - 86.94x}{100}$
Solve the equation to find *x*, the relative abundance of the lightest isotope.	$8547 = 84.95x + 8694 - 86.94x$ $8547 - 8694 = 84.95x - 86.94x$ $-147 = -1.99x$ $x = 73.87\%$
Determine the abundance of the heavier isotope.	Abundance of 86.94 isotope $= 100 - x$ $= 100 - 73.87$ $= 26.13\%$

Worked example: Try yourself 3.3.3

CALCULATING PERCENTAGE ABUNDANCES FROM RELATIVE ATOMIC MASS AND RELATIVE ISOTOPIC MASSES

The relative atomic mass of copper is 63.54. The relative isotopic masses of its two isotopes are 62.95 and 64.95. Calculate the relative abundances of the isotopes in naturally occurring copper.

RELATIVE MOLECULAR MASS

GO TO ➤ Section 5.3 page 162

Some elements and compounds exist as molecules, for example, oxygen (O_2), nitrogen (N_2) and carbon dioxide (CO_2). In Chapter 5, you will learn how non-metallic elements bond together to form covalent molecules. For these substances, a **relative molecular mass** can be determined.

The relative molecular mass, symbol M_r, is equal to the sum of the relative atomic masses of the atoms in the molecule (remember you can obtain relative atomic masses of elements from the periodic table). It is based (like relative atomic mass) on the scale in which the exact mass of an atom of carbon-12 is taken as 12 exactly.

ℹ The relative molecular mass is the mass of one molecule of that substance relative to the mass of a ^{12}C atom taken as 12 units exactly. It is numerically equal to the sum of the relative atomic masses of the elements in the formula for the molecule.

Worked example 3.3.4

CALCULATING THE RELATIVE MOLECULAR MASS OF MOLECULES

Calculate the relative molecular mass of carbon dioxide (CO_2).

Thinking	Working
Use the periodic table to find the relative atomic mass for the elements represented in the formula.	$A_r(C) = 12.01$ $A_r(O) = 16.00$
Determine the number of atoms of each element present, taking into consideration any brackets in the formula.	1 × C atom 2 × O atoms
Determine the relative molecular mass by adding the appropriate relative atomic masses.	$M_r = 1 \times A_r(C) + 2 \times A_r(O)$ $= 1 \times 12.01 + 2 \times 16.00$ $= 44.01$

Worked example: Try yourself 3.3.4

CALCULATING THE RELATIVE MOLECULAR MASS OF MOLECULES

Calculate the relative molecular mass of nitric acid (HNO_3).

GO TO ➤ Section 5.2 page 147

The relative formula mass is the mass of a formula unit relative to the mass of an atom of ^{12}C taken as 12 units exactly. It is numerically equal to the sum of the relative atomic masses of the elements in the formula.

Relative molecular mass and relative formula mass, like relative atomic mass, have no units, because they are based on the masses of atoms of elements compared with the mass of the carbon-12 isotope.

RELATIVE FORMULA MASS

You will learn in Chapter 5 that some compounds, such as sodium chloride (NaCl) and magnesium oxide (MgO), do not exist as molecules but rather as ionic networks. For ionic compounds, the term **relative formula mass** is used. Relative formula mass, like relative molecular mass, is calculated by taking the sum of the relative atomic masses of the elements in the formula.

Worked example 3.3.5

CALCULATING THE RELATIVE FORMULA MASS OF IONIC COMPOUNDS

Calculate the relative formula mass of magnesium hydroxide ($Mg(OH)_2$).

Thinking	Working
Use the periodic table to find the relative atomic mass for the elements represented in the formula.	$A_r(Mg) = 24.31$ $A_r(O) = 16.00$ $A_r(H) = 1.008$
Determine the number of atoms of each element present, taking into consideration any brackets in the formula.	1 × Mg atom 1 × 2 = 2 O atoms 1 × 2 = 2 H atoms
Determine the relative formula mass by adding the appropriate relative atomic masses.	Relative formula mass $= 1 \times A_r(Mg) + 2 \times A_r(O) + 2 \times A_r(H)$ $= 24.31 + 2 \times 16.00 + 2 \times 1.008$ $= 58.33$

Worked example: Try yourself 3.3.5

CALCULATING THE RELATIVE FORMULA MASS OF IONIC COMPOUNDS

Calculate the relative formula mass of copper(II) nitrate ($Cu(NO_3)_2$).

3.3 Review

SUMMARY

- Most elements consist of a mixture of isotopes.
- The most common isotope of carbon, carbon-12, is used as the reference standard against which to compare the masses of atoms.
- The carbon-12 isotope is assigned a mass of exactly 12 units.
- The relative isotopic mass, I_r, of an isotope is the mass of an atom of the isotope relative to the mass of an atom of carbon-12 taken as 12 units exactly.
- The relative isotopic mass and relative abundance of isotopes can be measured using a mass spectrometer.
- The relative atomic mass, A_r, of an element is a weighted average of its isotopic masses.
- The relative molecular mass of molecules, or relative formula mass of ionic compounds, is calculated from the sum of the relative atomic masses of its constituent elements.

KEY QUESTIONS

1 The isotopic composition of chlorine is shown in Table 3.3.3 on page 90. Select the correct statement about the relative atomic mass of chlorine.

- **A** The precise relative atomic mass of chlorine can be determined by adding the two relative isotopic masses and dividing by two.
- **B** The relative atomic mass of chlorine is based on the scale where carbon-12 has a mass of exactly 12.
- **C** The lighter isotope is less abundant.
- **D** The isotopes have different masses because they have different numbers of protons.

2 Use the data in Table 3.3.2 on page 88 to calculate the relative atomic mass of:

- **a** oxygen
- **b** silver
- **c** hydrogen.

3 The element lithium has two isotopes:
^{6}Li with a relative isotopic mass of 6.02
^{7}Li with a relative isotopic mass of 7.02.
The relative atomic mass of lithium is 6.94. Calculate the percentage abundance of the lighter isotope.

4 The mass spectrum of zirconium is shown in the graph on the right.

- **a** Measure the peak heights to calculate the percentage abundance of each zirconium isotope.
- **b** Use the percentage abundances calculated in part **a** to determine the relative atomic mass of zirconium. The mass number is a good approximation to the relative isotopic mass.

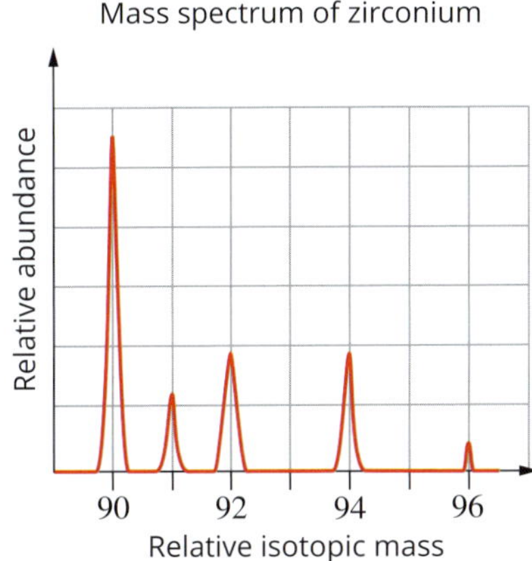

5 Calculate the relative molecular mass of:

- **a** sulfuric acid (H_2SO_4)
- **b** ammonia (NH_3)
- **c** ethane (C_2H_6).

6 Calculate the relative formula mass of:

- **a** potassium chloride (KCl)
- **b** sodium carbonate (Na_2CO_3)
- **c** aluminium sulfate ($Al_2(SO_4)_3$).

3.4 Electronic structure of atoms

When fireworks explode, they create a spectacular show of coloured lights (Figure 3.4.1). The light is produced by metal atoms that have been heated by the explosion. These coloured lights posed a significant problem for early scientists. The models the scientists were using could not explain the source of the light. However, the light was a clue that ultimately led to a better understanding of the arrangement of electrons in atoms.

FIGURE 3.4.1 The spectacular colours in this New Year's Eve fireworks display are emitted by metal atoms that have been heated to very high temperatures.

FLAME TESTS

A **flame test** is a simple method that can be used to determine the identity of a metal in a sample. A sample of the compound of the metal to be identified is inserted into a non-luminous Bunsen burner flame, as shown in Figure 3.4.2.

When the metal atoms are heated, they give off light of a characteristic colour. This means that the metal in an unknown sample can be identified by comparing the flame colour with the known characteristic colours produced by metals. Some examples of the flame colours produced by metals can be seen in Figure 3.4.3.

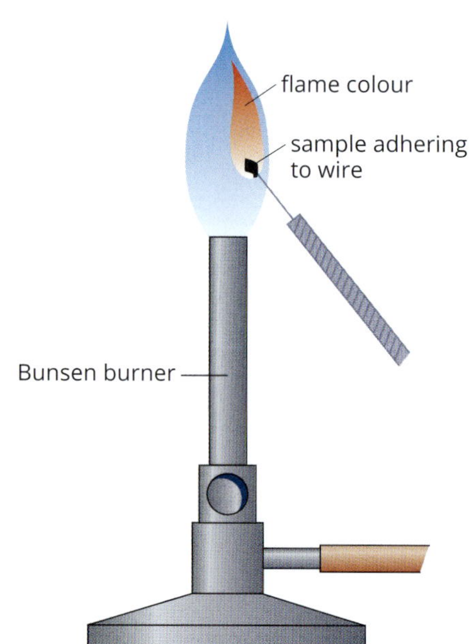

FIGURE 3.4.2 Performing a flame test. A moist wire has been dipped in the sample and then placed in the flame. A fine spray of solution from a spray bottle could be used instead.

FIGURE 3.4.3 The colour of each flame is caused by the different metal compounds present and can be used to identify the metals. The flame colours shown here are due to (from left to right): barium (yellow-green), lithium (crimson), strontium (scarlet), sodium (yellow), copper (green) and potassium (lilac).

EMISSION SPECTRA

When atoms are heated, they give off electromagnetic radiation or light. If the light passes through a prism, it produces a spectrum with a black background and a number of coloured lines. Figure 3.4.4 shows the apparatus used to produce these spectra.

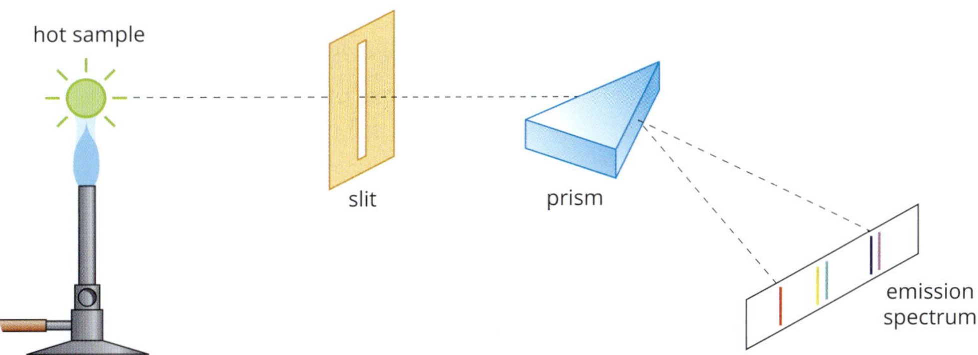

FIGURE 3.4.4 The apparatus used to analyse the light given out when an element is heated. The coloured lines are called an emission spectrum.

These spectra are known as line spectra or **emission spectra** and are related to the electronic structure within the atoms. Each emission spectrum is unique for a particular element and can be used to identify the elements present.

The line spectrum produced by helium is shown in Figure 3.4.6.

FIGURE 3.4.6 The emission spectrum of helium is made up of lines ranging from violet to red in colour.

Each line in the spectrum corresponds to light of a different energy. Violet lines, at the left of the spectrum, correspond to light with high energy. Moving to the right, the lines have successively lower energies. Red light, at the right of the spectrum, is the lowest-energy light visible to the human eye.

Information from emission spectra

Emission spectra give clues about the electronic structure of atoms. Two important clues are:

- atoms of the same element produce identical line spectra
- each element has a unique line spectrum and therefore a unique electronic structure.

CHEMFILE 

How helium got its name

Helium was discovered on the Sun before being discovered on Earth. The French astronomer Jules Janssen discovered helium in 1868 while studying the light from a solar eclipse in India similar to the solar eclipse shown in Figure 3.4.5. Although the spectrum showed the full range of colours, the bright yellow line in the helium spectrum stood out. The line could not be matched to the line spectra of any of the known elements. It was concluded that the line belonged to an unknown element. This element was named helium after the Greek Sun god, Helios.

FIGURE 3.4.5 The emission spectrum of helium was first detected in sunlight during a solar eclipse like the one shown here.

THE BOHR MODEL OF THE ATOM

In 1913, Danish physicist Niels Bohr developed a new model of the hydrogen atom that explained emission spectra. The **Bohr model** proposed the following:

- Electrons revolve around the nucleus in fixed, circular orbits.
- The electrons' orbits correspond to specific **energy levels** in the atom.
- Electrons can only occupy fixed energy levels and cannot exist between two energy levels.
- Orbits of larger radii correspond to higher energy levels.

In the Bohr model, it is possible for electrons to move between the energy levels by absorbing or emitting energy in the form of light. Bohr's model (Figure 3.4.7) gave close agreement between the calculated energies for lines in the hydrogen spectrum and the values observed in the spectrum.

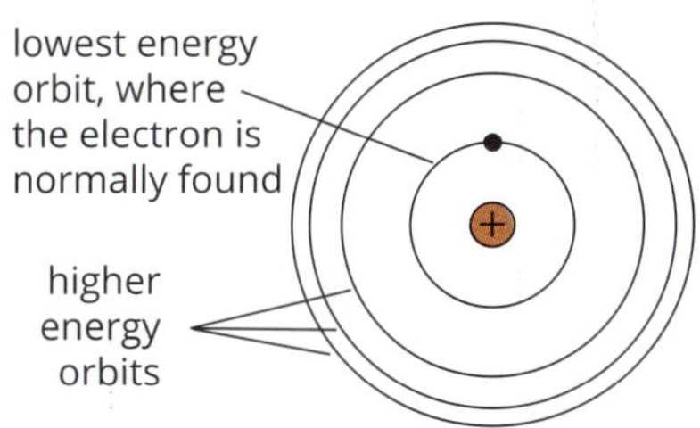

FIGURE 3.4.7 The Bohr model of a hydrogen atom. Bohr suggested that electrons move in orbits of particular energies.

Electron shells

Scientists quickly extended Bohr's model of the hydrogen atom to other atoms. They proposed that electrons are grouped into different energy levels, called **electron shells**. The electron shells are labelled with the number $n = 1, 2, 3 \ldots$, as shown in Figure 3.4.8.

The orbit in which an electron moves depends on the energy of the electron; electrons with low energy are in orbits close to the nucleus, while high-energy electrons are in outer orbits.

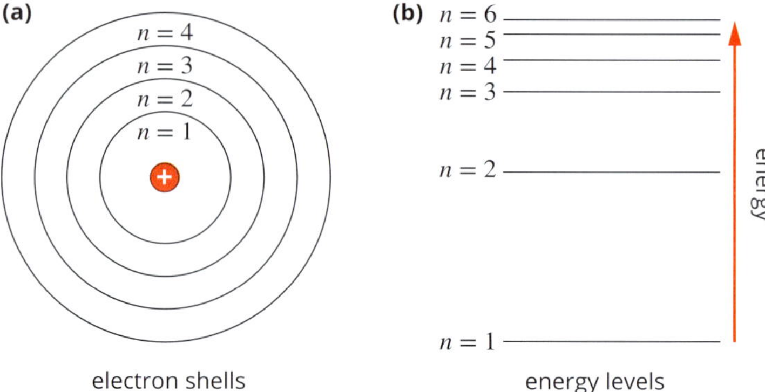

FIGURE 3.4.8 (a) The electron shells of an atom are labelled using integers $n = 1, 2, 3 \ldots$. The first shell is the shell closest to the nucleus, and the radii of the shells increase as the shell number increases. (b) Each shell corresponds to an energy level that electrons can occupy. The first shell has the lowest energy, and the energy increases as the shell number increases.

The lowest energy shell is the shell closest to the nucleus and is labelled $n = 1$. Shells with higher values of n correspond to higher energy levels. As the values of n increase, the energy levels get closer together.

Emission spectra and the shell model

Heating an element can cause an electron to absorb energy and jump to a higher energy level. Shortly afterwards, the electron returns to the lower energy level, releasing a fixed amount of energy as light. The electron can return in a number of different ways, some of which are shown in Figure 3.4.9 as coloured arrows. Each one of the particular pathways produces light of a particular colour in the emission spectrum.

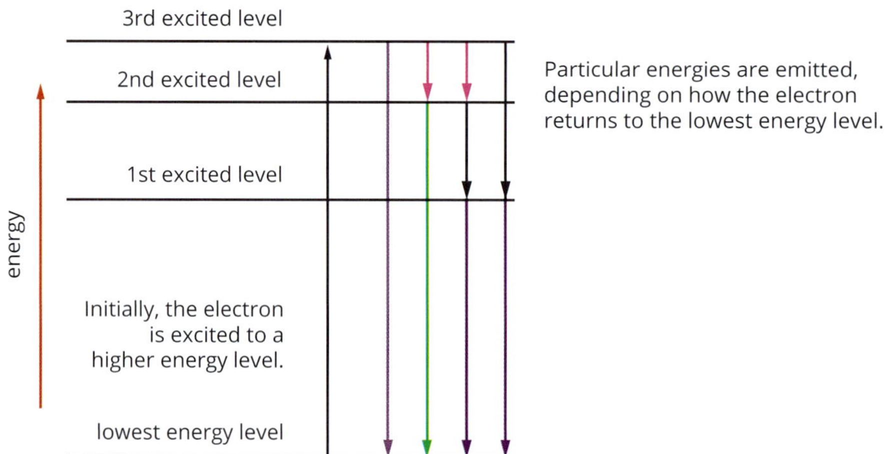

FIGURE 3.4.9 Emission of energy as electrons move from a higher to a lower energy level.

Consider a hydrogen atom with one proton and one electron. Usually, the electron exists in the $n = 1$ shell. This is the lowest energy state of the atom and is called the **ground state**.

When the hydrogen atom is heated, the electron absorbs energy and jumps to a higher energy level. This is known as an **excited state**.

Shortly afterwards, the electron returns to the ground state. The electron may return directly to the ground state or may move to other energy levels before returning to the ground state. For example, an electron in the $n = 4$ shell may move to the $n = 2$ shell before returning to the $n = 1$ ground state.

As the electron falls to a lower energy shell, it emits energy in the form of light. This energy is exactly equal to the energy difference between the two energy levels. Each transition corresponds to a specific energy of light and therefore a specific line in the line spectrum of hydrogen shown in Figure 3.4.10.

FIGURE 3.4.10 The emission spectrum of the hydrogen atom has four lines in the visible range: violet, blue, green and red.

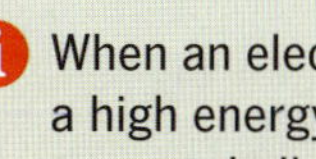

When an electron falls from a high energy shell to a lower energy shell, it emits energy in the form of light.

3.4 Review

SUMMARY

- When atoms are heated, they emit light.
- Flame tests are a simple technique that can be used to identify the metal in a sample.
- When light emitted by atoms is passed through a prism, it produces a spectrum made up of lines of different colours. These spectra are known as emission or line spectra.
- The emission spectra produced by atoms of the same element are identical.
- Each element has a unique emission spectrum.
- The Bohr model of the atom was the first atomic model to explain the origin of emission spectra.
- The Bohr model assumes that electrons can only exist in fixed, circular orbits of specific energies. These orbits later came to be known as energy levels or shells.
- When an electron absorbs energy (e.g. heat or light), the electron can jump from one shell to a higher energy shell.
- When an electron falls from a higher energy shell to a lower energy shell, it emits energy in the form of light.
- Each line in the emission spectrum corresponds to a specific electron transition between two shells.

KEY QUESTIONS

1 Select words from the following list to complete the sentences below. Not all of the words provided are required.

protons, higher, transition, electrons, lower, let out, emit, excited

When a sample containing copper is heated in the flame of a Bunsen burner, the flame turns a green colour. This is because the __________ in the copper atoms absorb energy and move to _________ energy levels and then _______ light that corresponds to a green colour as they return to __________ energy levels.

2 Explain what an emission spectrum is.

3 Explain how emission spectra are evidence for electron shells in the Bohr model.

4 What four assumptions did Bohr make about the electronic structure of the atom?

5 What form of energy is emitted when an electron moves from a higher energy shell to a lower energy shell?

3.5 Electronic configuration and the shell model

Bohr's model of the hydrogen atom gave close agreement between his calculated energies for the lines in hydrogen's emission spectra and the observed values. In Bohr's model, electrons are confined to specific energy levels or shells. In this section, you will look at how scientists were able to determine the electronic arrangement in other atoms.

IONISATION ENERGY

Evidence for the existence of energy levels in atoms was obtained from studies of successive **ionisation energies** in atoms of different elements. The ionisation energy is the energy needed to remove an electron from an atom. For example, a sodium atom contains 11 electrons. As each successive electron is removed, these ionisation energies can be measured. Figure 3.5.1 shows that the first electron to be removed has the lowest ionisation energy and is therefore the easiest to remove. The following eight electrons are more difficult to remove, and the last two require substantially more energy to remove.

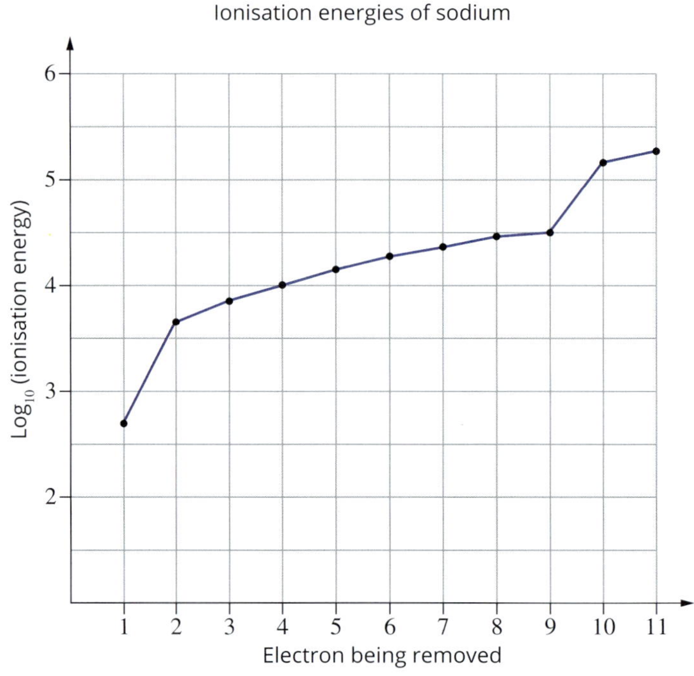

FIGURE 3.5.1 A graph of the ionisation energies of a sodium atom. (The logarithm of ionisation energy is used to provide a more convenient vertical scale.)

If an atom has six electrons, then each can be removed in turn. The electron that is least strongly attracted to the nucleus will be removed most easily. The amount of energy this requires is known as the first ionisation energy. Each one of the remaining five electrons will have a specific ionisation energy. The second ionisation energy will be greater than the first. The third will be greater than the second, and so on.

After scientists had determined the successive ionisation energies for a large number of elements, they concluded that electrons were grouped in different energy levels, which they called electron shells. Electrons in the same shell:

- are all about the same distance from the nucleus
- have about the same energy.

The different energy levels or shells can hold different numbers of electrons. The arrangement of these electrons around the nucleus is called the **electronic configuration**.

ELECTRONIC CONFIGURATION IN SHELLS

In all atoms, the electrons are as close to the nucleus as possible. This means that electrons will generally occupy inner shells before outer shells. For example, an atom of lithium has three electrons. Two electrons will occupy the first shell—that is all it can hold. The third electron is in the second shell.

A Bohr diagram is a simple diagram that shows the arrangement of electrons around the nucleus. In such diagrams, only the shells that are occupied are drawn. The Bohr diagram for lithium is shown in Figure 3.5.2.

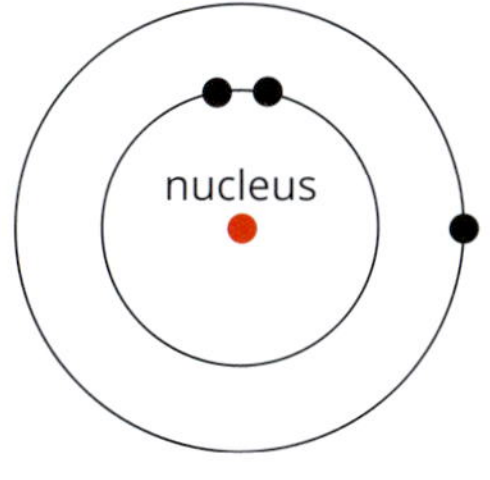

FIGURE 3.5.2 A Bohr diagram for the lithium atom, which has three electrons.

It is possible to determine the basic electronic configuration of any atom by applying the following three rules.

- *Rule 1*: Each shell can contain a maximum number of electrons. Table 3.5.1 shows the number of electrons that can occupy the first four shells.
- *Rule 2*: Lower energy shells fill before higher energy shells.

TABLE 3.5.1 Number of electrons held by each shell of an atom. Shell 1 is the shell closest to the nucleus.

Electron shell (n)	Maximum number of electrons
1	2
2	8
3	18
4	32
n	$2n^2$

Using these two rules, it is possible to accurately predict the electronic configuration of the first 18 elements.

Oxygen has eight electrons, so the first shell will hold two electrons and the second shell will hold six electrons. Its electronic configuration is therefore 2,6.

Figure 3.5.3 shows the electronic configuration of a sodium atom, which has 11 electrons. The first two electrons fill the first shell. Then the next eight electrons fill the second shell. The remaining electron occupies the third shell. The electronic configuration for this atom is written as 2,8,1 to indicate the arrangement of electrons in each shell.

- *Rule 3*: Electron shells fill in a particular order. The first two electrons go into the first shell. The next eight electrons go into the second shell. The third shell can hold 18 electrons, but once it contains eight electrons, the next two electrons go into the fourth shell. Only then does the third shell fill up.

A potassium atom contains 19 electrons. The first shell holds two electrons. The second shell holds eight and the third shell holds eight. The fourth shell will hold one electron. This arrangement can be seen in Figure 3.5.4.

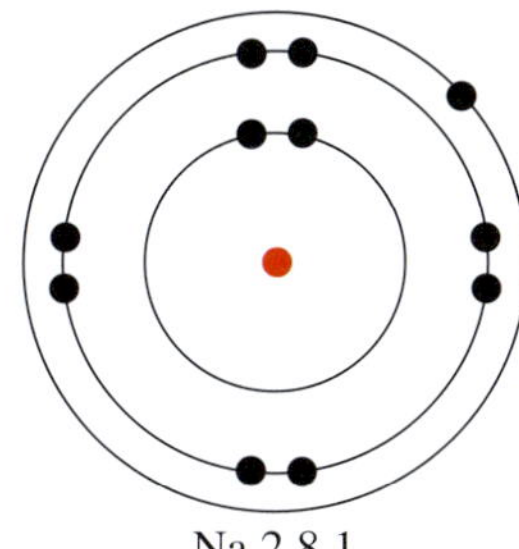

FIGURE 3.5.3 The electronic configuration of a sodium atom.

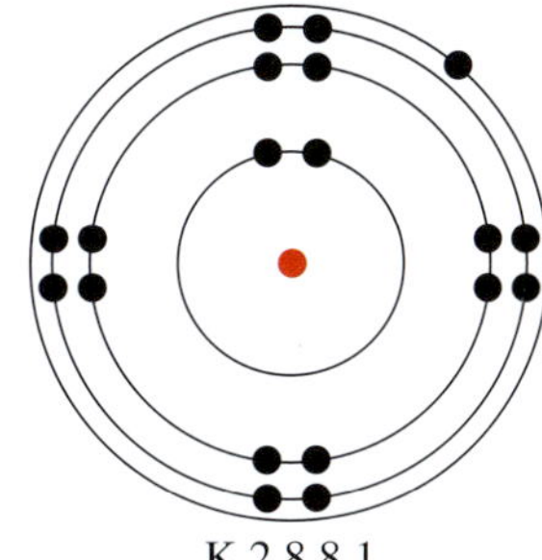

FIGURE 3.5.4 The electronic configuration of a potassium atom.

This pattern of shell filling continues until the third shell contains the maximum number of electrons it can hold (18). Therefore, an atom with 30 electrons has an electronic configuration of 2,8,18,2.

Worked example 3.5.1

CALCULATING ELECTRONIC CONFIGURATIONS FOR UP TO 36 ELECTRONS

Apply the rules of the shell model to determine the electronic configuration of an atom with 28 electrons.

Thinking	Working
Recall the maximum number of electrons that each shell can hold.	Shell (*n*) — Maximum number of electrons 1 — 2 2 — 8 3 — 18 4 — 32
Place the first 18 electrons in the shells from the lowest energy to the highest energy. Do not exceed the maximum number of electrons allowed.	Shell (*n*) — Electrons in atom 1 — 2 2 — 8 3 — 8 4 —
Place the next two electrons in the fourth shell.	Shell (*n*) — Electrons in atom 1 — 2 2 — 8 3 — 8 4 — 2
Continue filling the third shell until it holds up to 18 electrons. Put any remaining electrons in the fourth shell.	Shell (*n*) — Electrons in atom 1 — 2 2 — 8 3 — 16 4 — 2 The eight remaining electrons from the previous step have gone into the third shell.
Write the electronic configuration by listing the number of electrons in each shell separated by commas (with no space between).	The electronic configuration is 2,8,16,2.

Worked example: Try yourself 3.5.1

CALCULATING ELECTRONIC CONFIGURATIONS FOR UP TO 36 ELECTRONS

Apply the rules of the shell model to determine the electronic configuration of an atom with 34 electrons.

VALENCE ELECTRONS

The outermost shell of an atom is known as the atom's **valence shell**. The electrons in the outer shell are called **valence electrons**. These electrons require the least amount of energy to be removed. The valence electrons are involved in chemical reactions. Consequently, if you know the number of valence electrons in the atoms of an element, you can predict the chemical properties of the element.

In Chapter 5, you will learn about how atoms tend to lose, gain or share valence electrons in order to achieve eight electrons in their outer shell when they are involved in chemical reactions. This is known as the **octet rule**.

GO TO ➤ Section 5.2 page 147

Worked example 3.5.2

CALCULATING THE NUMBER OF VALENCE ELECTRONS IN AN ATOM

How many valence electrons are present in an atom of magnesium, which has an atomic number of 12?

Thinking	Working
Recall the maximum number of electrons that each shell can hold.	Shell (*n*) — Maximum number of electrons 1 — 2 2 — 8 3 — 18 4 — 32
Place 12 electrons in the shells from the lowest energy to the highest energy. Do not exceed the maximum number of electrons allowed.	Shell (*n*) — Electrons in atom 1 — 2 2 — 8 3 — 2 4
Write the electronic configuration by listing the number of electrons in each shell, separated by commas.	The electronic configuration is 2,8,2.
Determine the number of electrons in the outer shell or valence shell.	The number of electrons in the valence shell is 2.

Worked example: Try yourself 3.5.2

CALCULATING THE NUMBER OF VALENCE ELECTRONS IN AN ATOM

How many valence electrons are present in an atom of sulfur, which has an atomic number of 16?

3.5 Review

SUMMARY

- The electronic configuration of an atom indicates the arrangement of electrons in each shell.
- Bohr diagrams can be used to show the electronic configurations of atoms.
- Each shell can have a maximum number of electrons. This is summarised in the table below.

Electron shell (n)	Maximum number of electrons
1	2
2	8
3	18
4	32
n	$2n^2$

- The lowest energy shells fill first until they are full or have eight electrons. At that point, the next energy shell accepts two electrons before the unfilled shell with eight electrons continues filling.
- The outer shell cannot contain more than eight electrons.
- The outer shell is called the valence shell.
- Electrons in the outer shell are called valence electrons.

KEY QUESTIONS

1 Write the electronic configurations for atoms with:

a 5 electrons
b 12 electrons
c 20 electrons
d 35 electrons.

2 Write the electronic configuration of:

a Be
b S
c Ar
d Mg
e Ne.

3 Draw the Bohr diagram for one atom of scandium.

4 Write the name and symbol of the element with the electronic configuration:

a 2
b 2,7
c 2,8,3
d 2,5
e 2,8,7.

5 What is the number of valence electrons for an atom with 35 electrons?

3.6 The Schrödinger model of the atom

The shell model of the atom that developed from the work of Niels Bohr mathematically explained the lines in the emission spectrum of hydrogen atoms. However, there were some things that the model could not explain. The shell model:

- cannot accurately predict the emission spectra of atoms with more than one electron
- is unable to explain why electron shells can only hold $2n^2$ electrons
- does not explain why the fourth shell accepts two electrons before the third shell is completely filled.

These failings of the shell model indicate that the model is incomplete. Obtaining a better model of the atom required scientists to think about electrons in an entirely different way.

A QUANTUM MECHANICAL VIEW OF ATOMS

The Bohr model of the atom was revolutionary when it was proposed. Before Bohr's model, Rutherford and other scientists believed that electrons could orbit the nucleus at any distance from the nucleus. This picture of the electrons was based on how scientists observed the world around them. For example, planets can revolve at any distance around the Sun. However, Bohr's theory stated that electrons only occupy specific, circular orbits. This was the first suggestion that the physics inside atoms might be very different from the physics we experience in our daily lives.

Quantum mechanics

The word 'quantum' means 'a specific amount'. In the Bohr model, the electrons can only have specific amounts of energy, depending on which shell they are in. The energy of the electrons is said to be **quantised**.

In 1926, the German scientist Erwin Schrödinger proposed the idea that electrons behaved as waves around the nucleus. Using a mathematical approach and this wave theory, Schrödinger developed a model of the atom called **quantum mechanics**. This is the model of the atom that scientists use today.

Quantum mechanics describes the behaviour of extremely small particles like electrons. You rarely experience quantum mechanics in your everyday life. As a result, the predictions of quantum mechanics are often difficult to imagine. Nonetheless, quantum mechanics accurately predicts the behaviour of electrons in atoms.

The Schrödinger model and electron properties

The fundamental difference between the Bohr model and the **Schrödinger model** of the atom is in the way the electrons are considered. Bohr viewed electrons as tiny, hard particles that revolve around the nucleus in circular orbits. Schrödinger viewed electrons as having wave-like properties. In his model, the electrons occupy a three-dimensional space around the nucleus known as an **orbital**. Figure 3.6.1 compares the Bohr model with the Schrödinger model.

By assuming that electrons have wave-like properties, Schrödinger determined the following:

- There are major energy levels in an atom that, for historical reasons, are called shells.
- These shells contain separate energy levels of similar energy, called **subshells**, which he labelled *s*, *p*, *d* and *f*. Each subshell can only hold a certain number of electrons.
- The first shell ($n = 1$) contains only an *s*-subshell. The second shell contains *s*- and *p*-subshells. The third shell contains *s*-, *p*- and *d*-subshells and so on. The subshells for the first four shells are summarised in Table 3.6.1 on page 106.

(a)

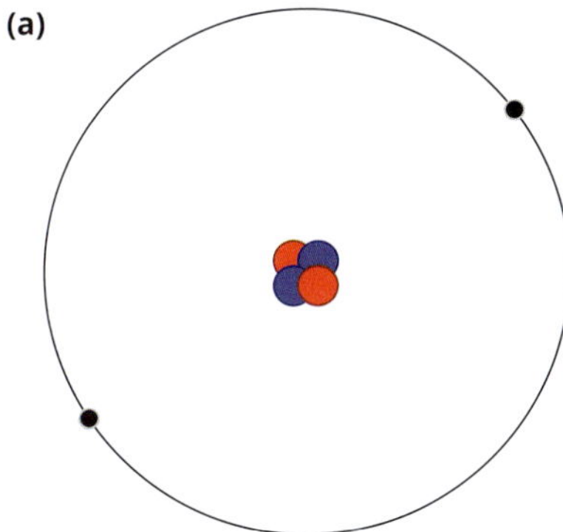

The Bohr model

(b)

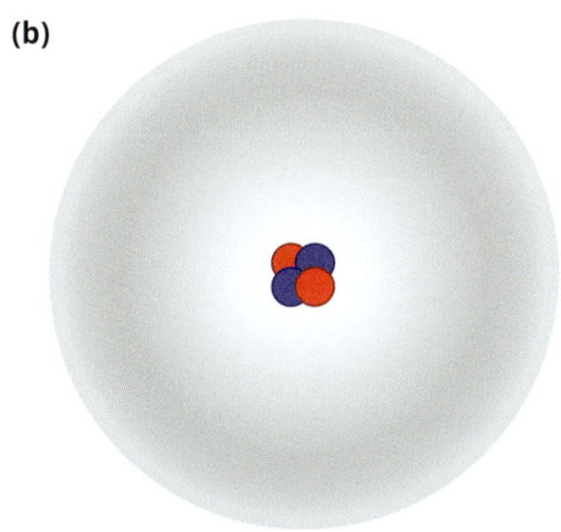

The Schrödinger model

FIGURE 3.6.1 (a) Bohr regarded electrons as particles that travel along a defined path in circular orbits. (b) In Schrödinger's quantum mechanical approach, the electrons behave as waves and occupy a three-dimensional space around the nucleus. The region occupied by electrons is known as an orbital.

TABLE 3.6.1 Energy levels within an atom

Shell number (n)	Subshells	Number of orbitals in subshell	Maximum number of electrons per subshell	Maximum number of electrons per shell
1	1s	1	2	2
2	2s	1	2	8
	2p	3	6	
3	3s	1	2	18
	3p	3	6	
	3d	5	10	
4	4s	1	2	32
	4p	3	6	
	4d	5	10	
	4f	7	14	

- Each subshell is made up of smaller components known as orbitals. Orbitals can be described as regions of space surrounding the nucleus of an atom in which electrons may be found. An *s*-subshell has just one orbital, a *p*-subshell has three orbitals, a *d*-subshell has five orbitals and an *f*-subshell has seven. The number of orbitals in each subshell for the first four shells is summarised in Table 3.6.1.

The Pauli exclusion principle

The **Pauli exclusion principle** states that each orbital can contain a maximum of two electrons. Each electron in an orbital must have a different quantum mechanical property called **spin**. The total number of orbitals in a shell is given by n^2; therefore, the total number of electrons per shell is given by $2n^2$. For example, the second shell contains *s*- and *p*-subshells and so contains a total of $1 + 3 = 4$ orbitals. Each orbital contains two electrons, so the second shell contains $2 \times 4 = 8$ electrons. The number of electrons in each subshell and shell for the first four shells is summarised in Table 3.6.1.

The Pauli exclusion principle states that each orbital can contain a maximum of two electrons, with each electron having a different spin.

Note that the Schrödinger model accurately predicts the maximum number of electrons that each shell can hold. This is something that the Bohr model could not explain.

ELECTRONIC CONFIGURATIONS AND THE SCHRÖDINGER MODEL

The electronic configurations for the Schrödinger model are more complicated than the electronic configurations of the shell model. This is because the Schrödinger model specifies subshells that electrons occupy. The electronic configuration of a sodium atom is shown in Figure 3.6.2.

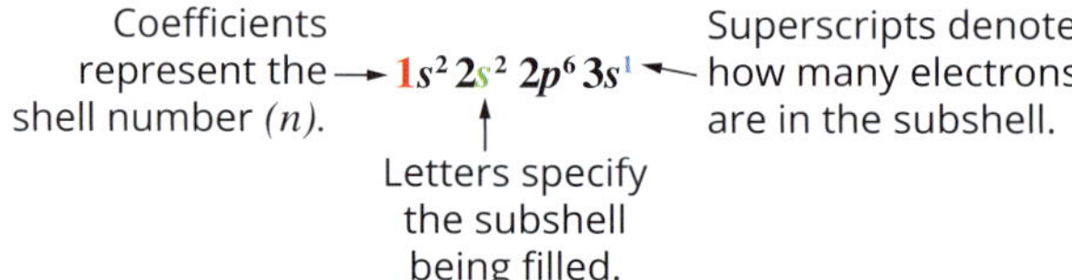

FIGURE 3.6.2 The electronic configuration of a sodium atom. It shows that there are 2 electrons in the 1*s*-subshell, 2 electrons in the 2*s*-subshell, 6 electrons in the 2*p*-subshell and 1 electron in the 3*s*-subshell.

Sodium has 11 electrons. The electronic configuration in Figure 3.6.2 indicates that, for sodium, there:

- are two electrons in the *s*-subshell in the first shell
- are two electrons in the *s*-subshell of the second shell
- are six electrons in the *p*-subshell of the second shell
- is one electron in the *s*-subshell of the third shell.

The Aufbau principle

Although the electronic configurations may look complicated, there are rules to help you construct them. The first rule is known as the **Aufbau principle**.

> The Aufbau principle states that the lowest energy orbitals are always filled with electrons first.

The order of energy levels of the subshells is:

$1s < 2s < 2p < 3s < 3p < 4s < 3d \ldots$

This is shown in Figure 3.6.3. In this diagram, each box represents an orbital that can hold two electrons.

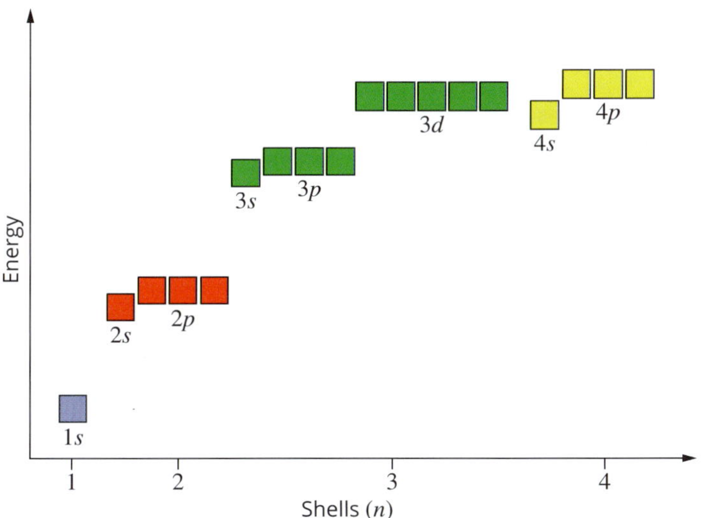

FIGURE 3.6.3 The relative energies of subshells. Each box represents an orbital in a subshell. Each orbital can contain two electrons.

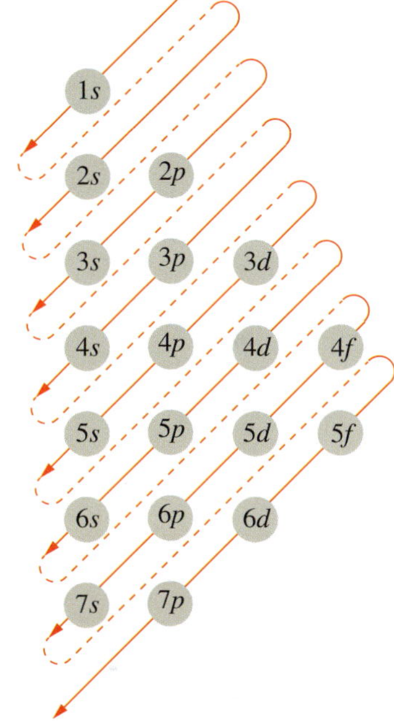

FIGURE 3.6.4 This geometric pattern shows the order in which the subshells are filled. Note that in this diagram, the fourth shell starts filling before the third shell is completely filled. This is because the 4*s*-orbital is lower in energy than the 3*d*-orbitals. As a result, the 4*s*-orbital accepts two electrons after the 3*s*- and 3*p*-orbitals are filled, but before the 3*d*-orbital begins filling.

The geometric pattern shown in Figure 3.6.4 is a commonly used and convenient way of remembering the order in which the subshells are filled.

Hund's rule

An **orbital diagram** (Figure 3.6.5) shows how the energy levels are filled in a boron atom, which has five electrons. The electrons are represented by arrows and the direction (up or down) represents the spin of the electron. The first two electrons fill the 1*s*-subshell, the second two electrons go into the next highest energy level, the 2*s*-subshell. The remaining electron then fills the next highest energy level, the 2*p*-subshell. The electronic configuration is written as $1s^22s^22p^1$.

The next element is carbon, which has six electrons in its atoms. Having one more electron than the boron atom means that carbon atoms have two electrons in the 2*p*-subshell. These two electrons could either occupy the same 2*p*-orbital or each electron could occupy a different 2*p*-orbital.

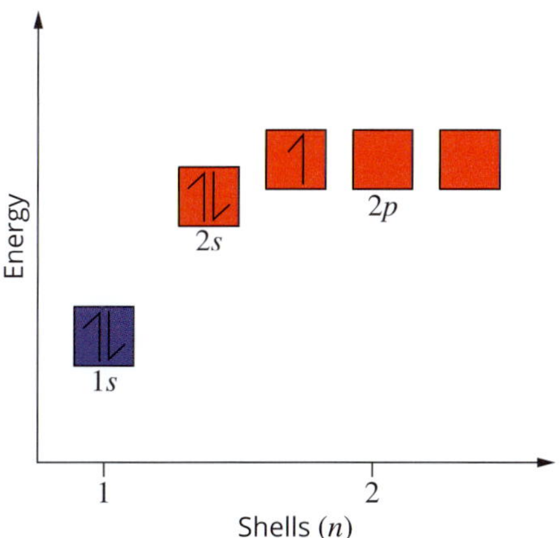

FIGURE 3.6.5 This orbital diagram shows how the orbitals in a boron atom are filled to give its electronic configuration.

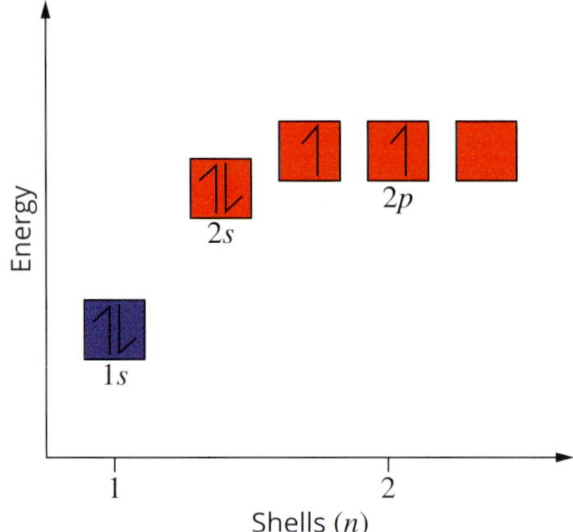

FIGURE 3.6.6 This orbital diagram shows how the orbitals in a carbon atom are filled to give its electronic configuration. Notice how Hund's rule applies: each of the 2*p*-subshell electrons with the same spin occupies different orbitals.

Hund's rule states that every orbital in a subshell must first be filled with one electron with the same spin before an orbital is filled with a second electron.

Using **Hund's rule**, in a carbon atom each electron would occupy a different 2*p*-orbital (Figure 3.6.6).

Figure 3.6.7 shows how to add electrons to atoms with 18 or more electrons. The first two electrons fill the 1*s*-orbital in the first shell. The next eight electrons fill the *s*- and *p*-orbitals in the second shell. The next eight electrons fill the *s*- and *p*-orbitals in the third shell. The 4*s*-orbital is then filled with two electrons because this orbital is lower in energy than the 3*d*-orbitals. Once the 4*s*-orbital is filled, the next 10 electrons are placed in the 3*d*-orbitals.

Finally, the remaining six electrons fill the 4*p*-orbitals.

Krypton has 36 electrons. According to the order of subshell filling, its electronic configuration is:

$1s^2 2s^2 2p^6 3s^2 3p^6 4s^2 3d^{10} 4p^6$

Although the 4*s*-subshell is filled before the 3*d*-subshell, the subshells of the third shell are usually grouped together. Therefore, the electronic configuration for a krypton atom is written as:

$1s^2 2s^2 2p^6 3s^2 3p^6 3d^{10} 4s^2 4p^6$

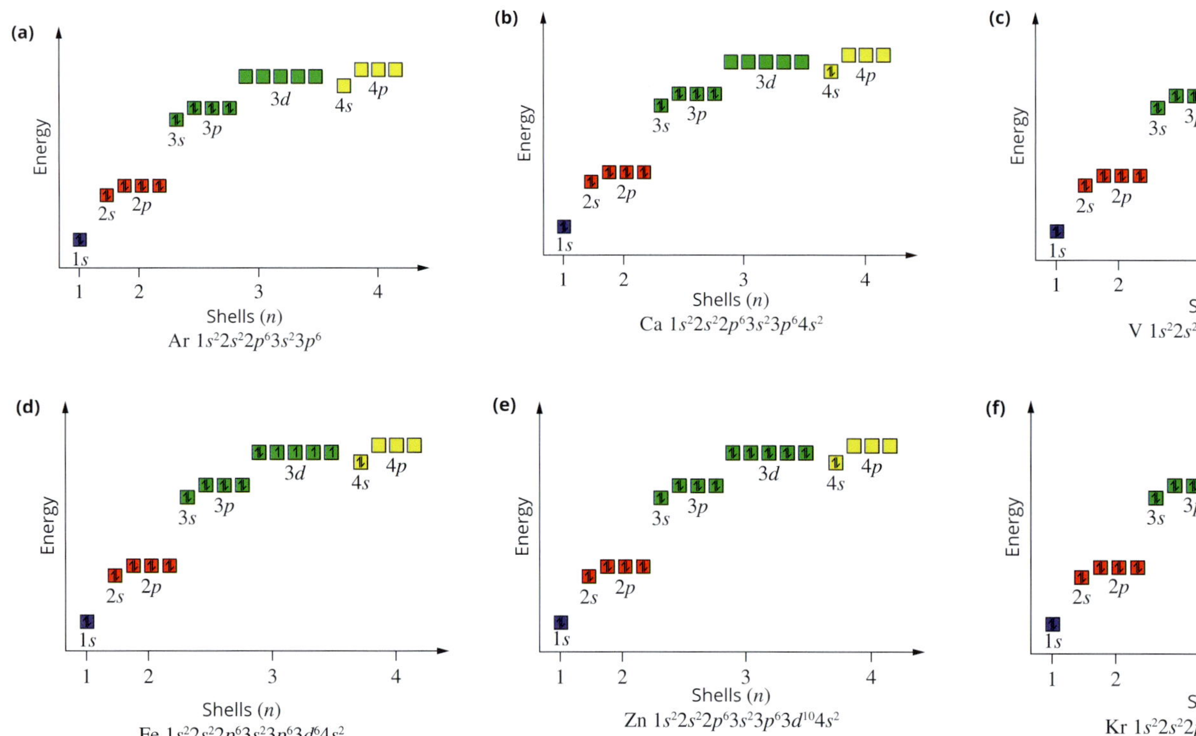

FIGURE 3.6.7 Orbital diagrams (showing how electrons are added to atoms with more than 18 electrons) for (a) argon (Ar), (b) calcium (Ca), (c) vanadium (V), (d) iron (Fe), (e) zinc (Zn) and (f) krypton (Kr).

Chromium and copper: exceptions

The electronic configurations for most elements follow the rules described above. There are two notable exceptions: element 24, chromium, and element 29, copper. Table 3.6.2 illustrates these exceptions.

TABLE 3.6.2 The electronic configurations for chromium and copper

Element	Electronic configuration predicted according to the rules	Actual electronic configuration
chromium (Cr)	$1s^22s^22p^63s^23p^63d^44s^2$	$1s^22s^22p^63s^23p^63d^54s^1$
copper (Cu)	$1s^22s^22p^63s^23p^63d^94s^2$	$1s^22s^22p^63s^23p^63d^{10}4s^1$

Each orbital can hold two electrons. In the *d*-subshell there are five orbitals and therefore 10 electrons that can be held within them. According to Hund's rule, as a subshell fills, a single electron is placed in each orbital first. Then a second electron is entered into the orbitals until the filling process is complete.

Chemists calculate that there is very little difference in energy between 3*d*- and 4*s*-orbitals, and the $3d^54s^1$ configuration for chromium is slightly more stable than the $3d^44s^2$ configuration. This is because each of the five *d*-orbitals is exactly half filled with one *d*-orbital empty.

Similarly for copper, the $3d^{10}4s^1$ arrangement with five completely filled *d*-orbitals is more stable than the $3d^94s^2$ configuration with a partially filled *d*-orbital.

Worked example 3.6.1

WRITING ELECTRONIC CONFIGURATIONS USING THE SCHRÖDINGER MODEL

Write the Schrödinger model of electronic configuration for a manganese atom with 25 electrons.

Thinking	Working
Recall the order in which the subshells fill by listing them from lowest energy to highest energy and the number of orbitals in each.	1*s*, 1 orbital 2*s*, 1 orbital 2*p*, 3 orbitals 3*s*, 1 orbital 3*p*, 3 orbitals 4*s*, 1 orbital 3*d*, 5 orbitals 4*p*, 3 orbitals
Fill the subshells by assigning two electrons per orbital, starting from the lowest energy subshells until you have reached the total number of electrons in your atom.	Subshell / Electrons in subshell / Progressive total of electrons 1*s* 2 2 2*s* 2 4 2*p* 6 10 3*s* 2 12 3*p* 6 18 4*s* 2 20 3*d* 5 25 4*p*
Write the electronic configuration by writing each subshell with the number of electrons as a superscript. Remember to group subshells from the same shell.	$1s^22s^22p^63s^23p^63d^54s^2$

Worked example: Try yourself 3.6.1

WRITING ELECTRONIC CONFIGURATIONS USING THE SCHRÖDINGER MODEL

Write the Schrödinger model of electronic configuration for a titanium atom with 22 electrons.

3.6 Review

SUMMARY

- The Bohr shell model of the atom was unable to fully explain the properties of atoms, and a new model was needed to describe the behaviour of electrons in atoms.
- The Schrödinger model proposed that electrons behave as waves and occupy a three-dimensional space around the nucleus.
- The Schrödinger model predicted that subshells are energy levels within the major shells. The subshells consist of orbitals.
- An orbital can be regarded as a region of space surrounding the nucleus in which an electron may be found.
- Orbitals of similar energy are grouped in subshells that are labelled *s*, *p*, *d* and *f*.
- The Pauli exclusion principle states that each orbital can hold a maximum of two electrons. Each electron in a given orbital must have a different quantum mechanical property called spin.
- Each subshell in an atom has a different energy.
- The Aufbau principle states that electrons fill the subshells from the lowest energy subshell to the highest energy subshell.
- The 4*s*-subshell is lower in energy than the 3*d*-subshell, so the fourth shell accepts two electrons before the third shell is completely filled.
- Electronic configurations of atoms in the Schrödinger model specify the number of electrons in each subshell.
- Hund's rule states that every orbital in a subshell must first be filled with one electron with the same spin before an orbital is filled with a second electron.

KEY QUESTIONS

1 Which subshell has the lowest energy?

A 3*d*

B 4*s*

C 4*p*

D 5*s*

2 Copy and complete the table to write the electronic configuration of each of the atoms listed.

Element (atomic number)	Electronic configuration using the shell model	Electronic configuration using the subshell model
boron (5)	2,3	$1s^2 2s^2 2p^1$
lithium (3)		
chlorine (17)		
sodium (11)		
neon (10)		
potassium (19)		
scandium (21)		
iron (26)		
bromine (35)		

3 In terms of energy levels, what is the essential difference between the shell model and the subshell model of the atom?

4 Draw orbital diagrams showing the electronic configuration of each of the following atoms.

a Be

b N

c Si

d Ni (only show the 3*d* and 4*s*-subshells for simplicity).

Chapter review

03

KEY TERMS

alpha particle
atom
atomic number
Aufbau principle
balanced nuclear reaction
band of stability
beta particle
Bohr model
carbon-12
electromagnetic radiation
electron
electronic configuration
electron shell
electrostatic attraction
element
emission spectrum
energy level
excited state
flame test
gamma radiation
ground state
Hund's rule
ionisation energy
ionise
isotope
mass number
mass spectrometer
mass spectrum
model
molecule
neutron
nucleon
nucleus
octet rule
orbital
orbital diagram
Pauli exclusion principle
periodic table
proton
quantised
quantum mechanics
radioactive
radioisotope
relative atomic mass
relative formula mass
relative isotopic abundance
relative isotopic mass
relative molecular mass
Schrödinger model
spin
subatomic particle
subshell
valence electron
valence shell

REVIEW QUESTIONS

1 Where would you find 99.97% of the mass of an atom?

2 How are protons, neutrons and electrons arranged in an atom?

3 Compare the mass and charge of protons, neutrons and electrons.

4 The radii of three elements were measured as being:

- **a** 0.000 000 000 120 m
- **b** 0.000 000 000 216 m
- **c** 0.000 000 000 348 m

Express these radii in scientific notation.

5 An atom of chromium can be represented by the notation $^{52}_{24}Cr$.

- **a** Determine its atomic number and mass number.
- **b** Determine the number of electrons, protons and neutrons in the chromium atom.

6 Two atoms both have 20 neutrons in their nucleus. The first has 19 protons and the other has 20 protons. Are they isotopes? Why or why not?

7 Explain why the number of electrons in an atom equals the number of protons.

8 Using the element bromine as an example, explain why elements are best identified by their atomic number and not their mass number.

9 Determine the relative molecular mass (M_r) of:

- **a** water (H_2O)
- **b** white phosphorus (P_4)
- **c** carbon monoxide (CO).

10 Determine the relative formula mass of:

- **a** zinc bromide ($ZnBr_2$)
- **b** barium hydroxide ($Ba(OH)_2$)
- **c** iron(III) carbonate ($Fe_2(CO_3)_3$).

11 Balance the equations for the following nuclear reactions:

- **a** $^{231}_{87}Fr \rightarrow ^{231}_{88}Ra + \_\_\_$
- **b** $^{239}_{94}Pu \rightarrow \_\_\_ + ^{4}_{2}He$
- **c** $\_\_\_ \rightarrow ^{227}_{89}Ac + ^{4}_{2}He$
- **d** $^{214}_{84}Po \rightarrow \_\_\_ + ^{4}_{2}He$
- **e** $^{149}_{62}Sm \rightarrow ^{145}_{61}Pm + \_\_\_ + \_\_\_$

12 The standard on which all relative masses are based is the ^{12}C isotope, which is given a mass of 12 exactly. Explain why in the table of relative atomic masses in the Appendix at the end of the book, the relative atomic mass of carbon is listed as 12.011.

13 When a sample of palladium is placed in a mass spectrometer, several peaks are recorded as the relative isotopic masses and corresponding percentage abundances given in the table.

Relative isotopic mass	Abundance (%)
101.9049	0.9600
103.9036	10.97
104.9046	22.23
105.9032	27.33
107.9039	26.71
109.9044	11.80

Calculate the relative atomic mass of palladium. Express your answer correct to 1 decimal place.

14 This table shows isotopic composition data for argon and potassium. Express your answers to part **a** and part **b** correct to 2 decimal places.

Element	Atomic number	Relative isotopic mass	Relative abundance (%)
argon	18	35.978	0.307
		37.974	0.060
		39.974	99.633
potassium	19	38.975	93.3
		39.976	0.011
		40.974	6.69

a Determine the relative atomic masses of argon and potassium.

b Explain why the relative atomic mass of argon is greater than that of potassium, even though potassium has a larger atomic number.

15 The mass spectrum of chromium is shown in the graph below.

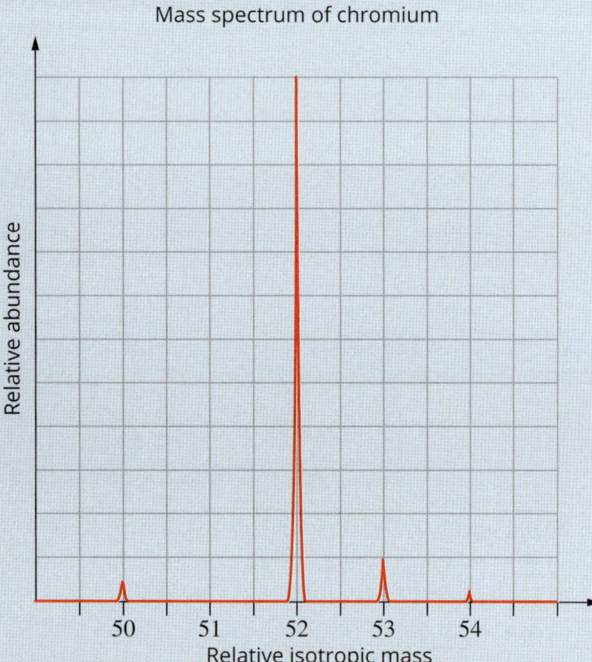

a Measure the peak heights to calculate the percentage abundance of each chromium isotope.

b Use the percentage abundances calculated in part **a** to determine the relative atomic mass for chromium. The mass number is a good approximation to the relative isotopic mass.

16 The relative atomic mass of europium is 151.96. The relative isotopic masses of its two isotopes are 150.92 and 152.92. Calculate the relative abundances of the isotopes in naturally occurring europium.

17 Determine the percentage abundance of the lighter isotope of:

a gallium: relative isotopic masses 68.95 and 70.95 respectively; $A_r = 69.72$

b boron: relative isotopic masses 10.02 and 11.01 respectively; $A_r = 10.81$.

18 In a hydrogen atom, which electron shell will the electron be in if the atom is in the ground state?

19 Determine which shell the 30th electron of an atom would go into, according to the rules for determining the electronic configuration of an atom.

20 What is the name of the element that has an electronic configuration of 2,8,2?

21 Write electronic configurations, using subshell notation, for the following elements. The atomic number of each element is shown in brackets.

a helium (2)
b carbon (6)
c fluorine (9)
d aluminium (13)
e argon (18)
f nickel (28)
g bromine (35)

22 Using a fluorine atom as an example, explain the difference between the terms 'shell', 'subshell' and 'orbital'.

23 Determine the Schrödinger model of electronic configuration that corresponds to the shell model electronic configuration 2,8,6.

24 Explain how the Schrödinger model of the atom explains why the fourth electron shell begins filling before the third shell is completely filled.

25 Explain why the 4*s*-subshell in chromium and copper is only half-filled compared with almost all other elements.

26 Draw orbital diagrams showing the electronic configuration of each of the following atoms.

a O

b Mg

c Mn (only show the 3*d* and 4*s*-subshells for simplicity)

d As (only show the 3*d*, 4*s* and 4*p*-subshells for simplicity)

27 Our model describes an atom as consisting of rapidly moving electrons at a relatively large distance from a very small central nucleus. What lies between those electrons and the nucleus?

28 New models for the atom are developed as scientists become aware of inconsistencies between current models and experimental data. Outline the problems with the existing model of the atom that led to the modifications suggested by:

a Bohr **b** Schrödinger.

29 Reflect on the Inquiry activity on page 78. Using your knowledge from this chapter, what is one limitation of this simple model?

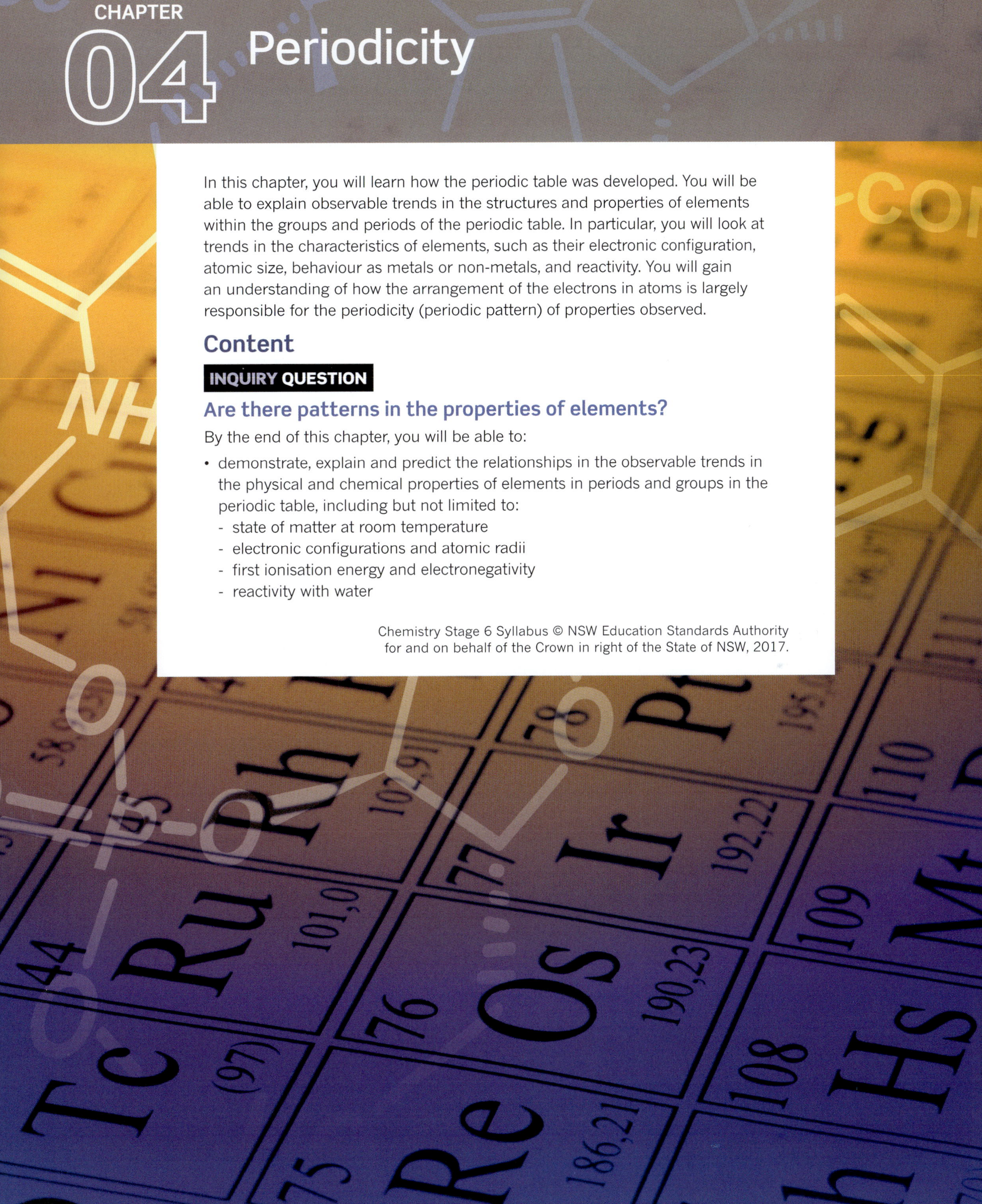

CHAPTER

04 Periodicity

In this chapter, you will learn how the periodic table was developed. You will be able to explain observable trends in the structures and properties of elements within the groups and periods of the periodic table. In particular, you will look at trends in the characteristics of elements, such as their electronic configuration, atomic size, behaviour as metals or non-metals, and reactivity. You will gain an understanding of how the arrangement of the electrons in atoms is largely responsible for the periodicity (periodic pattern) of properties observed.

Content

INQUIRY QUESTION

Are there patterns in the properties of elements?

By the end of this chapter, you will be able to:

- demonstrate, explain and predict the relationships in the observable trends in the physical and chemical properties of elements in periods and groups in the periodic table, including but not limited to:
  - state of matter at room temperature
  - electronic configurations and atomic radii
  - first ionisation energy and electronegativity
  - reactivity with water

4.1 The periodic table

CHEMISTRY INQUIRY 

Comparing the size of atoms

Are there patterns in the properties of elements?

COLLECT THIS ...

various sporting balls, measuring tape, camera

DO THIS ...

1. Draw up the results table below and record your predictions.
2. Arrange the balls in order of increasing diameter. You may need to measure the balls using the measuring tape.
3. Take four balls of various sizes. These will represent atoms of four elements.
   - **a** Place the balls down group 1 of the periodic table in the size order you think is correct.
   - **b** Place the balls across period 3 of the periodic table in the size order you think is correct.
4. Take a photo of your modelling and compare it with the Royal Society of Chemistry trends in the periodic table.
5. Rearrange the balls according to the data given by the Royal Society of Chemistry, and take a photo.
6. For a given type of ball/element, would the diameter increase or decrease if the element
   - **a** gained electrons
   - **b** lost electrons?

RECORD THIS ...

Describe the arrangement of the balls down group 1 and across period 3.
Present your observations in the results table.

Experiment	Prediction	Observations
2 Arrange the balls in order of increasing diameter.		
3a Place the balls down group 1 of the periodic table.		
3b Place the balls across period 3 of the periodic table.		
6a Would the diameter of the element increase or decrease if electrons were gained?		
6b Would the diameter of the element increase or decrease if electrons were lost?		

REFLECT ON THIS ...

Explain the meaning of your observations and any patterns you observed. What does this tell you about the rest of the groups and periods in the periodic table?

Chapter 2 looked at the observable properties of matter. Chapter 3 covered the nature of atoms, including the existence of subatomic particles, the charge and mass of these particles, the way subatomic particles are arranged in an atom and how they behave. As scientists' understanding of the atom improved and more elements were discovered, they needed a way to organise this knowledge.

The **periodic table** (Figure 4.1.1) is one of the most useful reference tools available to chemists. It underlies all of chemistry and is the organising principle for the entire discipline. The periodic table minimises the need to memorise isolated facts about different elements and provides a framework on which to organise our understanding. By knowing the properties of particular elements and their trends within the table, chemists are able to organise what would otherwise be an overwhelming collection of information. The arrangement of electrons in atoms is responsible for the **periodicity** (periodic patterns) in element properties. However, the periodic table was devised before scientists knew about electrons.

> The periodic table is so named because the properties of the elements vary periodically–that is, reoccur at regular intervals in elements ordered according to increasing mass of their atoms. These patterns were used in every design of the periodic table.

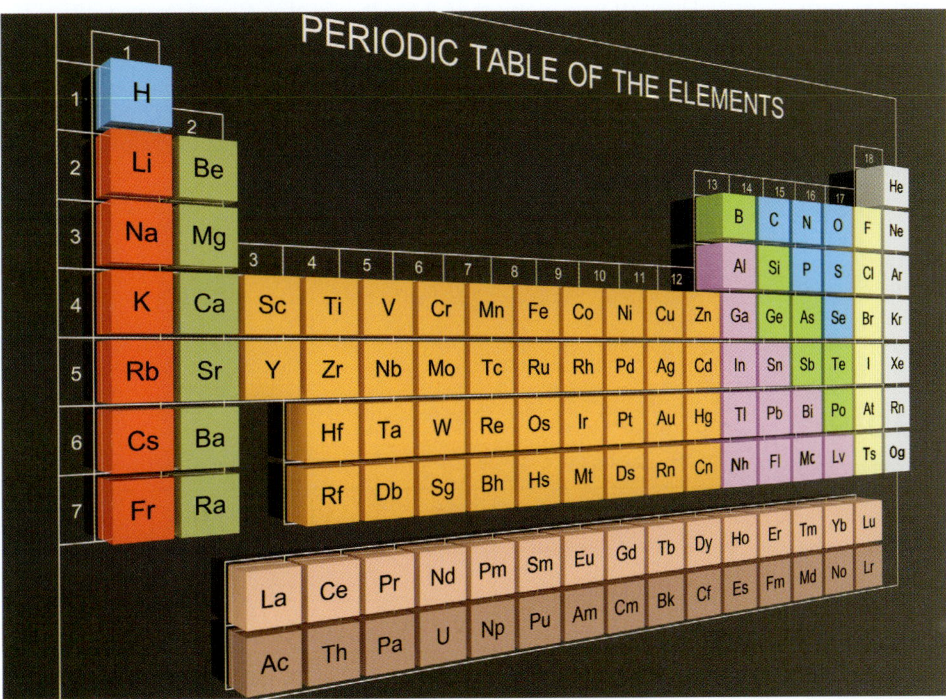

FIGURE 4.1.1 There are many versions of the modern periodic table of the elements. This is one of them.

DEVELOPMENT OF THE PERIODIC TABLE

In the 19th century, chemists wanted to use the information they had gathered about the elements to organise them into a useful and practical format. The work of these scientists led to what we now know as the periodic table. The early forms of the periodic table were very different from the one we use today; many elements had not been discovered and scientists only had limited information about some of the others. An early form of the periodic table created by a Russian chemist, Dmitri Mendeleev, is shown in Figure 4.1.2 (page 116).

During the time when the early periodic table was being constructed, the existence of subatomic particles was unknown, so the elements were placed in order of increasing mass of the atoms. Mendeleev recognised that the chemical properties of the elements varied periodically with increasing atomic mass. He arranged the elements with similar properties into vertical columns and proposed his **periodic law**: the properties of elements vary periodically with their atomic masses.

CHEMFILE IU L

Element names

The symbols for the chemical elements are used universally; however, the names of the elements differ between languages and cultures. Some elements, such as iron, tin and gold have been known since antiquity and were named independently within different cultures. For example, in ancient Rome, iron was known as *ferrum*, and several modern languages that are derived from Latin still use a word beginning with *fer-* for its name. Elements that were discovered in the last 200 years tend to have names that are similar in all cultures.

+ ADDITIONAL

Triads and octaves

In 1829, the German chemist Johann Wolfgang Döbereiner noticed that many of the known elements could be arranged in groups of three based on their chemical properties. He called these groups 'triads'. Within each of these triads, the properties of one element were intermediate between those of the other two. The intermediate element's relative atomic mass was almost exactly the average of the others.

One of Dobereiner's triads was lithium, sodium and potassium. Sodium is more reactive than lithium, but less reactive than potassium. Sodium's atomic mass is 23, which is the average of lithium's (atomic mass 7) and potassium's (atomic mass 39) atomic masses.

However, Dobereiner's theory was limited—not all elements could be included in triads. However, his work was quite remarkable, given he had fewer than 50 elements to work with at the time.

Decades later, English chemist John Alexander Newlands noticed a pattern in the atomic mass of elements. Newlands' law of octaves was published in 1865 and predicted properties of new elements such as germanium. His patterns worked well for the lighter elements, but did not fit for the heavier elements or allow for the discovery of new elements.

Four years later Mendeleev, working independently, published his periodic law, which, with a few modifications, was similar to Newlands' law of octaves.

+ ADDITIONAL

Mendeleev's periodic table

When Mendeleev published his table in 1869 (Figure 4.1.2), there were at least six other scientists working on finding the best way to order the elements. Mendeleev is remembered because of his amazing ability to predict the characteristics of elements that had not yet been discovered. For example, when gallium was discovered in 1875, its atomic mass, density and melting point were very close to what Mendeleev predicted in 1871. His predictions proved to be a very powerful guide for future chemists, helping them to understand things not yet discovered.

Period \ Group	I	II	III	IV	V	VI	VII	VIII
1	H = 1							
2	Li = 7	Be = 9.4	B = 11	C = 12	N = 14	O = 16	F = 19	
3	Na = 23	Mg = 24	Al = 27.3	Si = 28	P = 31	S = 32	Cl = 35.5	
4	K = 39	Ca = 40	? = 44	Ti = 48	V = 51	Cr = 52	Mn = 55	Fe = 56 Co = 59 Ni = 59
5	Cu = 63	Zn = 65	? = 68	? = 72	As = 75	Se = 78	Br = 80	
6	Rb = 85	Sr = 87	?Yt = 88	Zr = 90	Nb = 94	Mo = 96	? = 100	Ru = 104 Rh = 104 Pd = 106
7	Ag = 108	Cd = 112	In = 113	Sn = 118	Sb = 122	Te = 125	J = 127	
8	Cs = 133	Ba = 137	?Di = 138	?Ce = 140				
9								
10			?Er = 178	?La = 180	Ta = 182	W = 184		Os = 195 Ir = 197 Pt = 198
11	Au = 199	Hg = 200	Tl = 204	Pb = 207	Bi = 208			
12				Th = 231		U = 240		

FIGURE 4.1.2 A representation of Mendeleev's periodic table. Notice the spaces left for undiscovered elements.

THE MODERN PERIODIC TABLE

We now know that the number of protons (the **atomic number**) is what makes one element fundamentally different from another element. Therefore, the elements in the modern periodic table are arranged (in rows) in order of increasing atomic number.

The number of valence (outer-shell) electrons is used to organise the elements into columns. The form of the periodic table in common use is shown in Figure 4.1.3.

GO TO ➤ Section 3.5 page 100

period \ group	1	2	3	4	5	6	7	8	9	10	11	12	13	14	15	16	17	18
1	H 1																	He 2
2	Li 3	Be 4											B 5	C 6	N 7	O 8	F 9	Ne 10
3	Na 11	Mg 12											Al 13	Si 14	P 15	S 16	Cl 17	Ar 18
4	K 19	Ca 20	Sc 21	Ti 22	V 23	Cr 24	Mn 25	Fe 26	Co 27	Ni 28	Cu 29	Zn 30	Ga 31	Ge 32	As 33	Se 34	Br 35	Kr 36
5	Rb 37	Sr 38	Y 39	Zr 40	Nb 41	Mo 42	Tc 43	Ru 44	Rh 45	Pd 46	Ag 47	Cd 48	In 49	Sn 50	Sb 51	Te 52	I 53	Xe 54
6	Cs 55	Ba 56	La 57	Hf 72	Ta 73	W 74	Re 75	Os 76	Ir 77	Pt 78	Au 79	Hg 80	Tl 81	Pb 82	Bi 83	Po 84	At 85	Rn 86
7	Fr 87	Ra 88	Ac 89	Rf 104	Db 105	Sg 106	Bh 107	Hs 108	Mt 109	Ds 110	Rg 111	Cn 112	Nh 113	Fl 114	Mc 115	Lv 116	Ts 117	Og 118

lanthanoids

Ce	Pr	Nd	Pm	Sm	Eu	Gd	Tb	Dy	Ho	Er	Tm	Yb	Lu
58	59	60	61	62	63	64	65	66	67	68	69	70	71

actinoids

Th	Pa	U	Np	Pu	Am	Cm	Bk	Cf	Es	Fm	Md	No	Lr
90	91	92	93	94	95	96	97	98	99	100	101	102	103

FIGURE 4.1.3 This is the most common form of the periodic table in use.

In its current version:

- The periodic table is arranged in order of increasing atomic number.
- The horizontal rows are known as **periods** and are labelled 1–7.
- The vertical columns are known as **groups** and are labelled 1–18.
- **Main group elements** are elements in groups 1, 2 and 13–18.
- The elements in groups 3–12 are known as **transition metals**.

Some periodic tables you will see also indicate other properties of the elements, such as boiling point or whether the element is a solid, liquid or gas at room temperature.

GROUPS

Elements in the periodic table are arranged into vertical columns called groups. For main group elements, the group number can be used to determine the number of valence electrons in an atom of the element.

In groups 1 and 2, the number of valence electrons is equal to the group number. For example, magnesium is in group 2 and therefore has two valence electrons.

In groups 13–18, the number of valence electrons is equal to the group number minus 10. For example, oxygen is in group 16, so oxygen has six outer-shell electrons. Similarly, neon is in group 18, so neon has eight valence electrons. Helium is an important exception—it is in group 18 but only has two valence electrons. Helium is placed in group 18 because, like the other group 18 elements, it is unreactive. This information is summarised in Table 4.1.1.

TABLE 4.1.1 The number of valence electrons in elements belonging to each group

Group	Number of valence electrons
1	1
2	2
13	3
14	4
15	5
16	6
17	7
18	8*

*Helium has two valence electrons.

CHEMISTRY IN ACTION 

Element 117 and the Canberra scientists

Four atoms of a superheavy element known as number 117 have been created in a German laboratory as part of a collaboration with Professor David Hinde (Figure 4.1.4), director of the Heavy Ion Accelerator Facility in the Department of Nuclear Physics at the Australian National University.

Because the element is radioactive, the atoms disappeared within one tenth of a second, but the scientists were able to confirm the 2010 observation. This new element was given the temporary name of ununseptium ('one-one-seven' in Latin), but in November 2016 it was officially named tennessine (Ts) by the International Union of Pure and Applied Chemistry (IUPAC). This name was chosen because one of the teams working on its discovery was located at the Oakridge National Laboratory in Tennessee, USA. The mass of an atom of element 117 is equal to that of the heaviest atoms ever observed, and 40% heavier than an atom of lead.

Elements 113, 115 and 118 were also given permanent names in 2016. They are nihonium (Nh), moscovium (Mc) and oganesson (Og), respectively. All four names were chosen under the IUPAC rules. The scientists who discovered the elements were invited to propose names, and all four were named either for the geographical regions where the discoveries were made or in honour of scientists.

Professor Hinde said that because such an extremely small amount of tennessine has been synthesised, it will not be practical to use it. However, the Australian researchers will continue to concentrate on the quantum science behind the element's synthesis and attempt to produce the next superheavy element.

FIGURE 4.1.4 Professor David Hinde is director of the Heavy Ion Accelerator Facility in the Department of Nuclear Physics at the Australian National University in Canberra.

The electrons in the outer shell of an atom (the valence electrons) are the electrons that are involved in chemical reactions. As a consequence, the number of valence electrons determines many of the chemical properties that an element exhibits.

Table 4.1.2 shows the names of some groups in the periodic table.

Because elements in the same group have the same number of valence electrons, they also have similar properties. For example, the alkali metals are the elements in group 1 (with the exception of hydrogen). They are all relatively soft metals and are highly reactive with water and oxygen. Consider the electronic configurations of the atoms of the first three metals of this group:

Li $1s^2 2s^1$

Na $1s^2 2s^2 2p^6 3s^1$

K $1s^2 2s^2 2p^6 3s^2 3p^6 4s^1$

The valence shell of each atom of each element in group 1 contains one electron in an *s*-subshell. This similarity in the valence shell structure gives these elements similar chemical properties.

TABLE 4.1.2 Names of different groups in the periodic table

Group	Name
1	alkali metals
2	alkaline earth metals
17	halogens
18	noble gases

Fluorine, chlorine, bromine and iodine are halogens (group 17). They are all coloured and highly reactive (Figure 4.1.5). Their electronic configurations are:

F $1s^2 2s^2 2p^5$

Cl $1s^2 2s^2 2p^6 3s^2 3p^5$

Br $1s^2 2s^2 2p^6 3s^2 3p^6 3d^{10} 4s^2 4p^5$

I $1s^2 2s^2 2p^6 3s^2 3p^6 3d^{10} 4s^2 4p^6 4d^{10} 5s^2 5p^5$

Notice how all these elements have a highest-energy subshell electronic configuration of s^2p^5.

FIGURE 4.1.5 Three examples of halogens. These conical flasks contain, from left to right, chlorine (Cl, pale green), bromine (Br, red-brown) and iodine (I, purple).

The noble gases (group 18) are a particularly interesting group. The noble gases have a very stable electron arrangement: helium has a full outer shell of two electrons, and the other members of this group have a stable octet of valence electrons (eight electrons) in their outer shell. Chemical reactions involve the rearrangement of valence electrons to achieve a stable outer shell. Atoms with a stable electronic configuration do not tend to lose or gain electrons. Consequently, the noble gases have low reactivity.

The arrangement of electrons in atoms is also responsible for the periodicity (periodic pattern) of element properties.

PERIODS

The horizontal rows in the periodic table are called periods and are numbered 1–7. The number of a period gives information about the electronic configuration of an element. The period that an element is located within is equal to the number of occupied electron shells in the element's atoms. For example, the outer shell of magnesium and chlorine is the third shell, and both elements are in period 3:

Mg $1s^2 2s^2 2p^6 3s^2$

Cl $1s^2 2s^2 2p^6 3s^2 3p^5$

Similarly, the elements in period 5 all have outer shell electrons in the fifth shell.

BLOCKS

The periodic table has four main **blocks**–*s*, *p*, *d* and *f*. For every element in a block, the element's highest energy subshell is the same as the name of the block. For example, the highest energy subshell of an element in the *s*-block is the *s*-subshell.

The *s*-block contains the elements in group 1 (alkali metals), group 2 (alkaline earth metals), hydrogen and helium. These elements have a half-filled or fully-filled *s*-subshell, that is s^1 or s^2, as the highest energy subshell configuration. Table 4.1.3 and Figure 4.1.6 show which groups of elements fall into the four different blocks.

TABLE 4.1.3 Elements in the different blocks of the periodic table

Block	Elements	Highest energy subshell configurations
s-block	groups 1 and 2 and helium	s^1 or s^2
p-block	groups 13–18 (except helium)	s^2p^1 to s^2p^6
d-block	groups 3–12	d^1s^2 to $d^{10}s^2$
f-block	lanthanoids and actinoids	4*f*-subshell progressively being filled in the lanthanoids 5*f*-subshell progressively being filled in the actinoids

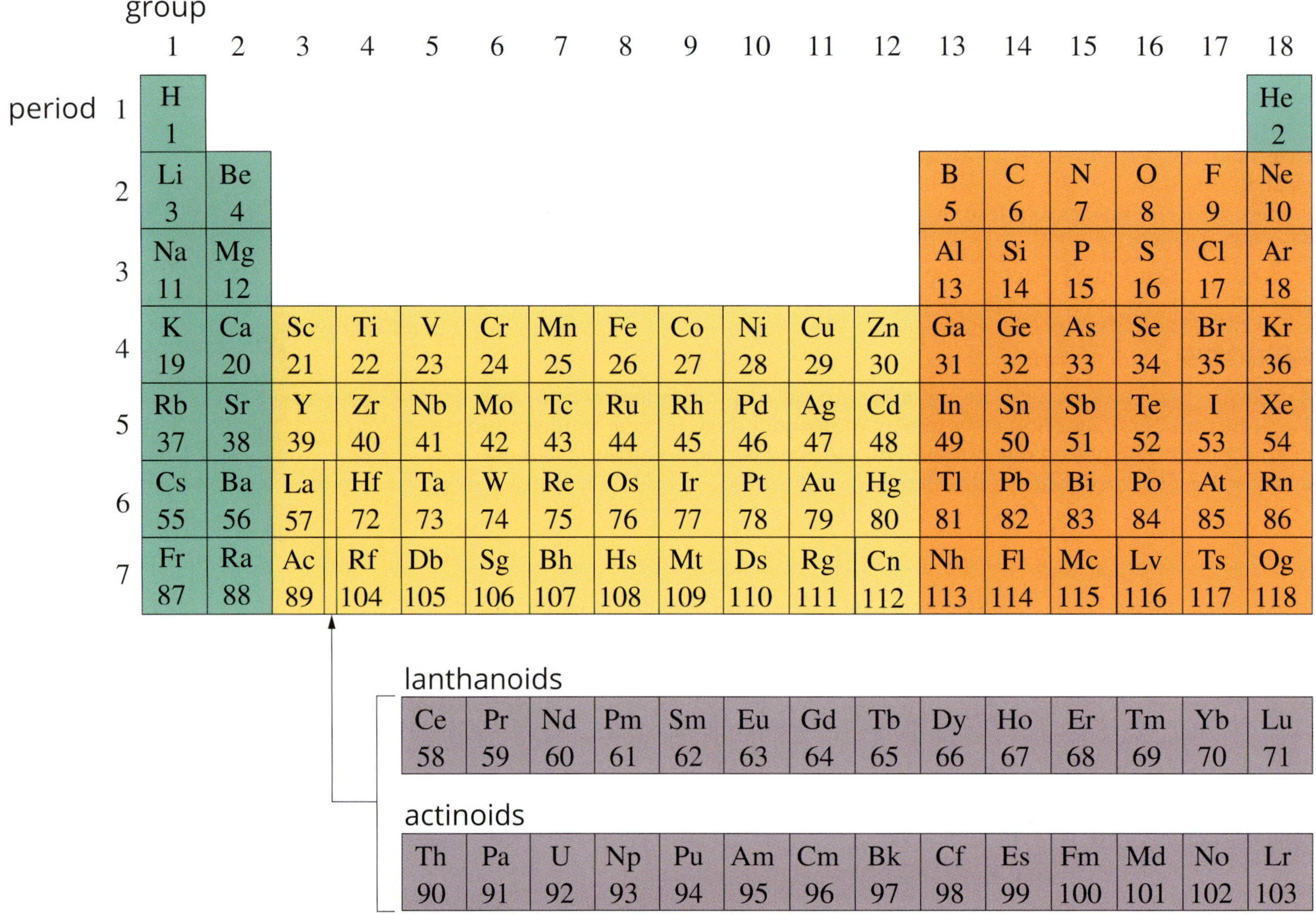

FIGURE 4.1.6 Colour is used to distinguish between the *s*, *p*, *d* and *f*-blocks of elements.

4.1 Review

SUMMARY

- The periodic table is a tool for organising elements according to their chemical and physical properties.
- The elements of the periodic table are arranged in order of increasing atomic number.
- Columns in the periodic table are known as groups and are numbered 1–18.
- The number of valence electrons in an atom of an element can be determined by the group in which it is located.
- Elements in groups 1, 2 and 13–18 in the periodic table are known as main group elements.
- The middle group of elements (groups 3–12) are known as the transition metals.
- Rows in the periodic table are known as periods and are numbered 1–7.
- The number of occupied electron shells in an atom of an element is equal to the number of the element's period.
- The periodic table has four main blocks of elements; the elements in each block all have the same type of subshell (*s*, *p*, *d* or *f*) as their highest energy subshell.

KEY QUESTIONS

1 In the periodic table, group is to column as period is to ________.

2 Main group elements are elements that belong to specific groups in the periodic table. Which groups are these?

3 How many valence electrons are in atoms of elements found in:

a group 1?
b group 15?
c group 17?
d group 2?

4 What is the electronic configuration of an atom of an element in period 3 and group 2?

5 Use the periodic table in Figure 4.1.3 on page 117 to answer these questions:

a In which group of the periodic table will you find the following?
i B
ii Cl
iii Na
iv Ar
v Si
vi Pb

b In which period of the periodic table will you find the following?
i K
ii F
iii He
iv H
v U
vi P

c What is the name, symbol and electronic configuration of the:
i second element in group 14?
ii second element in period 2?
iii element that is in group 18 and period 3?

6 Why are elements in the periodic table arranged in order of atomic number rather than relative atomic mass?

4.2 Trends in the periodic table: Part 1

The periodic table does not just provide information about an element's electronic configuration. It can also be used as a tool for summarising the relative properties of elements and explaining the trends observed in those properties.

You have already seen how the group number of an element indicates how many valence electrons an atom of that element has. The period indicates how many electron shells are occupied in an atom of an element. Properties such as atomic radii, electronegativity and ionisation energy show common trends in the periodic table.

Periodic trends were observed by Dmitri Mendeleev and formed the basis of the table of the elements that he first published in 1869. Mendeleev described the way the properties of the elements vary in the periodic law.

STATE OF MATTER AT ROOM TEMPERATURE

Some elements of the periodic table are gases at room temperature (25°C), many are solids and two are liquids. Melting points determine the state of matter of each element at room temperature. Trends in melting point can be seen in the first three periods. Melting points increase across periods from groups 1 to 14 and then drastically drop for groups 15–18. These trends are shown in Figure 4.2.1.

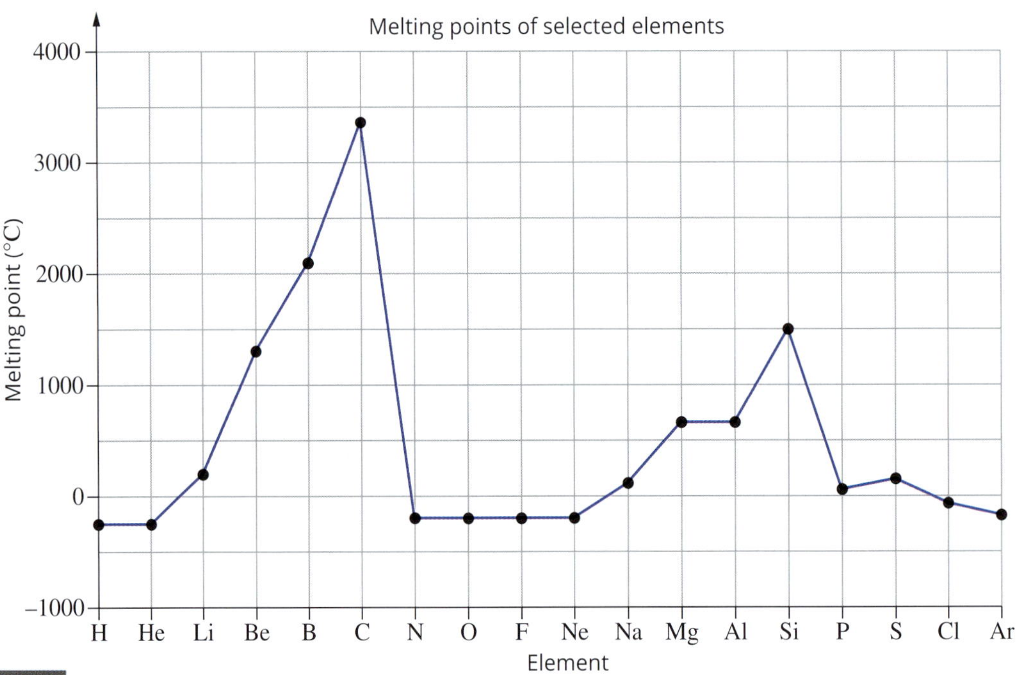

FIGURE 4.2.1 Melting points (in °C) of elements in periods 1–3.

FIGURE 4.2.2 Mercury is different to the rest of the *d*-block elements because it is liquid at room temperature (25°C), whereas the rest of the *d*-block elements are solid at room temperature.

There are other general trends in the state of elements at room temperature. Metals generally have high melting points, and most non-metals have low melting points. Elements in the *s*-block are all solids at room temperature, with the exception of hydrogen, which is a gas. The *d*-block elements are all solids at room temperature, except for mercury, which has a melting point of −39°C, making it a liquid at room temperature (Figure 4.2.2). The *p*-block has some elements that are solids at room temperature and some that are gases. All the noble gases (group 18) are gases at room temperature, but the only other gases in the *p*-block are nitrogen, oxygen, fluorine and chlorine. Bromine is a liquid at room temperature, as it has a melting point of −7°C.

These observed trends reflect similarities in the way that the atoms are bonded (joined together) in elements in each group of the periodic table. This bonding will be explained more fully in Chapter 5.

ELECTRONIC CONFIGURATION

To understand the reason for the periodicity of element properties, look at two groups of elements that Mendeleev recognised as being similar. The first group is the alkali metals (group 1). The elements in this group (lithium, sodium, potassium, rubidium and caesium) are all relatively soft metals and are highly reactive with water and oxygen.

Consider their electronic configurations:

Li $1s^2 2s^1$

Na $1s^2 2s^2 2p^6 3s^1$

K $1s^2 2s^2 2p^6 3s^2 3p^6 4s^1$

Rb $1s^2 2s^2 2p^6 3s^2 3p^6 3d^{10} 4s^2 4p^6 5s^1$

Cs $1s^2 2s^2 2p^6 3s^2 3p^6 3d^{10} 4s^2 4p^6 4d^{10} 5s^2 5p^6 6s^1$

These elements have similar valence shell electronic configurations—all have one electron in an *s*-subshell. This similarity in a group's arrangement of electrons gives elements similar properties and is responsible for the periodicity of element properties.

You can also see that the number of electron shells increases moving down the group. The increase in electron shells means that the valence electrons are in higher energy subshells and have a weaker attraction to the nucleus. The decrease in the attractive force between the valence electrons and the nucleus as you move down a group causes the trends in properties that are observed within a group.

CORE CHARGE

The **core charge** of an atom is a measure of the attractive force felt by the valence shell electrons towards the nucleus. Core charge can be used to predict the properties of elements and explain trends observed in the periodic table.

Consider an atom of lithium, which has an atomic number of three. It has three protons in its nucleus, two electrons in the first shell and one electron in the second shell (Figure 4.2.3).

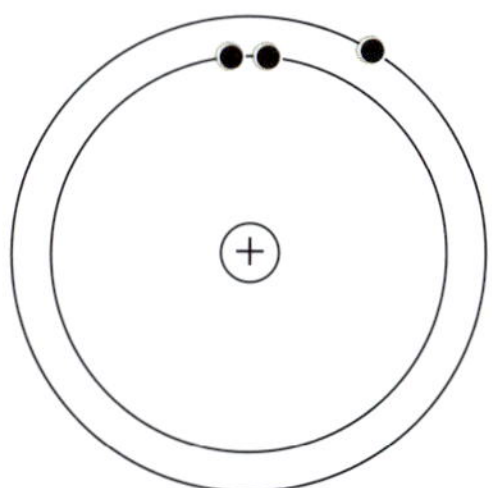

FIGURE 4.2.3 A Bohr diagram for the lithium atom, which has one valence electron and two electrons in the inner shell. The atom has a core charge of +1.

The valence shell electron is attracted to the three positive charges in the nucleus. This electron is also repelled by the two electrons in the inner shell. The electrons in the inner shell shield the valence shell electron from the attraction of the nucleus. The valence shell electron is effectively attracted to the nucleus as if there was a nuclear charge of +1. This atom is therefore said to have a core charge of +1.

In atoms with two or more shells filled with electrons, the attraction between the nucleus and the valence shell electrons is reduced by repulsion between the inner shell electrons and the valence shell electrons.

In general:

Core charge = number of protons in the nucleus − number of total inner-shell electrons

For example, an atom of chlorine (Figure 4.2.4) has 17 protons and 7 valence shell electrons; the number of electrons in the inner shells is 10. The core charge of a chlorine atom is 17 − 10 = +7.

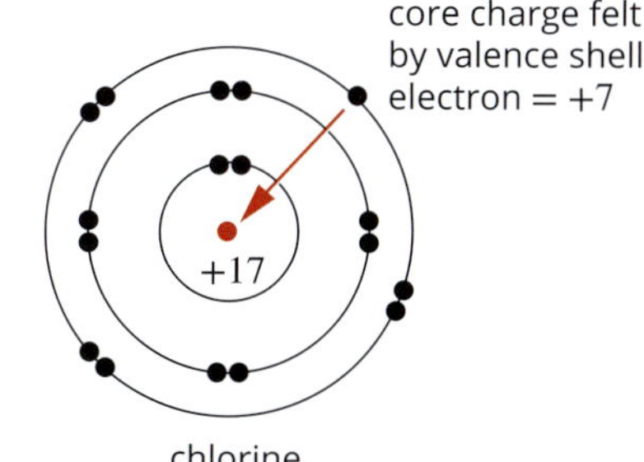

FIGURE 4.2.4 A Bohr diagram for the chlorine atom, showing a valence shell electron being shielded by 10 electrons (in shells $n = 1$ and $n = 2$).

Worked example 4.2.1

DETERMINING CORE CHARGE

Determine the core charge of an atom of aluminium.

Thinking	Working
Determine the number of electrons in an atom of the element, using the periodic table as a reference.	The atomic number of aluminium is 13. Therefore, an atom of aluminium has 13 protons and 13 electrons.
Use the number of electrons to determine the electronic configuration.	With 13 electrons, the electronic configuration is $1s^2 2s^2 2p^6 3s^2 3p^1$.
Determine the core charge. Core charge = number of protons – number of inner-shell electrons	The third shell is the valence shell in this atom. There are 10 inner-shell electrons, which in this atom are electrons in the first and second shells. Core charge = 13 – 10 = +3

Worked example: Try yourself 4.2.1

DETERMINING CORE CHARGE

Determine the core charge of an atom of fluorine.

Consider the atoms of two different elements in group 1, lithium and sodium, shown in Figure 4.2.5.

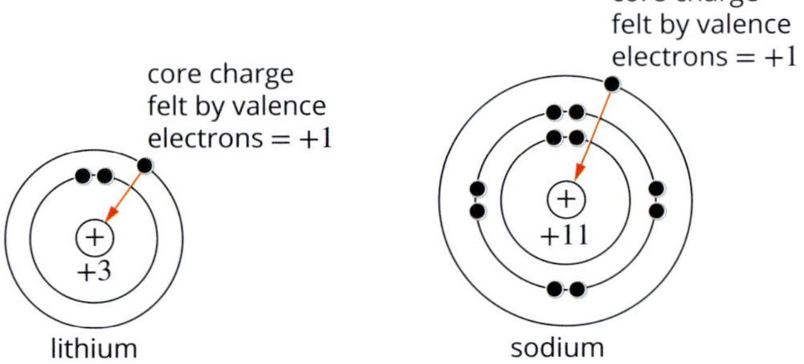

FIGURE 4.2.5 Lithium and sodium atoms both have a core charge of +1.

As for all group 1 elements, the valence electron of a lithium atom or a sodium atom experiences a core charge of +1.

As you look down a group, you can see that:

- the core charge remains constant, but the number of electron shells increases.
- the valence electrons become farther from the nucleus. Because of this, they will be pulled less strongly towards the nucleus.

Now consider sodium and chlorine. Both elements are in period 3 of the periodic table (Figure 4.2.6).

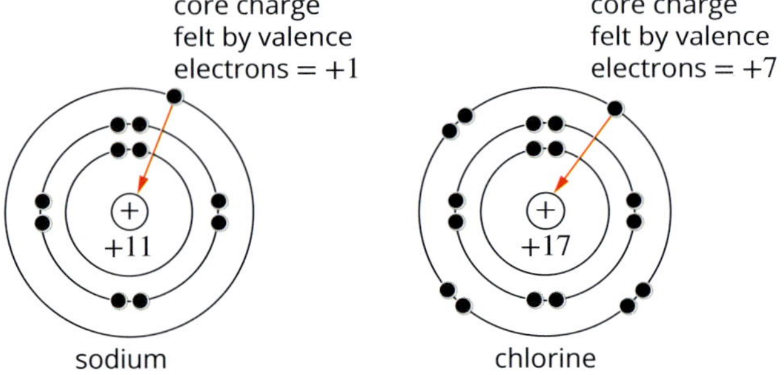

FIGURE 4.2.6 The core charges of two period 3 elements, sodium and chlorine, are +1 and +7 respectively.

The core charge experienced by the valence shell electrons in atoms of elements increases from left to right across a period, as demonstrated by sodium and chlorine. The core charge of an atom of a main group element is equal to the number of valence electrons in the atom, as summarised in Table 4.2.1.

Table 4.2.2 summarises how the attraction between the nucleus and valence electrons varies in the periodic table.

TABLE 4.2.2 The changes in attraction between the nucleus and valence electrons within groups and periods of the periodic table

	Trend in core charge	Trend in attraction between the nucleus and valence electrons
down a group	remains constant	Core charge stays constant down a group, but the valence electrons are held less strongly because they are farther from the nucleus (there are more shells in the atom).
left to right across a period	increases	The valence electrons are more attracted to the nucleus as the core charge increases.

TABLE 4.2.1 Core charges of main group elements

Group	Core charge
1	+1
2	+2
13	+3
14	+4
15	+5
16	+6
17	+7
18*	+8

*Helium has a core charge of +2.

ELECTRONEGATIVITY

Electronegativity is the ability of an atom to attract electrons towards itself. The more strongly the valence electrons of an atom are attracted to the nucleus of the atom, the greater the electronegativity. Therefore, the greater the core charge of an atom, the greater the electronegativity. Figure 4.2.7 shows the electronegativity of many of the main group elements.

Electronegativity increases across a period.

Electronegativity decreases down a group.

1	2	13	14	15	16	17
Li 1.0	Be 1.6	B 2.0	C 2.6	N 3.0	O 3.4	F 4.0
Na 0.9	Mg 1.3	Al 1.6	Si 1.9	P 2.2	S 2.6	Cl 3.2
K 0.8	Ca 1.0	Ga 1.8	Ge 2.0	As 2.2	Se 2.6	Br 3.0
Rb 0.8	Sr 1.0	In 1.8	Sn 2.0	Sb 2.1	Te 2.1	I 2.7
Cs 0.8	Ba 0.9	Tl 1.8	Pb 1.8	Bi 2.0	Po 2.0	At 2.2
Fr 0.7	Ra 0.9					

FIGURE 4.2.7 The electronegativity of elements generally decreases down a group and increases across a period, from left to right.

The trends observed in the electronegativity of the elements are summarised in Table 4.2.3.

TABLE 4.2.3 Trends in electronegativity in groups and periods of the periodic table

	Trend in electronegativity	Explanation
down a group	decreases	The core charge stays constant, and the number of shells increases down a group. Therefore, valence electrons are less strongly attracted to the nucleus when they are farther from the nucleus. As a result, electronegativity decreases.
left to right across a period	increases	The number of occupied shells in the atoms remains constant, but the core charge increases across a period. Therefore, the valence electrons become more strongly attracted to the nucleus. As a result, electronegativity increases.

ATOMIC RADIUS

Atomic radius is a measurement used for the size of atoms. It can be regarded as the distance from the nucleus to the valence shell electrons. It is usually measured by halving the distance between the nuclei of two atoms of the same element that are bonded together. Figure 4.2.8 depicts the atomic radii of many of the main group elements. Table 4.2.4 summarises the trends in atomic radii that are found in the periodic table. Within each group, the atomic radius tends to increase going down the group. Within each period, the atomic radius tends to decrease going from left to right.

Atoms do not have sharply defined boundaries, so it is not possible to measure their radii directly. One method of obtaining atomic radii, and therefore an indication of atomic size, is to measure the distance between nuclei of atoms in molecules. For example, in a hydrogen molecule (H_2) the two nuclei are 64 picometres (pm) apart. The radius of each hydrogen atom is assumed to be half that distance, i.e. 32 pm.

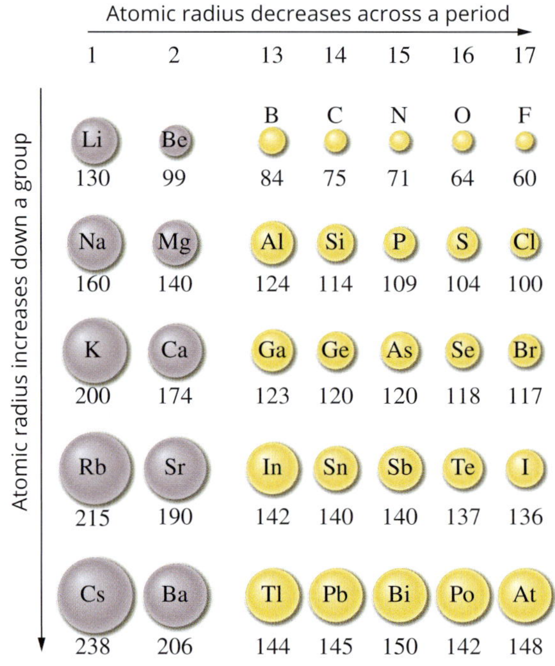

FIGURE 4.2.8 The relative sizes of atoms of selected main group elements. The atomic radii are given in picometres (pm). A picometre is 10^{-12} m.

TABLE 4.2.4 Trends in atomic radii in the periodic table

	Trend in atomic radii	Explanation
down a group	increases	Core charge stays constant, and the number of shells increases as you move down a group. As a result, atomic radius increases.
left to right across a period	decreases	As you move across a period, the number of occupied shells in the atoms remains constant, but the core charge increases. The valence electrons become more strongly attracted to the nucleus, so the atomic radius decreases across a period.

Converting between units

In science, you will often need to convert from one unit to another to complete a calculation. Knowing how to convert units is an important skill. As some units of measurement are difficult to visualise, knowing the size of different units in relation to one another will aid your understanding and help you to avoid errors in your calculations (Table 4.2.5).

It is important to give a symbol the correct case (upper or lower case): there is a big difference between 1 mm and 1 Mm.

TABLE 4.2.5 Prefixes and conversion factors

Multiplying factor	Power	Prefix	Symbol
1 000 000 000 000	10^{12}	tera	T
1 000 000 000	10^{9}	giga	G
1 000 000	10^{6}	mega	M
1000	10^{3}	kilo	k
0.01	10^{-2}	centi	c
0.001	10^{-3}	milli	m
0.000 001	10^{-6}	micro	µ
0.000 000 001	10^{-9}	nano	n
0.000 000 000 001	10^{-12}	pico	p

For example, atomic radii are given in picometres (pm). A hydrogen atom has an atomic radius of 32 pm. To put this into context, you can convert this value into a unit you can easily visualise, like millimetres (mm). The difference between the indices (the values that 10 is raised to) of mm and pm is nine, meaning you need to move the decimal place nine places to the left. So 32 pm becomes 0.000 000 032 mm. This is very small indeed!

4.2 Review

SUMMARY

- Elements can be solid, liquid or gas at room temperature, depending on their melting point. Trends in the state of the elements at room temperature reflect similarities in the way that the atoms are bonded (joined together) in elements in each group of the periodic table.
- The core charge of an atom is a measure of the attractive force felt by the valence electrons towards the nucleus.
- The core charge is calculated by subtracting the total number of inner-shell electrons from the number of protons in the nucleus.
- Electronegativity is the ability of an element to attract electrons towards itself.
- Atomic radius is a measurement used for the size of atoms. It can be regarded as the distance from the nucleus to the outermost electrons.
- The table below summarises how these properties have specific trends within the groups and periods of the periodic table.

Property	Down a group	Across a period (left to right)
core charge	no change	increases
electronegativity	decreases	increases
atomic radius	increases	decreases

KEY QUESTIONS

1 What is the core charge of an atom of carbon?

2 Explain the relationship between electronegativity and core charge.

3 Figure 4.2.7 gives electronegativity values for the elements in groups 1, 2 and 13–17 of the periodic table.

- **a** Give the name and symbol of the element that has the:
  - **i** highest electronegativity
  - **ii** lowest electronegativity.
- **b** In which group do you see the:
  - **i** greatest change in electronegativity as you go down the group?
  - **ii** smallest change in electronegativity as you go down the group?
- **c** Why are the elements of group 18 usually omitted from tables that give electronegativity values?

4 By referring to the periodic table, arrange the following atoms in order of smallest to largest size: Ca, F, Mg, O and P.

5 Which elements are gases at room temperature?

4.3 Trends in the periodic table: Part 2

You have learnt that the core charge and the number of electron shells in an atom of an element can be used to predict some properties of elements. You will remember that the core charge of an atom is the attractive force felt by the valence electrons towards the nucleus. Core charge increases as you move across a period. As you move down a group, the core charge stays constant, but atomic radius increases due to the additional electron shell in each period. The changes in core charge and atomic radii account for periodic trends in ionisation energy, reactivity and metallic character.

FIRST IONISATION ENERGY

When an element is heated, its electrons can move to higher energy shells. If an atom is given sufficient energy, an electron can be completely removed from the atom. If this occurs, the atom will now have one less electron than the number of protons in the nucleus, and will become a positively charged **ion**. A positively charged ion is called a **cation**. You will learn more about ions in Chapter 5.

GO TO ➤ Section 5.2 page 147

The process of removing an electron from an atom and forming an ion is called **ionisation**. The valence electrons are removed first because they are the farthest electrons from the nucleus and the least strongly held.

The energy required to remove one electron from an atom of an element in the gas phase is called the **first ionisation energy**. For example, the ionisation energy of sodium is 494 kJ per mole of sodium atoms.

An ion is any atom of an element with more or fewer electrons than protons. If an atom loses electrons, it becomes positively charged because there are now more protons than electrons. A positively charged ion is called a cation. If an atom gains electrons, it becomes negatively charged and is called an **anion**.

Figure 4.3.1 shows the first ionisation energies of most main group elements.

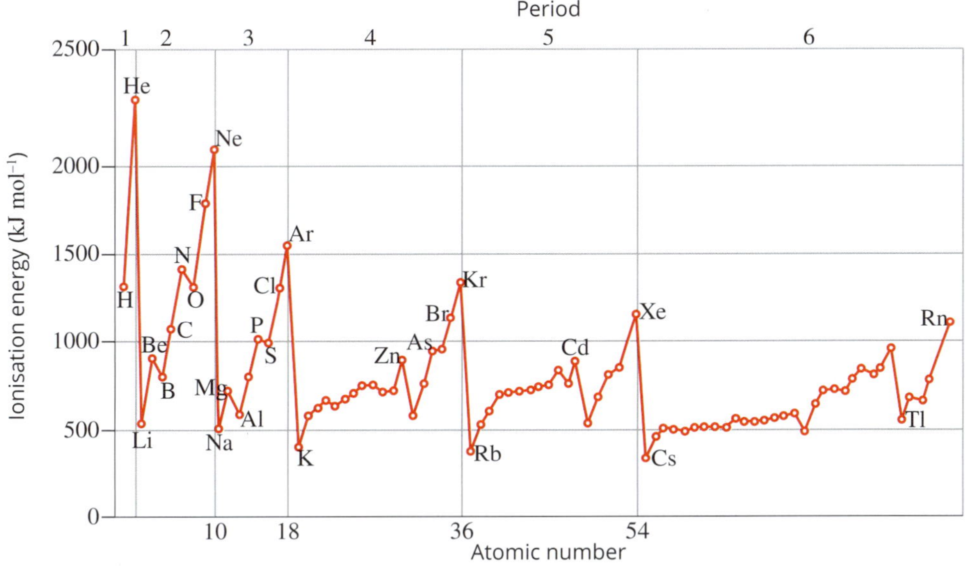

FIGURE 4.3.1 The first ionisation energy generally increases from left to right across a period.

The magnitude of the first ionisation energy reflects how strongly the valence electrons are attracted to the nucleus of the atoms. The more strongly the valence electrons are attracted to the nucleus, the greater the energy that is required to remove them from the atom and the higher the first ionisation energy. The trends in ionisation energy within the *s*- and *p*-block elements are described in Table 4.3.1 on page 130.

TABLE 4.3.1 The trends in ionisation properties in groups and periods of the periodic table for the *s*- and *p*-block elements

	Trend in ionisation energy	Explanation
down a group	decreases	Core charge stays constant, and the number of shells increases down a group. Therefore, the valence electrons are less attracted to the nucleus because they are farther from the nucleus. As a result, the energy required to overcome the attraction between the nucleus and the valence electron is less, and the first ionisation energy decreases down a group.
left to right across a period	increases	Core charge increases, and the number of occupied shells remains constant across a period. As a result, the valence electrons become more strongly attracted to the nucleus, and more energy is required to remove an electron. Therefore, first ionisation energy increases across a period.

In summary, as the core charge increases across a period, so too does the ionisation energy. As atomic radius increases down a group (due to the increasing number of occupied shells), the electrons become farther from the nucleus. Therefore, the valence electrons in elements lower in a group can be removed more easily, meaning the ionisation energy decreases.

Worked example 4.3.1

DETERMINING FIRST IONISATION ENERGY

Determine which of these atoms has the greatest first ionisation energy: Br, Cl, S, Se

Thinking	Working
Determine the group and period for each element.	Br is in group 17, period 4 Cl is in group 17, period 3 S is in group 16, period 3 Se is in group 16, period 4
Recall that first ionisation energies generally increase across a period and decrease down a group.	Of the four elements, Cl and Br are in group 17 and are expected to have greater first ionisation energies than the elements in group 16. Cl is in period 3, so is expected to have a greater first ionisation energy than Br, which is in period 4. Cl has the greatest first ionisation energy.

Worked example: Try yourself 4.3.1

DETERMINING FIRST IONISATION ENERGY

Determine which of these atoms has the greatest first ionisation energy: Al, B, C, Si.

+ ADDITIONAL

Trends in *d*- and *f*-block elements

The *d*-block elements are the transition metals. The *f*-block elements are the lanthanoids and actinoids (also called the lanthanides and actinides). The trends in the *d*-block and *f*-block elements are more complex than those in the first three periods. Looking at elements within the same group in periods 4, 5 and 6, a new pattern emerges.

Between periods 4 and 5 for *d*-block elements, the atomic radius increases as expected. When the 4*f*-subshell starts filling in period 6, the core charge increases. With this increase in core charge, the valence electrons are more closely attracted to the nucleus, and the atoms become denser and have smaller atomic radii.

The pattern is true for the transition metals in the sixth period. For example, in group 8, iron (period 4) has an atomic radius of 124 pm. Ruthenium (period 5) has a higher atomic radius at 136 pm because of the additional electron shell. Osmium (period 6) also has an atomic radius of 136 pm, even with an additional electron shell filled. This is because, as the electrons fill the 4*f*-subshell, the core charge increases.

This change in density and atomic radius also affects other properties of the transition metals. With increasing density, melting point increases. Tungsten has the highest melting temperature of any metal at 3414°C.

METALLIC CHARACTER

Metals conduct electricity and are usually solids at room temperature. Conversely, non-metallic elements usually do not conduct electricity and many are gases at room temperature.

The differences between the properties of metals and non-metals are related to the number of electrons in the outer shell of their atoms. In general, elements with atoms containing one, two or three valence electrons tend to behave as metals, whereas those with four or more valence electrons behave as non-metals. You will study this in more detail in Chapter 5.

Metals are located on the left side of the periodic table and non-metals on the right. Elements known as the **metalloids** are located between the metals and non-metals. The metalloids exhibit both metallic and non-metallic properties. Silicon is one of the most abundant metalloids (Figure 4.3.2). It is a brittle solid, which is a common property of non-metals. However, it is also a semiconductor, meaning it exhibits electrical conductivity, making it useful in many electronic devices such as computers and calculators.

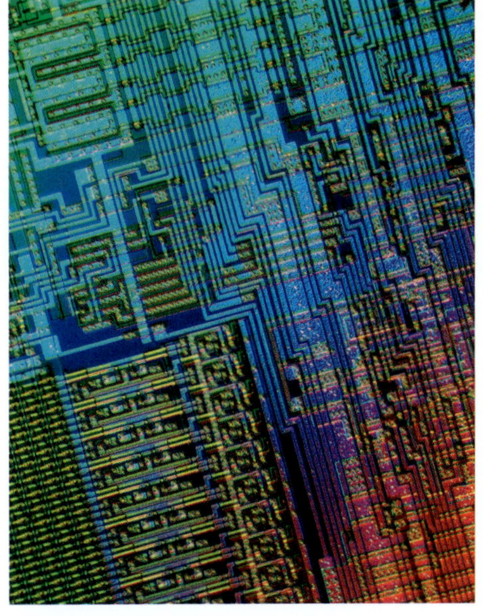

FIGURE 4.3.2 Light micrograph of the surface of an integrated circuit that uses silicon as a semiconductor.

REACTIVITY

The **reactivity** of an element is an indication of how easily an atom of that element loses or gains electrons.

Reactivity of metals

When metals react, they lose electrons. Therefore, the reactivity of metals is a measure of how easily an atom of a metallic element can lose electrons. The weaker the attraction of the valence electrons to the nucleus, the more easily the electrons can be lost.

Moving down a group, there is an increasing number of electron shells in the atoms and therefore a weaker attractive force between the nucleus and valence electrons. This means that reactivity of metals increases down a group. The rate at which metals react with water indicates their relative reactivity.

Table 4.3.2 on page 132 describes the reaction of some of the group 1 and 2 metals with water. In each case, a reaction results in the formation of hydrogen gas.

TABLE 4.3.2 Reactions of some group 1 and 2 metals with water.

Period	Group	Element	Reaction with water
3	1	sodium	vigorous, producing enough energy to melt the sodium, which fizzes and skates on the water surface
4	1	potassium	violent, making crackling sounds as the heat evolved ignites the hydrogen produced by the reaction
5	1	rubidium	violent explosion
3	2	magnesium	no reaction at room temperature but will react with steam
4	2	calcium	slow reaction at room temperature

For the five metals in Table 4.3.2:

- those in group 1 are more reactive in water than those in group 2
- the reactivity of the metal in water increases down a group.

These generalisations agree with the results of experiments with other metals.

From left to right across the periodic table, the core charge of the atoms increases and it becomes more difficult for the atoms of an element to lose electrons. Within the metals, this means there is a decrease in reactivity.

In summary, the reactivity of metals:

- increases down a group because it is easier for a metal atom with a greater number of shells to lose electrons
- decreases across the period because the increasing core charge makes it more difficult for a metal atom to lose electrons.

You will learn more about the reactivity of metals in Chapter 11.

Reactivity of non-metals

Atoms of non-metallic elements undergo chemical reactions to gain electrons and form a stable octet arrangement. The more easily a non-metal can attract or share electrons, the more reactive that non-metal is. Elements that have fewer electron shells and higher core charges will have a greater attractive force between the nucleus and the valence electrons and thus be more reactive.

Therefore, the reactivity of non-metals:

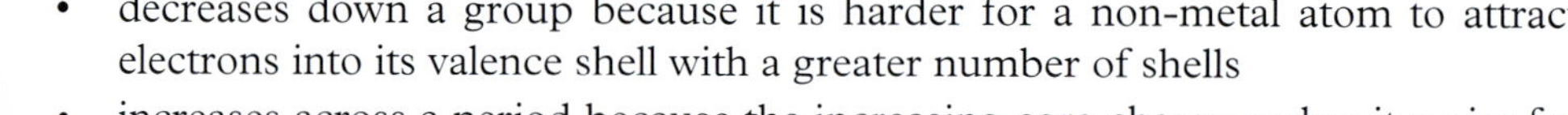

- decreases down a group because it is harder for a non-metal atom to attract electrons into its valence shell with a greater number of shells

- increases across a period because the increasing core charge makes it easier for a non-metallic atom to attract electrons.

4.3 Review

SUMMARY

- The first ionisation energy is the energy required to remove one electron from an atom of an element in the gas phase.
- First ionisation energy decreases down a group but increases across a period.
- Chemical reactivity is the ease with which an element undergoes a reaction. Metals tend to lose electrons, and non-metals tend to gain electrons. From left to right across a period in the periodic table, the elements change from metals to metalloids to non-metals.
- Chemical reactivity for metals increases down a group but decreases across a period. These trends are illustrated by the reactivity of metals with water.
- Chemical reactivity of non-metals decreases down a group but increases across a period.
- Many trends in the physical properties of elements in the periodic table can be explained using two key ideas:
  - From left to right across a period, the core charge of atoms increases, so the attractive force felt between the valence electrons and the nucleus increases.
  - Down a group, the number of shells in an atom increases, so the valence electrons are farther from the nucleus and are held less strongly.

KEY QUESTIONS

1. **a** Explain the meaning of first ionisation energy.
   b What factors need to be considered when predicting the trend in first ionisation energies across a period?
2. What is a metalloid?
3. Explain why first ionisation energy increases from left to right across a period.
4. Why is strontium more reactive with water than beryllium is?
5. **a** Select the most reactive non-metal from the following list: magnesium, sulfur, chlorine, fluorine, aluminium, oxygen.
   b Explain your reasoning.

Chapter review

KEY TERMS

anion
atomic number
atomic radius
block (periodic table)
cation
core charge
electronegativity
first ionisation energy
group (periodic table)
ion
ionisation
main group element
metalloid
period (periodic table)
periodicity
periodic law
periodic table
reactivity
transition metal

REVIEW QUESTIONS

1 Elements in the periodic table are arranged by increasing atomic number. What determines an atom's atomic number?

2 Use the periodic table to determine the period and block of the following elements:
  a hydrogen
  b carbon
  c phosphorus
  d copper
  e uranium

3 Determine the period and group of the elements with the following electronic configurations:
  a $1s^22s^2$
  b $1s^22s^22p^63s^23p^2$
  c $1s^22s^22p^63s^23p^63d^{10}4s^24p^1$
  d $1s^2$

4 In the periodic table, explain why there are:
  a two groups of elements in the *s*-block
  b six groups of elements in the *p*-block
  c 10 elements in each transition series
  d 14 elements in the actinoids and lanthanoids.

5 Name an element with properties similar to those of:
  a carbon
  b rubidium
  c iodine
  d phosphorus.

6 Describe the trend in melting points for the first 18 elements of the periodic table.

7 Which elements are liquid at room temperature?

8 Across a period, the number of subatomic particles in an atom increases, but the size of an atom decreases. Why?

9 Explain why it takes more energy to remove an electron from the outer shell of an atom of:
  a phosphorus than of magnesium
  b fluorine than of iodine.

10 a State the electronic configuration of nitrogen.
  b What period and group does nitrogen belong to in the periodic table?
  c How many valence electrons does nitrogen have?
  d What is nitrogen's core charge?

11 Explain why the radii of atoms do not increase uniformly as the atomic number of the atom increases.

12 The length of the bond between two fluorine atoms in F_2 is 120 pm. What is the atomic radius of fluorine?

13 Consider the elements in period 2 of the periodic table: lithium, beryllium, boron, carbon, nitrogen, oxygen and fluorine. Describe the changes that occur across the period. Consider:
  a the sizes of atoms
  b metallic character
  c electronegativity.

14 a Order the following elements from least reactive to most reactive: rubidium, sodium, lithium, potassium.
  b Explain your reasoning.

15 From each set of elements, select the element that has the largest first ionisation energy.
  a phosphorus, arsenic, nitrogen
  b silicon, chlorine, sulfur
  c bromine, chlorine, sulfur

16 How does the reactivity of elements change from left to right across period 3 in the periodic table?

17 Name some characteristics of metals.

18 Consider Figure 4.3.1 (page 129). First ionisation energies generally increase across periods. However, there is a slight decrease in first ionisation energy from Mg to Al. Explain this exception to the trend with reference to the electronic configurations of these elements.

19 Figures 4.2.7 (page 125), 4.2.8 (page 126) and 4.3.1 (page 129) show periodicity with respect to electronegativity, atomic radius and first ionisation energy. What does the term 'periodicity' mean?

20 What concepts or ideas are the two versions of the periodic table below trying to convey? Search the internet for other versions of the periodic table.

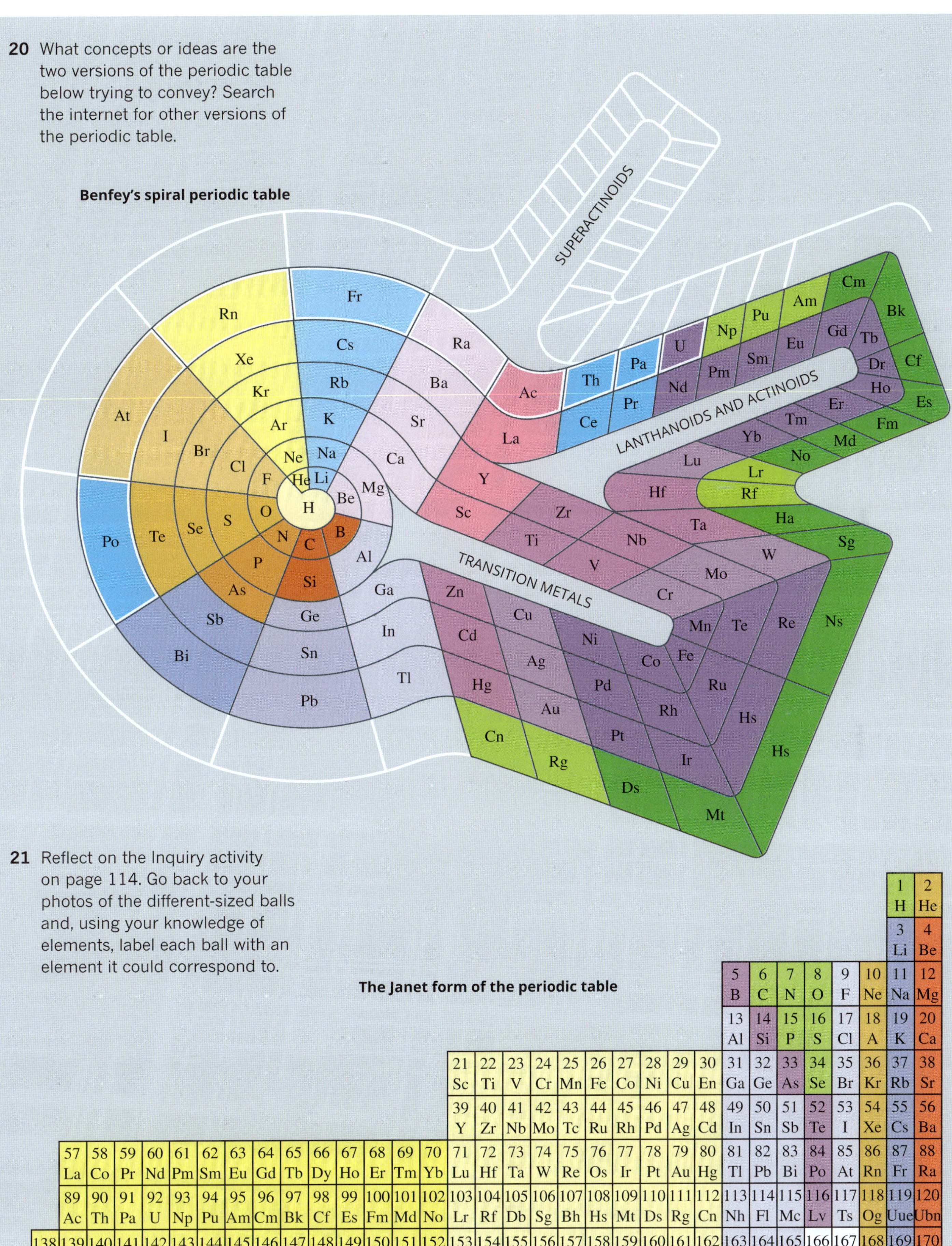

Benfey's spiral periodic table

21 Reflect on the Inquiry activity on page 114. Go back to your photos of the different-sized balls and, using your knowledge of elements, label each ball with an element it could correspond to.

The Janet form of the periodic table

																															1 H	2 He
																															3 Li	4 Be
																									5 B	6 C	7 N	8 O	9 F	10 Ne	11 Na	12 Mg
																									13 Al	14 Si	15 P	16 S	17 Cl	18 A	19 K	20 Ca
															21 Sc	22 Ti	23 V	24 Cr	25 Mn	26 Fe	27 Co	28 Ni	29 Cu	30 En	31 Ga	32 Ge	33 As	34 Se	35 Br	36 Kr	37 Rb	38 Sr
															39 Y	40 Zr	41 Nb	42 Mo	43 Tc	44 Ru	45 Rh	46 Pd	47 Ag	48 Cd	49 In	50 Sn	51 Sb	52 Te	53 I	54 Xe	55 Cs	56 Ba
	57 La	58 Co	59 Pr	60 Nd	61 Pm	62 Sm	63 Eu	64 Gd	65 Tb	66 Dy	67 Ho	68 Er	69 Tm	70 Yb	71 Lu	72 Hf	73 Ta	74 W	75 Re	76 Os	77 Ir	78 Pt	79 Au	80 Hg	81 Tl	82 Pb	83 Bi	84 Po	85 At	86 Rn	87 Fr	88 Ra
	89 Ac	90 Th	91 Pa	92 U	93 Np	94 Pu	95 Am	96 Cm	97 Bk	98 Cf	99 Es	100 Fm	101 Md	102 No	103 Lr	104 Rf	105 Db	106 Sg	107 Bh	108 Hs	109 Mt	110 Ds	111 Rg	112 Cn	113 Nh	114 Fl	115 Mc	116 Lv	117 Ts	118 Og	119 Uue	120 Ubn
138	139	140	141	142	143	144	145	146	147	148	149	150	151	152	153	154	155	156	157	158	159	160	161	162	163	164	165	166	167	168	169	170

CHAPTER
05 Bonding

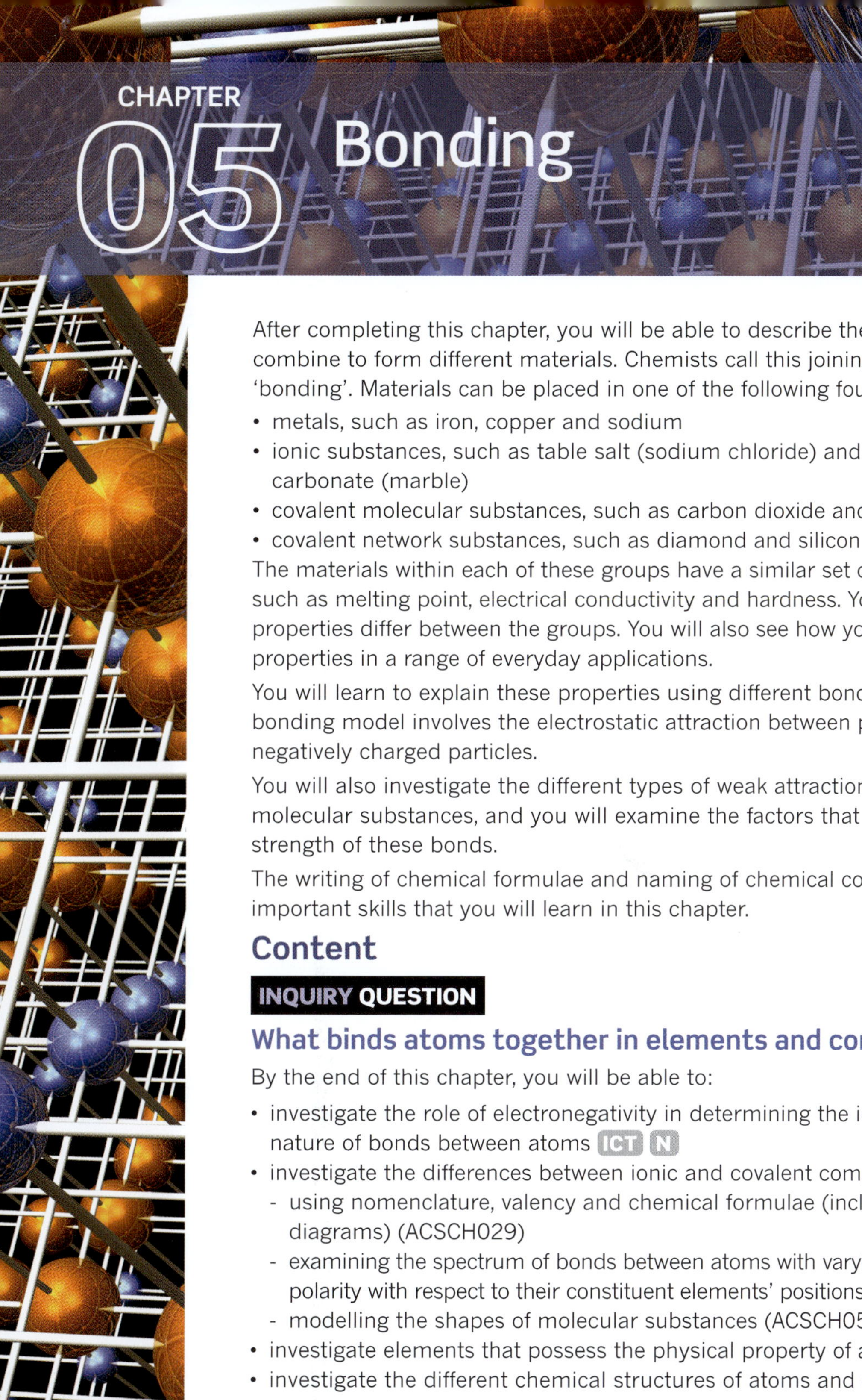

After completing this chapter, you will be able to describe the various ways atoms combine to form different materials. Chemists call this joining together of atoms 'bonding'. Materials can be placed in one of the following four groups:

- metals, such as iron, copper and sodium
- ionic substances, such as table salt (sodium chloride) and calcium carbonate (marble)
- covalent molecular substances, such as carbon dioxide and oxygen
- covalent network substances, such as diamond and silicon dioxide.

The materials within each of these groups have a similar set of physical properties such as melting point, electrical conductivity and hardness. You will see that the properties differ between the groups. You will also see how you can utilise these properties in a range of everyday applications.

You will learn to explain these properties using different bonding models. Each bonding model involves the electrostatic attraction between positively and negatively charged particles.

You will also investigate the different types of weak attractions that form between molecular substances, and you will examine the factors that influence the strength of these bonds.

The writing of chemical formulae and naming of chemical compounds are important skills that you will learn in this chapter.

Content

INQUIRY QUESTION

What binds atoms together in elements and compounds?

By the end of this chapter, you will be able to:

- investigate the role of electronegativity in determining the ionic or covalent nature of bonds between atoms ICT N
- investigate the differences between ionic and covalent compounds through:
  - using nomenclature, valency and chemical formulae (including Lewis dot diagrams) (ACSCH029)
  - examining the spectrum of bonds between atoms with varying degrees of polarity with respect to their constituent elements' positions on the periodic table
  - modelling the shapes of molecular substances (ACSCH056, ACSCH057)
- investigate elements that possess the physical property of allotropy ICT
- investigate the different chemical structures of atoms and elements, including but not limited to:
  - ionic networks
  - covalent networks (including diamond and silicon dioxide)
  - covalent molecular structures
  - metallic structure
- explore the similarities and differences between the nature of intermolecular and intramolecular bonds and the strength of the forces associated with each, in order to explain the:
  - physical properties of elements
  - physical properties of compounds (ACSCH020, ACSCH055, ACSCH058)

5.1 Metallic bonding

CHEMISTRY INQUIRY CCT

Comparing the properties of metals and ionic compounds

Is there a difference between the solubility of metals and ionic compounds in water?

COLLECT THIS ...

- piece of aluminium foil
- paper clip
- table salt
- bicarbonate of soda
- water
- 4 glasses of the same size and volume
- 4 teaspoons

DO THIS ...

1 Draw up the table below and predict what will happen when you fill each glass with warm tap water, add one of the following to each glass and stir for 5 minutes:
   a aluminium foil
   b a paper clip
   c a pinch of salt
   d a pinch of bicarb of soda.

RECORD THIS ...

Test your predictions by carrying out the four experiments.

Describe what happened in each condition by recording your observations.

Present a table of your results.

REFLECT ON THIS ...

Explain any patterns you observed.

Explain the meaning of your observations.

What could you do next time to improve your experiment?

Experiment	Prediction	Observations
a Add a piece of aluminium foil to a glass of warm water and stir.		
b Add a paper clip to a glass of warm water and stir.		
c Add a pinch of salt to a glass of warm water and stir.		
d Add a pinch of bicarbonate of soda to a glass of warm water and stir.		

PROPERTIES OF METALS

In this section, you will learn about the properties that all metals have in common. Metals have been important to human beings since early times. The development of civilisation can be measured by the way we have used metals. The Copper Age (5000–3000 BCE) was followed by the Bronze Age (3000–1000 BCE) and the Iron Age (from 1000 BCE).

Gold, silver and copper can be found on Earth in an almost pure form. These metals were employed by prehistoric humans to make ornaments, tools and weapons. As humans' knowledge of metallurgy (the science of modifying metals) has developed, metals have played a significant role in fields as diverse as construction, agriculture, art, medicine and transport (Figure 5.1.1).

FIGURE 5.1.1 The metal titanium has many uses, including in (a) the spectacular curved space museum building in Moscow and (b) the SR-71 Blackbird reconnaissance aircraft.

FIGURE 5.1.2 Iron is magnetic. These iron filings are aligned to the magnetic fields created by the two bar magnets.

The diverse properties of the different metals make them suitable for many purposes. Table 5.1.1 shows the uses of some metals. For example, titanium (Figure 5.1.4) is a very strong, relatively unreactive metal with a low density, close to that of bone. Consequently, it is used in surgical implants that can last up to 20 years with little effect on the body. Titanium is also used in the aerospace industry, in art and architecture, and in sporting products such as golf clubs.

TABLE 5.1.1 Properties and uses of some metals

Metal	Properties	Uses
iron (Figure 5.1.2)	soft, malleable, magnetic, good thermal and electrical conductor, fairly reactive, readily forms alloys	can corrode and is usually converted to more stable steel, which is used in buildings and bridges, automobiles, machinery and appliances
aluminium (Figure 5.1.3)	low density, relatively soft when pure, excellent thermal and electrical conductor, malleable and ductile, good reflector of heat and light, readily forms alloys	saucepans, frying pans, drink cans, cooking foil, food packaging, roofing, window frames, appliance trim, decorative furniture, electrical cables, aircraft and boat construction
titanium (Figure 5.1.4)	very strong, high melting point, low density, low reactivity, readily forms alloys	medical devices within the body, wheelchairs, computer cases; lightweight alloys used in high-temperature environments such as spacecraft and aircraft
gold (Figure 5.1.5)	shiny gold appearance, excellent thermal and electrical conductor, unreactive, readily forms alloys	electrical connections, jewellery, monetary standard, dentistry

FIGURE 5.1.3 Aluminium's properties make it ideal for use in soft drink cans.

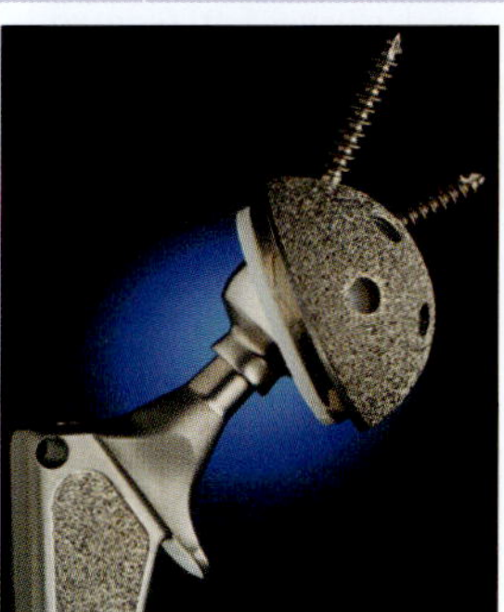

FIGURE 5.1.4 Titanium is a light but strong metal, making it ideal for use in medical implants, like this replacement hip joint.

FIGURE 5.1.5 The electrical conductivity of gold is utilised in this computer processor chip.

Table 5.1.2 lists the properties of some metals and non-metals. Despite the different properties of metals, most metals:

- have relatively high melting points and boiling points
- are good **conductors** of electricity
- are good conductors of heat
- have high **densities**.

TABLE 5.1.2 Properties of some metallic and non-metallic elements

Element	Melting point (°C)	Boiling point (°C)	Electrical conductivity ($MS\,m^{-1}$)*	Thermal conductivity ($J\,s^{-1}\,m^{-1}\,K^{-1}$)†	Density ($g\,mL^{-1}$)
metals					
gold	1063	2970	45	310	19.3
iron	1540	3000	9.6	78	7.86
mercury	−39	357	1	8.4	13.5
potassium	64	760	14	100	0.86
silver	961	2210	60	418	10.5
sodium	98	892	21	135	0.97
non-metals					
carbon (diamond)	3550	‡	10^{-17}	–	3.51
oxygen	−219	183	–	0.026	1.15 (liquid)

*$MS\,m^{-1}$ = megasiemens per metre.
†Thermal conductivity measures the conductance of heat.
‡Diamond sublimes (changes straight from a solid to a gas) when heated.

FIGURE 5.1.6 This power transmission tower relies on the strength of iron in steel for its structural integrity. The electricity cables are made from aluminium, utilising its ductility and electrical conductivity.

Generally, metals are good conductors of electricity and heat. They are malleable, ductile, have high tensile strength, low ionisation energies and low electronegativities.

Not all metals have all of these properties. Mercury is a liquid at room temperature—it has an unusually low melting point. The group 1 elements (the **alkali metals**) have some properties that make them different from most other metals—they are all soft enough to be cut with a knife, and they react vigorously with water to produce hydrogen gas. However, both mercury and the group 1 elements exhibit most of the other properties listed above and are classified as metals.

Metals also generally have the following characteristics in common. They:

- are **malleable**—they can be shaped by beating or rolling
- are **ductile**—they can be drawn into a wire
- are lustrous or reflective when freshly cut or polished
- are often hard, with high **tensile strength**
- have low **ionisation energies** and **electronegativities**.

These properties allow different metals to be used together to solve many engineering problems. The power transmission tower in Figure 5.1.6 is made of a range of metals to take advantage of their different properties.

Metals can be shaped for use in different applications by hammering, exploiting their malleability. Some metals, such as gold, copper and aluminium, are very malleable at room temperature. Other metals, such as iron, must be heated before they can be shaped.

Most metals are similar in appearance, being lustrous (reflective) and silvery-grey in colour. Gold and copper are notable exceptions. Gold is a yellow-coloured metal; copper is reddish.

CHEMFILE ICT

Beryllium

Beryllium is the fourth element in the periodic table. It is one of the lightest metals, its density being two-thirds the density of aluminium. Beryllium is non-magnetic and has six times the stiffness of steel. The Space Shuttle and the Spitzer Space Telescope both use beryllium due to its strength and lightness. NASA's next-generation James Webb Space Telescope (Figure 5.1.7), scheduled for launch in 2018, will depend on a 6.5 m mirror constructed using beryllium. It will be used to see objects 200 times fainter than those previously visible.

FIGURE 5.1.7 These are some of the 18 mirror segments of the James Webb Space Telescope. The mirrors are supported by beryllium ribs that will maintain the mirror's shape under extreme conditions.

METALS AND THE PERIODIC TABLE

More than 80% of the elements are metals. The location of metals in the periodic table is shown in Figure 5.1.8.

KEY
- non-metals
- metals
- metalloids

atomic number — 13; symbol — Al; name — aluminium

1	2	3	4	5	6	7	8	9	10	11	12	13	14	15	16	17	18
1 H hydrogen																	2 He helium
3 Li lithium	4 Be beryllium											5 B boron	6 C carbon	7 N nitrogen	8 O oxygen	9 F fluorine	10 Ne neon
11 Na sodium	12 Mg magnesium											13 Al aluminium	14 Si silicon	15 P phosphorus	16 S sulfur	17 Cl chlorine	18 Ar argon
19 K potassium	20 Ca calcium	21 Sc scandium	22 Ti titanium	23 V vanadium	24 Cr chromium	25 Mn manganese	26 Fe iron	27 Co cobalt	28 Ni nickel	29 Cu copper	30 Zn zinc	31 Ga gallium	32 Ge germanium	33 As arsenic	34 Se selenium	35 Br bromine	36 Kr krypton
37 Rb rubidium	38 Sr strontium	39 Y yttrium	40 Zr zirconium	41 Nb niobium	42 Mo molybdenum	43 Tc technetium	44 Ru ruthenium	45 Rh rhodium	46 Pd palladium	47 Ag silver	48 Cd cadmium	49 In indium	50 Sn tin	51 Sb antimony	52 Te tellurium	53 I iodine	54 Xe xenon
55 Cs caesium	56 Ba barium	57–71 lanthanoids	72 Hf hafnium	73 Ta tantalum	74 W tungsten	75 Re rhenium	76 Os osmium	77 Ir iridium	78 Pt platinum	79 Au gold	80 Hg mercury	81 Tl thallium	82 Pb lead	83 Bi bismuth	84 Po polonium	85 At astatine	86 Rn radon
87 Fr francium	88 Ra radium	89–103 actinoids	104 Rf rutherfordium	105 Db dubnium	106 Sg seaborgium	107 Bh bohrium	108 Hs hassium	109 Mt meitnerium	110 Ds darmstadtium	111 Rg roentgenium	112 Cn copernicium	113 Nh nihonium	114 Fl flerovium	115 Mc moscovium	116 Lv livermorium	117 Ts tennessine	118 Og oganesson

Lanthanoids	57 La lanthanum	58 Ce cerium	59 Pr praseodymium	60 Nd neodymium	61 Pm promethium	62 Sm samarium	63 Eu europium	64 Gd gadolinium	65 Tb trebium	66 Dy dysprosium	67 Ho holmium	68 Er erbium	69 Tm thulium	70 Yb ytterbium	71 Lu lutetium
Actinoids	89 Ac actinium	90 Th thorium	91 Pa protactinium	92 U uranium	93 Np neptunium	94 Pu plutonium	95 Am americium	96 Cm curium	97 Bk berkelium	98 Cf californium	99 Es einsteinium	100 Fm fremium	101 Md mendelevium	102 No nobelium	103 Lr lawrencium

FIGURE 5.1.8 About 80% of the elements are metals, represented in yellow in this periodic table.

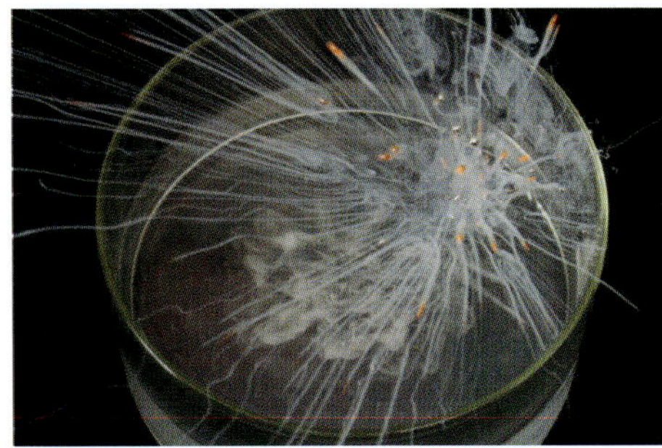

FIGURE 5.1.9 Potassium metal reacts violently with water.

As you will recall from Chapters 3 and 4, the metallic elements in group 1 have an unfilled *s*-subshell. The metals in group 2 have a filled *s*-subshell. Group 1 and 2 metals tend to have lower melting points and are softer than the metals in other groups. These metals have low ionisation energies and electronegativities; therefore, they are highly reactive and readily lose their valence *s*-subshell electrons to form positive ions. Group 1 and 2 metals will even react with water, in some cases violently (Figure 5.1.9).

Foods containing calcium, sodium, potassium and magnesium ions are essential for good health. Calcium is required for the production of hydroxyapatite ($Ca_5(PO_4)_3OH$), the structural component of your bones. Shellfish and the eggshells of birds are composed of calcium carbonate ($CaCO_3$). The transmission of nerve impulses involves the movement of sodium and potassium ions within nerve cells. Table salt (NaCl) is a widely used compound of sodium.

Magnesium plays an essential role in many chemical reactions that take place in your body. Chlorophyll, the green pigment in plants that is responsible for photosynthesis, contains magnesium.

These metals are also used in industry and agriculture. For example, sodium is used in the manufacture of glass. Lime (calcium oxide) is used to break down clay in soils and in the manufacture of cement. Magnesium is combined with aluminum to make lightweight alloys used in aircraft. Magnesium is also used in fireworks.

Between group 2 and group 13 in the periodic table is a block of elements known as the **transition metals**. These elements generally have unfilled *d*-subshells and are often referred to as the *d*-block elements (Figure 5.1.8 on page 141). They include metals such as iron and nickel that are used to build bridges, cars and railway lines, and precious metals such as silver and gold that have ornamental and economic uses. Most transition metals are silver-coloured and are similar in appearance, as can be seen in Figure 5.1.10.

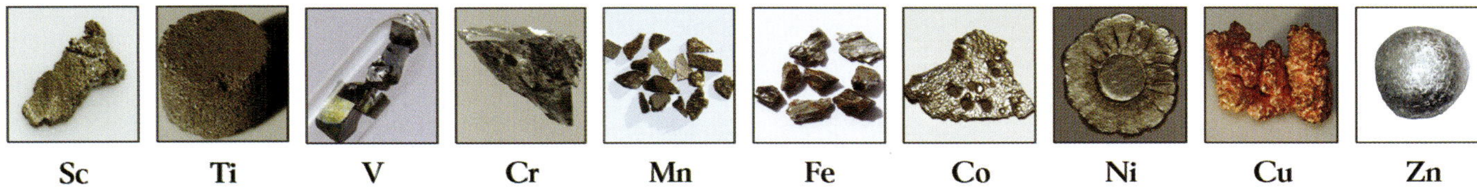

FIGURE 5.1.10 The first row of transition metals, found in period 4 of the periodic table.

All the transition metals in the first row (period 4), except scandium and titanium, are essential for animal life. Your body relies on the presence of trace elements to carry out certain biochemical reactions. For example, chromium, which you get from meat and bread, assists in the production of energy from glucose.

PROPERTIES OF TRANSITION METALS

Compared with the main group metals, transition metals have the following properties:

- They tend to be harder.
- They have higher densities.
- They have higher melting points.
- Some of them have strong magnetic properties.
- The high tensile strength of transition metals makes them suitable for use in the construction of buildings, cars, bridges and numerous other objects.

CHEMISTRY IN ACTION AHC

Transition metal compounds

Transition metal compounds display a wide range of colours. They are extensively used as pigments in paints, and to colour glass, ceramics and enamel. In Figure 5.1.11, the colours used by the artists are a result of the different transition metals present. 'Five Bells' was painted by John Olson in 1963, and the colours are still as vivid today as when they were painted.

Ochre is a type of hard clay that contains iron oxides and hydroxides. It occurs naturally, in many colours, including red, pink, white and yellow. Ground into a powder and mixed with liquids, ochre forms a paste that has been used for millennia by Indigenous Australian people for body decoration, cave painting, bark painting and other artwork.

FIGURE 5.1.11 (a) 'Five Bells' (1963) by the Sydney artist John Olson. (b) 'Untitled' (2010) by the Indigenous Australian artist Mavis Ngallametta.

Connecting properties and structure

Some of the properties of metals are listed in Table 5.1.3. Each of these properties gives some information about the structure and bonding of particles in metals.

TABLE 5.1.3 The properties of metals, and resulting conclusions about metallic structure and bonding

Property	What this indicates about structure
Metals are usually hard and tend to have high boiling points.	The forces between the particles must be strong.
Metals conduct electricity, both in the solid state and in the molten liquid state.	Metals have charged particles that are free to move.
Metals are malleable and ductile.	The attractive forces between the particles must be stronger than the repulsive forces between the particles when the layers of particles are moved.
Metals generally have high densities.	The particles are closely packed in a metal.
Metals are good conductors of heat.	There must be a way of quickly transferring energy throughout a metal object.
Metals are lustrous or reflective.	Free electrons are present, so metals can reflect light and appear shiny.
Metals tend to lose electrons when they are involved in chemical reactions.	Electrons must be relatively easily removed from metal atoms.

Chemists have developed a model for the structure of metals to explain all the properties that have been mentioned so far. You can deduce from the information in Table 5.1.3 that the model must include:

- charged particles that are free to move and conduct electricity
- strong forces of attraction between atoms throughout the metal structure
- some electrons that are relatively easily removed.

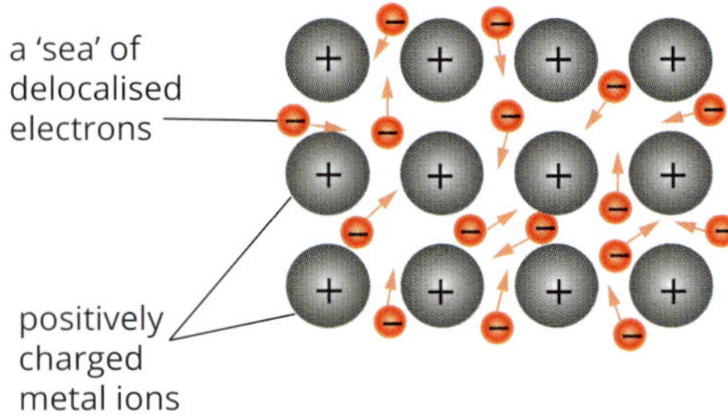

FIGURE 5.1.12 The metallic bonding model. Positive metal cations are surrounded by a mobile 'sea' of delocalised electrons. This diagram shows just one layer of metal ions.

> In the metallic bonding model, positive metal cations are surrounded by a 'sea' of delocalised electrons.

Positive sodium ions occupy fixed positions in the lattice.

'sea' of delocalised electrons

FIGURE 5.1.13 A representation of a sodium metal lattice. Each sodium atom loses its one valence electron. This electron is shared with all atoms in the lattice to form a 'sea' of delocalised electrons.

THE METALLIC BONDING MODEL

There are several models used to explain the bonding between the atoms of metals. One such model is the **metallic bonding model**. (Figure 5.1.12)

The basis for the metallic bonding model is that, in a solid sample of a metal:

- Positive ions or **cations** are arranged in a closely packed three-dimensional **network structure**, or lattice. They occupy fixed positions in the lattice.
- Negatively charged electrons move freely throughout the lattice. These electrons are called **delocalised electrons** because they belong to the lattice as a whole, rather than staying in the shell of a particular atom.
- The **delocalised** electrons come from the outer shells of the atoms. Inner-shell electrons are not free to move throughout the lattice and remain firmly bonded to individual cations.
- The positive cations are held in the lattice by the electrostatic force of attraction between these cations and the delocalised electrons. This attraction extends throughout the lattice and is called **metallic bonding**.

An example of how a metal such as sodium could be represented in this model is shown in Figure 5.1.13.

EXPLAINING THE PROPERTIES OF METALS

Table 5.1.4 shows how the metallic bonding model is consistent with the relatively high boiling point, electrical conductivity, malleability and ductility of metals.

TABLE 5.1.4 Physical properties of metals and explanations based on the metallic bonding model

Property	Explanation	Diagram
Metals are hard and have relatively high boiling points.	Strong electrostatic forces of attraction between the positive metal ions and the sea of delocalised electrons holds the metallic lattice together.	
Metals are good conductors of electricity.	Free-moving delocalised electrons will move towards a positive electrode and away from a negative electrode in an electric circuit.	
Metals are malleable and ductile.	When a force causes layers of metal ions to move past one another, the layers are still held together by their electrostatic attraction to the delocalised electrons between them.	

The strength of metallic bonds

Transition metals are harder, denser and have higher melting points than group 1 and 2 metals. This is due to the atoms of transition metals generally being a smaller size due to their greater **core charge**, which allows them to pack together more tightly with stronger bonds.

The outer *s*-subshell and inner *d*-subshell electrons in transition metals are delocalised and are involved in the formation of metallic bonds. The metallic bonding in group 1 and group 2 metals only involves the outer *s*-subshell electrons. These metals are softer and have lower melting points than the transition metals. The greater number of delocalised electrons in transition metals results in a greater strength of attraction between the delocalised electrons and the cations in the metal lattice, which in turn results in stronger metallic bonding. This makes sense when comparing the metallic solids iron and sodium. For the solids to melt, enough energy must be provided to allow the metal ions to break free from the metal lattice. Iron melts at 1535°C and sodium at 98°C. Clearly, more energy is required to overcome the electrostatic attraction between the positive ions and the delocalised electrons in the iron lattice than in the sodium lattice.

Other properties of metals

Metals generally have a high density. The cations in a metal lattice are closely packed. The density of a metal depends on the mass of the metal ions, their radius, and the way in which they are packed together in the lattice.

Metals are good conductors of heat (Figure 5.1.14). When the delocalised electrons bump into one another and into the metal ions, they transfer energy to their 'neighbours'. Heating a metal gives the ions and electrons more energy, so they vibrate more rapidly. The electrons, being free to move, transmit this energy rapidly throughout the lattice.

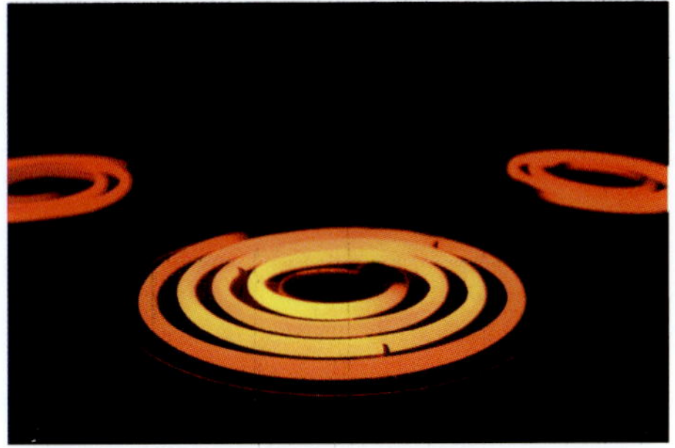

FIGURE 5.1.14 Metal heating elements are used in stove tops, ovens and kettles.

Metals are lustrous. Because of the presence of free electrons in the lattice, metals reflect light of all wavelengths and appear shiny.

Metals tend to lose electrons when they are involved in chemical reactions. The delocalised electrons in metals may participate in reactions anywhere on the metal's surface. The reactivity of a metal depends on how easily electrons can be removed from its atoms.

Limitations of the metallic bonding model

Although this model for metallic bonding explains many properties of metals, some cannot be explained so simply. These properties include:

- the range in melting points, hardness and densities of different metals
- the differences in electrical conductivities of metals
- the magnetic nature of metals such as cobalt, iron and nickel.

To explain these properties you need a more complex model of metallic bonding, and that is beyond the scope of this book.

Worked example 5.1.1

EXPLAINING THE ELECTRICAL CONDUCTIVITY OF ALUMINIUM

With reference to the electronic configuration of aluminium, explain why solid aluminium can conduct electricity.

Thinking	Working
Using the atomic number of the element, determine the electronic configuration of its atoms. (You may need to refer to a periodic table.)	Al has an atomic number of 13. This means that a neutral atom of Al has 13 electrons. The electronic configuration is $1s^22s^22p^63s^23p^1$.
From the electronic configuration, find how many outer-shell electrons are lost to form cations that have a stable octet of valence electrons. These electrons become delocalised.	Al has three electrons in its outer shell (the $3s^23p^1$ electrons). Al atoms will tend to lose these three valence electrons to form a cation with a charge of +3. The outer-shell electrons become delocalised and form the sea of delocalised electrons within the metal lattice.
An electric current occurs when there are free-moving charged particles.	If the Al is part of an electric circuit, the delocalised electrons can move through the lattice towards a positively charged electrode.

Worked example: Try yourself 5.1.1

EXPLAINING THE ELECTRICAL CONDUCTIVITY OF MAGNESIUM

With reference to the electronic configuration of magnesium, explain why solid magnesium can conduct electricity.

5.1 Review

SUMMARY

- Metals have the following characteristic properties:
  - high boiling point
  - good conductance of electricity in solid and liquid states
  - malleable and ductile
  - high density
  - good conductance of heat
  - lustrous
  - low electronegativity
  - low ionisation energy
  - chemically react by losing electrons.
- The main differences between the properties of main group and transition metals are:
  - transition metals are harder
  - transition metals are more dense
  - transition metals have higher melting points
  - some transition metals have strong magnetic properties
  - transition metal compounds tend to be brightly coloured.
- Metallic bonding is the electrostatic force of attraction between a lattice of positive ions (cations) and delocalised valence electrons. The lattice of cations is surrounded by a 'sea' of delocalised electrons.
- The metallic bonding model can be used to explain the properties of metals, including their malleability, thermal conductivity, generally high melting point and electrical conductivity.

KEY QUESTIONS

1. Determine the charge of the cations formed from the following metals if they lost all of their outer-shell electrons.
   - **a** Li
   - **b** Mg
   - **c** Ga
   - **d** Ba
2. **a** Potassium is classed as a metal. Which of its properties are similar to those of gold? In what ways is it different?
   - **b** Identify another element in Table 5.1.2 on page 140 that has similar properties to potassium.
   - **c** Identify another metal in Table 5.1.2 on page 140 that has similar properties to gold.
   - **d** Describe the locations of these four metals in the periodic table.
3. **a** Which metals from Table 5.1.2 on page 140 would you select if you wanted a good electrical conductor?
   - **b** What other factors might influence your choice?
4. Sodium and iron have very different physical properties. Explain why this is so based on where these metals are found in the periodic table.
5. Suggest some properties not included in Table 5.1.2 on page 140 that you would need to consider before choosing between aluminium and iron for building a bridge.
6. The properties of calcium mean that it is classed as a metal.
   - **a** Draw a diagram to represent a calcium metal lattice.
   - **b** Describe the forces that hold this lattice together.
7. Barium is an element in group 2 of the periodic table. It has a melting point of 850°C and conducts electricity in the solid state. Describe how the properties of barium can be explained in terms of its bonding and structure.
8. Graphite is a non-metallic substance that can be lustrous and conducts electricity and heat. It is not malleable, but breaks if a force is applied.
   - **a** What properties does graphite share with metals?
   - **b** What inferences can you make about the structure of graphite, given the properties it shares with metals?

5.2 Ionic bonding

In this section, you will study the structure and properties of a group of substances called **ionic compounds**. Ionic compounds are made by the chemical combination of metallic and non-metallic elements.

These materials are very common in the natural world because Earth's crust is largely made up of complex ionic compounds. Most rocks, minerals and **gemstones** (Figure 5.2.1) contain ionic bonds. Soil is made from weathered rocks mixed with decomposed organic material, so it contains large quantities of ionic compounds.

Ceramics, kitchen crockery and bricks are made from clays. Clays are produced by the weathering of rocks. Therefore, materials made from clays also contain ionic compounds. Kitchen crockery and bricks contain mixtures of ionic compounds. Table salt (sodium chloride) is a pure ionic compound.

FIGURE 5.2.1 Many gemstones are made from ionic compounds.

PROPERTIES OF IONIC COMPOUNDS

If you think about the characteristics of rocks, kitchen crockery and table salt, you will recognise that these materials, and therefore ionic compounds, have some properties in common.

Table 5.2.1 lists some typical ionic compounds and their properties. These compounds can be found in materials you might encounter in everyday life. Note that the compounds listed are simple ionic compounds, whereas rocks, ceramics and bricks contain a mixture of often complex ionic compounds.

TABLE 5.2.1 Properties of typical ionic compounds

Ionic compound	Melting point (°C)	Electrically conductive when solid	Electrically conductive when liquid	Electrically conductive when in aqueous solution (0.1 mol L^{-1})	Solubility in water at 25°C (g/100 g water)	Example of commercially available product containing the compound
copper(II) sulfate	decomposes at 110	no	yes	yes	22	bluestone spray (used to kill pathogens on fruit)
sodium chloride	801	no	yes	yes	36	table salt
calcium carbonate	1339*	no	yes	–	0.0013	main component in marble
zinc oxide	1975	no	yes	–	insoluble	zinc cream
sodium hydroxide	318	no	yes	yes	114	oven cleaner

*Melting point determined under pressure to prevent decomposition of compound.

Lists of data about ionic compounds such as those in Table 5.2.1 have enabled chemists to summarise their properties. Generally, ionic compounds:

- have high melting and boiling points. They are all solids at room temperature.
- are hard but **brittle**, shattering when given a hard blow
- do not conduct electricity in the solid state
- are good conductors of electricity in the liquid or **molten** state or when **dissolved** in water
- vary from very soluble to insoluble in water. They are not soluble in **non-polar solvents** such as oil.

Connecting properties and structure

The properties of ionic compounds suggest how the particles are arranged in the compounds.

Table 5.2.2 lists some observed properties of one particular ionic compound, sodium chloride (table salt). Beside each property is a description of the nature of the particles and of the forces acting between the particles. These descriptions can be inferred from the different properties of sodium chloride.

TABLE 5.2.2 Properties of sodium chloride and the information this provides about its structure

Property	What this indicates about structure
high melting point	The forces between the particles are strong.
hard, brittle crystals	The forces between the particles are strong.
does not conduct electricity in the solid state	There are no free-moving charged particles in solid sodium chloride.
conducts electricity in the molten state	Free-moving charged particles are present in molten sodium chloride.

THE IONIC BONDING MODEL

In summary, for ionic compounds generally:

- The forces between the particles are strong.
- There are no free-moving electrons present.
- There are ions present, but in the solid state they are not free to move.
- When an ionic compound melts, the ions are free to move, and then the compound will conduct electricity.

Now that you understand some of the details of the structure of ionic compounds, the next step is to work out how the particles in these compounds are arranged in the solid state.

When metallic and non-metallic atoms react to form ionic compounds, the following steps occur.

- The electronegativity of metals is less than the electronegativity and ionisation energy of non-metals. Consequently, metal atoms lose electrons to non-metallic atoms. They become positively charged metal ions (cations).
- The more electronegative non-metal atoms gain electrons from metal atoms and therefore become negatively charged non-metal ions (called **anions**).

GO TO ➤ Section 3.5 page 100

(You will remember from Chapter 3 how electrons arrange themselves into shells around the nucleus. Atoms are at their most stable when there are eight electrons in the valence shell. Therefore, an atom's ability to form a cation or an anion depends on how many electrons it needs to gain or lose to achieve this stable arrangement.)

Cations and anions then arrange themselves in the following way:

- Large numbers of cations and anions combine to form a three-dimensional network or lattice.
- The three-dimensional lattice is held together strongly by electrostatic forces of attraction between the oppositely charged ions. The electrostatic forces of attraction holding the ions together is called **ionic bonding**.
- In the case of sodium chloride, sodium (electronic configuration 2,8,1) loses one electron, so it has a stable eight electrons in the second electron shell (electronic configuration 2,8). Losing one electron makes it positively charged, and it is given the symbol Na^+. A chlorine atom normally has the electronic configuration of 2,8,7. Therefore, it only needs to gain one electron to become an anion, Cl^-, and its electronic configuration becomes 2,8,8.

Electrostatic forces of attraction in ionic compounds result from the attraction between positive and negative charges.

- In sodium chloride, each sodium ion is surrounded by six chloride ions, and each chloride ion is surrounded by six sodium ions, as shown in Figure 5.2.2. This arrangement maximises the forces of attraction between ions.
- Even though each chloride ion is close to another chloride ion (Figure 5.2.2), the attractive force between the chloride and the sodium ions outweighs the repulsive force from the other chloride ions, so the lattice is held together quite strongly.

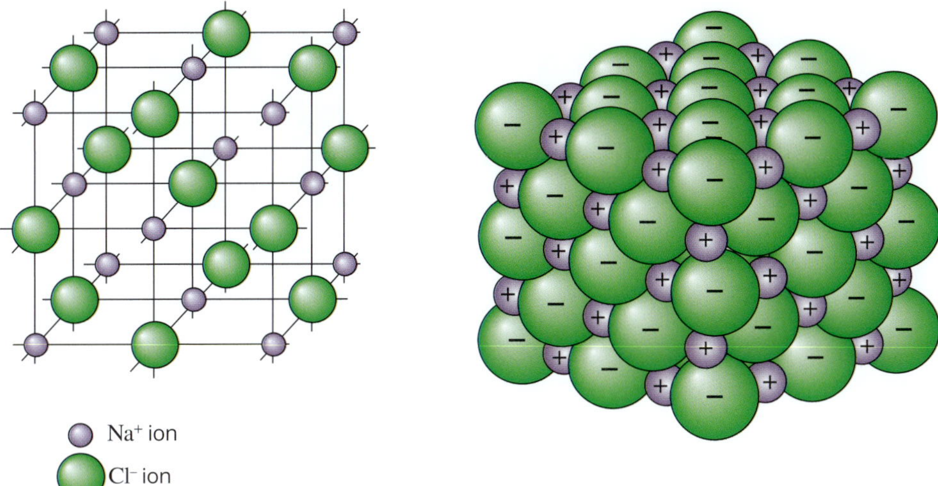

FIGURE 5.2.2 Two representations of part of a crystal lattice of an ionic compound sodium chloride. Forces of attraction between oppositely charged ions result in strong bonding.

The formula of sodium chloride

The **chemical formula** of sodium chloride is written as NaCl. However, it is important to note that in a solid sample of an ionic compound, such as sodium chloride, individual pairs of sodium and chloride ions do not exist. The solid is also not built up of discrete NaCl molecules.

Instead, the solid is made up of a continuous lattice of alternating Na^+ and Cl^- ions. All sodium ions are an equal distance from six chloride ions, and all chloride ions are an equal distance from six sodium ions. The overall ratio of sodium ions to chloride ions in the lattice is 1:1, and therefore the formula is written as NaCl.

EXPLAINING THE PROPERTIES OF IONIC COMPOUNDS

In this section, you will see how the ionic bonding model explains the properties of ionic compounds.

High melting points

To melt an ionic solid such as sodium chloride, you must provide energy to allow the ions to break free and move. Sodium chloride has a high melting point (801°C). This indicates that a large amount of energy is needed to overcome the electrostatic attraction between oppositely charged ions to allow them to move freely. The ionic bonds between the positive sodium ions and negative chloride ions must be strong, since a high temperature is required to melt solid sodium chloride.

The high melting point of ionic compounds is put to use in special bricks that line furnaces and kilns (Figure 5.2.3) and in the ceramic materials used to make brake discs for high-performance cars (Figure 5.2.4).

FIGURE 5.2.3 Bricks made from the ionic compound magnesium oxide are used to line furnaces and kilns.

FIGURE 5.2.4 Ceramic brake discs work more effectively than steel ones at high temperatures. A ceramic brake disc contains ionic compounds that have very high melting temperatures and are better than metals at withstanding the heat produced by braking.

Hardness and brittleness

There are strong electrostatic forces of attraction between ions in an ionic compound, so a strong force is needed to disrupt the **crystal lattice**. Therefore, one of the properties of ionic compounds is that they are hard. This also means that a sodium chloride crystal cannot be scratched easily.

The strength of house bricks, concrete bridges and cobbled streets can be attributed to the ionic bonding within their structures.

Although a salt crystal is hard, it will shatter under a strong force such as a hammer blow. This is because the force of the blow causes layers of ions to move relative to one another. During this movement, ions of like charge are shifted so that they are next to each other, as seen in Figure 5.2.5. The resulting repulsion between the similarly charged ions causes the crystals to shatter, giving ionic compounds their characteristic of brittleness.

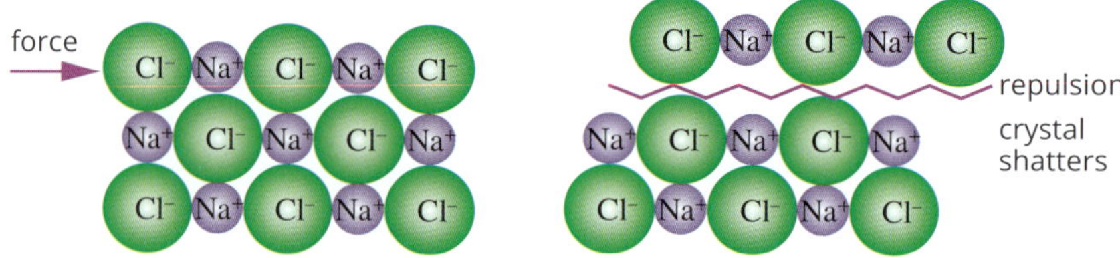

FIGURE 5.2.5 A lattice of an ionic compound shattering. Note that just before shattering, ions of like charge are shifted so that they are adjacent to each other. Repulsion between ions of like charge causes the crystal lattice to shatter.

FIGURE 5.2.6 A coffee cup being smashed.

Materials made from clay, such as kitchen crockery (Figure 5.2.6), ceramic tiles and bricks, are hard, but they are also brittle.

Electrical conductivity

In the solid form, ions in the crystal lattice of sodium chloride are not free to move, so solid sodium chloride does not conduct electricity. Remember that for a substance to conduct electricity, it must contain charged particles that are free to move. Figure 5.2.7 shows how the particles are arranged in an ionic compound in solid form.

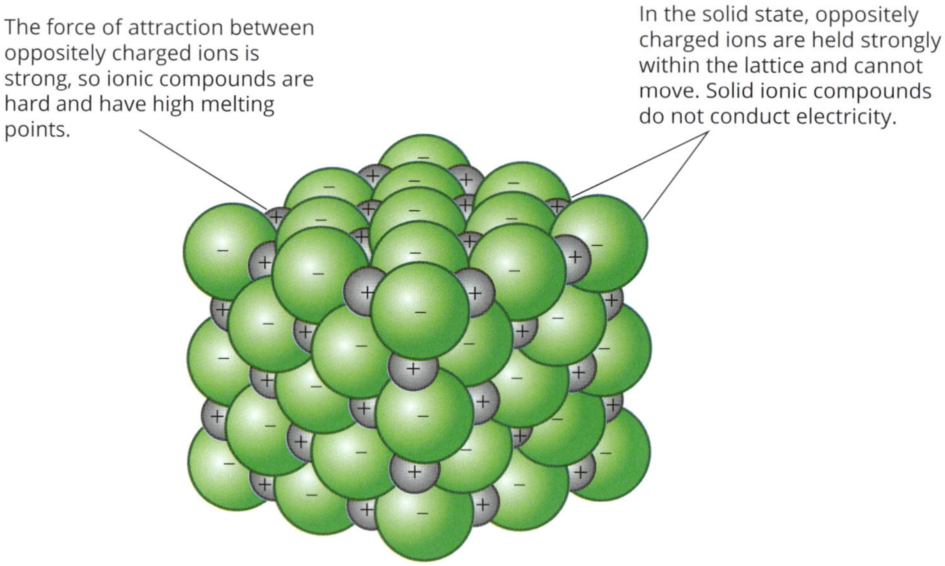

The force of attraction between oppositely charged ions is strong, so ionic compounds are hard and have high melting points.

In the solid state, oppositely charged ions are held strongly within the lattice and cannot move. Solid ionic compounds do not conduct electricity.

FIGURE 5.2.7 The arrangement of ions within a solid ionic compound takes the form of a crystal lattice.

The non-conducting property of ionic compounds is useful for ceramic insulators, which can be used to keep high-voltage power lines insulated from electricity poles and electric fence wires (Figure 5.2.8).

FIGURE 5.2.8 A ceramic insulator on the post of an electric fence.

When solid ionic compounds melt, the ions become free to move, enabling the cations and anions in the molten compound to conduct electricity.

Similarly, when ionic compounds dissolve in water, the ionic bonds in the lattice are broken, and the ions are separated and move freely in solution.

When an electric current is applied to either a molten the ionic compound or a solution of the compound in water, positive ions move towards the negatively charged electrode, and negative ions move towards the positively charged electrode, resulting in an electric current, as shown in Figure 5.2.9.

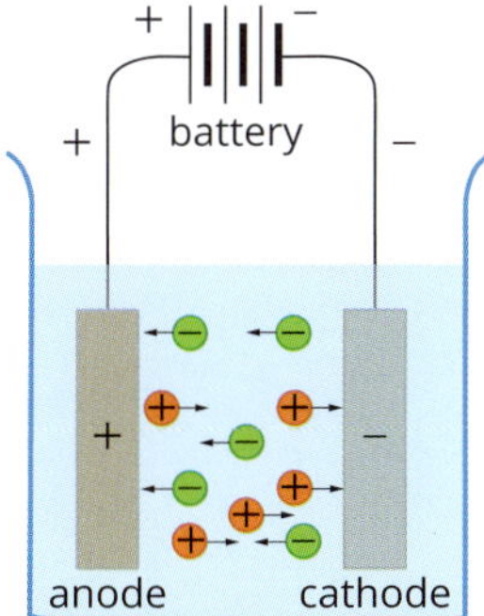

FIGURE 5.2.9 A molten ionic compound will conduct an electric current.

A solution or molten substance that conducts electricity by means of the movement of ions is called an **electrolyte**.

For an object to conduct electricity, it must contain charged particles that are free to move. In solid ionic compounds, the ions are arranged in a crystal lattice, so the ions are not free to move, and the solid cannot conduct electricity. When molten or dissolved in water, the ions are free to move, and the substance can conduct electricity.

Solubility

Some ionic compounds are very soluble in water, whereas others are very insoluble. When a soluble ionic compound is added to water, the ions break away from the ionic lattice and mix with the water molecules. If an insoluble compound is added to water, the ions remain bonded together in the ionic lattice and do not form a solution.

Whether an ionic compound is soluble or insoluble depends on the relative strength of the forces of attraction between:

1 the positive and negative ions in the lattice, and
2 the water molecules and the ions.

You will look at the solubility of ionic compounds in water and the reactions of ionic compounds in more detail in Chapter 11.

Connecting physical properties and uses

Ionic compounds have a wide variety of uses. Some of these uses are related directly to the physical properties of the compounds, such as hardness, high melting points and their ability to conduct electricity in solution. A number of ionic compounds and their uses in relation to their characteristic properties are described below.

Hardness

- Calcium phosphate is a constituent of bone tissue that gives it strength.
- Calcium sulfate in the form of gypsum is used to make plasterboard for lining the walls and ceilings of houses.
- Granite, limestone and sandstone are used as building stone.
- Bricks, tiles and crockery are made from clay, a material that contains particles held together by strong ionic bonding.

High melting point

- Magnesium oxide and other ionic compounds are used to line furnaces.
- Ceramics are materials that contain a mixture of strong ionic and covalent bonds. They are used in some engine parts.

Electrical conductivity

- Ammonium chloride is used as an electrolyte in dry-cell batteries. The ions contained in an electrolyte solution allow a current to flow in the battery.
- Potassium hydroxide is used as an electrolyte in the 'button' cells used in small electronic devices such as watches and calculators.

Other uses of ionic compounds

- Sodium hydrogen carbonate (bicarbonate of soda or baking soda) is used in baking to cause cakes to rise when placed in a hot oven. This occurs because the compound decomposes when heated to produce carbon dioxide gas.
- Sodium chloride has many uses, including as a flavouring agent and preservative in food.
- Sodium hypochlorite is used as a bleach and in swimming pools to kill microorganisms.

CHEMISTRY IN ACTION 

Salinity

Salinity is the presence of salt (sodium chloride) in groundwater and soil. Plants, like animals, can only tolerate a limited range of salt concentrations. Some plants growing on sand dunes tolerate higher amounts of salt than other plants. However, if the salt concentration in the soil is too high, even those plants die.

Increasing salinity is caused by increasing amounts of groundwater (water found under the surface of the Earth). As groundwater increases, the water table (water level) rises, bringing dissolved salts—in this case, sodium chloride—closer to the surface. Increased salinity of water near the soil surface leads to poor plant growth, and reduces the productivity of agricultural land that is used for grazing animals and growing crops.

The rise in the water table has been caused by the clearing of trees, which would otherwise have absorbed this water through their roots. This is a particular problem in irrigation areas due to excess irrigation water flowing down through the soil, raising the water table even further.

One property of salt that can lead to environmental damage is its ability to dissolve in water (Figure 5.2.10).

FIGURE 5.2.10 Land that is affected by salinity.

Reducing salinity

Many organisations and individuals are working to reduce the salinity of soil and water. Strategies include:

- using irrigation water more efficiently
- improving drainage on the surface of the land
- improving drainage under the surface
- growing trees to soak up excess water
- sealing irrigation channels to prevent leakage.

These are long-term projects; however, reduced salinity is already apparent in some areas.

FORMATION OF IONIC COMPOUNDS

Some of the reactions that occur between metals and non-metals to form ionic compounds are very vigorous. The reaction between sodium and chlorine to form sodium chloride produces a lot of heat. Sodium is very reactive. At high temperatures, the production of sodium chloride from sodium metal and chlorine gas is explosive, producing a flame and large amounts of energy.

In the previous sections, you saw that the positive and negative ions formed in this reaction are arranged in a three-dimensional lattice. In this section, you will learn how these positive and negative ions are formed from their respective atoms. You will also learn how the ratio of each type of atom in the compound is determined.

Forming ions

When metal atoms react with non-metal atoms to form an ionic compound, two things occur.

- Metal atoms lose electrons to form positively charged ions (cations).
- Non-metal atoms gain electrons to form negatively charged ions (anions).

From Chapter 4, you will remember that most metals have lower ionisation energies and lower electronegativities than non-metals. This means that non-metal atoms are usually more electronegative than metal atoms. In other words, non-metals have a stronger attraction for electrons than metals. In reactions that form ionic compounds, non-metal atoms take one or more electrons from the outer shell of metal atoms.

GO TO ➤ Section 4.3 page 129

GO TO ➤ Section 4.2 page 122

Most importantly, the ions that are formed usually have eight electrons in their outer shell—a stable electronic configuration. The tendency for elements to react in such a way that their atoms have eight electrons in their outer shell (valence shell) is known as the **octet rule**.

The noble gases (group 18) are elements that already have the most stable valence shell configuration. Hence, another way of thinking about the octet rule is that atoms tend to gain or lose electrons to obtain a stable electronic configuration identical to that of the noble gas nearest to them on the periodic table. The formation of stable ions is a powerful driving force in reactions between metals and non-metals when they produce ionic compounds.

For example, when sodium reacts with chlorine, each sodium atom loses one electron and each chlorine atom gains one electron.

After the reaction:

- The sodium ion has the stable electron shell configuration of 2,8 (the same as a neon atom)
- The chloride ion has the stable electron shell configuration of 2,8,8 (the same as an argon atom).

Note that, for simplicity, you will look at the arrangement of electrons just in shells of atoms (rather than in subshells).

Figure 5.2.11 illustrates how, when sodium reacts with chlorine, an electron is lost by a sodium atom and gained by a chlorine atom. A diagram of this type is called an **electron transfer diagram**.

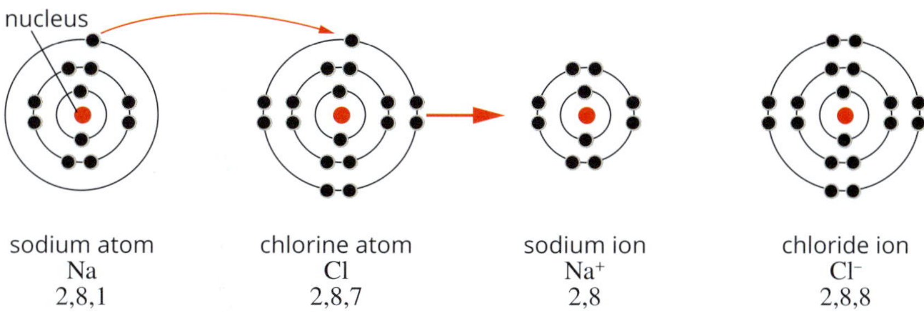

FIGURE 5.2.11 An electron transfer diagram showing the formation of sodium and chloride ions.

Atoms are at their most stable when they have a valence shell containing eight electrons. An exception to this is lithium. Lithium has only one electron in the second shell, and when it loses this, the first shell becomes the valence shell. The first shell can only hold two electrons, so a lithium ion is stable with an electronic configuration of two. This configuration is the same as that of the noble gas closest to it (helium).

The reaction between lithium and oxygen atoms is illustrated in Figure 5.2.12. In this reaction, an oxygen atom needs to gain two electrons to have eight electrons in its outer shell and form a stable ion. For this to happen, one oxygen atom will need to react with two lithium atoms, taking one electron from each lithium atom.

After the reaction, there are just two electrons in what is now the outer shell of the Li^+ ion, which is the same as the electronic configuration of a helium atom.

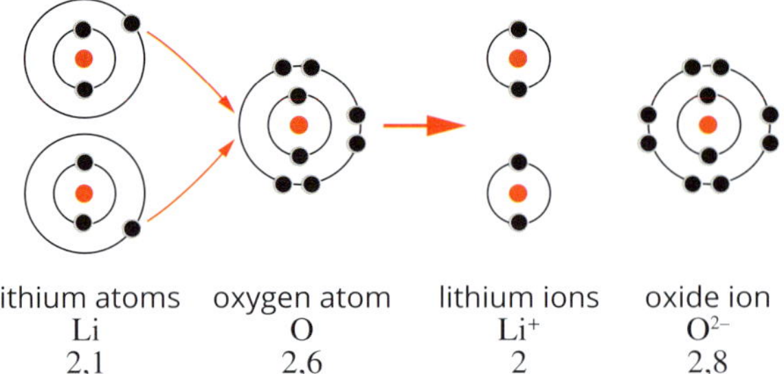

FIGURE 5.2.12 An electron transfer diagram showing the formation of lithium and oxide ions.

Electrovalencies and the periodic table

As shown in Figure 5.2.11 (page 153) and 5.2.12, ionic bonds form when there is a transfer of electrons from metals to non-metals. The loosely held **valence electrons** in metals are transferred to the more electronegative non-metal atoms.

The atoms of group 1 metals lose their single valence electron to form ions that have an **electrovalency** of +1. Group 2 metals form ions that have a +2 charge, while ions of group 3 metals have a +3 charge.

Magnesium is a metal in group 2 of the periodic table. An electron transfer diagram for magnesium is shown in Figure 5.2.13.

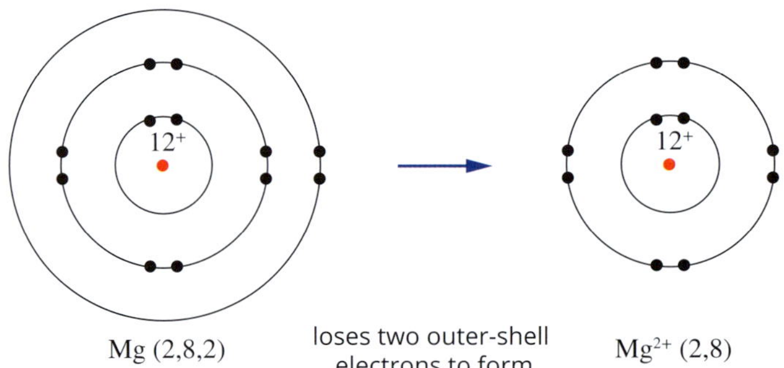

FIGURE 5.2.13 The electron transfer diagram for magnesium, showing the new electronic configuration when the ion Mg^{2+} is formed.

Some transition metals produce ions of various charges depending on the number of electrons lost. For example, iron can lose two electrons to form the iron(II) ion (Fe^{2+}). It can also lose three electrons to form iron(III) ions (Fe^{3+}). Metals lower down in groups 14 to 17 also have variable electrovalencies. For example, the group 4 metals tin and lead both form ions that have a +2 charge or a +4 charge.

Non-metals in group 16 have six electrons in their valence shells. Therefore, they gain two electrons to form anions with a charge of −2.

Group 17 non-metals have seven electrons in their valence shell. They readily gain one electron to fill the valence shell according to the octet rule. This means they form anions with a charge of −1.

Group 18 elements have a filled stable outer shell of electrons and do not form ions.

The electrovalencies of the main elements are summarised in the periodic table shown in Figure 5.2.14.

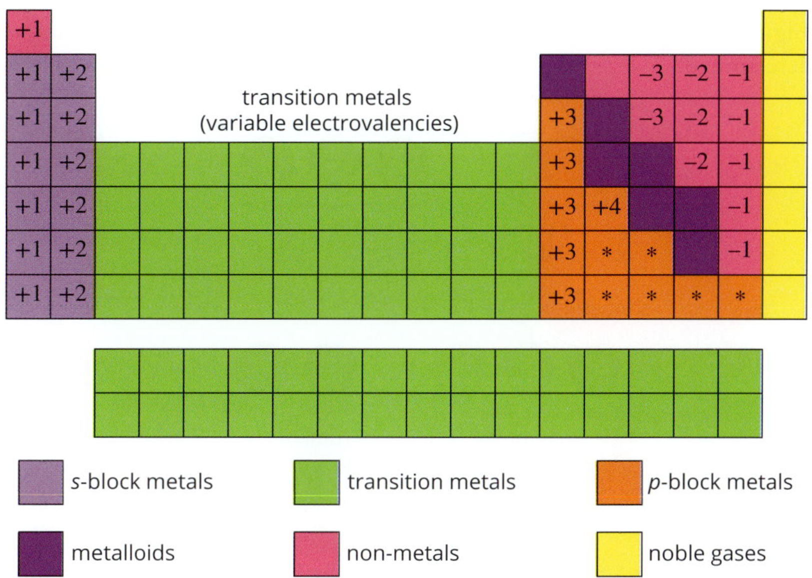

FIGURE 5.2.14 Electrovalencies of the elements. Note that some transition metals form ions of variable charge.

*The lower metals in groups 14 to 17 also produce ions that have variable electrovalencies.

When an atom loses electrons, it becomes more positively charged because the number of protons no longer equals the number of electrons. The ion formed is written with a superscript + sign indicating the charge. If two electrons are lost, it is written as 2+. When an atom gains electrons, becoming an anion, it becomes negatively charged. This is written with a superscript –. If three electrons are gained, the charge is written as 3– in superscript.

FORMULAE OF IONIC COMPOUNDS

You have seen that ionic compounds contain oppositely charged ions that are arranged in three-dimensional lattices. The ions can have different charges. For example, aluminium forms ions with a +3 charge, whereas oxygen forms an ion with a –2 charge.

In this section, you will learn how to use your knowledge of the charges on ions to write an overall formula for an ionic compound. You will also learn how to name ionic compounds.

Writing the formula of an ionic compound

Because ionic compounds are electrically neutral, the total number of positive charges on the metal ions must equal the total number of negative charges on the non-metal ions. This is the most important guiding principle when you are trying to work out the formula of an ionic compound.

This can be seen with the formula of the ionic compound sodium chloride:

- A sodium ion (Na^+) has a +1 charge.
- A chloride ion (Cl^-) has a –1 charge.
- Therefore, in a crystal of sodium chloride, the ratio of sodium ions to chloride ions is 1:1 and the formula of sodium chloride is NaCl.

Using the same steps, you can work out the formula of magnesium chloride:

- A magnesium ion (Mg^{2+}) has a +2 charge.
- A chloride ion (Cl^-) has a –1 charge.
- Thus, in a crystal of magnesium chloride, two chloride ions are needed to provide two negative charges to balance the +2 charge on every magnesium ion.

Therefore, in a crystal of magnesium chloride, the ratio of magnesium ions to chloride ions is 1:2 and the formula of magnesium chloride is $MgCl_2$.

Figure 5.2.15 illustrates how the formulae of some other ionic compounds can be determined.

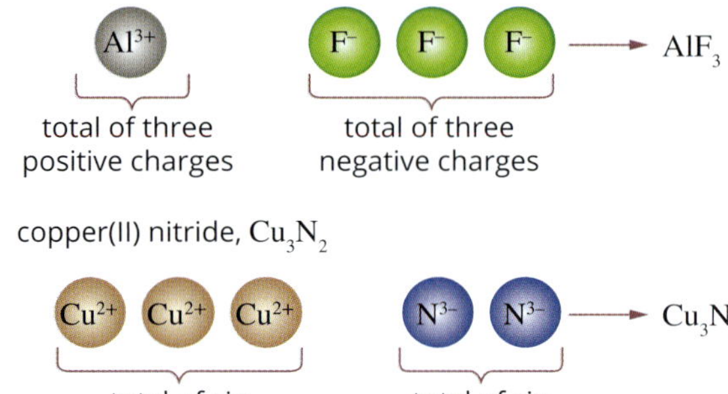

FIGURE 5.2.15 How to deduce chemical formulae from the charges on ions.

Tables 5.2.3 and 5.2.4 list some of the more common positively and negatively charged ions. These will be useful when you are writing formulae for ionic compounds.

TABLE 5.2.3 Names and formulae of some common cations

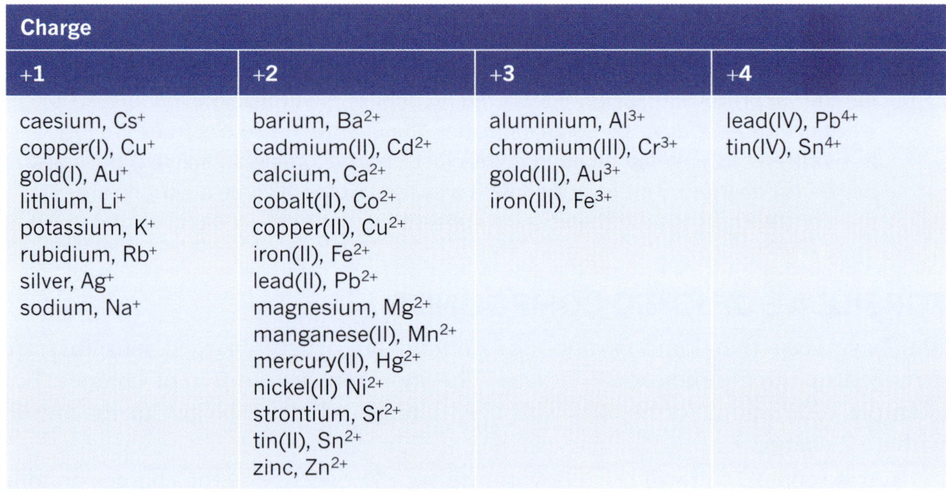

Charge			
+1	+2	+3	+4
caesium, Cs^+ copper(I), Cu^+ gold(I), Au^+ lithium, Li^+ potassium, K^+ rubidium, Rb^+ silver, Ag^+ sodium, Na^+	barium, Ba^{2+} cadmium(II), Cd^{2+} calcium, Ca^{2+} cobalt(II), Co^{2+} copper(II), Cu^{2+} iron(II), Fe^{2+} lead(II), Pb^{2+} magnesium, Mg^{2+} manganese(II), Mn^{2+} mercury(II), Hg^{2+} nickel(II) Ni^{2+} strontium, Sr^{2+} tin(II), Sn^{2+} zinc, Zn^{2+}	aluminium, Al^{3+} chromium(III), Cr^{3+} gold(III), Au^{3+} iron(III), Fe^{3+}	lead(IV), Pb^{4+} tin(IV), Sn^{4+}

TABLE 5.2.4 Names and formulae of some common anions

Charge		
–1	–2	–3
bromide, Br^- chloride, Cl^- fluoride, F^- iodide, I^-	oxide, O^{2-} sulfide, S^{2-}	nitride, N^{3-} phosphide, P^{3-}

Rules for writing ionic formulae

Here are some simple rules to follow when you are writing chemical formulae:

- Write the symbol for the positively charged ion first.
- Use subscripts to indicate the number of each ion in the formula. Write the subscripts after the ion they refer to.
- If there is just one ion present in the formula, omit the subscript '1'.
- Do not include the charges on the ions in the balanced formula.

Ionic formulae show the simplest ratio of the positive to negative ions present in an ionic compound.

These rules are illustrated in Figure 5.2.16.

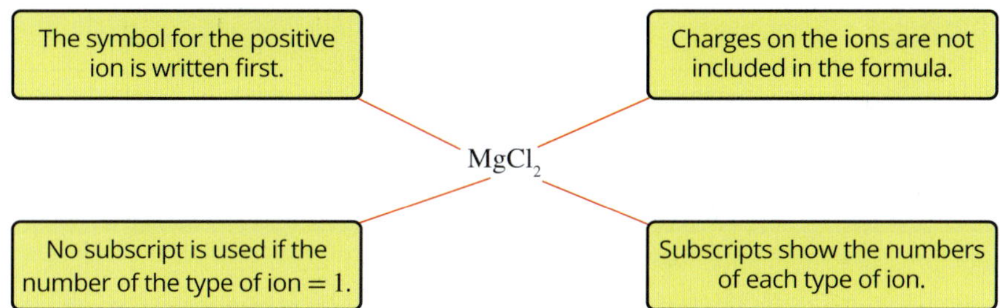

FIGURE 5.2.16 This diagram summarises the information provided by a chemical formula.

Worked example 5.2.1

WRITNG A CHEMICAL FORMULA

Determine the chemical formula of the ionic compound formed between zinc and nitride ions. You may need to refer to Tables 5.2.3 and 5.2.4.

Thinking	Working
Write the symbol and charge of the two ions forming the ionic compound.	Zn^{2+} and N^{3-}
Calculate the lowest common multiple of the two numbers in the charges of the ions.	$2 \times 3 = 6$
Calculate how many positive ions are needed to equal the lowest common multiple.	three Zn^{2+} ions
Calculate how many negative ions are needed to equal the lowest common multiple.	two N^{3-} ions
Use the answers from the previous two steps to write the formula for the ionic compound. Write the symbol of the positive ion first. (Note that 1 is not written as a subscript.)	Zn_3N_2

Worked example: Try yourself 5.2.1

WRITING A CHEMICAL FORMULA

Determine the chemical formula of the ionic compound formed between barium and fluoride ions. You may need to refer to Tables 5.2.3 and 5.2.4.

As you have seen above, simple ions contain only one atom. However, other ions, such as the sulfate ion (SO_4^{2-}), contain two or more atoms, which may be of different elements. These ions are called **polyatomic ions.**

In polyatomic ions:

- if different elements are present, then they are combined in a fixed ratio
- the group of atoms behaves as a single unit with a specific charge
- subscripts are used to indicate the number of each kind of atom in the ion.

For example, a carbonate ion (CO_3^{2-}) contains one carbon atom and three oxygen atoms combined together to form an ion. The carbonate ion has a charge of −2. Other polyatomic ions are nitrate (NO_3^-), hydroxide (OH^-) and phosphate (PO_4^{3-}). The formulae of a number of polyatomic ions can be seen in Table 5.2.5.

The alternative, systematic name for the acetate ion is ethanoate. The common name for the hydrogen carbonate ion is bicarbonate.

TABLE 5.2.5 Common polyatomic cations and anions

Charge			
+1	**−1**	**−2**	**−3**
ammonium, NH_4^+	acetate (ethanoate), CH_3COO^- cyanide, CN^- dihydrogen phosphate, $H_2PO_4^-$ hydrogen carbonate, HCO_3^- hydrogen sulfide, HS^- hydrogen sulfite, HSO_3^- hydrogen sulfate, HSO_4^- hydroxide, OH^- nitrite, NO_2^- nitrate, NO_3^- permanganate, MnO_4^-	carbonate, CO_3^{2-} chromate, CrO_4^{2-} dichromate, $Cr_2O_7^{2-}$ hydrogen phosphate, HPO_4^{2-} oxalate, $C_2O_4^{2-}$ sulfite, SO_3^{2-} sulfate, SO_4^{2-}	phosphate, PO_4^{3-}

Polyatomic ions are ions that are made up of two or more atoms that have an overall charge. The atoms are often of different types. They need to be written within brackets if there is more than one of a type of polyatomic ion present in an ionic compound. Subscripts are used to indicate the ratio of the ions in the crystal lattice.

If more than one polyatomic ion is required in a formula to balance the charge, then it is placed in brackets with the required number written as a subscript after the brackets. Some examples are:

- magnesium nitrate, $Mg(NO_3)_2$
- aluminium hydroxide, $Al(OH)_3$
- ammonium sulfate, $(NH_4)_2SO_4$.

Note that brackets are not required for the formula of sodium nitrate, $NaNO_3$, as there is only one NO_3^- ion present for each sodium ion.

The formulae of two other ionic compounds containing polyatomic ions are shown in Figure 5.2.17.

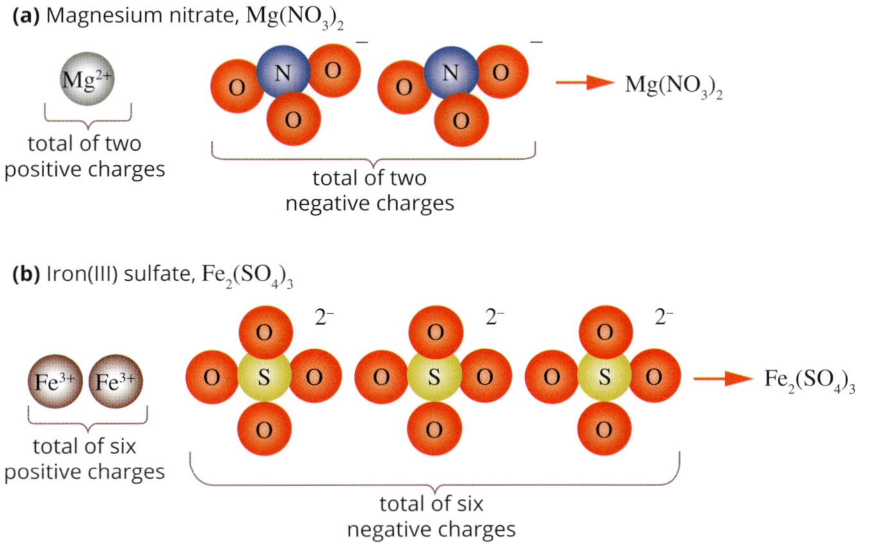

(a) Magnesium nitrate, $Mg(NO_3)_2$

Mg^{2+} — total of two positive charges

NO_3^-, NO_3^- — total of two negative charges

→ $Mg(NO_3)_2$

(b) Iron(III) sulfate, $Fe_2(SO_4)_3$

Fe^{3+} Fe^{3+} — total of six positive charges

SO_4^{2-}, SO_4^{2-}, SO_4^{2-} — total of six negative charges

→ $Fe_2(SO_4)_3$

FIGURE 5.2.17 The chemical formulae of the ionic compounds (a) magnesium nitrate and (b) iron(III) sulfate, which contain polyatomic ions.

NAMING IONIC COMPOUNDS

There are some basic conventions that chemists use when naming ionic compounds.

- The name of a positively charged metal ion (cation) is the same as the name of the metal. For example, the cation of a sodium atom is called a sodium ion; the cation of an aluminium atom is an aluminium ion.
- For simple non-metal ions (anions), the name of the ion is similar to that of the atom, but ends in '-ide'. For example, the anion of the chlorine atom is chloride; the anion of the oxygen atom is oxide.
- For polyatomic anions containing oxygen, the name of the ion will usually end in '-ite' or '-ate'. For example, the NO_2^- ion is called a nitrite ion; the NO_3^- ion is called a nitrate ion. (For two different ions of the same element with oxygen, the name of the ion with the smaller number of oxygen atoms usually ends in '-ite' and the one with the larger number of oxygen atoms ends in '-ate'.)

For compounds of metal with ions of variable electrovalencies, you need to specify the charge on the ion when naming the compound. This is done by placing a Roman numeral immediately after the metal in the name of the compound. For example:

- $FeCl_2$ contains the Fe^{2+} ion and is named iron(II) chloride
- $FeCl_3$ contains the Fe^{3+} ion and is named iron(III) chloride.

Some transition metals can form ions with variable electrovalencies. To indicate the charge on the metal, write a Roman numeral in brackets immediately after the metal in the name of the compound.

5.2 Review

SUMMARY

- Ionic compounds form a crystal lattice made up of metals and non-metals.
- The particles that make up the crystal lattice are cations and anions. Cations are positive ions and anions are negative ions.
- The three-dimensional lattice is held together strongly by electrostatic forces of attraction between the cations and the anions. The electrostatic forces of attraction are called ionic bonding.
- Ionic compounds are hard and have high melting and boiling points. They are solids at room temperature because of the strong forces of attraction between the positively and negatively charged ions in the ionic lattice. A lot of energy is required to overcome these forces of attraction.
- When an ionic compound is hit, the ions move within the lattice so that like-charged ions line up opposite each other and then repel, causing the lattice to be disrupted. This makes ionic compounds brittle.
- Ionic compounds do not conduct electricity in the solid state. Although the solid ionic lattice contains charged particles, the particles are locked in place by strong electrostatic forces of attraction and are not free to move.
- When ionic compounds are added to water or are in molten form, the charged particles are free to move, which means they can conduct electricity.
- In water, ionic compounds vary from very soluble to insoluble. The solubility depends on whether the forces between the water molecules and the ions in the lattice are strong enough to pull the ions out of the lattice.
- Ionic compounds are useful because of their physical properties such as hardness and high melting points.
- Metals have lower electronegativities than non-metals.
- During the formation of ionic compounds, metal atoms lose electrons to form positively charged ions (cations).
- During the formation of ionic compounds, non-metal atoms gain electrons to form negatively charged ions (anions).
- The ions present in an ionic compound have a stable electronic configuration identical to that of the noble gas nearest to them on the periodic table.
- The charge of an ion is called its electrovalency.
- When an ionic compound is formed from positively charged metal ions and negatively charged non-metal ions, the ions combine in proportions that produce an ionic compound with an overall zero charge.
- Ions that contain two or more atoms of different elements are called polyatomic ions.
- When writing formulae of ionic compounds:
  - the total number of positive charges on the metal ions must equal the total number of negative charges on the non-metal ions
  - the symbol for the positively charged ion is written first
  - subscripts are used to indicate the number of each ion in the formula
  - the charges on the ions are not included in the balanced formula
  - if a chemical formula contains more than one polyatomic ion, the formula of the ion is placed in brackets with the number of ions written as a subscript after the brackets
  - for metals that form ions with different charges, the charge on the ion is shown by placing a Roman numeral after the name of the metal.
- When naming ionic compounds, the following rules apply.
  - The name of the metal ion is the same as the name of the metal.
  - Simple non-metal ions take the name of the atom, but end in '-ide'.
  - Polyatomic anions containing oxygen usually end in '-ite' or '-ate'.

KEY QUESTIONS

1 Some properties of four different substances are described below. Which substance is most likely to be an ionic compound?

A Substance A has a melting point of 842°C and conducts electricity at 700°C.

B Substance B has a melting point of 308°C. It does not conduct electricity at 250°C but will conduct electricity at 350°C.

C Substance C has a melting point of 180°C and can be drawn into a wire.

D Substance D is a white solid that melts at 660°C and will not conduct electricity at 700°C.

2 Sodium chloride does not conduct electricity in the solid state, but does when molten (liquid).

a How could you use Figure 5.2.7 on page 150 to explain why solid sodium chloride does not conduct electricity?

b Explain why molten sodium chloride conducts electricity.

3 Why do ionic compounds have such high melting and boiling points?

4 Use diagrams similar to Figure 5.2.11 on page 153 to show the formation of ions in the reactions between:

a potassium and fluorine

b magnesium and sulfur

c aluminium and fluorine

d sodium and oxygen

e aluminium and oxygen.

5 Explain why potassium chloride has the formula KCl, whereas the formula of calcium chloride is $CaCl_2$.

6 Use the information in Tables 5.2.3 and 5.2.4 on page 156 to write formulae for the following ionic compounds:

a sodium chloride

b potassium bromide

c zinc chloride

d potassium oxide

e barium bromide

f aluminium iodide

g silver bromide

h zinc oxide

i barium oxide

j aluminium sulfide

7 Use the information in Tables 5.2.3 and 5.2.4 on page 156 to name the ionic compounds with the following formulae:

a KCl

b CaO

c MgS

d K_2O

e NaF

8 Use the information in Table 5.2.3 on page 156 and Table 5.2.5 on page 158 to write formulae for the following ionic compounds:

a sodium carbonate

b barium nitrate

c aluminium nitrate

d calcium hydroxide

e zinc sulfate

f potassium hydroxide

g potassium nitrate

h zinc carbonate

i potassium sulfate

j barium hydroxide

5.3 Covalent bonding

PROPERTIES OF COVALENT SUBSTANCES

In Section 5.1, you saw that the bonding between atoms in metallic elements is called metallic bonding. When metal atoms combine with atoms of non-metallic elements, the compounds contain another form of bonding, ionic bonding (Section 5.2). In this section, you will look at the chemical bonding that occurs when atoms of non-metals combine with each other. Non-metallic elements have higher electronegativities and ionisation energies than metallic elements. When non-metallic atoms combine, they achieve the stable electron configuration of noble gases by sharing electrons. The bond formed by electron sharing is called a **covalent bond**.

Examples of covalent compounds

Although there are fewer non-metals than metals in the periodic table, the atoms of non-metals form a much larger number of compounds than metals. The locations of non-metallic elements in the periodic table are shown in pink and yellow in Figure 5.3.1. The group 18 elements are inert and do not readily form bonds.

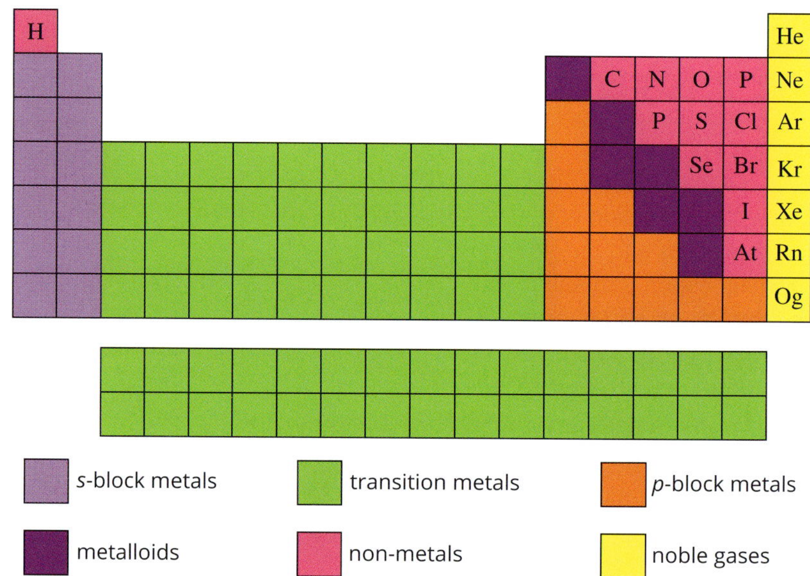

FIGURE 5.3.1 The location of non-metals in the periodic table is shown in pink and yellow. The group 18 elements are inert.

Water, carbon dioxide, caffeine, sugar and cooking oil are just a few examples of common compounds formed from non-metals. Most covalent substances consist of molecules. A **molecule** is a discrete (individually separate) group of atoms bonded together by covalent bonds. A molecule has a known formula that indicates the exact number of atoms of each element present in the molecule. Some elements exist as molecules. The element oxygen, for example, forms O_2 molecules consisting of two oxygen atoms held together by a covalent bond. Many compounds are also molecular. A water molecule (H_2O) contains covalent bonds joining two hydrogen atoms to an oxygen atom.

Properties of covalent elements and compounds

The properties of compounds formed by non-metallic elements can tell you a lot about their chemical structures. Table 5.3.1 lists some of the properties of a range of common covalent substances.

TABLE 5.3.1 Properties of some common covalent substances

Substance	Formula	Melting point (°C)	Boiling point (°C)	Conducts electricity as a solid?	Conducts electricity as a liquid?
ammonia	NH_3	–77	–33	no	no
laughing gas	N_2O	–90	–88	no	no
water	H_2O	0	100	no	no
paraffin wax	$C_{31}H_{64}$	37	370	no	no
oxygen	O_2	–219	–183	no	no

From Table 5.3.1, you can see that the covalent substances ammonia, laughing gas, water, paraffin wax and oxygen have low melting points and do not conduct electricity in either the solid or liquid states. Solid covalent compounds such as wax are soft and can easily be molded into different shapes.

Two conclusions that can be drawn from these properties about bonding in covalent compounds are shown in Figure 5.3.2.

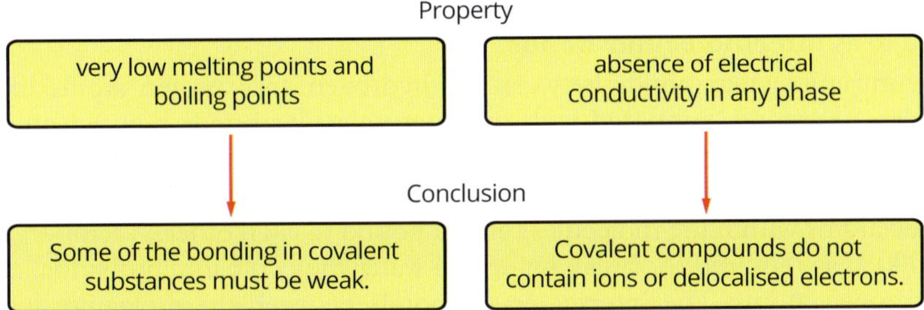

FIGURE 5.3.2 These conclusions can be drawn from the properties of covalent substances.

You can gain further insight into the chemical structure of covalent substances by considering what happens when you boil water in a beaker. Figure 5.3.3 shows that each water molecule contains two hydrogen atoms bonded to one oxygen atom. Between the water molecules are weak bonds holding the molecules close to one another.

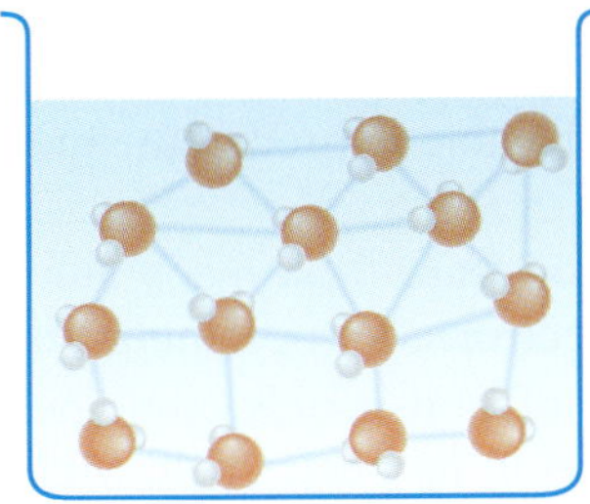

Water in a beaker contains two types of bonds:
1. the bonds between hydrogen and oxygen atoms within each water molecule
2. the bonds holding one molecule of water to another.

FIGURE 5.3.3 This representation shows the bonds between molecules of water in a beaker.

SKILLBUILDER 

Use of Latin and Greek words in science

Science is an international endeavour. To promote a common understanding of concepts, some language conventions are employed. For example, while the names of elements discovered many years ago have unique names in different languages (element 11 is sodium in English, but natrium in Dutch, from the original Latin), the element symbols are the same (element 11 is always Na).

As modern languages are evolving, science continues to use words from 'dead' languages like Latin and Greek (Tables 5.3.2 and 5.3.3). These words are not changed when they are used by speakers of different modern languages, meaning communication is clearer and easier as scientists have a common vocabulary. Sometimes the meanings of the Latin and Greek words are adapted for modern use, because the ancients obviously didn't understand all the things we do now.

TABLE 5.3.2 Examples of Latin terms in scientific use

Latin term	Meaning	Example
intra-	within	intramolecular bond
inter-	between, among	intermolecular bond
in vitro	in glass	in vitro fertilisation
in vivo	in life	The experiment was performed in vivo.

TABLE 5.3.3 Examples of Greek terms in scientific use

Greek term	Meaning	Example
mono-, mon-	one, single	carbon monoxide
di-	two, double	dipolar bond
helios	sun	heliocentric
sperma	seed	sperm cell

When water boils, the weak bonds between molecules are broken. However, the water molecules do not separate into hydrogen and oxygen atoms. Rather, as shown in Figure 5.3.4, the water vapour that is formed still contains molecules in which two hydrogen atoms are bonded to one oxygen atom. This indicates that liquid water must contain more than one form of bonding.

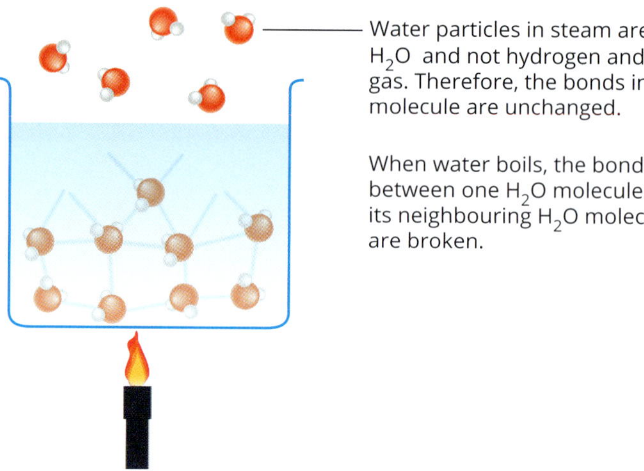

FIGURE 5.3.4 Changes occur to the bonding in water when it starts to boil.

The bonds between the hydrogen and oxygen atoms within water molecules are referred to as **intramolecular bonds**, and the bonds between water molecules are referred to as **intermolecular bonds**.

The intramolecular bonds between the hydrogen and oxygen atoms in water molecules are strong compared with the intermolecular bonds between the water molecules. It is the intermolecular bonds that are broken when molecular substances such as water boil or melt. This allows the molecules to separate from one another, while the atoms within the molecules remain bound to each other. A temperature of over 2000°C is required for breaking up the water molecule into individual oxygen and hydrogen atoms. The intermolecular bonds between the molecules in liquid water are broken when water boils at 100°C. You can see that it takes more energy to break the intramolecular bonds between the hydrogen and oxygen atom in a water molecule than it does to break the intermolecular bonds between water molecules.

COVALENT BONDING

This section examines a series of simple molecules to help you to understand the concept of a covalent bond, which is formed when non-metallic atoms share electrons. Using your knowledge of the valence shell electron arrangements of non-metallic atoms, you will be able to predict the molecules that different elements can form.

In the previous section, you learnt that many atoms become more stable if they gain an outer shell of eight electrons by combining with other atoms (the octet rule).

Commonly, when atoms of non-metals combine, electrons are shared so that each atom has eight electrons in its outer shell. Molecules formed in this way are more stable than the separate atoms.

Non-metallic atoms have a relatively high number of electrons in their outer shells, and they tend to share rather than to transfer electrons. Covalent bonding occurs when electrons are shared between atoms.

Single covalent bonds

When atoms share two electrons, one from each atom, the covalent bond formed is called a **single covalent bond**. Two examples of substances that contain single bonds are hydrogen and chlorine.

Example 1: Hydrogen

Hydrogen atoms have one electron. The valence shell for a hydrogen atom can hold a maximum of two electrons. A hydrogen atom can bond to another hydrogen atom to form a molecule of H_2, as shown in Figure 5.3.5.

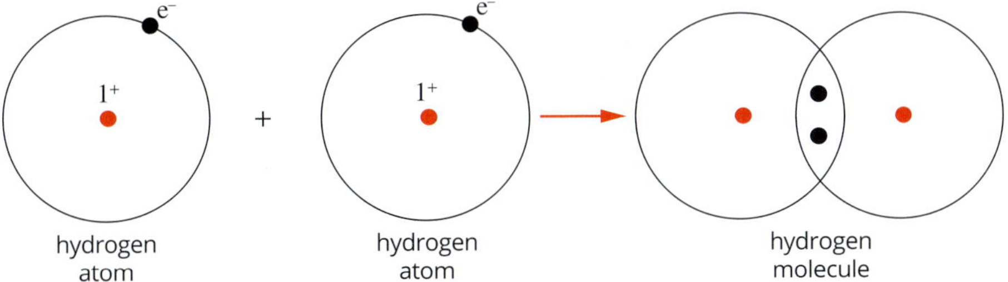

FIGURE 5.3.5 A covalent bond is formed when two hydrogen atoms share two electrons—one from each atom.

In the molecule that is formed:

- two hydrogen atoms share two electrons, one from each atom, to form a single covalent bond
- the atoms of hydrogen are now strongly bonded together by two electrons (an electron pair) in their outer shells.

The hydrogen molecule can be represented as H_2. Molecules that contain two atoms are called **diatomic molecules**.

Two alternative ways of representing a hydrogen molecule are shown in Figure 5.3.6.

FIGURE 5.3.6 A hydrogen molecule can be represented by (a) a valence structure or (b) a Lewis dot diagram.

In a hydrogen molecule, the electron in each atom is attracted to the proton within the neighbouring atom, as well as to their own proton. This means the two electrons will spend most of their time between the two nuclei instead of orbiting their own nuclei. Even though the protons in the two nuclei still repel each other (remember, like charges repel), the electrostatic attraction between the electrons and protons keeps the molecule held together.

This model of a hydrogen molecule is consistent with the observed properties of hydrogen gas. The covalent bonds in the molecules are strong, but the intermolecular forces in hydrogen (the attractions between one molecule and the surrounding molecules) are weak; therefore, hydrogen has a low melting temperature of −259°C. Hydrogen does not conduct electricity as it does not contain ions or delocalised electrons.

> A covalent bond is formed from the electrostatic attraction between shared electrons and the nuclei of the atoms involved in the bond.

Example 2: Chlorine

A chlorine atom has an electronic configuration of 2,8,7. It requires one more electron to achieve eight electrons in its outer shell. (Note that throughout this section, shell-model electronic configurations and diagrams are used to represent the electron arrangement in atoms as simply as possible. The inner shell electrons are not shown as only the outer shell electrons are involved in bond formation.) One chlorine atom can share an electron with another chlorine atom to form a molecule of chlorine with a single covalent bond. As a result, both atoms gain outer shells containing eight electrons as shown in Figure 5.3.7. This example illustrates the octet rule.

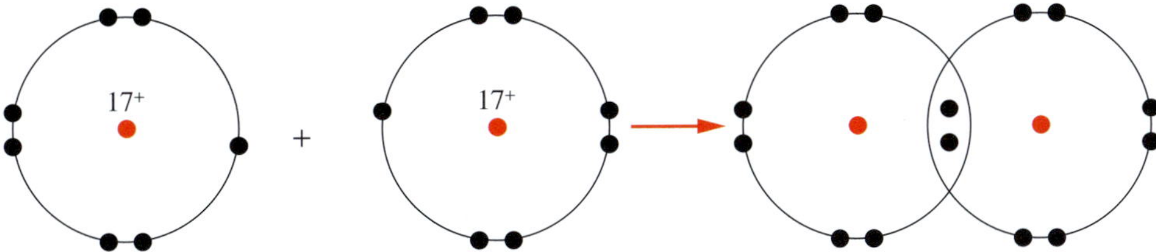

FIGURE 5.3.7 Two chlorine atoms share one electron each to form a chlorine molecule. Note that only outer-shell electrons are usually shown in these diagrams.

Lewis dot diagrams

Chemists often use **Lewis dot diagrams** (also known as electron dot diagrams) and **valence structures** to simplify the drawing of molecules.

Lewis dot diagrams show the valence shell electrons of an atom because only these electrons are involved in bonding. The Lewis dot diagram also allows you to distinguish between bonding electrons and non-bonding electrons. A chlorine molecule has one pair of bonding electrons. The outer-shell electrons that are not involved in bonding are called the **non-bonding electrons**. Each chlorine atom has six non-bonding electrons, present as three pairs of electrons. Pairs of non-bonding electrons are known as **lone pairs**.

In the valence structures used in this section, the non-bonding electrons are represented as dots, and each pair of electrons involved in the formation of a covalent bond is represented by a single line. Figure 5.3.8 shows the Lewis dot diagram and valence structure for a molecule of chlorine. It is easy to see that each chlorine atom in the molecule has eight electrons in its outer shell. The electrons in a Lewis dot diagram can be represented by either dots or crosses.

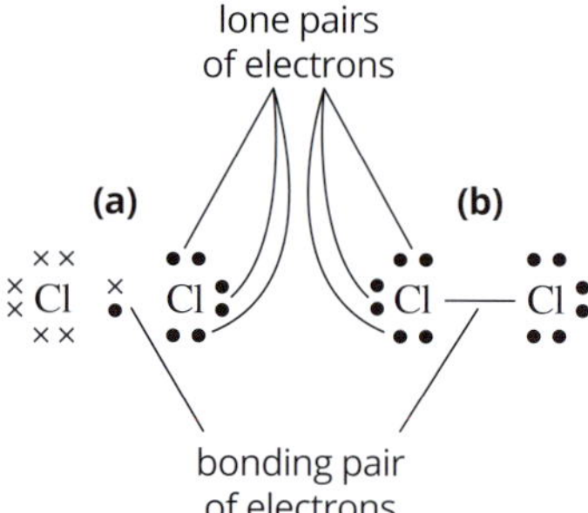

FIGURE 5.3.8 Two ways of representing the chlorine molecule. They show the outer-shell electrons only. (a) Lewis dot diagrams. The electrons can be represented by either dots or crosses. (b) In a valence structure, lines are used to represent pairs of electrons involved in bond formation.

Double covalent bonds

In a **double covalent bond**, two pairs of electrons (four electrons in total) are shared between the atoms, rather than just one pair.

The oxygen molecule contains a double covalent bond. The electronic configuration of an oxygen atom is 2,6. Each oxygen atom requires two electrons to gain a stable outer shell containing eight electrons. Therefore, when two oxygen atoms bond with each other, each atom shares two electrons.

In Figure 5.3.9, each oxygen atom in the molecule now has eight outer-shell electrons: four bonding electrons and four non-bonding electrons.

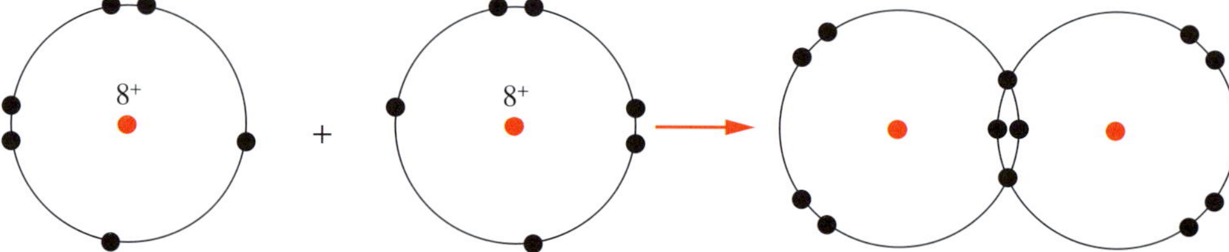

FIGURE 5.3.9 In oxygen molecules, each oxygen atom contributes two electrons to the bond.

Figure 5.3.10 shows two different ways of representing the oxygen molecule: the Lewis dot diagram and the valence structure of an oxygen molecule.

FIGURE 5.3.10 (a) The Lewis dot diagram shows that O_2 has a double covalent bond. Four electrons are shared, and each oxygen has two non-bonding electron pairs. (b) The valence structure shows the double bond as two parallel lines.

Triple covalent bonds

A **triple covalent bond** occurs when three electron pairs are shared between two atoms. The nitrogen molecule contains a triple bond. The electronic configuration of nitrogen is 2,5. A nitrogen atom requires three electrons to achieve eight electrons in its outer shell. When it bonds to another nitrogen atom, each atom contributes three electrons to the bond that forms (Figure 5.3.11) and two different ways of representing the molecule are shown in Figure 5.3.12.

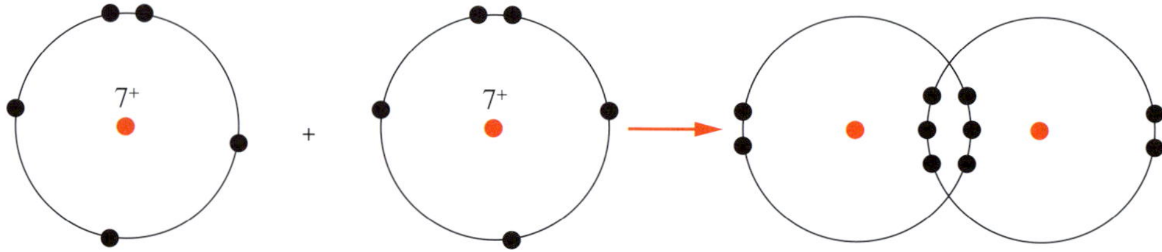

FIGURE 5.3.11 Nitrogen atoms contribute three electrons each to form a triple covalent bond in a molecule of N_2.

(a) N N (b) N≡N

FIGURE 5.3.12 (a) The Lewis dot diagram shows that N_2 has a triple covalent bond. (b) The valence structure shows the triple bond as three parallel lines.

The triple covalent bond in nitrogen gas (N_2) is relatively strong and not easily broken. This means that nitrogen gas is relatively unreactive.

MOLECULAR COMPOUNDS

A diatomic molecule contains two atoms. The molecules discussed so far have been diatomic molecules containing atoms of the same element.

Covalent bonds can also form between atoms of different elements. Hydrogen chloride (HCl) is a simple example (Figure 5.3.13). A hydrogen atom requires one electron to gain a stable outer shell, as does a chlorine atom. They can share an electron each and form a single covalent bond.

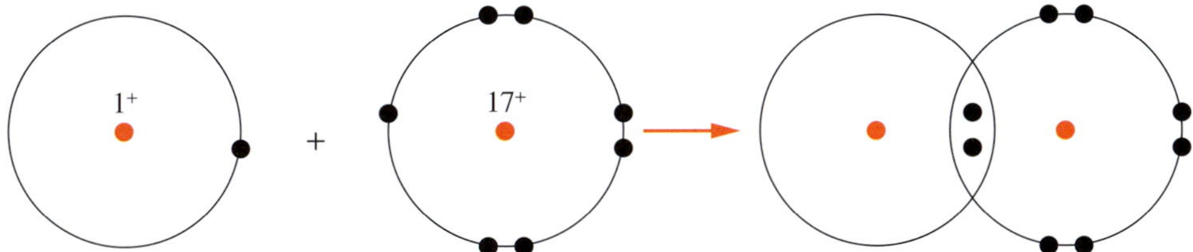

FIGURE 5.3.13 A pair of electrons is shared in the formation of a molecule of HCl.

Polyatomic molecules

If the atoms of one element have a different number of valence electrons from atoms of the element it is bonding with, the molecule that is formed may not be a diatomic molecule. Molecules made up of more than two atoms are called **polyatomic molecules**. Two examples of polyatomic molecules are water and methane.

Example 1: Water

When a compound forms between hydrogen and oxygen, an oxygen atom shares two electrons and a hydrogen atom shares one. To resolve this imbalance, two hydrogen atoms each share one electron with an oxygen atom.

As you can see in Figure 5.3.14, a water molecule contains:

- two single covalent bonds, each containing a shared electron pair
- four non-bonding electrons on the oxygen atom.

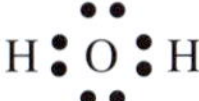

FIGURE 5.3.14 A water molecule has two single covalent bonds. Electrons are represented here simply by dots.

Example 2: Methane

When a compound forms between carbon and hydrogen, four hydrogen atoms are needed for providing the four electrons required to have eight electrons in the outer shell of a carbon atom (Figure 5.3.15). The molecule formed has a chemical formula of CH_4 and is called methane.

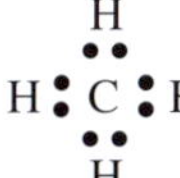

FIGURE 5.3.15 In a methane molecule, a carbon atom shares one electron with each of four hydrogen atoms to gain eight electrons in its outer shell.

Worked example 5.3.1

DRAWING LEWIS DOT DIAGRAMS

Draw a Lewis dot diagram of methane (CH_4).

Thinking	Working
Write the electronic configuration of the atoms in the molecule.	C—electronic configuration: 2,4 H—electronic configuration: 1
Determine how many electrons each atom requires for a stable outer shell.	C requires 4 electrons. H requires 1 electron.
Draw a Lewis dot diagram of the likely molecule, ensuring that each atom has a stable outer shell. Any electrons not involved in bonding will be in non-bonding (lone) pairs.	Draw a Lewis dot diagram of the molecule. H H : C : H H

Worked example: Try yourself 5.3.1

DRAWING LEWIS DOT DIAGRAMS

Draw a Lewis dot diagram of ammonia (NH_3).

NAMING COVALENT COMPOUNDS

To avoid confusion, chemists use a set of common rules when naming covalent compounds and writing their formulae. A simplified version of these rules is provided below.

- The least electronegative element is named first. These are found to the left in the periodic table and lower down the groups.
- The name of the second element in a covalent compound is modified by adding an 'ide' to the end of its name.

Table 5.3.4 lists the prefixes used to indicate the number of atoms of each element present in the molecule. The exception is the prefix mono- which is not used if the first element only has one atom in its molecule.

Some prefixes end in an 'a' or an 'o'. These letters are dropped when the name of the element starts with a vowel. For example, in the case of CO, carbon monoxide rather than carbon monooxide is used.

The compounds listed in Table 5.3.5 illustrate how these rules are used to name different oxides of nitrogen.

By convention, the common names of some compounds that contain hydrogen are used instead of the compound name. These compounds include water (H_2O), ammonia (NH_3), methane (CH_4) and hydrogen sulfide (H_2S). Hydrogen is always placed first when naming the hydrides of group 17 elements such as hydrogen chloride (HCl).

GO TO ➤ SkillBuilder page 164

TABLE 5.3.4 Prefixes used in the names of covalent compounds.

Number	Prefix
1	mono
2	di
3	tri
4	tetra
5	penta
6	hexa

TABLE 5.3.5 Naming the oxides of nitrogen

Formula	Name
NO	nitrogen monoxide
NO_2	nitrogen dioxide
N_2O	dinitrogen oxide
N_2O_3	dinitrogen trioxide
N_2O_4	dinitrogen tetroxide
N_2O_5	dinitrogen pentoxide

CHEMFILE 

Dihydrogen monoxide

There is a movement to ban dihydrogen monoxide—an odourless, colourless liquid that has been known to cause corrosion of metals. It is found in acid rain, industrial pollution and is even linked to global warming. Excessive consumption of dihydrogen monoxide can cause bloating, nausea and even death.

Looking at the name dihydrogen monoxide, you should be able to work out that its chemical formula is H_2O—plain old water! The movement to ban dihydrogen monoxide is a longstanding joke on the internet, and plays on the general population's lack of understanding of chemistry and scientific language.

So do not be fooled when someone asks you to sign a petition to ban dihydrogen monoxide. You might just be supporting the ban of something that is vital for your survival.

ALLOTROPES

Some elements can exist with their atoms in several different structural arrangements, called **allotropes**. In different allotropes, the atoms are bonded to each other in different, specific ways that give them a particular physical form and significantly different properties compared with other allotropes of the same element.

Allotropes are different forms of the same element.

Allotropes of oxygen

Oxygen forms allotropes. Oxygen gas consists of diatomic molecules with the formula O_2. Each oxygen atom in this arrangement is bound to one other oxygen atom. Ozone is another molecule containing only oxygen. Ozone molecules have the formula O_3 and consist of a central oxygen atom bound to two other oxygen atoms. Figure 5.3.16 shows the structure of these two molecules. Due to its different bonds, an ozone molecule is less stable than an oxygen molecule.

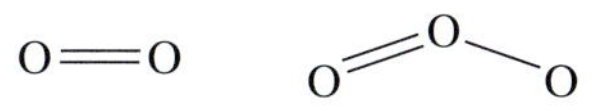

O=O

O=O–O

FIGURE 5.3.16 Oxygen and ozone are two molecules that contain only oxygen atoms.

Allotropes of phosphorus

Phosphorus has three main allotropes—white phosphorus, red phosphorus and black phosphorus (Figure 5.3.17).

FIGURE 5.3.17 The three allotropes of phosphorus—white, red and black phosphorus.

White phosphorus is made up of molecules with the formula P_4 (Figure 5.3.18). It has the appearance of a white, waxy solid, is highly flammable and can self-ignite when it comes into contact with air.

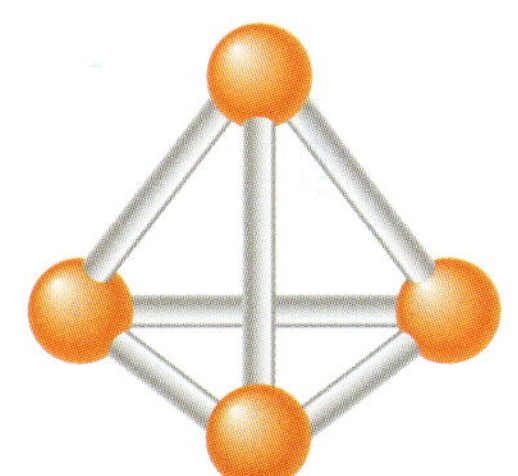

FIGURE 5.3.18 A molecule of white phosphorus.

The structure of red phosphorus consists of P_4 molecules joined together to form a chain of repeating units (Figure 5.3.19). In general, this type of structure is called a **polymer**. Red phosphorus is made by heating white phosphorus to 300°C. It is more stable than white phosphorus, so will not ignite until the temperature reaches above 240°C.

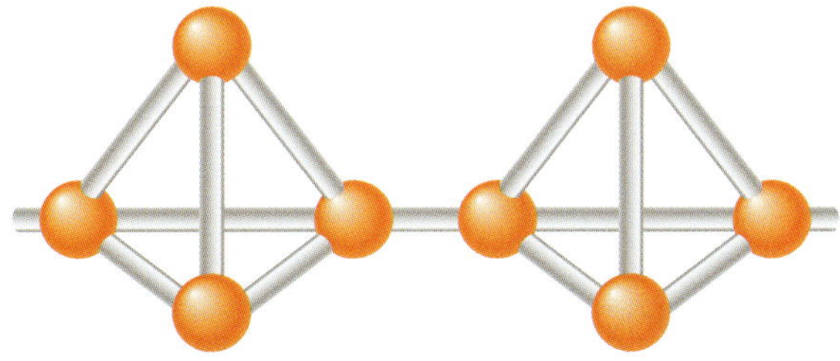

FIGURE 5.3.19 The structure of red phosphorus.

The most stable allotrope of phosphorus is black phosphorus. Black phosphorus is produced through heating white phosphorus under very high pressures. As shown in Figure 5.3.20, each phosphorus atom is bonded to three other phosphorus atoms, forming a covalent layer network (you will learn more about covalent layer networks in Section 5.5). The forces between layers are weak. Due to their similar structures, both black phosphorus, and graphite have similar properties, such as being hard but flaky solids.

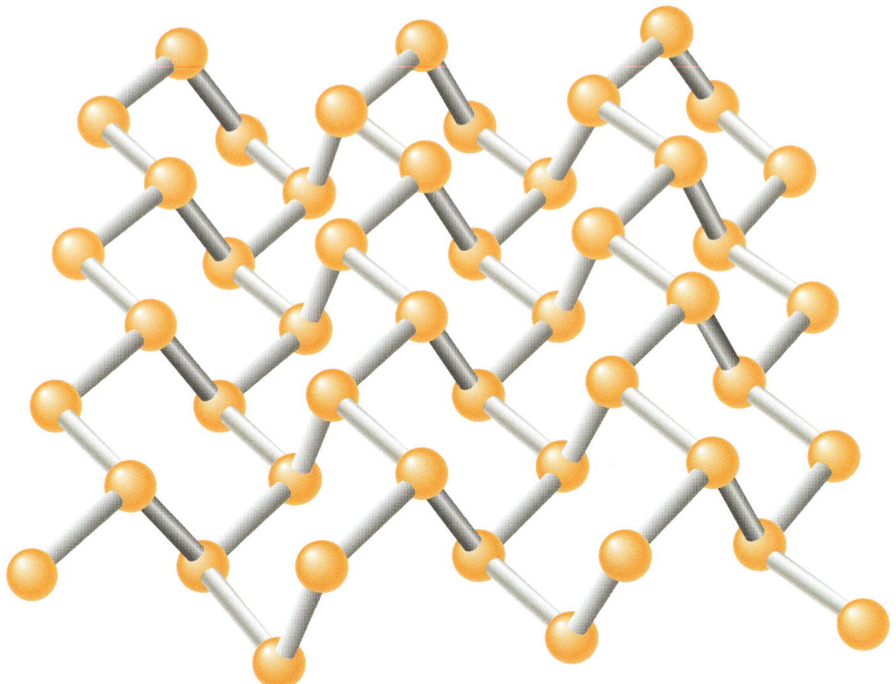

FIGURE 5.3.20 The structure of a layer of black phosphorus. Each phosphorus atom is strongly bonded to three other phosphorus atoms, forming a covalent layer network.

WS 1.7 WS 1.8

5.3 Review

SUMMARY

- Non-metallic elements and compounds usually have low boiling points and do not conduct electricity.
- In general, non-metallic elements and compounds do not conduct electricity because they do not contain free-moving charged particles (neither delocalised electrons nor ions).
- Many non-metallic elements and compounds are composed of molecules. Molecules are discrete groups of atoms of known formula, bonded together.
- The atoms in non-metallic molecules are held together by covalent bonds.
- A covalent bond involves the sharing of electrons.
- Covalent bonds form between non-metallic atoms, often enabling the atoms to gain outer shells containing eight electrons (except hydrogen, which gains an outer shell containing two electrons).
- A single covalent bond forms when two atoms share a pair of electrons.
- A double covalent bond forms when two atoms share four electrons.
- A triple covalent bond forms when two atoms share six electrons.
- Outer-shell electrons that are not involved in bonding are called lone pairs.
- Lewis dot diagrams show the valence electron arrangements of atoms in a molecule.
- Allotropes are different physical forms of the same element. Allotropes differ due to differences in the way atoms are bonded within their structures.
- A number of elements exist in different forms. Oxygen has two main allotropes—diatomic oxygen (O_2) and ozone (O_3).
- White, red and black phosphorus are the three main allotropes of phosphorus. White phosphorus is made up of molecules of P_4, red phosphorus is a polymer, and black phosphorus is a covalent layer network. The different structures result in the different properties of the three allotropes of phosphorus.

KEY QUESTIONS

1. How many covalent bonds are formed between atoms in these diatomic molecules?
   - **a** H_2
   - **b** N_2
   - **c** O_2
   - **d** F_2
2. Draw Lewis dot diagrams for each of the following molecules: fluorine (F_2), hydrogen fluoride (HF), water (H_2O), tetrachloromethane (CCl_4), phosphine (PH_3), butane (C_4H_{10}), carbon dioxide (CO_2).
3. What is the maximum number of covalent bonds an atom of each of the following elements can form?
   - **a** F
   - **b** O
   - **c** N
   - **d** C
   - **e** H
   - **f** Ne
4. When oxygen forms covalent molecular compounds with other non-metals, the valence structures that represent the molecules of these compounds all show each oxygen atom with two lone pairs of electrons. Why are there always two lone pairs?
5. Suggest the most likely formula and name of the compound formed between the following pairs of elements.
   - **a** C, Cl
   - **b** N, Br
   - **c** Si, O
   - **d** H, F
   - **e** P, F
6. Explain why covalent compounds do not conduct electricity in either the solid or liquid states.
7. **a** Describe how the octet rule can be used to explain the formation of bonds between non-metallic atoms.
   b Explain why hydrogen is an exception to the octet rule.
8. **a** What is meant by the term allotrope?
   b Name two allotropes of:
   - **i** oxygen
   - **ii** phosphorus.

5.4 Intermolecular forces

Covalent molecular substances have a much greater range of properties than ionic or metallic substances. Unlike ionic or metallic substances, they can be solids, liquids or gases at room temperature. Covalent molecular substances can be hard or soft, flexible or brittle, sticky or oily—almost any consistency. This broad range of properties means that they have a vast number of uses.

The properties of covalent molecular substances are the result of two main types of bonding—the strong covalent bonds inside the molecules, and the much weaker bonds between molecules. The bonds within a molecule are called intramolecular bonds, and the bonds between molecules are referred to as intermolecular bonds. In the previous section, you learned about the origin and nature of the strong intramolecular bonds in covalent molecules. In this section you will learn about the weaker intermolecular forces that exist between molecules and how these determine physical properties such as hardness, melting point and boiling point.

By the end of this section, you will know how intermolecular forces determine many of the physical properties of covalent molecular substances. You will become familiar with the **valence-shell electron-pair repulsion (VSEPR) theory** and use the theory to predict the shape and polarity of molecules. You will then use these predictions to draw conclusions about the nature of the bonding between molecules.

In addition, you will learn to identify intermolecular forces such as dipole–dipole forces, hydrogen bonding and dispersion forces. Finally, you will examine the factors that influence the strength of these intermolecular forces.

SHAPES OF MOLECULES

The shapes of molecules are critical in determining many physical properties of covalent molecular substances. In particular, molecular shape affects melting point, boiling point, hardness and solubility. This is because the shape of a molecule determines how it interacts with other molecules.

For very large molecules, such as DNA, proteins and enzymes, shape plays a key role in how they behave chemically and biologically. For example, the twisted double helix of DNA allows the molecule to coil up tightly so that it fits inside the nucleus of a cell.

To understand the shape of large molecules like DNA, scientists use complex techniques such as X-ray crystallography or powerful computer simulations. However, the shape of small molecules can be predicted using a relatively simple model known as the **valence-shell electron-pair repulsion (VSEPR) theory**. In this section, you will see how VSEPR theory can be used to predict the shape of molecules such as the water molecule shown in Figure 5.4.1.

FIGURE 5.4.1 Water (H_2O) molecules have a distinctive shape that is responsible for many of the properties of water. The valence-shell electron-pair repulsion theory accurately predicts the shape of water molecules.

VALENCE-SHELL ELECTRON-PAIR REPULSION THEORY

As the name suggests, the VSEPR theory uses our knowledge of the valence electrons in the atoms of a molecule to predict the shape of the molecule. The VSEPR theory is based on the principle that negatively charged electron pairs in the outer shell of an atom repel each other. As a consequence, these electron pairs are arranged as far away from each other as possible.

Electron-pair repulsion

In general, atoms in covalent molecules are most stable when they have eight electrons in their valence shell. This is known as the octet rule. These eight electrons are arranged into four pairs of electrons. For example, the carbon atom in methane is covalently bonded to four hydrogen atoms. In this arrangement, the carbon atom shares a pair of electrons with each hydrogen atom. The four pairs of electrons give the carbon atom a stable octet, as shown in Figure 5.4.2.

H
H C H
H

FIGURE 5.4.2 This Lewis dot diagram of a methane (CH_4) molecule shows the four electron pairs surrounding the central carbon atom.

Hydrogen atoms are an exception to the octet rule. Hydrogen atoms are stable with just two electrons because this gives them the stable electronic configuration of a helium atom.

The VSEPR theory states that the electron pairs in methane repel each other so that they are as far apart as possible. This repulsion between the electron pairs results in a **tetrahedral shape,** as shown in Figure 5.4.3. The tetrahedral shape ensures that the electron pairs are as far from each other as possible, with angles of 109.5° between all of the single bonds.

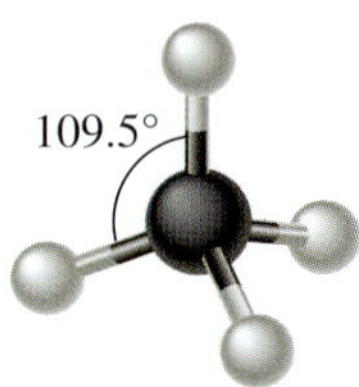

FIGURE 5.4.3 In a methane (CH_4) molecule, the electron pairs in the single covalent bonds repel each other. This repulsion forces the bonds as far apart as possible, pushing the hydrogen atoms into a tetrahedral shape with bonds at an angle of 109.5° to each other.

Lone pairs of electrons

Not all electron pairs in molecules exist as covalent bonds. From the previous section, you would have learnt that some electrons form a non-bonding pair of electrons known as a lone pair of electrons. In the VSEPR theory, lone pairs of electrons are treated in the same way as electron pairs in covalent bonds in order to determine the shape of a molecule.

In the ammonia molecule shown in Figure 5.4.4, the nitrogen atom has a stable octet made up of one lone pair of electrons and three single bonds. The four electron pairs repel each other to form a tetrahedral arrangement. However, the lone pair is ignored when describing the shape of the molecule. Instead, the three hydrogen atoms are described as forming a pyramidal arrangement with the nitrogen atom. The lone pair occupies slightly more space than the bonding electrons, so the three single covalent bonds are pushed closer together. The bond angle is therefore now slightly less than 109.5°.

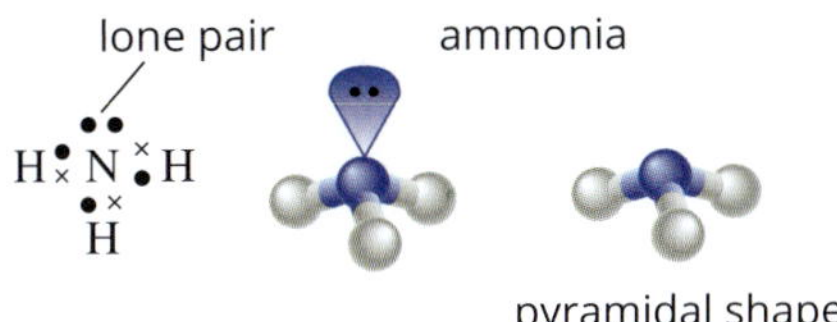

FIGURE 5.4.4 The Lewis dot diagram of an ammonia (NH_3) molecule shows that the nitrogen atom has one lone pair of electrons and three covalent bonds. These four electron pairs form a tetrahedral arrangement around the atom due to the repulsion of the electron pairs. The result is a pyramidal molecule.

In water molecules, the oxygen atom has a stable octet made up of two lone pairs and two single bonds. The four electron pairs repel each other to form a tetrahedral arrangement. This causes the two hydrogen atoms to form a V-shape or bent arrangement, with the oxygen atom as shown in Figure 5.4.5. With two lone pairs in the molecule, the two single covalent bonds are pushed closer together. In this case, the bond angle around the central atom is also slightly less than 109.5°.

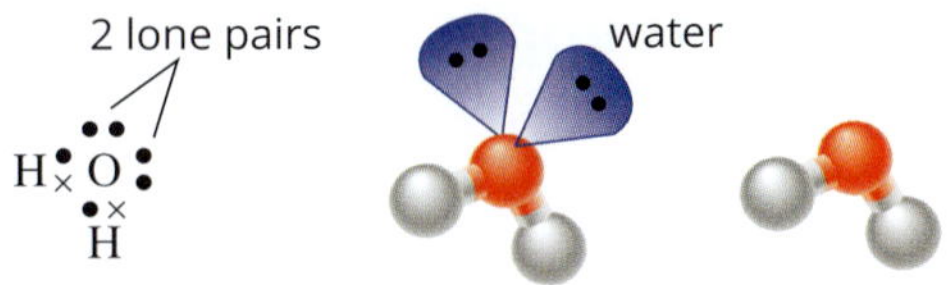

FIGURE 5.4.5 The Lewis dot diagram of a water (H_2O) molecule shows that the oxygen atom has two lone pairs of electrons and two covalent bonds. These four electron pairs form a tetrahedral arrangement around the atom due to the repulsion of the electron pairs. The result is a V-shaped or bent molecule.

In a hydrogen fluoride molecule, the fluorine atom has a stable octet made up of three lone pairs and one single bond. The four electron pairs repel each other to form a tetrahedral arrangement. Therefore, the hydrogen and fluorine atoms form a linear molecule, as you can see in Figure 5.4.6.

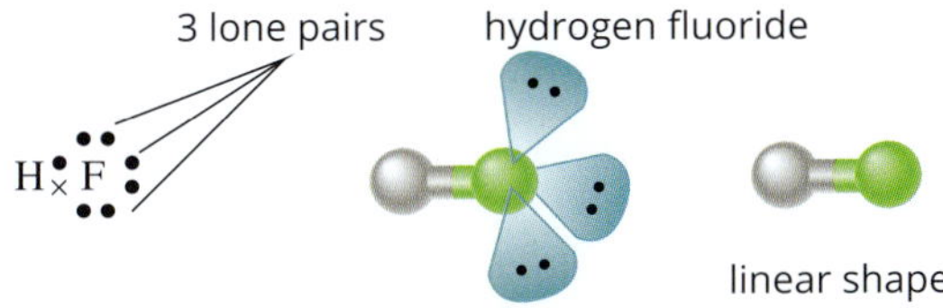

FIGURE 5.4.6 The Lewis dot diagram of a hydrogen fluoride (HF) molecule shows that the fluorine atom has three lone pairs of electrons and one covalent bond. These four electron pairs form a tetrahedral arrangement around the atom due to the mutual repulsion of the electron pairs. The result is a linear molecule.

Lone pairs of electrons influence a molecule's shape but are not considered a part of the shape.

Worked example 5.4.1

PREDICTING THE SHAPE OF MOLECULES

Predict the shape of a molecule of phosphine (PH_3).

Thinking	Working
Draw the electron dot diagram for the molecule.	H : P̤̈ : H H
Count the number of bonds and lone pairs on the central atom.	There are three bonds and one lone pair.
Determine how the groups of electrons will be arranged to get maximum separation.	Because there are four electron pairs, the groups will be arranged in a tetrahedral arrangement.
Deduce the shape of the molecule by considering the arrangement of just the atoms.	The phosphorus and hydrogen atoms are arranged in a pyramidal shape. P̈ H H H

Worked example: Try yourself 5.4.1

PREDICTING THE SHAPE OF MOLECULES

Predict the shape of a molecule of hydrogen sulfide (H_2S).

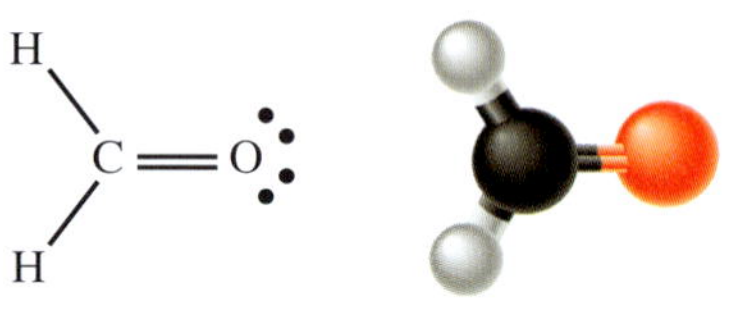

methanal, CH_2O
(trigonal planar)

FIGURE 5.4.7 Methanal has a central carbon atom that forms a double bond with an oxygen atom and single bonds with two hydrogen atoms. The bonds repel each other to form a trigonal planar arrangement.

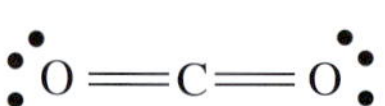

carbon dioxide, CO_2
(linear)

FIGURE 5.4.8 In a carbon dioxide molecule, the carbon atom forms double bonds with two oxygen atoms. The two double bonds repel each other. This results in a linear molecule.

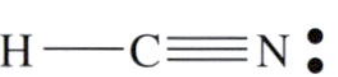

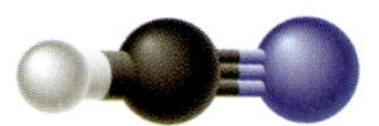

hydrogen cyanide, HCN
(linear)

FIGURE 5.4.9 The hydrogen cyanide molecule is linear. In this case, the central carbon atom forms a triple bond with the nitrogen atom and a single bond with the hydrogen atom.

Double bonds, triple bonds and valence-shell electron-pair repulsion theory

A double bond contains two pairs of electrons. A triple bond contains three pairs of electrons. The VSEPR theory treats double and triple bonds in the same way that it treats single bonds and lone pair electrons.

For example, if a central atom has two single bonds and one double bond, then the three sets of bonds will repel each other to get maximum separation. This results in a molecular shape known as trigonal planar because the atoms form a triangle in one plane. An example of this structure is the methanal (CH_2O) molecule shown in Figure 5.4.7. Trigonal planar molecules have bond angles of 120°.

If the central atom has two double bonds, then the two double bonds repel each other. This results in a linear molecule like carbon dioxide (CO_2), shown in Figure 5.4.8. A linear molecule has a bond angle of 180°.

Finally, if the central atom has a single bond and a triple bond, as in hydrogen cyanide (HCN), then the molecule is also linear (Figure 5.4.9).

INTERMOLECULAR FORCES

Intermolecular forces exist between molecules, and their effects are seen everywhere. If you have ever struggled to flatten out a sheet of cling wrap, then you've experienced the effects of intermolecular forces. Glue is an example of how intermolecular forces can be used to stick things together. The soap bubbles shown in Figure 5.4.10 are also held together by intermolecular forces.

If you take a closer look at liquids, you will notice further evidence of intermolecular bonding. Intermolecular bonds create the **surface tension** that allows insects to walk on water, as shown in Figure 5.4.11

FIGURE 5.4.10 These soap bubbles are held together by intermolecular forces.

FIGURE 5.4.11 The surface tension created by intermolecular forces allows insects to walk on water.

Strength of intermolecular forces

Intermolecular forces are 100 times weaker than the strong bonding found in ionic, metallic and covalent bonds.

The atoms in covalent molecules such as water (H_2O) and carbon dioxide (CO_2) are held together by strong covalent bonds. However, covalent molecular substances tend to have much lower melting and boiling points than ionic, metallic and covalent network substances. This is because the forces between the molecules are much weaker. It is the weak intermolecular forces that are broken when a covalent molecular substance is converted from a solid to a liquid or a liquid to a gas.

In general, covalent molecular substances tend to be softer than ionic or metallic substances. Wax, putty and butter are typical examples of soft, covalent molecular substances. These substances are soft because the intermolecular forces holding the molecules together are weak. Therefore, the bonds between the molecules can be readily broken, and the substances can be easily moulded, scratched or crushed.

Despite having weak intermolecular forces, covalent molecular substances can still form crystals. For example, sugar is a covalent molecular substance that forms crystals. Water also forms crystals of ice. However, most covalent molecular crystals are softer than typical metals and ionic crystals.

> When liquids change to gases, the intermolecular bonds are broken. For example, when liquid water changes to gaseous water, the intermolecular bonds holding the water molecules to other water molecules are broken, not the covalent bonds between the hydrogen and oxygen atoms.

ELECTRONEGATIVITY AND POLARITY

Intermolecular forces are an example of electrostatic forces. You have already seen many other examples of how electrostatic forces form bonds at the atomic level. The attraction between positive and negative ions in an ionic crystal lattice and the attraction between shared electrons and nuclei in a covalent bond are examples of electrostatic forces.

GO TO ➤ Section 5.2 page 147

GO TO ➤ Section 5.3 page 162

The electrostatic attraction between molecules works in a similar way. In intermolecular forces, the electrostatic attraction is between positive and negative charges in the molecules. These charges appear as a result of uneven electron distributions within the molecules. The following sections examine how the shape of the molecule and the electronegativity of its atoms can cause these uneven electron distributions.

GO TO ➤ Section 4.2 page 125

Polarity of diatomic molecules

Electronegativity is the key factor that determines the electron distribution in diatomic molecules. Electronegativity is the tendency of an atom in a covalent bond to attract electrons. Electronegativity increases from left to right across the periods of the periodic table and generally decreases down the groups of the table as shown in Figure 5.4.12. You will remember seeing these patterns in Chapter 4.

Electronegativity increases →

Electronegativity decreases ↓

Key: 2.20 – electronegativity; H – symbol

period \ group	1	2	3	4	5	6	7	8	9	10	11	12	13	14	15	16	17	18
1	2.20 H																	He
2	0.98 Li	1.57 Be											2.04 B	2.55 C	3.04 N	3.44 O	3.98 F	Ne
3	0.93 Na	1.31 Mg											1.61 Al	1.90 Si	2.19 P	2.58 S	3.16 Cl	Ar
4	0.82 K	1.00 Ca	1.36 Sc	1.54 Ti	1.63 V	1.66 Cr	1.55 Mn	1.83 Fe	1.88 Co	1.91 Ni	1.90 Cu	1.65 Zn	1.81 Ga	2.01 Ge	2.18 As	2.55 Se	2.96 Br	Kr
5	0.82 Rb	0.95 Sr	1.22 Y	1.33 Zr	1.6 Nb	2.16 Mo	2.10 Tc	2.2 Ru	2.28 Rh	2.20 Pd	1.93 Ag	1.69 Cd	1.78 In	1.96 Sn	2.05 Sb	2.1 Te	2.66 I	Xe
6	0.79 Cs	0.89 Ba	1.10 La*	1.7 Hf	1.5 Ta	1.7 W	1.9 Re	2.2 Os	2.2 Ir	2.2 Pt	2.4 Au	1.9 Hg	1.8 Tl	1.8 Pb	1.9 Bi	2.0 Po	2.2 At	Rn
7	0.7 Fr	0.9 Ra	1.1 Ac**	Rf	Db	Sg	Bh	Hs	Mt	Ds	Rg	Cn	Nh	Fl	Mc	Lv	Ts	Og

lanthanoids *	1.12 Ce	1.13 Pr	1.14 Nd	1.13 Pm	1.17 Sm	1.2 Eu	1.2 Gd	1.1 Tb	1.22 Dy	1.23 Ho	1.24 Er	1.25 Tm	1.1 Yb	1.27 Lu
actinoids **	1.3 Th	1.5 Pa	1.7 U	1.3 Np	1.3 Pu	1.13 Am	1.28 Cm	1.3 Bk	1.3 Cf	1.3 Es	1.3 Fm	1.3 Md	1.3 No	Lr

FIGURE 5.4.12 Table of electronegativity values. This periodic table shows the electronegativities of the atoms of each element. The electronegativities increase from left to right across the periods and generally decrease down the groups.

Non-polar diatomic molecules

When two atoms form a covalent bond, you can regard the atoms as competing for the electrons being shared between them. If the two atoms in a covalent bond are the same (i.e. have identical electronegativities), then the electrons are shared equally between the two atoms. This is the case for diatomic molecules such as chlorine (Cl_2), oxygen (O_2), hydrogen (H_2) and nitrogen (N_2).

Covalent bonds with an equal distribution of valence (outer shell) electrons are said to be **non-polar** because there is no charge on either end of the molecule.

> **Electron density** is the measure of the probability of an electron being present at a particular location within an atom. In molecules, areas of electron density are commonly found around the nuclei and within the bonds.

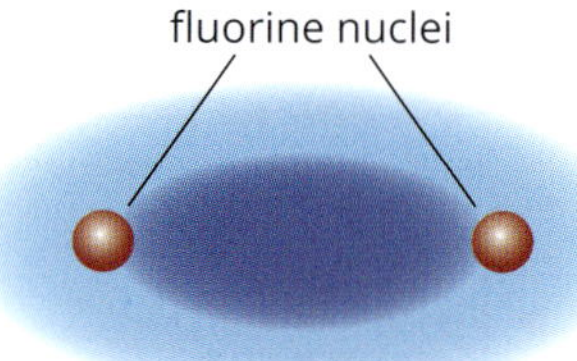

FIGURE 5.4.13 Fluorine molecules have a symmetric distribution of electrons and are therefore non-polar molecules with a non-polar bond.

Figure 5.4.13 shows the electron distribution in the non-polar fluorine molecule (F_2). The molecule has a high electron density between the two fluorine atoms, forming the covalent bond. The valence electrons are distributed evenly between the two atoms, making the bond non-polar.

Polar diatomic molecules

If the covalent bond is between atoms of two different elements, then the electrons will stay closer to the more electronegative atom because it has a stronger pull on the electrons in the bond. An example is the hydrogen fluoride (HF) molecule, shown in Figure 5.4.14. Molecules with an imbalanced electron distribution are said to be **polar**.

A fluorine atom is more electronegative than a hydrogen atom. Therefore, in a hydrogen fluoride molecule, the electrons tend to stay closer to the fluorine atom. This imbalance in the electron distribution means the fluorine atom is negatively charged and the hydrogen atom is positively charged. A molecule with two oppositely charged poles is said to be polar. The fluorine atom is described as having a partial negative charge, which is represented by the Greek letter delta and a minus sign (δ–). The hydrogen atom is described as having a partial positive charge, δ+. The separation of the positive and negative charges is known as an **electric dipole**, or simply a **dipole**, because they have two oppositely charged poles, one each end of the molecule.

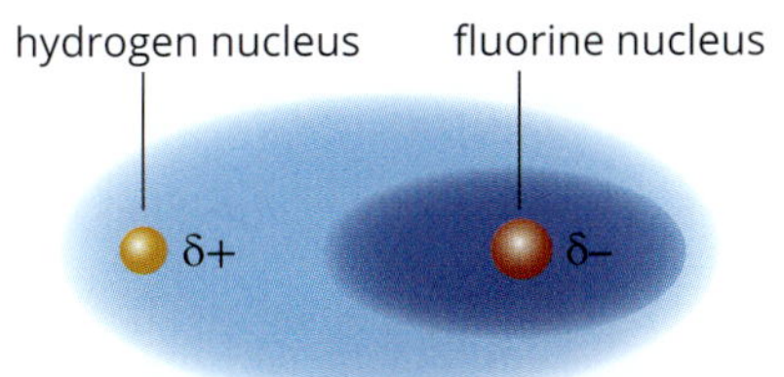

FIGURE 5.4.14 The electron distribution in hydrogen fluoride is asymmetric because of the different electronegativities of the hydrogen and fluorine atoms. Hydrogen fluoride is an example of a polar molecule with a polar bond.

Hydrogen fluoride is polar because it has a permanent dipole created by the different electronegativities of the two atoms. All diatomic molecules that are made up of atoms with different electronegativities are polar to some extent. The level of **polarity** will depend on the difference between the electronegativities of the two atoms. The greater the difference between the electronegativities, the greater the polarity of the molecule.

The partial charges on polar molecules are different from the charges on ions. The partial charges on a polar molecule will always add to give a total charge of zero since the charges on polar molecules are due to the unequal sharing of electrons between atoms. However, ions have an overall charge. The partial charges on a polar molecule are also a lot smaller than the charges on ions. That is why intermolecular bonds are much weaker than ionic bonds.

It is not just the covalent bonds in diatomic molecules that can be polar. Covalent bonds in larger molecules can have some degree of polarity. The polarity of any covalent bond can be compared by examining the difference in the electronegativities of the atoms involved in the bond.

As the difference in electronegativities of two atoms increases, the covalent bond formed between them increases in polarity. In other words, it becomes more like an ionic bond.

Worked example 5.4.2

COMPARING THE POLARITY OF COVALENT BONDS

Compare the polarity of the covalent bonds in hydrogen fluoride (HF) and carbon monoxide (CO).

Thinking	Working
Use the table of electronegativity values in Figure 5.4.12 on page 176 to find the electronegativities of the atoms in each molecule.	HF: hydrogen 2.20; fluorine 3.98 CO: carbon 2.55; oxygen 3.44
For each molecule, subtract the lowest electronegativity value from the highest value.	HF: 3.98 – 2.20 = 1.78 CO: 3.44 – 2.55 = 0.89
Determine which molecule has the biggest difference in electronegativity to determine the more polar molecule.	HF is more polar than CO.

Worked example: Try yourself 5.4.2

COMPARING THE POLARITY OF COVALENT BONDS

Compare the polarity of the covalent bonds in nitrogen monoxide (NO) and hydrogen chloride (HCl).

CHEMFILE CCT N

Mixing ethanol and water

Add 250 mL of ethanol to 250 mL of water in a measuring cylinder. You would be right to suppose that the combined volume would be 500 mL. However, the final volume is likely to be around 480 mL, as shown in Figure 5.4.17. Both liquids are highly polar and the interaction of the hydrogen bonds between the water and the ethanol molecules leads to the total volume being less than the combined individual volumes. This can make the preparation of solutions of known concentration difficult. Added to the problem is the fact that the size of the volume change varies with the temperature.

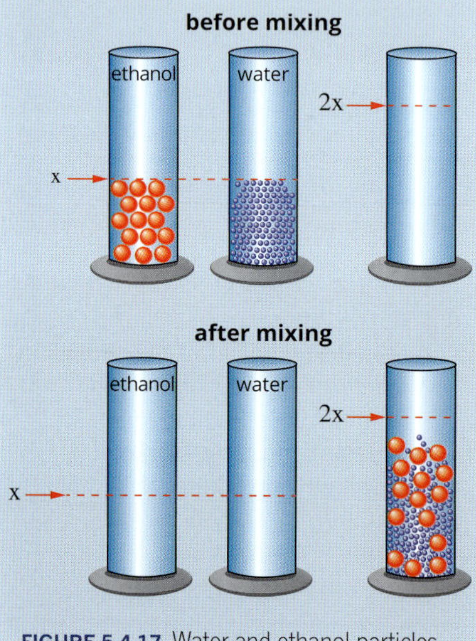

FIGURE 5.4.17 Water and ethanol particles bond strongly to each other. The volume when they are mixed is less than the combined individual volumes.

Polarity of polyatomic molecules

Determining the polarity of molecules with more than two atoms is a little more complicated. This is because the polarity of polyatomic molecules depends on the shape of the molecule as well as on the polarity of the covalent bonds. As a general rule:

- **symmetrical molecules** are non-polar
- **asymmetrical molecules** are polar.

Non-polar molecules

Even molecules with polar covalent bonds can be non-polar if the molecule is symmetrical.

In methane, the carbon atom is slightly more electronegative than the hydrogen atoms. Therefore, the carbon atom attracts the electrons more strongly than each hydrogen atom and has a partial negative charge, leaving hydrogen with a partial positive charge. However, the methane molecule has a tetrahedral shape and is therefore symmetrical. The symmetry of the molecule means that the individual dipoles of the covalent bonds cancel each other perfectly (Figure 5.4.14 on page 177). The result is a molecule with no overall dipole. It is non-polar.

The methane molecule shown in Figure 5.4.15 is an example of a non-polar molecule with polar covalent bonds.

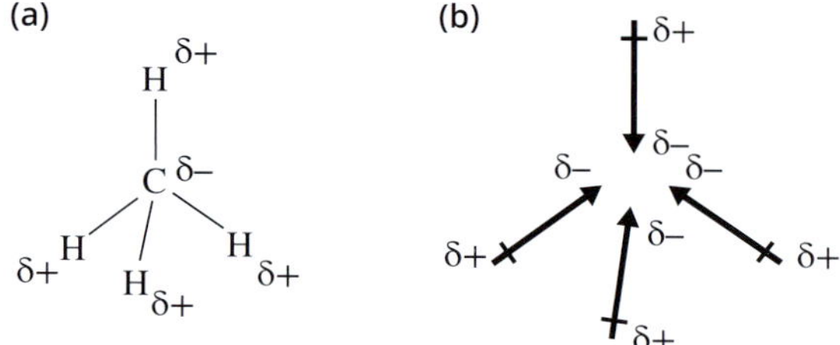

FIGURE 5.4.15 (a) Structure of a methane molecule showing the partial charges on the atoms. (b) The individual dipoles are distributed symmetrically around the molecule, resulting in a non-polar molecule.

Polar molecules

In asymmetrical molecules, the individual dipoles of the covalent bonds do not cancel one another. This results in a net dipole and the overall molecule is polar.

The chloromethane molecule shown in Figure 5.4.16 is an example of an asymmetric molecule. The chlorine atom is more electronegative than the carbon atom. Therefore, the chlorine atom attracts electrons from the carbon atom while the carbon atom attracts electrons from the hydrogen atoms. The individual dipoles of the covalent bonds are shown in the central image in Figure 5.4.16. These add to give the molecule a net or overall dipole.

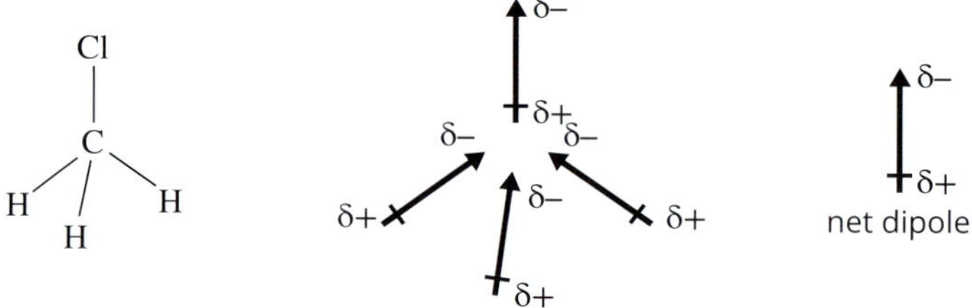

FIGURE 5.4.16 Chloromethane is a polar molecule because the individual dipoles of the covalent bonds add to give a net dipole.

Table 5.4.1 shows some more examples of how symmetry determines the polarity of covalent molecules.

TABLE 5.4.1 Structure, symmetry and polarity of selected covalent molecules

Molecule	Structural formula	Symmetrical/ asymmetrical	Polar/non-polar
methanal (formaldehyde)	$(H^{\delta+})_2C{=}O^{\delta-}$	asymmetrical	polar
carbon dioxide	$O^{\delta-}{=}C^{\delta+}{=}O^{\delta-}$	symmetrical	non-polar
tetrafluoromethane	$C^{\delta+}(F^{\delta-})_4$	symmetrical	non-polar
water	$H^{\delta+}{-}O^{\delta-}{-}H^{\delta+}$	asymmetrical	polar
ammonia	$N^{\delta-}(H^{\delta+})_3$	asymmetrical	polar

The polarity of a molecular substance has a significant impact on its solubility. In general, substances that are polar tend to dissolve in polar solvents. Non-polar substances tend to dissolve in non-polar solvents. You can remember this with the saying 'like dissolves like'.

Generally, like dissolves like: polar solvents dissolve polar substances and non-polar solvents dissolve non-polar substances.

THE SPECTRUM OF BOND TYPES

So far, this chapter has described the different types of bonding that can exist between atoms. The nature of a chemical bond can be predicted by considering the electronegativity difference between the atoms involved in bond formation.

There is a spectrum (continuum) of bonding types, ranging from non-polar covalent to polar covalent to ionic. The type of bonding depends upon the extent to which electrons are shared. This, in turn, depends upon the electronegativity difference between the atoms involved in the bond formation.

When the electronegativity difference between two atoms is zero, the bonding electrons are shared equally, and the bond formed between the two atoms is covalent and non-polar. For example, in molecular fluorine (F_2) both atoms have the same electronegativity and share the bonding electrons equally. As a result, the covalent bond between the two fluorine atoms is non-polar. If two different elements have the same electronegativity, the covalent bond between them will also be non-polar. For example, the electronegativity of both carbon and sulfur is 2.6. The covalent bonds between atoms of carbon and sulfur in carbon disulfide (CS_2) are non-polar because the electrons are shared equally (Figure 5.4.18).

S=C=S

FIGURE 5.4.18 The equal sharing of bonding electrons between the sulfur and carbon atoms results in a non-polar bond.

As you have learnt in this section, any difference in electronegativity between two bonded atoms will result in an unequal or partial sharing of electrons and the bond will be polar. The bonding pair of electrons in hydrogen fluoride (HF) is not shared equally and on average stays closer to the more electronegative fluorine atom. As a result, HF is a dipole, and the bonding in HF is described as polar covalent.

The polar nature of a bond between two atoms increases as the electronegativity difference between these atoms increases. The bonding in HF is more polar than in HCl because the fluorine atom is more electronegative than the chlorine atom.

Electrons will be transferred between two atoms if the electronegativity difference between them is great enough. As you discovered in Section 5.2, electron transfer between metallic and non-metallic atoms results in the formation of ionic bonds. The fluorine atom is much more electronegative than the sodium atom. When fluorine and sodium react, the sodium atom's valence electron is transferred to the outer shell of the fluorine atom. This results in the formation of Na^+ and F^- ions. The bonding between the ions is described as ionic. The spectrum of bond types is illustrated in Figure 5.4.19.

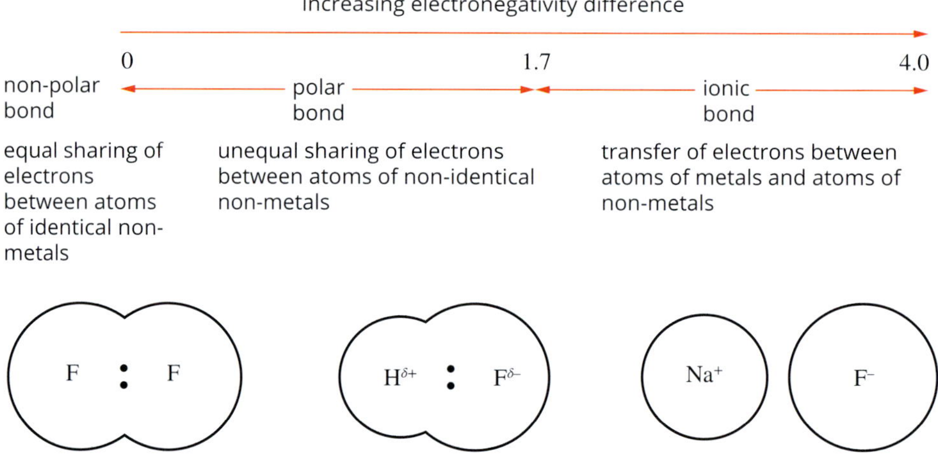

FIGURE 5.4.19 The spectrum of bond types. The polar and ionic nature of bonds increases with increasing electronegativity difference.

As a rule of thumb, an electronegativity difference of 1.7 is often used as the cut-off point between polar covalent bonding and ionic bonding (Table 5.4.2). In fact, the distinction between polar covalent and ionic bonding at this level of electronegativity is somewhat indistinct. Compounds that have an electronegativity difference between their atoms of around 1.7 should be considered as having both polar covalent and ionic bonding characteristics.

TABLE 5.4.2 Electronegativity difference and bond type

Electronegativity difference	Distribution of bonding electrons	Type of bond	Example
zero	electrons shared equally	non-polar covalent	H_2, Cl_2, CS_2
less than 1.7	electrons attracted to the more electronegative atom	polar covalent	NH_3, H_2O, HCl
greater than 1.7	electrons transferred to the more electronegative atom	ionic	NaCl, CaF_2

TYPES OF INTERMOLECULAR FORCES

Many factors determine the strength of intermolecular forces, including the size, shape and polarity of molecules. These factors not only determine the strength of the intermolecular forces in a substance, they also determine the types of intermolecular forces.

There are three main types of intermolecular forces:

- dipole–dipole forces
- hydrogen bonding
- dispersion forces.

In this section, you will examine the nature of these three types of intermolecular forces and their role in determining the physical properties of covalent molecular substances.

Dipole–dipole forces

Dipole–dipole forces only occur in polar molecules. These forces result from the electrostatic attraction between the positive and negative ends of the polar molecules, as shown in Figure 5.4.20.

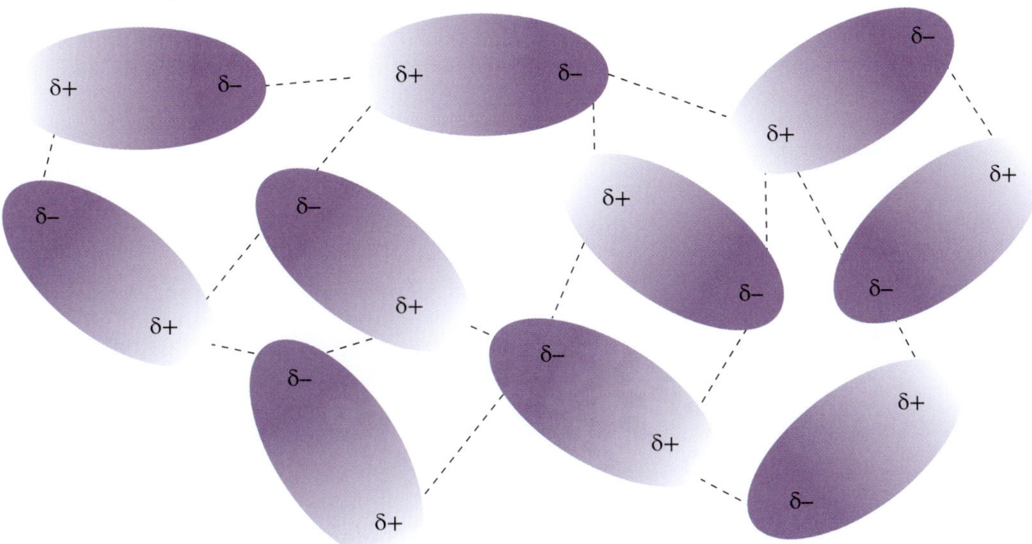

FIGURE 5.4.20 The positive and negative ends of polar molecules attract one another.

Dipole–dipole forces are relatively weak since the partial charges (the δ+ and δ–) on the molecules are small. However, the more polar a molecule is, the stronger the dipole–dipole forces are. The polarity will be larger when there is a large difference between the electronegativities of the atoms, or a large asymmetry in the shape of the molecule.

The strength of the dipole–dipole forces in molecules is directly related to the melting and boiling points of the substance. The stronger the dipole–dipole forces, the higher the melting and boiling points. This is because dipole–dipole forces bond the molecules together in the solid or liquid. Stronger dipole–dipole forces require more energy (i.e. higher temperatures) to break the bonds in order to change the solid to a liquid or the liquid to a gas.

As an example, we can compare the molecules of methanal (CH_2O) and ethane (CH_3CH_3), as shown in Figure 5.4.21. Methanal is asymmetrical and is therefore a polar molecule. This results in dipole–dipole forces between the molecules. Methanal's boiling point is –19°C. In contrast, the ethane molecule is symmetrical and therefore non-polar. This means there are no dipole–dipole forces and its boiling point is much lower at –88.5°C.

Molecules are more polar when there is a greater difference between the electronegativities of the atoms, or a large asymmetry in the shape of the molecule. Molecules that are more polar have stronger dipole–dipole forces between them.

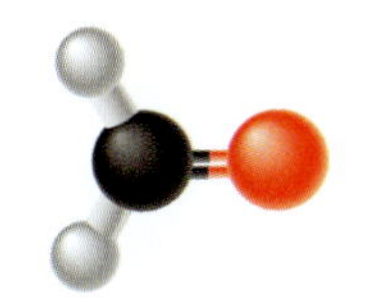

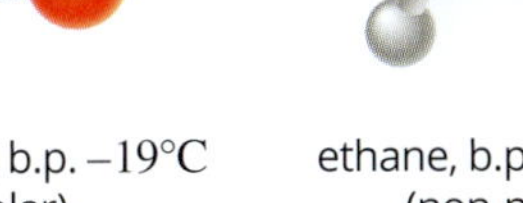

methanal, b.p. –19°C (polar) | ethane, b.p. –88.5°C (non-polar)

FIGURE 5.4.21 Molecular structures and boiling points (b.p.) of methanal and ethane.

Hydrogen bonding occurs in molecules where a hydrogen atom is bonded to an oxygen, a nitrogen or a fluorine atom.

Hydrogen bonding

Hydrogen bonding is a special form of dipole–dipole force. Hydrogen bonding only occurs between molecules in which a hydrogen atom is covalently bonded to an oxygen, a nitrogen or a fluorine atom.

Oxygen, nitrogen and fluorine atoms are small and highly electronegative. Therefore, they strongly attract the electron(s) within a covalent bond. This creates a significant partial positive charge on any hydrogen atoms bonded to these atoms. Remember that hydrogen only has one electron, and this electron is strongly attracted to the N, O or F atom in the covalent bond. The exposed positively charged hydrogen (proton) is attracted to a lone pair on a nitrogen, oxygen or fluorine atom in another molecule, as shown in Figure 5.4.22.

The result is a relatively strong intermolecular bond that is approximately ten times stronger than a dipole–dipole bond, but about one tenth the strength of an ionic or a covalent bond.

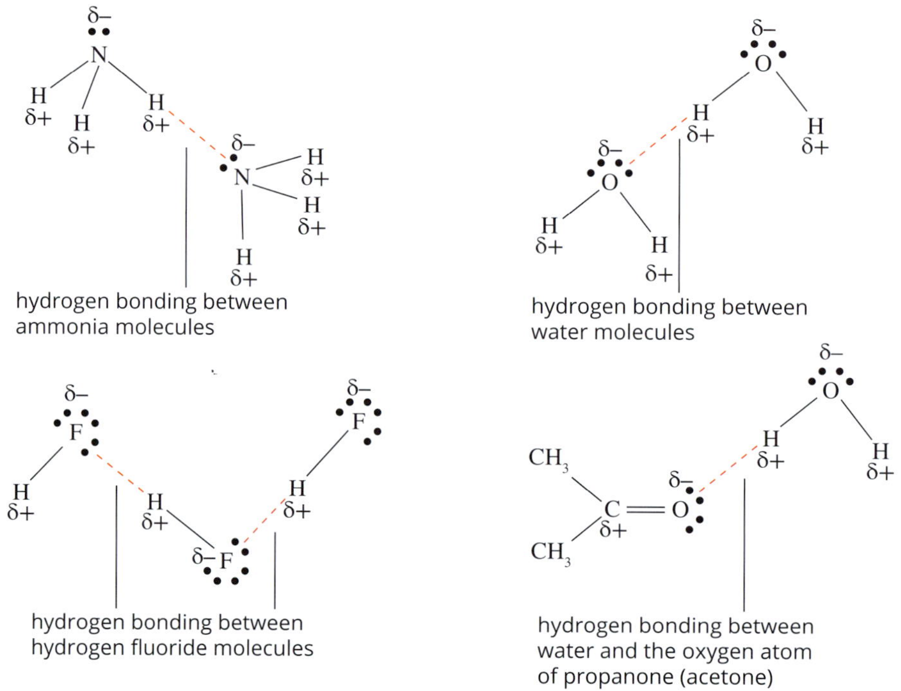

FIGURE 5.4.22 Examples of hydrogen bonding between molecules.

The presence of hydrogen bonds results in higher melting and boiling points. Figure 5.4.23 demonstrates the effect of hydrogen bonding on boiling point by comparing methanol with ethane or methanal. Recall that ethane is non-polar, so it has no dipole–dipole forces between molecules. Methanal has dipole–dipole forces but does not have hydrogen bonding. Methanol has hydrogen bonding between molecules because of the hydrogen atom attached to the oxygen atom. As a result, the boiling point of methanol (64.7°C) is significantly higher than the boiling points of methanal (−19°C) and ethane (−88.5°C).

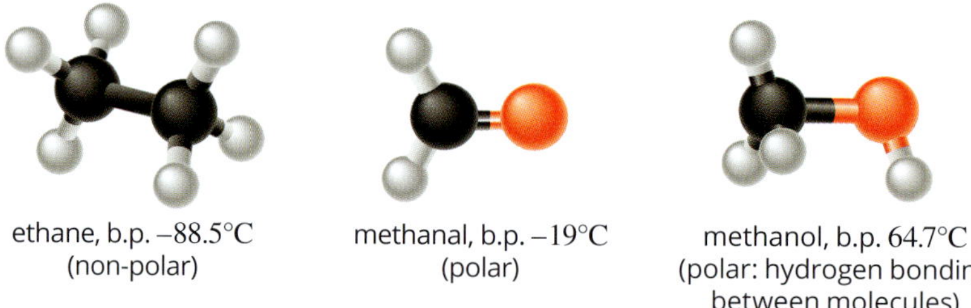

FIGURE 5.4.23 Methanol has a higher boiling point (b.p.) than ethane or methanal because hydrogen bonds can form between molecules of methanol.

The two key requirements for hydrogen bonding are:

- a hydrogen atom covalently bonded to an oxygen, a nitrogen or a fluorine atom
- a lone pair of electrons on the nitrogen, oxygen or fluorine atoms of neighbouring molecules.

It is both the very high electronegativities and the small atomic radii of oxygen, nitrogen and fluorine atoms that cause the formation of the hydrogen bonds. Chlorine atoms also have a very high electronegativity, but do not form hydrogen bonds because of having a larger atomic radius, which reduces the concentration of the negative charge around the atom.

Hydrogen bonding in water

Water is a very common substance, but it has some unusual properties. One unusual property of water is that ice floats. Ice floats because it is less dense than liquid water. For most other substances, the solid form is denser than the liquid.

The fact that ice is less dense than liquid water can be explained by hydrogen bonding. Water has two hydrogen atoms attached to an oxygen atom in a V-shape. Therefore, the hydrogen atoms can form hydrogen bonds with the lone pairs of electrons on the oxygen atoms of neighbouring molecules, as shown in Figure 5.4.24. In this way, a water molecule can form hydrogen bonds with four other water molecules.

The hydrogen bonding holds the water molecules in ice in a regular crystal network or lattice. In this lattice, the molecules are held further apart than in liquid water, as shown in Figure 5.4.25. As a result, ice is less dense and therefore floats on liquid water.

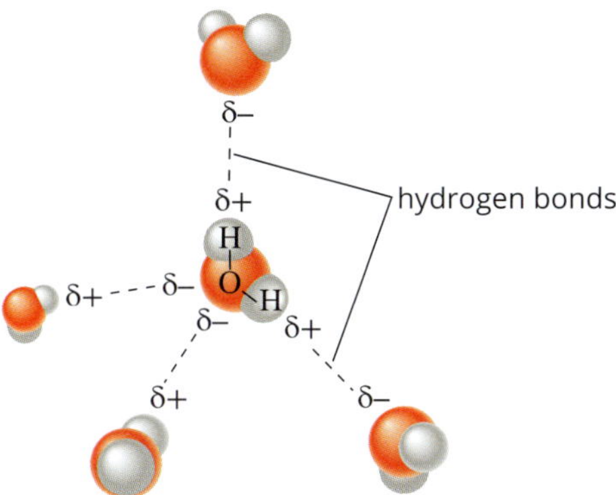

FIGURE 5.4.24 Each water molecule can form hydrogen bonds with four other water molecules. The two lone pairs of electrons on the molecule bond to hydrogen atoms on two neighbouring molecules. In addition, the two hydrogen atoms on the molecule form bonds with lone pairs of electrons of two neighbouring molecules.

FIGURE 5.4.25 Ice floats because the water molecules form a crystal lattice in which the molecules are spaced more widely apart than in liquid water. This arrangement means ice is less dense than liquid water.

When ice melts and forms liquid water, the density of the water increases. This is because the open crystal lattice collapses and the water molecules pack together more tightly.

However, as the temperature of the water increases further, water molecules in the liquid move and vibrate more rapidly. The movement causes the molecules to spread further apart. As the molecules move further apart, the liquid becomes less dense.

Dispersion forces

Dipole–dipole forces and hydrogen bonding explain the intermolecular forces in polar substances. However, they do not explain the existence of intermolecular forces in non-polar substances.

You now know that intermolecular forces are present in non-polar substances because non-polar substances form liquids and solids. Without intermolecular forces, there would be nothing to hold the molecules together and non-polar substances would only exist as gases. However, there are many non-polar compounds that are liquids at room temperature (such as vegetable oil), and even non-polar solids (such as candle wax). In fact, all non-polar substances form liquids or solids if cooled to a low enough temperature. Even hydrogen liquefies below −259°C.

The forces of attraction between non-polar molecules are known as **dispersion forces**. Dispersion forces are caused by **temporary dipoles** in the molecules that are the result of random movement of the electrons surrounding the molecule. These temporary dipoles are also known as **instantaneous dipoles**.

Dispersion forces are always present between molecules, no matter whether they are polar or non-polar, because electrons are constantly in motion within atoms (Figure 5.4.26).

(a) In a molecule, the electrons are constantly moving. In the case of a non-polar molecule, such as fluorine, the electrons spend an equal amount of time around each atom.

(b) Occasionally, the electrons gather more closely together at one end of the molecule, causing one end of the molecule to become negative and the other end to become positive. This is known as a temporary dipole.

(c) These temporary dipoles can then induce (create) dipoles in the neighbouring molecules.

(d) The neighbouring molecules then induce dipoles in their neighbours and so on. The temporary dipoles attract each other to create the intermolecular forces known as dispersion forces.

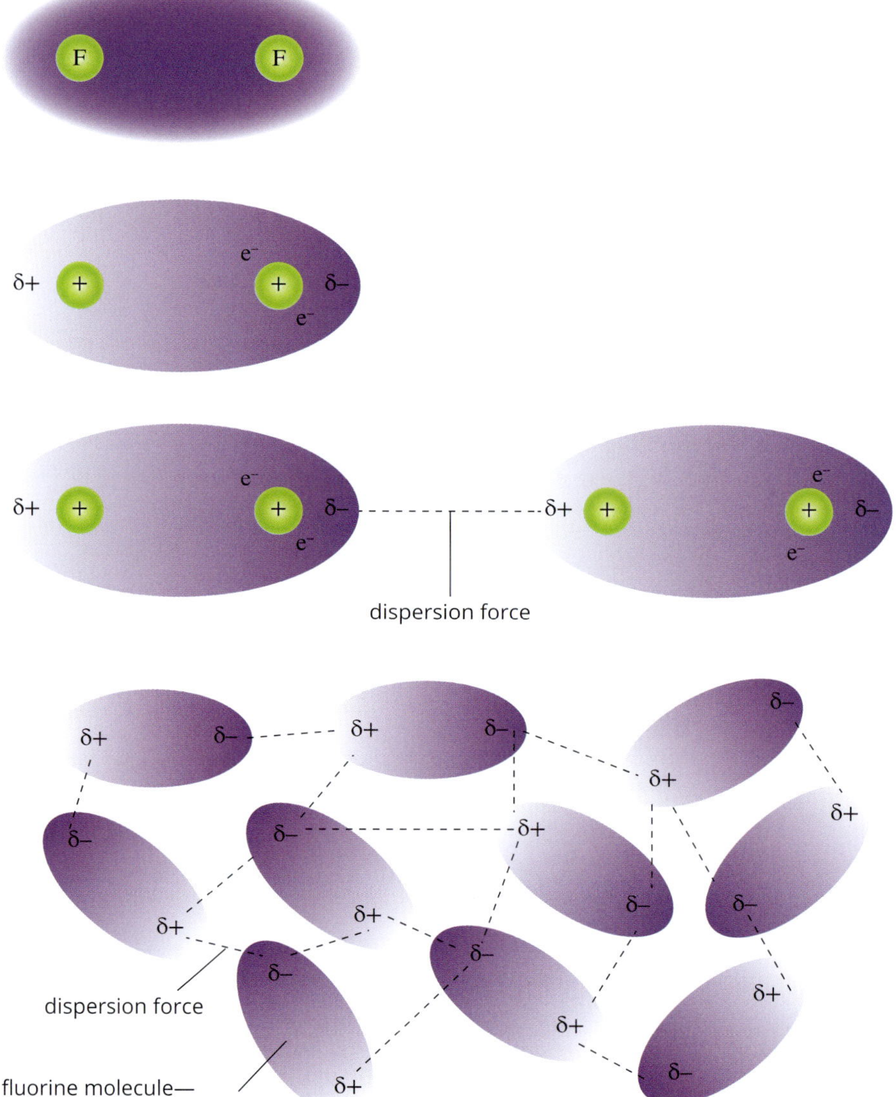

FIGURE 5.4.26 How dispersion forces form within non-polar substances. These diagrams show the forces forming between fluorine molecules.

Strength of dispersion forces

The strength of dispersion forces increases as the size of the molecule increases. This is because larger molecules have a larger number of electrons, and it is easier to produce temporary dipoles in molecules with large numbers of electrons. Since larger molecules have stronger dispersion forces, they have higher melting and boiling points.

Table 5.4.3 shows the boiling points of the halogens (group 17), which all form non-polar, diatomic molecules. The only forces between their molecules are dispersion forces. You can see that as the molecules increase in size and the dispersion forces become stronger, the boiling points of the substances increase.

TABLE 5.4.3 The effect of dispersion forces and the size of atoms on the boiling points of some molecules

Molecule	Molecular mass	Number of electrons	Boiling point (°C)
fluorine (F_2)	38.00	18	−188
chlorine (Cl_2)	70.90	34	−35
bromine (Br_2)	159.8	70	59
iodine (I_2)	253.8	106	184

The shape of a molecule also influences the strength of the dispersion forces. Molecules that form long chains will tend to have stronger dispersion forces than more compact molecules with similar numbers of electrons.

For example, butane and 2-methylpropane (Figure 5.4.27) both contain four carbon atoms and 10 hydrogen atoms. The boiling point of butane is −0.5°C, while the boiling point of 2-methylpropane is −11°C. The higher boiling point of butane is because of the different shapes of the two molecules: butane is a long molecule, whereas 2-methylpropane is compact. Being less compact and long means butane has more contact area over which to interact with its neighbouring molecules to form stronger dispersion forces.

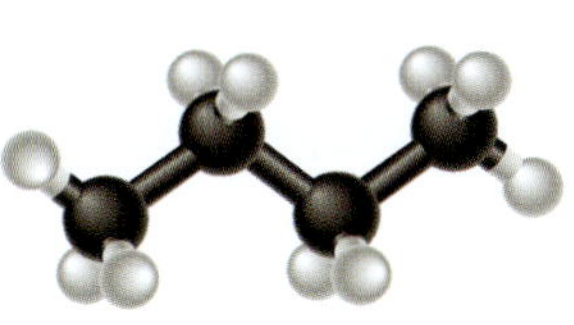

butane, C_4H_{10} (b.p. −0.5°C)

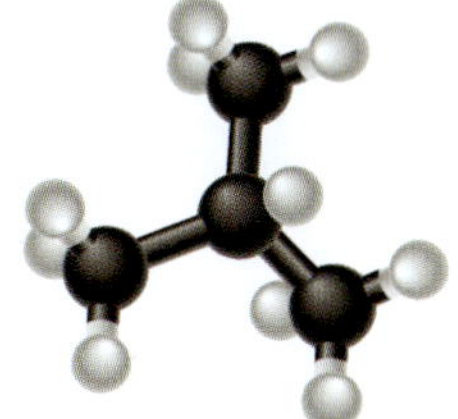

2-methylpropane, C_4H_{10} (b.p. −11°C)

FIGURE 5.4.27 Butane and 2-methylpropane have different boiling points (b.p.) because their molecules are different shapes.

It is important to remember that dispersion forces occur in both polar and non-polar molecular substances. For large molecules, the dispersion forces can even dominate over the dipole–dipole forces and hydrogen bonding.

Table 5.4.4 shows the boiling points of three polar molecules: hydrogen chloride (HCl), hydrogen bromide (HBr) and hydrogen iodide (HI). Hydrogen chloride is the most polar of these molecules and therefore has the strongest dipole–dipole forces. However, hydrogen iodide has the highest boiling point because it is a larger molecule and therefore has stronger dispersion forces.

TABLE 5.4.4 Comparison of the boiling points of three polar molecules

Substance	Molecular mass	Number of electrons	Boiling point (°C)
hydrogen chloride (HCl)	36.5	18	−85.1
hydrogen bromide (HBr)	81.0	36	−66.8
hydrogen iodide (HI)	127.9	54	−35.4

The boiling point of hydrogen fluoride (19.5°C) is much higher than any of these other compounds. This is because the hydrogen bonding between hydrogen fluoride molecules is much stronger than both the dispersion forces and the dipole–dipole forces between the other molecules listed in Table 5.4.4.

CHEMFILE AHC

Use of beeswax by Indigenous Australians

Indigenous Australians have traditionally used the properties of beeswax to make useful objects. They have used beeswax to make mouth pieces for didgeridoos, preserve string, make toys, preserve rock art and to attach stone tools such as axe heads to wooden handles. They have also used beeswax to seal woven bags (Figure 5.4.28). These conical bags are used for gathering food and in male ceremonies. Conical bags coated in beeswax are suited to collecting liquids, such as native honey.

FIGURE 5.4.28 Beeswax is used to seal this conical bag.

Beeswax is a soft substance, and it melts at about 65°C. It is made up of a mixture of long-chain molecules containing many carbon and hydrogen atoms. The typical formula of a molecule of beeswax is $C_{46}H_{92}O_2$.

Beeswax will soften when warmed in your hand, making it easy to mold into a desired shape. Dispersion forces are the main type of bonding between wax molecules. As the wax is warmed, the dispersion forces are broken, enabling the long-chain molecules to slide past one another. As the wax cools, new dispersion forces are formed between adjacent molecules, locking the piece of wax into a new shape.

5.4 Review

SUMMARY

- The shapes of molecules play an important role in determining the intermolecular forces between them, which are related to the physical properties of the substances.
- The shapes of simple molecules can be predicted by the valence-shell electron-pair repulsion (VSEPR) theory.
- The VSEPR theory is based on the principle that electron pairs in the outer shell of an atom repel each other. As a consequence, these electron pairs are arranged as far away from each other as possible.
- Electron pairs can exist in covalent bonds or as lone pairs of electrons. Lone pairs and covalent bonds repel each other.
- Lone pairs influence a molecule's shape, but are not considered when describing the shape.
- Many of the physical properties of covalent molecular substances are determined by intermolecular forces. These properties include melting point, boiling point and hardness.
- Intermolecular forces are 100 times weaker than ionic, covalent and metallic bonds.
- The shape and polarity of molecules are factors in determining the strength of intermolecular forces.
- As the difference in electronegativities of two atoms in a molecule increases, the covalent bond increases in polarity.
- Diatomic molecules containing the same type of atom are non-polar.
- In general, symmetrical molecules are non-polar and asymmetrical molecules with polar bonds are polar.
- Covalent, polar and ionic bonding are part of a spectrum of bond types. Each bond type is determined by the electronegativity difference between the atoms involved in the bond formation.
- The melting and boiling points of covalent molecular substances increase as the strength of the intermolecular bonding increases.
- There are three main types of intermolecular bonds: dipole–dipole forces, hydrogen bonds and dispersion forces.
- Dipole–dipole forces are only present between polar molecules and are the result of the attraction between the positive and negative ends of the molecules.
- The polarity of polyatomic molecules depends on the electronegativity of the atoms in the molecule and the symmetry of the molecule.
- The greater the polarity of a molecule, the stronger the dipole–dipole forces.
- Hydrogen bonding occurs between highly polar molecules in which hydrogen atoms are covalently bonded to an oxygen, a nitrogen or a fluorine atom.
- A hydrogen bond is formed between the hydrogen atom in one molecule and lone pairs of electrons in an oxygen, a nitrogen or a fluorine atom in another molecule.
- Hydrogen bonding is the reason why ice floats in liquid water. As a consequence of the hydrogen bonds between the water molecules, ice forms a crystal network or lattice in which the molecules are spaced further apart than in liquid water. The greater spacing of molecules makes ice less dense than liquid water, and so it floats on water.
- Dispersion forces occur between polar and non-polar molecules.
- Dispersion forces are the result of attraction between temporary dipoles that form in molecules.
- Temporary dipoles are due to random fluctuations in the distribution of electrons in molecules.
- Dispersion forces are stronger between larger molecules because it is easier to create temporary dipoles in molecules with a larger number of electrons.
- Hydrogen bonds are the strongest of the three main types of intermolecular forces.

KEY QUESTIONS

1 Describe the shape of each of the following molecules.
 a H_2S
 b HI
 c CCl_4
 d PH_3
 e CS_2

2 Covalent bonds can form between the following pairs of elements in a variety of compounds. Use the electronegativity values given in Figure 5.4.12 on page 176 to identify the atom in each pair that would have the greater share of bonding electrons.
 a S and O
 b C and H
 c C and N
 d N and H
 e F and O
 f P and F

3 Label the following diagram of a tetrachloromethane molecule to show the dipoles between the atoms.

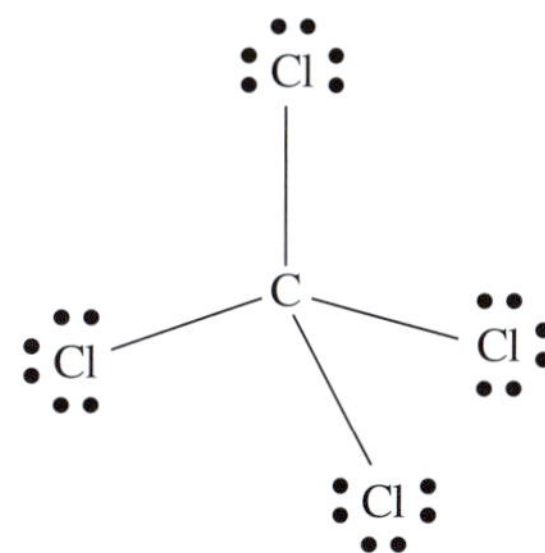

4 Examine the symmetry of each these general diagrams of molecular structures, and determine whether the molecules are likely to be polar or non-polar. The white, black and red circles represent atoms with different electronegativities.

a 

b, c 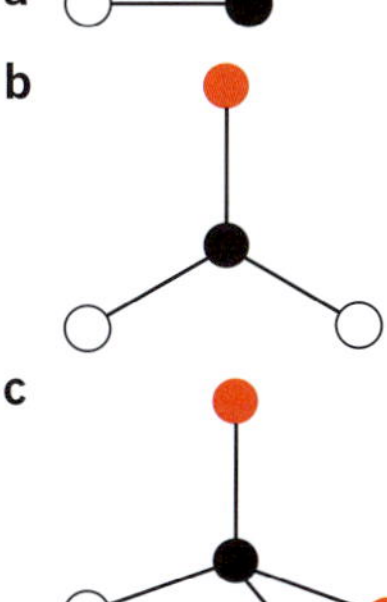

d 

e 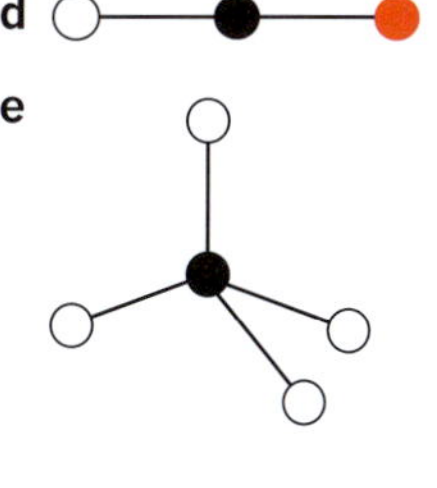

5 Identify the type of bonding that exists between atoms in
 a sodium bromide
 b hydrogen bromide
 c bromine (Br_2)

6 Identify which of the following substances would contain dipole–dipole forces between their molecules:
 A bromine (Br_2)
 B hydrogen chloride (HCl)
 C methane (CH_4)
 D tetrachloromethane (CCl_4)
 E chloromethane (CH_3Cl)

7 Refer to Figure 5.4.12 on page 176 for electronegativity values and select which one of the following substances would have the strongest dipole–dipole forces between the molecules in the liquid state.
 A fluorine (F_2)
 B carbon monoxide (CO)
 C hydrogen chloride (HCl)
 D methane (CH_4)

8 Consider the following substances. Identify whether the molecules are attracted to one another by dipole–dipole forces or hydrogen bonds.
 a NH_3
 b $CHCl_3$
 c CH_3Cl
 d F_2O
 e HBr
 f H_2S
 g HF
 h H_2O
 i H_2

5.5 Covalent network structures

While most covalent substances exist as individual molecules, there are some, such as diamond and graphite, that exist as continuous three-dimensional structures called **covalent network structures**. In general, substances that have a network structure have very high melting points or decomposition temperatures. They are also very hard because the atoms are held firmly in fixed positions in the network. The properties of some substances that have a network structure are shown in Table 5.5.1. As can be seen from the high melting points and hardness, the bonds between the particles are very strong. A great deal of energy is required to break these bonds. The formulae of covalent network compounds are written as the simplest ratio of the elements present in the network.

FIGURE 5.5.1 Diamonds are popular as a jewellery item. Their high refractive index makes diamonds sparkle.

TABLE 5.5.1 The properties of covalent networks

Material	Formula	Melting temperature (°C)	Hardness (Mohs scale)**
diamond*	C	>3350	10
graphite*	C	3974	1–2
silicon	Si	1410	6–7
silicon carbide*	SiC	2200	9.5
silicon dioxide	SiO_2	1700	7

* These materials sublime when heated at normal pressure.

** Mohs scale compares the hardness of materials on a scale from 1 to 10. Talc, a soft clay, is assigned a hardness value of 1; diamond, the hardest naturally occurring substance known, is rated as 10.

FIGURE 5.5.2 Diamond-tipped drills, which are used to drill through rock, mining for oil and natural gas.

ALLOTROPES OF CARBON

Diamonds (Figure 5.5.1) might be a 'girl's best friend', but it is unlikely that **graphite** (Figure 5.5.4) will ever be held in the same esteem. Both of these minerals are made of the same single element—carbon. Graphene and fullerenes are new materials that are also allotropes of carbon.

Diamond

The crystalline appearance of diamonds (Figure 5.5.1) and their high refractive index make them sparkle and has made them extremely popular as jewellery.

Diamond is the hardest naturally occurring material known. While the use of diamonds in jewellery is well known, many industrial cutting and drilling tools for working with tough materials are diamond-tipped. The drill tips in Figure 5.5.2 are used to drill through rock the mining for oil and natural gas. They contain small pieces of diamond that improve the hardness and durability of the tool.

Diamond does not contain small, discrete (individual) molecules. Instead, the carbon atoms bond with each other to form a continuous three-dimensional covalent network structure. There are no weak intermolecular forces present, only strong covalent bonds. This is what gives diamond its strength.

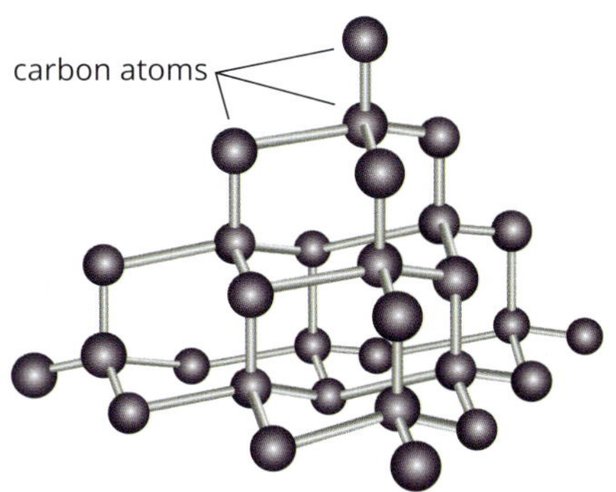

FIGURE 5.5.3 The structure of diamond, showing each carbon atom with four single covalent bonds to neighbouring atoms.

> Diamond is made up of carbon atoms that bond with four neighbouring carbon atoms, forming a covalent network structure. This structure makes diamond extremely hard.

As you saw in Section 5.4, when an atom has four electron pairs in its outer shell, the electron pairs position themselves as far away from each other as possible in a tetrahedral shape. In the covalent network structure for diamond shown in Figure 5.5.3, you can also see that individual atoms within diamond form single covalent bonds to four other carbon atoms in a tetrahedral arrangement.

The structure of diamond is directly related to its properties.

- Single covalent bonds between carbon atoms are strong bonds. The entire structure of diamond consists of a continuous network of these bonds, making diamond very hard and rigid.
- There are no small molecules in diamond, so there are no weak forces between the atoms. There are only strong covalent bonds between carbon atoms, and this makes the sublimation point very high (about 3500°C).
- The rigidity means that diamond is brittle, and it breaks rather than bends.
- Diamond does not conduct electricity because it does not contain any charged particles that are free to move.
- Because the atoms in diamond are held together very strongly, the thermal conductivity is extremely high. It is five times greater than that of copper, leading to some specialty electronic uses, in which diamond is used to transfer heat away from important electrical components.

Graphite

As you can see in Figure 5.5.4, graphite has a very different appearance to diamond. Figure 5.5.5, illustrates how the carbon atoms in graphite are in layers. There are strong covalent bonds between the carbon atoms in each layer. However, there are weak dispersion forces between the layers. As a consequence, graphite is hard in one direction, but quite slippery and soft in another direction. The structure of graphite is referred to as a **covalent layer network**.

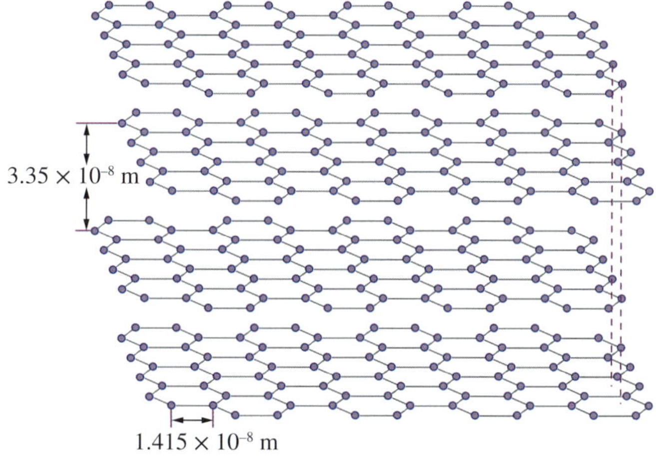

FIGURE 5.5.5 Graphite has a covalent layer network structure. The carbon atoms within each layer are covalently bonded to one another. Weak dispersion forces exist between the layers.

FIGURE 5.5.4 Natural graphite is soft and black.

In graphite, each carbon atom is covalently bonded to three other carbon atoms. The layered network structure contains delocalised electrons. The bonds within the layers are strong, but the bonds between layers are weak dispersion forces.

The covalent layer network structure of graphite also explains some of its other properties:

- The strong covalent bonds between the atoms in each layer explain graphite's resistance to heat. Graphite **sublimes** at a temperature of about 3600°C.
- Each carbon atom is bonded to three other carbon atoms. The fourth valence electron from each atom is able to move within the layer. Graphite can conduct electricity due to these delocalised electrons.

The conductivity of graphite makes it suitable for applications such as in battery electrodes where conductivity is required but a metal is not suitable.

Graphite can also be used as a lubricant. The weak dispersion forces between layers allow these layers to slide over one another and to reduce the friction between moving parts, such as in locks or machinery. For the same reason, graphite is also used in black lead pencils. Graphite is also used as an additive to improve the properties of rubber products, and it can be woven into a fibre that can help to reinforce plastics. Figure 5.5.6 shows spun graphite fibre, which can be used to make strong composite materials, such as those used in tennis racquets, fishing rods and racing car shells.

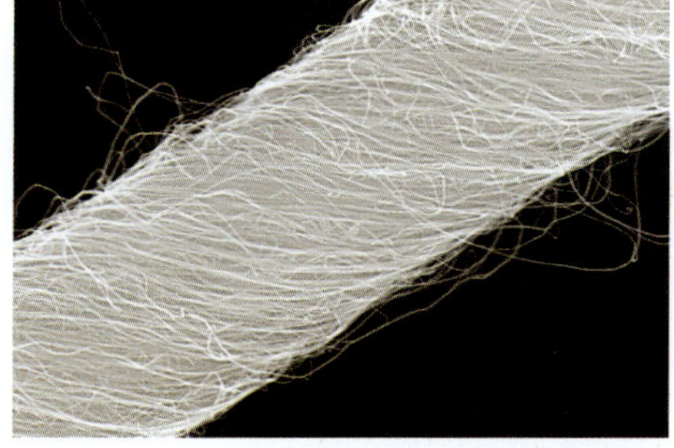

FIGURE 5.5.6 Graphite fibre can be used to reinforce plastics.

CHEMFILE 

Production of biochar

Most scientists agree that rising levels of carbon dioxide in the atmosphere, which are a result of human activities, are causing global warming. One area of research at the CSIRO to help combat rising carbon dioxide levels is the production of **biochar**.

Biochar is a high-carbon, porous and fine-grained residue produced by placing biomass (plant and forest waste) in a trench and covering it with soil. The biomass is allowed to smoulder, burning very slowly in the absence of oxygen. The product of this process is carbon and not carbon dioxide.

Biochar can increase the fertility of soils and agricultural production, using waste that would otherwise have become carbon dioxide. Figure 5.5.7 shows farmers adding biochar to the soil in a field.

FIGURE 5.5.7 Biochar added to soil improves the soil by supplying carbon and trapping nutrients.

CARBON NANOMATERIALS

Diamond and graphite have long been recognised as allotropes of carbon. However, since the 1980s, scientists have discovered how to make a new range of carbon allotropes that are examples of **nanomaterials**.

GO TO ➤ SkillBuilder page 127

Nanoparticles are between 1 and 100 nanometres wide. A nanometre (nm) is 10^{-9} metre. Nanomaterials are particularly interesting because they have a very high surface area to volume ratio, leading to some unique or enhanced properties.

Fullerenes

Fullerenes are an allotrope of carbon in which the atoms are arranged in a series of pentagons and hexagons.

Fullerenes, which are an allotrope of carbon, were discovered in the early 1980s. This allotrope was made up of molecules containing a roughly spherical group of carbon atoms arranged in a series of pentagons and hexagons, similar to the shape of a soccer ball, as you can see in Figure 5.5.8.

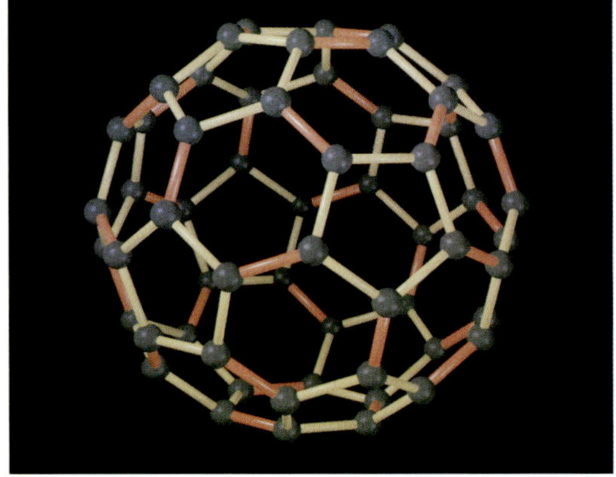

FIGURE 5.5.8 The structure of a fullerene molecule has a similar pattern to the surface of some soccer balls.

Scientists have since found further variations of this molecule. Fullerenes have three covalent bonds to each carbon atom and in some ways appear to be similar to graphite. This leaves delocalised electrons in the structure and the possibility of electrical conductivity. Although fullerenes were initially just a curiosity, scientists predict that they have a significant potential in a number of fields, such as in composite materials and **photovoltaic cells** (solar panels). The most stable fullerene molecule contains 60 carbon atoms bonded into an approximately spherical shape that is known as buckminsterfullerene or C_{60} or a **buckyball**.

CHEMFILE 

Fullerenes in flexible photovoltaic cells

Traditional photovoltaic cells (the solar panels you see on many rooftops) are made from highly refined and purified silicon crystals. Manufacturing these crystals is complex, and the cells produced are rigid and brittle.

Research into fullerenes has led to alternative types of photovoltaic cells, known as polymer solar cells. Polymer solar cells (Figure 5.5.9) use alternate layers of fullerene molecules instead of silicon. The cell produced is lighter than conventional cells and offers the advantage of being flexible. A flexible cell could match the curved shape of a caravan roof or the cabin of a boat. The initial problem with these cells was their low efficiency. Research into fullerenes and other aspects of the cells continues to improve their efficiency. These improvements will enable their increased use, resulting in a decreased reliance on the electricity generated by coal-fired power stations. Coal-fired power stations emit large amounts of the greenhouse gas carbon dioxide, thought to be responsible for increased global warming.

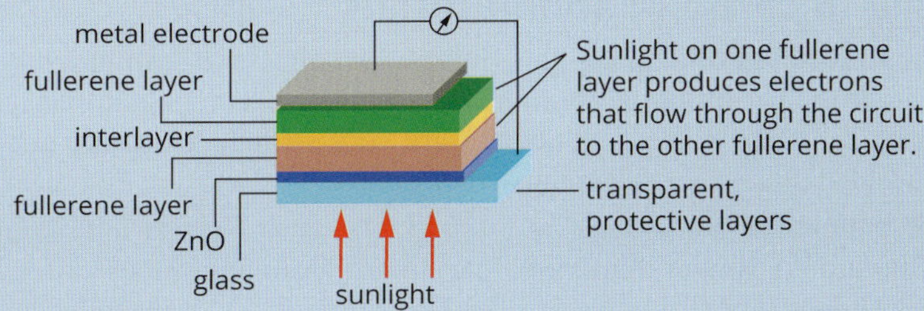

FIGURE 5.5.9 Polymer solar cells are constructed from transparent covering layers and alternate layers of fullerenes.

Graphene

You have seen that graphite has a layered structure. **Graphene** is best described as a single layer of graphite (Figure 5.5.10). Graphene is a single layer sheet with the same arrangement as the layers stacked in graphite. It is a very new material and was first isolated in 2004.

Graphite is soft, due to the weak dispersion forces between its layers. Graphene is only a single layer and retains the electrical conductivity of graphite, but it is an extremely strong and tough material.

Graphene has many potential uses. Graphene could:

- replace silicon as the basis for computer chips and circuits due to its high electrical conductivity
- be used in desalination plants—water under pressure can pass through the thin layer, but dissolved impurities cannot
- be used to construct electrodes, when it is an advantage for an electrode to be a non-metal
- be used in organic photovoltaic cells
- be used to reinforce composite materials because of its strength.

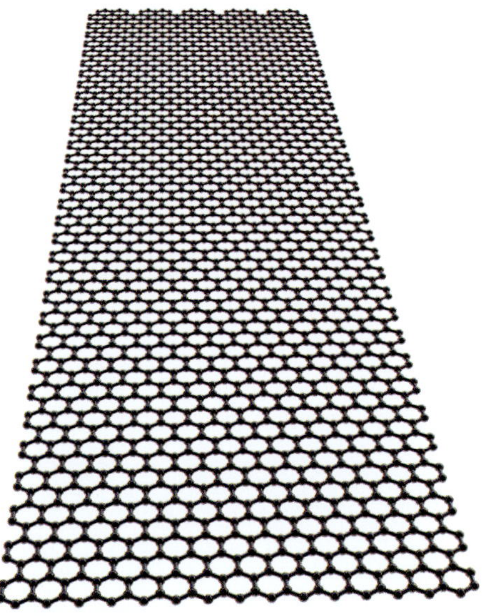

FIGURE 5.5.10 Graphene is a single layer sheet with the same arrangement as the layers stacked in graphite.

Nanotubes

Nanotubes are closely related to graphene. They have a long, hollow structure with walls formed from graphene. The diameter of these cylinders is very small, around 1 nanometre wide, while they can be millions of times longer. Each end of the cylinders can be capped by a half fullerene molecule, as shown in Figure 5.5.11.

Nanotubes can be single-walled or multiwalled. A multiwalled nanotube has smaller tubes sitting inside larger tubes.

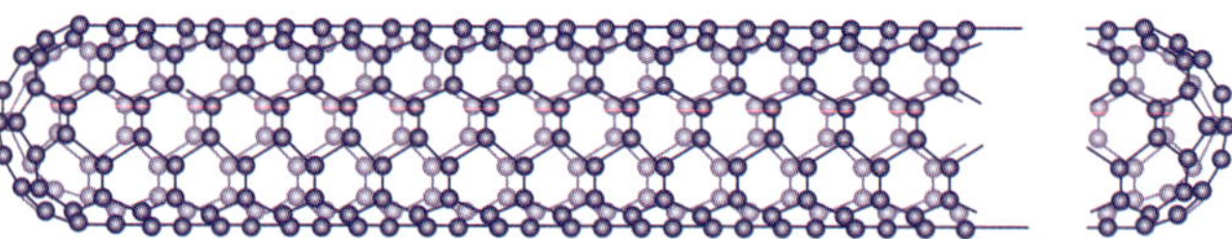

FIGURE 5.5.11 A carbon nanotube can be regarded as a sheet of graphene rolled into a cylinder and capped on each end by half a fullerene molecule.

Scientists are interested in nanotubes because of their:

- unique strength
- electrical conductivity (the conductivity of nanotubes depends on their shape; some are conductors, and others are semiconductors)
- thermal conductivity
- strong forces of attraction to one another.

Nanotubes hold great promise in fields such as optics, nanotechnology and electronics. Their extraordinary strength and thermal and electrical conductivity suggest they may be useful as additives in various structural materials.

Potential of carbon nanomaterials

Carbon nanomaterials offer huge gains in performance and properties over some other materials in current use and have a broad range of potential applications. The carbon–carbon bonds in these structures are very strong, and there are no weak points in a single layer of graphene or a nanotube.

Carbon nanotubes are:

- up to 300 times stronger than steel. Rope made from nanotubes with a diameter of 1 cm could support a mass of over 1000 tonnes (t). Nanotubes are already being used in high-performance sporting equipment.
- better conductors of electricity than silver. Since nanotubes are essentially 'wires' that are much narrower in diameter than metal wire, they offer the possibility for extreme miniaturisation of electrical circuits.
- better thermal conductors than diamond. Nanotubes could be used to transfer heat away from electrical components.
- stronger than Kevlar® fibres. Stain-resistant nanofabrics that never require washing are already available. You could even carry water in the pockets of a vest made from this material.
- capable of adsorbing more gas or impurities than activated charcoal.

OTHER COVALENT NETWORKS

Other substances such as silicon (Si) and silicon carbide (SiC) also form covalent network structures similar to diamond, with each silicon atom bonded to four other silicon atoms in a tetrahedral arrangement. Silicon is a semi-conductor that is widely used in the electronic industry. Silicon carbide is also known as **carborundum** and is commonly used as an abrasive for cutting and polishing objects. It can be heat-treated to form ceramic plates used in car brakes and bullet-proof vests. The carbon and silicon atoms in silicon carbide are linked by strong covalent bonds in a tetrahedral arrangement. Each carbon atom is bonded to four silicon atoms, and each silicon atom is bonded to four carbon atoms.

Silica (silicon dioxide, SiO_2), a major component of sand, also exists as a continuous three-dimensional covalent network structure (Figure 5.5.12). Each silicon atom is at the centre of a tetrahedron and is covalently bonded to four oxygen atoms. Each oxygen atom is bonded to two silicon atoms. As a result of the strong covalent bonding between the silicon and oxygen atoms, silicon dioxide is hard, has a high melting temperature and does not dissolve in water or organic **solvents**. It is widely used to make glass and ceramics.

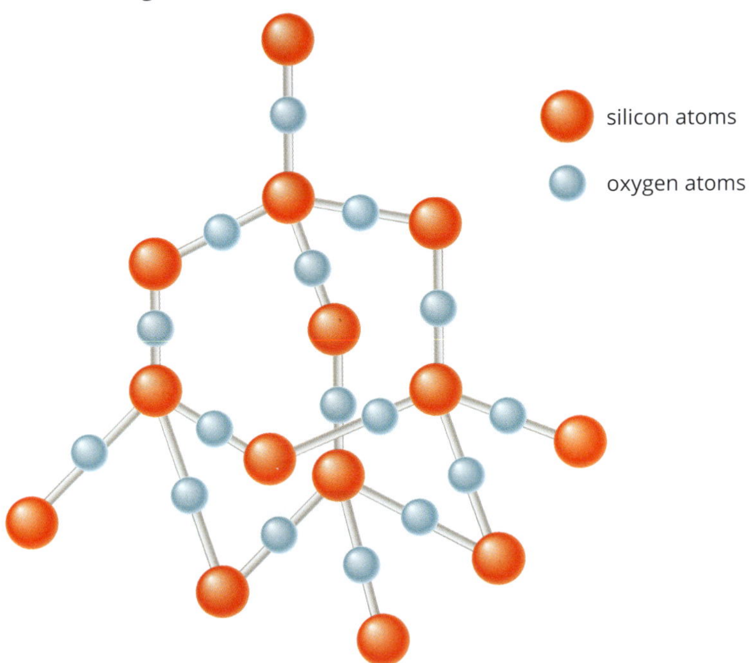

FIGURE 5.5.12 In silica (SiO_2) strong covalent bonds between silicon and oxygen atoms form a covalent network structure.

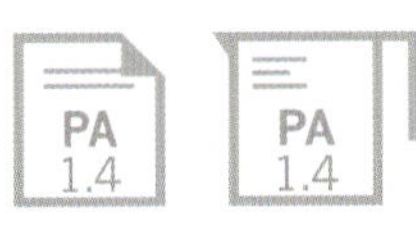

CHEMISTRY IN ACTION AHC

Stone tools

Indigenous Australians used a variety of stone tools for cutting and shaping wood, scraping animal skins, grinding seed, and as axe heads and spear points.

Rocks with a high silica (SiO_2) content, such as quartz, were used to make stone tools. Stone tools were made by precision flaking and grinding. For example, the rock to be shaped into an axe head was hit by another stone, causing small flakes to be chipped off. This process continued until the desired shape was achieved (Figure 5.5.13).

Other small flakes could be used as spearheads, scrapers, or to whittle wood (Figure 5.5.14).

Silica-rich stones were also used as grinding stones to grind seeds into flour, which was mixed with water and then cooked in the hot coals of a fire (Figure 5.5.15).

FIGURE 5.5.13 An Indigenous axe from Toorale.

FIGURE 5.5.14 These rock flakes have the potential for several uses.

FIGURE 5.5.15 Grinding stone from the Darling River region in NSW.

5.5 Review

SUMMARY

- Diamond, graphite and carbon nanomaterials such as fullerenes and graphene are different allotropes of carbon. Their structures and properties are very different.
- The atoms in a covalent network structure are held together by strong covalent bonds in a continuous three-dimensional network.
- In diamond, each carbon atom is covalently bonded to four other carbon atoms in a tetrahedral shape, forming a covalent network structure. Diamond sublimes at a high temperature, is extremely hard and has a sparkling crystalline appearance.
- In graphite, each carbon atom is covalently bonded to three other carbon atoms. The layered network structure contains delocalised electrons. The bonds within the layers are strong, but the bonds between layers are weak dispersion forces. Graphite is slippery, conducts electricity and sublimes at a high temperature.
- Nanomaterials are of interest to scientists because of the enhanced properties that their high surface area to volume ratio offers. Their high tensile strength and high electrical and heat conductivity are of particular interest.
- Carbon-based nanomaterials include fullerenes (buckyballs), nanotubes and graphene. In all of these allotropic forms, the carbon atoms are bonded to three other carbon atoms, as shown in the illustration to the right.

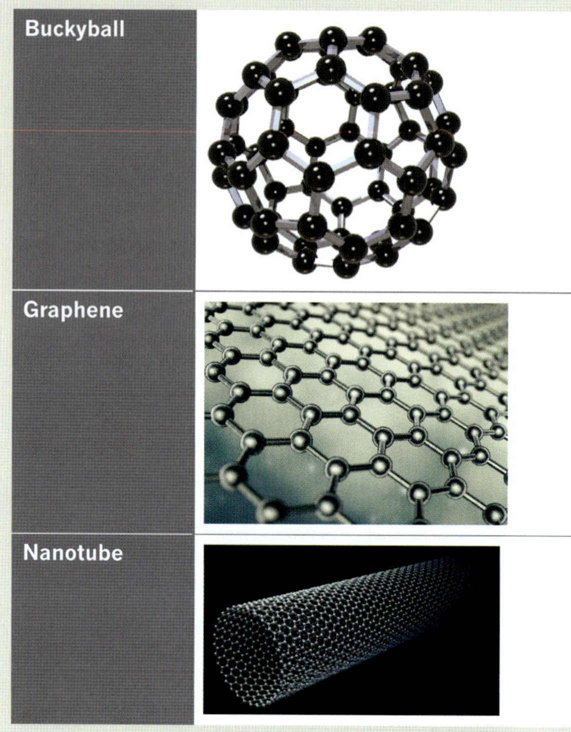

- Potential applications of fullerenes include fibres and fabrics, electrical circuits, photovoltaic cells and filtration systems.
- In silicon dioxide, each silicon atom is at the centre of a tetrahedron and is covalently bonded to four oxygen atoms. Each oxygen atom is bonded to two silicon atoms.

KEY QUESTIONS

1. Explain the following properties of diamond and graphite in terms of their respective structures:
   a. hardness or softness
   b. ability or inability to conduct electricity
2. Explain the following uses of graphite and diamond in terms of their structure.
   a. Graphite is used as a lubricant.
   b. Diamond is often used as an edge on saws and a tip on drills.
3. Describe how the structures and properties of fullerenes are both similar to and different from those of graphite.
4. Describe the bonding within a C_{60} buckyball.
5. Describe the bonding within a graphene sheet.
6. Briefly describe the nature of the structure and bonding in silicon dioxide.

Chapter review

05

KEY TERMS

alkali metal
allotrope
anion
asymmetrical molecule
biochar
brittle
buckyball
carborundum
cation
ceramic
chemical formula
conductor
core charge
covalent bond
covalent layer network
covalent network structure
crystal lattice
delocalise
delocalised electron
density
diamond
diatomic molecule
dipole
dipole–dipole force
dispersion force
dissolved
double covalent bond
ductile
electric dipole
electrolyte
electron density
electronegativity
electron transfer diagram
electrovalency
fullerenes
gemstone
graphene
graphite
hydrogen bond
ionic bonding
ionic compound
ionisation energy
instantaneous dipole
intermolecular bond
intermolecular force
intramolecular bond
Lewis dot diagram
lone pair
malleable
metallic bonding
metallic bonding model
molecule
molten
nanomaterial
nanotube
non-bonding electron
non-polar
non-polar solvent
octet rule
photovoltaic cell
polar
polarity
polyatomic ion
polyatomic molecule
polymer
salinity
single covalent bond
solvent
sublime
surface tension
symmetrical molecule
temporary dipole
tensile strength
tetrahedral shape
transition metal
triple covalent bond
valence electron
valence-shell electron-pair repulsion (VSEPR) theory
valence structure

REVIEW QUESTIONS

1 Which property most clearly distinguishes metals from non-metals?

2 Use the metallic bonding model to explain each of the following observations:
  a Copper wire conducts electricity.
  b A metal spoon used to stir a boiling mixture becomes too hot to handle.
  c Iron has a high melting point of 1540°C.
  d Lead has a high density (11.4 $g\,mL^{-1}$) compared with a non-metal such as sulfur.
  e Copper can be drawn out to form a wire.

3 Consider the metallic bonding model used to describe the bonding and structure of metals.
  a What is meant by the following terms?
    i delocalised electrons
    ii a lattice of cations
    iii metallic bonding
  b Which electrons are delocalised in a metal?

4 Describe an experiment you could carry out to demonstrate each of the following properties of the compounds given. In each case:
  i sketch the equipment you would use
  ii describe what you would expect to observe.
  a Solid magnesium chloride does not conduct electricity.
  b Molten sodium chloride does conduct electricity.
  c Solid sodium chloride is hard and brittle.

5 Use the ionic bonding model to explain the following properties of ionic compounds:
  a They generally have high melting points.
  b They are hard and brittle.
  c They do not conduct electricity in the solid state, but will conduct when molten or dissolved in water.

6 Use diagrams to show the electron transfer that occurs when:
  a lithium reacts with chlorine
  b magnesium reacts with fluorine
  c potassium reacts with sulfur
  d magnesium reacts with nitrogen.

7 Write the formula for the ionic compound formed in the reaction between:
  a potassium and bromine
  b magnesium and iodine
  c calcium and oxygen
  d aluminium and fluorine
  e calcium and nitrogen.

8 The electronic configurations of some metallic and non-metallic elements are given. (The symbols shown for the elements are not their real ones.) Write formulae for the compounds they are most likely to form if they react together. The first example has been done for you.

- **a** A: 2,1 and B: 2,6 → A_2B
- **b** C: 2,8,3 and D: 2,7
- **c** E: 2,8,8,2 and F: 2,8,6
- **d** G: 2,8,8,1 and H: 2,5
- **e** K: 2,8,2 and L: 2,6

9 Write the formula for each of the following ionic compounds:

- **a** copper(I) chloride
- **b** silver oxide
- **c** lithium nitride
- **d** potassium iodide
- **e** copper(II) nitrate
- **f** chromium(II) fluoride
- **g** potassium carbonate
- **h** magnesium hydrogen carbonate
- **i** nickel(II) phosphate

10 Name the ionic compounds with the following chemical formulae:

- **a** $(NH_4)_2CO_3$
- **b** $Cu(NO_3)_2$
- **c** $CrBr_3$

11 Hydrogen chloride (HCl) exists as a gas at room temperature. What can you conclude about the strength of the intermolecular bonding in pure hydrogen chloride?

12 Select the statement that best describes the way hydrogen atoms bond with one another:

- **A** One hydrogen atom donates an electron to another hydrogen atom to form a molecule.
- **B** Hydrogen atoms form a lattice with delocalised electrons.
- **C** Hydrogen atoms share electrons to obtain a complete outer shell of eight electrons.
- **D** Two hydrogen atoms share an electron each to form a hydrogen molecule.

13 In a molecule of nitrogen (N_2), how many of the following are there?

- **a** bonding electrons
- **b** covalent bonds
- **c** non-bonding electrons

14 Atom X has an electronic configuration of 2,6. Atom Y has a configuration of 2,7. What is the likely molecular formula of the compound that will form between X and Y?

15 Draw a Lewis dot diagram for each of the following molecules and identify the number of bonding and non-bonding electrons in each molecule.

- **a** HBr
- **b** H_2O_2
- **c** CF_4
- **d** C_2H_6
- **e** PF_3
- **f** Cl_2O
- **g** CH_4
- **h** H_2S

16 The atoms in molecules of nitrogen, oxygen and fluorine are held together by covalent bonds. How are the bonds in these molecules:

- **a** similar?
- **b** different?

17 Match the molecular formula to the correct molecular shape.

Molecular formula	Molecular shape
phosphorus trichloride (PCl_3)	tetrahedral
hypochlorous acid (HOCl)	linear
trichloromethane ($CHCl_3$)	V-shaped
hydrogen fluoride (HF)	pyramidal

18 Are the following molecules polar or non-polar? Draw structural formulae to help you decide.

- **a** CS_2
- **b** Cl_2O
- **c** SiH_4
- **d** CH_3Cl
- **e** CH_3CH_3

19 Order the following covalent bonds from most polar to least polar.

Si–O, N–O, F–F, H–Br, O–Cl

20 For each of the following structures, state whether:

- **i** the molecule is polar or non-polar
- **ii** the strongest intermolecular force of attraction between molecules of each type would be dispersion forces, hydrogen bonding or dipole–dipole attraction.

a

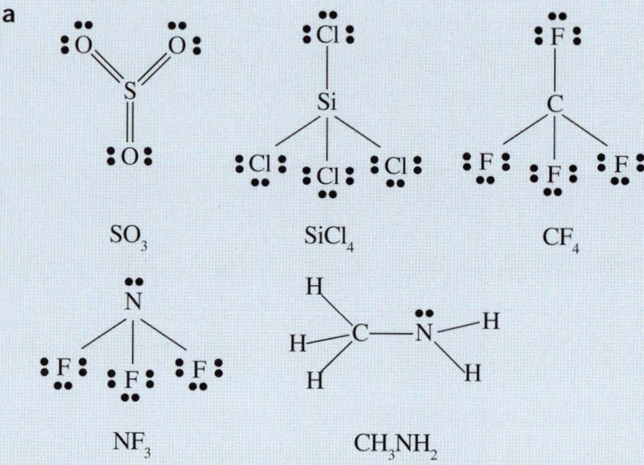

21 At room temperature, CCl_4 is a liquid, whereas CH_4 is a gas.

a Which substance has the stronger intermolecular attractive forces?

b Explain the difference in the strengths of the intermolecular forces.

22 When sugar ($C_{12}H_{22}O_{11}$), a covalent molecular substance, is gently heated, it turns into a clear liquid. If the liquid is heated strongly, it turns black and a gas is produced. Explain what is happening to the bonds in sugar when it is heated. Use the terms 'intermolecular bonds' and 'intramolecular bonds' in your answer.

23 Graphite has unusual properties. It sublimes at a high temperature, it conducts electricity, and it can be used as a lubricant. Use a diagram to explain why graphite has these properties.

24 The structures of methane and diamond are shown in the image below. Each carbon atom in methane (CH_4) has a tetrahedral arrangement of atoms around it. A carbon atom in diamond also has a tetrahedral arrangement. However, the two substances have very different properties.

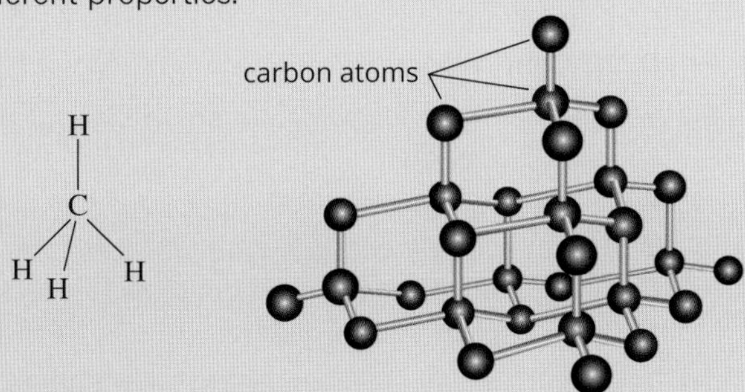

a Describe the type of bonding that would be present in each substance.

b Use the type of bonding present in each substance to explain the different properties you would expect each to have.

25 Diamond and silicon dioxide have some properties in common. They are both hard materials and do not conduct electricity. Silicon dioxide melts at 1600°C and diamond sublimes at 3640°C. Explain these properties in terms of the bonding and structure of diamond and silicon dioxide.

26 Classify the bonding in the following substances as metallic, ionic, polar covalent or non-polar covalent: $CuCl_2$, Ag, HCl, H_2O, Cu, CaS, NH_3, Cl_2
Explain the reasons for your classifications in terms of electronegativity and position of the elements in the periodic table.

27 State whether the following properties relate to metallic lattices only, ionic lattices only or both metallic and ionic lattices:

a They contain both positively and negatively charged particles.

b The lattice is held together by forces of attraction between positively and negatively charged particles.

c They are hard.

d They are brittle.

e They conduct electricity in the molten state.

28 Consider the following list of molecules:
N_2, Cl_2, O_2, NH_3, HCl, CH_4, H_2O, CO_2, CCl_4, $CHCl_3$

a From the list, identify and draw valence structures of molecules that contain the following features:

i one triple bond

ii one double bond

iii two double bonds.

b From the list, which substances contain:

i polar molecules?

ii symmetrical molecules?

iii molecules with hydrogen bonding between them?

29 Explain why covalent molecular substances generally have lower melting and boiling points than ionic and metallic substances.

---

30 The name, formula and melting point for a number of substances are provided in the following table.

a Complete the table by naming the type of bonding that exists between the particles of each substance when it is in its solid state.

Name	Nitrogen	Hydrogen bromide	Ammonia	Iron	Diamond	Potassium fluoride
formula	N_2	HBr	NH_3	Fe	C	KF
melting point (°C)	–210	–87	–78	1535	>3550 (sublimes)	858
bond type						

b Consider the following statement:
All bonding is an electrostatic attraction between positively and negatively charged particles.
Briefly discuss the nature of the electrostatic attraction between the particles in the solid state of each of the substances listed in the table above.

31 Reflect on the Inquiry activity on page 138. Explain, using concepts from this chapter, why the aluminium foil and paper clip did not dissolve in water but the salt and bicarbonate of soda did.

MODULE 1 • REVIEW

REVIEW QUESTIONS

Properties and structure of matter

Multiple choice

1 Matter in a ______ state has ______ volume and ______ shape.

A gaseous; no fixed; no fixed

B liquid; no fixed; a fixed

C solid; a fixed; no fixed

D solid; no fixed; a fixed

2 Which of the following is not a physical property of matter?

A volume

B corrosion resistance

C elasticity

D melting point

3 Zinc is an element. Therefore:

A zinc has no isotopes

B all zinc atoms are identical

C zinc atoms always contain the same number of protons

D zinc atoms contain equal numbers of protons and neutrons.

4 Copper has two isotopes, ^{63}Cu and ^{65}Cu. It has a relative atomic mass of 63.55. Therefore, any one particular copper atom would:

A contain either 63 or 65 protons

B contain either 34 or 36 neutrons

C have a mass number of 63.55

D have an atomic number of 29 or 31.

5 The electronic configuration of an atom of chromium (Cr) in its ground state is:

A $1s^2 2s^2 2p^6 3s^2 3p^6 3d^6$

B $1s^2 2s^2 2p^6 3s^2 3p^6 4s^2 4p^4$

C $1s^2 2s^2 2p^6 3s^2 3p^6 3d^4 4s^2$

D $1s^2 2s^2 2p^6 3s^2 3p^6 3d^5 4s^1$

6 The ground state electronic configuration for an ion of sulfur (S^{2-}) is:

A $1s^2 2s^2 2p^6 3s^2 3p^4$

B $1s^2 2s^2 2p^6 3s^2 3p^5$

C $1s^2 2s^2 2p^6 3s^2 3p^6$

D $1s^2 2s^2 2p^6 3s^2 3p^4 4s^2$

7 The 3*d*-subshell has:

A 3 orbitals and can hold up to 3 electrons

B 3 orbitals and can hold up to 6 electrons

C 5 orbitals and can hold up to 10 electrons

D 5 orbitals and can hold up to 15 electrons.

The following information relates to questions 8 and 9.

The atomic number, mass number and electron configuration of four particles, W, X, Y and Z, are given below:

Particle	Atomic number	Mass number	Electronic configuration
W	17	37	$1s^2 2s^2 2p^6 3s^2 3p^6$
X	19	39	$1s^2 2s^2 2p^6 3s^2 3p^6$
Y	20	40	$1s^2 2s^2 2p^6 3s^2 3p^6 4s^2$
Z	19	40	$1s^2 2s^2 2p^6 3s^2 3p^6 4s^1$

8 Which one of the following alternatives lists particles that are isotopes of the same element?

A W and X only

B X and Z only

C Y and Z only

D W, X and Y only

9 Which one of the following statements about particles W, X, Y and Z is correct?

A W is a noble gas.

B X is a positively charged ion.

C Y is in group 4 of the periodic table.

D Z is a negatively charged ion.

10 The study of the emission spectrum of hydrogen led the Danish physicist Niels Bohr to propose a model for an atom. An emission spectrum is produced when electrons in an atom:

A release energy as they move from higher to lower energy levels

B absorb energy as they move from higher to lower energy levels

C release energy as they move from lower to higher energy levels

D absorb energy as they move from lower to higher energy levels.

11 Identify the missing particle required to balance the following nuclear equation:

$$^{235}_{92}U \rightarrow \_\_\_ + {}^{4}_{2}He$$

A $^{0}_{-1}e$

B $^{231}_{90}Th$

C $^{239}_{94}Pu$

D $^{235}_{93}Np$

12 Which pair of elements are chemically most similar to each other?
A Li and Be
B Cl and Ar
C N and P
D Ne and Na

13 An ion that contains 11 protons, 12 neutrons and 10 electrons will have a mass number and charge corresponding to:

	Mass number	Charge
A	11	−1
B	11	+1
C	23	−1
D	23	+1

14 Main-group elements (groups 1, 2 and 13–18) are placed into groups in the periodic table according to each element's:
A mass number
B atomic number
C number of occupied electron shells
D number of electrons in the outer shell.

15 Which one of the following metals reacts most readily with oxygen?
A Na
B K
C Mg
D Ca

16 Predict which one of the following ionic compounds will have the highest melting point.
A MgO
B $MgCl_2$
C Na_2O
D NaCl

17 Which one of the following statements about ionic bonding is not correct?
A When molten, ionic compounds are conductors of electricity.
B An ionic lattice contains both cations and anions in fixed positions.
C Ionic bonding involves the sharing of electrons between two different atoms.
D Compounds held together by ionic bonding generally have high melting temperatures.

18 The formula for hydrogen cyanide is HCN. The number of electrons shared between the carbon and nitrogen atom is:
A 2
B 4
C 6
D 8

19 Which of the following gives the correct shape for each of the molecules listed?

	Linear	V-shaped	Tetrahedral
A	CO_2	H_2S	CH_4
B	H_2	CO_2	NH_3
C	HF	H_2O	NH_3
D	H_2O	NH_3	CH_4

20 Which one of the following correctly describes the intermolecular forces in pure samples of F_2, HF and CH_3F?

	F_2	HF	CH_3F
A	dispersion forces only	dispersion forces and hydrogen bonds	dispersion forces and dipole–dipole attraction
B	dispersion forces and hydrogen bonds	dispersion forces and hydrogen bonds	dispersion forces and hydrogen bonds
C	dispersion forces only	dispersion forces and hydrogen bonds	dispersion forces and hydrogen bonds
D	dispersion forces	dispersion forces and dipole–dipole attraction	dispersion forces and dipole–dipole attraction

Short answer

1 Magicians often use 'flash powder' in their acts. Flash powder contains 69.0% potassium chlorate (31.9% K, 28.9% Cl, 39.2% O) and the remainder is aluminium. Calculate the percentage composition by mass of each element in flash powder.

2 Iridium has two naturally occurring isotopes. Their relative abundances and masses are shown in the table:

Isotope	Relative isotopic mass	Percentage abundance
^{191}Ir	190.97	37.30%
^{193}Ir	192.97	62.70%

a What is the name of the instrument that is commonly used to experimentally obtain the abundances and relative isotopic masses data?
b Using the information in the table, determine the relative atomic mass of iridium. Show your working and give your answer to the appropriate number of significant figures.

3 a Using an atom of sodium as an example, clearly explain the difference between an electron shell, a subshell and an orbital.
  b Consider the following electronic configuration: $1s^22s^22p^63s^23p^63d^1$
  Does this electronic configuration represent a ground state or excited atom? Give a reason for your answer.

4 Give a concise explanation for each of the following.
  a The atomic radius of chlorine is smaller than that of sodium.
  b The first ionisation energy of fluorine is higher than that of lithium.
  c The reactivity of Be is less than that of Ba.
  d There are two groups in the *s*-block of the periodic table.

5 Write the formula of each of these compounds.
  a potassium phosphate
  b aluminium oxide
  c sodium nitrate
  d iron(III) sulfide

6 Ionic compounds have a variety of uses in everyday life.
  a Give the formula, and a use, of one ionic compound.
  b Use the ionic bonding model to explain the following properties of ionic compounds.
    i Solid ionic compounds cannot conduct electricity.
    ii Ionic compounds are generally brittle.
    iii Ionic compounds have high melting temperatures.

7 Draw the structure of each of the following compounds and include any non-bonding pairs of electrons. Give the name of the shape of the molecule and predict whether the molecule is polar or non-polar.
  a H_2S
  b PF_3
  c CCl_4
  d CS_2
  e SiH_4

8 Provide concise explanations for each of the following observations.
  a The melting temperature of ice (solid H_2O) is 0°C, but a temperature of over 1000°C is needed to decompose water molecules into hydrogen and oxygen gases.
  b Most substances are denser in their solid state than in their liquid state, yet solid water (ice) floats on liquid water.
  c A molecule of ethyne (C_2H_2) is linear, but a molecule of hydrogen peroxide (H_2O_2) is not.

9 Consider the following molecules: HCl, HF, H_2O, H_2S, CH_4, F_2, H_2, O_2.
  a List the molecules that are polar molecules.
  b List the molecules that would have dipole–dipole forces as their main intermolecular force.
  c List the molecules that have a linear shape.
  d List the molecules that are capable of hydrogen bonding.
  e List the molecules that contain a double bond.
  f Which substance in the list would have the lowest boiling temperature? Give a reason for your answer.

Extended response

1 Garden fertiliser supplies nutrients to plants. One common formulation is composed of moderately soluble urea derivatives to supply nitrogen, phosphate-containing rock to supply phosphorus, soluble potash to supply potassium, and coal powder to supply several other trace elements. Propose a strategy to separate these four components. (Hint: Coal powder is less dense than water.)

2 Magnesium is a commonly used structural metal.
  a Write the electronic configuration of a magnesium atom, using subshell notation.
  b i The radius of a magnesium atom is 140 pm (1 pm = 10^{-12} m). What is the radius of the magnesium atom in nanometres?
    ii Would you predict the radius of a sodium atom to be smaller or larger than 140 pm? Explain your answer.
  c i Describe the model commonly used to describe the structure of metals such as magnesium and the nature of the bonding between its particles. You may include a labelled diagram in your answer.
    ii Use the metallic bonding model to explain why magnesium is a good conductor of electricity.
    iii The metallic bonding model is useful for explaining many of the properties of metals. However, like most models, it has some limitations. Give one example of the limitations of the metallic bonding model.
  d Magnesium reacts readily with dilute hydrochloric acid. Identify one metal that would be expected to react more vigorously with hydrochloric acid than magnesium.
  e Magnesium is classified as a main-group metal and not as a transition metal.
  How does the electronic configuration of a transition metal differ from that of a main-group metal?

3 Ammonia (NH_3) is a constituent of many cleaning products for bathrooms.

a Draw a Lewis dot diagram of an ammonia molecule.

b Draw structural formulae for two ammonia molecules. Clearly show, and give the name of, the shape of these molecules. On your diagram, label the type of bonds that exist between:

- the atoms within each ammonia molecule
- two ammonia molecules.

c Draw a structural formula for a molecule of:

i nitrogen

ii carbon dioxide.

d i Explain why the bonds between nitrogen molecules and those between molecules of carbon dioxide are of the same type, even though the bonds inside these molecules differ in strength and polarity.

ii Explain why the bonds between ammonia molecules are different from those between nitrogen molecules or carbon dioxide molecules.

4 Diamond and graphite are allotropes of carbon. Although they are similar in some respects, they are very different in their structure and uses.

a Give a brief definition of the term 'allotrope'.

b Describe the similarity in the bonding of these two materials.

c Compare the structures of diamond and graphite.

d Graphite is an excellent conductor of electricity while diamond is unable to conduct electricity. Explain this difference in terms of the structure of the networks.

e Graphite is used as a 'lead' in pencils. With reference to the bonding in the network, explain why graphite can be used this way.

5 Phosphine is a hydride of phosphorus with the formula PH_3.

a Draw a Lewis dot diagram of a PH_3 molecule.

b Explain whether PH_3 is a polar or non-polar molecule.

c Explain why a PH_3 molecule does not have a trigonal planar shape.

d Determine the relative molecular mass of PH_3 and ammonia (NH_3).

e Explain why ammonia has a much higher melting point than phosphine.

f Phosphorus has one stable isotope and several radioactive isotopes. State the symbol, using the notation $^{A}_{Z}X$, for the radioactive phosphorus-32 isotope.

g Phosphorus-32 decays to produce a beta particle. Write a balanced nuclear reaction equation for this decay.

h Draw an orbital diagram showing the electronic configuration of a phosphorus atom.

Introduction to quantitative chemistry

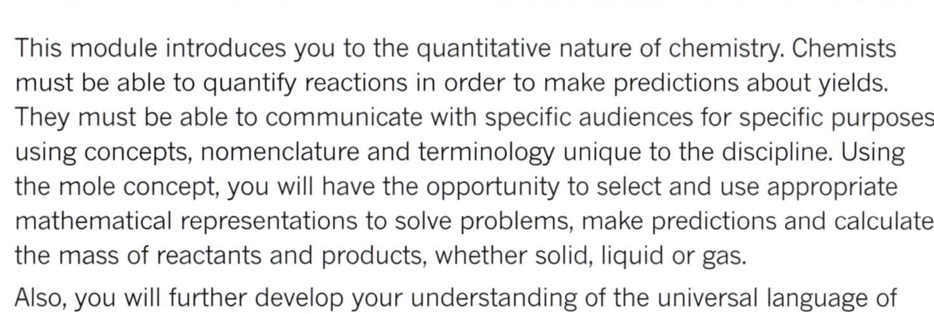

This module introduces you to the quantitative nature of chemistry. Chemists must be able to quantify reactions in order to make predictions about yields. They must be able to communicate with specific audiences for specific purposes, using concepts, nomenclature and terminology unique to the discipline. Using the mole concept, you will have the opportunity to select and use appropriate mathematical representations to solve problems, make predictions and calculate the mass of reactants and products, whether solid, liquid or gas.

Also, you will further develop your understanding of the universal language of chemistry. You will be introduced to the idea that science is a global enterprise that relies on clear communication, international conventions, peer review and reproducibility.

Outcomes

By the end of this module, you will be able to:

- design and evaluate investigations in order to obtain primary and secondary data and information (CH11-2)
- select and process appropriate qualitative and quantitative data and information using a range of appropriate media (CH11-4)
- solve scientific problems using primary and secondary data, critical thinking skills and scientific processes (CH11-6)
- describe, apply and quantitatively analyse the mole concept and stoichiometric relationships (CH11-9)

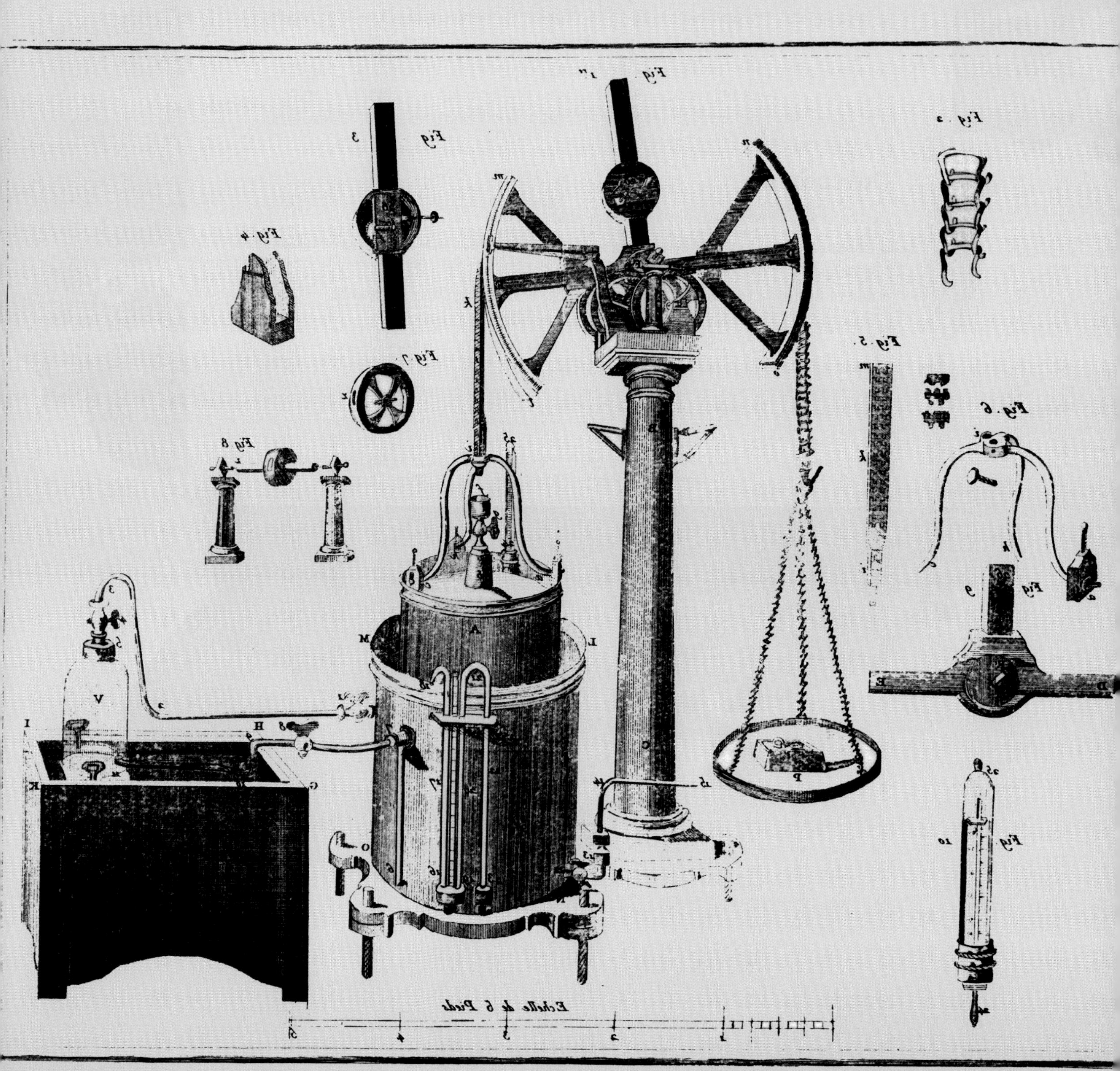

CHAPTER 06

Chemical reactions and stoichiometry

Chemical reactions are responsible for almost everything we see, smell and touch—even the process of obtaining energy from our food (respiration) involves chemical reactions. It is our need to better understand the processes in the world around us that drives the study of chemistry. The development of the law of conservation of mass in the late 18th century gave birth to many chemical advances, and the 19th century is known colloquially as the 'century of chemistry'. In the study of chemistry, it is vital not only to investigate chemical reactions qualitatively through descriptions of observable changes, but also to explore reactions in a quantitative manner. The law of conservation of mass gave rise to the term stoichiometry, which is a process that enables calculation of the quantities of chemical species involved in chemical reactions.

This chapter will explore the importance of the law of conservation of mass, how to balance chemical equations, and how to use the law of conservation of mass in calculations to determine the masses of solids and volumes of liquids or gases in chemical reactions.

Content

INQUIRY QUESTION

What happens in chemical reactions?

By the end of this chapter, you will be able to:

- conduct practical investigations to observe and measure the quantitative relationships of chemical reactions, including but not limited to:
  - masses of solids and/or liquids in chemical reactions
  - volumes of gases in chemical reactions (ACSCH046) ICT N
- relate stoichiometry to the law of conservation of mass in chemical reactions by investigating:
  - balancing chemical equations (ACSCH039) N
  - solving problems regarding mass changes in chemical reactions (ACSCH046) ICT N

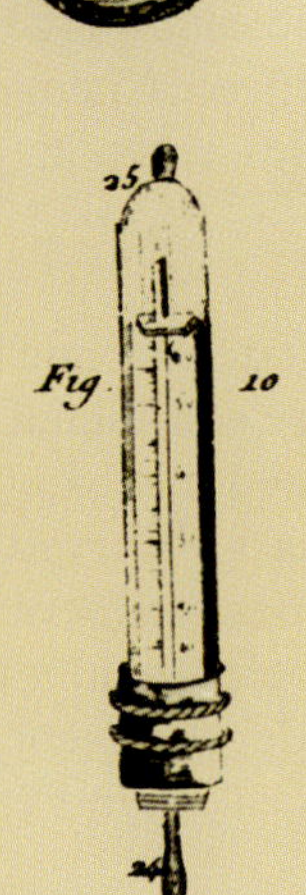

6.1 Writing chemical equations

Modelling conservation of mass: making a biopolymer

What happens in chemical reactions?

COLLECT THIS ...

- 50 mL measuring cylinder
- milk
- 250 mL conical flask
- hot water bath
- thermometer
- vinegar
- small test-tube
- electronic balance
- rubber stopper

DO THIS ...

1. Zero the electronic balance.
2. Measure approximately 50 mL of milk and pour it into the 250 mL conical flask.
3. Sit the conical flask in the hot water bath and place the thermometer in the milk. Heat until the milk temperature reaches approximately 50°C.
4. Take the conical flask out of the warm water and place it onto the electronic balance.
5. Measure approximately 10 mL of vinegar and pour into the test tube.
6. Place the test tube in the conical flask.
7. Stopper the conical flask and record the total mass.
8. Hold the stopper in place and invert the conical flask. Agitate carefully, allowing the vinegar and milk to mix.
9. Record the total mass again after the reaction.

RECORD THIS ...

Present your recorded masses.

Describe what happened during the mixing of the two liquids in a table.

REFLECT ON THIS ...

What happens in chemical reactions?

How does the rubber stopper help to model a closed system?

Compare the mass as recorded before and after the reaction.

Explain how this demonstrates the conservation of mass.

A closed system is a system where matter neither enters nor leaves.

Chemists need a way to communicate what occurs during a chemical reaction. From earlier years of science education, you may recall the observable indicators that a chemical reaction has occurred. These include things like a change in temperature, light being released, odours, bubbles and colour changes. However, these indicators usually do not give all of the information about what reacted or what was produced. Chemical equations are a symbolic representation of what is occurring in a chemical reaction.

THE LAW OF CONSERVATION OF MASS

FIGURE 6.1.1 Antoine Lavoisier (1743–1794). As a nobleman and unpopular tax collector, Lavoisier was executed by guillotine in 1794, during the French Revolution.

In 1774 the French chemist Antoine Lavoisier (Figure 6.1.1) conducted experiments investigating the oxidation of mercury in air. These experiments were conducted in sealed vessels— each an example of a **closed system**. He demonstrated that oxygen was the component of air responsible for the apparent increase in mass observed in combustion reactions. Perhaps, more importantly, through careful measurement of masses of the components of air, he concluded that the total mass before and after a reaction takes place is the same. This was a major contribution to development of the law of **conservation of mass**. 'Conservation' means 'the physical quantity of something remaining constant', applied to mass in this case.

It is because of this contribution that Lavoisier is regarded as one of the fathers of modern chemistry. In 1789, he published a chemistry textbook called *Traité élémentaire de chimie* (*Elementary Treatise of Chemistry*; Figure 6.1.2) in which he famously wrote 'Rien ne se perd, rien ne se crée, tout se transforme' ('Nothing is lost, nothing is created, everything is transformed').

> The law of conservation of mass refers to the fact that the total mass of the components of a chemical reaction in a closed system will remain constant throughout the reaction.

THE MEANING OF A CHEMICAL EQUATION

Chemical equations indicate several things. First, they show what **chemical species** are acting as **reactants** (also called reagents). Reactants are the chemical species that are used up in a chemical reaction. Second, chemical equations show what chemical species make up the **products** of the reaction. Products, as the name suggests, are the chemical species that are produced in a chemical reaction.

Coefficients

In many chemical reactions, more than one unit of the reactant species is required for the reaction to proceed. A coefficient is a number written before the formula of an element or compound in a chemical equation. The coefficient indicates the number of units of the species reacting or being produced in the reaction. The absence of a coefficient implies that there is only one unit of that chemical species present.

For example, $3H_2$ has a coefficient of 3, meaning there are three molecules of hydrogen gas (H_2), giving six hydrogen atoms in total. When the molecule H_2 has no written coefficient, this means the coefficient must be 1, so there are two hydrogen atoms in total.

Reaction arrow

In a mathematical equation, the two sides of the equation are separated by an '=' sign, but in a chemical equation an arrow is used instead. Consider this arrow as meaning 'moves towards' or 'to form' rather than 'equals'.

For example, in the synthesis of water from hydrogen and oxygen gas, two hydrogen molecules react with one oxygen molecule. This is shown on the left-hand side of the equation below, where you can also see a coefficient of 2 for the hydrogen molecules. On the product (right-hand) side of the equation, two water molecules are produced.

$$2H_2 + O_2 \rightarrow 2H_2O$$

State symbols

A fully balanced chemical equation includes the physical states or phases of the reactants. The four physical states of matter used in equations are:

- liquid (l)
- gas (g)
- solid (s)
- aqueous (aq), which refers to the solution formed when a substance dissolves in water.

The state symbols of (l), (g), (s) or (aq) are written in brackets immediately after the element or compound in a chemical equation.

For example, take the reaction Lavoisier conducted:

$$2Hg(l) + O_2(g) \rightarrow 2HgO(s)$$

You can read this reaction as 'two parts of mercury liquid reacts with one part of oxygen gas to form two parts of mercury(II) oxide'.

State symbols are important. One reason for this is because the states of the reactants and products affect the energy change of a reaction. Another reason is that it is useful to know the states of the products when designing the manufacture of a particular chemical. This may influence which chemical reactions are chosen for the production process.

TRAITÉ
ÉLÉMENTAIRE
DE CHIMIE,
PRÉSENTÉ DANS UN ORDRE NOUVEAU
ET D'APRÈS LES DÉCOUVERTES MODERNES;
Avec Figures :
Par M. LAVOISIER, de l'Académie des Sciences, de la Société Royale de Medecine, des Sociétés d'Agriculture de Paris & d'Orléans, de la Société Royale de Londres, de l'Inſtitut de Bologne, de la Société Helvétique de Baſle, de celles de Philadelphie, Harlem, Mancheſter, Padoue, &c.

TOME PREMIER.

A PARIS,
Chez CUCHET, Libraire, rue & hôtel Serpente.

M. DCC. LXXXIX.
Sous le Privilège de l'Académie des Sciences & de la Société Royale de Médecine.

FIGURE 6.1.2 The title page of Lavoisier's *Traité élémentaire de chimie*

+ ADDITIONAL

Reversible reactions

In chemistry, a double arrow ($\rightleftharpoons$) is used when writing a chemical equation to show a reversible process. In this way, you can show the phase changes associated with water, using the following equation:

$$H_2O(l) \rightleftharpoons H_2O(g)$$

Many reversible reactions are essential to our society, for example, the reactions that occur in rechargeable batteries (Figure 6.1.3).

FIGURE 6.1.3 Rechargeable batteries have a lower environmental impact than traditional single-use batteries. Rechargeable batteries use a reversible chemical reaction to gain and lose power.

BALANCING CHEMICAL EQUATIONS

There is conservation of mass, and also **conservation of elements**. The conservation of elements simply means that the total number and type of each element on each side of the chemical equation must be equal. Adding the correct coefficients will give you a **balanced chemical equation**. When balancing chemical equations, only change the coefficients of reactants and products, not the subscripts in the reactant and product formulae. Consider the difference between $2H_2O$ (two water molecules) and H_2O_2 (one molecule of hydrogen peroxide, a strong oxidising agent and the key component in products designed to bleach hair). Changing the subscript changes the chemical species (Figure 6.1.4).

(a)

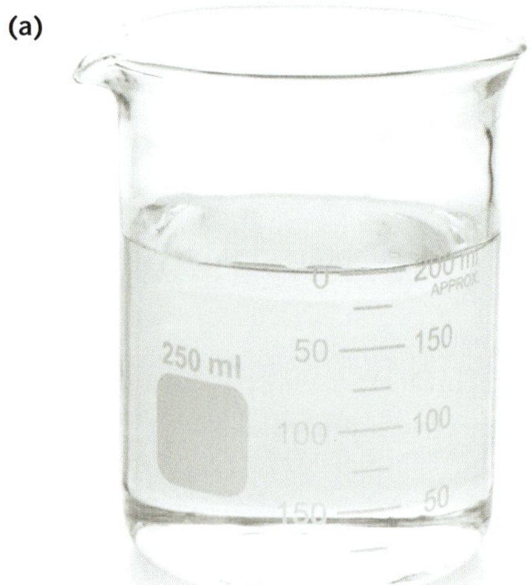

(b)

FIGURE 6.1.4 (a) Water and (b) hydrogen peroxide look very similar and have very similar chemical formulae—H_2O and H_2O_2. However, water is essential to human life while, whereas hydrogen peroxide can be dangerous in strong concentrations.

Worked example 6.1.1

BALANCING A SIMPLE CHEMICAL EQUATION

Balance the following chemical equation: $H_2(g) + O_2(g) \rightarrow H_2O(l)$

Thinking	Working
Underneath the arrow, write the letters R | E | P (Reactants, Elements, Products)	$H_2(g) + O_2(g) \rightarrow H_2O(l)$ R | E | P
List the elements under 'E' in the order in which they appear on the reactant side.	$H_2(g) + O_2(g) \rightarrow H_2O(l)$ R | E | P | H | | O |
Write the total number of each listed element on both the reactant and product sides.	$H_2(g) + O_2(g) \rightarrow H_2O(l)$ R | E | P 2 | H | 2 2 | O | 1 You can see that in this example, the hydrogen is currently balanced, but not the oxygen.
Balance by multiplying the coefficients until you have an equal number of each element on each side of the equation.	$H_2(g) + O_2(g) \rightarrow 2H_2O(l)$ R | E | P 2 | H | 2 × 2 = 4 2 | O | 1 × 2 = 2 On the products side: Multiplying the number of oxygen atoms by 2 requires doubling the number of H_2O molecules. This means the number of hydrogen atoms increases to 4. $2H_2(g) + O_2(g) \rightarrow 2H_2O(l)$ R | E | P 4 = 2 × 2 | H | 2 × 2 = 4 2 | O | 1 × 2 = 2 On the reactants side: Multiplying the number of hydrogen atoms by 2 requires doubling the number of H_2 molecules.
Write the equation with the correct coefficients.	$2H_2(g) + O_2(g) \rightarrow 2H_2O(l)$

Worked example: Try yourself 6.1.1

BALANCING A SIMPLE CHEMICAL EQUATION

Balance the following chemical equation: $H_2(g) + N_2(g) \rightarrow NH_3(g)$

BALANCING CHEMICAL EQUATIONS WITH A TRANSFER OF POLYATOMIC IONS

Chemical equations that appear more complex can still be solved using the method described in Worked example 6.1.1. If it can be recognised that there is an exchange of polyatomic ions, then the polyatomic ion can be treated as a single chemical species rather than as separate elements.

GO TO ➤ Section 5.2 page 147

Worked example 6.1.2

BALANCING A CHEMICAL EQUATION WITH TRANSFER OF POLYATOMIC IONS

Balance the following chemical equation: $Al_2(SO_4)_3(aq) + Ca(OH)_2(aq) \rightarrow Al(OH)_3(s) + CaSO_4(aq)$

<table>
<tr><th>Thinking</th><th>Working</th></tr>
<tr><td>Underneath the arrow, write the letters R | E | P (Reactants, Elements, Products)
Note: Polyatomic ions are treated like 'elements' when balancing chemical equations using this method.</td><td>$Al_2(SO_4)_3(aq) + Ca(OH)_2(aq) \rightarrow Al(OH)_3(s) + CaSO_4(aq)$
R | E | P</td></tr>
<tr><td>List the elements and polyatomic ions that do not change from reactant to product side under 'E' in the order in which they appear on the reactant side.</td><td>$Al_2(SO_4)_3(aq) + Ca(OH)_2(aq) \rightarrow Al(OH)_3(s) + CaSO_4(aq)$
R | E | P
| Al^{3+} |
| SO_4^{2-} |
| Ca^{2+} |
| OH− |</td></tr>
<tr><td>Write the total number of each listed element/ polyatomic ion on both the reactant and product sides.</td><td>$Al_2(SO_4)_3(aq) + Ca(OH)_2(aq) \rightarrow Al(OH)_3(s) + CaSO_4(aq)$
R | E | P
2 | Al^{3+} | 1
3 | SO_4^{2-} | 1
1 | Ca^{2+} | 1
2 | OH^- | 3</td></tr>
<tr><td>Balance by multiplying the coefficients until you have an equal number of each element/polyatomic ion on each side of the equation.</td><td>$Al_2(SO_4)_3(aq) + Ca(OH)_2(aq) \rightarrow 2Al(OH)_3(s) + 3CaSO_4(aq)$
R | E | P
2 | Al^{3+} | $1 \times 2 = 2$
3 | SO_4^{2-} | $1 \times 3 = 3$
1 | Ca^{2+} | $1 \times 3 = 3$
2 | OH^- | $3 \times 2 = 6$
On the products side:
Multiplying the number of Al^{3+} by 2 requires doubling the number of $Al(OH)_3$. This means the number of OH^- increases to 6.
Multiplying the number of Ca^{2+} by 3 requires tripling the number of $CaSO_4$. This means the number of SO_4^{2-} increases to 3.
$Al_2(SO_4)_3(aq) + 3Ca(OH)_2(aq) \rightarrow 2Al(OH)_3(aq) + 3CaSO_4(aq)$
R | E | P
2 | Al^{3+} | 2
3 | SO_4^{2-} | 3
$3 = 3 \times 1$ | Ca^{2+} | 3
$6 = 3 \times 2$ | OH^- | 6
On the reactants side:
Multiplying the number of Ca^{2+} by 3 requires tripling the number of $Ca(OH)_2$. This means the number of OH^- increases to 6.</td></tr>
<tr><td>Write the equation with the correct coefficients.</td><td>$Al_2(SO_4)_3(aq) + 3Ca(OH)_2(aq) \rightarrow 2Al(OH)_3(s) + 3CaSO_4(aq)$</td></tr>
</table>

Worked example: Try yourself 6.1.2

BALANCING A CHEMICAL EQUATION WITH TRANSFER OF POLYATOMIC IONS

Balance the following chemical equation:

$AgNO_3(aq) + K_3PO_4\ (aq) \rightarrow Ag_3PO_4(s) + KNO_3(aq)$

BALANCING MORE COMPLEX CHEMICAL EQUATIONS

As an alternative to the R|E|P method, more complex chemical equations can be balanced algebraically. This is where coefficients of chemical species are assigned a lower-case letter (so as to not confuse them with atomic symbols) and solved for those values accordingly.

Worked example 6.1.3

BALANCING A MORE COMPLEX CHEMICAL EQUATION

Balance the following chemical equation: $NaCl(aq) + SO_2(g) + H_2O(l) + O_2(g) \rightarrow Na_2SO_4(aq) + HCl(aq)$

Thinking	Working
In front of each chemical species, left to right, place the letters *a*, *b*, *c* etc. until all chemical species have a letter.	$a NaCl(aq) + b SO_2(g) + c H_2O(l) + d O_2(g) \rightarrow e Na_2SO_4(aq) + f HCl(aq)$
List the elements as they appear on the reactant side. Only list each element once, even if it appears a second (or third) time.	$a NaCl(aq) + b SO_2(g) + c H_2O(l) + d O_2(g) \rightarrow e Na_2SO_4(aq) + f HCl(aq)$ Na: Cl: S: O: H:
Look at each element and determine an algebraic expression to balance each side.	$a NaCl(aq) + b SO_2(g) + c H_2O(l) + d O_2(g) \rightarrow e Na_2SO_4(aq) + f HCl(aq)$ Na: $a = 2e$ Cl: $a = f$ S: $b = e$ O: $2b + c + 2d = 4e$ H: $2c = f$
As there is more Na on the product side, consider initially setting $e = 1$. Using this, solve for all values.	$a NaCl(aq) + b SO_2(g) + c H_2O(l) + d O_2(g) \rightarrow e Na_2SO_4(aq) + f HCl(aq)$ Na: $a = 2(1) = 2$ Cl: $a = f = 2$ S: $b = e = 1$ O: $2(1) + 1 + 2d = 4(1)$; $3 + 2d = 4$; $2d = 1$; $d = 0.5$ H: $2c = f$; $2c = 2$; $c = 1$ In summary: $a = 2$, $b = 1$, $c = 1$, $d = 0.5$, $e = 1$ and $f = 2$
In order to obtain a whole number solution, multiply all the coefficients by 2.	$a = 2 \times 2 = 4$ $b = 1 \times 2 = 2$ $c = 1 \times 2 = 2$ $d = 0.5 \times 2 = 1$ $e = 1 \times 2 = 2$ $f = 2 \times 2 = 4$
Write down the balanced equation. (Remember that coefficients of '1' do not need to be written.)	$4NaCl(aq) + 2SO_2(g) + 2H_2O(l) + O_2(g) \rightarrow 2Na_2SO_4(aq) + 4HCl(aq)$

Worked example: Try yourself 6.1.3

BALANCING A MORE COMPLEX CHEMICAL EQUATION

Balance the following chemical equation:

$Ca_3(PO_4)_2(s) + SiO_2(s) + C(s) \rightarrow CaSiO_3(s) + CO(g) + P(s)$

VISUALISING BALANCED CHEMICAL EQUATIONS

Chemical equations can also tell us how the atoms in the reactants are rearranged to make the products. This is best represented visually. Recall the reaction which Lavoisier conducted:

$$2Hg(l) + O_2(g) \rightarrow 2HgO(s)$$

Consider how this might appear visually. Figure 6.1.5 illustrates how the double bond in the reactant oxygen molecule is broken, the oxygen atoms split from the oxygen molecule, and new bonds are formed with each separate atom of mercury to give the products. Without the correct balancing, the correct number of atoms on each side of the equation would not be seen.

Hg + O=O → HgO
Hg HgO

FIGURE 6.1.5 Bonds in the oxygen molecule are broken and new bonds are formed between oxygen and mercury.

6.1 Review

SUMMARY

- Conservation of mass requires that the total mass of the components of a chemical reaction in a closed system remains constant.
- A coefficient is the number in front of a chemical species that indicates how many of that chemical species is present.
- Changing a subscript number changes the chemical species.
- The physical states in a chemical reaction are represented simply as: solid (s), liquid (l), gas (g) or aqueous (aq).
- Chemical equations are balanced by altering the coefficients until there is the same number of particles of each element on each of the reactant and product sides of the reaction arrow.

KEY QUESTIONS

1 Which of the following statements about the law of conservation of mass is correct?
  A The total mass of a closed system changes depending on the reaction conditions.
  B Mass cannot be conserved during a chemical reaction as a little bit of mass is always lost.
  C The mass of a closed system cannot change over time. Mass cannot be created nor destroyed.
  D The mass of a closed system always increases because mass is gained from the surroundings.

2 Write balanced chemical equations that correspond to the following word equations:
  a Gaseous nitrogen(II) oxide and hydrogen react to give ammonia gas and liquid water.
  b Solid carbon burns in a limited supply of oxygen gas to produce carbon monoxide gas.
  c Solid copper(II) oxide and ammonia gas react to give solid copper, gaseous nitrogen(II) oxide and liquid water.

3 Balance the following chemical equations.
  a $KClO_3(s) \rightarrow KCl(s) + O_2(g)$
  b $C_2H_2(g) + O_2(g) \rightarrow CO_2(g) + H_2O(g)$
  c $Na(s) + H_2O(l) \rightarrow NaOH(aq) + H_2(g)$

4 Calculate how many atoms of each different type of element are found in each of the following:
  a $8Ag_2S$
  b $2Mg_3(PO_4)_2$
  c $3NaClO_3$

5 Describe the following chemical equations using word equations.
  a $Zn(s) + CuSO_4(aq) \rightarrow Cu(s) + ZnSO_4(aq)$
  b $Mg(s) + CO_2(g) \rightarrow C(s) + MgO(s)$
  c $2KOH(aq) + H_2CO_3(aq) \rightarrow 2H_2O(l) + K_2CO_3(aq)$

6.2 Problems involving conservation of mass

In this section, you will explore the definitions of stoichiometry, SI units and density.

WHAT IS STOICHIOMETRY?

The term **stoichiometry** comes from the German word *stöchiometrie*, coined in 1792 by Jeremias Richter (Figure 6.2.1) from the Greek word 'stoicheion', meaning 'any first thing, from which others in a series take their rise'. Specifically, in chemistry, stoichiometry relates to the quantities and ratios of chemical species in a chemical reaction. It enables us to predict the **yield** of products (how much is produced), and whether a reactant is limiting or in excess. This will be further explored in Chapter 7.

GO TO ➤ Section 7.5 page 242

MEASURING MASSES OF SOLIDS IN CHEMICAL REACTIONS

The International System of Units is the most widely used system of measurement. It is abbreviated as SI as it is taken from the original French name, *Le Système International d'Unités*. **SI units** are the standard units of measurement for variables. The SI unit for mass is kilograms (abbreviated, kg). It is important to be conscious of any unit conversions required if a problem or real-world example is given to you in units other than SI units.

FIGURE 6.2.1 Jeremias Richter (1762–1807)

Mass–mass stoichiometry

Remember that mass is conserved, so the total mass in a chemical reaction must be the same before and after the reaction has taken place. Knowing this, you can work out the mass of product formed in a reaction, or how much mass of reactant has been used up in a chemical reaction.

Worked example 6.2.1

CALCULATING THE MASS OF REACTANT CONSUMED OR PRODUCT FORMED

Assuming that all reactants are consumed in the following reaction, calculate the mass of product formed. When 46.0 g of sodium metal (Na) reacts with 253.8 g of solid iodine (I), what mass of sodium iodide (NaI) is produced?

Thinking	Working
Write a balanced chemical equation for the process.	$2Na(s) + I_2(s) \rightarrow 2NaI(s)$
List the initial amounts of all chemical species before the reaction takes place.	$2Na(s) + I_2(s) \rightarrow 2NaI(s)$ Initial: 46.0 g 253.8 g 0 g
List the changes in mass of the chemical reactant species. As all reactants are consumed in this reaction, all of the mass is converted into the product.	$2Na(s) + I_2(s) \rightarrow 2NaI(s)$ Initial: 46.0 g 253.8 g 0 g Change: –46.0 –253.8 $+x$
The change in mass of the product will be equal to the sum of the mass lost by the reactants.	$x = 46.0 + 253.8$ $= 299.8$ g

Worked example: Try yourself 6.2.1

CALCULATING THE MASS OF REACTANT CONSUMED OR PRODUCT FORMED

Assuming that all reactants are consumed in the following reaction, calculate the mass of reactant consumed. When 84.6 g of solid iodine (I_2) reacts with sufficient sodium (Na) metal to produce 99.9 g of sodium iodide (NaI), what mass of sodium is consumed?

Mass–mass stoichiometry with excess reactants

GO TO ➤ Section 7.5 page 242

Sometimes you will have too much reactant in a chemical reaction; not all of it will be consumed in the reaction. This is known as a reactant that is in **excess**. This will be further explored in Chapter 7.

Worked example 6.2.2

CALCULATING THE MASS OF EXCESS REACTANT

One student measures 34.5 g of sodium (Na) and reacts it with 380.7 g of iodine (I_2), which is all consumed, producing 415.2 g of sodium iodide (NaI).

A second student measures the same amount of sodium and reacts it with 450.0 g of iodine and still only produces 415.2 g of sodium iodide. What mass of iodine is in excess?

Thinking	Working
Write a balanced chemical equation for the process.	$2Na(s) + I_2(s) \rightarrow 2NaI(s)$
List the initial amounts of all chemical species before the reaction takes place.	$2Na(s) + I_2(s) \rightarrow 2NaI(s)$ Initial: 34.5 g 450.0 g 0 g
List the changes in mass of the chemical reactant species. All reactants are consumed in the first student's reaction. The masses of sodium in each student's reaction are the same and the amount of iodine is more than what was required initially.	$2Na(s) + I_2(s) \rightarrow 2NaI(s)$ Initial: 34.5 g 380.7 g 0 g Change: –34.5 –380.7 +415.2 g
The mass of iodine remaining at the end of the second student's reaction will be the difference between the initial mass and the mass consumed in the reaction.	Excess iodine = 450.0 – 380.7 = 69.3 g

Worked example: Try yourself 6.2.2

CALCULATING THE MASS OF EXCESS REACTANT

One student measures 11.5 g of sodium (Na) and reacts it with 126.9 g of iodine (I_2), which is all consumed, producing 138.4 g of sodium iodide (NaI).

A second student measures 15.0 g of sodium and reacts it with the same amount of iodine and still produces 138.4 g of sodium iodide. What mass of sodium is in excess?

MEASURING LIQUIDS IN CHEMICAL REACTIONS

The SI unit for volume is litres (L), but it may be expressed in cubic centimetres (cm^3). Note that 1 cm^3 = 1 mL and 1000 mL = 1 L. At times it may be useful to measure a liquid as a volume. Volume and mass can be determined from one another by knowing the density of the liquid.

The **density** of a substance (d) is equal to the mass divided by the volume (typically in $g\,mL^{-1}$) and can be calculated using the following formula:

$$d = \frac{m}{V}$$

where:

d is the density in $g\,mL^{-1}$ or $g\,L^{-1}$

m is the mass in g

V is the volume in mL or L.

ℹ The density formula $d = \frac{m}{V}$ can be transposed to other useful forms:

$m = d \times V$ and $V = \frac{m}{d}$

For example, the density of ethanol is 0.789 g mL^{-1} at 25°C. Density is temperature dependent. Generally, the higher the temperature, the lower the density. For example, the density of water is typically taken as 1.00 g mL^{-1}, but this is only true at 4.0°C. At 20°C, the density of water is 0.998 g mL^{-1}, and at 65 °C it is 0.980 g mL^{-1}. The difference may appear subtle, but it is significant when calibrating precise pieces of equipment used for measuring volume, such as the pipettes used in analytical techniques. Lower densities at higher temperatures can be understood from particle theory: as particles gain kinetic energy, they spread further apart from one another.

Mass–volume stoichiometry

In stoichiometric calculations, if quantities are given as a volume, conversions into mass can be made by using the density formula. It is a common misconception that the 'thicker' or more viscous a liquid is, the greater is its density. **Viscosity** is a measure of how easily a substance flows, whereas density is how closely packed the particles are within a substance.

Worked example 6.2.3

CALCULATING MASS FROM VOLUME OF LIQUIDS

1.800 kg of glucose ($C_6H_{12}O_6$) from sugar cane is fermented in anaerobic conditions at 25°C to produce 1.167 L of ethanol (CH_3CH_2OH). The other product formed is carbon dioxide (CO_2).

Determine the mass of ethanol produced, and hence, using the law of conservation of mass, determine the mass of carbon dioxide produced. Take the density of ethanol to be 0.789 g mL^{-1} at 25°C.

Thinking	Working
Write a balanced chemical equation for the process.	$C_6H_{12}O_6(s) \rightarrow 2CH_3CH_2OH(l) + 2CO_2(g)$
Convert the volume of ethanol from L to mL.	$1.167 \times 1000 = 1167$ mL
Using the density formula, determine the mass of ethanol produced.	$d = \frac{m}{V}$ $m = d \times V$ $m(CH_3CH_2OH) = 0.789 \times 1167$ $= 920$ g
List the initial amounts of all chemical species before the reaction takes place.	$C_6H_{12}O_6(s) \rightarrow 2CH_3CH_2OH(l) + 2CO_2(g)$ Initial: 1800 g 0 g 0 g
List the changes in mass of the chemical reactant species. As all reactants are consumed in this reaction, all of the mass is converted into the product.	Initial: 1800 g 0 g 0 g Change: –1800 g +920 g +x g
The change in mass of the carbon dioxide will be equal to the difference between the mass of the glucose and the mass of the ethanol.	$x = 1800 - 920$ $= 880$ g $m(CO_2) = x = 880$ g

Worked example: Try yourself 6.2.3

CALCULATING MASS FROM VOLUME OF LIQUIDS

Hydrogen peroxide (H_2O_2) is a strong disinfectant and is also used in hair bleaching products. The production of hydrogen peroxide can be simplified to:

$$H_2(g) + O_2(g) \rightarrow H_2O_2(l)$$

0.04533 L of hydrogen peroxide is produced from the reaction between 4.0 g of hydrogen gas (H_2) and sufficient oxygen (O_2) to allow the reaction to be complete. Determine the mass of hydrogen peroxide produced, and hence, using the law of conservation of mass, determine the mass of oxygen gas consumed.

Take the density of hydrogen peroxide to be 1.50 g mL^{-1} at 25°C.

MEASURING GASES IN CHEMICAL REACTIONS

The mass of a gas is often difficult to measure, especially for a gas less dense than air. Volume can be a more useful way of measuring the amount of gas present in a chemical reaction.

Volume–volume stoichiometry

GO TO ➤ Section 9.3 page 289

Unlike mass, volume is not necessarily conserved in a chemical reaction. This is especially evident in gaseous reactions. This is further explored in Chapter 9 in relation to Avogadro's law. Using density and volume, the mass of reactants or products can be determined by applying our understanding that mass is conserved in chemical reactions.

As there are fewer gas particles present per unit of volume than in a liquid, the unit for the density of a gas is typically $g\,L^{-1}$. The density of a gas is both temperature and pressure dependent. For example, the density of methane at 25°C and one atmosphere (1 atm) of pressure is $0.656\,g\,L^{-1}$. At 0°C and 1 atm of pressure, the density is $0.716\,g\,L^{-1}$. This is further explored in Chapter 9.

Particle theory explains why the density of gases is pressure dependent: as gas particles are relatively widely spaced, compared with liquids or solids, gases are easily compressed when pressure is applied.

Worked example 6.2.4

CALCULATING MASS FROM VOLUME OF GASES

At 25°C and 1 atm of pressure, the density of methane (CH_4) is $0.656\,g\,L^{-1}$, the density of oxygen gas (O_2) is $1.31\,g\,L^{-1}$ and the density of carbon dioxide (CO_2) is $1.81\,g\,L^{-1}$.

Methane reacts with oxygen to form carbon dioxide and water according to the following equation:

$$CH_4(g) + 2O_2(g) \rightarrow CO_2(g) + 2H_2O(g)$$

If 12.2 L of methane reacts completely with 24.4 L of oxygen to produce 18.0 g of water, what mass of carbon dioxide is produced?

Thinking	Working
Write a balanced chemical equation for the process.	$CH_4(g) + 2O_2(g) \rightarrow CO_2(g) + 2H_2O(g)$
Using the density formula, calculate the mass of the reactant species.	$d = \frac{m}{V}$ $m = d \times V$ $m(CH_4) = 0.656 \times 12.2 = 8.00\,g$ $m(O_2) = 1.31 \times 24.4 = 32.0\,g$
List the initial amounts of all chemical species before the reaction takes place and list the changes in mass of the chemical reactant species. As all reactants are consumed, all of the mass is converted into the products.	$CH_4(g) + 2O_2(g) \rightarrow CO_2(g) + 2H_2O(g)$ Initial: 8.00 g, 32.0 g, 0 Change: −8.00, −32.0, $+x$, +18.0
The change in mass of the carbon dioxide will equal the difference between the total mass of the reactants and the mass of the water vapour produced, and thus will equal the mass of carbon dioxide produced.	$x = (8.00 + 32.0) - 18.0$ $= 22.0\,g$ $m(CO_2) = x = 22.0\,g$

CHEMFILE S CC

The Save Power campaign

Between 2009 and 2011, the Save Power campaign was a NSW Government initiative to deliver information to the wider community about their power usage and how to limit their consumption. The residential sector accounts for approximately 30% of all electricity use in NSW. As mass is difficult to visualise, each black balloon helps viewers 'see' 50 g of carbon dioxide (Figure 6.2.2). Save Power members pledged to save close to 2900 tonnes of carbon dioxide, equal to 58 million of the black balloons.

FIGURE 6.2.2 The Save Power campaign used a black balloon to represent 50 g of carbon dioxide produced.

Worked example: Try yourself 6.2.4

CALCULATING MASS FROM VOLUME OF GASES

At 15°C and 1 atm of pressure, the density of butane (C_4H_{10}) is $2.48\,g\,L^{-1}$, the density of oxygen gas (O_2) is $1.36\,g\,L^{-1}$ and the density of carbon dioxide (CO_2) is $1.87\,g\,L^{-1}$.

Butane reacts with oxygen to form carbon dioxide and water according to the following equation:

$$2C_4H_{10}(g) + 13O_2(g) \rightarrow 8CO_2(g) + 10H_2O(g)$$

If 23.4 L of butane reacts completely with 152.1 L of oxygen to produce 90.0 g of water, what mass of carbon dioxide is produced?

6.2 Review

SUMMARY

- Stoichiometry is an approach for determining how much of a chemical species has reacted or is expected to be produced in a chemical reaction.
- SI units refer to the standard units used internationally. The SI unit for mass is grams (g) and the unit for volume is litres (L).
- Excess reactants are those that are not consumed in a chemical reaction.
- Density is a measure of mass per unit of volume. The typical units are $g\,mL^{-1}$ for liquids and $g\,L^{-1}$ for gases.

KEY QUESTIONS

1 State whether the following statements are true or false. For those that are false, correct the statement.

- **a** The total mass in a chemical reaction is conserved.
- **b** Density is a measure of mass of a substance multiplied by the volume that it occupies.
- **c** The density of a gas is dependent on temperature and pressure.
- **d** Volume is conserved in a chemical reaction.

2 Magnesium reacts with oxygen according to the following equation:

$$2Mg(s) + O_2(g) \rightarrow 2MgO(s)$$

Calculate:

- **a** the mass of MgO produced when 5.00 g of Mg reacts to completion with 3.29 g of oxygen
- **b** the mass of O_2 consumed when 24.3 g of Mg reacts to produce 40.3 g of MgO
- **c** the mass of Mg that reacts with 12.05 g of O_2 to produce 30.35 g of MgO.

3 Consider the following reaction:

$$Fe_2O_3(s) + 2Al(s) \rightarrow Al_2O_3(s) + 2Fe(s)$$

If 200.0 g of Fe_2O_3 reacts completely with 67.6 g of Al to produce 127.8 g of Al_2O_3 and 139.8 g of Fe:

- **a** Check to see that this reaction follows the law of conservation of mass.
- **b** If 100.0 g of Al was used instead, what mass of Al would be in excess?
- **c** If 165.0 g of Fe_2O_3 was used instead, how much more Fe_2O_3, in g, would be required for this reaction to go to completion?
- **d** If you doubled all reactant quantities, what would the effect on the total mass of the products be?

4 In the decomposition reaction of solid calcium carbonate ($CaCO_3$), calcium oxide (CaO) and gaseous carbon dioxide (CO_2) form. At 25°C and 1 atm of pressure, the density of carbon dioxide is $1.81\,g\,L^{-1}$. When 10.00 g of calcium carbonate decomposes, 5.72 g of calcium oxide is produced.

- **a** What mass of carbon dioxide is formed?
- **b** Hence, determine what volume of carbon dioxide is produced.

Chapter review

KEY TERMS

balanced chemical equation
chemical species
closed system
conservation of elements
conservation of mass
density
excess reactant
product
reactant
SI unit
stoichiometry
viscosity
yield

REVIEW QUESTIONS

1 The law of conservation of mass can be best described as:
 A Mass in a closed system is free to change depending on the reaction.
 B The total mass after a reaction takes place will be less than at the beginning of a reaction.
 C The number of particles in a chemical reaction is conserved, but mass can change.
 D The total mass before and after a reaction takes place in a closed system is the same.

2 Which of the following shows a balanced chemical equation?
 A $CH_4(g) + O_2(g) \rightarrow CO_2(g) + 2H_2O(l)$
 B $F_2(g) + P(s) \rightarrow PF_3(g)$
 C $NaOH(aq) + H_2SO_4(aq) \rightarrow Na_2SO_4(aq) + H_2O(l)$
 D $2KF(s) + BaBr_2(s) \rightarrow 2KBr(s) + BaF_2(s)$

3 Which one of the following is a correctly balanced chemical equation for the reaction between a piece of copper metal and a solution of silver nitrate ($AgNO_3$)?
 A $CuNO_3(aq) + Ag(s) \rightarrow Cu(s) + AgNO_3(aq)$
 B $Cu(s) + 2AgNO_3(aq) \rightarrow Cu(NO_3)_2(aq) + 2Ag(s)$
 C $2Cu(s) + AgNO_3(aq) \rightarrow 2CuNO_3(aq) + Ag(s)$
 D $Cu(s) + Ag(NO_3)_2(aq) \rightarrow Cu(NO_3)_2(aq) + 2Ag(s)$

4 Balance the following unbalanced simple chemical equations:
 a $Li(s) + O_2(g) \rightarrow Li_2O(s)$
 b $Al(OH)_3(s) \rightarrow Al_2O_3(s) + H_2O(l)$
 c $Rb(s) + RbNO_3(aq) \rightarrow Rb_2O(s) + N_2(g)$
 d $Fe(OH)_2(aq) + O_2(g) \rightarrow Fe_2O_3(s) + H_2O(l)$
 e $NaOCl(aq) \rightarrow NaCl(aq) + O_2(g)$
 f $H_2SiCl_2(s) + H_2O(l) \rightarrow H_8Si_4O_4(aq) + HCl(aq)$

5 Balance the following unbalanced chemical equations involving the transfer of one or more polyatomic ions:
 a $HNO_3(aq) + Na_2CO_3(aq) \rightarrow NaNO_3(aq) + H_2O(l) + CO_2(g)$
 b $H_2SO_4(aq) + KOH(aq) \rightarrow K_2SO_4(aq) + H_2O(l)$
 c $Ba(NO_3)_2(aq) + H_2CO_3(aq) \rightarrow BaCO_3(s) + HNO_3(aq)$

6 Balance the following unbalanced chemical equations using the algebraic method:
 a $Cu(s) + HNO_3(aq) \rightarrow Cu(NO_3)_2(aq) + NO(g) + H_2O(l)$
 b $CO(g) + H_2(g) \rightarrow C_8H_{18}(l) + H_2O(l)$
 c $(NH_4)_2Cr_2O_7(s) \rightarrow N_2(g) + H_2O(g) + Cr_2O_3(s)$
 d $P_2I_4(s) + P_4(s) + H_2O(l) \rightarrow PH_4I\ (aq) + H_3PO_4(aq)$

7 Complete the following table by identifying the mass of the remaining chemical species.

Reaction	Reactant(s)			Product(s)	
a	$2H_2(g)$ +	$O_2(g)$	→	$H_2O(l)$	
	3.4 g	27.2 g			
b	$CH_4(g)$ +	$2O_2(g)$	→	$CO_2(g)$ +	$2H_2O(g)$
	12.2 g	48.8 g			20.0 g
c	$2HgO(s)$		→	$2Hg(l)$ +	$O_2(g)$
	23.6 g			21.9 g	

8 In Lavoisier's experiments, mercury and oxygen combined in the ratio that was always close to 25 g mercury to 2 g of oxygen. On the basis of this information, calculate the mass of oxygen that would combine with:
 a 5 g mercury
 b 4.2 g mercury
 c 0.27 g mercury.

9 Reflect on the Inquiry activity on page 206, and answer the following:
 a What evidence is there to determine that a chemical reaction has taken place?
 b What is happening to the atoms in the reaction between vinegar and milk?
 c If 30.0 g of milk happened to completely react with exactly 25.0 g of vinegar, what would the mass of the resultant products be?

10 a Balance the following chemical equation for the reaction of calcium with oxygen gas to form calcium oxide:

$$Ca(s) + O_2(g) \rightarrow CaO(s)$$

 b If 10.0 g of calcium reacts completely with exactly 4.0 g of oxygen gas, what mass of calcium oxide is formed?
 c If 10.0 g of calcium reacts with 10.0 g of oxygen gas instead, what mass of oxygen gas would be in excess?

11 A compound of tungsten and sulfur is useful as a solid lubricant. 1.84 g of tungsten reacts exactly with 0.64 g of sulfur. If 2.00 g of tungsten were to be used, how much would be in excess?

12 The density of pure acetic acid is 1.05 g mL^{-1} at 25°C. How much:
   a mass would you have in 30 mL?
   b volume would you have in 10.0 g?
   c mass would you have in 2.0 L?

13 The density of ammonia (NH_3) at 1 atm of pressure and 25°C is 0.860 g L^{-1}. How much:
   a volume would 50.0 g of ammonia occupy?
   b mass would there be in 15.0 L of ammonia?
   c volume would there be in 7.50 kg of ammonia occupy?

14 a Describe the trend in the density of a substance as temperature increases while the pressure remains constant.
   b Describe the trend in the density of a substance as pressures increases while the temperature remains constant.

15 At 25°C, 10.0 mL of ethanol (C_2H_5OH, density = 0.789 g mL^{-1}) completely reacts with 15.8 mL of butanoic acid (C_3H_7COOH, density = 0.953 g mL^{-1}) to form 3.09 g of water (H_2O) and ethyl butanoate ($C_6H_{12}O_2$, density = 0.879 g mL^{-1}). Ethyl butanoate is an artificial pineapple flavouring.
   a Write a balanced chemical equation to represent the reaction described.
   b What is the mass of ethanol in this reaction?
   c What is the mass of butanoic acid in this reaction?
   d Using the law of conservation of mass, determine the mass of ethyl butanoate produced.
   e Using the density for ethyl butanoate, determine the volume of ethyl butanoate produced.

16 Ammonium nitrate (NH_4NO_3) is an important fertiliser but readily forms explosive mixtures when combined with some other explosive compounds. When solid ammonium nitrate explodes, the products are all gases, including nitrogen, oxygen and water vapour. If 40.0 g of ammonium nitrate explodes, it produces 14.0 g of nitrogen (d = 1.25 g L^{-1} at 0°C and 1 atm of pressure) and 8.0 g of oxygen (d = 1.43 g L^{-1}), and the remaining mass is water vapour, which occupies a volume of 24.7 L.
   a Write a balanced chemical equation for the decomposition of ammonium nitrate.
   b What is the mass of the water vapour produced?
   c Determine the volume of nitrogen produced.
   d Determine the volume of oxygen produced.
   e Hence, determine the full volume of the products.
   f 40.0 g of ammonium nitrate only occupies a volume of 23.1 mL. Using your answer to part **e**, comment on the explosive nature of ammonium nitrate.

17 The following statements list a range of misconceptions in chemistry. Using your understanding of the law of conservation of mass and/or density, explain why these statements are not correct.
   a During a chemical reaction, the total number of product particles is less than the total number of reactant particles. Therefore, the total mass of the products is less than the total mass of the reactants.
   b Gases do not have mass.
   c A 'thick' liquid has a higher density than water.

18 When heated, mercury(II) oxide (HgO) produces mercury and oxygen. If these were the only three substances present, apply the law of conservation of mass to calculate:
   a the mass of mercury(II) oxide that would produce 2.00 g of liquid mercury and 0.16 g oxygen gas
   b the mass of mercury produced when 4.60 g of mercury(II) oxide is completely decomposed to produce 0.37 g of oxygen.

19 Solid sodium reacts with oxygen from the air, producing solid sodium oxide. The experimental data in the table below was obtained for the reaction between sodium and oxygen producing sodium oxide.

Mass of sodium reacting (g)	Mass of oxygen reacting (g)	Mass of sodium oxide produced (g)
2.00	0.70	2.70
3.00	1.04	4.04
4.00	1.39	5.39

   a Write a balanced chemical equation for this reaction.
   b How do the above results illustrate the law of conservation of mass?

20 Jessica heated some bright blue copper(II) nitrate crystals in a test tube. She noticed brown nitrogen dioxide gas being produced, and a glowing splint held at the top of the test tube relit (showing that oxygen gas was also produced). A fine black solid, copper(II) oxide, was left in the test tube.
   a Identify and list the reactants and products for this reaction.
   b Write a balanced chemical equation for the reaction that occurred.

21 David added some dilute hydrochloric acid to some solid limestone (calcium carbonate) in a beaker. When he weighed the products after the bubbling had stopped, he noticed that there had been a reduction in mass. Why did his results not agree with the law of conservation of mass?

CHAPTER

07 The mole concept

Chemists in fields as diverse as environmental monitoring, pharmaceuticals and fuel production routinely carry out chemical reactions as part of their work. It is important for them to be able to measure specific quantities of chemicals quickly and easily, in part because the amount of products formed depends on the amount of reactants.

By the end of this chapter, you will have a greater understanding of the way in which chemists measure quantities of chemicals, in particular of the way they can accurately count the number of particles in samples of elements and compounds simply by weighing them. This is essential for designing and producing materials, including cosmetics, fuels, fertilisers, pharmaceuticals and building materials.

Content

INQUIRY QUESTION

How are measurements made in chemistry?

By the end of this chapter, you will be able to:

- conduct a practical investigation to demonstrate and calculate the molar mass (mass of 1 mole) of:
  - an element
  - a compound (ACSCH046) ICT N
- conduct an investigation to determine that chemicals react in simple whole number ratios by moles ICT N
- explore the concept of the mole and relate this to Avogadro's constant to describe, calculate and manipulate masses, chemical amounts and numbers of particles in: (ACSCH007, ACSCH039) ICT N
  - moles of elements and compounds $n = \frac{m}{M}$ (n = number of moles, m = mass in g, M = molar mass in $g\,mol^{-1}$)
  - percentage composition calculations and empirical formulae
  - limiting reagent reactions

7.1 Introducing the mole

CHEMISTRY INQUIRY  

Moles of household items

How are moles measured in chemistry?

COLLECT THIS ...

- half a cup of each of:
  - table sugar (sucrose, $C_{12}H_{22}O_{11}$)
  - water
  - aluminium foil, tightly packed
  - table salt (NaCl)
  - bicarbonate of soda ($NaHCO_3$)
- digital kitchen scales
- measuring cup
- mobile phone or digital camera

DO THIS ...

1. Zero the scales with the measuring cup on the balance pan.
2. Measure out half a cup of each substance and weigh it. Record the masses in your table.
3. Take a photo as you weigh the substance.
4. If the substance came in a packet, write down the mass of the contents of the packet when it was unopened.

RECORD THIS ...

Draw up the table below in your workbook to record your results.

Use number of moles = $\frac{\text{mass (g)}}{\text{molar mass}}$ to calculate the number of moles in half a cup of each substance.

Calculate the number of molecules or particles in your sample using:

number of molecules = number of moles × 6.022×10^{23}.

REFLECT ON THIS ...

How do we measure the number of moles of substances?

Rank the substances from largest to smallest number of moles in half a cup.

Do liquids, ionic compounds, molecular substances or elements have the largest number of particles in half a cup?

Identify any patterns in your results and attempt to explain your observations.

What could you do next time to improve your experiment?

Substance	Mass (g)	Molar mass	Number of moles in half a cup	Mass of the full packet	Number of moles in the full packet
table sugar ($C_{12}H_{22}O_{11}$)		342.30			
water		18.02		-	-
aluminium		26.98			
table salt (NaCl)		55.44			
bicarbonate of soda ($NaHCO_3$)		84.01			

It is often essential for chemists to be able to measure the exact number of particles of an element or compound. However, the particles in elements and compounds (atoms, ions and molecules) are so small that it would be difficult to count them individually or even by the thousands of millions. If it were possible to count individual particles, the numbers in even very small samples would be huge and very inconvenient to work with.

The ice cubes shown in Figure 7.1.1 each contain more than 100 000 000 000 000 000 000 000 (or 10^{23}) water molecules (H_2O). As each water molecule is composed of two hydrogen atoms and one oxygen atom, the number of individual atoms in each ice cube is greater than 10^{23}. A quantity that allows chemists to measure accurate amounts of extremely small particles is required.

In this section, you will learn about the very convenient quantity used by chemists: the mole.

FIGURE 7.1.1 Each of these ice cubes contains more than 10^{23} water (H_2O) molecules. Molecules and atoms are so small, and the numbers of them in everyday samples are so large, that it would be very inconvenient to always count them individually.

THE CHEMIST'S COUNTING UNIT

A dozen is a convenient quantity for buying the eggs shown in Figure 7.1.2. For atoms, ions and molecules, which are much smaller than eggs, a quantity that describes a much larger number is needed.

The accepted quantity for chemists is the **mole**. The mole is the unit for **amount of substance** and is given the symbol *n* and the abbreviation 'mol'.

So n(glucose) = 2 mol is read as 'the amount of glucose molecules is 2 moles'.

Chemists use the mole as a counting measure. Figure 7.1.3 shows some quantities that you would be very familiar with, such as pair, dozen and ream. One dozen is equal to 12, two dozen equals 24, 20 dozen equals 240 and half a dozen equals 6. In the same way, chemists know that 1 mole is equivalent to a certain number and that 2 moles, 20 moles and half a mole are all multiples of that number.

FIGURE 7.1.2 One dozen is 12 eggs, two dozen is 24 eggs and half a dozen is 6 eggs.

FIGURE 7.1.3 Convenient quantities that you would be very familiar with: (a) a pair of shoes equals two, (b) a dozen roses equals 12, (c) a ream of paper equals 500 sheets.

Information provided by molecular formulae

The **molecular formula** of a substance indicates the number of atoms of each element in one molecule of the substance. In Chapter 5, you learnt that an oxygen molecule contains two oxygen atoms joined by a covalent bond.

GO TO ➤ Section 5.3 page 162

When referring to a mole of a substance, it is important to indicate which type of particle is being specified. The expression, '1 mol of oxygen' is ambiguous because it could describe 1 mol of oxygen atoms (O) or 1 mol of oxygen molecules (O_2). As there are two atoms in each oxygen molecule, 1 mol of oxygen molecules will contain 2 mol of oxygen atoms.

Some other examples of the use of the mole as a counting unit are provided in Table 7.1.1.

TABLE 7.1.1 Examples of the use of the mole as a counting unit

Number of moles of element or compound	Information that can be obtained about numbers of particles
1 mol of hydrogen atoms (H)	1 mol of hydrogen atoms (H)
1 mol of hydrogen molecules (H_2)	1 mol of hydrogen molecules (H_2) 2 mol of hydrogen atoms (H)
2 mol of aluminium atoms (Al)	2 mol of aluminium atoms (Al)
2 mol of calcium chloride ($CaCl_2$)	2 mol of Ca^{2+} ions 4 mol of Cl^- ions
10 mol of glucose ($C_6H_{12}O_6$) molecules	10 mol of glucose ($C_6H_{12}O_6$) molecules 60 mol of carbon atoms 120 mol of hydrogen atoms 60 mol of oxygen atoms

AVOGADRO'S CONSTANT

One mole of any substance is defined as the same number of particles as there are atoms in exactly 12 g of carbon-12. This number of atoms has been experimentally determined to be 6.022141×10^{23} atoms. That's 602 214 100 000 000 000 000 000 atoms!

GO TO ➤ SkillBuilder page 80

This number is commonly rounded to 6.022×10^{23} and is referred to as **Avogadro's constant** or Avogadro's number. It is given the symbol N_A.

Avogadro's constant is written in scientific notation.

$N_A = 6.022 \times 10^{23}\,mol^{-1}$

Therefore, 1 mol of particles contains 6.022×10^{23} particles.

FIGURE 7.1.4 One mole of sodium chloride and 1 mol of water contain the same number of particles, 6.022×10^{23}.

Avogadro's constant is an enormous number, but the extremely small size of atoms, ions and molecules means that 1 mol of most elements and compounds does not have a large mass or volume.

For example, in Figure 7.1.4 you can see that 1 mol of water, that is 6.022×10^{23} water molecules, has a volume of only 18 mL, and 1 mol of table salt (NaCl) has a mass of 58.44 g.

If you know that 1 mol of a substance contains 6.022×10^{23} particles, then:

- 2 mol of a substance contains $2 \times (6.022 \times 10^{23}) = 1.204 \times 10^{24}$ particles
- 0.3 mol of a substance contains $0.3 \times (6.022 \times 10^{23}) = 1.806 \times 10^{23}$ particles
- 4.70×10^{23} particles = $\frac{4.70 \times 10^{23}}{6.022 \times 10^{23}}$ particles = 0.781 mol
- 7.35×10^{24} particles = $\frac{7.35 \times 10^{24}}{6.022 \times 10^{23}}$ particles = 12.2 mol.

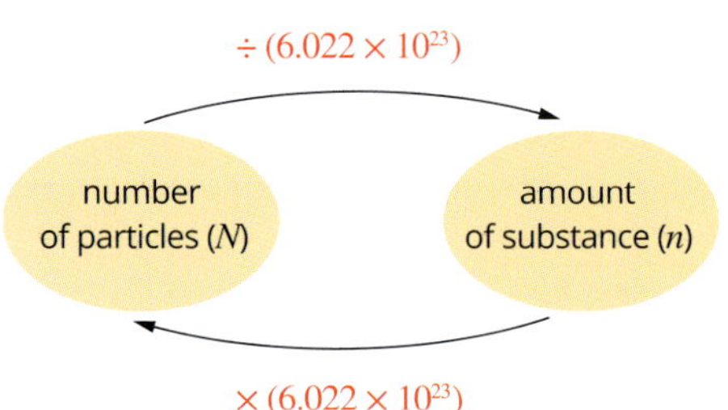

FIGURE 7.1.5 Relationship between number of particles and amount of substance in moles.

As you can see from Figure 7.1.5, a mathematical relationship exists between the number of particles, N, and the amount of substance in moles, n. This relationship can be written as:

$$n = \frac{N}{N_A}$$

Calculations using the mole and Avogadro's constant

Three quantities have been introduced so far:

- the mole, which is given the symbol n and the abbreviation mol
- Avogadro's constant, which is given the symbol N_A and has the value 6.022×10^{23}
- the actual number of particles (atoms, ions or molecules), which is given the symbol N.

The mathematical relationship that links the three quantities is $n = \frac{N}{N_A}$.

Worked example 7.1.1

CALCULATING THE NUMBER OF MOLECULES

Calculate the number of molecules in 3.5 mol of water (H_2O).

Thinking	Working
List the data given in the question next to the appropriate symbol. Include units.	The number of water molecules is the unknown, so: $N(H_2O) = ?$ $n(H_2O) = 3.5$ mol $N_A = 6.022 \times 10^{23}$
Rearrange the formula to make the unknown the subject.	$n = \frac{N}{N_A}$ so $N(H_2O) = n \times N_A$
Substitute in data and solve.	$N(H_2O) = n \times N_A$ $= 3.5 \times 6.022 \times 10^{23}$ $= 2.1 \times 10^{24}$ molecules

Worked example: Try yourself 7.1.1

CALCULATING THE NUMBER OF MOLECULES

Calculate the number of molecules in 1.6 mol of carbon dioxide (CO_2).

Worked example 7.1.2

CALCULATING THE NUMBER OF ATOMS

Calculate the number of oxygen atoms in 2.5 mol of oxygen gas (O_2).

Thinking	Working
List the data given in the question next to the appropriate symbol. Include units.	The number of oxygen atoms is the unknown, so: $N(O) = ?$ $n(O_2) = 2.5$ mol $N_A = 6.022 \times 10^{23}$
Calculate the amount, in mol, of oxygen atoms from the amount of oxygen and the molecular formula.	$n(O) = n(O_2) \times 2$ $= 2.5 \times 2$ $= 5.0$ mol
Rearrange the formula to make the unknown the subject.	$n = \frac{N}{N_A}$ so $N(O) = n \times N_A$
Substitute in data and solve.	$N(O) = n \times N_A$ $= 5.0 \times 6.022 \times 10^{23}$ $= 3.0 \times 10^{24}$ atoms

CHEMFILE N

Avogadro's constant

It is very difficult to imagine just how big Avogadro's constant really is, especially when atoms, ions and molecules are so small. Here are some examples to help.

- 6.022×10^{23} grains of sand, placed side by side, would stretch from Earth to the Sun and back about 7 million times.
- A computer counting 10 billion times every second would take 2 million years to reach 6.022×10^{23}.
- 6.022×10^{23} of the marshmallows shown in Figure 7.1.6 would cover Australia to a depth of 900 km!

FIGURE 7.1.6 One mole of (6.022×10^{23}) marshmallows would cover Australia to a depth of 900 km!

Worked example: Try yourself 7.1.2

CALCULATING THE NUMBER OF ATOMS

Calculate the number of hydrogen atoms in 0.35 mol of methane (CH_4).

Worked example 7.1.3

CALCULATING THE NUMBER OF MOLES, GIVEN THE NUMBER OF PARTICLES

Calculate the amount, in moles, of ammonia molecules (NH_3) represented by 2.5×10^{22} ammonia molecules.

Thinking	Working
List the data given in the question next to the appropriate symbol. Include units.	The number of mol of ammonia molecules is the unknown, so: $n(NH_3) = ?$ $N(NH_3) = 2.5 \times 10^{22}$ molecules $N_A = 6.022 \times 10^{23}$
Rearrange the formula to make the unknown the subject.	$n = \frac{N}{N_A}$ n is the unknown, so rearrangement is not required
Substitute in data and solve.	$n(NH_3) = \frac{N}{N_A}$ $= \frac{2.5 \times 10^{22}}{6.022 \times 10^{23}}$ $= 0.042$ mol

Worked example: Try yourself 7.1.3

CALCULATING THE NUMBER OF MOLES, GIVEN THE NUMBER OF PARTICLES

Calculate the amount, in mol, of magnesium atoms represented by 8.1×10^{20} magnesium atoms.

Worked example 7.1.4

CALCULATING THE NUMBER OF MOLES OF ATOMS, GIVEN THE NUMBER OF MOLES OF MOLECULES

Calculate the amount, in mol, of hydrogen atoms in 3.6 mol of sulfuric acid (H_2SO_4).

Thinking	Working
List the data given in the question next to the appropriate symbol. Include units.	The number of moles of hydrogen atoms is the unknown, so: $n(H) = ?$ $n(H_2SO_4) = 3.6$ mol
Calculate the number, in mol, of hydrogen atoms from the amount of sulfuric acid and the molecular formula.	$n(H) = n(H_2SO_4) \times 2$ $= 3.6 \times 2$ $= 7.2$ mol

Worked example: Try yourself 7.1.4

CALCULATING THE NUMBER OF MOLES OF ATOMS, GIVEN THE NUMBER OF MOLES OF MOLECULES

Calculate the amount, in mol, of hydrogen atoms in 0.75 mol of water (H_2O).

Identifying significant figures

When giving an answer to a calculation, it is important to take note of the number of significant figures that you use.

You should give an answer that is as accurate as possible. However, an answer can't be more precise than the data or the measuring device used to calculate it. For example, if a set of scales that measures to the nearest gram shows that an object has a mass of 56 g, then the mass should be recorded as 56 g, not 56.0 g. This is because you do not know whether it is 56.0 g, 56.1 g, 56.2 g or 55.8 g.

The value '56' has 2 significant figures. Recording it with 3 significant figures (e.g. 56.0 g or 55.8 g) would not be scientifically 'honest'. If the mass of 56 g was used to calculate another value, it would also not be 'honest' to give that answer with more than 2 significant figures.

The number of significant figures required in an answer depends on what kind of calculation you are doing.

If you are multiplying or dividing, use the smallest number of significant figures provided in the initial values.

If you are adding or subtracting, use the smallest number of decimal places provided in the initial values.

Working out the number of significant figures

The following rules should be followed to avoid confusion in determining how many significant figures are in a number.

- All non-zero digits are always significant. For example, 21.7 has three significant figures.
- All zeroes between two non-zero digits are significant. For example, 3015 has four significant figures.
- A zero to the right of a decimal point and following a non-zero digit is significant. For example, 0.5700 has four significant figures.
- Any other zero in a number less than one is not significant, as it is used only for locating decimal places. For example, 0.005 has just one significant figure.

7.1 Review

SUMMARY

- A mole is a convenient quantity (unit) for counting particles. The mole is given the symbol *n* and the abbreviation mol.
- One mole is defined as the amount of substance that contains the same number of 'specified' particles as there are atoms in 12 g of carbon-12.
- The number of particles in 1 mol is given the symbol N_A. This is known as Avogadro's constant and has the numerical value of 6.022×10^{23}.
- The formula $n = \frac{N}{N_A}$ can be used or rearranged to calculate the amount or number of specified particles in a sample.

KEY QUESTIONS

1 Calculate the number of:

  a atoms in 2.0 mol of sodium atoms (Na)

  b molecules in 0.10 mol of nitrogen molecules (N_2)

  c atoms in 20.0 mol of carbon atoms (C)

  d molecules in 4.2 mol of water molecules (H_2O)

  e atoms in 1.0×10^{-2} mol of iron atoms (Fe)

  f molecules in 4.62×10^{-5} mol of CO_2 molecules.

2 Calculate the amount of substance, in mol, represented by:

  a 3.0×10^{23} molecules of water (H_2O)

  b 1.5×10^{23} atoms of neon (Ne)

  c 4.2×10^{25} atoms of iron (Fe)

  d 4.2×10^{25} molecules of ethanol (C_2H_5OH).

3 Calculate the amount, in mol, of:

  a sodium atoms represented by 1.0×10^{20} sodium atoms

  b aluminium atoms represented by 1.0×10^{20} aluminium atoms

  c chlorine molecules represented by 1.0×10^{20} chlorine molecules.

4 Calculate the amount, in mol, of:

  a chlorine atoms in 0.4 mol of chlorine (Cl_2)

  b hydrogen atoms in 1.2 mol of methane (CH_4)

  c hydrogen atoms in 0.12 mol of ethane (C_2H_6)

  d oxygen atoms in 1.5 mol of sodium sulfate (Na_2SO_4).

7.2 Molar mass

Chemical laboratories always contain a balance like the one in Figure 7.2.1, which is used for weighing. If a chemist knows that a specific mass of a substance always contains a specific number of particles, it is possible to easily weigh a sample of the substance and calculate the exact number of particles present in the sample.

In this section, you will learn about how the amount of a substance, measured in moles, is related to the mass of the substance.

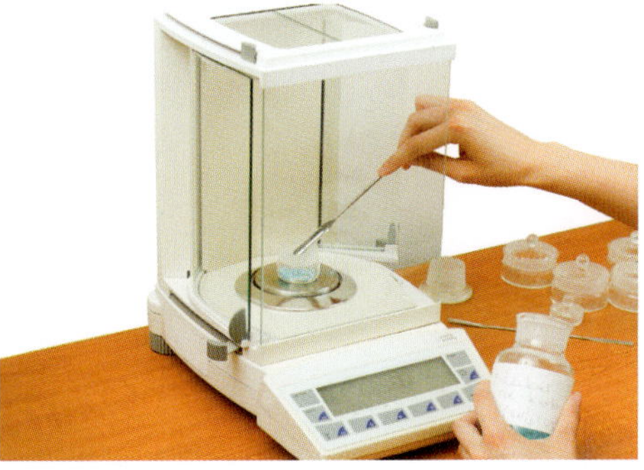

FIGURE 7.2.1 A digital balance is a simple piece of laboratory equipment used for weighing.

MOLAR MASS

Chemists have cleverly defined the mole so that you can determine the number of moles of a substance by simply measuring its mass.

The particles of different elements and compounds have different masses. Therefore, the masses of 1 mol of different elements or compounds will also be different. This is like saying that the mass of one dozen oranges will be greater than the mass of one dozen mandarins because one orange is heavier than one mandarin. The mass, in grams, of 1 mol of a particular element or compound is known as its **molar mass**. It is given the symbol M and the unit $g\,mol^{-1}$.

Remember that a mole is defined as the amount of substance that contains the same number of specified particles as there are atoms in 12 g of carbon-12. This is a very convenient definition because:

- 1 atom of ^{12}C has a **relative isotopic mass** (I_r) of 12 exactly
- 1 mol of atoms of ^{12}C has a mass of 12 g exactly.

Naturally occurring carbon is mainly composed of the ^{12}C isotope, so the molar mass of carbon is $12.01\,g\,mol^{-1}$.

Consider an atom of ^{12}C and an atom of ^{24}Mg. ^{12}C has been assigned a relative isotopic mass of 12 exactly. On that scale, the relative isotopic mass of ^{24}Mg is approximately 24. Since 1 mol of ^{12}C atoms weighs exactly 12 g, 1 mol of ^{24}Mg must weigh approximately twice as much, 24 g.

In general, the molar mass of an element is the mass of 1 mol of the element.

The molar mass of a compound is the mass of 1 mol of the compound. It is equal to the relative molecular or **relative formula mass** of the compound expressed in grams.

The molar mass is given the symbol, M, and the unit $g\,mol^{-1}$.

Table 7.2.1 shows you how to calculate the molar masses of some common substances.

TABLE 7.2.1 Calculating the molar mass of a substance by adding the relative atomic masses for each atom present in the substance based on the molecular or ionic formula

Substance	Relative atomic masses	Molar mass of substance
Na	Na: 22.99	$= 22.99\,g\,mol^{-1}$
O_2	O: 16.00	$= 2 \times 16.00$ $= 32.00\,g\,mol^{-1}$
H_2O	H: 1.008 O: 16.00	$= (2 \times 1.008) + 16.00$ $= 18.02\,g\,mol^{-1}$
CO_2	C: 12.01 O: 16.00	$= 12.01 + (2 \times 16.00)$ $= 44.01\,g\,mol^{-1}$
$NaNO_3$	Na: 22.99 N: 14.01 O: 16.00	$= 22.99 + 14.01 + (3 \times 16.00)$ $= 85.00\,g\,mol^{-1}$

From the calculations in Table 7.2.1 and the photograph of 1 mol of some common substances in Figure 7.2.2, you can see that 1 mol of each substance has a different mass.

FIGURE 7.2.2 One mole of each substance has a different mass.

Counting by weighing

A useful relationship links the amount of a substance (n) in moles, its molar mass (M) in grams per mole, and the given mass of the substance (m) in grams.

Mass of a given amount of substance (g) = amount of substance (mol) × molar mass ($g\,mol^{-1}$).

This can be written as $m = n \times M$ and rearranged to:

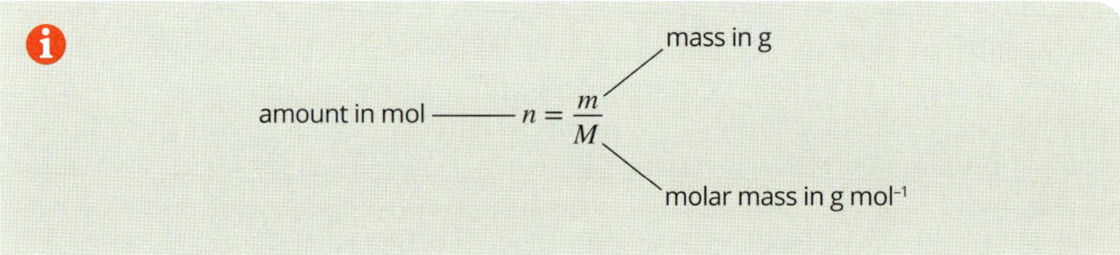

Worked example 7.2.1

CALCULATING THE MASS OF A SUBSTANCE

Calculate the mass of 0.35 mol of magnesium nitrate ($Mg(NO_3)_2$).

Thinking	Working
List the data given to you in the question. Remember that whenever you are given a molecular formula, you can calculate the molar mass.	$m(Mg(NO_3)_2)$ = ? $n(Mg(NO_3)_2)$ = 0.35 mol $M(Mg(NO_3)_2)$ = 24.31 + (2 × 14.01) + (6 × 16.00) = 148.33 $g\,mol^{-1}$
Calculate the mass of magnesium nitrate using: $n = \frac{m}{M}$	$n = \frac{m}{M}$, so $m = n \times M$ $m(Mg(NO_3)_2)$ = 0.35 × 148.33 = 52 g

Worked example: Try yourself 7.2.1

CALCULATING THE MASS OF A SUBSTANCE

Calculate the mass of 4.68 mol of sodium carbonate (Na_2CO_3).

You are now in a position to count atoms by weighing. When you use the mole, you are effectively counting the number of particles in a substance. The number of particles present in a substance is equal to the number of moles of the substance multiplied by 6.022×10^{23}.

Some calculations require you to use both of the formulae $n = \frac{m}{M}$ and $n = \frac{N}{N_A}$.

Worked example 7.2.2 is such a calculation.

Worked example 7.2.2

CALCULATING THE NUMBER OF MOLECULES

Calculate the number of CO_2 molecules present in 22 g of carbon dioxide.

Thinking	Working
List the data given to you in the question. Convert mass to grams if required. Remember that whenever you are given a molecular formula you can calculate the molar mass.	$N(CO_2) = ?$ $M(CO_2) = 12.01 + (2 \times 16.00) = 44.01\ g\ mol^{-1}$ $m(CO_2) = 22\ g$
Calculate the amount, in mol, of CO_2 using: $n = \frac{m}{M}$	$n(CO_2) = \frac{m}{M} = \frac{22}{44.01} = 0.50\ mol$
Calculate the number of CO_2 molecules using: $n = \frac{N}{N_A}$	$n = \frac{N}{N_A}$, so $N = n \times N_A$ $N(CO_2) = 0.50 \times 6.022 \times 10^{23} = 3.0 \times 10^{23}$ molecules

Worked example: Try yourself 7.2.2

CALCULATING THE NUMBER OF MOLECULES

Calculate the number of sucrose molecules in a teaspoon (4.2 g) of sucrose ($C_{12}H_{22}O_{11}$).

CHEMISTRY IN ACTION 

Big numbers in science and engineering

Astronomers and engineers also deal with very large numbers and have developed units to handle the huge distances and volumes they deal with. An astronomical unit (AU) is equal to the mean distance between the Sun and Earth, and it is used to compare the size of other planetary systems with our own. Distances between stars are measured in light-years: a light-year is the distance that light travels in a vacuum in 365.25 days. A more unusual Australian unit of measurement is a Sydharb, which is a volume defined in the Macquarie Dictionary as equal to the amount of water in Sydney Harbour at high tide (Figure 7.2.3).

News reports often discuss the size of a flood or the capacity of a reservoir in multiples of Sydney Harbour, and one Sydharb is approximately 5×10^{11} L. An astronomical unit is 1.49×10^{11} m and a light-year is 9.46×10^{15} m. In the same way that it is easier to talk about a mole than 6.022×10^{23} particles, these units are easy ways to visualise and discuss large numbers. These units also put the size of a mole into perspective. Even the numerical value of a light-year is 100 000 000 times smaller than a mole!

FIGURE 7.2.3 A Sydharb is an unusual unit of volume equal to the amount of water in Sydney Harbour at high tide. Floods and reservoir capacity are often measured in Sydharbs.

CHEMISTRY IN ACTION 

Measuring a molar mass of a compound without a mass spectrometer

When chemists were first determining the molar masses of molecules in the 19th century, a common method was to dissolve the molecule in a solvent and measure the decrease in the solvent's freezing point. This is called a freezing point depression experiment. It was a successful way to measure a molar mass because the freezing point of a solution depends on the number of particles dissolved in the solution, but not on their identity.

When water is the solvent, this relationship is written as:

T_f(solution) × m(water) = −1.86n

where:

T_f(solution) is the freezing point of the solution, in °C

m(water) is the mass of water, in kg

n is the number of moles of the dissolved substance.

Therefore, if you know the mass of the substance, you can use the formula $n = \frac{m}{M}$ to work out its molar mass.

7.2 Review

SUMMARY

- The molar mass of an element or compound is the mass, in grams, of 1 mol of that element or compound. Molar mass is given the symbol M and the unit g mol^{-1}.
- The molar mass of an element or compound has the same numerical value as the relative mass of the element or compound.
- The formula $n = \frac{m}{M}$ can be used or rearranged to calculate the mass, amount or molar mass of an element or compound.

KEY QUESTIONS

1 Calculate the molar mass of:
  - **a** nitrogen (N_2)
  - **b** ammonia (NH_3)
  - **c** sulfuric acid (H_2SO_4)
  - **d** iron(III) nitrate ($Fe(NO_3)_3$)
  - **e** acetic acid (CH_3COOH)
  - **f** sulfur atoms (S)
  - **g** vitamin C (ascorbic acid, $C_6H_8O_6$)
  - **h** hydrated copper(II) sulfate ($CuSO_4 \cdot 5H_2O$).

2 Calculate the mass of:
  - **a** 1.0 mol of sodium atoms (Na)
  - **b** 2.0 mol of oxygen molecules (O_2)
  - **c** 0.10 mol of methane molecules (CH_4)
  - **d** 0.25 mol of aluminium oxide (Al_2O_3).

3 Calculate the amount, in mol, of:
  - **a** H_2 molecules in 5.0 g of hydrogen (H_2)
  - **b** H atoms in 5.0 g of hydrogen (H_2)
  - **c** Al atoms in 2.7 g of aluminium (Al)
  - **d** CH_4 molecules in 0.40 g of methane (CH_4)
  - **e** O_2 molecules in 0.10 g of oxygen (O_2)
  - **f** O atoms in 0.10 g of oxygen (O_2)
  - **g** P_4 molecules in 1.2×10^{-3} g of phosphorus (P_4)
  - **h** P atoms in 1.2×10^{-3} g of phosphorus (P_4).

4 Calculate the number of atoms in:
  - **a** 23 g of sodium (Na)
  - **b** 4.0 g of argon (Ar)
  - **c** 0.243 g of magnesium (Mg)
  - **d** 10.0 g of gold (Au).

5 Calculate the:
  - **a** number of molecules in:
    - **i** 16 g of oxygen (O_2)
    - **ii** 2.8 g of nitrogen (N_2)
  - **b** number of oxygen atoms in 3.2 g of sulfur dioxide (SO_2)
  - **c** total number of atoms in 288 g of ammonia (NH_3).

7.3 Percentage composition and empirical formula

In Chapter 2, you learnt that compounds are substances that contain two or more different elements, and that the relative proportions of each element in a compound can be expressed as:

- a percentage in terms of the mass contributed by each element
- a formula, showing either the ratio of atoms contributed by each element in a compound or the actual numbers of atoms in a molecule.

In this section, you will learn how to use the **percentage composition** of elements in a compound to calculate a type of formula called an **empirical formula**.

PERCENTAGE COMPOSITION

In Chapter 2, you were introduced to the idea of the percentage composition of a given compound, which tells you the proportion by mass of the different elements in that compound. The proportion of each element is expressed as a percentage of the total mass of the compound.

GO TO ➤ Section 2.4 page 63

If the chemical formula of a compound is known, the percentage composition can be determined using the molar masses of the elements and compound. In general, this is written as:

$$\%\text{ by mass of an element in a compound} = \frac{\text{mass of the element in 1 mol of the compound}}{\text{molar mass of the compound}} \times 100$$

Worked example 7.3.1

CALCULATING PERCENTAGE COMPOSITION

Calculate the percentage by mass of aluminium in alumina (Al_2O_3).

Thinking	Working
Find the molar mass of the compound.	$M(Al_2O_3) = (2 \times 26.98) + (3 \times 16.00)$ $= 101.96\text{ g mol}^{-1}$
Find the total mass of the element in one mol of the compound.	mass of Al in 1 mol $= 2 \times M(\text{Al})$ $= 2 \times 26.98$ $= 53.96\text{ g}$
Find the percentage by mass of the element in the compound.	% by mass of Al in Al_2O_3 $= \frac{\text{mass of Al in 1 mol of } Al_2O_3}{\text{molar mass of } Al_2O_3} \times 100$ $= \frac{53.96}{101.96} \times 100$ $= 52.92\%$

Worked example: Try yourself 7.3.1

CALCULATING PERCENTAGE COMPOSITION

Calculate the percentage by mass of nitrogen in ammonium nitrate (NH_4NO_3).

EMPIRICAL FORMULA

Atoms or ions are present in compounds in fixed whole number ratios. The empirical formula of a compound gives the simplest whole number ratio of elements in that compound. See Table 7.3.1 for some examples.

TABLE 7.3.1 Empirical formulae of some common compounds

Compound	Empirical formula	Simplest whole number ratio of elements in the compound
water	H_2O	H:O 2:1
ethene	CH_2	C:H 1:2
calcium carbonate	$CaCO_3$	Ca:C:O 1:1:3

Determining empirical formulae

The empirical formula for a compound is determined from the mass of each element present in a given mass of the compound. These masses can be determined experimentally.

Once the masses of elements in a compound are known, the steps in Figure 7.3.1 are followed to convert these masses into a **mole ratio** (a ratio by number of atoms) and then into an empirical formula.

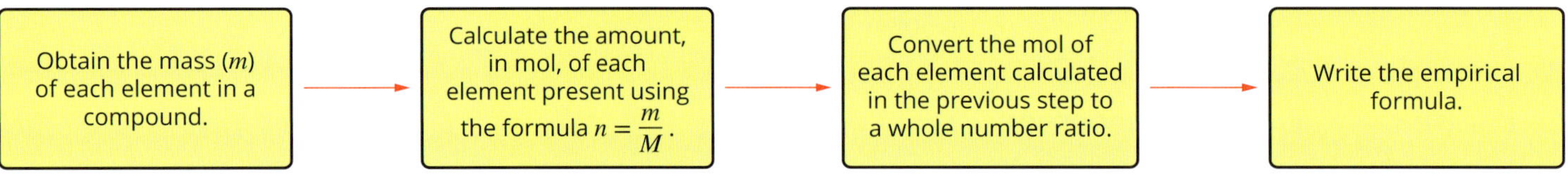

FIGURE 7.3.1 Follow these steps to calculate an empirical formula.

Worked example 7.3.2

DETERMINING THE EMPIRICAL FORMULA

A compound of carbon and oxygen contains 27.3% carbon and 72.7% oxygen by mass. Calculate the empirical formula of the compound.

Thinking	Working
Write down the mass, in g, of all elements present in the compound. If masses are given as percentages, assume that the sample weighs 100 g. The percentages then become masses in grams.	$m(C) = 27.3\ g$ $\quad m(O) = 72.7\ g$
Calculate the amount, in mol, of each element in the compound using: $n = \frac{m}{M}$	$n(C) = \frac{27.3}{12.01} = 2.28\ mol$ $\quad n(O) = \frac{72.7}{16.00} = 4.54\ mol$
Simplify the ratio by dividing each number of mol by the smallest number of mol calculated in the previous step. This gives you a ratio of the number of atoms of each element.	$\frac{2.28}{2.28} = 1$ $\quad \frac{4.54}{2.28} = 2$
Find the simplest whole number ratio.	1:2
Write the empirical formula.	CO_2

Worked example: Try yourself 7.3.2

DETERMINING THE EMPIRICAL FORMULA

0.50 g of magnesium is heated and allowed to completely react with chlorine. 1.96 g of white powder is formed. Determine the empirical formula of the compound.

MOLECULAR FORMULA

Molecular compounds have a molecular formula in addition to an empirical formula. The molecular formula gives the actual number of atoms of each element present in a molecule, rather than the simplest whole number ratio.

The molecular formula can be the same as or different from the empirical formula.

The empirical and molecular formulae of some common molecular compounds are shown in Table 7.3.2.

TABLE 7.3.2 Empirical and molecular formulae of some common molecular compounds

Molecule	Molecular formula	Empirical formula
water	H_2O	H_2O
ethane	C_2H_6	CH_3
carbon dioxide	CO_2	CO_2
glucose	$C_6H_{12}O_2$	CH_2O

Ionic compounds do not have molecular formulae because they do not exist as molecules. However, they do have empirical formulae that describe the fixed ratio of ions that exist in their lattices. The formula for calcium chloride ($CaCl_2$) is an example of an empirical formula of an ionic compound.

Determining molecular formulae

A molecular formula can be determined from the empirical formula of a compound if the molar mass of the compound is also known.

The molecular formula of a molecule is always a whole number multiple of the empirical formula. The number of the multiple is determined by the following formula.

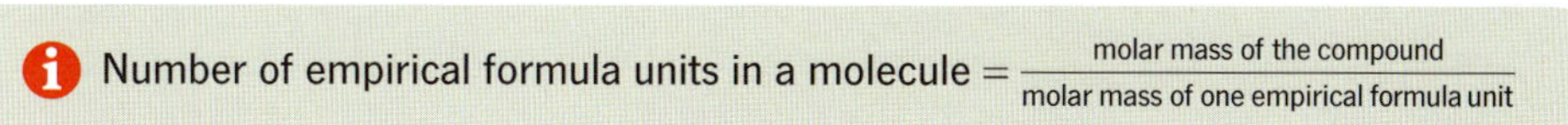

The general steps in the determination of a molecular formula are shown in Figure 7.3.2.

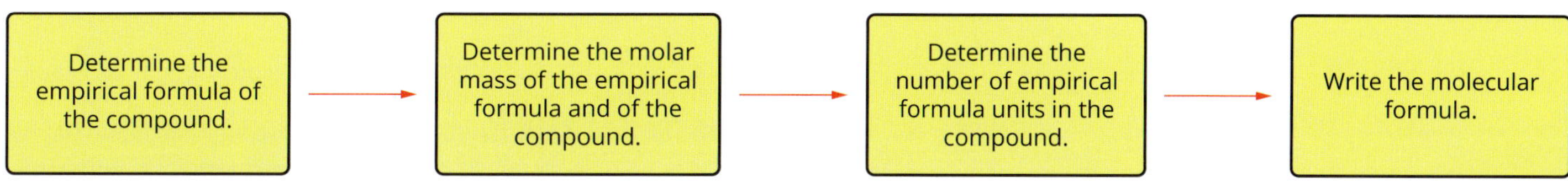

FIGURE 7.3.2 Steps for calculating a molecular formula.

Worked example 7.3.3

DETERMINING THE MOLECULAR FORMULA

A compound has the empirical formula CH. The molar mass of this compound is 78.11 g mol^{-1}. What is the molecular formula of the compound?

Thinking	Working
Calculate the molar mass of one unit of the empirical formula.	Molar mass of a CH unit = 12.01 + 1.008 = 13.02 g mol^{-1}
Determine the number of empirical formula units in the molecular formula.	number of CH units = $\frac{78.11}{13.02}$ = 6
Determine the molecular formula of the compound.	molecular formula = 6 × CH = C_6H_6

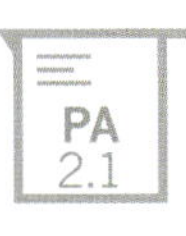

Worked example: Try yourself 7.3.3

DETERMINING THE MOLECULAR FORMULA

A compound has the empirical formula C_2H_5. The molar mass of this compound was determined to be 58.12 g mol^{-1}. What is the molecular formula of the compound?

7.3 Review

SUMMARY

- The percentage, by mass, of an element in a compound can be calculated from the mass of the element in 1 mol of the compound and the molar mass of the compound.
- The empirical formula of a compound gives the simplest whole number ratio of atoms or ions in the compound.
- Ionic compounds only have an empirical formula.
- Molecular compounds have a molecular formula that gives the actual number of atoms of each element in the molecule. It may be the same as, or different from, the empirical formula.

KEY QUESTIONS

1 Calculate the percentage by mass of:
  a iron in iron(III) oxide (Fe_2O_3)
  b uranium in uranium oxide (U_3O_8)
  c nitrogen in ammonium chloride (NH_4Cl)
  d oxygen in copper(II) nitrate ($Cu(NO_3)_2$).

2 Determine the empirical formulae of the compounds with the following compositions:
  a 2.74% hydrogen, 97.26% chlorine
  b 42.9% carbon, 57.1% oxygen
  c 10.0 g of a compound of magnesium and oxygen that contains 6.03 g of magnesium
  d 3.2 g of a hydrocarbon that contains 2.4 g of carbon

3 Determine the molecular formula of the following compounds.

	Empirical formula	Relative molecular mass
a	CH	78.11
b	HO	34.02
c	CH_2O	90.09
d	NO_2	46.01
e	CH_2	154.29

4 A hydrocarbon contains 85.7% carbon. Its relative molecular mass is 70. Determine the hydrocarbon's:
  a empirical formula
  b molecular formula.

5 A sample of the carbohydrate glucose contains 1.8 g carbon, 0.3 g hydrogen and 2.4 g oxygen.
  a Calculate the empirical formula of the compound.
  b Deduce its molecular formula, given that its relative molecular mass is 180.

7.4 Calculations based on the amount of a reactant or product

In this section, you will learn to calculate the amounts of reactants or products involved in chemical reactions. For example, in reactions such as that seen in Figure 7.4.1, you can use your knowledge of the mole concept to calculate the mass of products formed or reactants consumed.

FIGURE 7.4.1 When colourless aqueous solutions of mercury(II) acetate and sodium iodide are mixed, they produce a red solid, mercury(II) iodide. The mass of the solid from this reaction can be used to determine the amount of mercury present in the original sample.

STOICHIOMETRY

Chemists can apply the mole concept to chemical reactions to determine the quantities of reactants involved or products formed. This is useful in industry, as manufacturers need to know the quantities of raw materials required to produce a predicted amount of product. Although particles such as atoms, ions and molecules are so small that they cannot be counted individually, the mole concept allows us to determine the number of individual particles by simply measuring their mass.

Calculations based on the mole are used to make these predictions.

The study of ratios of moles of substances is called **stoichiometry**. **Stoichiometric calculations** are based on the law of conservation of mass.

The total mass of all products is equal to the total mass of all reactants in a chemical reaction.

In a chemical reaction, the total number of atoms of each element in the reactants is always equal to the total number of atoms of each element in the products. The atoms are just rearranged to form new compounds.

Mole ratios

Consider one of the important reactions in breadmaking, where table sugar (or sucrose, $C_{12}H_{22}O_{11}$) is fermented by baker's yeast to make carbon dioxide and ethanol (C_2H_5OH). The carbon dioxide makes the bread rise, and the ethanol evaporates in the baking process. The chemical equation is:

$$C_{12}H_{22}O_{11}(aq) + H_2O(l) \rightarrow 4CO_2(g) + 4C_2H_5OH(l)$$

The coefficients used to balance the equations show the ratios of reactants and products involved in the reaction. The equation indicates that 1 mol of $C_{12}H_{22}O_{11}$ reacts with 1 mol of water to form 4 mol of gaseous CO_2 and 4 mol of C_2H_5OH.

In more general terms, the amount, in moles, of CO_2 produced will always be four times the amount, in moles, of $C_{12}H_{22}O_{11}$ used. The amount, in moles, of C_2H_5OH produced will be equal to the amount, in moles, of CO_2 produced and four times the amount, in moles, of water consumed.

You can use the coefficients of this reaction to write relationships that show the mole ratios of any two chemicals involved in the reaction:

$$\frac{n(C_{12}H_{22}O_{11})}{n(H_2O)} = \frac{1}{1}, \quad \frac{n(CO_2)}{n(C_{12}H_{22}O_{11})} = \frac{4}{1} \text{ and } \frac{n(CO_2)}{n(C_2H_5OH)} = \frac{4}{4} = \frac{1}{1}$$

In the stoichiometric calculations that you will perform in this section, the number of mole or mass of one of the reactants or products will always be known (called the 'known chemical') and used to determine the number of mole or mass of one of the other reactants or products involved in the reaction (called the 'unknown chemical'). You can write the relationship between the known and the unknown chemicals using ratios:

$$\frac{n(\text{unknown chemical})}{n(\text{known chemical})} = \frac{\text{coefficient of unknown chemical}}{\text{coefficient of known chemical}}$$

Stoichiometric calculations allow you to use the mole ratio established in a chemical equation to predict the amount of a product that will be formed and how much reactant will be used.

When carrying out any stoichiometric calculation, you must always clearly state the mole ratio you are working with.

Worked example 7.4.1

USING MOLE RATIOS

0.375 mol of methanol (CH_3OH) is burnt in oxygen to make carbon dioxide and water. How many moles of carbon dioxide are formed?

Thinking	Working
Write a balanced equation for the reaction.	$2CH_3OH(l) + 2O_2(g) \rightarrow 2CO_2(g) + 2H_2O(l)$
Determine the number of mol of the 'known substance'. The known substance is the one that you are provided information about in the question.	$n(CH_3OH) = 0.375$ mol
Find the mole ratio: $\text{mole ratio} = \frac{\text{coefficient of unknown chemical}}{\text{coefficient of known chemical}}$ The 'unknown' substance is the one for which you need to calculate the number of mol	$\frac{\text{coefficient of } CO_2}{\text{coefficient of } CH_3OH} = \frac{1}{1}$
Calculate the number of mol of the unknown substance using: $n(\text{unknown}) = n(\text{known}) \times (\text{mole ratio})$	$n(CO_2) = 0.375 \times 1$ $= 0.375$ mol

Worked example: Try yourself 7.4.1

USING MOLE RATIOS

In the breadmaking reaction, baker's yeast ferments table sugar (sucrose, $C_{12}H_{22}O_{11}$) to form carbon dioxide and ethanol (C_2H_5OH). 0.050 mol of CO_2 is needed to make a loaf of bread rise properly. If the chemical reaction is $C_{12}H_{22}O_{11}(aq) + H_2O(l) \rightarrow 4CO_2(g) + 4CH_3CH_2OH(l)$, how much sucrose, in mol, is needed to form this amount of CO_2?

Mass–mass stoichiometry

When you carry out a reaction in the laboratory, you will measure quantities of chemicals in grams, not moles. For this reason, most calculations will require you to start and finish with mass rather than moles of a substance. To calculate the number of moles of the known substance from a mass, you can use the relationship:

$$\text{number of moles} = \frac{\text{mass (in g)}}{\text{molar mass (in g mol}^{-1})}$$

This can be written as:

$$n = \frac{m}{M}$$

To calculate a final answer as a mass, this formula is rearranged:

$$m = n \times M$$

Calculating the mass of a product

Stoichiometry can be used to calculate the mass of product that is expected from any reaction in which the mass of a reactant is known.

There are several steps involved in calculating the mass of a product.

1 Write a balanced equation for the reaction.
2 Calculate the number of moles of the reactant from its mass, using the formula:

$$n = \frac{m}{M}$$

3 Use the mole ratios in the equation to calculate the number of moles of the expected product.
4 Calculate the mass of the product using $m = n \times M$.

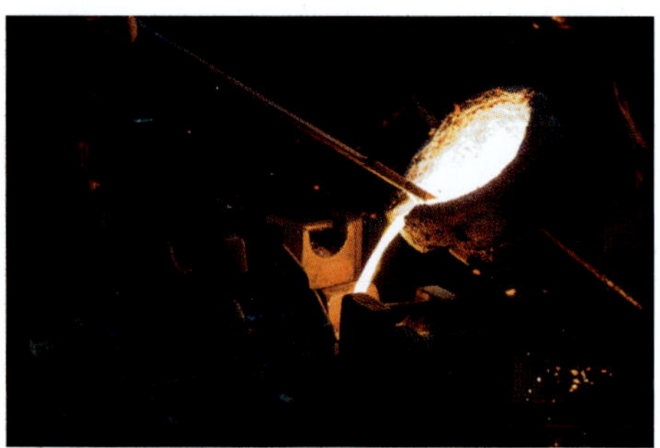

FIGURE 7.4.3 Iron metal is formed from iron(III) oxide reacting with carbon. Manufacturers use stoichiometry to calculate how much iron(III) oxide and carbon are needed to produce enough iron for their next order.

Figure 7.4.2 provides a flow chart that summarises this process, and Worked example 7.4.2 will help you to understand these steps.

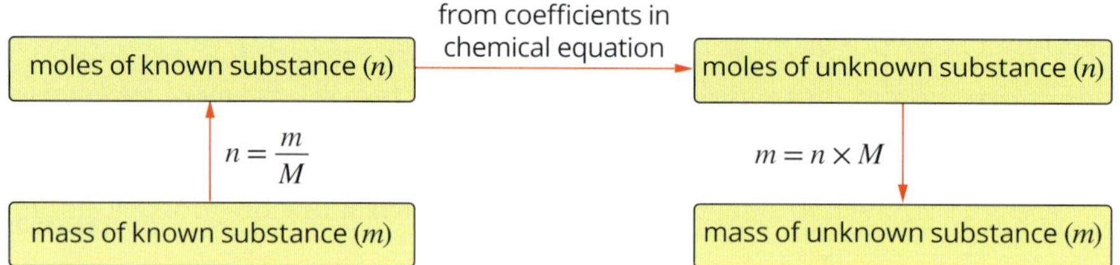

FIGURE 7.4.2 A flow chart for mass–mass stoichiometric calculations is helpful when trying to solve these problems.

Worked example 7.4.2

SOLVING MASS–MASS STOICHIOMETRIC PROBLEMS

When iron(III) oxide is reacted with carbon, iron metal is formed, along with carbon dioxide (Figure 7.4.3). Calculate the mass of iron that can be formed from 5.20 kg of iron(III) oxide.

Thinking	Working
Write a balanced equation for the reaction.	$2Fe_2O_3(s) + 3C(s) \rightarrow 4Fe(s) + 3CO_2(g)$
Calculate the number of mol of the known substance (the iron(III) oxide). Convert the mass to grams: $n = \frac{m}{M}$	$n(Fe_2O_3) = \frac{5.20 \times 10^3}{159.70}$ $= 32.6$ mol
Calculate the mole ratio: mole ratio $= \frac{\text{coefficient of unknown chemical}}{\text{coefficient of known chemical}}$	mole ratio $= \frac{\text{coefficient of Fe}}{\text{coefficient of } Fe_2O_3}$ $= \frac{4}{2}$
Calculate the number of mol of the unknown substance: $n(\text{unknown}) = n(\text{known}) \times (\text{mole ratio})$	$n(Fe) = 32.6 \times \frac{4}{2}$ $= 65.1$ mol
Calculate the mass of the unknown substance: $m = n(\text{unknown}) \times M$	$m(Fe) = 65.1 \times 55.85$ $= 3.64 \times 10^3$ g $= 3.64$ kg

Worked example: Try yourself 7.4.2

SOLVING MASS–MASS STOICHIOMETRIC PROBLEMS

A reaction between solutions of sodium sulfate and barium nitrate produces solid barium sulfate with a mass of 2.440 g. Aqueous sodium nitrate is also formed. Calculate the mass of sodium sulfate required to produce this amount of barium sulfate. The equation for this reaction is:

$$Na_2SO_4(aq) + Ba(NO_3)_2(aq) \rightarrow BaSO_4(s) + 2NaNO_3(aq)$$

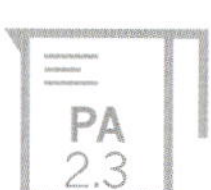

7.4 Review

SUMMARY

- A balanced equation shows the ratio of the amount, in moles, of reactants used and products formed in the reaction.
- Given the quantity of one of the reactants or products of a chemical reaction, such as in a precipitation reaction, the quantity of all other reactants and products can be predicted by working through the following steps.
  1. Write a balanced equation for the reaction.
  2. Calculate the amount, in moles, of the given substance.
  3. Use the mole ratios of reactants and products in the balanced equation to calculate the amount, in moles, of the required substance.
  4. Use the appropriate formula to determine the quantity needed of the required substance. Usually, this would be the mass, using the formula $m = n \times M$.

KEY QUESTIONS

1 A tree undergoing photosynthesis uses 120 g of carbon dioxide to make glucose ($C_6H_{12}O_6$). The equation for the reaction is:

$$6CO_2(g) + 6H_2O(l) \rightarrow C_6H_{12}O_6(aq) + 6O_2(g)$$

What mass of glucose is formed?

2 The mass of aluminium nitrate in a solution is determined by adding sodium carbonate solution to form aluminium carbonate and sodium nitrate. The equation for the reaction occurring is:

$$2Al(NO_3)_3(aq) + 3Na_2CO_3(aq) \rightarrow Al_2(CO_3)_3(s) + 6NaNO_3(aq)$$

In a particular reaction, 4.68 g of aluminium carbonate is obtained.

a Calculate the moles of aluminium carbonate produced.

b Determine the required mole ratio for the reaction.

c Calculate the mass of aluminium nitrate that reacted.

3 A solar panel is made from 45.9 g of silicon. What mass of silicon tetrachloride is needed to make this mass of silicon? The equation for the reaction is:

$$SiCl_4(g) + 2H_2(g) \rightarrow Si(s) + 4HCl(g)$$

4 The reaction between mercury(II) acetate and sodium iodide is represented by the following equation:

$$Hg(CH_3COO)_2(aq) + 2NaI(aq) \rightarrow HgI_2(s) + 2NaCH_3COO(aq)$$

Solid mercury(II) iodide of mass 4.82 g is formed when sodium iodide is added to a solution of mercury(II) acetate:

$M(Hg(CH_3COO)_2) = 318.69\ g\ mol^{-1}$

$M(HgI_2) = 454.4\ g\ mol^{-1}$

Calculate the mass of mercury(II) acetate that reacted to produce this solid.

7.5 Calculations based on the amounts of two reactants

In the previous section, you calculated the amount of product that could be formed given the amount of one reactant. You were also able to calculate the amount of a second reactant that would be required, given the amount of the first. In chemistry, once you know the mass of a reactant that is completely consumed in a chemical reaction, you can use that information to determine the amount of any other component based on the chemical equation.

In this section, you will learn how to perform stoichiometric calculations in which two reactants are involved, but the reactant that is completely consumed may not be obvious straight away. In this style of question, you are given sufficient information to allow you to calculate the amounts of both reactants present. But before you can calculate the amount of product that is formed, you will need to determine which reactant is completely consumed. You will then know which reactant will be the limiting factor in the reaction.

STOICHIOMETRY PROBLEMS INVOLVING EXCESS REACTANTS

When two reactants are mixed to create a chemical reaction, it is possible that they will be combined in just the right mole ratio, as indicated in the equation, for each to react completely. However, it is more likely, that they are not present in exactly the right mole ratio. In that case, one of the reactants will be used up before the other, and some of the other reactant will be left over at the end of the reaction.

To illustrate this idea, consider a problem that does not involve chemicals. Suppose that you have been given some skateboard decks and wheels, and you want to make as many complete skateboards as you can. As you can see in Figure 7.5.1, a complete skateboard is made up of one deck and four wheels.

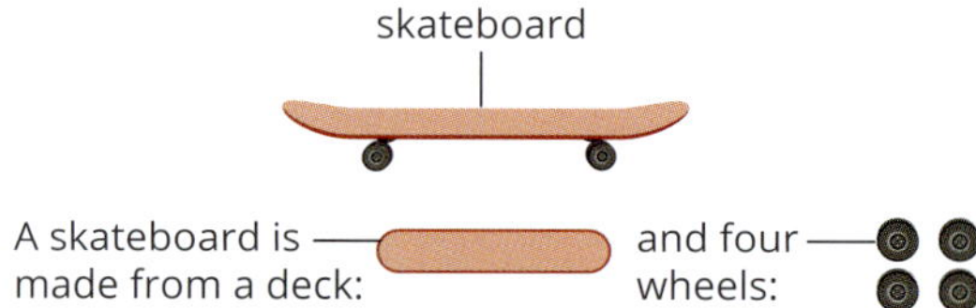

FIGURE 7.5.1 In order to make a complete skateboard, you must always use one deck and four wheels. This sets up the basic formula for a skateboard.

Now consider the situation shown in Figure 7.5.2. If you were given two decks and ten wheels, how many complete skateboards could you make from these materials?

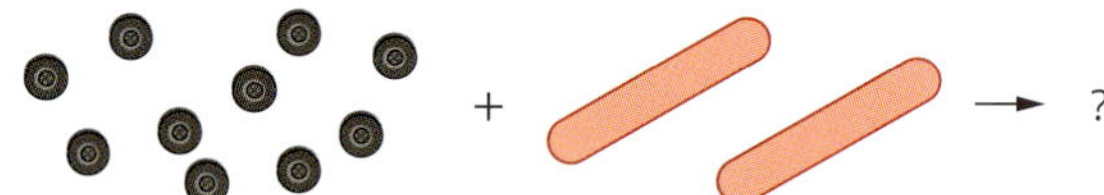

FIGURE 7.5.2 When provided with ten wheels and two skateboard decks, how many complete skateboards can be made?

The answer is that you could make two complete skateboards and there would be two wheels left over (Figure 7.5.3).

FIGURE 7.5.3 When supplied with two decks and ten wheels, the maximum number of skateboards that can be made is two. There will be two wheels that are not used.

In this example, you can say that the two wheels left at the end were in excess. Also, given two decks and ten wheels, the number of complete skateboards you could make was limited by the number of decks available.

A similar situation arises in chemical reactions when the quantities of the reactants supplied are not in the exact same ratio as that shown in the equation for the reaction.

When that happens, the reactant that is:

- completely consumed is the **limiting reactant**
- not completely consumed is the **excess reactant**.

Figure 7.5.4 shows three different scenarios for the reaction in which hydrogen gas and oxygen gas combine to form water, according to the equation:

$$2H_2(g) + O_2(g) \rightarrow 2H_2O(l)$$

Each of the diagrams provides examples to illustrate the concepts of limiting and excess reactants.

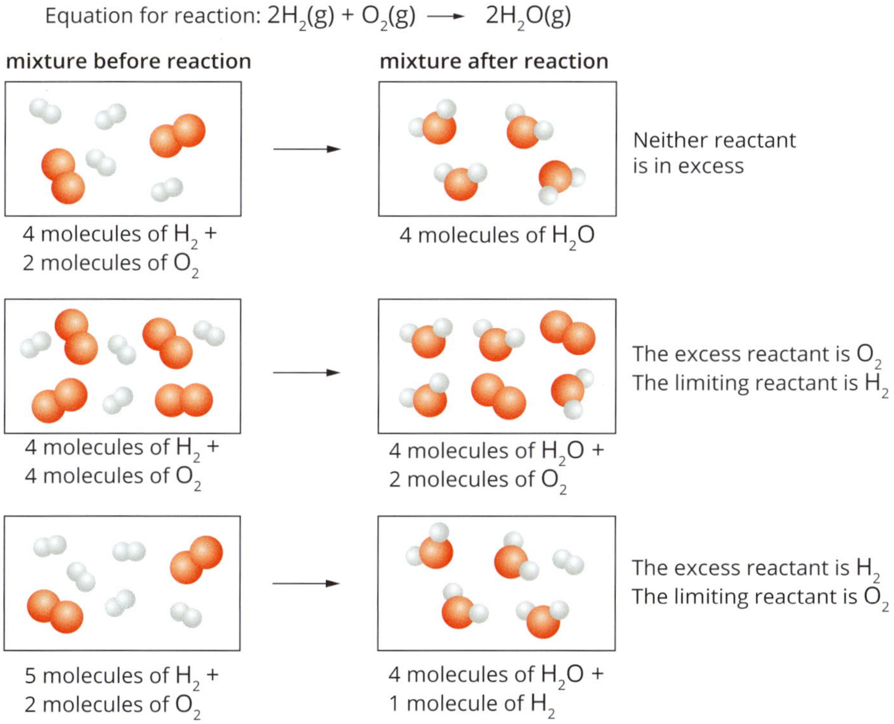

In a chemical reaction, the limiting reactant is completely used up.

Equation for reaction: $2H_2(g) + O_2(g) \longrightarrow 2H_2O(g)$

mixture before reaction		mixture after reaction	
4 molecules of H_2 + 2 molecules of O_2	→	4 molecules of H_2O	Neither reactant is in excess
4 molecules of H_2 + 4 molecules of O_2	→	4 molecules of H_2O + 2 molecules of O_2	The excess reactant is O_2 The limiting reactant is H_2
5 molecules of H_2 + 2 molecules of O_2	→	4 molecules of H_2O + 1 molecule of H_2	The excess reactant is H_2 The limiting reactant is O_2

FIGURE 7.5.4 Different scenarios showing the concept of a limiting reactant for the reaction $2H_2(g) + O_2(g) \rightarrow 2H_2O(l)$.

Note that in each of the examples shown in Figure 7.5.4, the amount of product formed in these types of reactions:

- is determined by the amount of the limiting reactant present in the reaction mixture
- cannot be determined from the amount of excess reactant.

In the skateboard example, it was the number of decks, not the number of wheels, that determined how many complete skateboards could be made.

You must always use the amount of the limiting reactant to determine the amount of product that will be formed.

Steps in solving stoichiometry problems involving excess reactants

When attempting to solve a limiting reactant problem in which you are required to work out the amount of product, there are three main steps:

1. Calculate the number of moles of each reactant.
2. Determine which reactant is in excess, and therefore which is the limiting reactant.
3. Use the amount of the limiting reactant to work out the amount of product formed.

Worked example 7.5.1

SOLVING MASS–MASS STOICHIOMETRY PROBLEMS WITH ONE REACTANT IN EXCESS

A solution containing 20.0 g of dissolved sodium hydroxide is added to a solution containing 25.0 g aluminium nitrate to form solid aluminium hydroxide and aqueous sodium nitrate. The equation for this reaction is:

$$3NaOH(aq) + Al(NO_3)_3(aq) \rightarrow Al(OH)_3(s) + 3NaNO_3(aq)$$

a Which reactant is the limiting reactant?	
Thinking	**Working**
Calculate the number of mol of each of the reactants in the equation using: $n = \frac{m}{M}$	Use the equation $n = \frac{m}{M}$. For NaOH: $n(NaOH) = \frac{20.0}{40.00} = 0.500$ mol For $Al(NO_3)_3$: $n(Al(NO_3)_3 = \frac{25.0}{213.01} = 0.117$ mol
Use the coefficients of the equation to find the limiting reactant.	The equation shows that 3 mol of NaOH react with 1 mol of $Al(NO_3)_3$. So to react all of the $Al(NO_3)$ you will require: $\frac{3}{1} \times n(Al(NO_3)_3$ of NaOH $\frac{3}{1} \times 0.117 = 0.352$ mol As there is 0.500 mol available, the NaOH is in excess. So, $Al(NO_3)_3$ is the limiting reactant. (It will be completely consumed.)

b What mass of aluminium hydroxide will be formed?	
Thinking	**Working**
Find the mole ratio of the unknown substance to the limiting reactant from the equation coefficients: $\text{mole ratio} = \frac{\text{coefficient of unknown chemical}}{\text{coefficient of known chemical}}$	From the equation coefficients: $\frac{\text{coefficient of } Al(OH)_3}{\text{coefficient of } Al(NO_3)_3} = \frac{1}{1}$
Calculate the number of mol of the unknown substance using the number of mol of the limiting reactant: $n(\text{unknown}) = n(\text{limiting reactant}) \times (\text{mole ratio})$	$n(Al(OH)_3) = n(Al(NO_3)_3) \times \frac{1}{1}$ $= 0.117 \times \frac{1}{1}$ $= 0.117$ mol
Calculate the mass of the unknown substance using: $m(\text{unknown}) = n(\text{unknown}) \times M$	Molar mass of $Al(OH)_3 = 78.00\ g\ mol^{-1}$ $m(Al(OH)_3) = 0.117 \times 78.00$ $= 9.15$ g

Worked example: Try yourself 7.5.1

SOLVING MASS–MASS STOICHIOMETRY PROBLEMS WITH ONE REACTANT IN EXCESS

A space shuttle is loaded with 1.06×10^8 g of liquefied H_2 and 6.29×10^8 g of liquefied O_2, which are burnt as fuels during take-off. The reaction is:

$$2H_2(l) + O_2(l) \rightarrow 2H_2O(g)$$

a Which reactant is the limiting reactant?

b What mass of water will be formed?

7.5 Review

SUMMARY

- If quantities of more than one reactant are given, the amount, in moles, of each reactant needs to be calculated.
- The limiting reactant needs to be determined; this will be the reactant that is consumed completely.
- The limiting reactant is used to predict the amount of product formed and the amount of the other reactant in excess.

KEY QUESTIONS

1 Aluminium reacts with oxygen gas to form aluminium oxide. The balanced equation for the reaction is:

$$4Al(s) + 3O_2(g) \rightarrow 2Al_2O_3(s)$$

In a particular reaction, 40 g of aluminium reacts with 35 g of oxygen gas. List the following steps in the order in which the calculations should be completed to determine the mass of aluminium oxide that will form.

A Use mole ratios to determine which reactant is limiting.

B Calculate the number of mol of aluminium and oxygen.

C Calculate the mass of aluminium oxide that forms.

D Refer to the balanced equation.

E Calculate the number of mol of aluminium oxide that forms.

2 In three separate experiments, different amounts of nitrogen and hydrogen reacted to form ammonia, according to the equation:

$$N_2(g) + 3H_2(g) \rightarrow 2NH_3(g)$$

This table shows the amounts of reactants and products in each experiment. Complete the table to indicate the amount of each product remaining at the end of the reaction.

Nitrogen molecules available	Hydrogen molecules available	Ammonia molecules produced	Nitrogen molecules in excess	Hydrogen molecules in excess
2	10			
879	477			
9 mol	6 mol			

3 Sodium metal can react with chlorine gas to form sodium chloride. 25.0 g of sodium is reacted with 50.0 g of chlorine gas.

a Write the balanced equation for the reaction between sodium and chlorine.

b Calculate the mass of sodium chloride that will form in the reaction.

4 Potassium iodide and lead(II) nitrate solutions react together to form solid lead(II) iodide and aqueous potassium nitrate:

$$2KI(aq) + Pb(NO_3)_2(aq) \rightarrow PbI_2(s) + 2KNO_3(aq)$$

In each of the following cases, carry out the calculations to determine the quantities required.

a If 1.0 mol of potassium iodide reacts with 1.0 mol of lead(II) nitrate, determine which reactant is in excess and by how many mol.

b If 0.50 mol of potassium iodide reacts with 2.0 mol of lead(II) nitrate, determine which reactant is in excess and by how many mol.

c If 1.00 g of lead(II) nitrate reacts with 1.50 g of potassium iodide, determine which reactant is in excess and the mass of lead(II) iodide that forms.

Chapter review

07

KEY TERMS

amount of substance
Avogadro's constant
empirical formula
excess reactant
limiting reactant
molar mass
mole
mole ratio
molecular formula
percentage composition
relative formula mass
relative isotopic mass
stoichiometric calculation
stoichiometry

REVIEW QUESTIONS

1 For each of the following numbers of molecules, calculate the amount of substance, in mol.
- **a** 4.50×10^{23} molecules of water (H_2O)
- **b** 9.00×10^{24} molecules of methane (CH_4)
- **c** 2.3×10^{28} molecules of chlorine (Cl_2)
- **d** 1 molecule of sucrose ($C_{12}H_{22}O_{11}$)

2 For each of the following amounts of molecular substances, calculate the:
- **i** number of molecules
- **ii** total number of atoms
- **iii** the number of significant figures to which the number of mol can be given.
- **a** 1.45 mol of ammonia (NH_3)
- **b** 0.576 mol of hydrogen sulfide (H_2S)
- **c** 0.0153 mol of hydrogen nitrate (HNO_3)
- **d** 2.5 mol of sucrose ($C_{12}H_{22}O_{11}$)

3 How would the molar mass (M) of a compound differ from its relative molecular mass (M_r)?

4 What is the molar mass (M) of each of the following?
- **a** iron (Fe)
- **b** sulfuric acid (H_2SO_4)
- **c** sodium oxide (Na_2O)
- **d** zinc nitrate ($Zn(NO_3)_2$)
- **e** glycine (H_2NCH_2COOH)
- **f** aluminium sulfate ($Al_2(SO_4)_3$)
- **g** hydrated iron(III) chloride ($FeCl_3 \cdot 6H_2O$)

5
- **a** If 6.022×10^{23} atoms of calcium have a mass of 40.08 g, what is the mass of one calcium atom?
- **b** If 1 mol of water molecules has a mass of 18.02 g, what is the mass of one water molecule?
- **c** What is the mass of one molecule of carbon dioxide?

6 For each of the following molecular substances, calculate the:
- **i** amount of substance in moles
- **ii** number of molecules
- **iii** total number of atoms.
- **a** 4.2 g of phosphorus (P_4)
- **b** 75.0 g of sulfur (S_8)
- **c** 0.32 g of hydrogen chloride (HCl)
- **d** 2.2×10^{-2} g of glucose ($C_6H_{12}O_6$).

7 For each of the following ionic substances, calculate the amount of:
- **i** substance, in moles
- **ii** each ion, in moles.
- **a** 5.85 g of NaCl
- **b** 45.0 g of $CaCl_2$
- **c** 1.68 g of $Fe_2(SO_4)_3$

8 Calculate the molar mass of a substance if:
- **a** 2.0 mol of the substance has a mass of 80 g
- **b** 0.10 mol of the substance has a mass of 9.8 g
- **c** 1.7 mol of the substance has a mass of 74.8 g
- **d** 3.50 mol of the substance has a mass of 371 g.

9 Which of the following metal samples has the greatest mass?
- **A** 100 g copper
- **B** 4.0 mol of iron atoms
- **C** 1.2×10^{24} atoms of silver

10 A new antibiotic has been isolated and only 2.0 mg is available. The molar mass is found to be 12.5 kg mol^{-1}.
- **a** Express the molar mass in g mol^{-1}.
- **b** Calculate the amount of antibiotic, in mol.
- **c** How many molecules of antibiotic have been isolated?

11 Calculate the percentage by mass of each element in:
- **a** Al_2O_3
- **b** $Cu(OH)_2$
- **c** $MgCl_2 \cdot 6H_2O$
- **d** $Fe_2(SO_4)_3$
- **e** perchloric acid ($HClO_4$).

12 A compound used as a solvent for dyes has the following composition by mass: 32.0% carbon, 6.7% hydrogen, 18.7% nitrogen and 42.6% oxygen. Find the empirical formula of the compound.

13 A clear liquid extracted from fermented lemons was found to consist of carbon, hydrogen and oxygen. Analysis showed it to be 52.2% carbon and 34.8% oxygen by mass.

a Find the empirical formula of the substance.

b If 2.17 mol of the compound has a mass of 100 g, find the molecular formula of the compound.

14 A hydrocarbon is a compound that contains carbon and hydrogen only. Determine the empirical formula of a hydrocarbon that is used as a specialty fuel and contains 90.0% carbon.

15 Find the relative atomic mass of nickel if 3.370 g nickel was obtained by reduction of 4.286 g of the oxide (NiO). You may need to refer to the periodic table at the end of this book.

16 4.150 g tungsten was burned in chlorine and 8.950 g tungsten chloride (WCl_6) was formed. Find the relative atomic mass of tungsten. You may need to refer to the periodic table at the end of this book.

17 Determine the molecular formulae of compounds with the following compositions and relative molecular masses:

a 82.75% carbon, 17.25% hydrogen; $M_r = 58.12$

b 43.66% phosphorus, 56.34% oxygen; $M_r = 283.88$

c 40.0% carbon, 6.7% hydrogen, 53.3% oxygen; $M_r = 180.16$

d 0.164 g hydrogen, 5.25 g sulfur, 9.18 g oxygen; $M_r = 178.16$

18 Using suitable examples, clearly distinguish between:

a relative isotopic mass

b relative atomic mass

c relative molecular mass

d relative formula mass

e molar mass.

19 Caffeine contains 49.48% carbon, 5.15% hydrogen, 28.87% nitrogen; the rest is oxygen.

a Determine the empirical formula of caffeine.

b If 0.200 mol of caffeine has a mass of 38.8 g, what is the molar mass of a caffeine molecule?

c Determine the molecular formula of caffeine.

d How many moles of caffeine molecules are in 1.00 g caffeine?

e How many molecules of caffeine are in 1.00 g caffeine?

f How many atoms all together are there in 1.00 g caffeine?

20 The empirical formula of a metal oxide can be found by experimentation, as shown in the figure below.

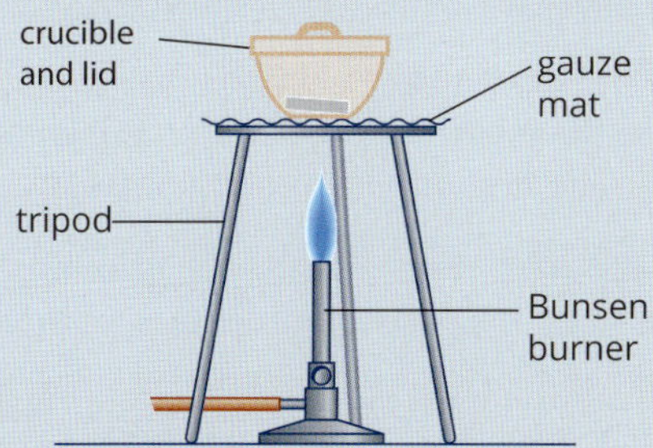

The mass of the oxygen that reacts with the mass of the metal must be determined. Steps A–F form the experimental method.

A Ignite a burner and heat the metal.

B Allow the crucible to cool, then weigh it.

C Continue the reaction until no further change occurs.

D Clean a piece of metal with emery paper to remove any oxide layer.

E Place the metal in a clean, weighed crucible and cover with a lid.

F Weigh the metal and record its mass.

a Place the steps in the correct order by letter.

b Wan and Eric collected the following data:

mass of the metal = 0.542 g

mass of the empty crucible = 20.310 g

mass of the crucible and metal oxide = 21.068 g

They found from this data that the metal oxide has a 1:1 formula, i.e. MO, where M = metal. Copy and complete the figure below with the data provided.

	Metal	Oxygen
mass (g)		
relative atomic mass		16.00
moles		
ratio		

c What metal was used in the experiment?

21 For each amount given, calculate the amounts, in mol, of the other reactants and products required for a complete reaction according to the following equation:

$$3Ca(NO_3)_2(aq) + 2Na_3PO_4(aq) \rightarrow Ca_3(PO_4)_2(s) + 6NaNO_3(aq)$$

$Ca(NO_3)_2$	Na_3PO_4	$Ca_3(PO_4)_2$	$NaNO_3$
27 mol			
	0.48 mol		
		0.18 mol	
			2.4 mol

22 When solutions of iron(II) sulfate and potassium hydroxide are mixed, the reaction that occurs can be represented as:

$FeSO_4(aq) + 2KOH(aq) \rightarrow Fe(OH)_2(s) + K_2SO_4(aq)$

Complete the following expressions based on this equation. The first one has been done for you.

$n(KOH) = \frac{2}{1} \times n(Fe(OH)_2)$

$n(FeSO_4) = \frac{(\ )}{(\ )} \times n(KOH)$

$n(KOH) = \frac{(\ )}{(\ )} \times n(K_2SO_4)$

$n(Fe(OH)_2) = \frac{(\ )}{(\ )} \times n(FeSO_4)$

23 A barbecue burns 50.2 g of propane (C_3H_8) to cook a meal with this reaction:

$$C_3H_8(l) + 10O_2(g) \rightarrow 3CO_2(g) + 4H_2O(g)$$

What mass of CO_2 is produced?

24 To make a mobile phone battery, molten lithium chloride is used to make lithium metal and chlorine gas in a reaction that is represented as:

$$2LiCl(l) \rightarrow 2Li(l) + Cl_2(g)$$

What mass of LiCl is needed to produce 8.22 g of Li for a battery?

25 In three separate experiments, different amounts of carbon and oxygen were reacted together to form carbon dioxide according to the equation:

$$C(s) + O_2(g) \rightarrow CO_2(g)$$

The table lists the amounts of reactants and products in each of the three experiments. Complete the table to indicate the amounts of the products formed and of the remaining reactants at the end of the reaction.

Carbon atoms available	Oxygen molecules available	Carbon dioxide molecules produced	Carbon atoms in excess	Oxygen molecules in excess
8	20			
1000	3000			
9 mol	6 mol			

26 Sodium can react with oxygen gas to form sodium oxide. The equation for this reaction is:

$$4Na(s) + O_2(g) \rightarrow 2Na_2O(s)$$

3.0 mol of sodium is reacted with 0.8 mol of oxygen gas.

a Determine which reactant is in excess.

b How many moles of sodium oxide is produced in the reaction?

27 16.0 g of hydrogen sulfide is mixed with 20.0 g of sulfur dioxide, and they react according to this equation:

$$2H_2S(g) + SO_2(g) \rightarrow 2H_2O(l) + 3S(s)$$

a Calculate the mass of sulfur produced.

b Calculate the mass of reactant left after the reaction.

28 4.40 g of P_4O_6 and 3.00 g of I_2 are mixed and allowed to react according to this equation:

$$5P_4O_6(s) + 8I_2(g) \rightarrow 4P_2I_4(s) + 3P_4O_{10}(s)$$

a Which reactant is in excess and by how much, in g?

b What mass of P_2I_4 forms?

c What mass of P_4O_{10} forms?

d What is the total mass of all the products? (Hint: Compare this with the mass of the reactants.)

29 Reflect on the Inquiry activity on page 222. Now that you understand the mole, explain any patterns you saw in your results.

CHAPTER

Concentration and molarity

We are surrounded by solutions—liquids that contain other substances dissolved in them. Earth's oceans, the plasma of human blood, the sap of trees and most beverages are typical examples of common solutions.

In fields such as medicine, pharmaceutical manufacturing and food preparation, it is important to know how much of a substance is present in a solution.

In this chapter, you will learn how to prepare and dilute solutions and to calculate the amount of a substance dissolved in a given solution.

Content

INQUIRY QUESTION

How are chemicals in solutions measured?

By the end of this chapter, you will be able to:

- conduct practical investigations to determine the concentrations of solutions and investigate the different ways in which concentrations are measured (ACSCH046, ACSCH063) ICT N
- manipulate variables and solve problems to calculate concentration, mass or volume using:
  - $c = \frac{n}{V}$ (molarity formula) (ACSCH063) N
  - dilutions (number of moles before dilution = number of moles of sample after dilution) ICT N
- conduct an investigation to make a standard solution and perform a dilution

8.1 Concentration of solutions

FIGURE 8.1.1 Both of these products have the same active ingredient: ethanoic acid (also known as acetic acid). Ethanoic acid from the container in (a) is safe to eat, whereas ethanoic acid from the container in (b) causes severe burns and blisters on skin. The difference between the two products is the concentration of the ethanoic acid—the concentration is lower in vinegar.

CHEMISTRY INQUIRY CCT

Comparing solutions

What does the 'concentration of a solution' mean?

COLLECT THIS ...

- vinegar
- clear plastic cups or beakers
- measuring cylinder
- baking soda (sodium bicarbonate)
- teaspoons or spatulas
- stopwatch

DO THIS ...

1. Prepare a set of vinegar solutions using the following guidelines:
   Solution 1: 50 mL of vinegar
   Solution 2: 40 mL of vinegar, 10 mL of water
   Solution 3: 30 mL of vinegar, 20 mL of water
   Solution 4: 20 mL of vinegar, 30 mL of water
   Solution 5: 10 mL of vinegar, 40 mL of water.
2. Rearrange the cups so that it is not obvious which cup is which—they should all be clear solutions.
3. Add a half of a teaspoon of baking soda to the first cup and time how long it takes for a reaction to stop.
4. Repeat step 3 for all cups.

RECORD THIS ...

Record the time taken for each reaction to stop.

Describe how the reaction in each cup differs slightly from the others.

REFLECT ON THIS ...

Are you able to arrange the cups in order of the percentage of vinegar in the cup?

What are the products of the reaction between baking soda and vinegar?

Can you think of products you use at home for which the 'concentration' of the solution can vary, for example concentrated detergent compared with regular detergent?

Can you think of any other ways of comparing the concentration of the vinegar in the cups?

A solution is formed when a substance (the **solute**) is dissolved in a liquid (the **solvent**). The **concentration** of a solution describes the relative amount of solute and solvent present in the solution. A solution in which the ratio of solute to solvent is high is said to be concentrated. Cordial that has not had any water added to it is an example of a **concentrated solution**. A solution in which the ratio of solute to solvent is low, such as a glass with 20 mL of cordial (the solute) and 200 mL of water (the solvent), is said to be a **dilute solution**. Figure 8.1.1 shows an example of a substance that is treated very differently when it is concentrated compared with when it is dilute.

'Concentrated' and 'dilute' are general terms. However, sometimes you need to know the actual concentration of a solution—the exact ratio of solute to solvent. The use of accurate solution concentrations is important in the prescription of medicines, chemical manufacturing and chemical analysis.

UNITS OF CONCENTRATION

Chemists use different measures of concentration, depending on the situation. The most common measures describe the amount of solute in a given amount of solution.

Some examples of units of concentration are:

- mass of solute per litre of solution (grams per litre, $g\,L^{-1}$)
- moles of solute per litre of solution (moles per litre, $mol\,L^{-1}$) (you will look more closely at this unit of concentration in Section 8.2)
- parts per million (ppm)
- percentage by mass.

Units of concentration measured in this way have two parts.

- The first part gives information about how much solute there is.
- The second part gives information about how much solution there is.

For example, if a solution contains sodium chloride (NaCl) with a concentration of $17\,g\,L^{-1}$, then 17 g of NaCl (first part) is dissolved in every 1 L of the solution (second part). You will now look at how to perform calculations of concentration using different units.

GO TO ➤ SkillBuilder page 127

Concentration in grams per litre ($g\,L^{-1}$)

The concentration of a solution in grams per litre ($g\,L^{-1}$) indicates the mass, in grams, of solute dissolved in one litre of the solution. For example, if the concentration of sodium chloride in seawater is $20\,g\,L^{-1}$, this means that in 1 L of seawater there is 20 g of sodium chloride. A formula used to calculate the concentration in $g\,L^{-1}$ is:

$$\text{concentration } (g\,L^{-1}) = \frac{\text{mass of solute (in g)}}{\text{volume of solution (in L)}}$$

Worked example 8.1.1

CALCULATING CONCENTRATION IN GRAMS PER LITRE ($g\,L^{-1}$)

What is the concentration, in $g\,L^{-1}$, of a solution containing 8.00 g of sodium chloride in 500 mL of solution?

Thinking	Working
Change the volume of solution so it is expressed in litres.	$500\,mL = \frac{500}{1000}$ $= 0.500\,L$
Calculate the concentration (c) in $g\,L^{-1}$.	$c = \frac{\text{mass of solute (in g)}}{\text{volume of solution (in L)}}$ $= \frac{8.00}{0.500}$ $= 16.0\,g\,L^{-1}$

Worked example: Try yourself 8.1.1

CALCULATING CONCENTRATION IN GRAMS PER LITRE ($g\,L^{-1}$)

What is the concentration, in $g\,L^{-1}$, of a solution containing 5.00 g of glucose in 250 mL of solution?

CHEMISTRY IN ACTION ICT

Blood alcohol concentration

Blood alcohol concentration (BAC) refers to the amount of alcohol present in the bloodstream. It is the concentration of ethanol, measured in grams of ethanol in every 100 millilitres of blood. This is different to the units of grams in every litre that you used in Worked example: Try yourself 8.1.1. A BAC of 0.05% means there is 0.05 g of ethanol in every 100 mL of blood (the same as 0.5 g of ethanol in every 1 L of blood).

Breathalysers or blood samples are commonly used to determine the BAC of drivers. Road safety authorities produce charts to educate drivers about the way their ability to drive safely deteriorates as their BAC rises. Figure 8.1.2a shows a portable breathalyser unit, and Figure 8.1.2b shows a typical chart outlining the impact of alcohol on a driver's capacity to drive.

(a)

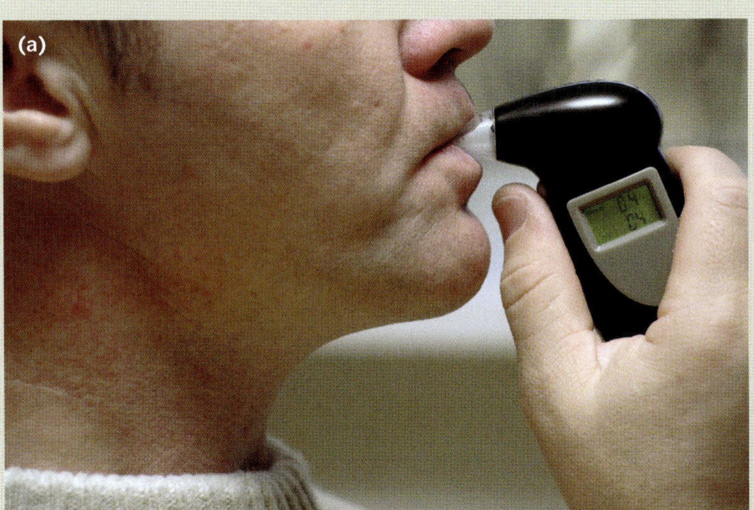

(b)

BAC	BAC and what driving skills are impaired
.10	
.09	
.08	concentrated attention, speed control
.07	
.06	information processing, judgement
.05	coordination
.04	eye movement control, standing steadiness, emergency responses
.03	tracking and steering
.02	divided attention, choice reaction time, visual function
.01	

FIGURE 8.1.2 (a) A portable random breath test device or breathalyser. (b) A chart illustrating the impact of alcohol on a person's ability to drive.

Concentration in parts per million (ppm)

FIGURE 8.1.3 Fish sold in Australia must have no more than 1 ppm of mercury according to the Australia New Zealand Food Standards Code.

When very small quantities of solute are dissolved to form a solution, the concentration is often measured in **parts per million (ppm)**. For example, the concentration of mercury in fish that are suitable for consumption is usually expressed in parts per million. The maximum concentration allowed for sale in Australia is 1 ppm (Figure 8.1.3).

In simple terms, the concentration in parts per million can be thought of as the mass in grams of solute that is dissolved in 1 000 000 g of solution. This can also be expressed as the mass in milligrams of solute dissolved in 1 kg of solution, because there are 1 million milligrams in a kilogram.

A solution of sodium chloride that has a concentration of 154 ppm contains 154 mg of sodium chloride dissolved in 1 kg of solution. A formula used to calculate the concentration of a solution in ppm is:

$$\text{concentration (ppm)} = \frac{\text{mass of solute (in mg)}}{\text{mass of solution (in kg)}}$$

Worked example 8.1.2

CALCULATING CONCENTRATION IN PARTS PER MILLION (PPM)

A saturated solution of calcium carbonate was found to contain 0.0198 g of calcium carbonate ($CaCO_3$) dissolved in 2000 g of solution. Calculate the concentration of calcium carbonate in the solution in parts per million (ppm).

Thinking	Working
Calculate the mass of solute in mg. Remember: mass (in mg) = mass (in g) × 1000	Mass of solute ($CaCO_3$) in mg: = 0.0198 × 1000 = 19.8 mg
Calculate the mass of solution in kg. Remember: mass (in kg) = $\frac{\text{mass (in g)}}{1000}$	Mass of solution in kg = $\frac{2000}{1000}$ = 2.000 kg
Calculate the concentration of the solution in mg kg^{-1}. This is the same as concentration in ppm.	Concentration of $CaCO_3$ in ppm = $\frac{\text{mass of solute (in mg)}}{\text{mass of solution (in kg)}}$ = $\frac{19.8}{2.000}$ = 9.90 mg kg^{-1} = 9.90 ppm

Worked example: Try yourself 8.1.2

CALCULATING CONCENTRATION IN PARTS PER MILLION (PPM)

A sample of tap water was found to contain 0.0537 g of sodium chloride (NaCl) per 250.0 g of solution. Calculate the concentration of NaCl in parts per million (ppm).

Concentration in milligrams per litre (mg L^{-1})

If, for example, the concentration of cadmium in a sample of wastewater was 5 mg L^{-1}, this means that in 1 L of the wastewater there will be 5 mg of cadmium. Concentrations in mg L^{-1} may be calculated using the formula:

$$\text{concentration (mg L}^{-1}) = \frac{\text{mass of solute (in mg)}}{\text{volume of solution (in L)}}$$

Concentrations in mg L^{-1} and ppm are equivalent

You saw earlier that a concentration of solution measured in units of ppm has the same value as concentration measured in mg kg^{-1}. Since 1 L of solution has a mass of almost exactly 1 kg, a concentration measured in mg L^{-1} can be considered as having the same value in units of ppm, i.e.:

1 ppm = 1 mg kg^{-1} = 1 mg L^{-1}

For example, the concentration, in mg L^{-1} and ppm, of 2.0 litres of a sugar solution containing 0.050 g of sugar can be calculated as follows:

0.050 g of sugar = 50 mg

$$\text{concentration (mg L}^{-1}) = \frac{50}{2.0}$$
$$= 25 \text{ mg L}^{-1}$$
$$= 25 \text{ ppm}$$

Concentration in parts per billion (ppb)

The amount of solute is always listed first: %(solute/solution)

For concentrations that are very low, the unit of **parts per billion (ppb)** can be used. If the concentration of a solution of a pesticide is 5 ppb, this means that in 1 billion grams of the solution, there will be 5 g of pesticide. Concentrations in units of ppb have the same value as concentrations in units of μg (microgram) per kg. A formula for calculating concentration in ppb is:

$$\text{concentration (ppb)} = \frac{\text{mass of solute (in μg)}}{\text{mass of solution (in kg)}}$$

Other units of concentration

You might have noticed symbols such as %(w/w), %(w/v) or %(v/v) on the labels of some foods, drinks and pharmaceuticals. These symbols represent other concentration units based on masses and volumes of solutes and solutions. These are useful in practical situations because people are familiar with these quantities (Figure 8.1.4).

FIGURE 8.1.4 Consumer products from hardware stores and supermarkets show a wide range of concentration units on their labels.

Percentage by mass (%(w/w))

Percentage by mass describes the mass of solute, measured in grams, present in 100 g of the solution.

Normal saline solution for washing contact lenses has a concentration of 0.9%(w/w). This can also be written as 0.9%(m/m). The abbreviation w/w indicates that the percentage is based on the weights or, more correctly, masses of both solute and solution. A concentration of 0.9% indicates that there is 0.9 g of sodium chloride dissolved in 100 g of solution.

FIGURE 8.1.5 The label from a bottle of Hunter Valley wine that shows the alcohol concentration in units of %(v/v).

Percentage by volume (%(v/v))

The abbreviation v/v indicates that the percentage is based on volumes of both solute and solution. The same units must be used to record both volumes. Percentage by volume is a more convenient unit to use than %(w/w) when the solute is a liquid.

Just like percentage by mass, percentage by volume is frequently expressed as volume per 100 mL of solution. For example, the wine label in Figure 8.1.5 shows 12.5% alcohol by volume. This means the wine contains 12.5% alcohol (ethanol) by volume (12.5%(v/v)). There will be 12.5 mL of alcohol in 100 mL of the wine.

For example, a 200 mL glass of champagne contains 28 mL of alcohol. The concentration expressed as %(v/v) of alcohol in this solution can be calculated as follows:

$$\begin{aligned}\text{concentration (\%v/v)} &= \frac{\text{volume of solute (in mL)}}{\text{volume of solution (in mL)}} \times 100\% \\ &= \frac{28}{200} \times 100\% \\ &= 14\%(\text{v/v})\end{aligned}$$

Percentage mass/volume (%(w/v))

Percentage mass/volume describes the mass of solute, measured in grams, present in 100 mL of the solution.

For example, if a solution of plant food contains a particular potassium compound at a concentration of 3%(w/v), this indicates that there is 3 g of potassium compound in 100 mL of solution.

CHEMFILE N

Saline drip

A 'saline drip' is sometimes used during medical procedures to replenish body fluids in patients (Figure 8.1.6). The solution used contains sodium chloride, commonly at a concentration of 0.9%(w/v). That means that 100 mL of the solution contains 0.9 g of sodium chloride.

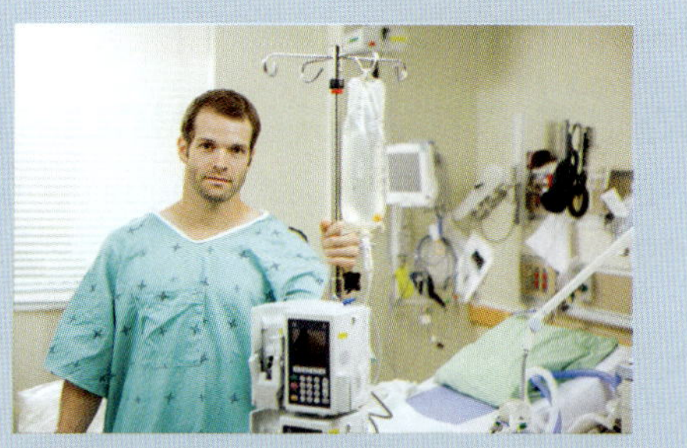

FIGURE 8.1.6 A saline drip is a solution of sodium chloride.

8.1 Review

SUMMARY

- The concentration of a solution is defined as the quantity of solute dissolved in a quantity of solution.
- Concentration units have two parts. The first part provides information about the quantity of solute. The second part provides information about the quantity of solution.
- The concentration of a solution can be expressed in different units, including $g\,L^{-1}$, $mg\,L^{-1}$, %(w/w), %(w/v), %(v/v), ppm and ppb.

KEY QUESTIONS

1 What is the concentration, in $g\,L^{-1}$, of a 60 mL solution that contains 5.0 g of sugar?

A 300 **B** 0.83 **C** 12 **D** 83

2 If 4.22 g of Li_2SO_4 is dissolved in water to make 380 mL of solution, what is the concentration in:

a %(w/v)?

b $g\,L^{-1}$?

3 **a** A 750 mL bottle of whisky contains 244 mL of ethanol. What is the concentration of ethanol in %(v/v)?

b The concentration of ethanol in a 350 mL bottle of beer is 4.9%(v/v). Calculate the volume of ethanol in the bottle.

4 Calculate the concentration of the following solutions in ppm:

a 25 mg of $CaCl_2$ dissolved in 5.0 kg of solution

b 1.25 g of lead(II) nitrate dissolved in 2000 g of solution

c 4.0×10^{-3} g of $MgSO_4$ dissolved in 150 g of solution

5 The nutrition panel on a flavoured milk drink states that it contains 35.0 g of sugar and 7.5 g of fat in every 250 mL serving. What are the concentrations of sugar and fat expressed as %(w/v)?

6 Ciguatoxin is a dangerous toxin that is found in some fish in Australia's tropical waters. The amount of this chemical that can kill a 70 kg human is approximately 15 mg. What is the lethal concentration of the toxin in a human body in parts per billion (ppb)?

7 Fluorine is added to toothpaste in the form of sodium fluoride (NaF). The permissible level of fluorine in toothpaste is 0.22%(w/w). A 175 g toothpaste tube contains 0.42 g of fluorine. Calculate the fluoride concentration in %(w/w) to see if it is above or below the legal level.

8.2 Molar concentration

In the previous section, you saw that units of concentration such as $g\,L^{-1}$, %(w/w) and %(v/v) are commonly used on the labels of consumer products found in supermarkets, pharmacies and hardware stores. Another unit of concentration that is commonly used by chemists is $mol\,L^{-1}$.

In this section, you will learn how to carry out calculations using this unit of concentration.

CONCENTRATION IN MOLES PER LITRE ($mol\,L^{-1}$)

GO TO ➤ Section 7.1 page 222

Concentrations expressed in moles per litre ($mol\,L^{-1}$) allow chemists to compare the number of atoms, molecules or ions present in a given volume of solution.

For example, a $0.1\,mol\,L^{-1}$ solution of hydrochloric acid would contain 0.1 mol of HCI in 1 L of the solution. A bottle of HCl of this concentration is shown in Figure 8.2.1.

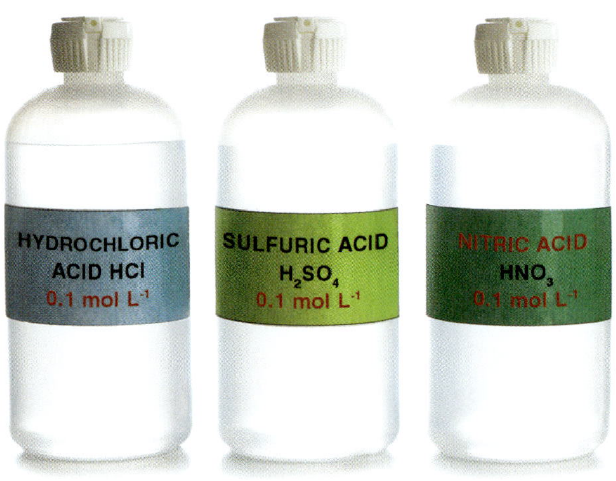

FIGURE 8.2.1 The concentrations of the solutions in these containers are shown in units of $mol\,L^{-1}$.

The concentration in $mol\,L^{-1}$ of a solution can be calculated as follows:

$$\text{concentration } (mol\,L^{-1}) = \frac{\text{amount of solute (in mol)}}{\text{volume of solution (in L)}}$$

This can be written using variables:

$$c = \frac{n}{V}$$

where

c is the concentration in moles per litre ($mol\,L^{-1}$)

n is the amount in moles (mol)

V is the volume in litres (L).

The concentration of a solution in moles per litre is often referred to as the **molarity**, or molar concentration of the solution, and is given the unit $mol\,L^{-1}$.

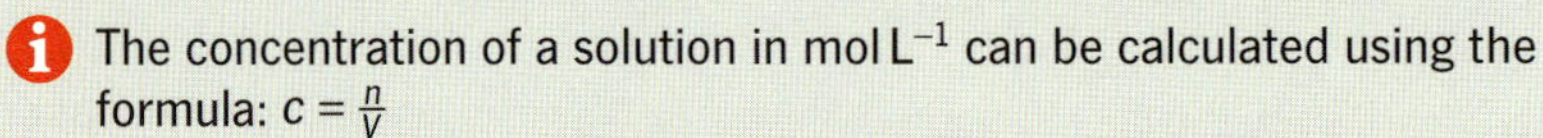
The concentration of a solution in $mol\,L^{-1}$ can be calculated using the formula: $c = \frac{n}{V}$

A solution containing 1 mol of solute dissolved in 1 L of solution can therefore be described in several different ways. You can say that the solution:

- has a concentration of 1 mole per litre
- has a concentration of $1\,mol\,L^{-1}$
- is 1 molar.

Worked example 8.2.1

CALCULATING MOLAR CONCENTRATIONS (MOLARITY)

Calculate the molar concentration of a solution that contains 0.105 mol of potassium nitrate (KNO_3) dissolved in 200 mL of solution.

Thinking	Working
Convert the given volume to litres.	$V(KNO_3) = \frac{200}{1000}$ $= 0.200\ L$
Calculate the molar concentration using the formula: $c = \frac{n}{V}$	$c(KNO_3) = \frac{n}{V}$ $= \frac{0.105}{0.200}$ $= 0.525\ mol\ L^{-1}$

Worked example: Try yourself 8.2.1

CALCULATING MOLAR CONCENTRATIONS (MOLARITY)

Calculate the molar concentration of a solution that contains 0.24 mol of glucose dissolved in 500 mL of solution.

Calculating the number of moles of solute in a solution

The formula used to calculate the molarity of a solution is $c = \frac{n}{V}$. If you rearrange the formula, you can use it to calculate the number of moles of solute in a solution of given concentration and volume:

$n = c \times V$

where

n is the amount of solute in moles (mol)

c is the concentration in moles per litre ($mol\ L^{-1}$)

V is the volume in litres (L).

Worked example 8.2.2

CALCULATING THE NUMBER OF MOLES OF SOLUTE IN A SOLUTION

Calculate the amount, in mol, of ammonia (NH_3) in 25.0 mL of a 0.3277 mol L^{-1} ammonia solution.

Thinking	Working
Convert the given volume to litres.	$V(NH_3) = \frac{25.0}{1000}$ $= 0.0250\ L$
Calculate the amount of compound, in mol, using the formula: $n = c \times V$	$n(NH_3) = c \times V$ $= 0.3277 \times 0.0250$ $= 8.19 \times 10^{-3}\ mol$

Worked example: Try yourself 8.2.2

CALCULATING THE NUMBER OF MOLES OF SOLUTE IN A SOLUTION

Calculate the amount, in mol, of potassium permanganate ($KMnO_4$), in 100 mL of a 0.0250 mol L^{-1} solution of the compound.

Calculating molarity given the mass of solute

Sometimes you will know the mass of a solute and volume of solution and want to calculate the molarity. Two main calculations are involved:

1 Calculate the number of moles of solute from its mass.

2 Calculate the concentration using the number of moles and the volume (in litres).

Worked example 8.2.3

CALCULATING MOLARITY GIVEN THE MASS OF SOLUTE

Calculate the concentration, in mol L^{-1}, of a solution that contains 16.8 mg of silver nitrate ($AgNO_3$) dissolved in 150 mL of solution.

Thinking	Working
Convert the volume to litres.	$V(AgNO_3) = \frac{150}{1000}$ $= 0.150\ L$
Convert the mass to grams.	$m(AgNO_3) = \frac{16.8}{1000}$ $= 0.0168\ g$
Calculate the molar mass of the solute.	$M(AgNO_3) = 107.9 + 14.01 + (3 \times 16.00)$ $= 169.9\ g\ mol^{-1}$
Calculate the number of moles of solute using the formula: $n = \frac{m}{M}$	$n(AgNO_3) = \frac{m}{M}$ $= \frac{0.0168}{169.9}$ $= 9.89 \times 10^{-5}\ mol$
Calculate the molar concentration using the formula: $c = \frac{n}{V}$	$c(AgNO_3) = \frac{n}{V}$ $= \frac{9.89 \times 10^{-5}}{0.150}$ $= 6.59 \times 10^{-4}\ mol\ L^{-1}$

Worked example: Try yourself 8.2.3

CALCULATING MOLARITY GIVEN THE MASS OF SOLUTE

Calculate the concentration, in mol L^{-1}, of a solution that contains 4000 mg of ethanoic acid (CH_3COOH) dissolved in 100 mL of solution.

8.2 Review

SUMMARY

- Molarity is defined as the number of moles of solute per litre of solution.
- Molarity is calculated using the formula: $c = \frac{n}{V}$
- The formula $n = c \times V$ can be used to calculate the number of moles of solute in a given volume of solution.

KEY QUESTIONS

1 Hydrogen peroxide solutions for hair bleaching are sold as 120 mL solutions containing 5.00 g of H_2O_2 dissolved in water. What is the molar concentration of hydrogen peroxide in this bleaching product?

A 0.81%(w/v)

B 1.23 mol L^{-1}

C 41.7 g L^{-1}

D 1.23×10^{-3} mol L^{-1}

2 Calculate the molar concentration of the following solutions:

a a 25.0 mL solution that contains 2.0×10^{-3} mol of NaCl

b a 4.1 L solution that contains 1.23 mol of CH_3COOH

c a 9.3×10^3 L solution that contains 1.8×10^3 mol of KCl

3 Calculate the molar concentration of the following solutions.

a an 8.0 L solution that contains 2.0 mol of NaCl

b a 500 mL solution that contains 0.25 mol of $MgCl_2$

c a 200 mL solution that contains 0.0876 mol of sucrose

4 A 750 mL bottle of vinegar contains 0.98 mol of ethanoic acid. Calculate the molar concentration of ethanoic acid.

5 Calculate the amount, in mol, of solute in each of the following solutions.

a 0.10 L of 0.22 mol L^{-1} KOH solution

b 10 mL of 0.64 mol L^{-1} NaI solution

c 15.6 mL of 0.0150 mol L^{-1} $CuSO_4$ solution

d 1.5×10^{-1} mL of 5.2 mol L^{-1} HCl solution

6 Calculate the molar concentration of a 250 mL solution that contains 5.09 g of $AgNO_3$.

7 Calculate the molar concentration of a 1.55 L solution that contains 1.223 g of $CaCl_2$.

8 What is the molar concentration of a sodium carbonate solution prepared by dissolving 1.32 g of sodium carbonate in a volumetric flask that is made up to 250.0 mL?

8.3 Dilution

FIGURE 8.3.1 Pesticide solutions used in aerial spraying are prepared by diluting a concentrated solution of the pesticide compound.

Many commercially available domestic and industrial products come in the form of concentrated solutions. Examples are pesticides (Figure 8.3.1), fertilisers, detergents, fruit juices, acids and other chemicals. A major reason for using concentrates is to save on transportation costs. Diluted solutions contain a lot of water and that extra mass has to be transported, which increases costs. It is also more convenient to buy concentrated products and dilute with water, whether at home or in the workplace.

Everyday examples of dilution are:

- adding water to cordial
- a laboratory technician making a 1 $mol L^{-1}$ solution of hydrochloric acid from a bottle of concentrated hydrochloric acid
- a home gardener diluting fertiliser concentrate to spray on the lawn
- a farmer diluting weedkiller concentrate to spray on a wheat crop
- an assistant in a commercial kitchen diluting concentrated detergent solution before using it to wash dishes.

In this section, you will learn how to calculate the concentrations of solutions before and after they have been diluted.

CALCULATING CONCENTRATIONS WHEN SOLUTIONS ARE DILUTED

The process of adding more solvent to a solution is known as **dilution**. When a solution is diluted, its concentration is decreased.

For example, if 50 mL of water is added to 50 mL of 0.10 $mol L^{-1}$ sugar solution, the amount of sugar remains unchanged but the volume of the solution in which it is dissolved doubles. As Figure 8.3.2 shows, this means the sugar molecules are spread further apart during the dilution process, so the concentration of the sugar solution is decreased (it will become 0.050 $mol L^{-1}$ in this instance).

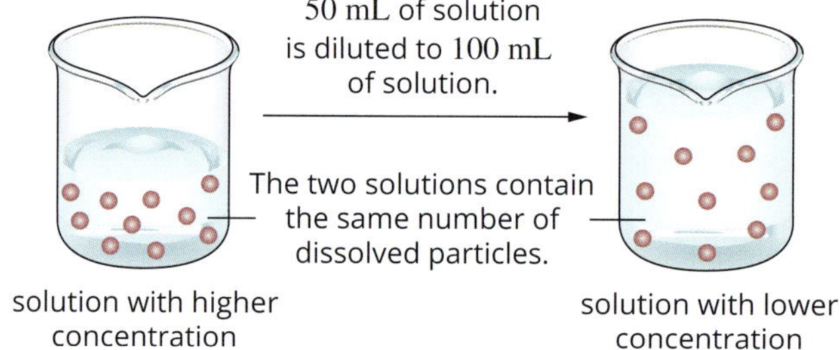

FIGURE 8.3.2 Dilution does not change the number of solute particles, but the concentration of the solute decreases.

It is important to recognise that diluting a solution (by adding more solvent) does not change the amount of solute present.

Suppose you had V_1 litres of a solution and the concentration was c_1 $mol L^{-1}$. The amount of solute, in moles, is given by:

$$n_1 = c_1 \times V_1$$

Suppose water was added to make a new volume, V_2, and change the concentration to c_2. The amount of solute, n_2, in this diluted solution is given by:

$$n_2 = c_2 \times V_2$$

The number of moles of solute has not changed, $n_1 = n_2$; therefore:

$$c_1 V_1 = c_2 V_2$$

This formula is useful when solving problems involving diluted solutions. (Note that this formula can also be used with different concentration and volume units, as long as they are the same on both sides.)

Worked example 8.3.1

CALCULATING THE CONCENTRATION OF A DILUTED SOLUTION

Calculate the concentration of the solution formed when 10.0 mL of water is added to 5.00 mL of 1.2 mol L^{-1} HCl.

Thinking	Working
Write down the value of c_1 and V_1. Note: c_1 and V_1 refer to the original solution, before water was added.	$c_1 = 1.2\ \text{mol L}^{-1}$ $V_1 = 5.00\ \text{mL}$
Write down the value of V_2. Note: V_2 is the total volume of the original solution plus the added water.	$V_2 = 10.0 + 5.00$ $= 15.0\ \text{mL}$
Transpose the equation $c_1V_1 = c_2V_2$ to calculate the concentration (c_2) of the new solution.	$c_1V_1 = c_2V_2$ $c_2 = \frac{c_1V_1}{V_2}$
Calculate the concentration of the diluted solution.	$c_2 = \frac{1.2 \times 5.00}{15.0}$ $= 0.40\ \text{mol L}^{-1}$

Worked example: Try yourself 8.3.1

CALCULATING THE CONCENTRATION OF A DILUTED SOLUTION

Calculate the concentration of the solution formed when 95.0 mL of water is added to 5.00 mL of 0.500 mol L^{-1} sulfuric acid (H_2SO_4).

CHEMISTRY IN ACTION PSC

Diluting strong acids

When schools purchase acid supplies for their science departments, they usually choose concentrated solutions. If the solution is concentrated, the volume of acid to transport is less. The school can prepare solutions of varying concentration by diluting this acid. A laboratory assistant diluting the strong acid needs to be aware of the following basic principles:

- Dilutions should be conducted in a fume cupboard.
- A spill plan must be ready.
- The laboratory assistant must wear goggles, gloves and a lab coat (Figure 8.3.3).
- Acid is added to water; water is not added to acid.
- It is easier to manage the release of heat energy if small amounts of acid are added gradually.
- Rinse all equipment thoroughly after use.
- Label and store the solutions appropriately.

FIGURE 8.3.3 Correct procedures need to be followed to safely dilute strong acids.

CHEMFILE N S CC

Threshold limit values of common substances

Sodium fluoride is added to the water supplies of most Australian cities. At low concentrations, the fluoride ions help to prevent cavities in teeth. If, however, you were to swallow a tablespoon of sodium fluoride, it might prove fatal. There are many substances like sodium fluoride that the human body can safely ingest in dilute solutions but not in concentrated solutions. It is important for society to have guidelines to follow and for companies to monitor the levels of exposure of their workers to chemicals. Threshold limit values (TLVs) are used for this purpose.

The TLV of a substance is the level to which it is believed a worker or consumer can be exposed day after day without adverse health effects. Three examples of TLVs are:

- sulfur dioxide (SO_2) at 5 ppm. Sulphur dioxide is produced when coal and other fossil fuels are burnt. It is also used as a preservative in foods.
- ozone (O_3) at 0.1 ppm. Ozone is produced in car engines. Figure 8.3.4 shows an example of an ozone warning sign.
- octane at 30 ppm. Octane is a component of petrol. Workers in garages and service stations are exposed to petrol fumes.

FIGURE 8.3.4 An ozone exposure warning.

CHANGING THE UNITS OF CONCENTRATION

At times it is useful to change, or 'convert', concentration values from one unit to another unit. One way to do this is in a series of simple steps. Worked example 8.3.2 shows how this can be done.

Worked example 8.3.2

CONVERTING CONCENTRATION UNITS

What is the concentration, in ppm, of a 0.00 200 mol L^{-1} solution of NaCl? Remember that concentration in ppm is the same as mg L^{-1}.

Thinking	Working
Calculate the number of mol of solute in 1.00 L of the solution.	$n(\text{NaCl}) = c \times V$ $= 0.00200 \times 1.00$ $= 0.00200$ mol
Calculate the mass, in g, of solute in 1.00 L of the solution.	$M(\text{NaCl}) = 35.45 + 22.99$ $= 58.44$ g mol^{-1} $m(\text{NaCl}) = n \times M$ $= 0.00200 \times 58.44$ $= 0.117$ g
Calculate the mass, in mg, of solute in 1.00 L of the solution.	$m(\text{NaCl}) = 0.117 \times 1000$ $= 117$ mg
Express the concentration of the solute in ppm.	$c(\text{NaCl}) = 117$ ppm

Worked example: Try yourself 8.3.2

CONVERTING CONCENTRATION UNITS

What is the concentration, in ppm, of a 0.0100 mol L^{-1} solution of NaOH? Remember that concentration in ppm is the same as mg L^{-1}.

8.3 Review

SUMMARY

- Numerical problems involving dilution can be solved using the formula:

$$c_1V_1 = c_2V_2$$

- When using this formula, c_1 and c_2 must be in the same units of concentration, and V_1 and V_2 must be in the same units of volume.
- One type of concentration unit can be converted to another unit in simple steps.

KEY QUESTIONS

1 Water is added to a 50 mL sample of 2.0 mol L^{-1} NaCl solution until the final volume is 250 mL. Calculate the molar concentration of the diluted solution.

2 Calculate the final concentration of each of the following diluted solutions:

a 10.0 mL of water is added to 5.0 mL of 1.2 mol L^{-1} HCl

b 1.0 L of water is added to 3.0 L of 0.10 mol L^{-1} HCl

c 5.0 mL of 0.50 mol L^{-1} HCl is added to 95.0 mL of water

3 What volume of 10 mol L^{-1} hydrochloric acid would be required to prepare 250 mL of a 0.30 mol L^{-1} HCl solution?

A 7.5 L

B 133 L

C 0.133 L

D 0.0075 L

4 The concentration of a solution of ammonia (NH_3) is 1.5%(w/v). What is the molar concentration of a solution produced by diluting 25.0 mL of this solution with 250 mL of water?

5 The concentration of a solution of NaOH is 17.0%(w/v). What is the concentration of this solution in mol L^{-1}?

6 The concentration of sulfuric acid sold to schools is usually 18 mol L^{-1}. Explain how a laboratory assistant would prepare 2.0 L of 0.10 mol L^{-1} sulfuric acid.

8.4 Standard solutions

There are many circumstances where the concentration of a solution needs to be accurately known. The levels of contaminants in water supplies, the concentrations of active ingredients in pharmaceuticals, and impurities in blood are examples where the exact concentration is significant.

A solution of accurately known concentration is called a **standard solution.** This section focuses on the preparation of standard solutions. Some of the equipment used to prepare a standard solution is shown in Figure 8.4.1.

FIGURE 8.4.1 Apparatus and materials needed to prepare a standard solution. The measured mass of solid will be dissolved in water in a volumetric flask to make an accurate volume of solution.

PRIMARY STANDARDS

Pure substances are widely used in the laboratory to prepare solutions of accurately known concentrations. Substances that are so pure that the amount of substance, in moles, can be calculated accurately from their mass are called **primary standards**.

A primary standard should:

- be readily obtainable in a pure form
- have a known chemical formula
- be easy to store without deteriorating or reacting with the atmosphere
- have a high molar mass, which will help to minimise the effect of errors in weighing
- be inexpensive.

Examples of primary standards are:

- bases: anhydrous sodium carbonate (Na_2CO_3) and hydrated sodium borate ($Na_2B_4O_7{\cdot}10H_2O$)
  (The term **anhydrous** indicates there is no water present in the compound.)
- acids: hydrated oxalic acid ($H_2C_2O_4{\cdot}2H_2O$) and potassium hydrogen phthalate ($KH(C_8H_4O_4)$).

PREPARING STANDARD SOLUTIONS

A standard solution can be prepared by dissolving an accurate mass of a primary standard in an accurately measured volume of water.

Standard solutions are prepared by dissolving an accurately measured mass of a primary standard in an accurately measured volume of water.

Digital balances are used in analytical laboratories to accurately weigh primary standards. A top-loading balance can weigh to an accuracy of between 0.1 g and 0.001 g, depending on the model. Analytical balances can weigh to an accuracy of between 0.0001 g and 0.000 01 g. The two types of balances commonly used can be seen in Figure 8.4.2.

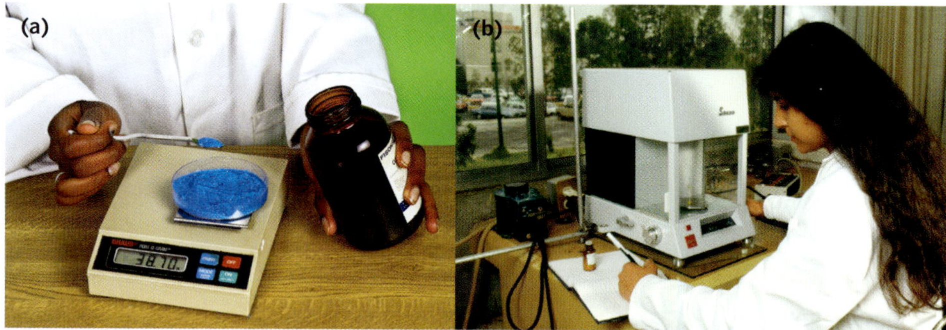

FIGURE 8.4.2 (a) A top-loading balance and (b) an analytical balance can be used to accurately weigh substances used in chemical analysis.

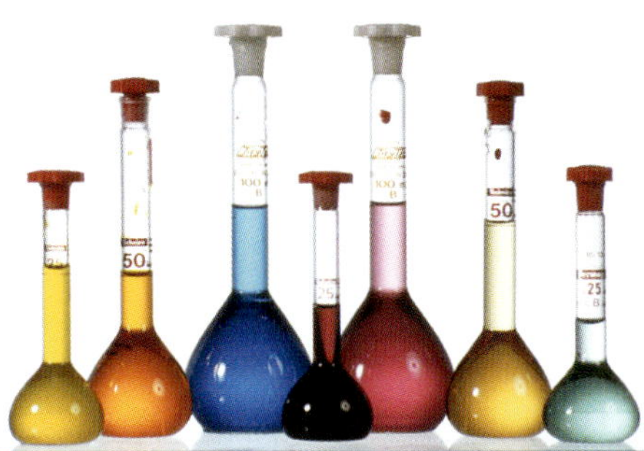

FIGURE 8.4.3 Volumetric flasks of various sizes are used to prepare standard solutions.

A **volumetric flask** or standard flask (Figure 8.4.3) is used to prepare a solution that has an accurately known volume. Volumetric flasks of 50.00 mL, 100.00 mL and 250.0 mL are frequently used in the laboratory.

A standard flask is filled so that the bottom of the meniscus is level with the graduation line on the neck of the flask (Figure 8.4.4). When checking this, your eye should be level with the line to avoid **parallax errors**.

To prepare a standard solution from a primary standard, you need to dissolve an accurately known amount of the substance in deionised water (water that has been purified by removing dissolved ions) to produce a solution of known volume. The steps in this process are shown in Figure 8.4.5 (page 266).

The molar concentration of the standard solution is then found using these formulae:

$$\text{amount in mol, } n = \frac{\text{mass of solute (in g)}}{\text{molar mass (in g mol}^{-1}\text{)}} = \frac{m}{M}$$

$$\text{concentration, } c = \frac{\text{amount of solute (in mol)}}{\text{volume of solution (in L)}} = \frac{n}{V} = \frac{m}{M \times V}$$

In practice, making a standard solution directly from a primary standard is only possible for a few of the chemicals found in the laboratory. Many chemicals are impure because they decompose or react with chemicals in the atmosphere. For example:

- strong bases, such as sodium hydroxide (NaOH) and potassium hydroxide (KOH), absorb water and react with carbon dioxide in the air
- many hydrated salts, such as hydrated sodium carbonate ($Na_2CO_3 \cdot 10H_2O$), lose water to the atmosphere over time.

In addition, the concentrations of commercial supplies of strong acids (HCl, H_2SO_4 and HNO_3) cannot be accurately specified.

Solutions such as HCl(aq), H_2SO_4(aq), NaOH(aq) and KOH(aq) must be standardised to determine their concentration. To do this, an accurately measured volume of the solution is reacted with a known amount of a standard solution such as Na_2CO_3(aq) or $KH(C_8H_4O_4)$(aq).

SKILLBUILDER WE

Reading volume levels

Correctly reading the volume of a liquid in a flask is an important skill. The following steps will help you to be as accurate as possible.

1. Ensure the flask is on a level surface.
2. Bring your eye level down to the surface of the liquid. This avoids parallax errors (incorrectly reading a different volume due to the position of your eyes). Do not hold the flask up to your eye level, as the liquid will not be level, and it is potentially hazardous.
3. Know what to read. Read the volume that is equal to the bottom of the meniscus (the concave curve of the surface of the water; Figure 8.4.4).
4. Hold up a piece of coloured card or tile to provide a contrasting background. This makes reading the volume easier.

FIGURE 8.4.4 This close-up view of the neck of a standard flask shows that the bottom of the meniscus is level with the white graduation line.

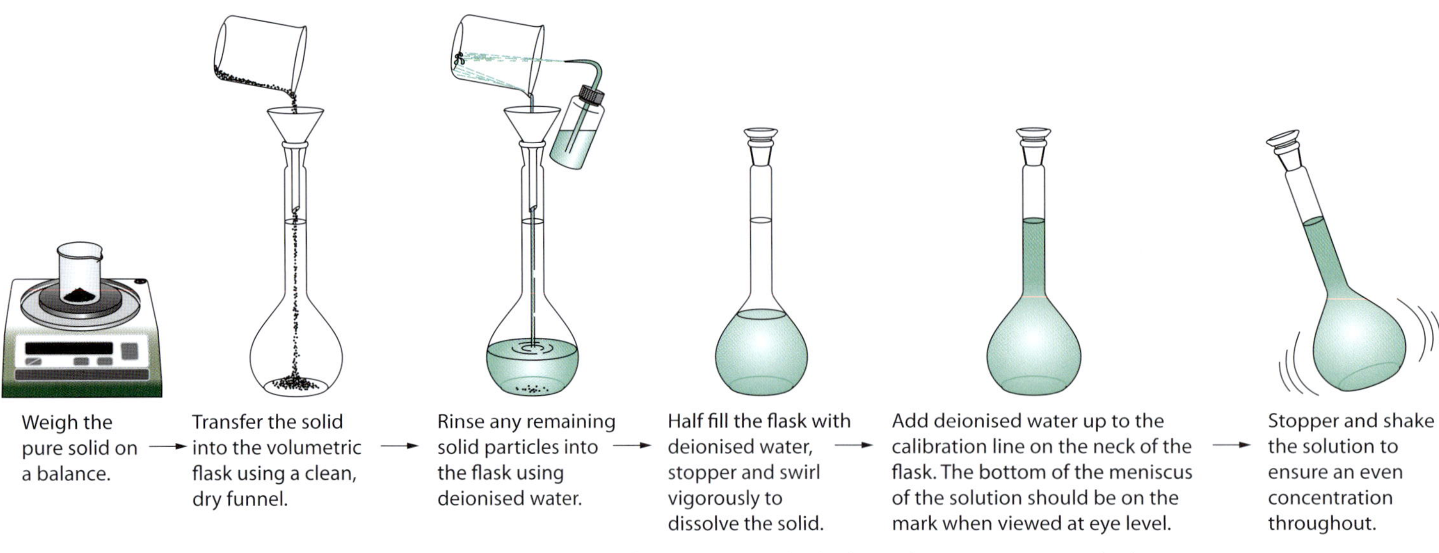

FIGURE 8.4.5 Preparing a standard solution from a primary standard.

Concentration of standard solutions

Earlier in this chapter you looked at calculating the concentration of a solution and determining how much solute is required to prepare a solution of a specific concentration.

Figure 8.4.6 summarises the steps involved in calculating the concentration of a standard solution from the accurately measured mass of a primary standard.

Use the chemical formula of the primary standard to determine the molar mass of the compound.

↓

Use the mass and molar mass of the primary standard to determine the amount, in mol, dissolved in the volumetric flask.

↓

Use the amount, in mol, of the primary standard and the volume of the flask to determine the concentration of the standard solution.

FIGURE 8.4.6 A summary of the steps used in the calculation of the concentration of a standard solution prepared from a primary standard.

Worked example 8.4.1

CALCULATING THE CONCENTRATION OF A STANDARD SOLUTION PREPARED FROM A PRIMARY STANDARD

Calculate the concentration of a standard solution prepared from 29.22 g of sodium chloride (NaCl) dissolved in a 1.00 L volumetric flask.

Thinking	Working
Use the chemical formula to determine the molar mass (M) of the compound.	$M(\text{NaCl}) = 22.99 + 35.45$ $= 58.44\ \text{g mol}^{-1}$
Use the mass (m) and molar mass (M) of the compound and the formula $n = \frac{m}{M}$ to determine the amount, in mol.	$n = \frac{29.22}{58.44}$ $= 0.500\ \text{mol}$
Use the amount, in mol, to determine the concentration of the solution using the formula $c = \frac{n}{V}$.	$c = \frac{0.500}{1.00}$ $= 0.500\ \text{mol L}^{-1}$

CHEMFILE WE

Purchasing standard solutions

Preparing standard solutions can be very challenging for some industries, especially if the accuracy of the concentration is important to the outcome. An alternative is to purchase the solution from a certified laboratory that guarantees the accuracy of their solutions. The hydrochloric acid solution shown in Figure 8.4.7 is guaranteed as 1.0 mol L^{-1}.

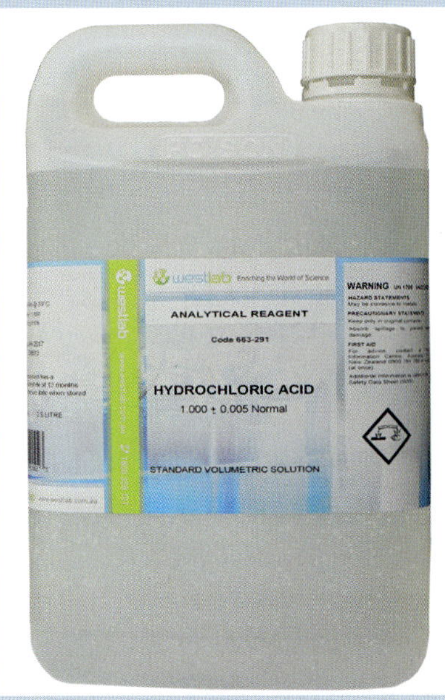

Figure 8.4.7 An alternative to preparing standard solutions is to purchase them from certified laboratories.

Worked example: Try yourself 8.4.1

CALCULATING THE CONCENTRATION OF A STANDARD SOLUTION PREPARED FROM A PRIMARY STANDARD

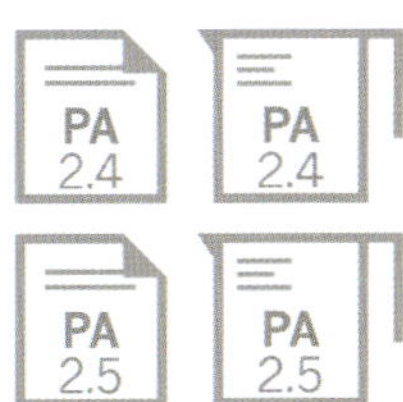

Calculate the concentration of a standard solution prepared from 117.0 g of NaCl dissolved in a 500.0 mL volumetric flask.

8.4 Review

SUMMARY

- A substance is suitable for use as a primary standard if it:
  - is readily obtainable in a pure form
  - has a known chemical formula
  - is easy to store without deteriorating or reacting with the atmosphere
  - has a high molar mass to minimise the effect of errors in weighing
  - is inexpensive.
- A standard solution is a solution of an accurately known concentration.
- The concentration, in mol L^{-1}, of a prepared standard solution can be determined by measuring the mass of solid dissolved and the volume of solution prepared.

KEY QUESTIONS

1 Which substance can be used as a primary standard?

A KOH

B H_2SO_4

C Na_2CO_3

D NaOH

2 Describe the steps required for the correct preparation of a standard solution.

3 Potassium hydrogen phthalate ($KH(C_8H_4O_4)$) is used as a primary standard for the analysis of bases. Calculate the concentration of a standard solution prepared in a 50.00 mL volumetric flask by dissolving 2.042 g of potassium hydrogen phthalate in deionised water. The molar mass of $KH(C_8H_4O_4)$ is 204.22 g mol^{-1}.

4 Calculate the mass of anhydrous sodium carbonate (Na_2CO_3) required to prepare 250.0 mL of a 0.500 mol L^{-1} standard solution.

5 Hydrated oxalic acid ($H_2C_2O_4{\cdot}2H_2O$) can be used as a primary standard. Calculate the mass of hydrated oxalic acid required to prepare 500 mL of a 0.200 mol L^{-1} standard solution. The molar mass of hydrated oxalic acid = 126.08 g mol^{-1}.

Chapter review

KEY TERMS

anhydrous
concentrated solution
concentration
dilute solution
dilution
molarity
parallax errors
ppb (parts per billion)
ppm (parts per million)
primary standard
solute
solvent
standard solution
volumetric flask

REVIEW QUESTIONS

1 A 250 mL antiseptic solution contains 28 mL of ethanol. The most convenient unit to use to describe the ethanol concentration is:

A %(w/v)
B %(v/v)
C %(w/w)
D $mg\,L^{-1}$

2 **a** What is the concentration, in ppm, of lead in a 6.0 kg solution that contains 12 mg of lead?
b What is the concentration as percentage by mass?

3 What is the concentration, in %(w/w), of a 300 g solution of potassium nitrate (KNO_3) that contains 15 g of solute?

4 In each 5.0 mL dose of a cold and flu medicine, there is 2.00 mg of the active ingredient. What is the concentration of this chemical, in $mg\,L^{-1}$ and %(w/v)?

5 A brand of disinfectant comes in a 1.25 L container, and the concentration of the active ingredient, benzalkonium chloride, is 1.50%(w/w). The mass of the solution is 1.30 kg.
a What is the mass of benzalkonium chloride in the solution?
b What mass of water is required for the solution?

6 The saline solution shown is for eye drops and other medical purposes. The 15 mL dispenser contains 1.2 mg of sodium chloride (NaCl).

Calculate the sodium chloride concentration in the eye drops in
a $mg\,L^{-1}$
b %(w/v)

7 160 mL of a solution contains 0.380 mol of NaBr. What is the molar concentration of the solution?

8 What is the molar concentration of a 2.0 L solution containing 30 g of NaOH?

9 Calculate the molar concentration of a sodium sulfate solution that has a concentration of 4.26%(w/v).

10 How many moles of solute are present in the following solutions?
a 12 mL of 0.22 $mol\,L^{-1}$ NaI
b 150 mL of 0.0250 $mol\,L^{-1}$ $KMnO_4$
c 7.2 L of 3.15×10^{-3} $mol\,L^{-1}$ KBr

11 What is the mass, in grams, of solute present in the following solutions?
a 100 mL of 1.20 $mol\,L^{-1}$ NH_3
b 20 mL of 0.50 $mol\,L^{-1}$ $AgNO_3$

12 The HCl solution shown is a commercially available standard solution. The 2.50 L container has a concentration of 1.00 $mol\,L^{-1}$.

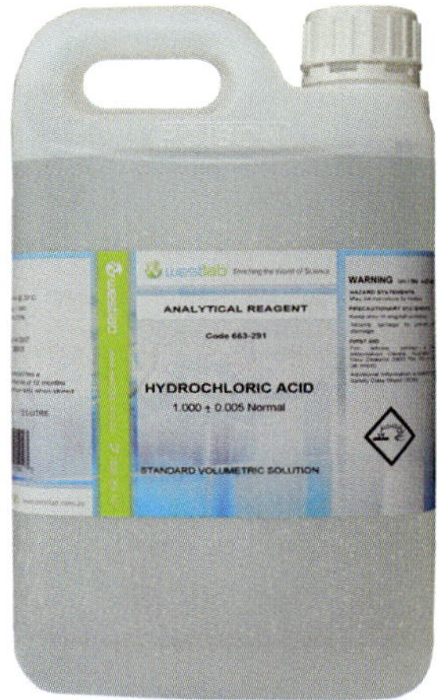

a Calculate the mass of HCl required to prepare this solution.
b What is the concentration of the solution in units of %(w/v)?

13 Select the correct alternative statement about the impact of dilution of a solution.
A The number of mole of solvent is unchanged.
B If the volume is doubled, the concentration will be doubled.
C The volume of solvent is unchanged.
D The number of mole of solute is unchanged.

14 Water is added to 50.0 mL of 5.00 mol L^{-1} HCl until the total volume is 250 mL. What is the molar concentration of HCl in the diluted solution?

15 What volume of water must be added to 25 mL of a 4.0 mol L^{-1} solution of potassium carbonate solution to dilute it to a concentration of 1.6 mol L^{-1}?

16 1.50 mL of a 0.0500 mol L^{-1} solution of calcium chloride ($CaCl_2$) was diluted with water to a volume of 10.0 L. What was the concentration of Cl^- ions in the diluted solution in ppm?

17 Concentrated sulfuric acid (H_2SO_4) is sold as an 18 mol L^{-1} solution. A company uses this concentrated solution to prepare 2.0 L bottles of 1.0 mol L^{-1} sulfuric acid. How many 2.0 L bottles can be prepared from each litre of concentrated solution?

18 Sodium borate ($Na_2B_4O_7{\cdot}10H_2O$) is used as a primary standard in volumetric analysis. Some properties of $Na_2B_4O_7{\cdot}10H_2O$ are listed below. Which one of these properties is not important in its use as a primary standard?

A It is highly soluble in water.
B Its purity is greater than 99.5%.
C It is a soft, white crystalline solid.
D It has a molar mass of 381 g mol^{-1}.

19 Calculate the concentration of a standard solution of hydrated oxalic acid ($H_2C_2O_4{\cdot}2H_2O$) prepared by dissolving 25.21 g of hydrated oxalic acid in 250.0 mL of deionised water.

20 Calculate the mass of Na_2CO_3 required to make a 500 mL standard solution of 0.400 mol L^{-1} Na_2CO_3.

21 Which of the following contains the greatest mass of sulfuric acid? (Molar mass = 98 g mol^{-1})

A 500 mL of 2.0 mol L^{-1} H_2SO_4
B 5 L of 0.1 mol L^{-1} H_2SO_4
C 100 L of H_2SO_4 of concentration 100 mg L^{-1}
D 50 g of H_2SO_4 added to 2 L of water

22 A student has three substances at her disposal, hydrated sodium carbonate ($Na_2CO_3{\cdot}10H_2O$), anhydrous sodium carbonate (Na_2CO_3) and sodium hydroxide (NaOH). Of these three substances, anhydrous sodium carbonate is the only one that is suitable as a primary standard.

a Explain why hydrated sodium carbonate is not suitable as a primary standard.
b Explain why sodium hydroxide is not suitable as a primary standard.
c Anhydrous sodium carbonate is often heated before it is weighed. Why do you think it is heated?
d What mass of sodium carbonate is needed to prepare a 250 mL standard solution of molar concentration 0.200 mol L^{-1}?

23 Iodine can be used as an antiseptic. A chemist shop sells two different iodine solutions:
Product A: 25 mL concentration 28%(w/v) iodine, cost $4.10
Product B: 500 mL concentration 4.0%(w/v) iodine, cost $12.80

a Which product contains the greatest mass of iodine?
b Which product offers better value for money?

24 Copy and complete the diagram below by inserting into the white boxes the processes required to convert between the quantities in the yellow boxes. The answer to the top left box is given to you.

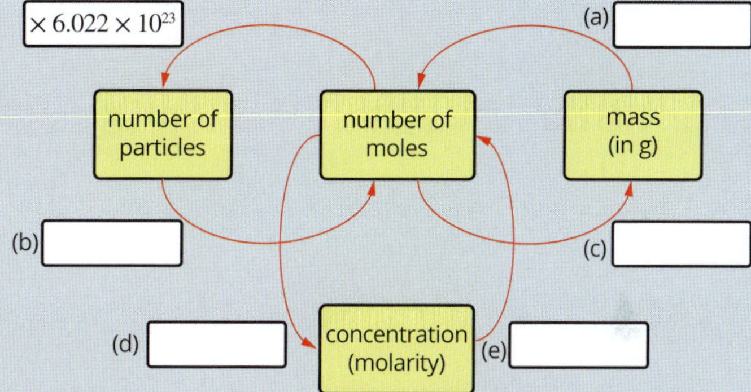

25 What is the molar concentration of a saturated $CuSO_4{\cdot}5H_2O$ solution that contains 20 g of solute in 100 g of water at 20°C? (Assume that 1 g of water gives 1 mL of solution.)

26 A cleaning product distributor has a warehouse storing 180 containers of oven-cleaning solution. The volume of each container is 25.0 L and the concentration of sodium hydroxide (NaOH) in each is 1.65 mol L^{-1}.

a Calculate the mass of NaOH in each container.
b Calculate the total mass of NaOH in the containers.
c Determine the concentration of NaOH in each container in g L^{-1}.
d The volume of solution remaining in one container that has been in use is 8.0 L. A worker adds deionised water to the container to make it back up to 25.0 L. What is the molar concentration of this diluted solution?

27 Reflect on the Inquiry activity on page 250. The chemical name of baking soda is sodium hydrogen carbonate ($NaHCO_3$). The chemical formula of ethanoic (acetic) acid, the active ingredient in vinegar, is CH_3COOH.
The reaction between vinegar and baking soda produces a gas.

a Use the chemical formulae provided to deduce the identity of the gas formed.
b Write an equation in words for the reaction occurring.
c Why do you think baking soda is added to a cake mix?

CHAPTER

09 Gas laws

Our knowledge of the behaviour of gases is crucial to our understanding of many phenomena, from hot air balloons rising through the air to car tyres bursting in the heat, expanding exhaust gases and explosions.

Our modern understanding of gases began with the experiments of J.B. van Helmont, a 17th century Flemish chemist, who identified carbon dioxide and recognised that the atmosphere was made of discrete elements. He coined the term 'gas' from the Greek word χάος (chaos). The scientific understanding of gases continued to develop over the next 200 years through the work of scientists such as Robert Boyle, Jacques Charles, John Dalton, Joseph Gay-Lussac and Amedeo Avogadro on a variety of gases in many different settings. Their detailed investigations of the properties of gases—pressure, volume, number of particles (moles) and temperature—ultimately led to the development of the ideal gas law, a mathematical model that links these properties together.

Content

INQUIRY QUESTION

How does the ideal gas law relate to all other gas laws?

By the end of this chapter, you will be able to:

- conduct investigations and solve problems to determine the relationship between the ideal gas law and:
  - Gay-Lussac's law (temperature)
  - Boyle's law
  - Charles' law
  - Avogadro's law (ACSCH060) 

CHEMISTRY INQUIRY 

Absolute zero 

How cold can it get?

COLLECT THIS ...

- fine capillary tube 5 cm in length, sealed at one end
- ruler
- approx. 450 mL water
- 600 mL beaker
- rubber band
- thermometer
- Bunsen burner

DO THIS ...

1 Invert the capillary tube so the sealed end is at the top, and strap it and the thermometer to the ruler using the rubber band.
2 Add the water to the beaker and heat it until it boils. Remove it from the heat.
3 Position the ruler, thermometer and capillary tube in the hot water.
4 Allow the beaker of water to cool.

RECORD THIS ...

Describe what happened.

Immediately record the temperature and length of the air column in the capillary tube and then at each decrease of 10°C.

Present your results in a spreadsheet (volume versus temperature). Create a scatter plot of the data with a trend line.

REFLECT ON THIS ...

Describe the relationship seen in the graph.

How could this graph be used to identify absolute zero temperature?

Is it possible for a gas to have zero volume?

9.1 Introducing gases

Every day you observe the behaviour of gases—such as those shown in Figure 9.1.1. Such examples can tell you a great deal about the physical properties of gases—those properties that can be observed and measured without changing the nature of the gas itself.

In this section you will learn about the properties and behaviour of gases.

FIGURE 9.1.1 (a) Air is used to inflate vehicle tyres. Air is a mixture of gases and is easily compressed. When the car goes over a bump in the road, the air compresses slightly and absorbs the impact of the bump. (b) The gases that cause the smell of a freshly brewed cup of coffee rapidly fill an entire room. Gases mix readily and, unlike solids and liquids, occupy all the available space. (c) This weather balloon is only partially inflated when released. Its volume increases because of pressure changes as it ascends into the atmosphere, where it will collect data.

PROPERTIES OF GASES

Each of the examples shown in Figure 9.1.1 can be explained in terms of the properties of gases. Table 9.1.1 summarises some of the properties of gases and compares them with the properties of solids and liquids. These observations can be used to develop a particle model of gas behaviour.

TABLE 9.1.1 Some properties of the three states of matter

	Gases	Liquids	Solids
density	low	high	high
volume and shape	fill the space available, because particles move independently of one another	fixed volume; adopt the shape of their container because particles are affected by attractive forces	fixed volume and shape because particles are affected by attractive forces
compressibility	compress easily	almost incompressible	almost incompressible
ability to mix	gases mix together rapidly	liquids mix together slowly unless stirred	solids do not mix unless finely divided

The low density of gases relative to that of liquids and solids suggests that the particles in a gas are spaced much further apart. The mass of any gas in a given **volume** is less than the mass of a liquid or solid in the same volume. The theory that gas particles are widely spaced can also explain the observation that gases are easily compressed.

The fact that gases spread to fill the space available, as shown in Figure 9.1.2, suggests that the particles of a gas move independently of one another.

The wide spacing and independent movement of particles explains why different gases mix rapidly.

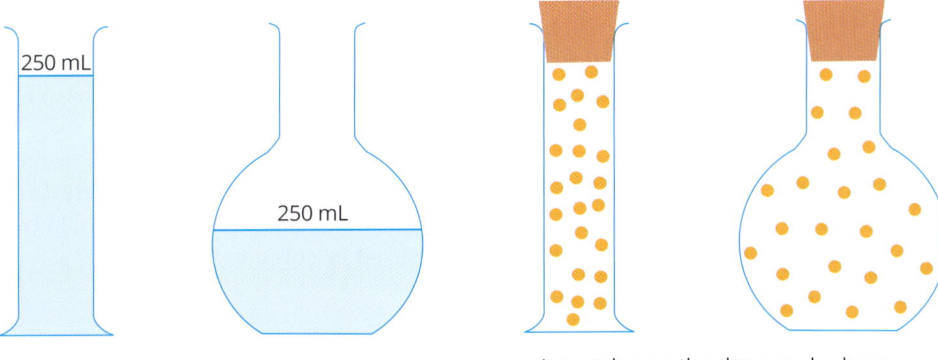

FIGURE 9.1.2 Both liquids and gases take the shape of the container they are in. However, a liquid has a fixed volume, whereas the volume of a gas expands to fill all the available space in a container.

KINETIC MOLECULAR THEORY

Scientists have developed a model to explain gas behaviour based on the behaviour of the particles of a gas. This model is known as the **kinetic molecular theory** of gases. According to this theory:

- Gases are composed of small particles, either atoms or molecules.
- The volume of the particles in a gas is negligible compared with the volume they occupy; consequently, most of the volume occupied by a gas is empty space.
- Gas particles move rapidly in random, straight-line motion.
- Particles collide with one another and with the walls of the container.
- The forces between particles are negligible.
- In a gas, when particles collide, the kinetic energy of the gas particles can be transferred from one particle to another, but the total kinetic energy remains constant. Therefore, collisions between gas particles are described as elastic collisions—kinetic energy is conserved.
- The average kinetic energy of the particles increases as the temperature of the gas increases.

FIGURE 9.1.3 Pumping more air into a tyre increases the pressure in the tyre because more particles are being pumped into a nearly fixed volume.

Kinetic energy (KE) is the energy of motion. The kinetic energy of a particle is calculated from the formula:

$$\text{KE} = \frac{1}{2}mv^2$$

where

KE is the kinetic energy in joules (J)

m is the mass in grams (g)

v is the **velocity** in metres per second (m s^{-1}).

THE NATURE OF PRESSURE AND VOLUME

When you blow up a balloon, the balloon expands because you put more air into it, and the rubber of the balloon stretches. Pumping more air into a tyre, as shown in Figure 9.1.3, increases the pressure inside the tyre. Figure 9.1.4 shows a scuba diver exhaling. The air bubbles get larger (increase in volume) as they rise to the surface of the ocean and the external pressure on them is reduced. In Section 9.3 you will examine the mathematical relationship between the amount (number of moles), volume, pressure and temperature of a gas. This requires an understanding of the nature of pressure and volume.

FIGURE 9.1.4 The air bubbles produced by a scuba diver expand as they rise.

Volume

Volume is the quantity used to describe the space that a substance occupies. Because a gas occupies the whole container that it is in, the volume of a gas is equal to the volume of its container.

There are several different units used for volume, some of which are represented in Figure 9.1.5. The common units are litre (L), millilitre (mL), cubic metre (m^3) and cubic centimetre (cm^3). Small volumes of gas are usually measured in millilitres (mL) or litres (L). Very large samples are measured in cubic metres (m^3).

- $1\,mL = 1\,cm \times 1\,cm \times 1\,cm = 1\,cm^3$
- $1\,L = 1 \times 10^3\,ml$ or $1000\,mL$
- $1\,m^3 = 1\,m \times 1\,m \times 1\,m = 100\,cm \times 100\,cm \times 100\,cm = 1 \times 10^6\,cm^3$
- $1\,m^3 = 1 \times 10^6\,cm^3 = 1 \times 10^6\,mL = 1000\,L$

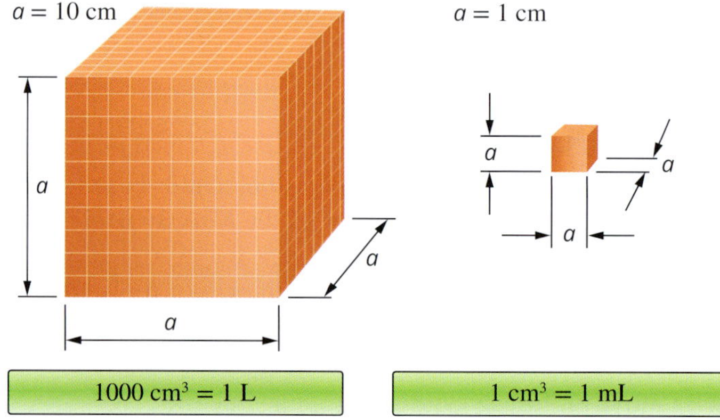

FIGURE 9.1.5 Cubic centimetres, millilitres and litres are the most commonly used units of volume.

> There are seven base SI units. Volume is not one of these units, so it is derived (created) from the base unit of length, the metre. The derived SI base unit for volume is cubic metres (m^3).

Worked example 9.1.1

CONVERTING VOLUME UNITS

A gas has a volume of 255 mL.

a What is its volume in cubic centimetres (cm^3)? **b** What is its volume in litres (L)? **c** What is its volume in cubic metres (m^3)?	
Thinking	**Working**
Recall the conversion factors for each of the units of volume. Apply the correct conversion to each situation.	**a** The units of mL and cm^3 are equivalent. $1\,mL = 1\,cm^3$ $255\,ml = 255\,cm^3$ **b** $1000\,ml = 1\,L$ Divide volume in mL by 1000 to convert to L. $255\,mL = \frac{255}{1000}$ $= 0.255\,L$ **c** $1 \times 10^6\,mL = 1\,m^3$ Divide volume in mL by 1×10^6 to convert to m^3. $255\,mL = \frac{255}{1 \times 10^6}$ $= 2.55 \times 10^{-4}\,m^3$

Worked example: Try yourself 9.1.1

CONVERTING VOLUME UNITS A GAS HAS A VOLUME OF 700 ML.

A gas has a volume of 700 mL.

a What is its volume in cubic centimetres (cm^3)?
b What is its volume in litres (L)?
c What is its volume in cubic metres (m^3)?

Pressure

People often talk about exerting pressure on something as if it is a type of force. In terms of the kinetic theory of gases, **pressure** is defined as the force exerted on a unit area of a surface by the particles of a gas as they collide with the surface.

The more a gas is compressed, the more frequently the gas particles collide with each other and the walls of their container. The increased frequency of collisions with the walls of the container increases the force on the walls, such as on the inside of a tyre. The force per unit area is described as pressure.

Air is a mixture of gases including nitrogen, oxygen, carbon dioxide and argon. In air, the nitrogen molecules collide with the walls of a container, exerting a pressure. In a similar way, the oxygen molecules exert a pressure, as do molecules of each gas present in the mixture. In the gaseous mixture of nitrogen and oxygen shown in Figure 9.1.6, the measured pressure is the sum of the **partial pressure** (or individual pressure) of oxygen and the partial pressure of nitrogen.

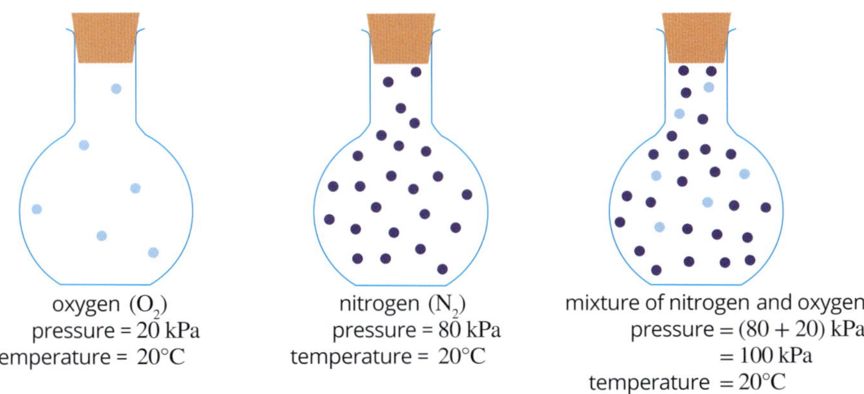

FIGURE 9.1.6 The total pressure of a mixture of gases is the sum of the partial pressures (individual pressures) of each of the gases in the mixture.

This relationship is often known as Dalton's law of partial pressures. It is expressed as:

$$P_{tot} = P_1 + P_2 + P_3 \ldots$$

where the total gas pressure within a container is the sum of the pressures of the constituent gas species, each of which exert a pressure independent of any other gas species.

Units of pressure

Because pressure is the force exerted on a unit area of a surface, the relationship can be written as:

$$\text{pressure} = \frac{\text{force}}{\text{area}}$$

$$P = \frac{F}{A}$$

The units of pressure depend on the units used to measure force and area. Over the years, scientists in different countries have used different units to measure force and area, so there are a number of different units of pressure.

The SI unit for force is the newton (N) and the SI unit for area is the square metre (m^2). Therefore, the SI units for pressure are newtons per square metre ($N\,m^{-2}$). One newton per square metre is equivalent to a pressure of one **pascal** (Pa).

In 1982, the International Union of Pure and Applied Chemistry (IUPAC), the organisation responsible for naming chemicals and setting standards, adopted a standard for pressure equivalent to 100 000 Pa or 100 kPa. This gave rise to a new pressure unit called the **bar**, where 1 bar equals 100 kPa.

The use of mercury **barometers** resulted in pressure often being measured in millimetres of mercury (mmHg).

Another unit for pressure is the **standard atmosphere (atm)**. One standard atmosphere (1 atm) is the pressure required to support 760 millimetres of mercury (760 mmHg) in a mercury barometer at 25°C. This is the average atmospheric pressure at sea level. One atmosphere equals 101.3 kPa.

Summary of pressure units

There are four commonly used units of gas pressure: pascal, bar, millimetres of mercury and atmosphere (Table 9.1.2).

TABLE 9.1.2 Commonly used units of gas pressure

Name of unit	Symbol for unit	Conversion to $N\,m^{-2}$
newtons per square metre	$N\,m^{-2}$	
pascal	Pa	$1\,Pa = 1\,N\,m^{-2}$
kilopascal	kPa	$1\,kPa = 1 \times 10^3\,Pa = 1 \times 10^3\,N\,m^{-2}$
atmosphere	atm	$1\,atm = 101.3\,kPa = 1.013 \times 10^5\,N\,m^{-2}$
bar	bar	$1\,bar = 100\,kPa = 1 \times 10^5\,N\,m^{-2}$
millimetres of mercury	mmHg	$760\,mmHg = 1\,atm = 1.013 \times 10^5\,N\,m^{-2}$

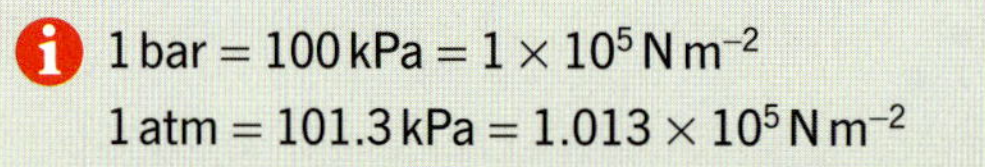

$1\,bar = 100\,kPa = 1 \times 10^5\,N\,m^{-2}$

$1\,atm = 101.3\,kPa = 1.013 \times 10^5\,N\,m^{-2}$

These relationships can be used to convert pressure from one unit to another.

CHEMFILE N

Torricelli's barometer

In the 17th century, the Italian physicist Evangelista Torricelli invented the earliest barometer, an instrument used to measure atmospheric pressure. It was a straight glass tube, closed at one end, containing mercury. The tube was inverted so that the open end was below the surface of the mercury in a bowl, as seen in Figure 9.1.7.

The column of mercury in Torricelli's barometer was supported by the pressure of the gas particles in the atmosphere colliding with the surface of the mercury in the open bowl.

At sea level, the top of the column of mercury was about 760 mm above the surface of the mercury in the bowl. Torricelli found that the height of the mercury column decreased when he took his barometer to higher altitudes in the mountains.

At higher altitudes, there are fewer air particles and therefore less frequent collisions on the surface area of the mercury. The reduced pressure supports a shorter column of mercury.

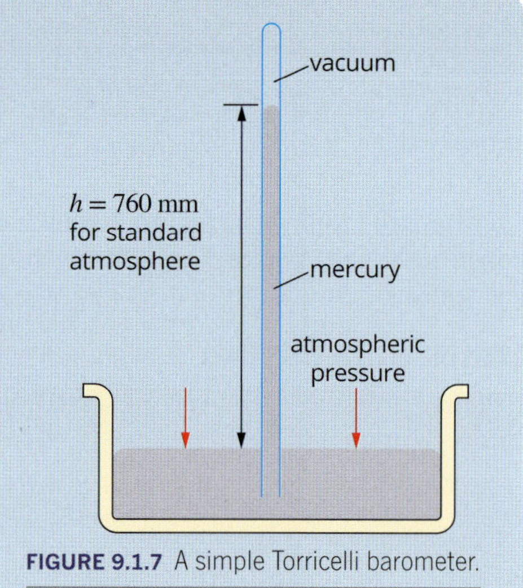

FIGURE 9.1.7 A simple Torricelli barometer.

Worked example 9.1.2

CONVERTING PRESSURE UNITS

Mount Everest (Figure 9.1.8) is the highest mountain on Earth.

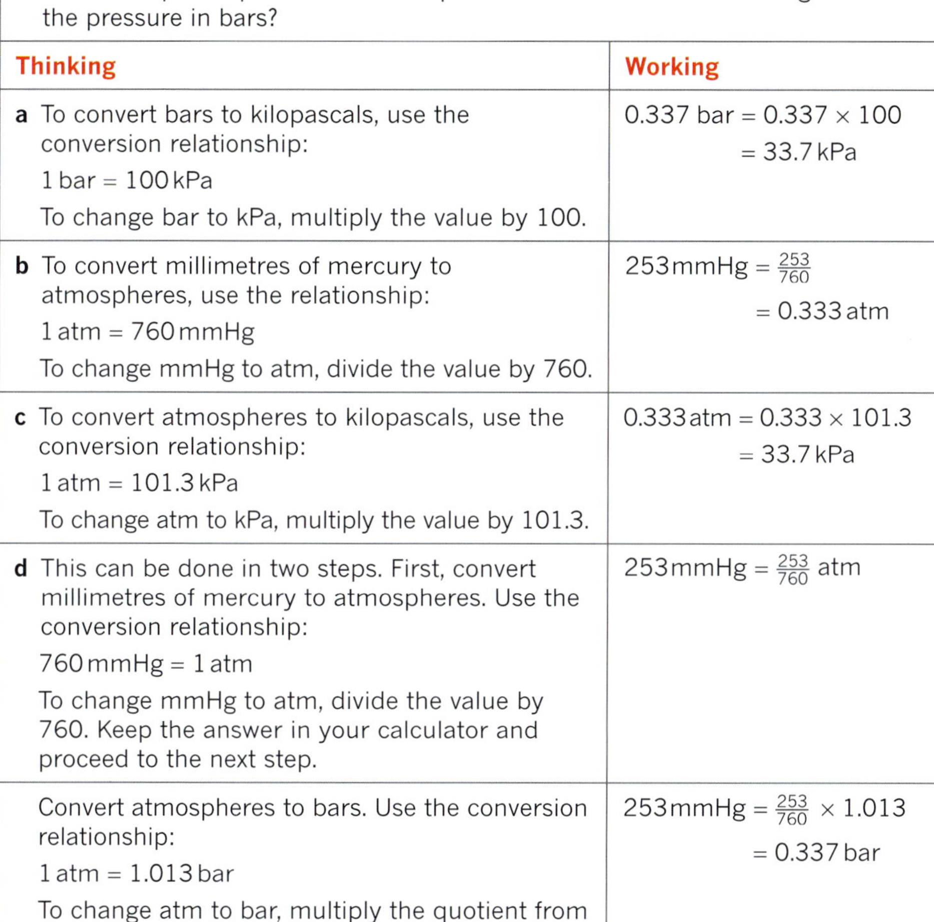

a The atmospheric pressure at the top of Mount Everest is 0.337 bar. What is the pressure in kilopascals (kPa)? **b** The atmospheric pressure at the top of Mount Everest is 253 mmHg. What is the pressure in atmospheres (atm)? **c** The atmospheric pressure at the top of Mount Everest is 0.333 atm. What is the pressure in kilopascals (kPa)? **d** The atmospheric pressure at the top of Mount Everest is 253 mmHg. What is the pressure in bars?	
Thinking	**Working**
a To convert bars to kilopascals, use the conversion relationship: 1 bar = 100 kPa To change bar to kPa, multiply the value by 100.	0.337 bar = 0.337 × 100 = 33.7 kPa
b To convert millimetres of mercury to atmospheres, use the relationship: 1 atm = 760 mmHg To change mmHg to atm, divide the value by 760.	253 mmHg = $\frac{253}{760}$ = 0.333 atm
c To convert atmospheres to kilopascals, use the conversion relationship: 1 atm = 101.3 kPa To change atm to kPa, multiply the value by 101.3.	0.333 atm = 0.333 × 101.3 = 33.7 kPa
d This can be done in two steps. First, convert millimetres of mercury to atmospheres. Use the conversion relationship: 760 mmHg = 1 atm To change mmHg to atm, divide the value by 760. Keep the answer in your calculator and proceed to the next step.	253 mmHg = $\frac{253}{760}$ atm
Convert atmospheres to bars. Use the conversion relationship: 1 atm = 1.013 bar To change atm to bar, multiply the quotient from the previous step by 1.013.	253 mmHg = $\frac{253}{760}$ × 1.013 = 0.337 bar

FIGURE 9.1.8 The atmospheric pressure at the top of the world's highest mountain, Mount Everest, is different to the pressure at the foot of the mountain, 8848 m below.

Worked example: Try yourself 9.1.2

CONVERTING PRESSURE UNITS

Cyclone Debbie was one of the biggest cyclones in Australian history.

a The atmospheric pressure in the eye of Cyclone Debbie was measured as 0.959 bar. What was the pressure in kilopascals (kPa)?

b The atmospheric pressure in the eye of Cyclone Debbie was 720 mmHg. What was the pressure in atmospheres (atm)?

c The atmospheric pressure in the eye of Cyclone Debbie was 0.947 atm. What was the pressure in kilopascals (kPa)?

d The atmospheric pressure in the eye of Cyclone Debbie was 720 mmHg. What was the pressure in bars?

9.1 Review

SUMMARY

- Gas properties can be explained by the kinetic molecular theory.
- According to the kinetic molecular theory:
  - The volume of the particles in a gas is very small compared with the distance between the particles.
  - The average kinetic energy of the particles in a gas is proportional to its temperature.
  - Gas particles are in rapid random motion, colliding with each other and with the container wall.
  - The forces between the particles in a sample of gas are negligible.
- To convert between different volume units, use these relationships:

  $1\,L = 1000\,mL = 1000\,cm^3$

  $1\,m^3 = 1 \times 10^6\,cm^3 = 1 \times 10^6\,mL = 1000\,L$
- Pressure is defined as the force per unit area.
- The pressure exerted by a gas depends on the collisions of gas particles with the wall of the container.
- To convert between different pressure units, use these relationships:

  $1\,bar = 100\,kPa = 1.00 \times 10^5\,Pa$

  $1\,atm = 760\,mmHg = 1.013 \times 10^5\,Pa$

  $= 101.3\,kPa = 1.013\,bar$
- The total pressure exerted by a mixture of gases is the sum of each gas's partial pressure:

  $$P_{tot} = P_1 + P_2 + P_3 \ldots$$

KEY QUESTIONS

1. Use the kinetic molecular theory to explain the following observed properties of gases:
   - **a** Gases occupy all the available space in a container.
   - **b** Gases can be easily compressed, compared with their corresponding liquid forms.
   - **c** A given volume of a gaseous substance has a smaller mass than the same volume of the substance in the liquid state.
   - **d** Gases readily mix together.
   - **e** The total pressure of a mixture of gases is equal to the sum of the pressures exerted by each of the gases in the mixture.
2. Use the ideas of the kinetic molecular theory of gases to explain the following observations:
   - **a** Tyre manufacturers recommend a maximum pressure for tyres.
   - **b** The pressure in a car's tyres will increase if a long distance is travelled on a hot day.
   - **c** You can smell dinner cooking as you enter your house.
   - **d** A balloon will burst if you blow it up too much.
3. In the kinetic molecular theory, pressure is described as the force per unit area of surface. Explain what happens to the pressure in each of the following situations:
   - **a** The temperature of a filled aerosol can is increased.
   - **b** A gas in a syringe is compressed.
4. Convert each of the following pressures to the units specified:
   - **a** 140 kPa to Pa
   - **b** 92 000 Pa to kPa
   - **c** 4.24 atm to mmHg and Pa
   - **d** 120 kPa to mmHg, atm and bar
   - **e** 1400 mmHg to atm, Pa and bar
   - **f** 80 000 Pa to atm, mmHg and bar
5. Convert the following volumes to the unit specified:
   - **a** 2 L to mL
   - **b** 4.5 L to m^3
   - **c** 2250 mL to L
   - **d** 120 mL to L
   - **e** 5.6 mL to L
   - **f** 3.7 m^3 to L
   - **g** 285 mL to m^3
   - **h** $4.70 \times 10^{-3}\,m^3$ to L and cm^3
6. The total pressure of a mixture of gases is the sum of each gas's partial pressure: $P_{tot} = P_1 + P_2 + P_3 \ldots$
   - **a** A container of air at a pressure of 100 kPa contains water vapour at a partial pressure of 2.5 kPa. If the water vapour is removed, calculate the pressure of the remaining dry air.
   - **b** The partial pressure of oxygen in dry air is 160 mmHg. If the dry air is at a pressure of 760 mmHg, calculate the percentage of oxygen in dry air.

9.2 The gas laws

The behaviour of gases has been described **qualitatively** in Section 9.1. In this section, you will learn about the mathematical relationships that link the volume, pressure, temperature and number of particles of a gas. In particular, you will learn how to use these relationships to calculate the volumes of gases.

These mathematical relationships were developed over a period of several hundred years by various scientists who performed experiments on gases. In 1662, Robert Boyle (Figure 9.2.1) experimentally determined the relationship between the pressure and volume of a gas. French chemist Jacques Charles identified the relationship between the volume and temperature of a gas. In 1787, Charles determined that the volume of a fixed amount of gas is directly proportional to temperature, provided the pressure remains constant. The relationship between gas volume and temperature is known as Charles' law. In 1802, another French chemist, Joseph Louis Gay-Lussac, published his investigations into the relationship between temperature and pressure.

These relationships have become known as the gas laws. The gas laws are used to describe the behaviour of all gases, regardless of their chemical composition. The three gas laws are combined mathematically to form a combined gas law.

FIGURE 9.2.1 Robert Boyle (1627–1691). In 1662, Boyle showed by experiment that for a given amount of gas at constant temperature, the volume of a gas is inversely proportional to its pressure. The relationship between gas volume and pressure is known as Boyle's law.

BOYLE'S LAW: VOLUME AND PRESSURE

Changing the volume of a fixed amount of gas at a constant temperature causes a change in the pressure of the gas. The pressure of the gas in the syringe shown in Figure 9.2.2 increases as the plunger is pushed in, and decreases as the plunger is pulled out.

For a given amount of gas at constant temperature, the volume of the gas is inversely proportional to its pressure. This relationship is seen in the changing volume of a weather balloon as it rises to altitudes with much lower pressure than at ground level. A weather balloon filled with helium gas to a volume of 40 L at a pressure of 1 atm increases in volume to 200 L by the time it reaches an altitude with a pressure of 0.2 atm (Figure 9.2.3) (page 280).

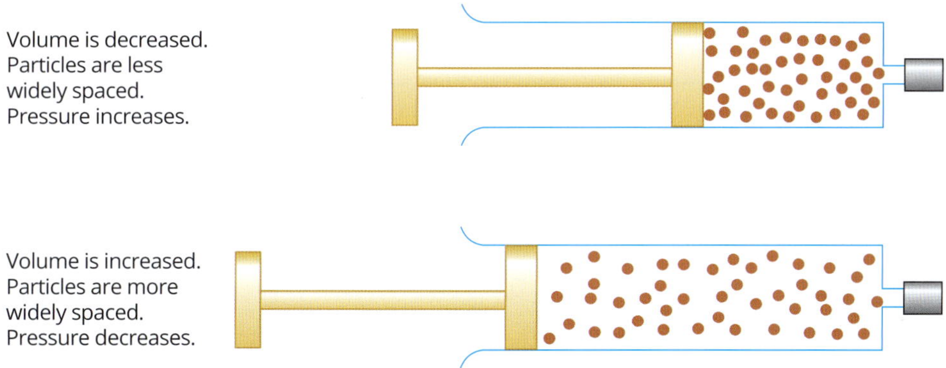

FIGURE 9.2.2 The pressure of the gas in the syringe is affected by a change in volume.

> The mathematical relationship between the pressure, P, exerted by a gas and the volume, V, it occupies can be written as:
>
> $$P \propto \frac{1}{V}$$
>
> This relationship can also be expressed as $PV = k$, where k is a constant at a given temperature.

CHEMISTRY IN ACTION 

Weather balloons

Figure 9.2.3 shows a weather balloon launch. Weather balloons measure temperature, humidity and atmospheric pressure at an altitude of around 25 km. Tracking the balloon during its journey, using a global positioning system (GPS), allows measurement of wind speed and direction. The balloon increases in volume as it rises through the atmosphere due to the decreasing atmospheric pressure. Ultimately, it bursts and the instruments fall to Earth, suspended from a parachute.

FIGURE 9.2.3 A researcher launching a radiosonde weather balloon, which is carrying a package of instruments. Study of the atmosphere allows more accurate weather forecasting and enables environmental change to be monitored. Photographed at the Halley Research Station, Antarctica.

SKILLBUILDER 

Understanding mathematical symbols

An important part of the language of science is the use of symbols. These are used to represent quantities or to give meanings. For example, the four symbols $<$, $>$, $\leq$ and $\geq$ are known as 'inequalities'.

The following mathematical symbols are commonly used in science.

Symbol	Meaning	Example	Explanation
$<$	less than	$2 < 3$	2 is less than 3
$>$	greater than	$6 > 1$	6 is greater than 1
$\leq$	less than or equal to	$2x \leq 10$	$2x$ is less than or equal to 10
$\geq$	greater than or equal to	$3y \geq 12$	$3y$ is greater than or equal to 12
$\sqrt{}$	square root	$\sqrt{4} = 2$	The square root of 4 is 2.
Δ	change in (difference between)	Δt	change in t (time)
$\approx$	approximately equal to	$\pi \approx 3.14$	π is approximately equal to 3.14
Σ	summation	$\sum_{i=1}^{4} i$	the sum of consecutive integers from 1 to 4, i.e. $1 + 2 + 3 + 4 = 10$
$\propto$	proportional to	$P \propto \frac{1}{V}$	pressure is proportional to 1 divided by volume
$\therefore$	therefore	$x + 1 = 6$ $\therefore x = 5$	x plus 1 equals 6 therefore x is equal to 5

Worked example 9.2.1

USING BOYLE'S LAW TO PREDICT CHANGES IN PRESSURE AND VOLUME

A helium balloon at 1.0 atm pressure has a volume of 5.0 L. The external pressure on the balloon is decreased to 0.30 atm. If the temperature remains constant, what will be the new volume of the balloon?

Thinking	Working
State Boyle's law.	$PV = k$
Determine the relationship between the two states.	As the amount of gas has not changed and the temperature has been held constant, the relationship between the two states is: $P_1V_1 = k = P_2V_2$
State the known and unknown variables.	$P_1 = 1.0$ atm $V_1 = 5.0$ L $P_2 = 0.30$ atm $V_2 = ?$
Rearrange the equation to make V_2 the subject.	$P_1V_1 = P_2V_2$ $V_2 = \frac{P_1V_1}{P_2}$

Substitute in the known values and solve.	$V_2 = \frac{P_1V_1}{P_2}$ $= \frac{1.0 \times 5.0}{0.30}$ $= 17\text{L}$
State the new volume.	The new volume is 17 L.

Worked example: Try yourself 9.2.1

USING BOYLE'S LAW TO PREDICT CHANGES IN PRESSURE AND VOLUME

The swim bladder of a fish has a volume of 30 mL near the surface of the water, where the pressure acting on it is 1.00 bar. Assuming no changes to the gas composition of the bladder, and no change in the temperature of the water, what will be the volume of the bladder at a depth of 10 m, where the pressure on the swim bladder is 2.00 bar?

CHARLES' LAW: VOLUME AND TEMPERATURE

If a balloon is anchored in a beaker of water, as shown in Figure 9.2.4, and the water is heated with a Bunsen burner, the volume of air in the balloon increases with the rising temperature. This is an example of Jacques Charles' law (Figure 9.2.5).

FIGURE 9.2.4 Demonstration of the thermal expansion of gas, and the relationship between volume and temperature.

The kinetic molecular theory states that increasing the temperature of a gas increases the average kinetic energy of the molecules. The molecules move more rapidly and collide with the walls of the container more frequently and with greater force. This can cause:

- the volume of gas to increase if the pressure on the gas remains much the same, such as for a gas in a syringe or balloon (Figure 9.2.5)
- the pressure to increase if the volume of the gas container is fixed, such as in a sealed flask or a gas cylinder.

Table 9.2.1 shows the results of an experiment that investigates the relationship between temperature and the volume of a given amount of gas. In this experiment, the gas in a syringe was heated slowly in an oven. The pressure on the plunger of the syringe was held constant.

TABLE 9.2.1 Variation of volume with temperature

Temperature (°C)	20	40	60	80	100	120	140
Volume (mL)	60.0	64.1	68.2	72.3	76.4	80.5	84.6

FIGURE 9.2.5 (a) Jacques Charles (1746–1823) was a French chemist, physicist and aeronaut. (b) Charles was famous for making the first flight in a hydrogen balloon on 1 December 1783. Charles showed that the volume of a gas is directly proportional to its temperature.

CHEMFILE IU

William Thomson

The Kelvin scale is named after Irish mathematical physicist and engineer William Thomson (1824–1907), also known as Lord Kelvin (Figure 9.2.7).

Born in Belfast, Thomson attended Glasgow University from the age of ten and Cambridge University from 1841 to 1845. In 1846, he returned to Glasgow to become a Professor of Natural Philosophy, remaining in this position for 53 years. He is most famous for formulating the second law of thermodynamics, working to install telegraph cables under the Atlantic Ocean and proposing the absolute temperature scale. He correctly determined the value of absolute zero to be −273°C.

A marine engineering enthusiast, Thomson invented several marine instruments to improve navigation and safety, including a mariner's compass and a deep-sea sounding apparatus. In 1892, Thomson became the first scientist to join the House of Lords in England. By the time of his death in 1907, he was an international celebrity. He headed an international commission responsible for the design of the Niagara Falls power station, and he became a lecturer at Johns Hopkins University in Baltimore, USA. On his death he was buried beside Sir Isaac Newton in Westminster Abbey.

FIGURE 9.2.7 Lord Kelvin (1824–1907) proposed the absolute temperature scale.

The graph of these results is linear, as shown in Figure 9.2.6. When the graph is extrapolated to a volume of 0 L, it crosses the temperature axis at −273°C. If the origin is reset at −273°C, this graph passes through the origin. This has led scientists to develop a new temperature scale, known as the **Kelvin scale** or **absolute temperature scale**. On the Kelvin scale, each temperature increment is equal to one temperature increment on the Celsius scale, and 0°C is equal to 273 K.

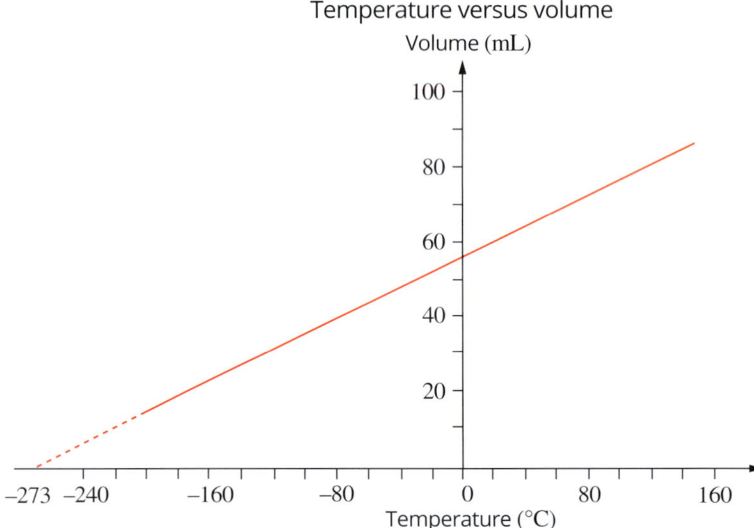

FIGURE 9.2.6 The variation of volume with temperature for a fixed amount of gas at constant pressure.

The relationship between temperature on the Celsius scale and temperature on the Kelvin scale is given by the equation:

$$T \text{ (in K)} = T \text{ (in °C)} + 273$$

The temperature 0 K (−273°C) is the lowest temperature theoretically possible. For this reason, 0 K is known as **absolute zero**. At this temperature, all molecules and atoms have minimum kinetic energy. Therefore, a temperature in degrees Celsius is converted to absolute temperature in kelvin by adding 273. For example, the boiling point of water at sea level is converted to absolute temperature by adding 273, so 100°C is equivalent to 373 K (Figure 9.2.8).

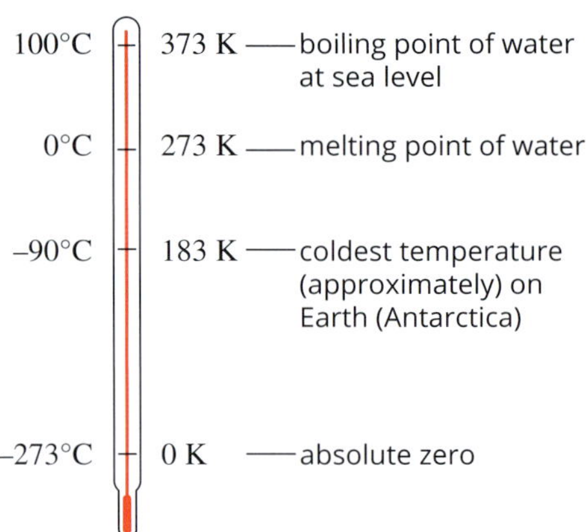

FIGURE 9.2.8 The Celsius and Kelvin temperature scales.

To convert degrees Celsius to kelvin:

$$T \text{ (in K)} = T \text{ (in °C)} + 273$$

Extremely low temperatures, within 1×10^{-10} K of absolute zero, have been reached in the laboratory. Note that the Kelvin scale has no degrees sign—it is written as just K.

Worked example 9.2.2

CONVERTING TEMPERATURES FROM CELSIUS TO KELVIN

What is 300°C on the Kelvin temperature scale?

Thinking	Working
T (in K) = T (in °C) + 273	T (in K) = T (in °C) + 273 = 300 + 273 = 573 K

Worked example: Try yourself 9.2.2

CONVERTING TEMPERATURES FROM CELSIUS TO KELVIN

What is 100°C on the Kelvin temperature scale?

The ascending hot-air balloon in Figure 9.2.9 demonstrates the relationship between volume and temperature. Heating the air causes the balloon to expand, which reduces the density of the air inside and allows the balloon to rise.

FIGURE 9.2.9 Hot-air balloonists take advantage of the relationship between the volume of a gas and temperature.

Worked example 9.2.3

USING CHARLES' LAW TO PREDICT CHANGES IN TEMPERATURE AND VOLUME

A hot air balloon at 273 K has a volume of 5.0 L. The temperature is increased to 365 K. If the pressure remains constant, what will be the new volume of the balloon?

Thinking	Working
State Charles' law.	$V = kT$
Determine the relationship between the two states.	As the amount of gas has not changed and the pressure between the two states has been held constant, the relationship between the two states is: $\frac{V_1}{T_1} = \frac{V_2}{T_2}$
State the known and unknown variables.	$V_1 = 5.0$ L $T_1 = 273$ K $V_2 = ?$ $T_2 = 365$ K
Rearrange the equation to make V_2 the subject.	$\frac{V_1}{T_1} = \frac{V_2}{T_2}$ $V_2 = \frac{V_1 T_2}{T_1}$
Substitute in the known values and solve.	$V_2 = \frac{V_1 T_2}{T_1}$ $= \frac{5.0 \times 365}{273}$ = 6.68 L
State the new volume.	The new volume is 6.68 L.

As the temperature, T, of a gas increases, the volume, V, of the gas increases. Therefore, V and T are directly proportional:

$$V \propto T$$

This relationship can also be expressed as $\frac{V}{T} = k$, where k is a constant at a given pressure. Note that the value of T must be expressed in kelvin, not degrees Celsius.

Worked example: Try yourself 9.2.3

USING CHARLES' LAW TO PREDICT CHANGES IN TEMPERATURE AND VOLUME

A hot air balloon at 273 K has a volume of 355 L. The temperature is increased to 395 K. If the pressure remains constant, what will be the new volume of the balloon?

GAY-LUSSAC'S LAW: PRESSURE AND TEMPERATURE

In 1802, Joseph Louis Gay-Lussac (Figure 9.2.10) published the findings of his investigations into the behaviour of gases. Gay-Lussac confirmed Boyle's law and Charles' law. He also found a direct relationship between temperature and pressure, as shown in Figure 9.2.11. This relationship is now referred to as Gay-Lussac's law.

FIGURE 9.2.10 Joseph Louis Gay-Lussac (1778–1850) was a French chemist. In 1804 he made balloon ascents to measure changes in the earth's magnetism and in the composition of the air with altitude. Gay-Lussac showed that the pressure of a gas is directly proportional to its temperature.

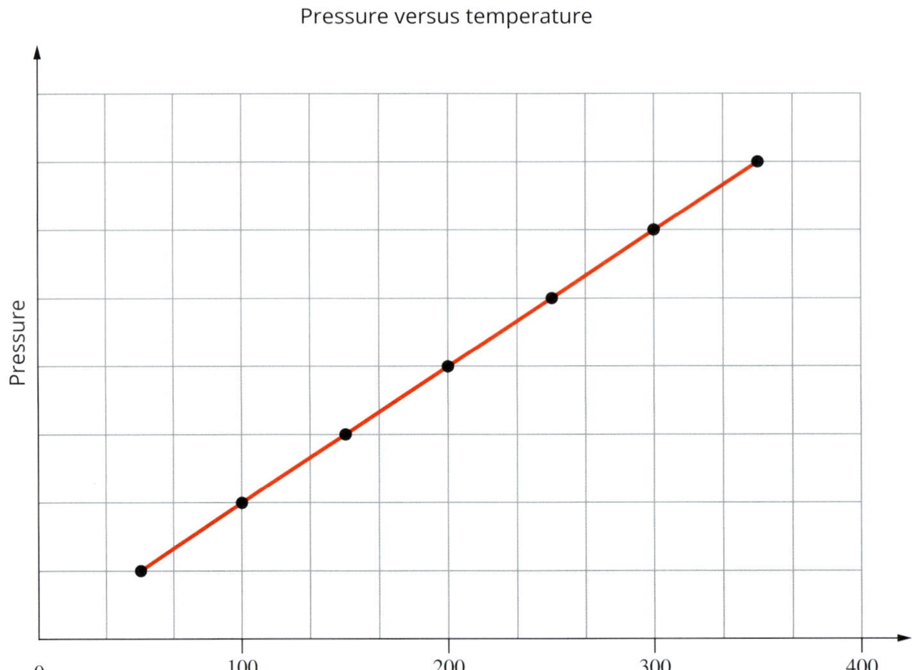

FIGURE 9.2.11 The variation of pressure with temperature for a fixed amount of gas of constant volume.

> As the temperature, T, of a gas increases, the pressure, P, of the gas increases. Therefore, P and T are directly proportional:
>
> $$P \propto T$$
>
> This relationship can also be expressed as $\frac{P}{T} = k$, where k is a constant at a given pressure. Note that the value of T must be expressed in kelvin, not degrees Celsius.

Worked example 9.2.4

USING GAY-LUSSAC'S LAW TO PREDICT CHANGES IN TEMPERATURE AND PRESSURE

The temperature of the air inside a car tyre (Figure 9.2.12) is initially 20°C. The car tyre is left in the hot sun and the pressure inside the tyre rises from 1.00 bar to 1.15 bar. What is the new temperature in the tyre?

Thinking	Working
State Gay-Lussac's law.	$\frac{P}{T} = k$
Determine the relationship between the two states.	As the amount of gas has not changed and the volume between the two states has been held constant, the relationship between the two states is: $\frac{P_1}{T_1} = k = \frac{P_2}{T_2}$
Convert temperature from degrees Celsius to kelvin.	$T_1 = 273 + 20$ $= 293\text{ K}$
State the known and unknown variables.	$P_1 = 1.00\text{ bar}$ $T_1 = 293\text{ K}$ $P_2 = 1.15\text{ bar}$ $T_2 = ?$
Rearrange the equation to make T_2 the subject.	$T_2 = \frac{P_2 T_1}{P_1}$
Substitute in the known values.	$T_2 = \frac{P_2 T_1}{P_1}$ $= \frac{1.15 \times 293}{1.00}$ $= 337\text{K}$
Convert temperature from kelvin to degrees Celsius.	$T_2 = 337 - 273$ $= 64°\text{C}$
State the new temperature.	The final temperature is 64°C.

FIGURE 9.2.12 The pressure of the air inside a car tyre is dependent on the temperature.

Worked example: Try yourself 9.2.4

USING GAY-LUSSAC'S LAW TO PREDICT CHANGES IN TEMPERATURE AND PRESSURE

A gas in a sealed container at 25°C is heated in an oven. If the initial gas pressure was 1.0 bar and the final gas pressure is 1.6 bar, calculate the new temperature of the gas.

COMBINED GAS LAW

The gas laws can be combined together.

Boyle's law states that the pressure–volume product is constant:

$$PV = k_1 \quad (1)$$

Charles's law states that the volume is proportional to the absolute temperature:

$$\frac{V}{T} = k_2 \quad (2)$$

Gay-Lussac's law states that the pressure is proportional to the absolute temperature:

$$P = k_3T \quad (3)$$

where

P is the pressure in kilopascals (kPa)

V is the volume in litres (L)

T is the absolute temperature of an ideal gas in kelvin (K).

By combining (1) and either of (2) or (3), a new equation is derived with P, V and T.

$$\frac{PV}{T} = k$$

The following equation allows for the comparison of a gas under two different sets of conditions:

$$\frac{P_1V_1}{T_1} = \frac{P_2V_2}{T_2}$$

The combined gas law is used to compare a gas under two different sets of conditions.

$$\frac{P_1V_1}{T_1} = \frac{P_2V_2}{T_2}$$

where

P_1 and P_2 are in any units, as long as they are consistent

V_1 and V_2 are in any units, as long as they are consistent

T_1 and T_2 are the absolute temperatures in kelvin.

Worked example 9.2.5

USING THE COMBINED GAS LAW

A student measured the pressure and volume of carbon dioxide at 25°C to be 148 kPa and 15.0 mL. The gas was then heated to 200°C. Calculate the new pressure if the volume increased to 30.0 mL.

Thinking	Working
Determine the relationship between the two states.	As the amount of gas has not changed, the relationship between the two states is: $\frac{P_1V_1}{T_1} = \frac{P_2V_2}{T_2}$
Convert temperature from degrees Celsius to kelvin.	$T_1 = 273 + 25 = 298$ K $T_2 = 273 + 200 = 473$ K
State the known and unknown variables.	$P_1 = 148$ kPa $V_1 = 15.0$ mL $T_1 = 298$ K $P_2 = ?$ $V_2 = 30.0$ mL $T_2 = 473$ K
Rearrange the equation to make P_2 the subject.	$P_2 = \frac{P_1V_1T_2}{T_1V_2}$
Substitute in the known values.	$P_2 = \frac{148 \times 15.0 \times 473}{298 \times 30.0}$ $= 117$ kPa
State the new gas pressure.	The pressure of the gas at 200°C with a volume of 30 ml is 117 kPa.

Worked example: Try yourself 9.2.5

USING THE COMBINED GAS LAW

A student measured the pressure and volume of carbon dioxide at 25°C to be 96.0 kPa and 500 mL. The gas was then heated to 450°C. Calculate the new pressure if the volume decreased to 338 mL.

9.2 Review

SUMMARY

- The following three laws describe the behaviour of gases under different conditions:

  Boyle's law: $V \propto \frac{1}{P}$ (for constant T)

  Charles' law: $V \propto T$ (for constant P)

  Gay-Lussac's law: $P \propto T$ (for constant V)
- Absolute zero is a temperature of −273°C or 0 K. Molecules and atoms have minimal kinetic energy at this temperature.
- The combined gas law is used to compare a gas under two different sets of conditions:

$$\frac{P_1V_1}{T_1} = \frac{P_2V_2}{T_2}$$

KEY QUESTIONS

1 Convert the following temperatures from degrees Celsius to kelvin.

a 0°C

b 25°C

c 100°C

d 175°C

e −145°C

2 Each part of this question refers to a constant mass of gas at a fixed temperature. Complete the following table.

	Initial condition		Final condition	
	Pressure	Volume	Pressure	Volume
a	1 atm	10 L	5 atm	
b	160 mmHg	235 mL	760 mmHg	
c	101.3 kPa	1.6 L	693 kPa	
d	1.00 bar	20 L	3.00 bar	
e	1 atm	100 L	200 atm	

3 Each part of this question refers to a constant mass of gas at a fixed pressure. Complete the following table.

	Initial condition		Final condition	
	Volume	Temperature	Volume	Temperature
a	100 mL	300 K		900 K
b	500 mL	500°C		250°C
c	10.0 L	25°C		300°C
d	10.0 L	300 K	25.0 L	
e	3.2 m^3	27°C	6.9 m^3	

4 Each part of this question refers to a constant mass of gas at a fixed volume. Complete the following table.

	Initial condition		Final condition	
	Pressure	Temperature	Pressure	Temperature
a	5 atm	300 K		900 K
b	760 mmHg	500°C		250°C
c	693 kPa	25°C	1582 kPa	
d	300 bar	300 K	500 bar	
e	200 atm	300 K		650 K

9.2 Review *continued*

5 Each part of this question refers to a constant mass of gas. Complete the following table:

	Initial condition			Final condition		
	Pressure	**Volume**	**Temperature**	**Pressure**	**Volume**	**Temperature**
a	101.3 kPa	10.0 L	25°C	53.2 kPa		–56°C
b	1 atm	3.2 L	3000 K		1.0 L	273 K
c	5.00 bar	116 mL	–25°C		100 mL	75°C
d	10 atm	375 mL	298 K	1 atm	7500 mL	
e	760 mmHg	10.0 L	298 K	480 mmHg		273 K

6 An upper atmosphere balloon is filled on the ground (where the temperature is 25°C and the pressure is 1 atm) to a volume of 5.0 L. It rises through the atmosphere to a region where the temperature is –50°C and its volume increases to 10.0 L. Calculate the pressure in the balloon.

7 A tyre is inflated to 30 kPa pressure at 15°C. Assuming a constant tyre volume of 33 L, what will the pressure in the tyre be at a temperature at 47°C?

8 A helium airship has a gasbag of volume 10 000 m^3 at sea level (1.00 bar) and at 20°C. At higher altitudes, the air pressure falls at a rate of 0.10 bar for every 1000 m increase in altitude.

a What will the volume of the gasbag be at a height of 1500 m, when the pilot's thermometer registers a temperature of 5°C?

b The airship rises through the atmosphere until the volume of its gasbag is 10 500 m^3. The temperature is 0°C.

i What is the pressure indicated by the pilot's barometer?

ii Calculate the height of the airship.

9.3 The ideal gas law

By the beginning of the 19th century, the fluid behaviour of gases had been successfully described by the relationships that had been discovered between the properties of pressure, volume and temperature. Meanwhile, experiments with gases had led John Dalton to propose a modern theory of the atom. However, gas behaviour was yet to be linked to atomic theory. Italian scientist Amedeo Avogadro (Figure 9.3.1) was able to provide this link. His work led to a universal theory of gases—the **ideal gas law**.

FIGURE 9.3.1 Amedeo Avogadro, 1776–1856.

AVOGADRO'S LAW: VOLUME AND AMOUNT OF GAS

In 1811, Avogadro published the results of his investigations into the relative masses of elements and molecules. In the paper, he stated that equal volumes of gases under the same conditions of temperature and pressure will contain equal numbers of molecules. In other words, the volume, V, occupied by a gas also depends directly on the amount of gas, n, in moles.

This relationship is shown in Figure 9.3.2. Both syringes show a gas at a constant temperature and pressure. The volume doubles with twice the number of molecules of gas in the syringe.

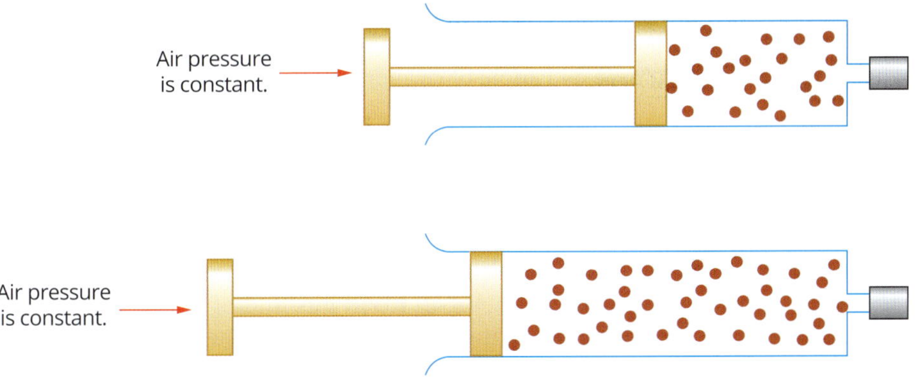

FIGURE 9.3.2 When the amount of gas in the syringe is doubled, the volume doubles, provided the pressure on the plunger and the temperature of the gas remain constant.

The mathematical relationship between the volume, V, occupied by a gas and the amount of gas, n, at constant temperature and pressure can be written as:

$$V \propto n$$

This relationship can also be expressed as $\frac{V}{n} = k$, where k is a constant at a given temperature and pressure.

MOLAR VOLUME OF A GAS

If the amount of gas is fixed at 1 mole as shown in Figure 9.3.3, the volume it occupies depends almost entirely on its temperature and pressure. We define this volume as the **molar volume**, V_m, of a gas. Molar volume is the amount of space, or volume, occupied by 1 mole of any gas at a particular pressure and temperature.

The volume of 1 mole of gas, V_m, is equal to its total volume, V, divided by the number of moles, n, of gas present. This can be represented by the relationship:

$$V_m = \frac{V}{n} \text{ (at a given temperature and pressure)}$$

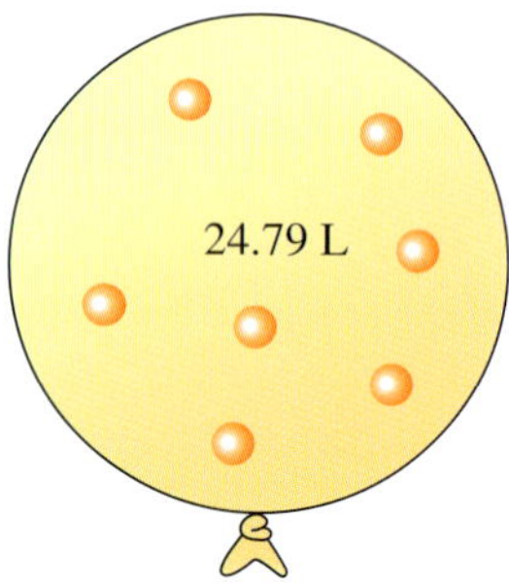

24.79 L

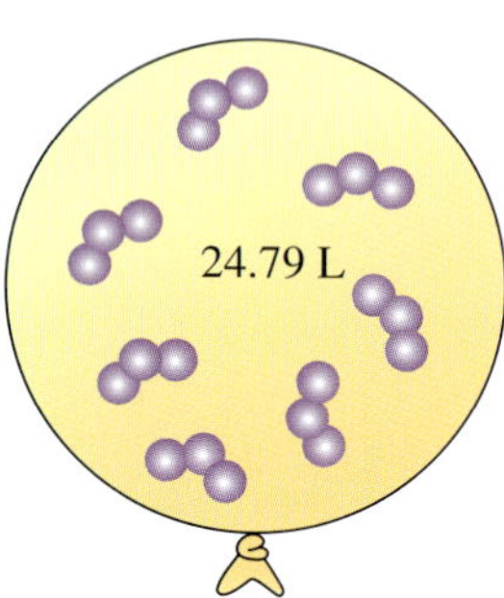

1 mole of neon, Ne
$P = 100$ kPa
$T = 25°C$
$V = 24.79$ L
number of molecules = 6.022×10^{23}
mass = 20.18 g

1 mole of nitrogen, N_2
$P = 100$ kPa
$T = 25°C$
$V = 24.79$ L
number of molecules = 6.022×10^{23}
mass = 28.01 g

1 mole of ozone, O_3
$P = 100$ kPa
$T = 25°C$
$V = 24.79$ L
number of molecules = 6.022×10^{23}
mass = 48.00 g

FIGURE 9.3.3 One mole of any gas occupies the same volume, given the same temperature and pressure. In this example, 1 mole of each gas occupies 24.79 L at 100 kPa and 25°C (298 K).

By rearranging this expression:

 $n = \frac{V}{V_m}$ (at a given temperature and pressure)

The molar volume of a gas varies with temperature and pressure. However, molar volume does not vary with the identity of the gas. For example, 1 mole of hydrogen gas occupies the same volume as 1 mole of oxygen gas at the same temperature and pressure.

Standard conditions

A temperature of 25°C (298 K) and a pressure of 100 kPa is typical of the conditions you will encounter when you are working in a laboratory. These conditions are known as **standard laboratory conditions (SLC)**.

Table 9.3.1 shows the molar volume of an **ideal gas** and of helium at SLC. An ideal gas is a theoretical gas composed of particles that do not interact except through elastic collisions. At SLC, most gases behave very like an ideal gas and therefore have a molar volume very close to that of an ideal gas.

TABLE 9.3.1 Molar volume at SLC

Gas	Formula	Molar volume at SLC ($L\,mol^{-1}$)
ideal gas	–	24.79
helium	He	24.83

It is usual to assume that the molar volume of a gas is $24.79\,L\,mol^{-1}$ at SLC. From this value, you can calculate the amount, in moles, of a gas, given its volume at SLC.

Another set of commonly used standard conditions is **standard temperature and pressure (STP)**. STP refers to a temperature of 0°C (273 K) and a pressure of 100 kPa. The molar volume of an ideal gas at STP is $22.71\,L\,mol^{-1}$.

In Section 9.4, you will see that the concept of molar volume is often used to determine the volume of gases required by or produced in a chemical reaction.

In Chapters 15 and 16, you will learn about another set of standard conditions, standard state conditions—conditions relevant to the standard state of a substance, where the temperature is 25°C (298K) and the pressure is 1 atm.

$V_m = 24.79\,L\,mol^{-1}$ at SLC (25°C or 298 K and 100 kPa)

$V_m = 22.71\,L\,mol^{-1}$ at STP (0°C or 273 K and 100 kPa)

GO TO ➤ Chapter 15 page 477

GO TO ➤ Chapter 16 page 501

Worked example 9.3.1

CALCULATING THE VOLUME OF A GAS FROM ITS AMOUNT (IN MOL)

Calculate the volume, in L, occupied by 0.24 mol of nitrogen gas at SLC. Assume that nitrogen behaves as an ideal gas.

Thinking	Working
Rearrange $n = \frac{V}{V_m}$ to make volume the subject.	$n = \frac{V}{V_m}$ $V = n \times V_m$
Substitute in the known values, including $V_m = 24.79\,L\,mol^{-1}$ (at SLC), and solve.	$V = n \times V_m$ $= 0.24 \times 24.79$ $= 5.9496\,L$
Consider the units and significant figures.	$V = 5.9\,L$

Worked example: Try yourself 9.3.1

CALCULATING THE VOLUME OF A GAS FROM ITS AMOUNT (IN MOL)

Calculate the volume, in L, occupied by 3.5 mol of oxygen gas at SLC. Assume that oxygen behaves as an ideal gas.

THE IDEAL GAS LAW

The following three laws describe the behaviour of gases under different conditions:

- $V \propto \frac{1}{P}$ (for constant T and n)
- $V \propto T$ (for constant P and n)
- $V \propto n$ (for constant P and T)

We can combine all three laws by writing:

$$V \propto \frac{nT}{P}$$

This relationship can be expressed as $V = \frac{RnT}{P}$, where R is a proportionality constant.

The equation is known as the ideal gas law and is more usually written in the form:

$$PV = nRT$$

where R is called the ideal gas constant, universal gas constant, or simply, the **gas constant**.

This constant can be determined experimentally by measuring the volume occupied by a known amount of gas at a known temperature and pressure.

> The value of R depends on the units of pressure and volume used. It has a value of 8.314 J mol^{-1} K^{-1} when:
>
> P is measured in kilopascals (kPa)
>
> V is measured in litres (L)
>
> n is measured in moles (mol)
>
> T is measured on the Kelvin scale (K).

You will use the ideal gas law in Section 9.4 to calculate the volume of gases consumed or produced during chemical reactions.

Worked example 9.3.2

USING THE IDEAL GAS LAW TO CALCULATE THE VOLUME OF A GAS

Calculate the volume, in L, occupied by 2.24 mol of oxygen gas (O_2) if the pressure is 200 kPa at 50°C.

Thinking	Working
Convert units, if necessary. Pressure is in kPa and temperature in K.	P = 200 kPa (no conversion required) $T = 50 + 273$ $= 323$ K
Rearrange the ideal gas law so that volume, V, is the subject.	$PV = nRT$ $V = \frac{nRT}{P}$
Substitute in values for pressure, amount, temperature and the gas constant, R, then solve for V.	$V = \frac{2.24 \times 8.314 \times 323}{200}$ $= 30.1$ L

Worked example: Try yourself 9.3.2

USING THE IDEAL GAS LAW TO CALCULATE THE VOLUME OF A GAS

Calculate the volume, in L, occupied by 13.0 mol of carbon dioxide gas (CO_2) if the pressure is 250 kPa at 75.0°C.

CHEMISTRY IN ACTION 

Decompression chambers

Scuba diving and water pressure

If you swim at the water's surface, your body experiences a pressure, due to the surrounding air, of about 1 atm. Below the surface, your body experiences an additional pressure due to the water. This additional pressure amounts to about 1 atm for every 10 m of depth. Therefore, at 20 m the pressure on your body is about 3 atm.

As the pressure on your body increases, the volume of your body cavities such as your lungs and inner ears decreases. This squeezing effect makes diving well below the water's surface without scuba equipment very uncomfortable. Scuba equipment overcomes this problem by supplying air to the mouth from tanks at the same pressure as that produced by the underwater environment.

Scuba diving and gas solubility

As the pressure in a diver's lungs increases during a dive, more gas dissolves in the blood. Nitrogen (N_2) is one of these gases. When a diver ascends the pressure drops, and the nitrogen becomes less soluble in the blood and comes out of solution. If a diver ascends too quickly, the rapid pressure drop causes the nitrogen to come out of the blood as tiny bubbles (Figure 9.3.4), a bit like the bubbles of carbon dioxide you observe when you open a bottle of soft drink.

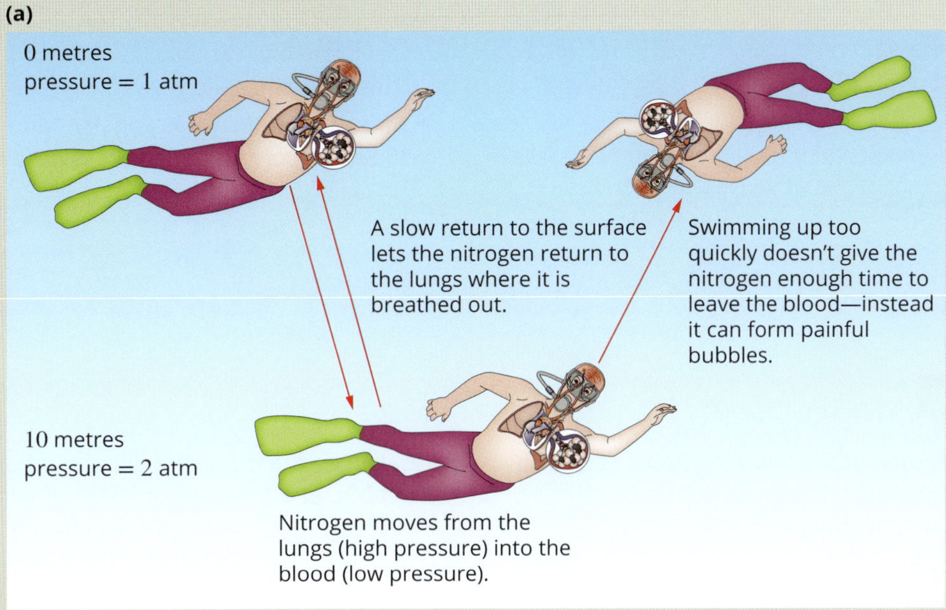

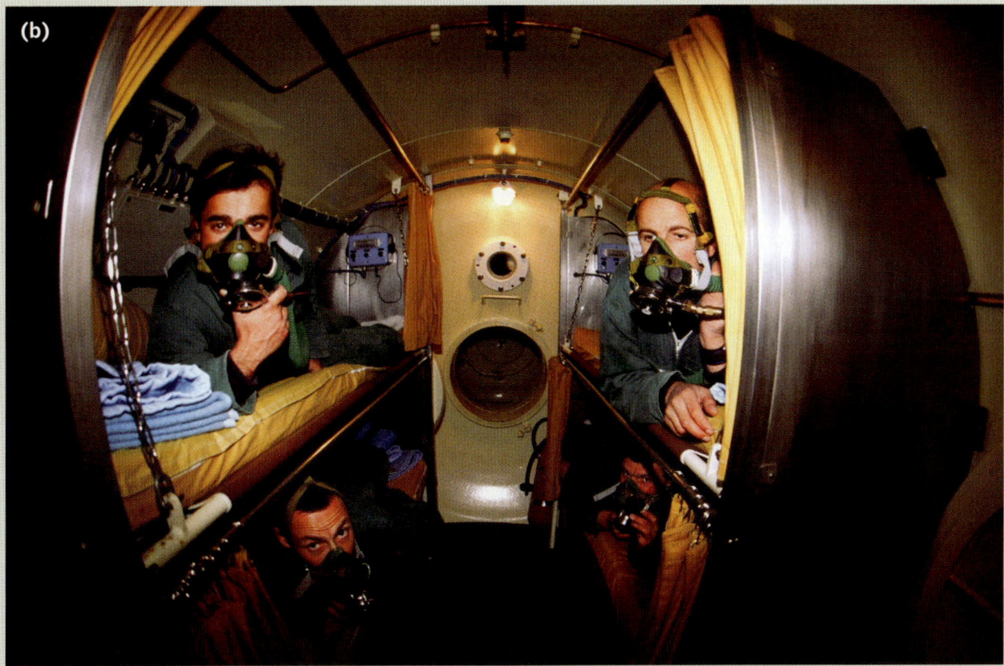

FIGURE 9.3.4 (a) Scuba diving and the bends. (b) A decompression chamber used to treat divers with the bends.

These bubbles cause pain in joints and muscles. If they form in the spinal cord, brain or lungs, they can cause paralysis or death. The treatment for divers suffering from this effect (the bends) involves time in a decompression chamber similar to the one shown in Figure 9.3.4b.

The chamber increases the pressure surrounding the diver's body, forcing any nitrogen bubbles to dissolve in the blood. It then slowly reduces the pressure back to 1 atm. The duration of treatment depends on the severity of the symptoms, the dive history and the patient's response to treatment.

9.3 Review

SUMMARY

- Standard laboratory conditions (SLC) refers to a temperature of 25°C (298 K) and a pressure of 100 kPa.
- Standard temperature and pressure (STP) refers to a temperature of 0°C (273 K) and a pressure of 100 kPa.
- The molar volume, V_m, of a gas is the volume occupied by 1 mol of gas at a given temperature and pressure:

$$n = \frac{V}{V_m}$$

- The value of V_m is 24.79 L mol^{-1} at SLC.
- The value of V_m is 22.71 L mol^{-1} at STP.
- $PV = nRT$ is the ideal gas law. It can be used to calculate one variable (P, V, n or T) when the other three variables are known.
- R is the universal gas constant and its value is 8.314 J mol^{-1} K^{-1} when pressure is in kPa, volume is in L, amount of gas is in mol and temperature is in K.

KEY QUESTIONS

1 Calculate the volume of the following gases at SLC.

a 1.4 mol of chlorine

b 1.0×10^{-3} mol of hydrogen

c 1.4 g of nitrogen

2 Calculate the mass of the following gas samples. All volumes are measured at SLC.

a 2.80 L of neon (Ne)

b 50.0 L of oxygen (O_2)

c 140 mL of carbon dioxide (CO_2)

3 0.25 mol of nitrogen is placed in a flask of volume 5.0 L at a temperature of 5°C. What is the pressure in the flask?

4 What volume of gas, in litres, is occupied by:

a 0.20 mol of hydrogen at 115 kPa and 40°C?

b 12.5 mol of carbon dioxide at 5.00 atm and 150°C?

c 8.50 g of hydrogen sulfide (H_2S) at 100 kPa and 27°C?

5 Calculate the mass of helium in a balloon if the volume is 100 L at a pressure of 95 000 Pa and a temperature of 0°C.

6 At a given temperature, a sample of nitrogen of mass 11.3 g exerts a pressure of 102 kPa in a gas cylinder of volume 10.0 L. Calculate the temperature of the gas.

7 Which sample of gas contains the greater amount, in mol, of gas: 3.2 L of nitrogen at 25°C and a pressure of 1.2 bar or 2.5 L of helium at 23°C and a pressure of 1.2 atm?

9.4 Stoichiometric calculations involving gases

In Chapter 6 you learned how to use stoichiometric calculations to find the mass of chemicals consumed or produced in a chemical reaction. In this section, you will use **stoichiometry** to calculate the volumes of gaseous chemicals consumed or produced during chemical reactions. For example, carbon dioxide emissions from power plants such as the one shown in Figure 9.4.1 can be quantified using stoichiometry.

GO TO ➤ Section 6.2 page 213

FIGURE 9.4.1 It is useful to be able to calculate the volume of carbon dioxide emissions produced by the combustion of a particular fuel.

MASS–VOLUME STOICHIOMETRY

Some stoichiometric calculations require you to determine the volume of a gas that reacts with, or is produced from, a given mass of a reactant. For these calculations, you need to determine the number of moles of the reactant, and use the mole ratio from the balanced chemical equation.

The volume of the gas, in litres, is determined from its amount in moles using one of the following formulae:

1 At standard laboratory conditions (SLC) of 25°C and 100 kPa, the molar volume equation is used:

$$n = \frac{V}{V_m}$$

This formula can be rearranged to make volume the subject:

$$V = n \times V_m$$

Remember that at SLC, the accepted volume of one mole of any gas is 24.79 L mol^{-1}.

2 At non-standard conditions, the ideal gas law is used:

$$PV = nRT$$

The ideal gas law can also be rearranged to make volume the subject:

$$V = \frac{nRT}{P}$$

Remember that for use in the ideal gas law, pressure must be in kPa, temperature must be in K and the gas constant is equal to 8.314 J mol^{-1} K^{-1}.

Worked examples 9.4.1 and 9.4.2 show how to calculate the volume of carbon dioxide that is produced from a known mass of fuel at SLC and at non-standard conditions.

Worked example 9.4.1

SOLVING MASS–VOLUME STOICHIOMETRY PROBLEMS AT STANDARD LABORATORY CONDITIONS

Calculate the volume of carbon dioxide, in L, produced when 2.00 kg of propane is burned completely in oxygen to produce carbon dioxide and water. The gas volume is measured at SLC.

Thinking	Working
Write a balanced equation for the reaction.	$C_3H_8(g) + 5O_2(g) \rightarrow 3CO_2(g) + 4H_2O(l)$
Calculate the number of moles of the known substance using: $n = \frac{m}{M}$	$n(C_3H_8) = \frac{2000}{44.09}$ $= 45.36$ mol
Find the mole ratio: $\frac{\text{coefficient of unknown}}{\text{coefficient of known}}$	$\frac{\text{coefficient of } CO_2}{\text{coefficient of } C_3H_8} = \frac{3}{1}$
Calculate the number of moles of the unknown substance using: $n(\text{unknown}) = \text{mole ratio} \times n(\text{known})$	$n(CO_2) = \frac{3}{1} \times 45.36$ $= 136.1$ mol
Calculate the volume of the unknown substance using: $V = n \times V_m$	$V(CO_2) = 136.1 \times 24.79$ $= 3373$ L

Worked example Try yourself: 9.4.1

SOLVING MASS–VOLUME STOICHIOMETRY PROBLEMS AT STANDARD LABORATORY CONDITIONS

Calculate the volume of carbon dioxide, in L, produced when 300 g of butane is burned completely in oxygen to produce carbon dioxide and water. The gas volume is measured at SLC.

Worked example 9.4.2

SOLVING MASS–VOLUME STOICHIOMETRY PROBLEMS AT NON-STANDARD CONDITIONS

Calculate the volume of carbon dioxide, in L, produced when 800 g of propane is burned completely in oxygen to produce carbon dioxide and water. The gas volume is measured at 60°C and 200 kPa.

Thinking	Working
Write a balanced equation for the reaction.	$C_3H_8(g) + 5O_2(g) \rightarrow 3CO_2(g) + 4H_2O(l)$
Calculate the number of moles of the known substance using: $n = \frac{m}{M}$	$n(C_3H_8) = \frac{800}{44.09}$ $= 18.1$ mol
Find the mole ratio: $\frac{\text{coefficient of unknown}}{\text{coefficient of known}}$	$\frac{\text{coefficient of } CO_2}{\text{coefficient of } C_3H_8} = \frac{3}{1}$

Calculate the number of moles of the unknown substance using: n(unknown) = mole ratio × n(known)	$n(CO_2) = \frac{3}{1} \times 18.1$ $= 54.3$ mol
Express the pressure and temperature in the required units.	$P = 200$ kPa $T = 60 + 273$ $= 333$ K
Calculate the volume of the unknown substance using: $V = \frac{nRT}{P}$	$V(CO_2) = \frac{54.4 \times 8.314 \times 333}{200}$ $= 753$ L

Worked example: Try yourself 9.4.2

SOLVING MASS–VOLUME STOICHIOMETRY PROBLEMS AT NON-STANDARD CONDITIONS

Calculate the volume of carbon dioxide, in L, produced when 5.00 kg of butane is burned completely in oxygen to produce carbon dioxide and water. The gas volume is measured at 40°C and 400 kPa.

GAS VOLUME–VOLUME CALCULATIONS

For chemical reactions where both the reactants and products are in the gaseous state, it is often convenient to measure volumes, rather than masses.

For example, the reaction between propane gas and oxygen can be represented by the equation:

$$C_3H_8(g) + 5O_2(g) \rightarrow 3CO_2(g) + 4H_2O(g)$$

This equation tells us that when 1 mol of propane reacts with 5 mol of oxygen gas, 3 mol of carbon dioxide and 4 mol of water vapour are produced.

You saw in Section 9.3 that equal amounts, in moles, of all gases occupy equal volumes measured at the same temperature and pressure.

Therefore, the mole ratios in the balanced equation become volume ratios at the same temperature and pressure. In the above reaction, this means that when 1 L of propane gas reacts with 5 L of oxygen gas, 3 L of carbon dioxide and 4 L of water vapour are produced.

Worked example 9.4.3

SOLVING GAS VOLUME–VOLUME STOICHIOMETRY PROBLEMS

Methane gas (CH_4) is burned in a gas stove according to the following equation:

$$CH_4(g) + 2O_2(g) \rightarrow CO_2(g) + 2H_2O(g)$$

If 50 mL of methane is burned, calculate the volume of O_2 gas required for complete combustion of the methane under constant temperature and pressure conditions.

Thinking	Working
Use the balanced equation to find the mole ratio of the two gases involved.	1 mol of CH_4 gas reacts with 2 mol of O_2 gas.
The temperature and pressure are constant, so volume ratios are the same as mole ratios.	1 volume of CH_4 reacts with 2 volumes of O_2 gas, so 50 mL of CH_4 reacts with 100 mL of O_2.

Worked example: Try yourself 9.4.3

SOLVING GAS VOLUME–VOLUME STOICHIOMETRY PROBLEMS

Methane gas (CH_4) is burned in a gas stove according to the following equation:

$$CH_4(g) + 2O_2(g) \rightarrow CO_2(g) + 2H_2O(g)$$

If 50 mL of methane is burned in air, calculate the volume of CO_2 gas produced under constant temperature and pressure conditions.

Always use the number of moles of the limiting reactant to determine the amount of product that will be formed.

Calculations involving excess reactants

GO TO ➤ Section 7.5 page 242

Stoichiometry calculations become more complex if the reactants are not present in their stoichiometric ratio. In these cases, you must determine which reactant is completely consumed in the reaction—the limiting reactant—and which one is present in excess. The amount of limiting reactant determines how much product is formed.

Worked example 9.4.4 introduces a strategy that can be used to determine the limiting reactant in a reaction.

Worked example 9.4.4

SOLVING EXCESS REACTANT STOICHIOMETRY PROBLEMS

25.0 g of hydrogen gas and 100.0 g of oxygen gas are mixed and ignited. The water produced is collected and weighed. The equation for the reaction is:

$$2H_2(g) + O_2(g) \rightarrow 2H_2O(g)$$

a Which reactant is the limiting reactant?

b What is the volume of water vapour formed at SLC?

Thinking	Working
a Calculate the number of moles of each reactant using $n = \frac{m}{M}$, $n = \frac{V}{V_m}$ or $n = \frac{PV}{RT}$, as appropriate.	$n(H_2) = \frac{m}{M}$ $= \frac{25.0}{2.016}$ $= 12.4$ mol $n(O_2) = \frac{m}{M}$ $= \frac{100.0}{32.00}$ $= 3.125$ mol
Use the coefficients of the equation to find the limiting reactant.	The equation shows 2 mol of H_2 reacts with 1 mol of O_2. So to react all of the O_2 will require $2 \times n(O_2)$ of H_2 $= 2 \times 3.125$ $= 6.250$ mol As there is 12.4 mol of H_2, the H_2 is in excess. The O_2 is the limiting reactant. It will be completely consumed.

b Find the mole ratio using: $\frac{\text{coefficient of unknown}}{\text{coefficient of known}}$ The limiting reactant is the known substance.	$\frac{\text{coefficient of } H_2O}{\text{coefficient of } O_2} = \frac{2}{1}$
Calculate the number of moles of the unknown substance using: $n(\text{unknown}) = \text{mole ratio} \times n(\text{known})$	$n(H_2O) = 2 \times 3.125$ $= 6.250\ \text{mol}$
Calculate the volume of the unknown using: $V = n \times V_m$	$V(H_2O) = 6.250 \times 24.79$ $= 154.9\ \text{L}$

Worked example: Try yourself 9.4.4

SOLVING EXCESS REACTANT STOICHIOMETRY PROBLEMS

65.0 g of butane is burned completely in 200 L of oxygen. The gas volume is measured at SLC. The equation for the reaction is:

$$2C_4H_{10}(g) + 13O_2(g) \rightarrow 8CO_2(g) + 10H_2O(l)$$

a Which reactant is the limiting reactant?

b What is the volume of carbon dioxide formed?

CHEMFILE CCT

The carburettor

Car engines such as the one used in the Morris Minor (Figure 9.4.2) work by burning petrol to create enormous gas pressure, and then turning that pressure into motion. The engine uses about 10 mg of petrol during each combustion cycle.

FIGURE 9.4.2 Older cars use a carburettor to control the fuel–air mixture to the engine.

The function of the carburettor is to turn the fuel into a fine mist, and then to mix the fuel and air in just the right proportion. Too much petrol leads to a 'rich' mixture, where the fuel is in excess. This leads to incomplete combustion. Consequently, the car runs poorly, creates a sooty exhaust and uses fuel inefficiently. Too much air leads to a 'lean' mixture, where there is an excess of air (oxygen). This runs the engine hot, potentially leading to the combustion of nitrogen, which damages the engine and produces exhaust gases containing toxic, polluting nitrous oxides.

The carburettor (Figure 9.4.3) should get the fuel–air mixture just right. Unfortunately, carburettors were never perfect at getting the mix right. Since the 1990s, they have been superseded by fuel injectors.

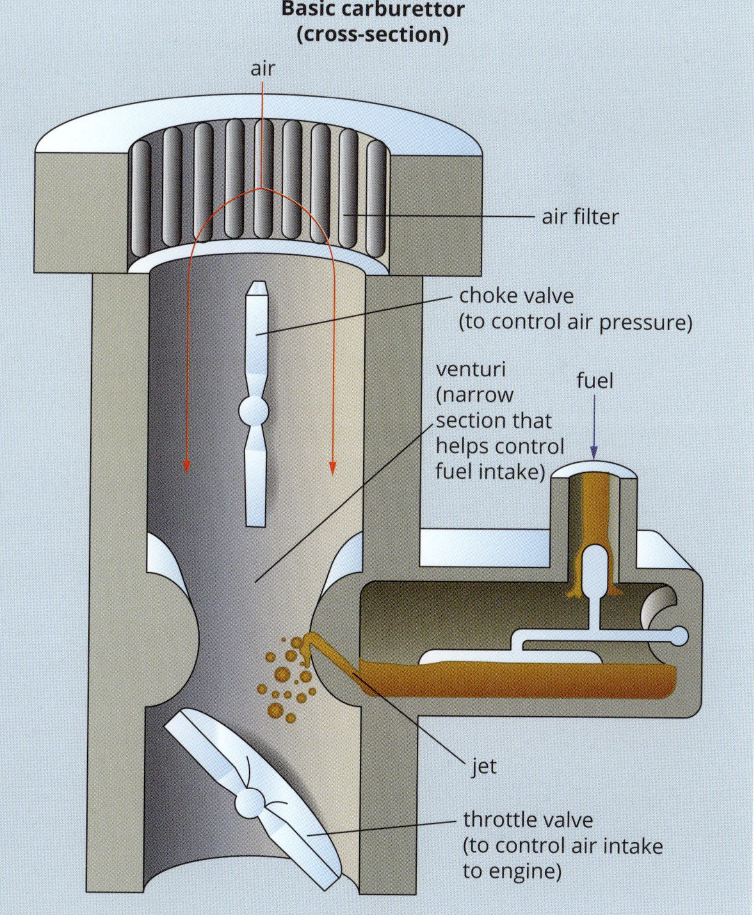

FIGURE 9.4.3 A schematic diagram of a carburettor. Its job is to control the mixture of fuel and air going into the engine.

9.4 Review

SUMMARY

- A balanced equation shows the ratio of the amounts, in moles, of reactants and products in the reaction.
- Stoichiometric calculations follow the general steps:
  1. Calculate the amount, in moles, of a known substance from the data given, using

     $$n = \frac{m}{M},\ n = \frac{V}{V_m} \text{ or } n = \frac{PV}{RT}$$

  2. Use the mole ratio from a balanced chemical equation to determine the amount, in moles, of the unknown substance.

     $$\frac{n(\text{unknown chemical})}{n(\text{known chemical})} = \frac{\text{coefficient of unknown chemical}}{\text{coefficient of known chemical}}$$

  3. Find the desired quantity of the unknown substance from its amount, in moles, using

     $m = n \times M$, $V = n \times V_m$ or $PV = nRT$
- The mole ratio in a balanced equation is also a volume ratio if all reactants and products are in the gaseous state and the temperature and pressure are kept constant.
- Stoichiometric calculations can be used to calculate the volume of gases involved in chemical reactions.

KEY QUESTIONS

1 Propane (C_3H_8) burns in oxygen according to the equation:

$$C_3H_8(g) + 5O_2(g) \rightarrow 3CO_2(g) + 4H_2O(l)$$

When the following masses of propane react completely with excess oxygen, calculate the volume of:

i oxygen used at SLC

ii carbon dioxide produced at SLC.

a 22 g

b 16.5 g

c 3.40 kg

2 Octane (C_8H_{18}) is one of the main constituents of petrol. It burns according to the equation:

$$2C_8H_{18}(g) + 25O_2(g) \rightarrow 16CO_2(g) + 18H_2O(g)$$

What mass of octane must have been used if 50.0 L of carbon dioxide, measured at 120°C and 1.10 atm, was produced?

3 Hydrogen peroxide (H_2O_2) decomposes according to the equation:

$$2H_2O_2(aq) \rightarrow 2H_2O(l) + O_2(g)$$

Calculate the volume of oxygen, collected at 30.0°C and 91.0 kPa, that is produced when 10.0 g of hydrogen peroxide decomposes.

4 What volume of NO_2 is produced when 0.5 L of nitrogen(II) oxide (NO) reacts with excess oxygen? (All volumes are measured at 25°C and 100 kPa.) The equation for the reaction is:

$$2NO(g) + O_2(g) \rightarrow 2NO_2(g)$$

5 Calculate the volume of oxygen needed to completely react with 150 mL of carbon monoxide according to the following equation. Assume all volumes are measured at the same temperature and pressure.

$$2CO(g) + O_2(g) \rightarrow 2CO_2(g)$$

6 Calculate the pressure produced in a sealed tube of volume 100 mL when 1.00 g of liquid water is vaporised at 220°C.

7 23 g of methane gas (CH_4) is reacted with 10 L of oxygen at SLC. Calculate the volume of carbon dioxide produced in this reaction.
The equation for the reaction is:

$$CH_4(g) + 2O_2(g) \rightarrow CO_2(g) + 2H_2O(g)$$

Chapter review

KEY TERMS

absolute temperature scale
absolute zero
bar
barometer
gas constant
ideal gas
ideal gas law
Kelvin scale
kinetic energy
kinetic molecular theory
molar volume
partial pressure
pascal
pressure
qualitatively
standard atmosphere
standard laboratory conditions (SLC)
standard temperature and pressure (STP)
stoichiometry
velocity
volume

REVIEW QUESTIONS

1 Which one of the following volumes is equal to 4.5 L?

A 4.5×10^2 mL

B 4.5×10^3 mL

C 4.5×10^{-3} mL

D 4.5×10 mL

2 Which one of the following best describes the effect of an increase in temperature on gas particles?

A Both the average kinetic energy and the average speed of the particles increase.

B Both the average kinetic energy and the average speed of the particles decrease.

C The average kinetic energy of the particles increases and the average speed of the particles decreases.

D The average kinetic energy of the particles decreases and the average speed of the particles increases.

3 Select the correct answers from the pairs of words to complete the paragraph about gases.
The volume occupied by the atoms or molecules in a gas is much **smaller/larger** than the total volume occupied by the gas. The particles move in rapid, **straight-line/curved** paths and collide with each other and with the walls of the container. The forces between particles are extremely **weak/strong**. The collisions between particles are **elastic/rigid**. The average kinetic energy of the particles is **directly/inversely** proportional to the temperature of the gas, in units of **K/°C**.

4 Use the kinetic theory of gases to explain the following statements:

a The pressure of a gas increases if its volume is reduced at constant temperature.

b The pressure of a gas decreases if its temperature is lowered at a constant volume.

c In a mixture of gases, the total pressure is the sum of the partial pressures of each gas.

d The pressure of a gas, held at constant volume and temperature, will increase if more gas is added to the container.

5 **a** If a container of gas is opened and some of the gas escapes, what happens to the pressure of the remaining gas in the container?

b Use the kinetic molecular theory to explain what happens to the gas pressure in part **a**.

6 A sample of propane in a 9.00 L container is heated and its temperature and pressure recorded.

P (kPa)	*T* (K)
106.6	295
107.6	298
108.7	301
109.8	304
112.0	310
113.8	315

a Use the data to determine whether propane is behaving as an ideal gas over this range. Explain your answer.

b Give one reason for gases deviating from ideal gas behaviour.

7 The graph shows the relationship between volume and absolute temperature for an ideal gas and two real gases.

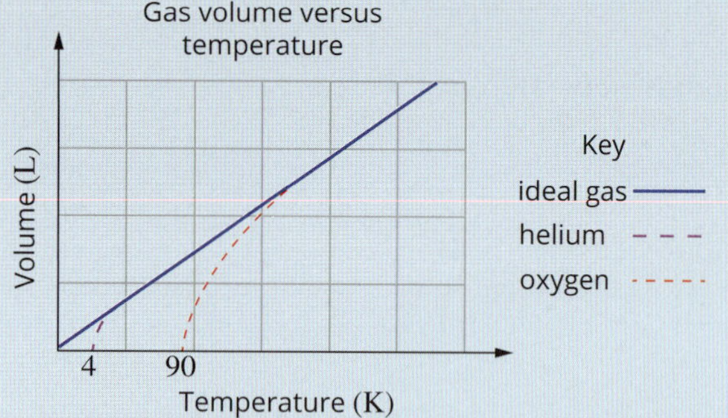

- **a** Describe the relationship for the ideal gas shown in this graph.
- **b** What conditions are necessary for this relationship to be true?
- **c** Explain what happens to the oxygen at 90 K.
- **d** Why does helium behave more like an ideal gas than does oxygen?

8 2.00 L of a gas is collected at 65°C and 720 mmHg pressure. Calculate the gas volume at SLC.

9 24 L of air at SLC in a cylinder is compressed with a piston until its volume is halved and its is temperature increased by 20 K. Calculate the new pressure, in bar, of the air in the cylinder.

10 A bike pump contains 500 mL of air at SLC. The air is sealed within the pump, and then the piston is compressed so that the volume of air is reduced to just 220 mL. If the pressure in the pump is now 2.5 bar, calculate the temperature of the air inside the pump.

11 Gases are often collected 'over water' by bubbling the gas into an inverted graduated cylinder filled with water, which is sitting in a water bath. As the gas is produced, it displaces the water from the cylinder. The result is a 'wet gas' that contains both the gas and water vapour.

CO_2 from a chemical reaction is collected over water. It has a volume of 300 mL at a pressure of 98.0 kPa and a temperature of 29°C. Calculate the volume of dry CO_2 at SLC. The vapour pressure of $H_2O(g)$ at 29°C is 4.0 kPa.

12 Calculate the mass of oxygen present in a 50.0 L container of oxygen at SLC.

13 Use the molar volume of a gas at SLC to find:

- **a** the volume occupied by 8.0 g of oxygen at SLC
- **b** the mass of nitrogen dioxide (NO_2) present in 10 L at SLC.

14 If 64.0 g of oxygen gas occupies a volume of 25.0 L when the temperature is 30.0°C, then the pressure of the gas, in kPa, is closest to:

- **A** 20.0
- **B** 200
- **C** 400
- **D** 6.40×10^3

15 Calculate the volume, in litres, occupied by 10.0 g of carbon dioxide at 25°C and 101.3 kPa.

16 What is the mass of oxygen present in a 10.0 L container of oxygen at a pressure of 105 kPa and at 20°C?

17 At what temperature will 0.20 g of helium exert a pressure of 80 kPa in a container with a fixed volume of 4.0 L?

18 Carbon dioxide gas is a product of the complete combustion of fuels.

- **a** Calculate the mass of 1.00 mol of carbon dioxide.
- **b** What is the volume occupied by 1.00 mol of carbon dioxide at SLC?
- **c** Given that density is defined as $\frac{\text{mass}}{\text{volume}}$, calculate the density of carbon dioxide at SLC in $g\,L^{-1}$.
- **d** Would you expect the density of carbon dioxide at STP to be less than, equal to or greater than its density at SLC? Justify your answer.

19 A sample of gas of mass 10.0 g occupies a volume of 5.4 L at 27°C and 100×10^3 Pa.

- **a** Calculate the amount, in mol, of gas in the sample.
- **b** Determine the molar mass of the gas.

20 Which container holds more molecules of oxygen gas: container A of volume 40.0 L at 25°C and 770 mmHg or container B of volume 0.10 L at 45°C and 390 mmHg?

21 A room has a volume of 220 m^3.

- **a** Calculate the amount, in mol, of air particles in the room at 23°C and a pressure of 100 kPa.
- **b** Assume that 20% of the molecules in the air are oxygen molecules and the remaining molecules are nitrogen. Calculate the mass of air in the room.

22 Octane (C_8H_{18}) is one of the main constituents of petrol. It burns according to the equation:

$$2C_8H_{18}(g) + 25O_2(g) \rightarrow 16CO_2(g) + 18H_2O(g)$$

Calculate the mass of octane used if 50.0 L of carbon dioxide, measured at 120°C and 1.10 atm, was produced.

23 What volume of oxygen gas (in L) is required for the complete combustion of 10.0 L of ethane gas (C_2H_6) at constant pressure and a temperature of 150°C?

The balanced equation for the reaction is:

$$2C_2H_6(g) + 7O_2(g) \rightarrow 4CO_2(g) + 6H_2O(g)$$

24 Methane (CH_4) burns in excess oxygen according to the equation:

$$CH_4(g) + 2O_2(g) \rightarrow CO_2(g) + 2H_2O(g)$$

This reaction produces 5.0 L of carbon dioxide at 200°C and 100 kPa. Assuming all volumes are measured at the same temperature and pressure, calculate:

a the volume of methane used

b the volume of oxygen used

c the mass of water vapour produced.

25 Propane (C_3H_8) undergoes complete combustion as follows:

$$C_3H_8(g) + 5O_2(g) \rightarrow 3CO_2(g) + 4H_2O(g)$$

All volumes are measured at 120°C and 102 kPa. When 80 mL of propane and 500 mL of oxygen are reacted:

a one of the gases does not react completely. Which gas is it and what volume of it is unreacted?

b what volumes of carbon dioxide and water are produced in the reaction?

c what change in the total volume of all the gases has occurred as a result of the reaction?

26 There are many scientists investigating possible fuels to replace fossil fuels. A group of Japanese chemists is investigating the following reaction as a source of methane (CH_4):

$$CaCO_3(s) + 4H_2(g) \rightarrow CH_4(g) + Ca(OH)_2(s) + H_2O(g)$$

At 400°C, 100 kPa and under suitable reaction conditions, what:

a volume of methane is produced if 100 L of hydrogen is completely reacted?

b mass of calcium carbonate is used in part **a**?

27 An indoor gas heater burns propane (C_3H_8) at a rate of 12.7 g per min. Calculate the minimum volume of oxygen per min, in L (at SLC) that needs to be available for the complete combustion of propane.

The balanced equation for the reaction is:

$$C_3H_8(g) + 5O_2(g) \rightarrow 3CO_2(g) + 4H_2O(g)$$

28 The Liddell Power Station in Maitland consumes about 4.9 million tonnes (1 tonne = 10^6 g) of coal in 1 year. The coal used in the power station is composed of approximately 75% carbon by mass. Calculate the volume of the greenhouse gas carbon dioxide released each year by the power station at SLC.

The balanced equation for the reaction is:

$$C(s) + O_2(g) \rightarrow CO_2(g)$$

29 Consider two containers of equal size. One contains oxygen and the other carbon dioxide. Both containers are at 23°C and at a pressure of 1.0 atm. Answer each of the following questions about the two gases and give a reason for your answers.

a Which of the two samples of gas contains more molecules?

b Which of the two samples of gas contains the greater number of atoms?

c Which of the two gases has the greater density?

30 Ethanol can be used as an alternative fuel for cars. When ethanol (C_2H_5OH) undergoes complete combustion by reacting with oxygen, it forms carbon dioxide and water.

a Write a balanced equation that represents the complete combustion of ethanol.

b **i** What mass of carbon dioxide would be formed if 1.00 kg of ethanol reacts?

ii What volume of carbon dioxide would be formed from 1.00 kg of ethanol at SLC?

c If the density of ethanol is 0.785 g mL^{-1}, calculate the mass of ethanol in a 50.0 L tank of the fuel.

31 Reflect on the Inquiry activity on page 272. Which of the gas laws most applies to this experiment?

MODULE 2 • REVIEW

Introduction to quantitative chemistry

Multiple choice

1 Determine the missing coefficient for O_2 in the following chemical equation:

$2C_4H_{10}(g) + \_\_O_2(g) \rightarrow 8CO_2(g) + 10H_2O(l)$

A 5
B 8
C 13
D 16

2 A 12.2 g sample of X reacts with a sample of Y to form 78.9 g of XY. What is the mass, in g, of Y that reacted?

A 12.2
B 66.7
C 78.9
D 91.1

3 Which of the following chemical equations is balanced correctly?

A $2H_2O(l) \rightarrow H_2(g) + O_2(g)$
B $Zn(s) + HCl(aq) \rightarrow ZnCl_2(aq) + H_2(g)$
C $Al_4C_3(s) + 3H_2O(l) \rightarrow CH_4(g) + 4Al(OH)_3(aq)$
D $CH_4(g) + 2O_2(g) \rightarrow CO_2(g) + 2H_2O(l)$

4 A stone has a mass of 4.50 g and occupies a volume of 1.45 mL. What is the stone's density, in $g\,mL^{-1}$?

A 0.32
B 3.10
C 4.50
D 6.53

5 What is the volume, in L, occupied by a gas that has a mass of 3.20 kg and a density of $0.535\,g\,L^{-1}$?

A 5.98
B 1.71
C 1712
D 5980

6 The mass, in g, of one molecule of carbon dioxide is closest to:

A 7.3×10^{-23}
B 44
C 6.022×10^{23}
D $44 \times 6.022 \times 10^{23}$

7 A particular brand of plant fertiliser contains 58.5% urea ($CO(NH_2)_2$) as the only nitrogen-containing compound. The mass of nitrogen, in g, in 125 g of this fertiliser is:

A 34.1
B 40.0
C 50.0
D 73.1

8 The mass, in g, of calcium chloride that contains 6.022×10^{23} chloride ions is:

A 35.45
B 55.49
C 74.53
D 110.98

9 Which sample contains the largest amount, in mol, of oxygen atoms?

A 0.20 mol NO_3
B 0.30 mol HCOOH
C 0.40 mol $KMnO_4$
D 0.70 mol H_2O_2

10 2.0 mol of $Fe_2O_3(s)$ reacts with excess CO(g) according to the following equation:

$Fe_2O_3(s) + 3CO(g) \rightarrow 2Fe(s) + 3CO_2(g)$

What mass of Fe(s), in g, is produced?

A 7.2×10^{-2}
B 4.0
C 1.1×10^{2}
D 2.2×10^{2}

11 Hydrochloric acid (HCl) reacts with solid sodium carbonate according to the following equation:

$2HCl(aq) + Na_2CO_3(s) \rightarrow 2NaCl(aq) + H_2O(l) + CO_2(g)$

In which mixture is HCl the limiting reagent?

A 0.20 mol HCl + 0.10 mol Na_2CO_3
B 0.30 mol HCl + 0.05 mol Na_2CO_3
C 0.10 mol HCl + 0.20 mol Na_2CO_3
D 0.40 mol HCl + 0.20 mol Na_2CO_3

12 What volume of $3.0\,mol\,L^{-1}$ Na_2CO_3, in mL, is required to prepare 750 mL of $0.15\,mol\,L^{-1}$ solution?

A 38
B 45
C 113
D 250

13 What volume of water, in mL, must be added to 20.0 mL of $0.50\,mol\,L^{-1}$ NaCl in order to change its concentration to $0.20\,mol\,L^{-1}$?

A 30.0
B 50.0
C 200
D 500

14 Cadmium is a metal used in photography, in nickel–cadmium batteries and solar cells, and for metal plating. Drinking water contaminated with cadmium can have harmful health effects. What is the concentration of cadmium, in ppb, in a reservoir containing 500 ML (1 ML = 10^6 L) of water if 1.75 kg of cadmium is discharged into it?

A 3.5
B 875
C 1750
D 3000

15 Excess silver nitrate ($AgNO_3$) is added to a 250.0 mL sample of river water. The precipitate of silver chloride (AgCl) was dried and weighed. Its mass was found to be 20.37 g. What is the concentration, in $g\,L^{-1}$, of chloride ions in the river water?

A 0.1425
B 0.5691
C 20.14
D 81.52

16 Which of the following points of the kinetic molecular theory can be used to explain that the volume of a fixed mass of gas is inversely proportional to pressure at a constant temperature?

A Collisions between particles are elastic.
B Forces between particles are extremely weak.
C The kinetic energy of the particles is proportional to the temperature of the gas.
D The volume of gas particles is very small compared with the volume occupied by the gas.

17 As a diver comes to the surface, the pressure inside her lungs changes from 200 to 100 kPa. Assuming that the volume of the gas in the lungs was initially 3.0 L, at the surface the volume of the same amount of gas, in litres, would be:

A 1.5
B 2.0
C 4.5
D 6.0

18 4.0 L of hydrogen gas is collected in a syringe. Then the pressure and the temperature (in kelvin) of the gas inside the syringe are both doubled. When this happens, the volume of the hydrogen gas, in litres, in the syringe would be:

A 1.0
B 4.0
C 6.0
D 16

19 Given that the density, *d*, of a substance is given by the formula $d = \frac{m}{V}$ and that *M* represents molar mass, then the density of any gas is given by which formula?

A $d = \frac{R \times T}{P \times M}$

B $d = \frac{P}{R \times T \times M}$

C $d = \frac{R \times T \times M}{P}$

D $d = \frac{P \times M}{R \times T}$

20 Methane burns in oxygen according to the following equation:

$$CH_4(g) + 2O_2(g) \rightarrow CO_2(g) + 2H_2O(g)$$

If all measurements were made at 200°C and 1 atm pressure, what would be the total volume of gases, in mL, after the reaction if 20 mL of methane burned in 20 mL of oxygen?

A 20
B 30
C 40
D 50

Short answer

1 Balance the following chemical equations:

a $C_3H_6(g) + O_2(g) \rightarrow CO_2(g) + H_2O(g)$
b $Al(OH)_3(aq) + HBr(aq) \rightarrow AlBr_3(aq) + H_2O(l)$
c $MgCl_2(aq) + AgNO_3(aq) \rightarrow AgCl(s) + Mg(NO_3)_2(aq)$
d $HNO_3(aq) + FeSO_4(aq) + H_2SO_4(aq) \rightarrow Fe_2(SO_4)_3(aq) + H_2O(l) + NO(g)$

2 Write balanced chemical equations that correspond to the following word equations:

a Carbon monoxide gas reacts with solid iron(III) oxide to form solid iron and carbon dioxide gas.
b A piece of sodium reacts with water to form a solution of sodium hydroxide and hydrogen gas.
c When molten calcium bromide is electrolysed, it forms liquid calcium metal and bromine gas.

3 In the following reaction:

$$CaCO_3(s) \rightarrow CaO(s) + CO_2(g)$$

calculate the mass of:

a $CaCO_3$ that reacted to produce 10.00 g of CaO and 7.85 g of CO_2.
b CO_2 produced when 100.1 g of $CaCO_3$ reacts to produce 56.1 g of CaO.
c CaO produced when 75.07 g of $CaCO_3$ reacts to produce 33.01 g of CO_2.

4 A sample of aluminium nitrate ($Al(NO_3)_3$) has a mass of 30.5 g.
   a Calculate the amount, in mol, of aluminium nitrate in the sample.
   b Calculate the amount, in mol, of nitrogen atoms present.
   c Calculate the total number of nitrogen atoms present in the sample.
   d Determine the percentage by mass of oxygen in aluminium nitrate.

5 A compound of tungsten and sulfur is a useful solid lubricant. Deduce the empirical formula of this compound if a particular sample is formed when 1.84 g of tungsten reacts exactly with 0.64 g of sulfur.

6 Copper forms two oxides, copper(I) oxide (Cu_2O) and copper(II) oxide (CuO). The following experiment was carried out to determine whether a particular oxide of copper was Cu_2O or CuO.
   A sample of the oxide was dissolved in sulfuric acid. An excess of zinc powder was added to displace the copper as copper metal. The copper metal was collected, washed, thoroughly dried and then weighed. The results were:
   m(oxide) = 2.127 g
   m(dried copper) = 1.704 g
   a Determine the empirical formula of this oxide of copper.
   b If the sample analysed was a mixture of the two oxides instead of one of the oxides only, would the mass of copper obtained be greater than or less than 1.704 g? Explain your answer.

7 Metal X burns in air and forms a compound with the empirical formula X_2O.
   When 0.753 g of the metal burns completely, 0.907 g of X_2O forms.
   a Calculate the molar mass of X.
   b Identify metal M.

8 Potassium nitrate (KNO_3) and ammonium sulfate ($(NH_4)_2SO_4$) can both be used in fertilisers as a source of nitrogen for plants.
   What mass of ammonium sulfate contains the same mass of nitrogen as 65.0 g of potassium nitrate?

9 Deep-sea fish can build up high levels of mercury from contaminated water and by eating other contaminated fish. A 3.0 g sample of fish is ground up and mixed with water. The mixture is made up to 100 mL with water. A sample was found to have a mercury concentration of 20 mg L^{-1}.
   Determine the:
   a percentage composition by mass of mercury in the fish
   b concentration in ppm of mercury in the fish.

10 A helium cylinder for the inflation of party balloons holds 25.0 L of gas, and is filled to a pressure of 16 500 kPa at 15°C.
   a Express the pressure of helium gas inside the cylinder in units of:
      i atm
      ii mmHg
   b What mass of helium does the cylinder contain when full?
   c What volume would the helium occupy at standard laboratory conditions (SLC)?
   d How many balloons can be inflated from a single cylinder at 30°C if the volume of one balloon is 6.5 L and each needs to be inflated to a pressure of 108 kPa?

Extended response

1 Barium chloride ($BaCl_2$) is an inexpensive, soluble salt of barium that has wide application in the laboratory, including in the gravimetric determination of the presence of sulfate ions. Solid barium chloride can be produced through the decomposition reaction of solid barium chlorate ($Ba(ClO_3)_2$), which also produces oxygen gas. When 7.25 g of barium chlorate decomposes completely, 4.96 g of barium chloride is produced.
   a Write a balanced chemical equation for the decomposition of barium chlorate.
   b Using your understanding of the conservation of mass, determine how much oxygen is produced.
   c If 5.00 g of barium chlorate was used instead, how much more barium chlorate would be required to produce the same amount of barium chloride?
   d If you tripled the amount of barium chlorate, what would be the effect on the total mass of products?
   e If the decomposition of 7.25 g of barium chlorate produced only 3.72 g of barium chloride and 1.72 g of oxygen, how much barium chlorate did not decompose?

2 A particular rock is known to contain magnesium carbonate. There is no other metal carbonate present in the rock. A sample of the rock weighing 12.8 g is finely ground then added to hydrochloric acid (HCl(aq)) in a reaction vessel. The magnesium carbonate in the rock dissolves in the acid to produce a solution of magnesium chloride, water and 2.60 L of carbon dioxide gas at SLC.
   a Write an equation for the reaction between the magnesium carbonate in the rock and the hydrochloric acid.
   b Calculate the amount, in mol, of carbon dioxide produced in this reaction.
   c Determine the percentage composition by mass of magnesium carbonate in the rock.

d The hydrochloric acid has a concentration of 0.50 mol L^{-1}. Determine the minimum volume of acid required for this reaction.

e Carbon dioxide can be detected by bubbling it through limewater ($Ca(OH)_2(aq)$). The reaction produces a precipitate of calcium carbonate and water.

  i Write an equation for the reaction of carbon dioxide with limewater.

  ii Determine the mass of calcium carbonate that would be produced if the carbon dioxide produced in the first reaction was bubbled through limewater.

3 A solution of lead(II) nitrate is prepared by dissolving 9.80 g of solid lead(II) nitrate in water. The total volume of the solution formed was 50.0 mL.

a Calculate the molar concentration of this solution.

b The lead(II) nitrate solution is diluted by adding 30.0 mL of water. What is the new concentration of the solution?

c The diluted lead(II) nitrate solution from part **b** is mixed with 50.0 mL of 0.650 mol L^{-1} sodium iodide. The products of this reaction are a bright yellow precipitate of lead(II) iodide and a solution of sodium nitrate.

  i Write a balanced chemical equation for this reaction.

  ii Determine the amount, in mol, of lead(II) nitrate and sodium iodide that were mixed together.

  iii Determine whether lead(II) nitrate or sodium iodide was limiting in this reaction.

  iv Determine the mass of precipitate that forms.

4 The volume of a quantity of carbon dioxide was measured at different pressures. The experiment was carried out at a constant temperature of 25°C. The results were recorded and graphed.

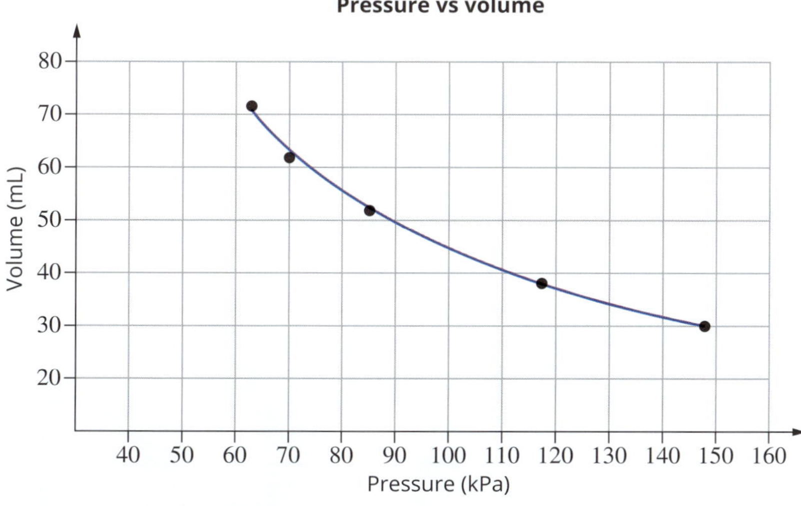

a Describe what happens to the volume of the gas as the pressure changes.

b What relationship between pressure and volume can be derived from these results?

c Calculate the volume of the carbon dioxide gas if the pressure was increased to 160 kPa. Show your working.

d If the sample of carbon dioxide was subjected to extremely high pressures, would the relationship in part **b** still hold? Explain your answer.

e Find the number of mol of carbon dioxide used in the experiment.

f If carbon dioxide were replaced with an equal amount of hydrogen and the experiment repeated, what differences would you expect to find in this new set of results? Explain your answer.

5 A mass of 5.948 g of white phosphorus was used to make phosphine (PH_3) according to the following unbalanced equation:

$$P_4(s) + KOH(aq) + H_2O(l) \rightarrow PH_3(g) + KH_2PO_2(aq)$$

a Balance the equation for the reaction.

b Calculate the amount, in mol, of white phosphorus that reacted.

c The phosphorus reacted with 4.500 g of potassium hydroxide. Determine whether phosphorus or potassium hydroxide was limiting.

d Determine the amount, in mol, of the other reactant in excess.

e Determine the volume of phosphine produced at SLC.

f If the 4.500 g of potassium hydroxide was dissolved in water to produce 250.0 mL of solution, what is the molar concentration of the solution?

3 Reactive chemistry

All chemical reactions involve the creation of new substances and associated energy transformations. These are often detectable as changes in the temperature of the surroundings and/or the emission of light. The reactions are controlled by chemists to produce substances for the development of useful products.

Chemicals can react at many different speeds and in many different ways, but the reactions all involve the breaking and making of chemical bonds. In this module, you will study how chemicals react, the changes in matter and energy that take place during these reactions, and how these chemical reactions and changes relate to the chemicals used in everyday life.

Outcomes

By the end of this module, you will be able to:

- design and evaluate investigations in order to obtain primary and secondary data and information (CH11-2)
- conduct investigations to collect valid and reliable primary and secondary data and information (CH11-3)
- select and process appropriate qualitative and quantitative data and information using a range of appropriate media (CH11-4)
- explore the many different types of chemical reactions, in particular the reactivity of metals, and the factors that affect the rate of chemical reactions (CH11-10)

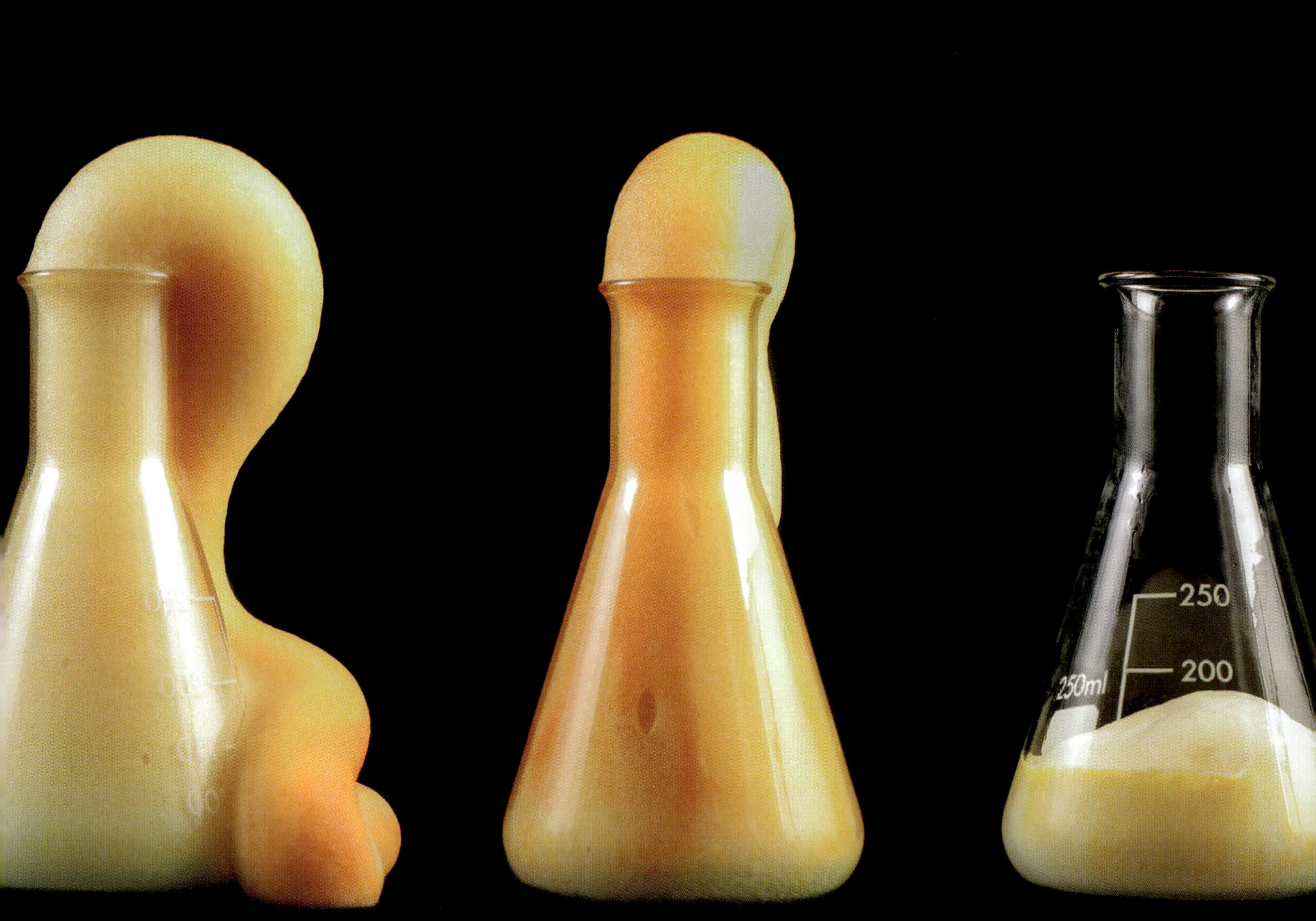
250
200
250ml

CHAPTER

10 Chemical reactions

One of the most important functions of chemists is to change substances. Crude oil in Earth's crust is of little use in its natural form, but once separated into components such as petrol, diesel and oil, it forms a vital supply of energy for most modern societies. Many of the common metals we use, such as iron, aluminium and lithium, are not found as elements in nature; chemists have to design processing plants that can separate the metals from the non-metals they are bonded to.

Changes to matter also occur in the natural environment. The decomposition of a dead animal involves changes to many of the substances in the animal.

In this chapter, you will learn about different types of chemical reactions and be able to predict and identify the reactants and products in these reactions.

Content

INQUIRY QUESTION

What are the products of a chemical reaction?

By the end of this chapter, you will be able to:

- investigate a variety of reactions to identify possible indicators of a chemical change
- use modelling to demonstrate
  - the rearrangement of atoms to form new substances ICT
  - the conservation of atoms in a chemical reaction (ACSCH042, ACSCH080)  N
- conduct investigations to predict and identify the products of a range of reactions, for example:
  - synthesis
  - decomposition
  - combustion
  - precipitation
  - acid base reactions
  - acid carbonate reactions (ACSCH042, ACSCH080)
- investigate the chemical processes that occur when Aboriginal and Torres Strait Islander Peoples detoxify poisonous food items AHC
- construct balanced equations to represent chemical reactions ICT N

10.1 Chemical change

CHEMISTRY INQUIRY 

Colourful chemical changes

What are the products of a chemical reaction?

COLLECT THIS ...

- flower petals
- methylated spirits
- vinegar
- washing soda (sodium carbonate) or baking soda (sodium bicarbonate)
- test-tubes or clear plastic cups
- flask
- hot plate

DO THIS ...

1. Add 30 mL of water and 30 mL of methylated spirits to a flask.
2. Tear flower petals into small pieces and add to the flask. Use one type of flower only.
3. Heat the liquid to around 60 °C.
4. Allow to cool.
5. Pour liquid off to separate it from the petals.
6. Add some washing soda and water to a test-tube. Add about 10 drops of your flower petal solution.
7. Add vinegar slowly, observing any signs of a reaction and any colour change of the flower petal solution.
8. Add washing soda to see if you can reverse the colour change.

RECORD THIS ...

Describe the colour changes you have seen.

Record any signs of a chemical reaction.

REFLECT ON THIS ...

What are the products of the reaction between washing soda and vinegar?

What is causing the colour changes?

Will all parts of a flower work?

Do petals from other flowers work?

Could this process help you tell how concentrated an acid is?

Changes to matter can be classified as **chemical changes** or **physical changes**. Extracting aluminium from bauxite ore is an example of a chemical change (Figure 10.1.1). Chemical changes are those in which new substances with different compositions and properties are formed. The covalent bonds in carbon dioxide and water are broken and new bonds are made when glucose is formed in photosynthesis.

GO TO ➤ Section 2.2 page 54

You learnt back in Chapter 2 that physical changes are changes in properties (such as volume or density) or changes in state (such as a liquid boiling and becoming a gas). Physical changes occur without altering the composition of the substance. Table salt can be obtained from the evaporation of sea water. This is an example of a physical change, because the salt was always present in the water.

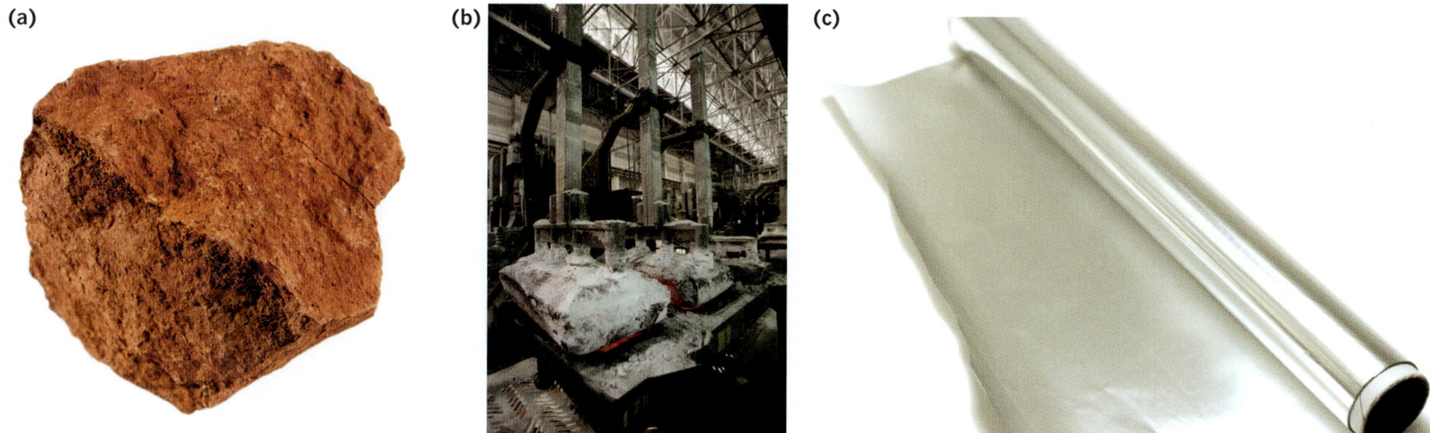

FIGURE 10.1.1 (a) The production of aluminium from bauxite is an example of a chemical change. (b) Smelters built for this task are large and expensive. The bauxite is heated to high temperatures and an electric current is passed through it. (c) The chemical change is complete when the bauxite has been converted to shiny aluminium foil.

CHEMISTRY IN ACTION 

Salt harvesting

Many Australian lakes and rivers have a high salt concentration. The salt can be separated from the water by allowing the salt water to sit in shallow ponds. Energy from the sun evaporates the water, leaving the salt to be scooped up and purified.

Salt harvesting is an example of a physical change. The salt particles are separated from water in the process, but are themselves unaltered. The photos in Figure 10.1.2 show some of the operations of Australian Saltworks, a company with several distribution centres in NSW.

FIGURE 10.1.2 (a) Sections of an inland lake partitioned off for salt harvesting. (b) A front-end loader scraping up the dried salt from the lake floor.

EXAMPLES OF CHANGE

Some examples of physical and chemical changes are given in Table 10.1.1.

TABLE 10.1.1 Examples of physical and chemical changes

Physical changes	Chemical changes
water vapour condensing on a cold window	vinegar reacting with baking soda
iceberg melting	rusting of nail
stirring copper(II) sulfate into water to form a blue solution	burning of wood in a fire
separation of milk into curds and whey	lighting a match

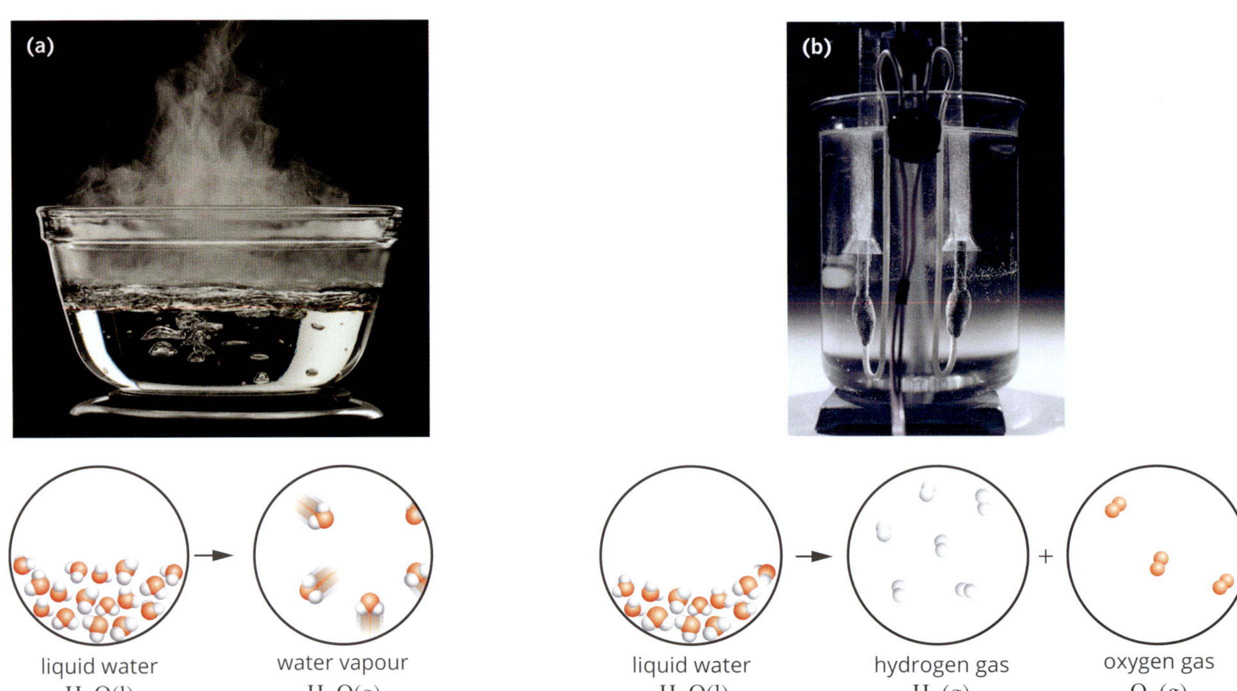

FIGURE 10.1.3 (a) Boiling water is a physical change, whereas (b) the electrolysis of water is a chemical change.

Figure 10.1.3 contrasts the boiling of water to form steam (a physical change) with the electrolysis of water to form hydrogen gas and oxygen gas (a chemical change). In the physical change from water to steam, the covalent bonds in the water molecules are unchanged. Only the intermolecular bonds between water molecules are broken. The water molecules are still present in the steam, and the steam can be condensed back to liquid water. However, when an electric current is passed through water, the covalent bonds in the water molecules are broken and hydrogen molecules and oxygen molecules are formed.

Signs of change

When a chemical change occurs, the process is referred to as a **chemical reaction**. Bonds in the reactants break and products with different properties are formed. There are usually signs that a reaction has occurred. When an egg cooks, as in Figure 10.1.4a, the change of colour of the egg white is evidence of the chemical reaction. Burning paper leaves a pile of black ash. Other common signs of a chemical reaction occurring are:

- a significant temperature change. When a candle burns, thermal energy is released.
- formation of a solid (**precipitate**) in a solution. When carbon dioxide bubbles through limewater, a white precipitate of calcium carbonate forms.
- production of an odour. If sugar is left too long on a hotplate, the distinctive toffee smell indicates that the sugar has been broken down by the heat.
- production of a gas. When baking soda is added to vinegar, a reaction is evident from the stream of bubbles released (Figure 10.1.4b).

FIGURE 10.1.4 (a) The colour changes occurring when an egg cooks are due to chemical changes. (b) The stream of gas erupting from a bottle of vinegar after baking soda has been added is also due to a chemical change.

GO TO ➤ Section 6.1 page 206

CHEMICAL EQUATIONS

In Chapter 6, you learnt that chemical reactions can be represented by **chemical equations**. Three examples of chemical equations are shown in Table 10.1.2. It is sometimes helpful to first write the equation in words, then to write the correct formulae, and finally to balance the equation. It is important that you are able to write balanced chemical equations to represent the types of reactions that you will learn about in this chapter.

TABLE 10.1.2 Examples of chemical equations

	Reactants			→	Products		
Example 1	methane	+	oxygen	→	carbon dioxide	+	water
	$CH_4(g)$	+	$2O_2(g)$	→	$CO_2(g)$	+	$2H_2O(g)$
Example 2	magnesium	+	oxygen	→	magnesium oxide		
	$2Mg(s)$	+	$O_2(g)$	→	$2MgO(s)$		
Example 3	hydrogen	+	chlorine	→	hydrogen chloride		
	$H_2(g)$	+	$Cl_2(g)$	→	$2HCl(g)$		

Worked example 10.1.1

BALANCING EQUATIONS

Barium nitrate solution reacts with sodium sulfate solution to form a solution of sodium nitrate and a precipitate of barium sulfate. Write a balanced equation for this reaction.

Thinking	Working
Write an equation in words.	barium nitrate + sodium sulfate → barium sulfate + sodium nitrate
Write the correct formulae for each substance and add states.	$Ba(NO_3)_2(aq) + Na_2SO_4(aq) \rightarrow BaSO_4(s) + NaNO_3(aq)$
Balance the equation by adjusting coefficients.	$Ba(NO_3)_2(aq) + Na_2SO_4(aq) \rightarrow BaSO_4(s) + 2NaNO_3(aq)$

Worked example: Try yourself 10.1.1

BALANCING EQUATIONS

Aluminium metal reacts in hydrochloric acid (a hydrogen chloride solution) to form aluminium chloride solution and hydrogen gas. Write a balanced equation for this reaction.

FIGURE 10.1.5 Magnesium ribbon and the oxygen in the air are reactants. The white powder forming on the tongs is the product, magnesium oxide.

GO TO ➤ Section 6.1 page 206

Conservation of atoms

During a chemical reaction, the atoms in the reactants are rearranged. Bonds are broken and new bonds form. However, no atoms are created or destroyed. All the atoms present in the reactants will still be there after the reaction. Some or all of the bonds will have changed, but not the number of atoms of each element. As the number of atoms is unchanged, the mass of the products must equal the mass of the reactants.

This is the **law of conservation of mass**, which states that mass is neither created nor destroyed in a chemical reaction (you learnt about this concept back in Chapter 6). The law of conservation of mass is the reason chemists insist on balancing equations. Balancing adjusts the ratio of each substance until the number of reactant atoms equals the number of product atoms.

The reaction between magnesium and oxygen (Figure 10.1.5) has already been mentioned in this section:

$$2Mg(s) + O_2(g) \rightarrow 2MgO(s)$$

Figure 10.1.6 illustrates that there are the same number of reactant atoms as product atoms; they are just arranged differently. Therefore, the mass of the products is the same as that of the reactants.

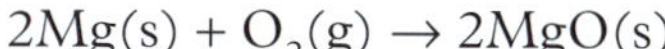

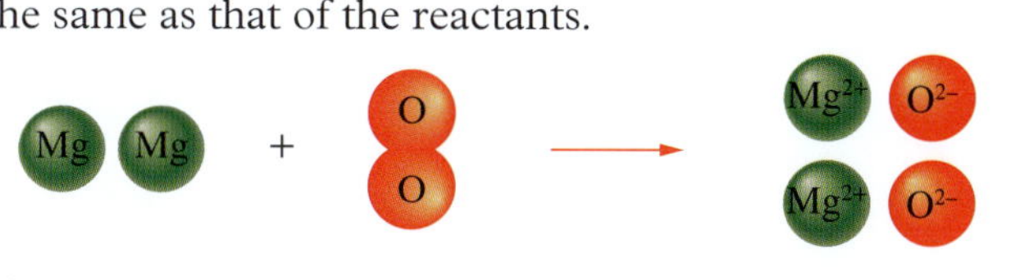

FIGURE 10.1.6 The number of atoms of each element does not change in a chemical reaction. The coefficients in a balanced equation reflect the ratios needed of each substance present for mass to be conserved.

10.1 Review

SUMMARY

- A physical change is a change in the physical properties of a substance or a change in state.
- A chemical change leads to the formation of new substances with different compositions and different properties.
- There are often signs that a chemical change has occurred, such as the evolution of a gas, production of an odour, formation of a precipitate, a change in temperature or a change in colour.
- Chemical change can be represented by a chemical equation. A chemical equation contains information about the ratios and the physical state of each substance.
- Mass is conserved in a chemical reaction, because the number of atoms of each element present does not change. This is the law of conservation of mass.

KEY QUESTIONS

1 Select the alternative that represents a physical change.
  A lighting a match
  B digestion of food
  C melting of gold
  D fermentation of sugar

2 Classify each of the following as physical or chemical changes:
  a candle burning
  b dissolving sugar in tea
  c separating milk into cream and skim milk
  d rusting of a nail
  e gold panning
  f browning of a banana.

3 Identify the reactants and products in the following reactions:
  a During photosynthesis, glucose and oxygen are formed from carbon dioxide and water.
  b Hydrogen gas and oxygen gas can be formed when an electric current is passed through water.
  c Heat can be used to break calcium carbonate down into calcium oxide and carbon dioxide.
  d The burning of coal produces carbon dioxide.

4 For each of the reactions described below, write an equation using words and then a balanced chemical equation.
  a Sulfur solid can react with oxygen gas to form sulfur dioxide gas (SO_2).
  b A silver nitrate solution mixed with potassium iodide solution forms solid silver iodide and potassium nitrate solution.

5 A sample of sugar is added to a crucible and heated in two stages. In the first stage, a clear, sweet liquid forms. In the second stage, after strong heating, a black residue is formed and gases are evolved.
  a Describe what type of change is occurring during stage one. Explain your answer.
  b Describe what type of change is occurring during stage two. Explain your answer.

6 36 g of carbon can react with oxygen to form 132 g of carbon dioxide. What mass of oxygen was required for this reaction?

7 Magnesium metal reacts in hydrochloric acid to form magnesium chloride and hydrogen gas.
  a 30 mL of hydrochloric acid is added to a plastic bag and a piece of magnesium is dropped in. The top of the plastic bag is left open.
  Describe how the mass of the plastic bag will change over the next few minutes.
  b 30 mL of hydrochloric acid is added to a plastic bag and a piece of magnesium is dropped in, then a knot is quickly tied in the top of the plastic bag.
  Describe how the mass of the plastic bag will change over the next few minutes.

8 For each of the following reactions, suggest a possible sign that a chemical reaction has occurred:
  a a dead animal in the bush
  b a candle burning
  c tarnishing of a metal trophy.

10.2 Synthesis reactions

In order to make sense of the huge array of known chemical reactions, chemists often classify reactions based on how they proceed. Five different classes of reaction will be discussed in this chapter: synthesis, decomposition, precipitation, combustion and acid base.

Many of our pharmaceuticals have complex chemical structures. The molecules do not exist in nature, and chemists have to build the molecules from elements or smaller molecules. The production of a complex pharmaceutical is an example of a **synthesis** reaction.

History shows that many synthesised compounds came into production because a previous naturally occurring source was diminishing or limited. Most of our fruit flavours (like banana flavouring, Figure 10.2.1) are now synthesised from molecules isolated from crude oil, rather than from fruits themselves. Kevlar, nylon and neoprene are examples of the wide range of plastics to choose from in modern society. They are produced in **synthesis reactions**.

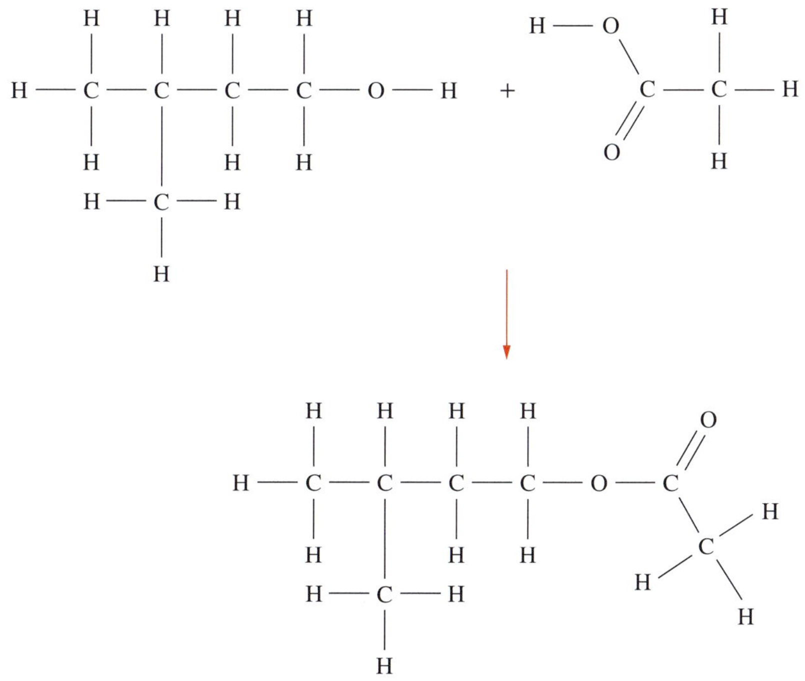

A synthesis reaction involves forming a compound from elements or from other, usually smaller, compounds.

FIGURE 10.2.1 Artificial banana flavour (3-methylbutyl ethanoate) can be synthesised from the two molecules shown. The flavouring formed has the same formula as one of the components of real bananas.

Before the 19th century, farmers relied on animal manure to fertilise their crops. As the world's population has grown rapidly, the need for fertiliser has also grown rapidly. This demand has led to the production of artificial fertilisers such as ammonium sulfate ($(NH_4)_2SO_4$). This can be synthesised by first producing ammonia (NH_3) and sulfuric acid (H_2SO_4). These two products are then reacted together to form ammonium sulfate. Figure 10.2.2 illustrates this production.

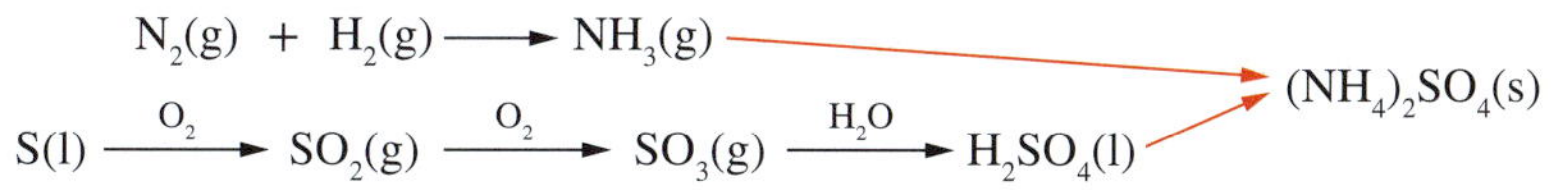

$$N_2(g) + H_2(g) \longrightarrow NH_3(g)$$

$$S(l) \xrightarrow{O_2} SO_2(g) \xrightarrow{O_2} SO_3(g) \xrightarrow{H_2O} H_2SO_4(l)$$

$$NH_3(g) + H_2SO_4(l) \longrightarrow (NH_4)_2SO_4(s)$$

FIGURE 10.2.2 The synthesis of ammonium sulfate ($(NH_4)_2SO_4$) requires the synthesis of ammonia (NH_3) and the synthesis of sulfuric acid (H_2SO_4). These two products are then combined to complete the formation of the fertiliser.

CHEMFILE WE

Mauve

Eighteen-year-old laboratory worker William Henry Perkin was trying in 1859 to prepare a drug to be used as a cure for malaria. Perkin did not succeed in this endeavour, but he did notice a residue with an unusual colour. Rather than dismiss the residue as a waste product, Perkin had the initiative to market the residue as a clothing dye. This dye, now known as mauve, was the first synthetic dye ever produced, and it launched a range of similar dyes that soon replaced most of the natural dyes previously in use.

The production of mauve made Perkin a wealthy man, and he spent most of his working life researching and marketing other synthetic dyes. Perkin can be seen in Figure 10.2.3 with a sample of cloth dyed mauve.

FIGURE 10.2.3 Sir William Henry Perkin holding a cloth that he dyed mauve.

10.2 Review

SUMMARY

- In a synthesis reaction, a compound is formed from elements or from other, usually smaller compounds.

KEY QUESTIONS

1 Select the synthesis reaction from the following alternatives.

A Plants can produce glucose from carbon dioxide and water during photosynthesis.

B An electric current is passed through molten sodium chloride to produce sodium and chlorine.

C Lead metal can be extracted from rocks containing lead(II) sulfide.

D During digestion, starch molecules can be broken down into glucose molecules.

2 Aluminium metal can be produced by passing an electric current through molten aluminium oxide. The equation for this reaction is:

$$2Al_2O_3(l) \rightarrow 4Al(l) + 3O_2(g)$$

Aluminium is produced in this reaction, yet it is not considered to be an example of a synthesis reaction. Explain why.

3 Hydrogen chloride gas can be formed from the reaction between hydrogen gas and chlorine gas.

a Write a word equation for this reaction.

b Write a balanced chemical equation for this reaction.

c How many molecules of reactants are required to manufacture 40 molecules of HCl?

d Will the mass of chlorine gas required for this reaction be the same as the mass of hydrogen gas required? Explain your answer.

4 Sulfur trioxide gas (SO_3) can be formed when sulfur dioxide (SO_2) reacts with oxygen gas.

a Write a word equation for this reaction.

b Write a balanced chemical equation for this reaction.

c How will the mass of SO_3 formed compare with the mass of the reactants?

d If 100 g of SO_2 is reacted with 100 g of O_2, will 200 g of SO_3 be formed? Explain your answer.

10.3 Decomposition reactions

An example of a **decomposition** reaction is the 'elephant's toothpaste' reaction shown in Figure 10.3.1. The equation for this reaction is:

$$2H_2O_2(l) \rightarrow 2H_2O(l) + O_2(g)$$

Notice that hydrogen peroxide is the only reactant, but that there are two products: water and oxygen gas. Another interesting aspect to this reaction is the variety of substances that can act as a catalyst. Typically, manganese dioxide is used, but iodine or potassium iodide will produce the same result. Crushed liver is also a catalyst. Liver contains the enzyme catalase, and it is the catalase that speeds up this decomposition. Potatoes and apple also contain catalase.

Looking at this reaction in more detail, some of the covalent bonds in the hydrogen peroxide are broken, and new bonds between oxygen atoms are formed (Figure 10.3.2). The bonds that break when hydrogen peroxide decomposes are covalent bonds, but decomposition of other types of substances (such as those that contain ionic bonds) can also occur.

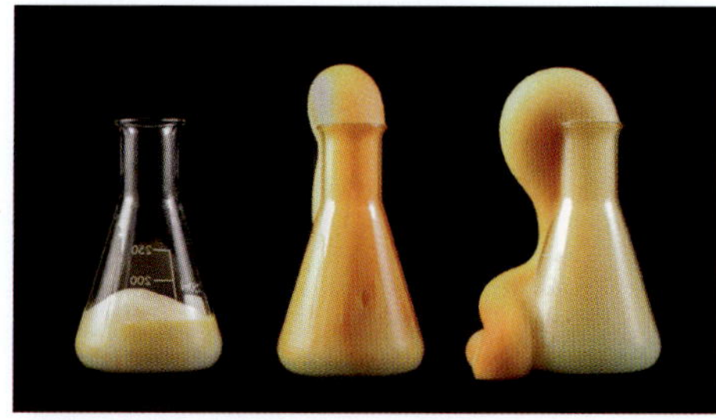

FIGURE 10.3.1 The elephant's toothpaste decomposition reaction begins with a hydrogen peroxide sample. A catalyst causes rapid decomposition, producing what is called 'elephant's toothpaste'.

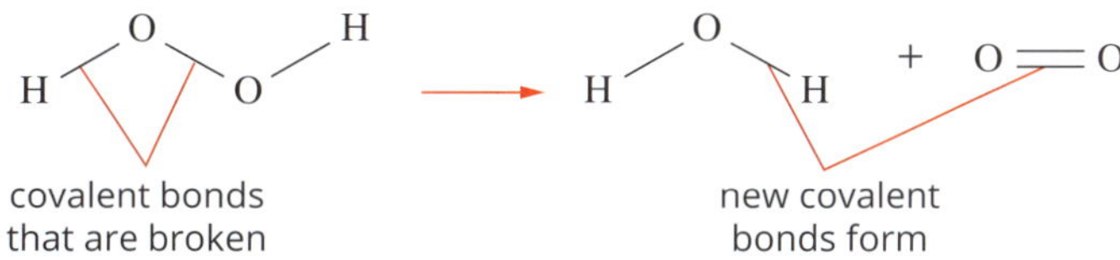

FIGURE 10.3.2 Decomposition reactions involve the breaking of one reactant and the formation of two or more products.

A decomposition reaction is a reaction in which one reactant is broken down into two or more products.

CHEMISTRY IN ACTION CCT

Discovery of oxygen

A decomposition reaction led to the isolation of the first pure samples of oxygen gas. Scientists had struggled to understand and study gases because their colourless nature and a lack of airtight equipment made them very difficult to work with. It was only in the 18th century that scientists proved that there were different gases and that air was not a pure, unreactive substance. In 1774, English scientist Joseph Priestley focused sunlight from a large convex lens onto a lump of reddish mercuric oxide in an airtight container (Figure 10.3.3). The red solid formed a grey liquid and he found the gas emitted to be 'five or six times as good as common air'. Priestley arrived at this conclusion because a mouse stayed alive over five times longer when added to a container that had been heated with mercuric oxide than in a similar container of air. The gas that was produced from the decomposition reaction is oxygen. The equation for the reaction is:

$$2HgO(s) \rightarrow 2Hg(l) + O_2(g)$$

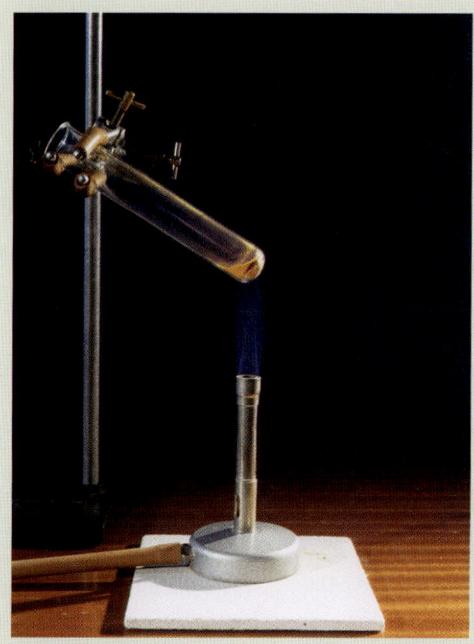

FIGURE 10.3.3 A modern version of Priestley's experimental set-up.

THERMAL DECOMPOSITION

Decomposition reactions often require energy. Heating is an obvious way to supply that energy. Priestley's decomposition of mercuric oxide described in the Chemistry in Action box on page 319 is an example. The process by which heat breaks compounds down into simpler substances is known as **thermal decomposition**.

Most carbonate compounds will decompose when heated, releasing carbon dioxide gas. An example is calcium carbonate, which decomposes into calcium oxide and carbon dioxide:

$$CaCO_3(s) \rightarrow CaO(s) + CO_2(g)$$

The calcium oxide formed, also known as lime, has many uses, ranging from marking out football pitch lines to reducing the acidity of soil.

The decomposition of sodium azide (NaN_3) is an example of a thermal decomposition that saves lives every day by inflating vehicle airbags, as shown in Figure 10.3.4. A collision in a modern vehicle triggers rapid heating of sodium azide. The decomposition produces large volumes of nitrogen gas, inflating the airbag to protect the occupants of the vehicle.

$$2NaN_3(s) \rightarrow 2Na(s) + 3N_2(g)$$

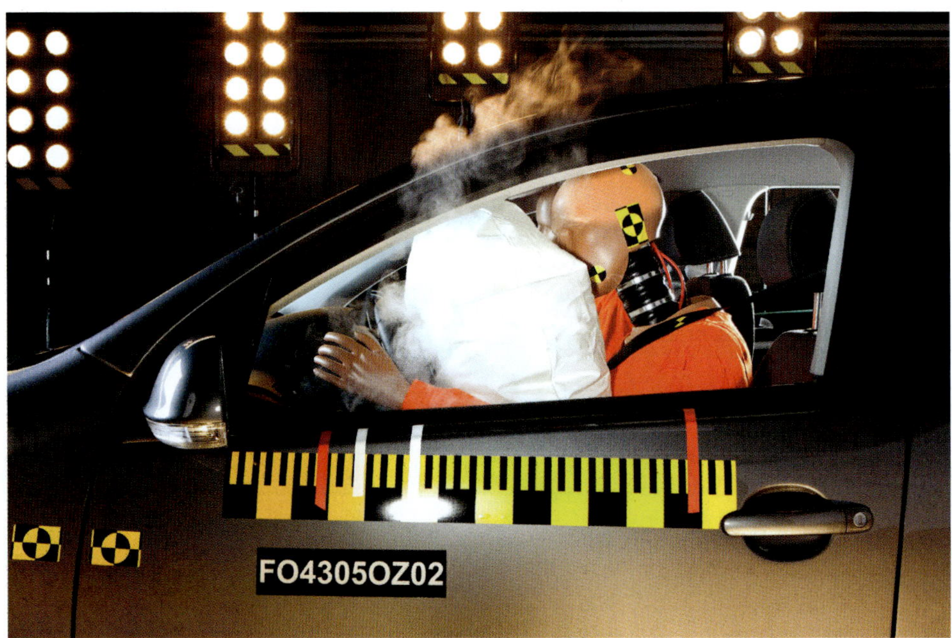

FIGURE 10.3.4 A collision triggers the release of heat energy. This heat energy leads to the decomposition of sodium azide and the formation of nitrogen gas, which inflates the airbag.

CHEMFILE PSC

Decomposition in cooking

Anyone baking a cake is probably harnessing thermal decomposition. Baking soda is added to a cake mix so that, when heated, it decomposes and releases carbon dioxide gas (Figure 10.3.5). The carbon dioxide is trapped in the cake mixture, giving it a soft, light texture. The equation for this reaction is shown below. Notice that three products are formed: decomposition reactions do not necessarily form only two products.

$$2NaHCO_3(s) \rightarrow Na_2CO_3(s) + H_2O(g) + CO_2(g)$$

FIGURE 10.3.5 Baking soda decomposes during cooking, releasing carbon dioxide gas.

DECOMPOSITION BY ELECTRICAL ENERGY

Other forms of energy such as electricity and light may also bring about the decomposition of compounds. Electrical energy can be used to decompose water to hydrogen and oxygen (Figure 10.3.6). Using electrical energy to cause a decomposition reaction is called **electrolysis**. The equation for this example is:

$$2H_2O(l) \rightarrow 2H_2(g) + O_2(g)$$

The process of electrolysis is also used to produce commercial supplies of reactive metals. In Australia, aluminium smelters electrolyse bauxite (Al_2O_3) to produce aluminium metal in large quantities.

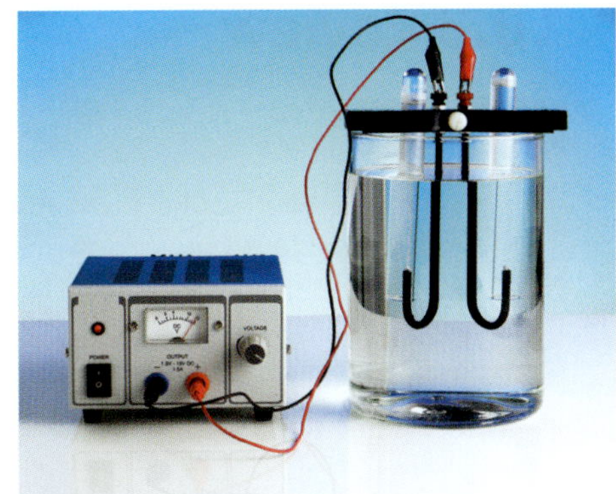

FIGURE 10.3.6 Electrical energy from a power supply is causing water to decompose to hydrogen gas at one electrode and oxygen gas at the other. The gases can be collected by placing a container over the electrodes.

DECOMPOSITION BY LIGHT ENERGY

Light energy can also cause decomposition of some compounds. This process is known as **photolysis**. The hydrogen peroxide mentioned at the start of this section is usually sold in an opaque container to improve its shelf life (Figure 10.3.7). Similarly, cooking oils and other food items are often sold in dark glass containers to limit the decomposition caused by light passing through the liquid.

For many years, photographers harnessed the decomposition of silver compounds to produce permanent images. Silver bromide on photographic film decomposes to silver when light enters the camera.

$$2AgBr(s) \rightarrow 2Ag(s) + Br_2(g)$$

The silver atoms form an image that is enhanced when the film is developed.

FIGURE 10.3.7 Opaque containers protect the contents of this bottle from light, extending the shelf life of the product by reducing the amount of decomposition.

CHEMFILE S

Biodegradation

Biodegradation is nature's way of decomposing organic substances. A fallen tree in a forest will slowly decompose over the ensuing years. Nature uses microorganisms such as bacteria and fungi to bring about decomposition of once-living things. The number of products formed in a biodegradation reaction can be very high, depending on the type of organism being biodegraded.

The sewage treatment process shown in Figure 10.3.8 uses bacteria to break down human waste into harmless products.

FIGURE 10.3.8 Decomposition of human waste at a sewage treatment plant.

Scientists look for ways to both slow down and speed up biodegradation. When the Egyptians prepared mummies, they were using techniques to slow the biodegradation of a corpse. Modern scientists also develop techniques to slow the decomposition of human bodies in a morgue. On the other hand, scientists work to speed up the rate of biodegradation of plastic waste, sewage and garden compost. The faster these waste products are converted to harmless alternatives, the better for all.

10.3 Review

SUMMARY

- Decomposition reactions are reactions in which a single reactant breaks down to form two or more products.
- Heat energy is often used to bring about decomposition. This is referred to as thermal decomposition.
- Electrical energy can be used to bring about decomposition. This is called electrolysis.
- Light energy can be used to bring about decomposition. This is called photolysis.

KEY QUESTIONS

1. Select the reaction that is an example of decomposition.
   A $AgNO_3(aq) + NaCl(aq) \rightarrow AgCl(s) + NaNO_3(aq)$
   B $S(s) + O_2(g) \rightarrow SO_2(g)$
   C $CuCO_3(s) \rightarrow CuO(s) + CO_2(g)$
   D $CaCO_3(s) + 2HCl(aq) \rightarrow CuCl_2(aq) + H_2O(l) + CO_2(g)$
2. Solid potassium chlorate ($KClO_3$) can decompose when heated to form solid potassium chloride and oxygen gas.
   Write a balanced chemical equation for this reaction.
3. When water boils there is only one reactant, yet this is not considered a decomposition reaction. Explain why.
4. Magnesium metal can be produced from the electrolysis of molten magnesium chloride ($MgCl_2$). Write a balanced equation for this reaction.
5. Soft drink contains dissolved carbonic acid (H_2CO_3). When the soft drink is opened, decomposition of the carbonic acid occurs. One of the products is a gas, providing the 'fizz' we are familiar with.
   a Write a balanced equation for the decomposition of carbonic acid.
   b Explain why soft drink goes 'flat' after being opened.
6. Magnesium carbonate ($MgCO_3$) will decompose when heated to form a metal oxide and carbon dioxide.
   a Write a balanced equation for this reaction.
   b Write a general equation in words for the decomposition of a metal carbonate.

10.4 Combustion reactions

Fire heats our homes, powers our cars and entertains us with firework displays. Bushfires can destroy homes and lives, devastating huge areas of bush and damaging the habitat of many animals. However, a bushfire can also be an agent of regrowth and renewal in land management.

In ancient cultures, such as the Ancient Greeks and Ancient Japanese, fire was considered to be one of the fundamental elements, along with air, earth and water (Figure 10.4.1).

Fire results from the **combustion** of substances. Combustion reactions need three things:

- fuel to burn
- oxygen for the fuel to burn in
- energy to get the process started.

Fire can be understood and controlled by applying our knowledge of chemistry. In this section, you will look at combustion reactions and their importance to the world in more detail.

COMBUSTION AS A CHEMICAL PROCESS

In combustion reactions, the reactant combines with oxygen to produce oxides. This type of reaction is often referred to as an **oxidation** reaction. Thermal energy is released during combustion.

In a combustion reaction, the reactant combines with oxygen to produce oxides.

The combustion of a **hydrocarbon** (a compound containing only carbon and hydrogen) produces carbon dioxide and water, provided there is enough oxygen present. An example is the combustion of propane (C_3H_8), a major component of LPG (liquefied petroleum gas), shown in the following equation:

$$C_3H_8(g) + 5O_2(g) \rightarrow 3CO_2(g) + 4H_2O(l)$$

Combustion reactions release energy. Water may be formed as a gas rather than as a liquid if the temperature is elevated.

FIGURE 10.4.1 According to the Ancient Greeks, the classical four elements were earth, fire, air and water.

Complete and incomplete combustion

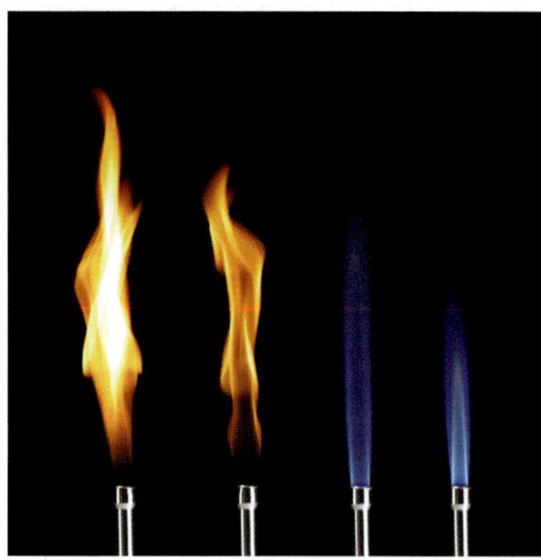

FIGURE 10.4.2 The yellow flame of a Bunsen burner is due to incomplete combustion and produces carbon. The blue flame is the hotter flame that occurs when the collar hole is open and more oxygen is allowed into the reaction.

Combustion reactions can be described as complete or incomplete. The type of combustion reaction that occurs will depend on the amount of oxygen available to react with the fuel.

Complete combustion occurs when oxygen is plentiful. The only products are carbon dioxide and water.

An example is the complete combustion of methane:

$$CH_4(g) + 2O_2(g) \rightarrow CO_2(g) + 2H_2O(l)$$

When the oxygen supply is limited, **incomplete combustion** occurs. As less oxygen is available, not all of the carbon can be converted into carbon dioxide. Carbon monoxide and/or carbon are produced instead. The hydrocarbon burns with a yellow, smoky or sooty flame, due to the presence of glowing carbon particles. Figure 10.4.2 shows the appearance of the different flames of a Bunsen burner due to incomplete and complete combustion.

The equation for the incomplete combustion of methane to form carbon monoxide is:

$$2CH_4(g) + 3O_2(g) \rightarrow 2CO(g) + 4H_2O(l)$$

CHEMFILE PSC

Carbon monoxide poisoning

Carbon monoxide is a highly poisonous gas. It combines readily with haemoglobin, the oxygen carrier in blood. When attached to carbon monoxide, haemoglobin cannot transport oxygen around the body, which leads to oxygen starvation of tissues.

Even at concentrations as low as 10 parts per million (ppm), carbon monoxide can cause drowsiness, dizziness and headaches. At about 200 ppm, carbon monoxide can lead to death. The average carbon monoxide concentration in large cities, mostly due to incomplete combustion of fuels in cars (Figure 10.4.3), is now 7 ppm, but it can be as high as 120 ppm at busy intersections in heavy traffic.

FIGURE 10.4.3 Car exhaust gases can contain high levels of carbon monoxide as a result of incomplete combustion of fuels.

The complete combustion of hydrocarbons occurs when there is sufficient oxygen for the fuel to burn. The products of complete combustion are carbon dioxide and water. When oxygen is not plentiful, incomplete combustion occurs. The products of incomplete combustion are carbon monoxide and/or carbon and water.

WRITING EQUATIONS FOR COMPLETE COMBUSTION OF FUELS

It is important to write chemical equations correctly because they tell you a lot about chemical reactions. Writing equations for the complete combustion reactions of fuels containing carbon and hydrogen is relatively straightforward, because the products are always carbon dioxide and water.

Our society has made extensive use of combustion reactions involving the burning of petrol. Petrol is a mixture of hydrocarbons, including octane.

The combustion reactions of octane (C_8H_{18}) and the other hydrocarbons in petrol power the internal combustion engines in most of Australia's 17.6 million motor vehicles.

Worked example 10.4.1

WRITING EQUATIONS FOR COMPLETE COMBUSTION OF HYDROCARBON FUELS

Write the equation, including state symbols, for the complete combustion of butane gas (C_4H_{10}).

Thinking	Working
Add oxygen as a reactant, and carbon dioxide and water as the products.	$C_4H_{10} + O_2 \rightarrow CO_2 + H_2O$
Balance carbon and hydrogen atoms, based on the formula of the hydrocarbon.	$C_4H_{10} + O_2 \rightarrow 4CO_2 + 5H_2O$
Find the total number of oxygen atoms on the product side.	Total O = $(4 \times 2) + 5$ $= 13$
If this is an odd number, multiply all of the coefficients in the equation by two, except for the coefficient of oxygen.	$2C_4H_{10} + O_2 \rightarrow 8CO_2 + 10H_2O$
Balance oxygen by adding the appropriate coefficient to the reactant side of the equation.	$2C_4H_{10} + 13O_2 \rightarrow 8CO_2 + 10H_2O$
Add state symbols.	$2C_4H_{10}(g) + 13O_2(g) \rightarrow 8CO_2(g) + 10H_2O(l)$

Worked example: Try yourself 10.4.1

WRITING EQUATIONS FOR COMPLETE COMBUSTION OF HYDROCARBON FUELS

Write the equation, including state symbols, for the complete combustion of liquid hexane (C_6H_{14}).

A similar series of steps can also be used to write the combustion equations for other carbon-based fuels that contain oxygen, such as alcohols.

Worked example 10.4.2

WRITING EQUATIONS FOR COMBUSTION REACTIONS OF ALCOHOLS

Write the equation, including state symbols, for the complete combustion of liquid ethanol (C_2H_5OH) (Figure 10.4.4).

Thinking	Working
Add oxygen as a reactant, and carbon dioxide and water as the products.	$C_2H_5OH + O_2 \rightarrow CO_2 + H_2O$
Balance carbon and hydrogen atoms, based on the formula of the alcohol.	$C_2H_5OH + O_2 \rightarrow 2CO_2 + 3H_2O$
Find the total number of oxygen atoms on the product side. Then subtract the one oxygen atom in the alcohol molecule from the total number of oxygen atoms on the product side.	Total O on product side $= (2 \times 2) + (3 \times 1)$ $= 7$ Total O on product side − 1 in alcohol $= 7 - 1 = 6$
If this is an odd number, multiply all the coefficients in the equation by two, except for the coefficient of oxygen.	$C_2H_5OH + O_2 \rightarrow 2CO_2 + 3H_2O$
Balance oxygen by adding the appropriate coefficient to the reactant side of the equation.	$C_2H_5OH + 3O_2 \rightarrow 2CO_2 + 3H_2O$
Add state symbols.	$C_2H_5OH(l) + 3O_2(g) \rightarrow 2CO_2(g) + 3H_2O(l)$

Worked example: Try yourself 10.4.2

WRITING EQUATIONS FOR COMBUSTION REACTIONS OF ALCOHOLS

Write the equation, including state symbols, for the complete combustion of liquid methanol (CH_3OH).

FIGURE 10.4.4 Alcohol burners or spirit lamps are used in some science laboratories. They burn liquid alcohols, such as ethanol, and produce carbon dioxide and water. Spirit lamps produce a smaller, cooler flame than Bunsen burners.

WRITING EQUATIONS FOR INCOMPLETE COMBUSTION OF FUELS

When the supply of oxygen is insufficient, incomplete combustion of fuels occurs. Equations can also be written to represent this. In general, for the incomplete combustion of hydrocarbons, as well as carbon-based fuels that contain oxygen, the products are carbon monoxide and/or carbon and water.

Worked example 10.4.3

WRITING EQUATIONS FOR INCOMPLETE COMBUSTION OF FUELS

Write an equation, including state symbols, for the incomplete combustion of ethane gas (C_2H_6) to form carbon monoxide and water vapour.

Thinking	Working
Add oxygen as a reactant, and carbon monoxide and water as the products.	$C_2H_6 + O_2 \rightarrow CO + H_2O$
Balance the carbon and hydrogen atoms, based on the formula of the hydrocarbon.	$C_2H_6 + O_2 \rightarrow 2CO + 3H_2O$
Balance oxygen by adding the appropriate coefficient to the reactant side of the equation.	$C_2H_6 + \frac{5}{2}O_2 \rightarrow 2CO + 3H_2O$
If oxygen gas has a coefficient that is half of a whole number, multiply all of the coefficients in the equation by two.	$2C_2H_6 + 5O_2 \rightarrow 4CO + 6H_2O$
Add state symbols.	$2C_2H_6(g) + 5O_2(g) \rightarrow 4CO(g) + 6H_2O(g)$

Worked example: Try yourself 10.4.3

WRITING EQUATIONS FOR INCOMPLETE COMBUSTION OF FUELS

Write an equation, including state symbols, for the incomplete combustion of liquid methanol (CH_3OH) to form carbon monoxide and water vapour.

+ ADDITIONAL

Additional products in incomplete combustion reactions

When the amount of oxygen is less than 50% of the required amount for the reaction to combust completely, methane (CH_4) can be produced from larger hydrocarbons.

Also, as carbon monoxide reacts with water according to the following equation, it can be seen that small amounts of hydrogen can also be produced.

$$CO(g) + H_2O(g) \rightarrow CO_2(g) + H_2(g)$$

10.4 Review

SUMMARY

- In combustion reactions, the reactant combines with oxygen to produce oxides. Combustion reactions release energy.
- The products of the complete combustion of hydrocarbons and carbon-based fuels containing oxygen are carbon dioxide and water.
- Incomplete combustion, resulting in the production of carbon monoxide and/or carbon and water, occurs when hydrocarbons and carbon-based fuels containing oxygen undergo combustion in a limited supply of oxygen.

KEY QUESTIONS

1 The most reliable process to follow when writing a combustion equation is to:
   A ensure the number of oxygen atoms is balanced first
   B adjust the formula of the fuel to balance the equation
   C keep the numbers of carbon and hydrogen atoms equal to one another
   D balance the carbon atoms, then the hydrogen atoms and finally the oxygen atoms.

2 Write a balanced equation for the complete combustion of liquid benzene (C_6H_6).

3 Write a balanced equation for the incomplete combustion of liquid ethanol (C_2H_5OH), in which carbon monoxide is formed.

4 Give concise explanations for each of the following:
   a A blacksmith uses a bellows to increase the temperature of burning coal.
   b A fire in a caravan can often flare dangerously if a rescuer opens the door to check for occupants.
   c Kerosene lamps often have a thick black film on the glass of the lamp.
   d The centre of a haystack can smoulder for weeks without bursting into flames.

5 The combustion of diesel (a hydrocarbon fuel) in a portable generator can be used to produce electrical energy.
   a Explain why diesel generators should only be operated in well-ventilated areas.
   b List the reactants and products for the reaction that occurs in a diesel generator.
   c Explain how the law of conservation of mass applies to the reaction occurring in a diesel generator.

10.5 Precipitation reactions

Sometimes, when two clear solutions are mixed together, they react to form a solid. The insoluble compound formed in such a reaction is called a **precipitate**, and this category of reaction is referred to as a **precipitation reaction**. For example, when a lead(II) nitrate solution is mixed with potassium iodide solution, a bright yellow solid forms, as shown in Figure 10.5.1. Many precipitates will collect on the bottom of the container.

FIGURE 10.5.1 The mixing of two clear solutions can be seen to produce a yellow precipitate.

> In a precipitation reaction, an insoluble compound is formed when two clear solutions are mixed together.

Precipitation reactions are used to remove minerals from drinking water, to remove heavy metals from wastewater, and in the purification plants of reservoirs.

In this section, you will look at what takes place during precipitation reactions.

Precipitation reactions occur naturally in undersea hydrothermal vents. The vents release superheated solutions containing sulfides, which then combine with metal ions to form precipitates of mineral sulfides, creating the chimney-like structures seen in Figure 10.5.2. The areas around these chimneys are biologically rich, often hosting complex communities fuelled by the chemicals dissolved in the vent fluids.

FIGURE 10.5.2 Undersea hydrothermal vents release superheated water containing sulfides, which form precipitates with metal ions.

DEDUCING THE IDENTITY OF A PRECIPITATE

When a colourless solution of silver nitrate is mixed with a colourless solution of sodium chloride, a white solid precipitate is formed, as shown in Figure 10.5.3.

FIGURE 10.5.3 Mixing aqueous solutions of sodium chloride and silver nitrate produces a solid, which is an example of a precipitate.

To understand what happens in the reaction between silver nitrate and sodium chloride, you need to identify the ions present in the reactant solutions and work out how they interact with each other.

- In the silver nitrate solution, there are silver ions (Ag^+) and nitrate ions (NO_3^-).
- In the sodium chloride solution, there are sodium ions (Na^+) and chloride ions (Cl^-).
- When one solution is added to the other, the mixture formed will contain all of the ions.

In both of these solutions, all the ions are moving around independently. As the ions move in the solution, they will collide with one another. When positive and negative ions collide, they may join together to form a new, insoluble precipitate.

SKILLBUILDER WE

Mixing solutions to form a precipitate

When mixing two solutions to form a precipitate, you need to remember the following points.

- Always wear safety goggles. Your teacher or lab technician will tell you if you also need to wear gloves.
- Place the beaker you will be pouring into on a flat surface.
- Always stand up when pouring the solution.
- Slowly trickle the solution down the inside of the beaker to form the best precipitate. If you add the solution quickly or don't trickle it down the inside of the beaker, the precipitate will have trapped bubbles and will not be as clear.
- Always bring your eyes down to the beaker to observe the precipitate; never hold the beaker up to your eye level.
- Always ask your teacher or lab technician how you need to dispose of the remaining solutions and precipitate.

Two new combinations of positive and negative ions are possible:

- sodium and nitrate ions (to form sodium nitrate)
- silver and chloride ions (to form silver chloride).

The solubility tables (Tables 10.5.1 and 10.5.2 pages 333 and 334) indicate that sodium nitrate is soluble in water, but that silver chloride is insoluble. Therefore, the precipitate must be silver chloride. Solubility tables will be discussed later in this section.

The process for the precipitation reaction between sodium chloride and silver nitrate is shown in Figure 10.5.4. When the hydrated Ag^+ and Cl^- ions come into contact, the attraction between the ions is greater than that of the ions for the water molecules. An ionic lattice of AgCl is formed.

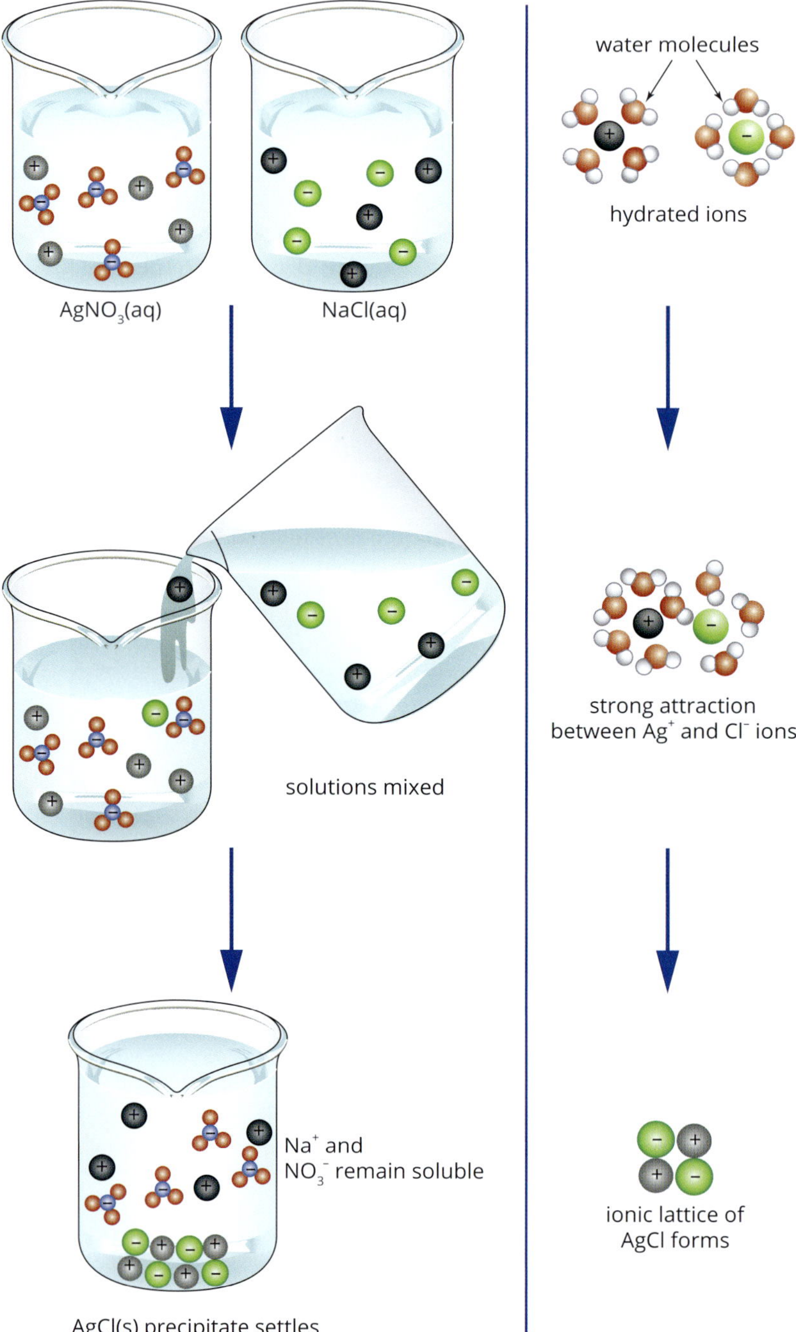

FIGURE 10.5.4 Pictorial representation of mixing aqueous solutions of sodium chloride and silver nitrate to produce a precipitate of silver chloride.

There is a simple way that allows you to work out which compound will form the precipitate in a reaction between two ionic solutions:

1 Write down the formula for the positive ion of one of the compounds, followed by its negative ion. Repeat the process for the second compound. For example: Ag^+, NO_3^-, Na^+, Cl^-.
2 Draw two lines. The first line joins the positive ion of the first solution to the negative ion of the second. The second line joins the negative ion of the first solution to the positive ion of the second (Figure 10.5.5).
3 Use solubility tables to work out which of the two combinations of ions will result in an insoluble compound. This will be the precipitate. The other ions will remain in solution.

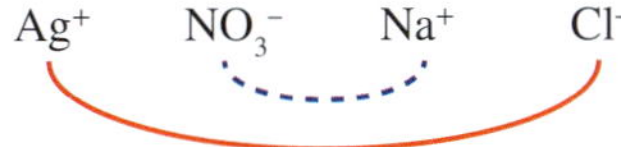

FIGURE 10.5.5 Working out which compound is the precipitate in a reaction.

It is important for chemists to be able to predict whether a precipitate will form in a reaction and what this precipitate will be. Worked example 10.5.1 takes you through the process of predicting the products of a precipitation reaction.

Worked example 10.5.1

PREDICTING THE PRODUCTS OF A PRECIPITATION REACTION

What precipitate, if any, will be produced when solutions of potassium hydroxide and lead(II) nitrate are added together? You will need to refer to the solubility tables (Tables 10.5.1 and 10.5.2 pages 333 and 334) to complete this question.

Thinking	Working
Identify which ions are produced by each of the ionic compounds in the mixture.	$K^+(aq)$, $OH^-(aq)$, $Pb^{2+}(aq)$ and $NO_3^-(aq)$
Identify which two new combinations of positive and negative ions are possible in the mixture of the solutions.	$K^+(aq)$ and $NO_3^-(aq)$ $Pb^{2+}(aq)$ and $OH^-(aq)$
Use the solubility tables to check which, if any, of these combinations will produce an insoluble compound.	Compounds containing potassium ions are soluble, so potassium nitrate will not form a precipitate. Lead(II) hydroxide is insoluble, so lead(II) hydroxide will form a precipitate.

Worked example: Try yourself 10.5.1

PREDICTING THE PRODUCTS OF A PRECIPITATION REACTION

What precipitate, if any, will be produced when solutions of sodium sulfide (Na_2S) and copper(II) nitrate ($Cu(NO_3)_2$) are added together? You will need to refer to the solubility tables (Tables 10.5.1 and 10.5.2 pages 333 and 334) to complete this question.

CHEMFILE S

Limescale accumulation

Have you ever wondered where that flaky white build-up on the element of your kettle comes from?

When you boil water in the kettle, ions present in the water can precipitate out, leaving a white coating called limescale on the element (Figure 10.5.6).

The amount of build-up depends on the type of water treatment in your area. Areas that have hard water (i.e. high levels of calcium and magnesium ions) have a bigger problem with limescale.

Limescale mostly consists of calcium carbonate ($CaCO_3$), which precipitates out as a crystalline solid when the water is boiled. High levels of limescale can build up in pipes and eventually restrict or even block the flow of water (Figure 10.5.7).

Limescale can be a big problem in the home. A coating as thin as 1.5 mm over a heating element can reduce its efficiency by as much as 12%.

Many modern houses use ion filters or water conditioners to remove dissolved ions from the water and reduce the accumulation of limescale.

FIGURE 10.5.6 The accumulation of limescale on domestic kettle elements is the result of the precipitation of calcium carbonate when water is repeatedly boiled.

FIGURE 10.5.7 Accumulation of limescale in pipes is a result of precipitation of $CaCO_3$.

WRITING EQUATIONS FOR PRECIPITATION REACTIONS

Now that you can identify the precipitate that forms in a reaction, the next step is to show the complete reaction by writing a chemical equation.

The reaction between silver nitrate and sodium chloride solutions can be summarised in words as:

silver nitrate + sodium chloride → silver chloride + sodium nitrate
(solution) (solution) (solid) (solution)

An alternative representation is an equation that uses formulae; this type of equation is called a **full equation**. The complete formulae of all the reagents and products are shown in the reaction. The state symbols for each of the species in the chemical reaction must also be shown:

$$AgNO_3(aq) + NaCl(aq) \rightarrow AgCl(s) + NaNO_3(aq)$$

Although 'sodium nitrate' or '$NaNO_3(aq)$' is written as a product in these equations, the sodium ions and nitrate ions are not combined with each other. They move freely through the solution. They are present at the start of the reaction and they are still there, as separate ions, at the end of the reaction. Because the ions have not been involved in forming a precipitate, they are said to be **spectator ions**.

GO TO ➤ Section 6.1 page 206

Spectator ions do not undergo a chemical change in the reaction. In a precipitation reaction, they will always start as aqueous (aq) ions, and they will remain as aqueous ions after the reaction is complete.

Worked example 10.5.2 looks at the process of writing full equations for precipitation reactions and identifying spectator ions.

Worked example 10.5.2

WRITING EQUATIONS FOR PRECIPITATION REACTIONS

Write a full equation for the reaction between iron(III) nitrate and sodium sulfide, in which the precipitate is iron(III) sulfide. Identify the spectator ions in this reaction.

Thinking	Working
Write an incomplete, unbalanced equation showing the reactants and the precipitate product. Include symbols of state.	$Fe(NO_3)_3(aq) + Na_2S(aq) \rightarrow Fe_2S_3(s)$
Add to this equation the formula of the other compound formed in the reaction.	$Fe(NO_3)_3(aq) + Na_2S(aq) \rightarrow Fe_2S_3(s) + NaNO_3(aq)$
Balance the equation.	$2Fe(NO_3)_3(aq) + 3Na_2S(aq) \rightarrow Fe_2S_3(s) + 6NaNO_3(aq)$
Write the formulae of the ions that do not form a precipitate in the reaction. These are the spectator ions.	$Na^+(aq)$ and $NO_3^-(aq)$ are spectator ions.

Worked example: Try yourself 10.5.2

WRITING EQUATIONS FOR PRECIPITATION REACTIONS

Write a full equation for the reaction between copper(II) sulfate and sodium hydroxide, in which the precipitate is copper(II) hydroxide. Identify the spectator ions in this reaction.

Writing ionic equations for precipitation reactions

The essential feature of the reaction between silver nitrate and sodium chloride is the combination of silver ions and chloride ions to form a precipitate. This reaction can be summarised in an **ionic equation**:

$$Ag^+(aq) + Cl^-(aq) \rightarrow AgCl(s)$$

Note that spectator ions are not included in an ionic equation. Only the species that change are included.

A simple way to write an ionic equation for a precipitation reaction is shown in the following example of the reaction between solutions of aluminium nitrate and sodium sulfide. The full equation for the precipitation reaction is:

$$2Al(NO_3)_3(aq) + 3Na_2S(aq) \rightarrow Al_2S_3(s) + 6NaNO_3(aq)$$

To write an ionic equation:

1 Write down the formula of the precipitate on the right-hand side of the page. Include a symbol of state. Place an arrow to the left of it.

$$\rightarrow Al_2S_3(s)$$

2 To the left of this formula, add the formulae of the ions that form the precipitate, in the ratio shown by the formula of the precipitate. Include state symbols.

$$2Al^{3+}(aq) + 3S^{2-}(aq) \rightarrow Al_2S_3(s)$$

3 Check that the equation is balanced.

SOLUBILITY TABLES AND RULES

It is important to have ways of knowing which ionic solids are soluble in water and which ones are not. Different forms of solubility tables are used for this purpose.

The SNAAP rule

A handy way to remember many of the soluble salts is with the initials SNAAP. Salts that contain one or more of the following ions are soluble:

- **S**odium (Na^+)
- **N**itrate (NO_3^-)
- **A**mmonium (NH_4^+)
- **A**cetate (CH_3COO^-)
- **P**otassium (K^+)

Solubility tables

A **solubility table** can be used to determine whether common ionic compounds are soluble in water. **Anions** and **cations** are listed with indications of their solubility in different compounds. Table 10.5.1 is a sample solubility table listing ionic compounds that are soluble in water. Note that for some ions there are exceptions. Insoluble compounds are given in Table 10.5.2.

TABLE 10.5.1 Relative solubilities of soluble ionic compounds

Soluble in water (>0.1 mol dissolves per L at 25°C)	Exceptions: insoluble (<0.1 mol dissolves per L at 25°C)	Exceptions: slightly soluble (0.01–0.1 mol dissolves per L at 25°C)
most chlorides (Cl^-), bromides (Br^-) and iodides (I^-)	AgCl, AgBr, AgI, PbI_2	$PbCl_2$, $PbBr_2$
all nitrates (NO_3^-)	no exceptions	no exceptions
all ammonium (NH_4^+) salts	no exceptions	no exceptions
all sodium (Na^+) and potassium (K^+) salts	no exceptions	no exceptions
all acetates (CH_3COO^-)	no exceptions	no exceptions
most sulfates (SO_4^{2-})	$SrSO_4$, $BaSO_4$, $PbSO_4$	$CaSO_4$, Ag_2SO_4

TABLE 10.5.2 Relative solubilities of insoluble ionic compounds

Insoluble in water (>0.1 mol dissolves per L at 25°C)	Exceptions: soluble (<0.1 mol dissolves per L at 25°C)	Exceptions: slightly soluble (0.01–0.1 mol dissolves per L at 25°C)
most hydroxides (OH^-)	$NaOH$, KOH, $Ba(OH)_2$, NH_4OH*, $AgOH$†	$Ca(OH)_2$, $Sr(OH)_2$
most carbonates (CO_3^{2-})	Na_2CO_3, K_2CO_3, $(NH_4)_2CO_3$	no exceptions
most phosphates (PO_4^{3-})	Na_3PO_4, K_3PO_4, $(NH_4)_3PO_4$	no exceptions
most sulfides (S^{2-})	Na_2S, K_2S, $(NH_4)_2S$	no exceptions

*NH_4OH does not exist in significant amounts in an ammonia solution. Ammonium and hydroxide ions readily combine to form ammonia and water.

†AgOH readily decomposes to form a precipitate of silver oxide and water.

Using solubility tables to predict solubility

The solubility tables (Tables 10.5.1 and 10.5.2) can be used to predict whether a particular ionic substance will be soluble, slightly soluble or insoluble in water.

Worked example 10.5.3 shows you a systematic way of using solubility tables to determine whether an ionic compound will be soluble or insoluble in water.

Worked example 10.5.3

DETERMINING WHETHER IONIC COMPOUNDS ARE SOLUBLE OR INSOLUBLE IN WATER

Is barium sulfide (BaS) soluble or insoluble in water? You will need to refer to the solubility tables to complete this question (Tables 10.5.1 and 10.5.2 on page 333 and 334).

Thinking	Working
Identify the ions that are present in the ionic compound.	Barium (Ba^{2+}) and sulfide (S^{2-})
Check the solubility tables to see if compounds containing the cation are usually soluble or insoluble in water.	Barium ions do not appear in the table. This is no help, so look for the other ion.
Check the solubility tables to see if compounds containing the anion are usually soluble or insoluble in water.	Compounds containing sulfide ions are usually insoluble in water. So, barium sulfide will be insoluble in water.

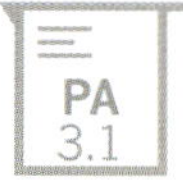

Worked example: Try yourself 10.5.3

DETERMINING WHETHER IONIC COMPOUNDS ARE SOLUBLE OR INSOLUBLE IN WATER

Is ammonium phosphate ($(NH_4)_3PO_4$) soluble or insoluble in water? You will need to refer to the solubility tables to complete this question (Tables 10.5.1 and 10.5.2 on pages 333 and 334).

10.5 Review

SUMMARY

- A precipitation reaction occurs when two solutions of compounds are mixed and a solid product is formed. The solid product is called a precipitate.
- The SNAAP rule can be used to predict the solubility of an ionic compound.
- Solubility tables can be used to predict which compound, if any, will precipitate in a reaction.
- Ions that are not directly involved in the formation of the precipitate are called spectator ions.
- Full and ionic equations can be written for precipitation reactions.
- Ionic equations do not include spectator ions.

KEY QUESTIONS

1 Which of the following ionic compounds is insoluble in water?
   A calcium nitrate
   B calcium carbonate
   C calcium chloride
   D calcium acetate

2 Which of the following substances would you expect to be soluble in water? Refer to Tables 10.5.1 and 10.5.2 (pages 333 and 334) to complete this question.
   A sodium carbonate
   B lead(II) nitrate
   C magnesium carbonate
   D ammonium sulfate
   E iron(II) sulfate
   F magnesium phosphate
   G zinc carbonate
   H sodium sulfide
   I silver chloride
   J barium sulfate

3 Which of the following compounds would you expect to be insoluble in water? Refer to Tables 10.5.1 and 10.5.2 (pages 333 and 334) to complete this question.
   A silver carbonate
   B zinc nitrate
   C copper(II) carbonate
   D silver chloride
   E lead(II) bromide
   F magnesium hydroxide
   G barium nitrate
   H aluminium sulfide

4 Use the solubility tables to identify the precipitate formed, if any, when the following solutions are mixed.
   a silver nitrate and potassium carbonate
   b potassium hydroxide and lead(II) nitrate
   c magnesium chloride and sodium sulfide
   d sodium chloride and iron(II) sulfate

5 **a** Name the precipitate formed when aqueous solutions of the following compounds are mixed.
      i K_2S and $MgCl_2$
      ii $CuCl_2$ and $AgNO_3$
      iii KOH and $AlCl_3$
      iv $MgSO_4$ and NaOH
   b Write a fully balanced chemical equation for each reaction.

6 For each of the following precipitation reactions, write:
      i a full equation
      ii an ionic equation.
   a $AgNO_3(aq) + NaCl(aq) \rightarrow$
   b $CuSO_4(aq) + Na_2CO_3(aq) \rightarrow$
   c $(NH_4)_2SO_4(aq) + BaCl_2(aq) \rightarrow$
   d $K_2S(aq) + Pb(NO_3)_2(aq) \rightarrow$
   e $CaCl_2(aq) + Na_3PO_4(aq) \rightarrow$
   f $NaOH(aq) + Pb(NO_3)_2(aq) \rightarrow$

7 For each of the reactions in Question **6**, identify the spectator ions.

10.6 Reactions of acids and bases

Acids were originally grouped together because of their similar chemical behaviour. Chemists use indicators, such as litmus, to identify acidic solutions. Acids and bases react readily with many other chemicals, and some of the early definitions of acids and bases were derived from these reactions.

In this section, you will learn to use the patterns in the reactions of acids and bases to predict the products that are formed.

GENERAL REACTION TYPES INVOLVING ACIDS AND BASES

Acids and bases react with many substances. However, it is possible to group some reactions together according to the similarity of the reactants involved and the products formed. While the identification of products should be based on experimental data, these groups, or reaction types, can be useful. The reaction types you will be studying are the reactions of acids with reactive metals and with the following bases:

- metal hydroxides
- metal carbonates and
- metal hydrogen carbonates.

Common acids used in school laboratories are:

- hydrochloric acid (HCl)
- sulfuric acid (H_2SO_4)
- nitric acid (HNO_3)
- acetic acid (CH_3COOH).

ACIDS AND REACTIVE METALS

When dilute acids are added to main-group metals and some transition metals, bubbles of hydrogen gas are released and a salt is formed. In general, the equation for the reaction is:

$$\text{acid} + \text{reactive metal} \rightarrow \text{salt} + \text{hydrogen}$$

Reactive metals include calcium, magnesium, iron and zinc, but not copper, silver or gold. For example, the reaction between dilute hydrochloric acid and zinc metal can be seen in Figure 10.6.1 and represented by the equation:

$$2HCl(aq) + Zn(s) \rightarrow ZnCl_2(aq) + H_2(g)$$

This reaction can also be represented by an ionic equation. In an aqueous solution, the hydrochloric acid and the salt, zinc chloride (a soluble ionic compound), are **dissociated**. The ionic equation can be determined as shown in Worked example 10.6.1.

FIGURE 10.6.1 Hydrogen gas is produced from the reaction of zinc with dilute hydrochloric acid.

Worked example 10.6.1

WRITING IONIC EQUATIONS FOR REACTIONS BETWEEN ACIDS AND REACTIVE METALS

Write an ionic equation for the reaction that occurs when dilute hydrochloric acid is added to a sample of zinc metal.

Thinking	Working
Write the general reaction. Identify the products formed.	acid + reactive metal → salt + hydrogen Hydrogen gas and zinc chloride solution are produced.
Identify the reactants and products. Indicate the state of each, i.e. (aq), (s), (l) or (g).	Reactants: zinc is a solid, Zn(s). Hydrochloric acid is dissociated, forming $H^+(aq)$ and $Cl^-(aq)$ ions. Products: hydrogen gas, $H_2(g)$. Zinc chloride is dissociated into $Zn^{2+}(aq)$ and $Cl^-(aq)$ ions.
Write the equation showing all reactants and products. (There is no need to balance the equation yet.)	$H^+(aq) + Cl^-(aq) + Zn(s) \rightarrow Zn^{2+}(aq) + Cl^-(aq) + H_2(g)$
Identify the spectator ions.	$Cl^-(aq)$
Rewrite the equation without the spectator ions. Balance the equation with respect to number of atoms of each element and charge.	$2H^+(aq) + Zn(s) \rightarrow Zn^{2+}(aq) + H_2(g)$

Worked example: Try yourself 10.6.1

WRITING IONIC EQUATIONS FOR REACTIONS BETWEEN ACIDS AND REACTIVE METALS

Write an ionic equation for the reaction that occurs when aluminium is added to a dilute solution of hydrochloric acid.

ACIDS AND METAL HYDROXIDES

Soluble metal hydroxides, such as NaOH, dissociate in water to form metal cations and hydroxide ions, $OH^-(aq)$. The products of the reaction of an acid with a metal hydroxide are an ionic compound, called a **salt**, and water.

The general rule for the reaction between acids and metal hydroxides can be expressed as:

$$\text{acid} + \text{metal hydroxide} \rightarrow \text{salt} + \text{water}$$

For example, solutions of sulfuric acid and sodium hydroxide react to form sodium sulfate and water. This can be represented by the equation:

$$H_2SO_4(aq) + 2NaOH(aq) \rightarrow Na_2SO_4(aq) + 2H_2O(l)$$

The salt formed in the reaction between sulfuric acid and sodium hydroxide is sodium sulfate. If water was evaporated from the reaction mixture, solid sodium sulfate would be left behind.

Salts

Salts consist of the positive ion or cation from the base and the negative ion or anion from the acid. In Chapter 5, you saw that the names of anions that contain oxygen often end in '-ate': for example, sulfate, phosphate. The names of anions that do not contain oxygen end in '-ide': for example, chloride, bromide.

GO TO ➤ Section 5.2 page 147

Table 10.6.1 lists the names of salts formed from some **neutralisation reactions** of acids with metal hydroxides. Note that the name of the positive ion is listed first, and the name of the negative ion from the acid is listed second.

TABLE 10.6.1 Salts formed from some common neutralisation reactions

Reactants (acid + metal hydroxide)	Name of salt formed	Formulae of ions present in the salt solution
hydrochloric acid + potassium hydroxide	potassium chloride	$K^+(aq) + 2Cl^-(aq)$
hydrochloric acid + magnesium hydroxide	magnesium chloride	$Mg^{2+}(aq) + 2Cl^-(aq)$
nitric acid + sodium hydroxide	sodium nitrate	$Na^+(aq) + NO_3^-(aq)$
sulfuric acid + zinc hydroxide	zinc sulfate	$Zn^{2+}(aq) + SO_4^{2-}(aq)$
phosphoric acid + potassium hydroxide	potassium phosphate	$3K^+(aq) + PO_4^{3-}(aq)$
acetic acid + calcium hydroxide	calcium acetate	$Ca^{2+}(aq) + 2CH_3COO^-(aq)$

Neutralisation reactions

If a solution of a metal hydroxide is added to a solution of an acid, the hydroxide ions will react with the hydrogen ions. The acid and base are said to have been **neutralised** at the point when all the hydroxide ions have reacted with all the hydrogen ions, forming water (H_2O).

Ionic equations

The hydroxide ions from metal hydroxides, such as sodium hydroxide (NaOH), calcium hydroxide ($Ca(OH)_2$) and magnesium hydroxide ($Mg(OH)_2$), react readily with the hydrogen ion ($H^+(aq)$) from acids.

The reaction between an acid and a metal hydroxide can be represented by an **ionic equation** as well as by an overall equation.

When writing ionic equations for neutralisation reactions, you need to remember that:

- strong acids such as HCl, H_2SO_4 and HNO_3 dissociate completely in solution and are written as ions: for example, HCl in solution is written as $H^+(aq) + Cl^-(aq)$.
- metal hydroxides and salts are ionic and, if soluble, dissociate in solution and are written as ions: for example, KOH dissolving in water is written as

$$KOH(s) \rightarrow K^+(aq) + OH^-(aq)$$

- water is written as $H_2O(l)$.

Figure 10.6.2 is a diagrammatic representation of the ions in solutions of HCl and NaOH when mixed in a neutralisation reaction.

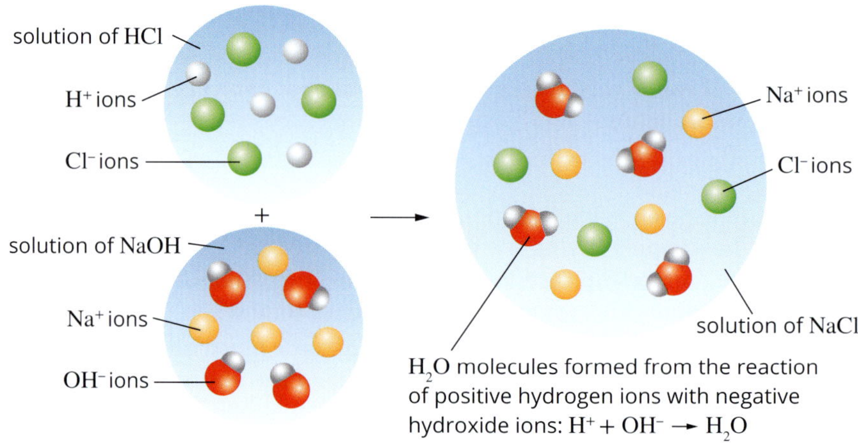

FIGURE 10.6.2 A representation of the reaction between $H^+(aq)$ and $OH^-(aq)$ ions that occurs when solutions of HCl and NaOH are mixed.

Worked example 10.6.2 indicates the steps to follow when writing ionic equations for the reactions of acids.

Worked example 10.6.2

WRITING AN IONIC EQUATION FOR AN ACID BASE REACTION

Write an ionic equation for the reaction that occurs when hydrochloric acid is added to sodium hydroxide solution. A representation of this reaction is shown in Figure 10.6.2.

Thinking	Working
Write the general reaction. Identify the products formed.	acid + metal hydroxide → salt + water A solution of sodium chloride and water is formed.
Identify the reactants and products. Indicate the state of each, i.e. (aq), (s), (l) or (g).	Reactants: HCl(aq) is dissociated in solution, forming $H^+(aq)$ and $Cl^-(aq)$. NaOH(aq) is dissociated in solution, forming $Na^+(aq)$ and $OH^-(aq)$. Products: sodium chloride is dissociated and exists as $Na^+(aq)$ and $Cl^-(aq)$. Water is a molecular compound and its formula is $H_2O(l)$.
Write the equation showing all reactants and products, in ionised form where possible. (There is no need to balance the equation yet.)	$H^+(aq) + Cl^-(aq) + Na^+(aq) + OH^-(aq) \rightarrow Na^+(aq) + Cl^-(aq) + H_2O(l)$
Identify the spectator ions: the ions that have an (aq) state both as a reactant and as a product.	$Na^+(aq)$ and $Cl^-(aq)$
Rewrite the equation without the spectator ions. Balance the equation with respect to the number of atoms of each element and charge.	$H^+(aq) + OH^-(aq) \rightarrow H_2O(l)$

Worked example: Try yourself 10.6.2

WRITING AN IONIC EQUATION FOR AN ACID BASE REACTION

Write an ionic equation for the reaction that occurs when sulfuric acid is added to potassium hydroxide solution.

CHEMISTRY IN ACTION PSC

Benefits of neutralisation

A neutralisation reaction enables the adjustment of the acidity of a solution. If excess acid is harmful, it can be neutralised by adding a base. Conversely, an environment that is too basic can be neutralised by adding an acid. Common examples of this include:

- Methanoic acid (otherwise known as formic acid) is released from the sting of ants, bees and nettles (Figure 10.6.3). If affected skin is rinsed with limewater or a dilute solution of ammonia, the acid becomes neutralised and is no longer painful to the person affected.
- The venom from wasps is basic. A common treatment for wasp bites is to rinse the site of the bite with vinegar (acetic acid, CH_3COOH) because this neutralises the base within the venom.
- Excess acid in the stomach is the cause of indigestion. This condition is treated with substances that neutralise acids. Antacid tablets are bases that neutralise the excess acid in the stomach.
- The bacteria occurring on tooth enamel feed on the sugars present in food. The products of their metabolism are acids that attack the enamel, and this leads to tooth decay. This is why toothpastes (Figure 10.6.4) are weak bases.

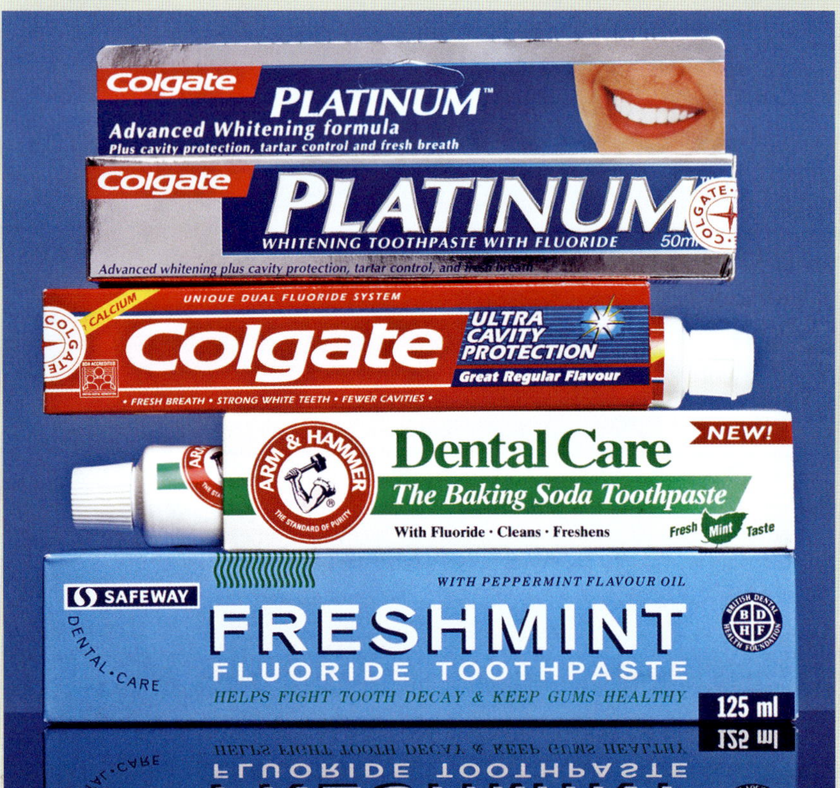

FIGURE 10.6.4 A selection of toothpastes. These products are advertised as providing protection against tooth cavities, gum disease and tooth decay.

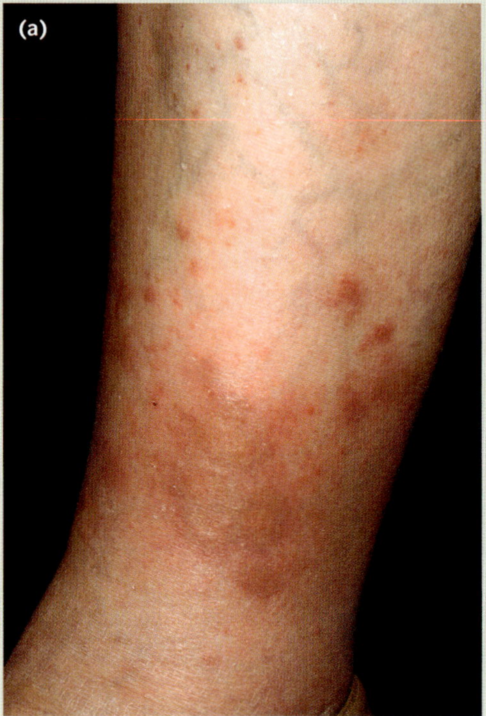

FIGURE 10.6.3 (a) Irritation on the leg of a person bitten by ants. (b) Scanning electron microscope image of a nettle (*Urtica* sp.) leaf surface, showing stinging, hair-like structures (colourised).

ACIDS AND METAL CARBONATES

The weathering of buildings and statues (Figure 10.6.5) is due in part to the reaction between acid rain and the carbonate minerals in the stone.

FIGURE 10.6.5 This statue has been heavily eroded by acid rain, which reacts with carbonate salts in limestone. Acid rain is formed when gases, such as sulfur dioxide and nitrogen oxides, dissolve in atmospheric water to form acidic solutions.

Acids reacting with metal carbonates and metal hydrogen carbonates produce carbon dioxide gas, together with a salt and water. Metal carbonates include sodium carbonate (Na_2CO_3), magnesium carbonate ($MgCO_3$) and calcium carbonate ($CaCO_3$).

The general reaction for metal carbonates with acids can be summarised as:

acid + metal carbonate → salt + water + carbon dioxide

For example, a solution of hydrochloric acid reacting with sodium carbonate solution produces a solution of sodium chloride, water and carbon dioxide gas. The reaction is represented by the equation:

$$2HCl(aq) + Na_2CO_3(aq) \rightarrow 2NaCl(aq) + CO_2(g) + H_2O(l)$$

The reaction between hydrochloric acid and sodium carbonate is represented in Figure 10.6.6.

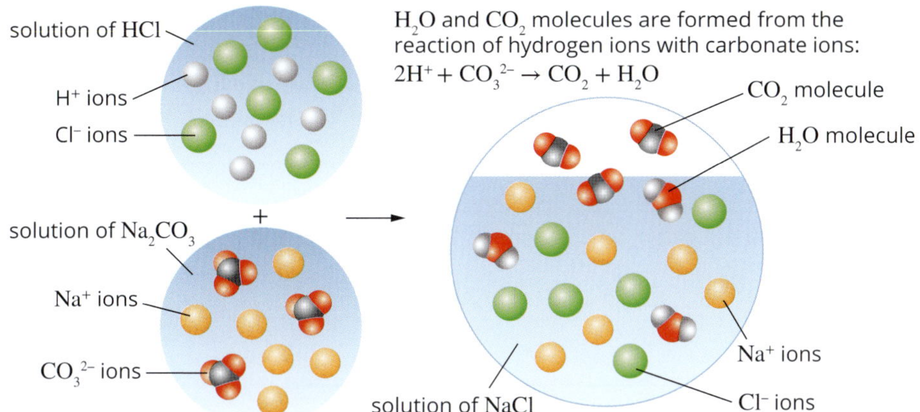

FIGURE 10.6.6 The reaction between solutions of sodium carbonate and hydrochloric acid.

Metal hydrogen carbonates (also known as bicarbonates) include sodium hydrogen carbonate ($NaHCO_3$), potassium hydrogen carbonate ($KHCO_3$) and calcium hydrogen carbonate ($Ca(HCO_3)_2$). Acids added to metal hydrogen carbonates also produce carbon dioxide, together with a salt and water. The general reaction is:

acid + metal hydrogen carbonate → salt + water + carbon dioxide

For example, when solutions of hydrochloric acid and sodium hydrogen carbonate are mixed, the following reaction occurs:

$$HCl(aq) + NaHCO_3(s) \rightarrow NaCl(aq) + H_2O(l) + CO_2(g)$$

The reactions between acids and metal carbonates, and the reactions between acids and metal hydrogen carbonates, can also be represented by ionic equations, following steps similar to the steps for writing reactions between acids and bases.

CHEMFILE PSC

Bicarbonate of soda

Self-raising flour contains tartaric acid and some sodium hydrogen carbonate (bicarbonate of soda) (Figure 10.6.7). It is used in baking cakes because on heating in the oven, the acid and hydrogen carbonate react. Carbon dioxide is released, which causes the cake mixture to rise.

FIGURE 10.6.7 In self-raising flour, bicarbonate of soda reacts with tartaric acid to produce carbon dioxide, and that reaction has caused these muffins to rise.

Worked example 10.6.3

WRITING IONIC EQUATIONS FOR REACTIONS BETWEEN ACIDS AND METAL CARBONATES

What products are formed when a dilute solution of nitric acid is added to solid magnesium carbonate? Write an ionic equation for this reaction.

Thinking	Working
Write the general reaction. Identify the products.	acid + metal carbonate → salt + water + carbon dioxide The products of this reaction are magnesium nitrate in solution, water and carbon dioxide gas.
Identify the reactants and products. Indicate the state of each, i.e. (aq), (s), (l) or (g).	Reactants: nitric acid is dissociated in solution, forming $H^+(aq)$ and $NO_3^-(aq)$ ions. Magnesium carbonate is an ionic solid, $MgCO_3(s)$. Products: magnesium nitrate is dissociated into $Mg^{2+}(aq)$ and NO_3^- (aq) ions. Water has the formula $H_2O(l)$. Carbon dioxide has the formula $CO_2(g)$.
Write the equation showing all reactants and products. (There is no need to balance the equation yet.)	$H^+(aq) + NO_3^-(aq) + MgCO_3(s) \rightarrow Mg^{2+}(aq) + NO_3^-(aq) + H_2O(l) + CO_2(g)$
Identify the spectator ions.	$NO_3^-(aq)$
Rewrite the equation without the spectator ions. Balance the equation with respect to the number of atoms of each element and charge.	The balanced equation with the spectator ions omitted is: $2H^+(aq) + MgCO_3(s) \rightarrow Mg^{2+}(aq) + H_2O^+(l) + CO_2(g)$

FIGURE 10.6.8 Limewater test. Carbon dioxide is bubbled through limewater, turning the limewater cloudy.

Worked example: Try yourself 10.6.3

WRITING IONIC EQUATIONS FOR REACTIONS BETWEEN ACIDS AND METAL CARBONATES

What products are formed when a solution of hydrochloric acid is added to a solution of sodium carbonate? Write an ionic equation for this reaction.

Testing for carbonate salts

Acids can be used to detect the presence of carbonate salts. Carbon dioxide is produced when an acid is added to a carbonate.

The **limewater test** is a simple laboratory test used to confirm the presence of carbon dioxide gas. Limewater is a saturated solution of calcium hydroxide ($Ca(OH)_2$). When carbon dioxide is bubbled through this solution, it will turn 'milky' or 'cloudy' in appearance (Figure 10.6.8) due to the precipitation of calcium carbonate:

$$Ca(OH)_2(aq) + CO_2(g) \rightarrow CaCO_3(s) + H_2O(l)$$

10.6 Review

SUMMARY

- Generalisations can be made about the likely products of reactions involving acids and bases:
  - acid + reactive metal → salt + hydrogen
  - acid + metal hydroxide → salt + water
  - acid + metal carbonate → salt + water + carbon dioxide
  - acid + metal hydrogen carbonate → salt + water + carbon dioxide
- Each of these reactions can be represented by full and ionic equations.
- An ionic equation only shows those ions, atoms or molecules that take part in the reaction.
- Spectator ions (ions that do not take part in the reaction) are not included in ionic equations.
- Ionic equations are balanced with respect to both the number of atoms of each element and the charge.

KEY QUESTIONS

1 The venom from a wasp is alkaline. Which of the following substances will relieve the pain of a wasp sting by neutralising the venom?

A water
B concentrated sodium hydroxide solution
C vinegar (acetic acid)
D olive oil

2 Write full and ionic chemical equations for the reactions between:

a magnesium and sulfuric acid
b calcium and hydrochloric acid
c zinc and acetic acid
d aluminium and hydrochloric acid.

3 Name the salt produced in each of the reactions in Question **2**.

4 For each of the following reactions write:

i a full equation to represent the reaction (remember to include states)
ii an ionic equation.

a solid zinc oxide and sulfuric acid
b solid calcium and nitric acid
c solid copper(II) hydroxide and nitric acid
d solid magnesium hydrogen carbonate and hydrochloric acid
e solid tin(II) carbonate and sulfuric acid

5 Predict the products of the following reactions and write full and ionic chemical equations for each.

a A solution of sulfuric acid is added to a solution of potassium hydroxide.
b Nitric acid solution is mixed with sodium hydroxide solution.
c Hydrochloric acid solution is poured on to some solid magnesium hydroxide.
d Blue copper(II) carbonate powder is added to dilute sulfuric acid.
e Dilute hydrofluoric acid (HF) is mixed with a solution of potassium hydrogen carbonate.
f Dilute nitric acid is added to a spoon coated with solid zinc.
g Hydrochloric acid solution is added to some marble chips (calcium carbonate).
h Solid bicarbonate of soda (sodium hydrogen carbonate) is mixed with vinegar (a solution of acetic acid).

6 The products of a reaction between a metal and an acid are zinc chloride solution and hydrogen gas.

a Name the reactants used.
b Write a full equation for the reaction occurring.

7 The salt lithium sulfate is formed in a neutralisation reaction.

a Name the reactants used.
b Write a full equation for the reaction.

10.7 Removing toxins from food

Carrying out chemical reactions is not limited to modern chemists. Chemical processes such as the fermentation of alcohol, the manufacture of iron and the production of glass have been in use for many centuries. In this section, you will read about examples of chemical processes that Indigenous Australians have employed for many years to purify and prepare foods.

CHEMFILE AHC

Tea-tree

One of the first plants of interest to the landing party of Lieutenant James Cook in 1770 was a particular tree with sticky, aromatic leaves (Figure 10.7.1). The English visitors called it 'tea-tree' because they liked the spicy tea that could be brewed from its leaves. The visitors also noticed how the Indigenous Dharawal people used this oil for a range of medicinal purposes. Recent studies rate tea-tree oil to be the most powerful natural antiseptic obtained from any vegetation.

FIGURE 10.7.1 An illustration of a tea-tree plant published in 1777 after Lieutenant Cook returned to England.

INDIGENOUS FOOD PREPARATION

When European colonists settled in Australia in the late 18th century, they considered the Indigenous people to be very primitive. It became apparent, however, that Indigenous Australians were skilled practitioners of bush medicines, ink production and food preparation.

Cycad

The fruit of the cycad (*Macrozamia*) was exploited as an important food source by Indigenous peoples on the east coast, despite it being highly toxic and carcinogenic (Figure 10.7.2a). Indigenous Australians developed two different methods for removing the toxins to allow the fruit to be consumed safely.

- **Leaching**: The kernels are cut open, ground (increasing the surface area) and allowed to soak in water. The toxins are relatively soluble and can wash out of the fruit. The fruit can then be ground further and baked into a type of bread.
- **Fermentation**: The kernels are stored for months in a moist environment that allows them to ferment. They are no longer toxic after this process.

The existence of stands of cycads did not happen by chance. Indigenous Australians deliberately burnt the cycads regularly to increase seed production and ripening (Figure 10.7.2b).

FIGURE 10.7.2 (a) Cone and seeds of the cycad. (b) Fire was used to increase the rate of seeding of cycads.

The example of the cycad demonstrates how Indigenous Australians

- used heat to kill germs, destroy toxins or improve texture
- used water to leach soluble toxins from food
- ground food to facilitate the removal of impurities
- grated vegetables on rough bark to remove the skin.

Indigenous Australians also removed impurities by placing foods in dilly bags and putting them in streams, or pouring water over the food to strain out toxins.

CHEMISTRY IN ACTION AHC

Scientists research Indigenous Australian diet

Figure 10.7.3 shows a very interesting headline from the *Cairns Post*.

Scientists eat poisonous rainforest nuts at Atherton near Cairns for Aboriginal diet insight

FIGURE 10.7.3 Headline from the *Cairns Post*.

Since 2014, La Trobe University researchers, with assistance from local elders, have been researching the diets of Indigenous people living in the Atherton Tablelands rainforests. Part of their investigation has been to eat poisonous nuts and seeds, but not before the nuts have been through traditional purification processes.

The black beans, cycads and yellow walnuts trialled by the researchers are considered to be significantly toxic. However, a lengthy process of cooking, scraping and leaching in local water supplies is being found to be successful in removing toxins. Cooking the ground up food leads to an increase in the levels of digestible starch in most of the nuts.

No food additives

Indigenous Australians modified their food less than European settlers did, and they used very few if any food additives. Figure 10.7.4 shows the label on a typical modern food bar. The list of ingredients includes many additives as well as the natural components of the product. Nutritionists regularly express concerns about the impact on the body of these additives. Indigenous Australian diets, however, consisted of meat and vegetables, with absolutely no additives. As such, the possible impact of food additives was not an issue.

Nutrition information (average)

Servings per package: 6

Average serving size: 22g (1 bar)

	quantity per serving	% daily intake ▲ per serving	quantity per 100g
Energy	380 kJ	**4%**	1720 kJ
Protein	0.9 g	**2%**	4.1 g
Fat, total	2.2 g	**3%**	9.9 g
- Saturated	1.4 g	**6%**	6.4 g
Carbohydrate	16.5 g	**5%**	75.2 g
- Sugars	6.9 g	**8%**	31.4 g
Dietary fibre	0.2 g	**0.6%**	0.8 g
Sodium	56 mg	**2%**	255 mg

Ingredients

Puffed rice (36%) (whole white rice, sugar, salt, barley malt extract), glucose, choc chips (12%) (sugar, vegetable fat, milk solids, cocoa powder, emulsifiers [soy lecithin, 476, 492], flavour, salt), fructose, hydrogenated soyabean oil (antioxidants [304, 307b, 320]), sugar, glucose solids, invert sugar, humectant (glycerol), gelatin, natural flavour, salt, emulsifiers (472e, 472a), skim milk powder.

FIGURE 10.7.4 Modern foods include a variety of additives that were not present in the diet of Indigenous Australians.

10.7 Review

SUMMARY

- Indigenous Australians have long utilised a range of chemical processes to remove toxins from native nuts and berries.
- These techniques have included leaching, fermentation, grinding, heating and grating.

KEY QUESTIONS

1 Select the correct explanation as to why Indigenous Australians can often consume native foods that are considered toxic to humans.

A The foods have been incorrectly designated as toxic.

B Indigenous Australians have a digestive system that can tolerate a range of toxins.

C Indigenous Australians have developed processes that can remove the toxins.

D Cooking foods always destroys the toxins in them.

2 Indigenous Australians used more than one technique to remove toxins from foods. Suggest an explanation as to why each of the following processes might be effective in removing toxins:

a grinding food to a powder

b sitting the food in a dilly bag in water for several days

c cooking the food.

3 Indigenous Australians often set fire to cycad bushes. Explain why they might deliberately damage a reliable food source.

4 Coffee beans contain caffeine, an addictive substance that some people choose to avoid. Caffeine is a partially polar molecule, so is soluble in both water and non-polar solvents. Suggest a possible way to remove caffeine from coffee beans.

Chapter review

10

KEY TERMS

anion
biodegradation
cation
chemical change
chemical equation
chemical reaction
combustion
complete combustion
decomposition
dissociate
electrolysis
fermentation
full equation
hydrocarbon
incomplete combustion
ionic equation
law of conservation of mass
leaching
limewater test
neutralisation reaction
neutralise
oxidation
photolysis
physical change
precipitate
precipitation reaction
salt
solubility table
spectator ion
synthesis
synthesis reaction
thermal decomposition

REVIEW QUESTIONS

1 Suggest a sign of a chemical change for each of the following processes:

a ripening of fruit
b milk that is past its use-by date
c raw egg added to vinegar
d starting a car engine
e an air-bag triggered in a car collision.

2 Three substances undergo changes when treated as described in the following table:

Substance	Treatment	Observation
gold nuggets	heated to 1400°C in a crucible	Gold turns to a liquid and is poured into a mould to form a gold bar.
mercury(II) oxide	heated strongly	A silver liquid is formed and a gas is evolved.
aluminium	heated strongly in a flame	Metal burns, producing a white powder.

Identify each change as a physical change or a chemical change. Describe what you think is happening to the substance.

3 **a** Give one example of a chemical change that occurs during a car trip.
b Give one example of a chemical change that occurs during the cooking of food.

4 Summarise in words the information available from the following equation:

$$4Al(s) + 3O_2(g) \rightarrow 2Al_2O_3(s)$$

5 For each of the following reactions, write a word equation and then a balanced chemical equation.

a Magnesium metal can react with sulfur powder to form solid magnesium sulfide.
b Zinc metal reacts with copper(II) sulfate solution to form copper metal and zinc sulfate solution.
c Solid magnesium oxide and carbon dioxide gas are produced when solid magnesium carbonate is heated.

6 Which alternative represents a synthesis reaction?

A $X + Y \rightarrow Z$
B $Z \rightarrow X + Y$
C $A + B \rightarrow C + D$
D $A + B \rightarrow C + D + E$

7 Nitrogen gas in the air can combine with oxygen gas to form nitrogen monoxide gas (NO) when petrol burns in a car engine. The nitrogen monoxide can react with more oxygen gas to form nitrogen dioxide gas (NO_2).

a Write an equation in words and then a balanced chemical equation for the formation of NO.
b Write an equation in words and then a balanced chemical equation for the formation of NO_2.

8 When solid potassium chlorate ($KClO_3$) is heated, it decomposes to solid potassium chloride and oxygen gas.

a What is the name given to decomposition reactions that are caused by heat energy?
b Write a balanced chemical equation for the decomposition of potassium chlorate.

9 Molten aluminium metal is manufactured by passing an electric current through molten aluminium oxide (Al_2O_3). The other product is oxygen gas.

a What is the name given to decomposition reactions that are caused by electrical energy?
b Write a balanced chemical equation for the decomposition of molten aluminium oxide.

10 When light hits solid silver chloride (AgCl), it decomposes into the elements it is made from.

a What is the name given to decomposition reactions that are caused by light energy?
b Write a balanced chemical equation for the decomposition of silver chloride.

11 Identify the correct statement about combustion reactions.

A Combustion reactions will always have two products only.

B Combustion reactions will always have one product only.

C There is an overall release of energy during combustion reactions.

D There is an overall absorption of energy during combustion reactions.

12 Natural gas is mainly methane (CH_4).

a Write a word equation for the complete combustion of methane.

b Write a balanced chemical equation for the complete combustion of methane.

c Write a balanced chemical equation for the incomplete combustion of methane to form carbon monoxide and water.

d Methane is a common fuel for a Bunsen burner. Explain how you can change the level of combustion when using a Bunsen burner.

13 The photo below shows a modern wood heater, which has a flue (chimney), side-opening door and an air intake valve. Heat is produced in this type of heater from the combustion of wood.

a Explain how the level of combustion is controlled in this type of heater.

b What are the disadvantages and advantages to the home owner of running the heater on a setting that limits the degree of combustion?

c Two heavy logs are added to a stove one night, and the next morning all that is left is a pile of light ash. Explain how the law of conservation of mass still holds in this situation.

14 Consider the reaction represented by the following equation:

$NaCl(aq) + AgNO_3(aq) \rightarrow AgCl(s) + NaNO_3(aq)$

Indicate whether the following statements about the reaction and equation are true or false.

a The equation represents a precipitation reaction.

b Ionic compounds that contain sodium or nitrate ions are always soluble.

c The precipitate in this reaction is $AgNO_3$.

d The equation is an example of an ionic equation.

e Silver chloride is insoluble in water.

f The sodium and chlorine ions can be described as spectator ions.

15 Copy and complete the following table. Identify which reaction mixtures will produce precipitates and write their formulae.

	NaOH	KBr	NaI	$MgSO_4$	$BaCl_2$
$Pb(NO_3)_2$					
KI					
$CaCl_2$					
Na_2CO_3					
Na_2S					

16 Consider the following ions: SO_4^{2-}, PO_4^{3-}, Br^- and S^{2-}. Which of these ions would combine with $Fe^{2+}(aq)$ to give a precipitate?

17 What precipitate will be formed (if any) when the following solutions are mixed?

a barium nitrate and sodium sulfate

b sodium chloride and copper(II) sulfate

c magnesium sulfate and lead(II) nitrate

d potassium chloride and barium nitrate

18 Write balanced chemical equations and ionic equations for each of the following precipitation reactions:

a $NH_4Cl(aq) + AgNO_3(aq) \rightarrow$

b $FeCl_2(aq) + Na_2S(aq) \rightarrow$

c $Fe(NO_3)_3(aq) + KOH(aq) \rightarrow$

d $CuSO_4(aq) + NaOH(aq) \rightarrow$

e $Ba(NO_3)_2(aq) + Na_2SO_4(aq) \rightarrow$

19 Write a balanced chemical equation and ionic equation for the reaction that takes place when solutions of the following compounds are mixed. In each case, name the spectator ions. Refer to the solubility tables on pages 333 and 334 to help identify the precipitate for each reaction.

a copper(II) sulfate and sodium carbonate

b silver nitrate and potassium chloride

c sodium sulfide and lead(II) nitrate

d iron(III) chloride and sodium hydroxide

e iron(III) sulfate and potassium hydroxide

20 Complete and balance the following chemical equations:

a $HNO_3(aq) + KOH(aq) \rightarrow$

b $H_2SO_4(aq) + K_2CO_3(aq) \rightarrow$

c $H_3PO_4(aq) + Ca(HCO_3)_2(s) \rightarrow$

d $HF(aq) + Zn(OH)_2(s) \rightarrow$

21 Which one of the following correctly identifies all the products formed when magnesium hydroxide reacts with hydrochloric acid?

A water

B chloride ions

C magnesium ions

D magnesium chloride precipitate

E water, magnesium ions and chloride ions

22 Write an ionic equation for the reaction between an acid (HA) and a soluble metal carbonate (MCO_3).

23 Dilute hydrochloric acid is added to a white solid in a test-tube. A colourless gas is produced. The gas turns limewater cloudy. What class of compound might the white solid be?

24 Hydrogen is produced when dilute sulfuric acid reacts with aluminium metal.

a Write a balanced chemical equation for this reaction.

b Write an ionic equation for this reaction.

25 Classify each of the following general equations for reactions as synthesis, decomposition, combustion, precipitation or acid base:

a $A \rightarrow B + C$

b $A + B \rightarrow C$

c $A + B \rightarrow \text{salt} + H_2O$

d $A(aq) + B(aq) \rightarrow C(s) + D(aq)$

e $A + O_2 \rightarrow CO_2 + H_2O$

26 Classify each of the following reactions as synthesis, decomposition, combustion, precipitation or acid base:

a $C_3H_8(g) + 5O_2(g) \rightarrow 3CO_2(g) + 4H_2O(g)$

b $Ba(NO_3)_2(aq) + CuSO_4(aq) \rightarrow BaSO_4(s) + Cu(NO_3)_2(aq)$

c $H_2(g) + F_2(g) \rightarrow 2HF(g)$

d $2NaOH(s) \rightarrow Na_2O(s) + H_2O(g)$

e $HCl(aq) + NaOH(aq) \rightarrow NaCl(aq) + H_2O(l)$

27 The photo shows a flask in which sodium is burning in chlorine gas.

a Write a word equation for the reaction that is occurring.

b Write a balanced chemical equation for the reaction.

c Which category of reaction does this belong to?

28 Predict the products of the following reactions to complete and balance the equations:

a $HCl(aq) + KOH(aq) \rightarrow$

b $HCl(aq) + Ca(s) \rightarrow$

c $HNO_3(aq) + CaCO_3(s) \rightarrow$

d $Ca(s) + O_2(g) \rightarrow$

e $C_2H_6(g) + O_2(g) \rightarrow$

29 Reflect on the Inquiry activity on page 312. What type of chemical reaction did you observe when you added the vinegar to the washing soda and flower petal solution?

CHAPTER 11

Predicting reactions of metals

The structure, bonding and physical properties of metals were discussed in Chapter 5. In this chapter, you will investigate how the chemical properties of various metals can be predicted.

Some metals are highly reactive. For example, rubidium and caesium (both group 1 metals) will react explosively in water. Other metals (such as copper) are far less reactive. Platinum and gold hardly react at all.

You will explore the reactivities of a range of metals with oxygen, water and dilute acids. The reactions between metals and metal compounds in solution will also be considered. You will see how observation of the relative reactivity of metals enables us to place metals into an activity series. Such an activity series can be used to predict the reactions of metals.

Finally, the factors that determine the reactivity of metals will be examined and linked to other properties such as ionisation energy, atomic radius and electronegativity.

Content

INQUIRY QUESTION

How is the reactivity of various metals predicted?

By the end of this chapter, you will be able to:

- conduct practical investigations to compare the reactivity of a variety of metals in:
  - water
  - dilute acid (ACSCH032, ACSCH037)
  - oxygen
  - other metal ions in solution
- construct a metal activity series using the data obtained from practical investigations and compare this series with that obtained from standard secondary-sourced information (ACSCH103)
- analyse patterns in metal activity on the periodic table and explain why they correlate with, for example:
  - ionisation energy (ACSCH045)
  - atomic radius (ACSCH007)
  - electronegativity (ACSCH057)

11.1 Reactions of metals

CHEMISTRY INQUIRY 

Comparing the reactivity of metals in water

Can you predict the reactions of metals in water?

COLLECT THIS ...

- iron nail
- galvanised nail
- 2 plastic containers of the same size
- table salt (sodium chloride)
- water
- teaspoon
- camera

DO THIS ...

1 Draw up the table and predict what will happen when you complete the following steps.
2 Fill the plastic containers with water.
3 Add to each a teaspoon of table salt and stir until the salt is dissolved.
4 Add one nail to each container, and label which type of nail is in the container.
5 Take a photo of each nail.
6 Leave undisturbed for 5 days.
7 Take another photo of each nail.

RECORD THIS ...

Record your predictions in the table.

Description of experiment	Prediction	Observations
iron nail in salty water		
galvanised nail (coated in zinc) in salty water		

Describe your observations of what happened to the nails after 5 days.

REFLECT ON THIS ...

Explain your observations.

What could you do next time to improve your experiment?

FIGURE 11.1.1 A nugget of pure gold found in a river bed.

Some metals are more reactive than others. An iron pipe will corrode, but a copper pipe will not. One of the reasons why gold is used to make jewellery is that it does not combine readily with other elements. Gold exists in nature as the pure metal and can be found as nuggets in river beds or alluvial soils (Figure 11.1.1). It is also found in quartz veins in rocks.

In nature, silver is found as either pure nuggets or chemically combined with other elements such as sulfur or chlorine. Copper is found as either pure nuggets or chemically combined with other elements in copper ores such as azurite ($Cu_3(CO_3)_2(OH)_2$) (Figure 11.1.3) and calcocite (Cu_2S).

Metals such as sodium and calcium are too reactive to be found in their elemental state. They combine readily with other elements. Table salt or sodium chloride (NaCl) is a common compound of sodium and chlorine. Calcium occurs as calcium carbonate ($CaCO_3$) in the form of marble, an attractive rock used in sculpture and to decorate buildings (Figure 11.1.4).

In this section, you will see how the reactivity of different metals can be experimentally determined by observing their reactions with oxygen, water, dilute acid, and solutions of different metal salts. The metals we will use as a sample set to investigate the reactivity of metals are shown in the periodic table in Figure 11.1.2.

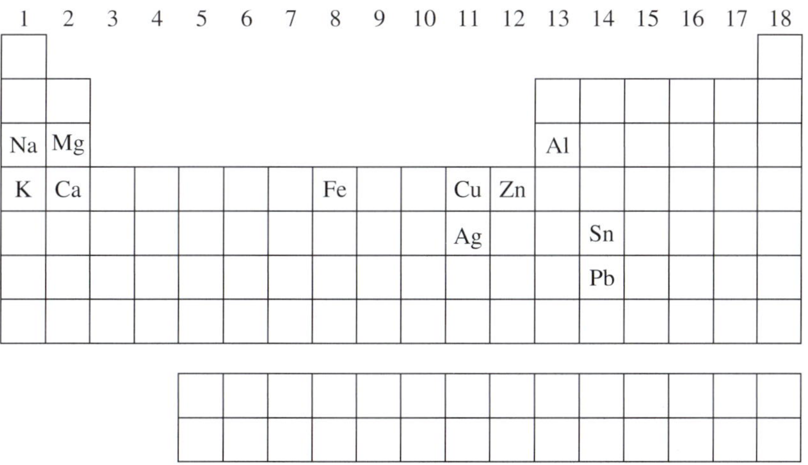

FIGURE 11.1.2 The sample set of metals that will form the basis of the discussion on the reactivity of metals.

FIGURE 11.1.3 The copper ore azurite contains the copper compound $Cu_3(CO_3)_2(OH)_2$.

REACTIONS OF METALS WITH OXYGEN

There is great variation in the **reactivity** of metals with oxygen. Sodium is so reactive that it is stored under paraffin oil to prevent exposure to oxygen and water in the air. The rusting of iron is a slow reaction between iron, oxygen and water. Gold, silver and platinum do not react with atmospheric oxygen and water in this way.

When metals react with oxygen, they form a metal oxide:

$$\text{metal} + \text{oxygen} \rightarrow \text{metal oxide}$$

Sodium and potassium metals are quickly tarnished in air, forming an oxide layer on the surface of the metal. The equation for the reaction of sodium with oxygen is:

$$4Na(s) + O_2(g) \rightarrow 2Na_2O(s)$$

Calcium and magnesium both burn vigorously in air, producing a large amount of energy in the form of heat and light. Magnesium burns with an intense white flame, as shown in Figure 11.1.5. Magnesium oxide powder is formed in this reaction.

Aluminium, zinc and iron will burn in air only if the metal is in a form with a large surface area, as shown in Figure 11.1.6. (You will learn about how surface area influences the rate of reaction in Section 13.3.) Iron reacts with oxygen to form iron(III) oxide (Fe_2O_3):

$$4Fe(s) + 3O_2(g) \rightarrow 2Fe_2O_3(s)$$

FIGURE 11.1.4 Calcium carbonate ($CaCO_3$) in the form of marble was used to build the Taj Mahal in the 17th century by the Mughal Emperor as a tomb for his wife.

Some metals react with oxygen, forming metal oxides.

GO TO ➤ Section 13.3 page 421

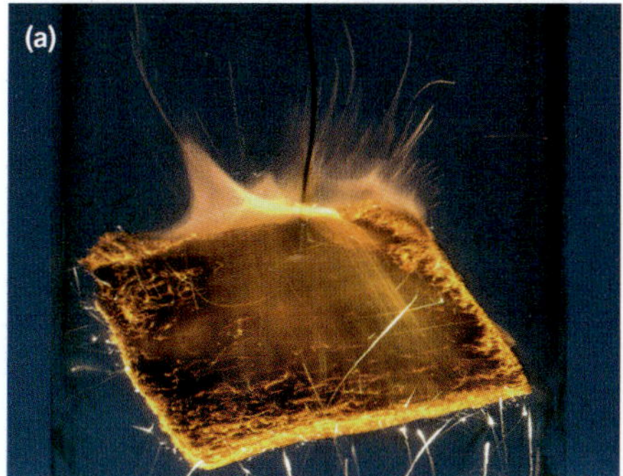

FIGURE 11.1.6 Some finely divided metals will burn in air: (a) iron wool consists of fine iron fibres that ignite when strongly heated, and (b) powdered aluminium burns in the flame of a Bunsen burner. Powdered aluminium is often used in fireworks.

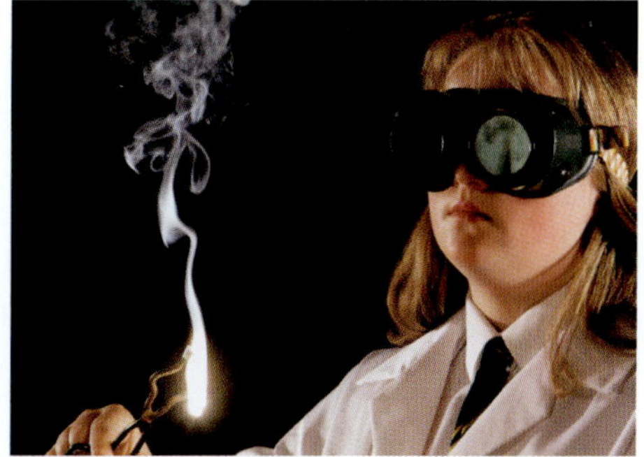

FIGURE 11.1.5 Magnesium ribbon burns brightly when heated in air to form a white powder, magnesium oxide.

Pure aluminium reacts readily with atmospheric oxygen, forming a thin layer of aluminium oxide on the metal surface. This layer protects the underlying metal from further reaction with oxygen. The equation for this reaction is:

$$4Al(s) + 3O_2(g) \rightarrow 2Al_2O_3(s)$$

Zinc burns in air, forming zinc oxide:

$$2Zn(s) + O_2(g) \rightarrow 2ZnO(s)$$

Tin and lead react with oxygen when heated, forming tin(II) oxide and lead(II) oxide, respectively:

$$2Sn(s) + O_2(g) \rightarrow 2SnO(s)$$
$$2Pb(s) + O_2(g) \rightarrow 2PbO(s)$$

Copper reacts more slowly with oxygen than tin or lead when heated:

$$2Cu(s) + O_2(g) \rightarrow 2CuO(s)$$

Silver does not react with atmospheric oxygen, even when heated.

The reactivity of metals with oxygen can be used to rank them in order from most reactive to least reactive (Table 11.1.1).

TABLE 11.1.1 The order of reactivity of various metals with oxygen

Metal	Reactivity with oxygen
potassium sodium calcium magnesium	react readily with oxygen
aluminium zinc iron	react readily with oxygen if the metal is in a powdered form
tin lead copper	react slowly when heated in oxygen
silver	does not react with oxygen

The metal oxides that are formed are ionic solids consisting of a lattice of positive metal **cations** and negative **anions**. The reaction of metals with oxygen involves a transfer of electrons from metal atoms to oxygen atoms.

For example, when magnesium metal reacts with oxygen, the ionic solid magnesium oxide is formed. This consists of an ionic lattice in which each Mg^{2+} **ion** is surrounded by O^{2-} ions, and each O^{2-} ion is surrounded by Mg^{2+} ions. Ionic lattices were discussed in greater detail in Section 5.2. The Mg^{2+} ions are formed by the loss of two electrons from the **valence shell** of each Mg atom, as represented by the following equation. The electronic configurations of a magnesium atom and ion are also shown:

GO TO ➤ Section 5.2 page 147

$$\begin{array}{ccc} Mg(s) & \rightarrow & Mg^{2+}(s) + 2e^- \\ 2,8,2 & & 2,8 \end{array}$$

The two electrons are transferred to the valence shell of each oxygen atom to form O^{2-} ions. Since oxygen is a diatomic molecule, a total of four electrons from two magnesium atoms are transferred.

$$\begin{array}{ccc} O_2(g) + 4e^- & \rightarrow & 2O^{2-}(s) \\ 2,6 & & 2,8 \end{array}$$

Since the number of electrons lost must equal the number of electrons gained, this means that two magnesium atoms will react with one oxygen molecule. The overall equation can be written as:

$$2Mg(s) + O_2(g) \rightarrow 2MgO(s)$$

CHEMISTRY IN ACTION ICT

Use of gold in communication satellites

Gold is one of the most unreactive of all the metals. This lack of reactivity makes it ideal for use in the advanced technologies used by the aerospace and satellite telecommunications industries. Gold-coated plastic sheets are used as a radiation shield in space satellites (Figure 11.1.7) and in astronauts' space suits. Gold reflects both infrared and ultraviolet radiation, which can damage the delicate instruments in a satellite. Gold is softer and more malleable than other metals, making it easier to work with. It also requires less maintenance, because it will not react with oxygen. Microprocessors in the satellite are made of gold because it is an excellent conductor of electricity. An ultra-thin layer of gold is sprayed onto the mirrors used in space telescopes to improve the mirror's reflective properties.

FIGURE 11.1.7 A gold-coated radiation shield is installed on a section of a satellite.

REACTIONS OF METALS WITH WATER

The sample set of metals listed in Table 11.1.1 will have different reactivities with water.

A vigorous reaction occurs when a piece of sodium is added to a beaker of water (Figure 11.1.8). The sodium moves in a random manner on the surface of the water, producing a spitting sound. Hydrogen gas (H_2) is produced, and the water in the beaker turns pink when the indicator phenolphthalein is added, indicating the presence of hydroxide ions (OH^-). The amount of heat generated by the reaction is sufficient to ignite the hydrogen gas.

FIGURE 11.1.8 Sodium reacts vigorously with water.

The reaction between sodium metal and water can be described by the following ionic equation:

$$2Na(s) + 2H_2O(l) \rightarrow 2Na^+(aq) + 2OH^-(aq) + H_2(g)$$

The 'pop' test is a simple test for hydrogen gas and involves collecting some of the gas in a test-tube. A characteristic 'popping' sound is heard when a lit match is held over the mouth of the test-tube. A flame can sometimes be seen in the test-tube as the hydrogen gas ignites.

Potassium metal reacts even more vigorously with water. The heat produced by the reaction ignites the hydrogen, producing a flame above the floating potassium. Hydrogen gas and hydroxide ions are produced:

$$2K(s) + 2H_2O(l) \rightarrow 2K^+(aq) + 2OH^-(aq) + H_2(g)$$

Calcium metal will also react with water, but not as violently as sodium or potassium. Hydrogen gas and a white precipitate of calcium hydroxide are formed:

$$Ca(s) + 2H_2O(l) \rightarrow Ca(OH)_2(s) + H_2(g)$$

The other metals in our list do not react with cold water. However, magnesium will react with hot water to produce hydrogen gas and a white precipitate of magnesium hydroxide:

$$Mg(s) + 2H_2O(l) \rightarrow Mg(OH)_2(s) + H_2(g)$$

Aluminium, zinc and iron will react with steam to produce hydrogen and the metal oxide. Aluminium reacts more vigorously than zinc. Iron reacts slowly with steam.

Hydrogen gas and the metal oxide are formed in these reactions:

$$2Al(s) + 3H_2O(g) \rightarrow Al_2O_3(s) + 6H_2(g)$$
$$Zn(s) + H_2O(g) \rightarrow ZnO(s) + H_2(g)$$
$$2Fe(s) + 3H_2O(g) \rightarrow Fe_3O_4(s) + 3H_2(g)$$

On the basis of these experiments, it is possible to place some of the metals in our list in order of their reactivity with water and steam, as shown in Table 11.1.2.

Since tin, lead, copper and silver do not react with either water or steam, it is not possible to place them in an order of reactivity based on this test.

In each of the reactions described above, the metal removes or displaces the hydrogen from water. These types of reactions are called **displacement** reactions. In these reactions, electrons are transferred from the metal atom to a water molecule.

TABLE 11.1.2 The order of reactivity of various metals with water

Metal	Reactivity with water
potassium sodium calcium	react with cold water to produce hydrogen and hydroxide ions
magnesium	reacts with hot water to produce hydrogen and hydroxide ions
aluminium zinc iron	react with steam to produce hydrogen and the metal oxide

> Some metals displace hydrogen from water or steam, forming either the metal hydroxide or metal oxide, and hydrogen gas.

REACTIONS OF METALS WITH ACIDS

The ranking of metals in order of reactivity can be further refined by testing the reactivity of metals in dilute hydrochloric acid (HCl). Dilute hydrochloric acid contains hydrogen ions (H^+). You can conduct simple experiments to test the ease with which various metals displace hydrogen ions from hydrochloric acid. Back in Chapter 10, you learnt that hydrogen gas is produced when a reactive metal reacts with a dilute acid. Metal ions are also produced:

$$\text{acid} + \text{metal} \rightarrow \text{metal ion} + \text{hydrogen gas}$$

GO TO ➤ Section 10.6 page 336

The 'pop' test can be used to detect the presence of hydrogen gas. There are specific tests that can be used to test for the presence of metal ions.

The reactivity of sodium, potassium and calcium metals is not tested with dilute acids. You will recall that these metals react vigorously with water. Since acids are diluted with water, we cannot be certain whether any displaced hydrogen comes from reaction with the water or with the acid. Since the reactions of these metals with water are vigorous, it can be assumed that they will be even more reactive with acids. For safety reasons, the reactivity of sodium, potassium and calcium with acids should not be tested.

There is great variation in the reactivity of metals with acids. Magnesium reacts vigorously with dilute hydrochloric acid, forming magnesium ions and hydrogen gas:

$$Mg(s) + 2H^+(aq) \rightarrow Mg^{2+}(aq) + H_2(g)$$

Zinc reacts less vigorously than magnesium with dilute hydrochloric acid, forming zinc ions and hydrogen gas:

$$Zn(s) + 2H^+(aq) \rightarrow Zn^{2+}(aq) + H_2(g)$$

Iron reacts very slowly in cold, dilute hydrochloric acid. It reacts more vigorously in warm acid, producing iron(II) ions and hydrogen gas:

$$Fe(s) + 2H^+(aq) \rightarrow Fe^{2+}(aq) + H_2(g)$$

Tin and lead react very slowly with warm acid, forming tin(II) and lead(II) ions, respectively, and hydrogen gas:

$$Sn(s) + 2H^+(aq) \rightarrow Sn^{2+}(aq) + H_2(g)$$

$$Pb(s) + 2H^+(aq) \rightarrow Pb^{2+}(aq) + H_2(g)$$

Copper and silver do not react with dilute hydrochloric acid.

FIGURE 11.1.9 Some metals react with dilute acid to form a salt and hydrogen gas. The reaction between magnesium and dilute acid is extremely vigorous. The reaction between zinc and dilute acid is less vigorous. The reaction between iron and dilute acid is very slow. There is no reaction between lead and cold dilute acid. Based on this information, the order of metal reactivity from most reactive to least reactive is: magnesium, zinc, iron, lead.

The relative reactivities of magnesium, zinc, iron and lead with dilute acids are shown in Figure 11.1.9.

The outcomes of the reactions of all the metals in the sample set with dilute hydrochloric acid are summarised in Table 11.1.3. They are arranged in order of the ease with which they displace hydrogen from dilute hydrochloric acid.

TABLE 11.1.3 The order of reactivity of various metals with dilute hydrochloric acid

Metal	Reactivity with dilute hydrochloric acid
potassium sodium calcium	not tested
magnesium	reacts vigorously in cold acid, producing magnesium ions and hydrogen gas
aluminium*	reacts slowly at first, then vigorously, producing aluminium ions and hydrogen gas
zinc	reacts less vigorously than magnesium, producing zinc ions and hydrogen gas
iron	reacts slowly in cold acid (vigorously in warm acid), producing iron(II) ions and hydrogen gas
tin	reacts very slowly in warm acid, producing tin(II) ions and hydrogen gas
lead	reacts very slowly in warm acid, producing lead(II) ions and hydrogen gas
copper silver	no reaction with cold or warm dilute acid

*Aluminium reacts slowly at first as the protective layer of aluminium oxide dissolves. The reaction is more rapid once the bare metal has been exposed to the acid.

When metals react with dilute acids, there is a transfer of electrons from metal atoms to hydrogen ions. For example, when magnesium reacts with dilute hydrochloric acid, each magnesium atom loses two electrons, and two hydrogen ions from the acid gain two electrons to form hydrogen gas. The chloride ions from the hydrochloric acid are not involved in the reaction.

$$Mg(s) \rightarrow Mg^{2+}(aq) + 2e^-$$
$$2H^+(aq) + 2e^- \rightarrow H_2(g)$$

The ionic equation for this reaction is:

$$Mg(s) + 2H^+(aq) \rightarrow Mg^{2+}(aq) + H_2(g)$$

The full K(overall) equation is:

$$Mg(s) + 2HCl(aq) \rightarrow MgCl_2(aq) + H_2(g)$$

Some metals react with dilute acids to produce metal ions and hydrogen gas.

REACTIONS OF METALS WITH OTHER METALLIC IONS IN SOLUTION

When a coil of copper metal is placed in a solution of silver nitrate, the copper becomes coated with a black deposit, as shown in the flask on the right in Figure 11.1.10. After a while, the solution turns a pale blue colour. When the deposit is tested, it can be shown to be silver. The blue colour of the solution is due to the presence of copper(II) ions (Cu^{2+}). The copper metal has displaced the Ag^+ ions from the solution of silver nitrate. The ionic equation for this reaction is:

$$Cu(s) + 2Ag^+(aq) \rightarrow Cu^{2+}(aq) + 2Ag(s)$$

A similar reaction occurs when a strip of zinc is placed into a silver nitrate solution:

$$Zn(s) + 2Ag^+(aq) \rightarrow Zn^{2+}(aq) + 2Ag(s)$$

If a strip of zinc is placed into a solution of copper(II) sulfate, a brown deposit forms on the zinc, and the blue colour of the solution slowly fades (Figure 11.1.11 page 358). Tests reveal that the solution now contains Zn^{2+} ions. The zinc metal has displaced the Cu^{2+} ions from solution:

$$Zn(s) + Cu^{2+}(aq) \rightarrow Cu(s) + Zn^{2+}(aq)$$

However, no reaction occurs when a strip of silver metal is placed into solutions containing either Ag^+ or Zn^{2+} ions.

Copper metal will not react with solutions containing Zn^{2+} ions.

These reactions are summarised in Table 11.1.4.

FIGURE 11.1.10 When a coil of copper wire is suspended in a solution of silver nitrate (as in the flask on the left), long crystals of silver metal start to form. In the flask on the right, the copper metal has displaced the silver ions from the solution.

TABLE 11.1.4 Reactions of copper, silver and zinc with solutions containing their ions

Metal	Metal ion in solution		
	Zn^{2+}	Cu^{2+}	Ag^+
Zn	no reaction	brown deposit	black deposit
Cu	no reaction	no reaction	black deposit
Ag	no reaction	no reaction	no reaction

From Table 11.1.4, you can see that zinc is more reactive than copper and silver, and that copper is more reactive than silver. The results of similar tests carried out using the sample set of metals and solutions containing their metal ions are shown in Table 11.1.5 page 358. Sodium, potassium and calcium metal were not used because they react with water. Aluminium was not tested since the protective oxide coating on the surface of aluminium makes it difficult to judge its reactivity relative to that of the other metals.

FIGURE 11.1.11 A brown deposit of copper metal is observed forming on the zinc, and the blue copper(II) sulfate solution gradually becomes colourless as the concentration of Cu^{2+} ions decreases.

TABLE 11.1.5 Reactivity of metals with solutions of metal ions

Metal	Metal ion in solution						
	Mg^{2+}	Zn^{2+}	Fe^{2+}	Sn^{2+}	Pb^{2+}	Cu^{2+}	Ag^{+}
Mg	X	✔	✔	✔	✔	✔	✔
Zn	X	X	✔	✔	✔	✔	✔
Fe	X	X	X	✔	✔	✔	✔
Sn	X	X	X	X	✔	✔	✔
Pb	X	X	X	X	X	✔	✔
Cu	X	X	X	X	X	X	✔
Ag	X	X	X	X	X	X	X

✔ a reaction occurred; X no reaction occurred

From these results you can see that the order of reactivity, from highest to lowest, is:

magnesium > zinc > iron > tin > lead > copper > silver

For example, Table 11.1.5 indicates that magnesium metal is more reactive than copper metal, so magnesium metal will displace copper(II) ions from a solution of copper(II) sulfate. In this reaction, each magnesium atom will lose two electrons, and each copper(II) ion will gain two electrons:

$$Mg(s) \rightarrow Mg^{2+}(aq) + 2e^-$$

$$Cu^{2+}(aq) + 2e^- \rightarrow Cu(s)$$

The ionic equation for this reaction is:

$$Mg(s) + Cu^{2+}(aq) \rightarrow Mg^{2+}(aq) + Cu(s)$$

Note that the reactivity of the metal ions is in the reverse order. The more reactive the metal, the less reactive its ion. The less reactive the metal, the more reactive its ion.

GO TO ➤ Section 11.3 page 366

The reactivity of metals in terms of the ease with which they lose electrons is explained in detail later in the chapter.

11.1 Review

SUMMARY

- Most metals combine with oxygen to form a metal oxide.
- The reaction between a metal and oxygen involves a transfer of electrons from the metal atoms to the oxygen atoms.
- There is considerable variety in the reactivity of metals with oxygen. Group 1 and 2 metals react readily with oxygen. Most transition metals react slowly with oxygen at room temperature, but more quickly when heated. Silver and gold do not react with oxygen, even when heated.
- Potassium, sodium and calcium displace hydrogen from cold water. Magnesium displaces hydrogen from hot water. A solution containing metal ions and hydroxide ions is also formed. Aluminium, zinc and iron displace hydrogen from steam. Tin, lead, copper and silver do not react with either water or steam.
- Hydrogen ions are displaced from dilute acids by magnesium, aluminium, zinc, iron, tin and lead (in decreasing order of reactivity). Hydrogen gas and solutions containing the metal ions are formed by these reactions.
- A more reactive metal will displace the ion of a less reactive metal from solution.

KEY QUESTIONS

1 Name the product formed when each of these metals reacts with oxygen.

a Na
b Mg
c Al
d Sn

2 You are provided with the following set of metals: calcium, iron, potassium, silver and zinc.

a Arrange these metals in decreasing order of reactivity with oxygen.
b Write a balanced chemical equation for the reaction of each of these metals with oxygen. Write 'no reaction' if the metal does not react with oxygen.

3 **a** Identify one metal that will:

i react with cold water
ii react with steam
iii not react with either cold water or steam.

b Identify the general products formed when a metal reacts with:

i cold water
ii steam.

4 Write ionic equations for the reaction of the following metals with dilute acid:

a zinc
b aluminium
c tin

5 Explain why the reactions between metals and dilute acids are called displacement reactions.

6 **a** Identify whether a reaction occurs when zinc metal is placed into a solution containing:

i silver ions
ii magnesium ions.

b Write ionic equations for any reactions that occur in part **a**.

11.2 The activity series of metals

In Section 11.1, the reactivity of a representative set of metals with oxygen, water, dilute acids and solutions containing metal ions was examined. You will recall that some metals:

- react with oxygen to form metal oxides
- displace hydrogen from water and steam
- displace hydrogen from dilute hydrochloric acid
- displace metal ions of less reactive metals from solution.

In each set of these reactions, you will have noticed the similarity in the ranking of metals from most reactive to least reactive. The reactivity of other metals can be tested using the same experiments as those described in the previous section.

The reactions of metals are summarised in Table 11.2.1. The metals have been ranked in order of their reactivity, from most reactive to least reactive. This ranking is called the **activity series of metals**.

TABLE 11.2.1 Ranking of the reactivity of metals with oxygen, water, dilute acid and metal ions

Metal	Reaction with			
	oxygen	water	dilute acid	solution of metal ions
K	react with oxygen to form a metal oxide	displace hydrogen from cold water	not tested	not tested
Na				
Li				
Ca				
Mg		displace hydrogen from steam	displace hydrogen from dilute acids	will displace the ion of a metal lower in the series from a solution of its salt
Al				
Mn				
Zn				
Cr				
Fe				
Co		do not displace hydrogen from either cold water or steam		
Ni				
Sn				
Pb				
Cu			do not displace hydrogen from dilute acids	
Ag	do not react with oxygen			
Au				

A metal that is higher in the activity series will displace the ion of a less reactive metal from solution.

You will have noted that the various reactions of metals involve the loss of electrons. Table 11.2.2 shows the **half-equations** representing the loss of electrons by metal atoms. The metals are ranked in order of reactivity, as in the activity series. Metals at the top of the activity series lose electrons more readily than metals lower down the series, and consequently they are more reactive. The reason for this will be outlined later in the chapter.

GO TO ➤ Section 11.3 page 366

TABLE 11.2.2 The activity series of metals with their half-equations

Order of reactivity of the metal	Half-equation representing loss of electrons
decreasing metal reactivity ↓	$K(s) \rightarrow K^+(aq) + e^-$
	$Na(s) \rightarrow Na^+(aq) + e^-$
	$Li(s) \rightarrow Li^+(aq) + e^-$
	$Ca(s) \rightarrow Ca^{2+}(aq) + 2e^-$
	$Mg(s) \rightarrow Mg^{2+}(aq) + 2e^-$
	$Al(s) \rightarrow Al^{3+}(aq) + 3e^-$
	$Mn(s) \rightarrow Mn^{2+}(aq) + 2e^-$
	$Zn(s) \rightarrow Zn^{2+}(aq) + 2e^-$
	$Fe(s) \rightarrow Fe^{2+}(aq) + 2e^-$
	$Cr(s) \rightarrow Cr^{3+}(aq) + 3e^-$
	$Co(s) \rightarrow Co^{2+}(aq) + 2e^-$
	$Ni(s) \rightarrow Ni^{2+}(aq) + 2e^-$
	$Sn(s) \rightarrow Sn^{2+}(aq) + 2e^-$
	$Pb(s) \rightarrow Pb^{2+}(aq) + 2e^-$
	$Cu(s) \rightarrow Cu^{2+}(aq) + 2e^-$
	$Ag(s) \rightarrow Ag^+(aq) + e^-$
	$Au(s) \rightarrow Au^+(aq) + e^-$

The activity series of metals ranks metals in order of their reactivity from highest to lowest. Reactive metals lose electrons more readily than less reactive metals.

PREDICTING THE REACTION OF METALS

The activity series of metals can be used to predict whether or not a reaction will take place between a metal and the ion of another metal.

Worked example 11.2.1

PREDICTING METAL DISPLACEMENT REACTIONS

Predict whether zinc will displace copper from a solution containing copper(II) ions and, if appropriate, write the ionic equation for the reaction.

Thinking	Working
Locate the metal and the metal ion in the activity series.	Metals are found in the right-hand column of the activity series. $Al(s) \rightarrow Al^{3+}(aq) + 3e^-$ $Zn(s) \rightarrow Zn^{2+}(aq) + 2e^-$ $Cr(s) \rightarrow Cr^{3+}(aq) + 3e^-$ $Fe(s) \rightarrow Fe^{2+}(aq) + 2e^-$ $Co(s) \rightarrow Co^{2+}(aq) + 2e^-$ $Ni(s) \rightarrow Ni^{2+}(aq) + 2e^-$ $Sn(s) \rightarrow Sn^{2+}(aq) + 2e^-$ $Pb(s) \rightarrow Pb^{2+}(aq) + 2e^-$ $Cu(s) \rightarrow Cu^{2+}(aq) + 2e^-$ $Ag(s) \rightarrow Ag^+(aq) + e^-$
Determine whether the metal ion is below, and to the right of the metal in the table. If this is the case, there will be a reaction.	You can see from the activity series that Cu^{2+} is in the right-hand column and below Zn. This means that Cu^{2+} is the ion of a less reactive metal, so there will be a reaction.

The more reactive metal atom will lose electrons to form a metal ion. Write a half-equation for this reaction.	$Zn(s) \rightarrow Zn^{2+}(aq) + 2e^-$
The ion of the metal ion lower down the activity series will gain electrons, forming the metal atom. Write a half-equation for this reaction.	$Cu^{2+}(aq) + 2e^- \rightarrow Cu(s)$
Combine the two half-equations, balancing electrons, to give the ionic equation for the reaction.	$Zn(s) \rightarrow Zn^{2+}(aq) + 2e^-$ $Cu^{2+}(aq) + 2e^- \rightarrow Cu(s)$ $Zn(s) + Cu^{2+}(aq) \rightarrow Zn^{2+}(aq) + Cu(s)$

Worked example: Try yourself 11.2.1

PREDICTING METAL DISPLACEMENT REACTIONS

Predict whether copper will displace gold from a solution containing gold ions and, if appropriate, write the ionic equation for the reaction.

LIMITATIONS OF THE ACTIVITY SERIES

GO TO ➤ Chapter 13 page 409

The activity series of metals is based on **qualitative** experimental observations, using judgements about how fast the reactions occur. There are a number of factors, other than reactivity, that influence how fast a reaction occurs. They are the:

- temperature at which the experiment is carried out
- surface area of the solid reactants
- concentration of the reactants in solution
- pressure of the gaseous reactants and products.

Whether an observed reaction is fast or slow depends in part on the reactivity of the metal and in part on the conditions under which the reaction takes place.

The use of terms such as vigorous, fast, slow, or very slow is an imprecise way of describing the reactivity of metals. These terms are difficult to interpret. For example, one observer may describe a reaction as very fast, and another may describe it as violent. The activity series is not based on precise measurements or numerical data. Because of these factors, there may be slight variations in the ranking of metals by different textbooks or websites.

GO TO ➤ Section 12.4 page 396

Another ranking of chemical reactivity, called the **table of standard reduction potentials**, ranks both metals and non-metals in order of reactivity (you will learn more about this in Chapter 12). This ranking is based on more sophisticated **quantitative** measurements. Since the reaction of metals involves a transfer of electrons, it is possible to measure the electrical voltage or potential produced by the transfer. This electrical potential is measured under specified conditions of temperature (25°C), concentration ($1\,mol\,L^{-1}$) and pressure (100 kPa). In the table of standard reduction potentials, metals and non-metals are ranked in order of this measured electrical potential.

A section of the table of standard reduction potentials, ranking the metals whose reactivity was examined earlier (Section 11.1), is shown in Table 11.2.3. In the table of standard reduction potentials used in this course, the metals are placed in the right-hand column, with the most reactive metals at the top. The table of standard reduction potentials can be used to predict the reactions of metals with metal ions and with non-metals.

TABLE 11.2.3 A section of the table of standard reduction potentials (containing various metals and hydrogen)

Half-equation	Reduction potential (V)	
$K^{+}(aq) + e^{-} \rightarrow K(s)$	–2.94	decreasing metal reactivity ↓
$Ca^{2+}(aq) + 2e^{-} \rightarrow Ca(s)$	–2.87	
$Na^{+}(aq) + e^{-} \rightarrow Na(s)$	–2.71	
$Mg^{2+}(aq) + 2e^{-} \rightarrow Mg(s)$	–2.36	
$Al^{3+}(aq) + 3e^{-} \rightarrow Al(s)$	–1.68	
$Zn^{2+}(aq) + 2e^{-} \rightarrow Zn(s)$	–0.76	
$Fe^{2+}(aq) + 2e^{-} \rightarrow Fe(s)$	–0.44	
$Sn^{2+}(aq) + 2e^{-} \rightarrow Sn(s)$	–0.14	
$Pb^{2+}(aq) + 2e^{-} \rightarrow Pb(s)$	–0.13	
$H^{+}(aq) + e^{-} \rightarrow \frac{1}{2}H_2(g)$	0.00	
$Cu^{2+}(aq) + 2e^{-} \rightarrow Cu(s)$	+0.34	
$Ag^{+}(aq) + e^{-} \rightarrow Ag(s)$	+0.80	

With the exceptions of sodium and calcium, there is a very close agreement in the order of reactivity of metals between the two series. The activity series ranks sodium as being more reactive than calcium, whereas the table of standard reduction potentials ranks calcium as being more reactive.

A comparison of the activity series with the table of standard reduction potentials is provided in Table 11.2.4.

TABLE 11.2.4 Comparison of the activity series with the table of standard reduction potentials

Activity series	Table of standard reduction potentials
Metals are listed in the left-hand column of the table, in order of decreasing reactivity from top to bottom.	Metals are listed in the right-hand column of the table, with standard reduction potential increasing from top to bottom.
Metals are ranked on their ability to react with other chemicals.	Metals, non-metal and polyatomic ions are ranked on the basis of measured standard reduction potential.
Reaction conditions are not specified.	Reaction conditions are specified.
The activity series can be used to predict the reactions of metals with metal ions.	The table of standard reduction potentials can be used to predict the reactions of metals with metal ions and with non-metals. It can also be used predict the reactions between non-metals.
The order of reactivity of metals from most reactive to least reactive is: K > Na > Ca > Mg > Al > Zn > Fe > Sn > Pb > Cu > Ag Note that Na is placed before Ca.	The order of reactivity of metals from most reactive to least reactive is: K > Ca > Na > Mg > Al > Zn > Fe > Sn > Pb > Cu > Ag Note that Ca is placed before Na.

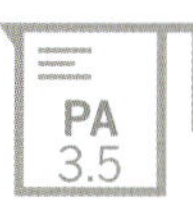

SKILLBUILDER N

Qualitative and quantitative variables

Variables are the factors that are changed or controlled during an experiment. There are three types of variables.

- The independent variable is the variable that is changed by the researcher.
- The dependent variable is the variable that may change in response to a change in the independent variable. This is the variable that you will measure or observe.
- Controlled variables are all the variables that must be kept constant during the investigation.

You should only test one independent variable at a time; otherwise, you cannot be sure that the changes in the dependent variable are the result of changes in the independent variable.

Variables can be either quantitative or qualitative, depending on whether or not they are measured. Figure 11.2.1 shows examples of quantitative and qualitative variables.

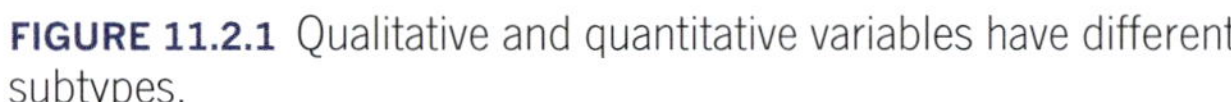

FIGURE 11.2.1 Qualitative and quantitative variables have different subtypes.

Quantitative data can be mathematically processed. For example, a mean (average) can be calculated, or the data can be represented as a continuous line graph.

A mixture of qualitative and quantitative observations can be made in an investigation. For example, an investigation into the acidity of solutions may involve the classification of samples as acidic, basic or neutral, depending on the colour change qualitatively observed when they react with an indicator such as litmus. A quantitative measure of the acidity of the samples can be made using a pH meter (Figure 11.2.2).

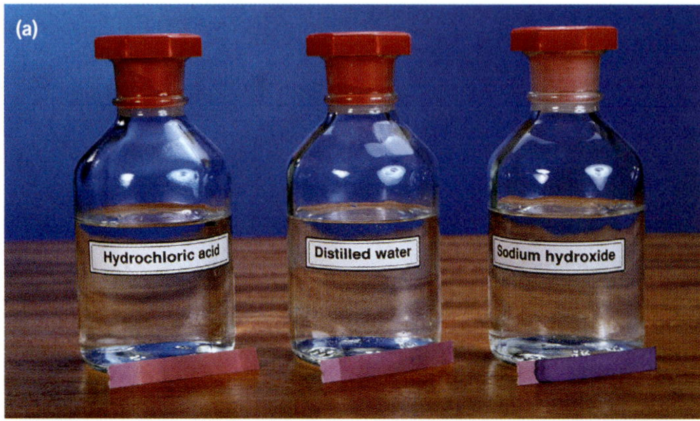

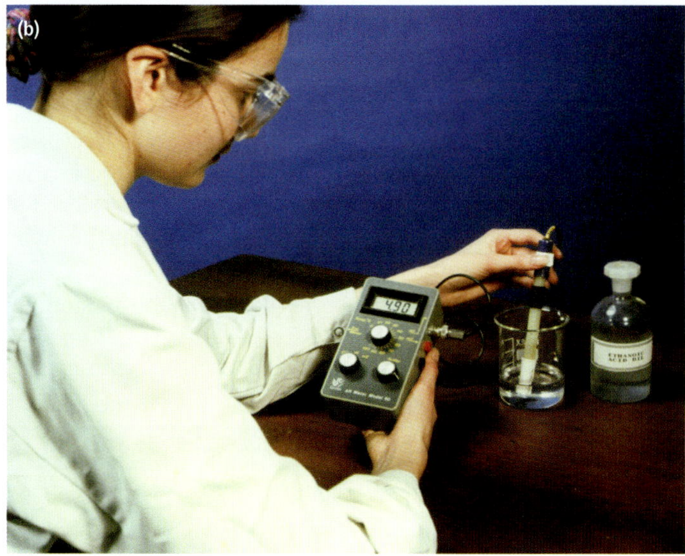

FIGURE 11.2.2 Determining whether a solution is acidic or basic can be qualitative: (a) litmus paper enables a qualitative assessment of the nature of the solution. (If the paper turns purple, the solution is basic; if the paper turns pink, the solution is acidic.) On the other hand, a calibrated pH meter (b) enables a quantitative measurement of the degree of acidity or alkalinity of the solution ($pH < 7$, the solution is acidic; $pH > 7$, the solution is basic).

11.2 Review

SUMMARY

- The activity series ranks metals in order of their relative reactivity.
- Observations made of reactions of different metals are used to rank metals in order of most reactive to least reactive.
- Metal atoms lose electrons in a reaction.
- The atoms of more reactive metals lose electrons more readily than the atoms of less reactive metals.
- The activity series can be used to predict the reactions of metals.
- A metal will displace the ion of a less reactive metal from solution.
- During a displacement reaction, the metal atom loses electrons and the metal ion gains electrons.
- The loss and gain of electrons can be represented by half-equations.
- Metal displacement reactions can be represented by ionic equations derived by combining half-equations.
- The activity series of metals is based on observations of the relative vigour of the reactions of various metals. Terms such as 'violent', 'fast', 'slow' and 'barely reacts' are used to describe the vigour of the reaction. Reaction conditions are not standardised.
- The table of standard reduction potentials ranks both metals and non-metals in order of reactivity. Their reactivity is assigned a numerical value, the standard reduction potential, based on precise measurements made under standardised conditions.

KEY QUESTIONS

1 A reddish brown deposit forms on a strip of manganese metal when it is placed in a solution of copper(II) sulfate. No reaction is observed when a strip of manganese is placed in a solution of magnesium nitrate. Arrange the metals copper, magnesium and manganese in order of:
  - **a** their reactivity, from highest to lowest
  - **b** the ease, from highest to lowest, with which the metal atom loses electrons.

2 Write the half-equations and the ionic equation for the reaction that occurs between a piece of nickel and a solution of copper(II) sulfate.

3 Refer to the activity series (Table 11.2.2, page 361) and predict whether a reaction will occur when:
  - **a** silver metal is placed in a copper(II) nitrate solution
  - **b** a strip of aluminium is placed in a sodium chloride solution
  - **c** magnesium is added to a solution of iron(II) sulfate
  - **d** the element zinc is placed in a tin(II) sulfate solution
  - **e** a piece of tin is placed in a silver nitrate solution
  - **f** lead(II) nitrate solution is poured into a beaker containing zinc granules
  - **g** silver foil is added to a lead(II) nitrate solution.

4 Use the reactivity series to predict whether a reaction will occur in each of the following situations. Write an ionic equation for each reaction that you predict will occur.
  - **a** Copper(II) sulfate solution is stored in an aluminium container.
  - **b** Sodium chloride solution is stored in a copper container.
  - **c** Silver nitrate solution is stored in a zinc container.

5 Solutions of zinc nitrate, tin(II) nitrate and copper(II) nitrate have been prepared in a laboratory, but have inadvertently been left unlabelled. Name two metals that could be used to identify each solution.

6 Identify one difference between the metal activity series and the table of standard reduction potentials.

11.3 Metal activity and the periodic table

As you saw in Section 11.2, the activity series places metals in order of their relative reactivity. You learnt that metals lose electrons when they react, and that a reactive metal loses electrons more readily than less reactive metals. In this section, you will analyse the trends in the reactivity of metals across the periodic table.

GO TO ➤ Section 4.2 page 122

The reactivity of metals is in part dependent on their **atomic radius**, **ionisation energy** and **electronegativity**. The trends in these properties in the periodic table were explained in Chapter 4. You will recall the following definitions:

- Atomic radius, the distance from the nucleus to the outermost electrons, is a measure of the size of an atom.
- Ionisation energy is the energy required to remove one electron from an atom.
- Electronegativity is the ability of atoms to attract electrons.

Variation in these factors within the periodic table can be understood in terms of **core charge,** which is a measure of the force of attraction between the valence electrons and the nucleus of the atom.

The relative reactivity of some of the metals in groups 1, 2 and 3 was outlined in Section 11.1. On the basis of experimental observation, aluminium, calcium, magnesium, sodium and potassium can be ranked in order of most reactive to least reactive as follows:

$$K > Na > Ca > Mg > Al$$

This ranking can be explained in terms of atomic radius, ionisation energy and electronegativity.

Sodium and potassium are both group 1 metals, with potassium being more reactive. Since both metals are in the same group, they have the same core charge. Some properties of sodium and potassium are listed in Table 11.3.1.

TABLE 11.3.1 Selected properties of sodium and potassium

Property	Sodium	Potassium
atomic number	11	19
electronic configuration	2,8,1	2,8,8,1
core charge	+1	+1
atomic radius (pm)	160	200
first ionisation energy (kJ mol^{-1})	496	419
electronegativity	0.9	0.8

Potassium has one more electron shell than sodium, and its atomic radius is consequently greater. Because potassium has a greater atomic radius, the electrostatic force of attraction between its valence electron and the nucleus is weaker. As a result, the ionisation energy and electronegativity of potassium are less than those of sodium. The potassium atom is more reactive, since it will lose the outer shell or valence electron more readily than will a sodium atom. The relative reactivities of the group 2 metals calcium and magnesium can be explained in a similar way.

Sodium, magnesium and aluminium are sequential elements in the third period of the periodic table, and are respectively found in groups 1, 2 and 13. Some properties of sodium, magnesium and aluminium are listed in Table 11.3.2.

TABLE 11.3.2 Selected properties of sodium, magnesium and aluminium

Property	Sodium	Magnesium	Aluminium
atomic number	11	12	13
electronic configuration	2,8,1	2,8,2	2,8,3
core charge	+1	+2	+3
atomic radius (pm)	160	140	124
ionisation equation	$Na \rightarrow Na^+ + e^-$	$Mg \rightarrow Mg^{2+} + 2e^-$	$Al \rightarrow Al^{3+} + 3e^-$
energy required to form ion ($kJ\,mol^{-1}$)	500	2200	5137
electronegativity	0.9	1.3	1.6

The electronic configuration of sodium is 2,8,1, that of magnesium is 2,8,2 and that of aluminium is 2,8,3. The number of occupied or partially occupied shells stays constant, but the core charge increases with increasing atomic number. Thus, the strength of attraction between valence electrons and the nucleus increases from sodium to magnesium to aluminium. As a consequence, the atomic radius decreases across the period from left to right. The ionisation energy and electronegativity increases due to the combined effects of increased core charge and decreased atomic radius.

The energy required to form an ion also depends upon the number of electrons that are lost in formation of the metal cation. The sodium atom only loses a single electron when it forms the Na^+ ion. The formation of a magnesium ion involves the loss of two electrons. A greater amount of energy is required to remove the second electron than the first, and more than four times the energy is required to form the Mg^{2+} ion than is needed to form the Na^+ ion. The formation of the aluminium ion Al^{3+} involves the loss of three electrons, requiring about ten times more energy than is required to remove a single electron from sodium.

As a result of the factors described above, the ability of the three metals to form ions by losing electrons decreases across the period. This is reflected in the order of their reactivities: Na > Mg > Al.

Similar trends also occur in the transition metals in groups 3 to 10. As core charges increase, atomic radii, ionisation energy and electronegativity decrease. Generally, the reactivity of the transition metals decreases with increasing atomic number, although there are a number of exceptions. A more detailed explanation of the trends in the reactivity of the transition metals is beyond the scope of this course.

Table 11.3.3 summarises the relationship between the reactivity of metals and core charge, atomic radius, electronegativity and ionisation energy.

TABLE 11.3.3 A summary of the periodic table trends

Property	Down a group	Across a period (left to right)
core charge	no change	increases
electronegativity	decreases	increases
atomic radius	increases	decreases
ionisation energy	decreases	increases
reactivity of metals	increases	decreases

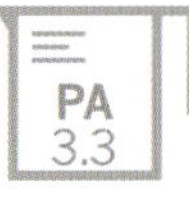

11.3 Review

SUMMARY

- The chemical reactivity of metals increases down a group and decreases across a period of the periodic table.
- The atoms of metals lose electrons when they react.
- A reactive metal loses its valence electrons more readily than a less reactive metal.
- The relative reactivity of metals can be explained in terms of atomic radius, ionisation energy and electronegativity.

KEY QUESTIONS

1 Describe the periodic trend in reactivity of metals down the groups and across the periods.

2 Explain, in terms of atomic radius, ionisation energy and electronegativity, why calcium is more reactive than magnesium.

3 Explain why the reactivity of potassium is greater than that of both sodium and calcium.

4 Some properties of three metals (I, II and III) are provided in the following table. Which of these metals is the most reactive? Justify your selection in terms of each of the three properties listed in the table.

Property	Metal		
	I	II	III
atomic radius (pm)	118	215	174
first ionisation energy ($kJ\,mol^{-1}$)	760	403	590
electronegativity	1.9	0.8	1.0

Chapter review

KEY TERMS

activity series of metals
anion
atomic radius
cation
core charge
displace
electronegativity
half-equation
ion
ionisation energy
qualitative
quantitative
reactivity
table of standard reduction potentials
valence shell
variable

REVIEW QUESTIONS

1 Name the products formed when:
- **a** calcium is burnt in oxygen
- **b** potassium reacts with water
- **c** zinc reacts with a dilute acid.

2 Which one of the following alternatives describes what happens when sodium and oxygen react?
- **A** Each sodium atom gains one electron.
- **B** Each sodium atom loses one electron.
- **C** Each oxygen atom gains one electron.
- **D** Each oxygen atom loses one electron.

3 Which one of the following metals will displace hydrogen from cold water?
- **A** lead
- **B** potassium
- **C** silver
- **D** zinc

4 Identify the metal from the following list that will not displace hydrogen from a dilute acid.
Al, Cu, Fe, Mg, Zn

5 Which pair of chemicals will not react with each other in a displacement reaction?
- **A** calcium and water
- **B** magnesium and steam
- **C** zinc and dilute hydrochloric acid
- **D** tin and a solution of magnesium nitrate

6 Describe how you could recover copper from a mixture of zinc and copper metals. Explain your reasoning. Include a relevant ionic equation with your explanation.

7 Which one of the following ionic equations has not been balanced correctly?
- **A** $Cu^{2+}(aq) + Zn(s) \rightarrow Cu(s) + Zn^{2+}(aq)$
- **B** $Fe^{2+}(aq) + Zn(s) \rightarrow Fe(s) + Zn^{2+}(aq)$
- **C** $Ag^{+}(aq) + Cu(s) \rightarrow Ag(s) + Cu^{2+}(aq)$
- **D** $Zn^{2+}(aq) + Mg(s) \rightarrow Zn(s) + Mg^{2+}(aq)$

8 Balance these half-equations:
- **a** $Ag^{+}(aq) \rightarrow Ag(s)$
- **b** $Cu(s) \rightarrow Cu^{2+}(aq)$
- **c** $Zn(s) \rightarrow Zn^{2+}(aq)$

9 List the following in order of decreasing reactivity:
Mg, Ag, Al, Sn, Li, Cu, Fe, Ca, Na

10 Explain what is meant by the following statement:
The activity series of metals is derived from qualitative not quantitative data.

You may need to refer to the activity series of metals in Table 11.2.2 (page 361) to answer Questions **11–16**.

11 Predict whether the following mixtures would result in reactions. Write an ionic equation for each reaction that you predict will occur.
- **a** Zinc metal is added to a solution of silver nitrate.
- **b** Copper metal is placed in an aluminium chloride solution.
- **c** Tin(II) sulfate is placed in a copper container.
- **d** Magnesium metal is added to a solution of lead(II) nitrate.
- **e** Silver metal is added to nickel(II) chloride solution.
- **f** Solutions of potassium chloride and copper(II) chloride are mixed.
- **g** Potassium nitrate solution is added to a silver container.
- **h** Lead metal is placed in a solution of silver nitrate.

12 If a piece of zinc metal is added to solutions of the following compounds, will a reaction be expected? If so, write a balanced chemical equation for the reaction.
- **a** $AlCl_3$
- **b** $AgNO_3$
- **c** $SnCl_2$
- **d** $CuSO_4$

13 You are given three colourless solutions (A, B and C) known to be sodium nitrate, silver nitrate and lead(II) nitrate, but not necessarily in that order. You also have some pieces of magnesium ribbon and copper wire. Describe how you could identify each of the solutions, using only the chemicals supplied.

14 An unknown metal is placed in solutions of aluminium nitrate and iron(II) sulfate. After a period of time, the metal is found to have reacted with the iron(II) sulfate solution, but not with the aluminium nitrate solution. Is the unknown metal zinc or cobalt?

15 Iron nails are placed into solutions of 1 mol L^{-1} $CuSO_4$, $MgCl_2$, $Pb(NO_3)_2$ and $ZnCl_2$. In which solution or solutions would you expect a coating of another metal to appear on the nail? Explain your answer.

16 Account for the fact that it takes more energy to remove an electron from the outer shell of an atom of:

a magnesium than sodium

b lithium than sodium.

17 **a** Select the least reactive metal from the following list: magnesium, sulfur, chlorine, fluorine, aluminium, oxygen.

b Explain your reasoning.

18 From each set of elements, select the element that has the largest first ionisation energy. Explain your reasoning.

a aluminium, gallium, indium

b aluminium, magnesium, sodium

19 The following half-equations form part of the activity series of metals. The metals are ranked in order of decreasing reactivity, as shown:

$Mg(s) \rightarrow Mg^{2+}(aq) + 2e^-$

$Al(s) \rightarrow Al^{3+}(aq) + 3e^-$

$Zn(s) \rightarrow Zn^{2+}(aq) + 2e^-$

$Fe(s) \rightarrow Fe^{2+}(aq) + 2e^-$

$Pb(s) \rightarrow Pb^{2+}(aq) + 2e^-$

$Ag(s) \rightarrow Ag^{+}(aq) + e^-$

a Which species loses its valence electrons most readily?

b Which species gains electrons most readily?

c Lead rods are placed in solutions of silver nitrate, iron(II) sulfate and magnesium chloride. In which solutions would you expect to see a coating of another metal form on the lead rod?

d Which of the metals silver, zinc or magnesium might be coated with lead when immersed in a solution of lead(II) nitrate?

20 Consider the following information relating to three metals ('R', 'S' and 'T') and solutions of their salts ($T(NO_3)_2$, RNO_3 and $S(NO_3)_2$).
Three experiments were carried out with these metals to determine their order of reactivity. Equations representing what happened in these experiments are listed below.

Experiment 1: $T(s) + 2RNO_3(aq) \rightarrow T(NO_3)_2(aq) + 2R(s)$

Experiment 2: $T(NO_3)_2(aq) + S(s) \rightarrow$ no reaction

Experiment 3: $S(s) + RNO_3(aq) \rightarrow$ no reaction

Determine the order of reactivity of the three metals, from most reactive to least reactive.

21 Would elements from List A be more likely to gain or lose electrons when they take part in chemical reactions with elements from List B? Give reasons for your answers.

List A: sodium, calcium and magnesium

List B: sulfur, oxygen and chlorine

22 Metal X has the electronic structure $1s^22s^22p^63s^23p^63d^{10}4s^24p^65s^2$. On the basis of metal X's ability to displace hydrogen from cold water, suggest where it should be placed in the following activity series: K, Na, Ca, Mg, Al, Zn, Cu, Ag.
Justify your answer.

23 Use the information provided in the following table to rank metals I, II, III and IV in order of reactivity from most reactive to least reactive. Justify your ranking in terms of each of the three properties listed in the table.

Property	Metal			
	I	II	III	IV
atomic radius (pm)	122	200	140	160
first ionisation energy (kJ mol^{-1})	745	419	738	496
electronegativity	1.9	0.8	1.3	0.9

24 Reflect on the Inquiry activity on page 352. Using the activity series of metals, explain why the iron nail corroded but the galvanised nail did not.

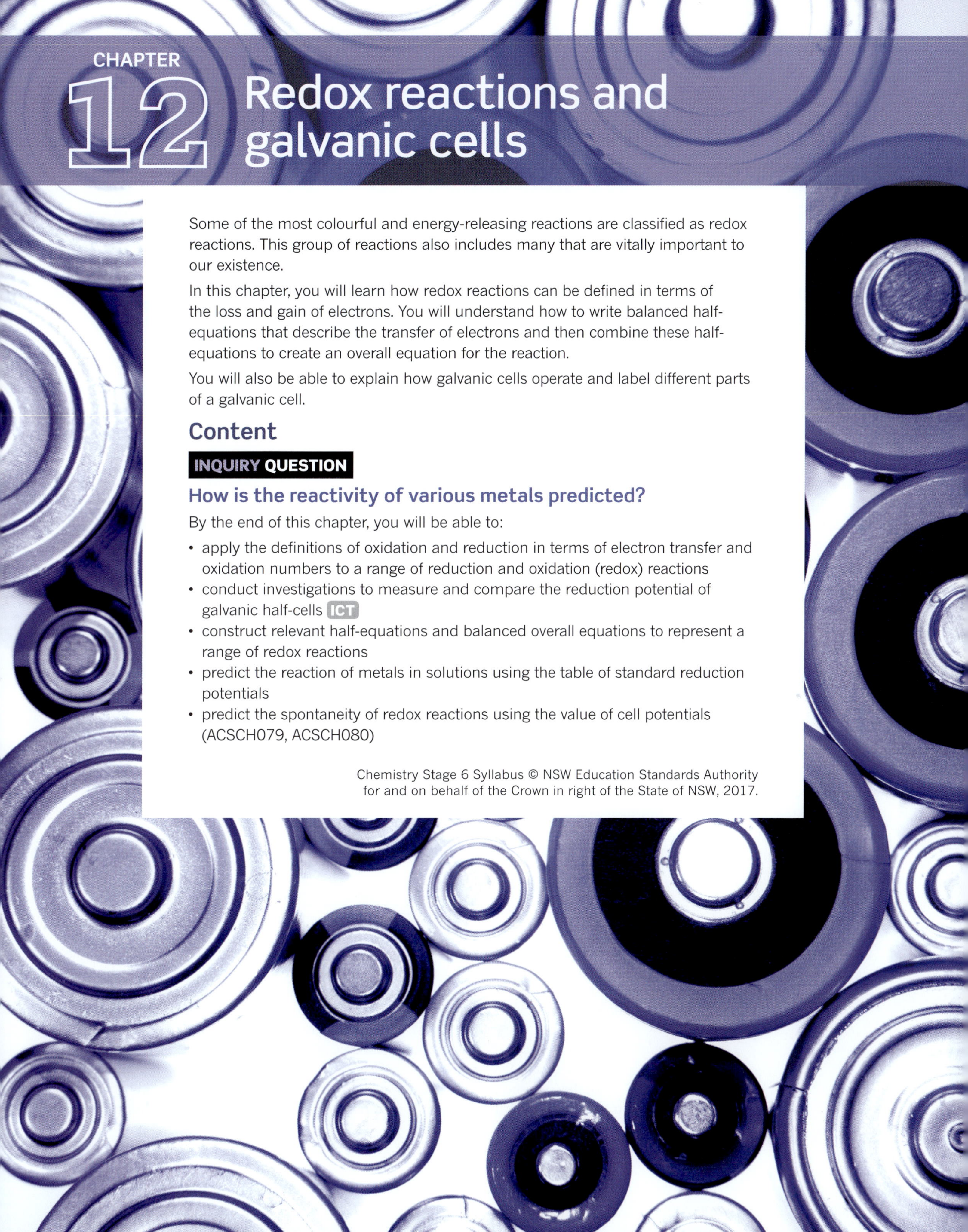

CHAPTER 12

Redox reactions and galvanic cells

Some of the most colourful and energy-releasing reactions are classified as redox reactions. This group of reactions also includes many that are vitally important to our existence.

In this chapter, you will learn how redox reactions can be defined in terms of the loss and gain of electrons. You will understand how to write balanced half-equations that describe the transfer of electrons and then combine these half-equations to create an overall equation for the reaction.

You will also be able to explain how galvanic cells operate and label different parts of a galvanic cell.

Content

INQUIRY QUESTION

How is the reactivity of various metals predicted?

By the end of this chapter, you will be able to:

- apply the definitions of oxidation and reduction in terms of electron transfer and oxidation numbers to a range of reduction and oxidation (redox) reactions
- conduct investigations to measure and compare the reduction potential of galvanic half-cells ICT
- construct relevant half-equations and balanced overall equations to represent a range of redox reactions
- predict the reaction of metals in solutions using the table of standard reduction potentials
- predict the spontaneity of redox reactions using the value of cell potentials (ACSCH079, ACSCH080)

12.1 Introducing redox reactions

CHEMISTRY INQUIRY 

Observing the corrosion of iron

Can you accelerate the corrosion of iron?

COLLECT THIS ...

- white vinegar
- 5 pieces of steel wool (non-soapy type)
- water
- camera
- 5 plastic containers or disposable plastic cups (all of equal shape and volume)

DO THIS ...

1 Draw up the table below and predict what will happen when you:

   a place a piece of steel wool into a container and leave it undisturbed for 2 days

   b place a piece of steel wool into a container, add enough water to just cover the steel wool and leave it undisturbed for 2 days

   c place a piece of steel wool into a container, add enough vinegar to just cover the steel wool and leave it undisturbed for 2 days

   d place a piece of steel wool into a container, cover with water, empty the container and squeeze out most of the water, replace the steel wool and leave it undisturbed for 2 days

   e place a piece of steel wool into a container, cover with vinegar, empty the container and squeeze out most of the vinegar, replace the steel wool and leave it undisturbed for 2 days

2 Photograph your experiments when initially set up and after 2 days.

RECORD THIS ...

Draw up this table to record your predictions and results.

Description of experiment	Prediction	Observations
a dry steel wool placed in plastic cup for 2 days		
b steel wool immersed in water for 2 days		
c steel wool immersed in vinegar for 2 days		
d steel wool soaked in water, excess water squeezed out, placed in cup for 2 days		
e steel wool soaked in vinegar, excess vinegar squeezed out, placed in cup for 2 days		

Test your predictions.

Describe what happened by recording your observations.

Present a table of your results.

REFLECT ON THIS ...

Explain the meaning of your observations.

What could you do next time to improve your experiment?

Your everyday life depends on a large number of chemical reactions. Many of these are **redox reactions** (*red*-uction; *ox*-idation). From the respiration reactions that enable your cells to produce energy and the combustion reactions that warm your home, to the reactions in the batteries that keep your mobile phone working, redox reactions are occurring within you and around you all the time. The spectacular reaction between potassium and chlorine gas shown in Figure 12.1.1 is also an example of a redox reaction.

In this section, you will learn how redox reactions are defined in terms of electron transfer and how to represent this transfer of electrons using half-equations.

REDOX REACTIONS

Early understandings of redox reactions

When chemistry evolved from the ancient study of alchemy, many of the reactions known to early chemists involved air. French chemist Antoine Lavoisier identified the reactive component of air and named it oxygen. As a result, reactions in which oxygen was a reactant were described as **oxidation** reactions. In air, the combustion of an element such as carbon, sulfur or iron produces an oxide:

$$C(s) + O_2(g) \rightarrow CO_2(g)$$

$$S(s) + O_2(g) \rightarrow SO_2(g)$$

$$4Fe(s) + 3O_2(g) \rightarrow 2Fe_2O_3(s)$$

Because elemental iron reacts readily with oxygen, iron is generally found in nature in ores containing minerals such as haematite (Fe_2O_3) and magnetite (Fe_3O_4). The iron metal used extensively for construction has been extracted from iron ore in a blast furnace.

The extraction of iron from iron ore in a blast furnace can be represented by the equation:

$$Fe_2O_3(s) + 3CO(g) \rightarrow 2Fe(l) + 3CO_2(g)$$

In this reaction, the iron(III) oxide has lost oxygen and the carbon monoxide has gained oxygen. The iron(III) oxide is described as having been **reduced** and the carbon monoxide is described as having been **oxidised**.

Oxidation and **reduction** always occur simultaneously, hence the term 'redox reaction'.

FIGURE 12.1.1 Potassium burning in chlorine gas—a spectacular example of a redox reaction.

CHEMFILE 

Origins of the words 'oxidation' and 'reduction'

Scientists first used the term 'oxidation' in the late 18th century following on from the work of Antoine Lavoisier. Lavoisier showed that the 'burning' of metals, such as mercury, involved a combination with oxygen.

The term 'reduction' was used long before this to describe the process of extracting metals from their ores. The word 'reduction' comes from the Latin *reduco*, meaning to restore. The process of metal extraction was seen as restoring the metal from its compounds, such as iron from iron oxide or copper from copper(II) oxide. The reduction of copper(II) oxide to form copper powder occurs when copper(II) oxide is heated in the presence of hydrogen or methane gas, as shown in Figure 12.1.2. Some fine particles of copper escape with the gas, causing the green flame.

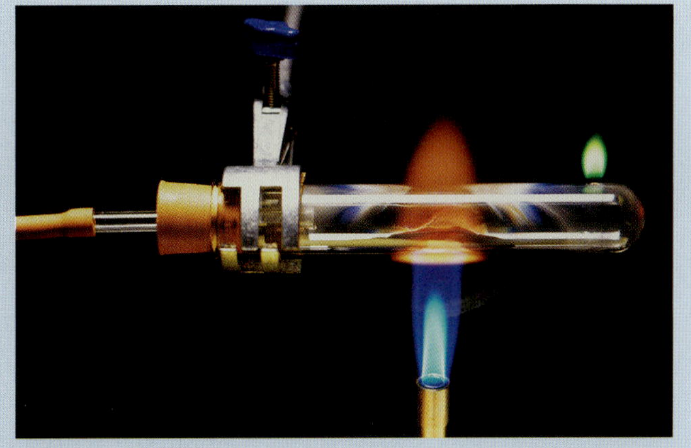

FIGURE 12.1.2 The reduction of copper(II) oxide to form copper powder occurs when it is heated in the presence of hydrogen or methane gas.

Transfer of electrons

If you have ever heated a piece of magnesium ribbon in an experiment, as shown in Figure 12.1.3, you will remember that it burns with a brilliant white flame. Magnesium oxide powder is formed.

This reaction can be represented by the equation:

$$2Mg(s) + O_2(g) \rightarrow 2MgO(s)$$

The reaction involves an electron transfer between the reactants, which can be represented by two **half-equations**.

Each magnesium atom loses two electrons to form a magnesium (Mg^{2+}) ion. The half-equation for this part of the overall reaction is written as shown. The electronic configurations are also shown:

$$Mg(s) \rightarrow Mg^{2+}(s) + 2e^-$$

2,8,2   2,8

FIGURE 12.1.3 Magnesium ribbon burns brightly when heated in air to form a white powder, magnesium oxide.

GO TO ➤ Chapter 11 page 351   GO TO ➤ Section 4.2 page 122

Notice that when electrons are lost by one of the reactants, they appear as products in the half-equation.

At the same time, each oxygen atom in the oxygen molecule (O_2) gains two electrons (i.e. four electrons per oxygen molecule):

$$O_2(g) + 4e^- \rightarrow 2O^{2-}(s)$$
$$2,6 \qquad\qquad 2,8$$

Oxidation is defined as the loss of electrons.

Reduction is defined as the gain of electrons.

Notice that when electrons are gained by a reactant, they appear as reactants in the half-equation. The electrons that are gained by the oxygen have come from the magnesium atoms.

The burning of magnesium involves the transfer of electrons from magnesium atoms to oxygen atoms. The transfer of electrons between reactants occurs in many other reactions, and this transfer of electrons provides a widely used definition of oxidation and reduction.

The definitions of oxidation and reduction can be recalled using the mnemonic (memory aid) OIL RIG, as shown in Figure 12.1.4.

Note that there is no overall loss of electrons, but there is a transfer of electrons from one atom to another. If an atom loses electrons, there must be another atom that gains electrons. Therefore, oxidation and reduction always occur simultaneously.

FIGURE 12.1.4 The mnemonic OIL RIG is a useful way to remember that Oxidation Is the Loss of electrons and Reduction Is the Gain of electrons.

Other examples of redox reactions

Many redox reactions do not involve a reaction with oxygen. The reaction between potassium and chlorine shown in Figure 12.1.1 on page 373 is an example:

$$2K(s) + Cl_2(g) \rightarrow 2KCl(s)$$

Oxidation half-equation: $K(s) \rightarrow K^+(s) + e^-$

Reduction half-equation: $Cl_2(g) + 2e^- \rightarrow 2Cl^-(s)$

Another example is the reaction between copper and sulfur:

$$Cu(s) + S(l) \rightarrow CuS(s)$$

Oxidation half-equation: $Cu(s) \rightarrow Cu^{2+}(s) + 2e^-$

Reduction half-equation: $S(l) + 2e^- \rightarrow S^{2-}(s)$

Worked example: 12.1.1

RECOGNISING OXIDATION AND REDUCTION

Write the oxidation and reduction half-equations for the reaction with the overall equation:

$$2Li(s) + Br_2(l) \rightarrow 2LiBr(s)$$

Thinking	Working
Identify the ions in the product.	LiBr is made up of Li^+ and Br^- ions.
Write the half-equation for the oxidation of the reactant that forms positive ions and balance the equation with electrons.	$Li(s) \rightarrow Li^+(s) + e^-$
Write the half-equation for the reduction of the reactant that forms negative ions and balance the equation with electrons.	$Br_2(l) + 2e^- \rightarrow 2Br^-(s)$

Worked example: Try yourself 12.1.1

RECOGNISING OXIDATION AND REDUCTION

Write the oxidation and reduction half-equations for the reaction with the overall equation:

$$2Na(s) + Cl_2(g) \rightarrow 2NaCl(s)$$

Oxidising agents and reducing agents

Just as an employment agent enables (or causes) a client to become employed, an **oxidising agent,** or **oxidant,** enables or causes another chemical to be oxidised. Similarly, a **reducing agent**, or **reductant**, enables or causes another chemical to be reduced. Redox reactions always involve an oxidising agent and a reducing agent that react together.

In the reaction between magnesium and oxygen shown in Figure 12.1.5, magnesium is being oxidised by oxygen. Therefore, oxygen is the oxidising agent. In turn, oxygen is gaining electrons from magnesium. It is being reduced by the magnesium, so magnesium is the reducing agent.

> Reducing agents cause another chemical to be reduced. In the reaction, they are oxidised. Oxidising agents cause another chemical to be oxidised. In the reaction, they are reduced.

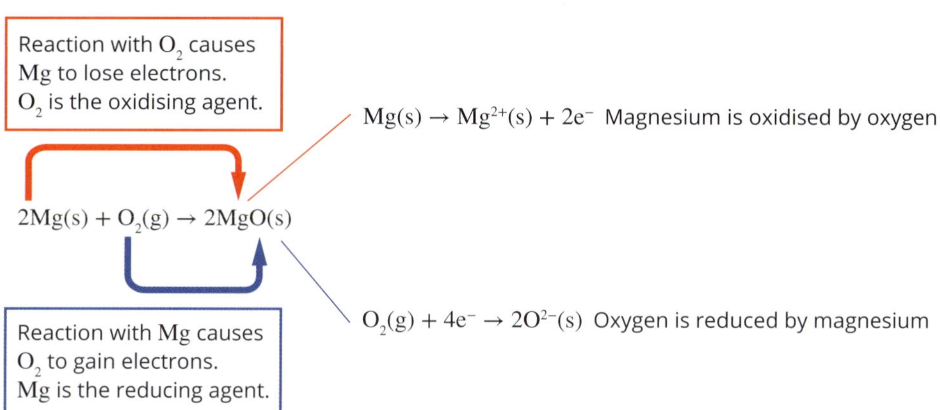

FIGURE 12.1.5 In the reaction between magnesium and oxygen, magnesium is the reducing agent and oxygen is the oxidising agent.

Since metals tend to lose electrons, they act as reducing agents.

Figure 12.1.6 summarises the list of redox terms introduced in this section.

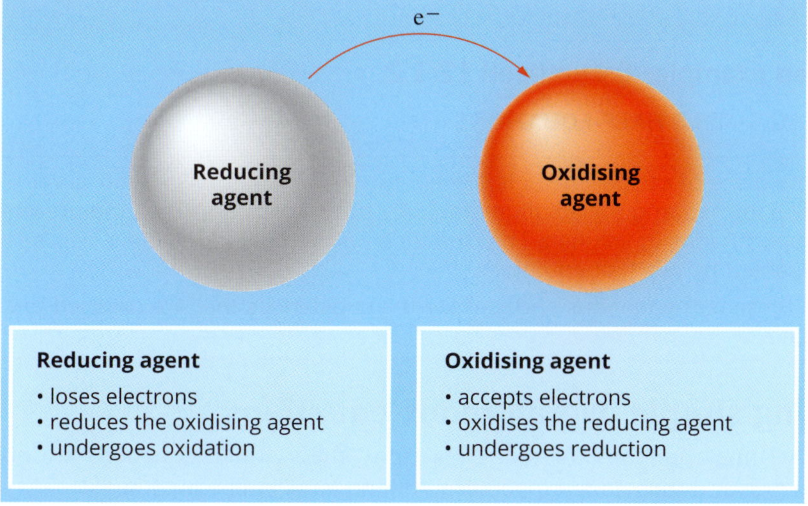

FIGURE 12.1.6 Summary of redox reaction terms.

SIMPLE REDOX EQUATIONS

Writing simple half-equations

Half-equations enable you to see the detail of what is happening in a redox reaction. Like other chemical equations, half-equations must be balanced so there is the same number of atoms of each element on each side of the arrow. Similarly, charge must also be balanced. Half-equations and overall equations should indicate the states of all the species in the reaction as well.

Worked example 12.1.2

WRITING SIMPLE HALF-EQUATIONS

When sodium metal reacts with fluorine gas (F_2), solid sodium fluoride is formed. The oxidation and reduction reactions can be represented by two half-equations.

Write these half-equations and identify the substances that are oxidised and reduced.

Thinking	Working
Identify one reactant and the product it forms and write them on each side of an equation. Balance the equation for the element.	$F_2(g) \rightarrow 2F^-(s)$
Add electrons to balance the equation for charge.	$F_2(g) + 2e^- \rightarrow 2F^-(s)$
To decide whether the reactant is oxidised or reduced, remember that oxidation is loss of electrons and reduction is gain of electrons.	Electrons are gained, so this is reduction. The $F_2(g)$ is being reduced.
Identify the second reactant and the product it forms, and write them on each side of an equation. Balance the equation for the element.	$Na(s) \rightarrow Na^+(s)$
Add electrons to balance the equation for charge.	$Na(s) \rightarrow Na^+(s) + e^-$
To decide whether the reactant is oxidised or reduced, remember that oxidation is loss of electrons and reduction is gain of electrons.	Electrons are lost, so this is oxidation. The Na(s) is being oxidised.

Worked example: Try yourself 12.1.2

WRITING SIMPLE HALF-EQUATIONS

When a piece of copper metal is placed into a silver nitrate solution, silver metal is formed and the solution gradually turns blue, indicating the presence of copper(II) ions in solution. The oxidation and reduction reactions can be represented by two half-equations.

Write these half-equations and identify the substances that are oxidised and reduced.

Writing simple overall redox equations

When writing equations for redox reactions, the two half-equations are normally written first and then added together to obtain an overall equation.

An overall equation does not show any of the electrons transferred; all of the electrons lost in the oxidation reaction are gained in the reduction reaction. One, or perhaps both, of the half-equations may need to be multiplied by a factor to ensure that the electrons balance and can be cancelled out in the overall equation.

Consider the reaction of sodium and chlorine gas. In this reaction, each Na atom is oxidised and loses one electron. Each Cl_2 molecule is reduced and gains two electrons:

$$Na(s) \rightarrow Na^+(s) + e^-$$
$$Cl_2(g) + 2e^- \rightarrow 2Cl^-(s)$$

Two Na atoms must therefore be oxidised to provide the electrons required by each chlorine molecule that is reduced. To write an overall equation for this reaction, the half-equation involving the oxidation of Na is multiplied by a factor of two before combining it with the half-equation for reduction of Cl_2:

$$2Na(s) \rightarrow 2Na^+(s) + 2e^-$$

You can now write the two half-equations and add them to find the overall equation:

$$2Na(s) \rightarrow 2Na^+(s) + 2e^-$$
$$Cl_2(g) + 2e^- \rightarrow 2Cl^-(s)$$

When the electrons have been cancelled, the overall equation is:

$$2Na(s) + Cl_2(g) \rightarrow 2NaCl(s)$$

When writing an overall redox equation from two half-equations, you need to balance the number of electrons.

Worked example 12.1.3

WRITING OVERALL REDOX EQUATIONS FROM HALF-EQUATIONS

Sodium metal is oxidised by oxygen gas in air to form solid sodium oxide.
Write the half-equations for the reaction, and hence write the balanced overall equation.

Thinking	Working
Identify one reactant and the product it forms, and write them on either side of the equation. Balance the equation for the element.	$O_2(g) \rightarrow 2O^{2-}(s)$
Add electrons to balance the equation for charge.	$O_2(g) + 4e^- \rightarrow 2O^{2-}(s)$
Identify the second reactant and the product it forms, and write them on either side of the equation. Balance the equation for the element.	$Na(s) \rightarrow Na^+(s)$
Add electrons to balance the equation for charge.	$Na(s) \rightarrow Na^+(s) + e^-$
Multiply one equation by a suitable factor to ensure that the number of electrons is balanced.	$(Na(s) \rightarrow Na^+(s) + e^-) \times 4$ $4Na(s) \rightarrow 4Na^+(s) + 4e^-$
Add the oxidation and the reduction half-equations together, cancelling out electrons so that none are in the final equation. Combine ions to create the formula of the product.	$O_2(g) + 4e^- \rightarrow 2O^{2-}(s)$ $4Na(s) \rightarrow 4Na^+(s) + 4e^-$ When the electrons have been cancelled, the overall equation is: $4Na(s) + O_2(g) \rightarrow 2Na_2O(s)$

Worked example: Try yourself 12.1.3

WRITING OVERALL REDOX EQUATIONS FROM HALF-EQUATIONS

Potassium metal is oxidised by oxygen gas in air to form solid potassium oxide.
Write the half-equations for the reaction, and hence write the balanced overall equation.

Not all oxidation and reduction half-equations involve simple ions and their elements. Many complex redox reactions involve reactants and products that have oxygen and hydrogen in their formulae.

COMPLEX REDOX EQUATIONS

Writing complex half-equations

Half-equations that involve atoms or simple ions can be written quite easily. For example, knowing that magnesium metal is oxidised to form Mg^{2+} ions in solution, you can readily write the half-equation as:

$$Mg(s) \rightarrow Mg^{2+}(aq) + 2e^-$$

However, complex half-equations involving polyatomic ions are usually less obvious. The anaesthetic nitrous oxide, or laughing gas (N_2O), can be prepared by the reduction of nitrate ions in an acidic solution:

$$2NO_3^-(aq) + 10H^+(aq) + 8e^- \rightarrow N_2O(g) + 5H_2O(l)$$

Such equations can be deduced from the following steps. The reduction of nitrate ions will be used to illustrate this process.

1. Balance all elements except hydrogen and oxygen in the half-equation.

$$2NO_3^- \rightarrow N_2O$$

2. Balance the oxygen atoms by adding water molecules.

$$2NO_3^- \rightarrow N_2O + 5H_2O$$

3. Balance the hydrogen atoms by adding H^+ ions.

$$2NO_3^- + 10H^+ \rightarrow N_2O + 5H_2O$$

4. Balance the charge in the equation by adding electrons.
   In this case, the total charge on the left-hand side is $(2 \times -1) + (10 \times +1) = +8$. The total charge on the right-hand side is 0. Make the charges equal by adding 8 electrons to the left-hand side.

$$2NO_3^- + 10H^+ + 8e^- \rightarrow N_2O + 5H_2O$$

5. Add states to complete the half-equation.

$$2NO_3^-(aq) + 10H^+(aq) + 8e^- \rightarrow N_2O(g) + 5H_2O(l)$$

> The steps outlined are for balancing redox half-equations in acidic solutions only.

When writing half- and overall equations, it is important that they are fully balanced (Figure 12.1.7). The number of each element on each side must be equal, just as with any other chemical equation. The charge on each side of the equation must also be equal. It is important to remember that the charges being equal does not mean they must be zero.

FIGURE 12.1.7 In balanced half-and overall equations, the number of atoms of each element on each side is equal, as is the total charge.

Worked example 12.1.4

BALANCING COMPLEX HALF-EQUATIONS

Write the half-equation for the reduction of an acidified solution of $Cr_2O_7^{2-}$ to aqueous Cr^{3+}.

Thinking	Working
Balance all elements except hydrogen and oxygen in the half-equation.	There are 2 Cr atoms in $Cr_2O_7^{2-}$, so 2 Cr atoms are needed on the right-hand side (RHS). $Cr_2O_7^{2-} \rightarrow 2Cr^{3+}$

Balance the oxygen atoms by adding water.	There are 7 O atoms in $Cr_2O_7^{2-}$, so 7 H_2O molecules are added to the RHS. $Cr_2O_7^{2-} \rightarrow 2Cr^{3+} + 7H_2O$
Balance the hydrogen atoms by adding H^+ ions. Acids provide a source of H^+ ions.	There are now 14 H atoms on the RHS and none on the left-hand side (LHS), so 14 H^+ ions are added to the LHS. $Cr_2O_7^{2-} + 14H^+ \rightarrow 2Cr^{3+} + 7H_2O$
Balance the charge in the equation by adding electrons.	The charge on the LHS is (–2) + (+14) = +12 and on the RHS is 2 × +3 = +6, so 6 electrons are added to the LHS to make the charges equal. $Cr_2O_7^{2-} + 14H^+ + 6e^- \rightarrow 2Cr^{3+} + 7H_2O$
Add states to complete the half-equation.	All states are (aq) except for water, which is (l). $Cr_2O_7^{2-}(aq) + 14H^+(aq) + 6e^- \rightarrow 2Cr^{3+}(aq) + 7H_2O(l)$

Worked example: Try yourself 12.1.4

BALANCING COMPLEX HALF-EQUATIONS

Write the half-equation for the reduction of an acidified solution of MnO_4^- to solid MnO_2.

Writing overall redox equations involving hydrogen and oxygen

To write an overall redox equation from complex half-equations, you still add the oxidation half-equation to the reduction half-equation, making sure that the number of electrons used in reduction equals the number of electrons released during oxidation. In addition, H^+ ions and H_2O molecules may also be present as reactants and products, which will need to be cancelled out.

Worked example 12.1.5

COMBINING COMPLEX HALF-EQUATIONS

Write balanced oxidation and reduction half-equations for the reaction in which ethanol ($C_2H_5OH(aq)$) and dichromate ion ($Cr_2O_7^{2-}(aq)$) react to form acetic acid ($CH_3COOH(aq)$) and chromium(III) ion ($Cr^{3+}(aq)$). Then write the overall equation for the reaction.

Thinking	Working
Identify one reactant and the product it forms, and write the balanced half-equation.	$C_2H_5OH(aq) + H_2O(l) \rightarrow CH_3COOH(aq) + 4H^+(aq) + 4e^-$
Identify the second reactant and the product it forms, and write the balanced half-equation.	$Cr_2O_7^{2-}(aq) + 14H^+(aq) + 6e^- \rightarrow 2Cr^{3+}(aq) + 7H_2O(l)$
Multiply one or both equation(s) by a suitable factor to ensure that the number of electrons on both sides of the arrow is equal.	Lowest common multiple = 12 $3 \times [C_2H_5OH(aq) + H_2O(l) \rightarrow CH_3COOH(aq) + 4H^+(aq) + 4e^-]$ $2 \times [Cr_2O_7^{2-}(aq) + 14H^+(aq) + 6e^- \rightarrow 2Cr^{3+}(aq) + 7H_2O(l)]$ This gives: $3C_2H_5OH(aq) + 3H_2O(l) \rightarrow 3CH_3COOH(aq) + 12H^+(aq) + 12e^-$ $2Cr_2O_7^{2-}(aq) + 28H^+(aq) + 12e^- \rightarrow 4Cr^{3+}(aq) + 14H_2O(l)$
Add the oxidation and the reduction half-equations together, cancelling electrons so that none appear in the final equation. Also cancel H_2O and H^+ if these occur on both sides of the arrow.	$3C_2H_5OH(aq) + \cancel{3H_2O(l)} \rightarrow 3CH_3COOH(aq) + \cancel{12H^+(aq)} + \cancel{12e^-}$ $2Cr_2O_7^{2-}(aq) + \overset{16H^+}{\cancel{28H^+}}(aq) + \cancel{12e^-} \rightarrow 4Cr^{3+}(aq) + \overset{11H_2O}{\cancel{14H_2O}}(l)$ $3C_2H_5OH(aq) + 2Cr_2O_7^{2-}(aq) + 16H^+(aq) \rightarrow 3CH_3COOH(aq) + 4Cr^{3+}(aq) + 11H_2O(l)$

Worked example: Try yourself 12.1.5

COMBINING COMPLEX HALF-EQUATIONS

Write balanced oxidation and reduction half-equations for the reaction in which SO_3^{2-}(aq) and ClO^-(aq) react to form H_2S(g) and ClO_3^-(aq). Then write the overall equation for the reaction.

Conjugate redox pairs

When a half-equation is written for an oxidation reaction, the reactant is a reducing agent. The product is an oxidising agent. The reactant and the product that it forms are known as a **conjugate redox pair**.

Consider the half-equation in which zinc is oxidised:

$$Zn(s) \rightarrow Zn^{2+}(aq) + 2e^-$$

The zinc metal is a reducing agent because it would cause another reactant to gain electrons. In the half-equation, Zn^{2+}(aq) is formed. This is the conjugate oxidising agent. Therefore, Zn^{2+}(aq) and Zn(s) are a conjugate redox pair.

As another example, consider the half-equation in which Cu^{2+}(aq) is reduced:

$$Cu^{2+}(aq) + 2e^- \rightarrow Cu(s)$$

The copper(II) ion is the oxidising agent because it would cause another reactant to lose electrons. In the half-equation, Cu(s) is formed. This is the conjugate reducing agent. Thus, Cu^{2+}(aq) and Cu(s) are a conjugate redox pair.

There are two conjugate redox pairs in a redox reaction, as shown in the two examples in Table 12.1.1.

An oxidising agent and its corresponding reducing agent are known as a conjugate redox pair.

TABLE 12.1.1 Identifying conjugate redox pairs in a redox reaction

	Example 1	Example 2
Overall equation	$Fe(s) + Sn^{2+}(aq) \rightarrow Fe^{2+}(aq) + Sn(s)$	$Cu(s) + 2Ag^+(aq) \rightarrow Cu^{2+}(aq) + 2Ag(s)$
Oxidation half-equation	$Fe(s) \rightarrow Fe^{2+}(aq) + 2e^-$	$Cu(s) \rightarrow Cu^{2+}(aq) + 2e^-$
Reduction half-equation	$Sn^{2+}(aq) + 2e^- \rightarrow Sn(s)$	$Ag^+(aq) + e^- \rightarrow Ag(s)$
Conjugate redox pairs	Fe^{2+}(aq)/Fe(s) Sn^{2+}(aq)/Sn(s)	Cu^{2+}(aq)/Cu(s) Ag^+(aq)/Ag(s)

12.1 Review

SUMMARY

- Redox (*red*-uction; *ox*-idation) reactions involve the transfer of electrons from one species to another.
- Oxidation and reduction always occur at the same time.
- Half-equations are used to represent oxidation and reduction.
- Oxidation is defined as the loss of electrons, e.g. $Mg(s) \rightarrow Mg^{2+}(aq) + 2e^-$
- Reduction is defined as the gain of electrons, e.g. $Br_2(aq) + 2e^- \rightarrow 2Br^-(aq)$
- The reducing agent (reductant) donates electrons to another substance, causing that substance to be reduced. The reductant is itself oxidised.
- Metals can act as reducing agents.
- The oxidising agent (oxidant) accepts electrons from another substance, causing that substance to be oxidised. The oxidant is itself reduced.
- Half-equations are added together to determine the overall redox equation. It may be necessary to multiply one or both half-equations by a factor to balance the electrons.
- To balance redox half-equations under acidic conditions:
  1. Balance all atoms except hydrogen and oxygen.
  2. Balance the oxygen atoms by adding water molecules.
  3. Balance the hydrogen atoms by adding H^+ ions.
  4. Balance the charge by adding electrons.
  5. Add states.
- When combining complex oxidation and reduction half-equations to write an overall equation, any $H^+(aq)$ and $H_2O(l)$ that appear on both sides of the arrow should be cancelled out.
- In a redox reaction, a conjugate redox pair is made up of a reducing agent and the oxidising agent that is formed, or an oxidising agent and the reducing agent that is formed.

KEY QUESTIONS

1 Identify each of the following half-equations as involving either oxidation or reduction.

a $Na(s) \rightarrow Na^+(aq) + e^-$
b $Cl_2(g) + 2e^- \rightarrow 2Cl^-(aq)$
c $S(s) + 2e^- \rightarrow S^{2-}(aq)$
d $Zn(s) \rightarrow Zn^{2+}(aq) + 2e^-$

2 Balance the following half-equations and then identify each as an oxidation or a reduction reaction.

a $Fe(s) \rightarrow Fe^{3+}(aq)$
b $K(s) \rightarrow K^+(aq)$
c $F_2(g) \rightarrow F^-(aq)$
d $O_2(g) \rightarrow O^{2-}(aq)$

3 Iron reacts with hydrochloric acid according to the ionic equation:

$$Fe(s) + 2H^+(aq) \rightarrow Fe^{2+}(aq) + H_2(g)$$

a What has been oxidised in this reaction? What is the product?
b Write a half-equation for the oxidation reaction.
c Identify the oxidising agent.
d What has been reduced in this reaction? What is the product?
e Write a half-equation for the reduction reaction.
f Identify the reducing agent.
g Identify the two conjugate redox pairs in this reaction.

4 When a strip of magnesium metal is placed in a blue solution containing copper(II) ions ($Cu^{2+}(aq)$), crystals of copper appear and the solution soon becomes paler in colour.

a Show that this reaction is a redox reaction by identifying the substance that is oxidised and the one that is reduced.
b Write a half-equation for the oxidation reaction.
c Write a half-equation for the reduction reaction.
d Write an overall redox equation.
e Identify the oxidising agent and the reducing agent.
f Explain why the solution loses some of its blue colour as a result of the reaction.

12.1 Review *continued*

5 Write complex half-equations for the:

a reduction of $MnO_2(s)$ to $Mn^{2+}(aq)$

b reduction of $MnO_4^-(aq)$ to $MnO_2(s)$

c reduction of $SO_4^{2-}(aq)$ to $H_2S(g)$

d oxidation of $SO_2(g)$ to $SO_4^{2-}(aq)$

e oxidation of $H_2S(g)$ to $S(s)$

f oxidation of $SO_3^{2-}(aq)$ to $SO_4^{2-}(aq)$.

6 Calcium metal that is exposed to the air forms an oxide coating.

a What is the formula of calcium oxide?

b What has been oxidised in this reaction?

c Write a half-equation for the oxidation reaction.

d What has been reduced in this reaction?

e Write a balanced half-equation for the reduction reaction.

f Write an overall equation for this redox reaction.

g Copy the following statement and fill in the blank spaces with the appropriate words.

Calcium has been ____________ by ____________ to calcium ions. The ______________ has gained electrons from the ______________. The oxygen has been ______________ by ______________ to oxide ions. The ____________ has lost electrons to the ____________.

12.2 Oxidation numbers

In this section, you will learn the set of rules that chemists have devised to allow a wider range of reactions to be classified as redox reactions. This involves assigning oxidation numbers to the atoms in a reaction.

You will see how oxidation numbers can be used to determine whether a reaction that does not involve the formation of ions could be classified as a redox reaction and which substances in a redox reaction have been oxidised or reduced.

You will also discover that many transition metals have multiple oxidation states and that many of these compounds can be coloured, such as in the reactions shown in Figure 12.2.1.

FIGURE 12.2.1 Many redox reactions involve colour changes. It is convenient to assign individual oxidation numbers to the atoms involved in redox reactions such as this.

OXIDATION NUMBERS

In Section 12.1, you looked at redox reactions that involved the production of ionic compounds from their elements. In these cases, it was relatively easy to deduce which element gained or lost electrons by considering the charge on the ions produced in the reaction. A species that was reduced gained electrons, becoming less positive, whereas the oxidised species lost electrons, becoming more positive.

For some redox reactions, it is harder to identify the species that are being oxidised and reduced. For example, the reaction that occurs during the wet corrosion of iron metal can be represented by the half-equations:

$$Fe(s) \rightarrow Fe^{2+}(aq) + 2e^-$$

$$O_2(g) + 2H_2O(l) + 4e^- \rightarrow 4OH^-(aq)$$

The first half-equation shows that the iron loses electrons and is oxidised. The second half-equation represents a reduction reaction because electrons have been gained. However, it is not obvious whether oxygen or hydrogen has gained these electrons.

It is possible to determine which element has been oxidised and which has been reduced in this reaction by looking at **oxidation numbers**. Oxidation numbers are also called oxidation states.

Oxidation numbers have no physical meaning—they do not indicate a formal charge or the physical or chemical properties of the substance. However, they are a useful tool for identifying which atoms have been oxidised and which atoms have been reduced.

Oxidation number rules

Oxidation numbers are assigned to elements involved in a reaction by following a specific set of rules. In applying these rules, we regard all compounds and polyatomic ions as though they are individual ions.

Table 12.2.1 (page 384) describes the rules for determining oxidation numbers. In the examples, the oxidation number of an element is placed above its symbol. The plus or minus sign precedes the number, distinguishing the oxidation number from the charge on an ion, where the sign is generally placed after the number. For example, the oxide ion (O^{2-}) has a charge of 2– and an oxidation number of –2. While the values are the same in this instance, it is important to remember that oxidation states do not always indicate the charge on the species.

TABLE 12.2.1 Rules for determining oxidation numbers of elements in compounds

Rule	Examples
1 The oxidation number of a free element is zero.	$\overset{0}{Na}$, $\overset{0}{C}$, $\overset{0}{Cl_2}$, $\overset{0}{P_4}$
2 The oxidation number of a simple ion is equal to the charge on the ion.	$\overset{+1}{Na^+}$, $\overset{-1}{Cl^-}$, $\overset{+2}{Mg^{2+}}$, $\overset{-2}{O^{2-}}$, $\overset{+3}{Al^{3+}}$, $\overset{-3}{N^{3-}}$
3 In compounds, some elements have oxidation numbers that are regarded as fixed, except in a few exceptional circumstances.	
a Main group metals have an oxidation number equal to the charge on their ions.	Ionic compounds: $\overset{+1}{K}Cl$, $\overset{+2}{Mg}SO_4$
b Hydrogen has an oxidation number of +1 when it forms compounds with non-metals. Exception: In metal hydrides the oxidation number of hydrogen is −1.	Compounds of H: $\overset{+1}{H_2}O$ Metal hydrides: $Na\overset{-1}{H}$, $Ca\overset{-1}{H_2}$
c Oxygen usually has an oxidation number of −2. Exceptions: In compounds with fluorine, oxygen has a positive oxidation number. In peroxides, oxygen has an oxidation number of −1.	Compounds of O: $H_2\overset{-2}{O}$ Peroxides: $H_2\overset{-1}{O_2}$, $Ba\overset{-1}{O_2}$
4 The sum of the oxidation numbers in a neutral compound is zero.	$\overset{+4}{C}\overset{-2}{O_2}$ Note that in CO_2, the oxidation number of each oxygen atom is written as −2. It is not written as −4 for two O atoms.
5 The sum of the oxidation numbers in a polyatomic ion is equal to the charge on the ion.	$\overset{+6}{S}\overset{-2}{O_4}{}^{2-}$, $\overset{-3}{N}\overset{+1}{H_4}{}^+$
6 The most electronegative element is assigned the negative oxidation number.	$\overset{+2}{O}\overset{-1}{F_2}$

Common oxidation numbers of the first 36 elements in their compounds are shown in the periodic table in Figure 12.2.2. Transition metals and some non-metals can have a range of oxidation numbers; these are usually calculated after applying the rules for all other elements in the compound.

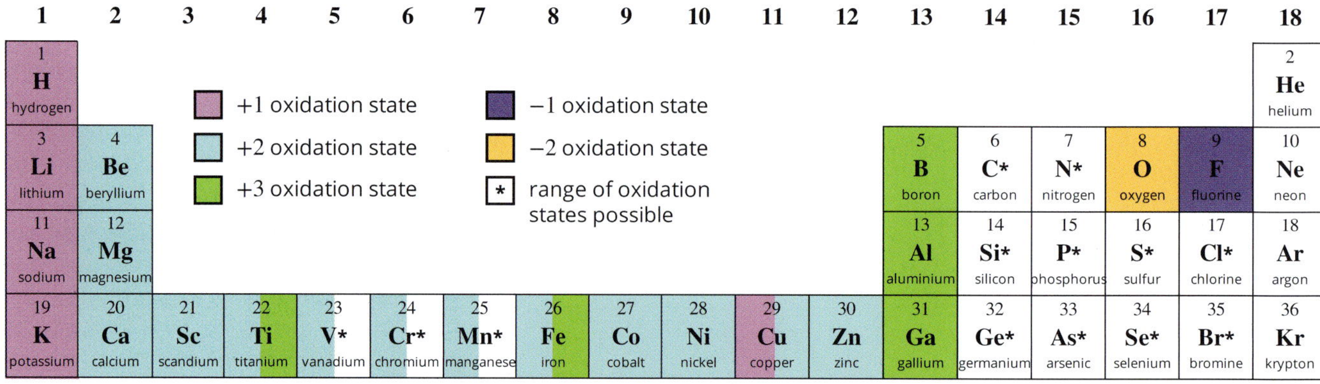

FIGURE 12.2.2 Part of the periodic table showing the most common oxidation numbers of some elements.

Calculating oxidation numbers

For a compound containing several elements, you can use algebra and the rules given in Table 12.2.1 to calculate the oxidation number of an element.

For example, to find the oxidation number of sulfur in H_2SO_4, there is a rule for hydrogen and for oxygen, which leaves sulfur as the only unknown. If you let the oxidation number of S equal x, the following expression can be written to solve for x:

$$(2 \times +1) + x + (4 \times -2) = 0$$
$$2 + x - 8 = 0$$
$$x - 6 = 0$$
$$x = +6$$

Later in this section, you will learn how to use oxidation numbers to determine whether a substance has been oxidised or reduced.

Worked example 12.2.1

CALCULATING OXIDATION NUMBERS

Use the rules in Table 12.2.1 to determine the oxidation number of each element in $KClO_4$.

Thinking	Working
Identify an element that has a set value.	K is a main group metal in group 1. Applying rule 3a, the oxidation number of potassium is +1.
Identify any other elements that have set values.	According to rule 3c, oxygen has an oxidation number of –2 unless attached to fluorine or in a peroxide.
Use algebra to work out the oxidation number of other elements.	Let the oxidation number of chlorine in $KClO_4$ be x. Solve the sum of the oxidation numbers for x: $+1 + x + (4 \times -2) = 0$ $+1 + x - 8 = 0$ $x - 7 = 0$ $x = +7$
Write oxidation numbers above the elements in the formula.	$\overset{+1}{K}\overset{+7}{Cl}\overset{-2}{O}_4$ Note that the oxidation number of oxygen is written as –2 (not as –8), even though there are four oxygen atoms in the formula.

Worked example: Try yourself 12.2.1

CALCULATING OXIDATION NUMBERS

Use the rules in Table 12.2.1 to determine the oxidation number of each element in $NaNO_3$.

USING OXIDATION NUMBERS

Using oxidation numbers to name chemicals

Transition elements can form ions with a number of different charges. This means that many transition metals have variable oxidation numbers. For example, there are two compounds that can be called iron chloride: $FeCl_2$ and $FeCl_3$.

Using the rules in Table 12.2.1 (on page 384), you can see that the chloride ion has an oxidation number of −1. In $FeCl_2$ this means the oxidation number of iron is +2, whereas in $FeCl_3$ the oxidation number is +3.

To distinguish between the two iron chlorides, insert Roman numerals representing the appropriate oxidation number in the name.

- $FeCl_2$ is named iron(II) chloride.
- $FeCl_3$ is named iron(III) chloride.

For elements that can have variable oxidation states, the use of a Roman numeral in the name indicates the specific oxidation state of the element.

When naming non-metal compounds, you can also use Roman numerals to show the oxidation number of an element such as nitrogen that has several possible oxidation states. Nitrogen dioxide (NO_2) is also called nitrogen(IV) oxide, while nitric oxide (NO) is nitrogen(II) oxide. This method of naming makes it much easier to determine the formula from the name of the oxide.

Using oxidation numbers to identify oxidation and reduction

You can use the concept of oxidation numbers to extend the definition of oxidation and reduction.

In this new definition, a change in oxidation numbers indicates that a redox reaction has taken place. This can be used as an alternative definition of oxidation and reduction, instead of our earlier definition involving loss and gain of electrons. It is particularly useful for non-ionic compounds, for which it is difficult to determine whether electrons have been transferred.

It can now be stated that:

- oxidation involves an increase in oxidation number
- reduction involves a decrease in oxidation number.

Remember that oxidation and reduction always occur together in a redox reaction. One process cannot happen without the other.

Unless there is a change in the oxidations number of all elements in a reaction, the reaction is not a redox reaction.

Oxidation numbers can be used to analyse the equation of a reaction and determine whether it represents a redox process.

An increase in oxidation number indicates the element was oxidised. A decrease in oxidation number indicates the element was reduced.

Many everyday processes involve redox reactions. Combustion reactions, where fuel is burned to produce heat while also producing CO_2 and water, are an example of redox reactions that are very important in our society. The equation for the burning of carbon in excess oxygen is:

$$C(s) + O_2(g) \rightarrow CO_2(g)$$

At first glance, the reaction may not seem like a redox reaction. Because none of the species are ionic compounds, it is not clear which reactant is losing or gaining electrons. Using oxidation numbers though, you can identify both an oxidation process and a reduction process for the reaction.

$$\overset{0}{C}(s) + \overset{0}{O}_2(g) \rightarrow \overset{+4}{C}\overset{-2}{O}_2(g)$$

In this reaction, the carbon is oxidised because its oxidation number increases from 0 to +4; the oxygen is reduced because its oxidation number decreases from 0 to −2.

Worked example 12.2.2

USING OXIDATION NUMBERS TO IDENTIFY OXIDATION AND REDUCTION IN AN EQUATION

Use oxidation numbers to determine which element has been oxidised and which has been reduced in the following equation:

$$CH_4(g) + 2O_2(g) \rightarrow CO_2(g) + 2H_2O(l)$$

Thinking	Working
Determine the oxidation numbers of one of the elements on each side of the equation.	Choose C as the first element. $\overset{-4}{C}H_4(g) + 2O_2(g) \rightarrow \overset{+4}{C}O_2(g) + 2H_2O(l)$
Assess whether the oxidation number has changed. If so, identify whether it has increased (oxidation) or decreased (reduction).	The oxidation number of C has increased from –4 to +4, so the carbon in CH_4 has been oxidised.
Determine the oxidation numbers of a second element on the left-hand and the right-hand side of the equation.	Choose oxygen as the second element. $CH_4(g) + 2\overset{0}{O}_2(g) \rightarrow C\overset{-2}{O}_2(g) + 2H_2\overset{-2}{O}(l)$
Assess whether the oxidation number has changed. whether so, identify if it has increased (oxidation) or decreased (reduction).	The oxidation number of O has decreased from 0 to –2, so O_2 has been reduced.
Continue this process until the oxidation numbers of all elements have been determined.	Determine the oxidation numbers of hydrogen. $C\overset{+1}{H}_4(g) + 2O_2(g) \rightarrow CO_2(g) + 2\overset{+1}{H}_2O(l)$ The oxidation number of H has not changed.

Worked example: Try yourself 12.2.2

USING OXIDATION NUMBERS TO IDENTIFY OXIDATION AND REDUCTION IN AN EQUATION

Use oxidation numbers to determine which element has been oxidised and which has been reduced in the following equation:

$$CuO(s) + H_2(g) \rightarrow Cu(s) + H_2O(l)$$

Using oxidation numbers to identify conjugate redox pairs

When a half-equation is written for an oxidation reaction, the reactant that loses electrons is a reducing agent. The product that gains electrons is an oxidising agent. In Section 12.1, you learnt that the reactant and the product of a half-equation form a conjugate redox pair.

For example, in the half-equation for the oxidation of zinc:

$$Zn(s) \rightarrow Zn^{2+}(aq) + 2e^-$$

Zinc metal (Zn) is a reducing agent and it forms $Zn^{2+}(aq)$, an oxidising agent. Zn(s) and $Zn^{2+}(aq)$ form a conjugate redox pair: $Zn^{2+}(aq)/Zn(s)$.

In the $Zn^{2+}(aq)/Zn(s)$ half-equation, the oxidation number of zinc increases from 0 to +2. The increase in the oxidation number of zinc indicates that it is an oxidation half-reaction.

In a reduction half-equation, the reactant is an oxidising agent and will gain electrons. The product formed is a reducing agent. Therefore, another conjugate redox pair is present in the redox reaction.

When listing conjugate redox pairs, it is best to include the states of both the oxidising and reducing agents.

For example, consider the half-equation for the reduction of $Ag^+(aq)$:

$$Ag^+(aq) + e^- \rightarrow Ag(s)$$

$Ag^+(aq)$ is an oxidising agent and forms Ag(s), which is a reducing agent. $Ag^+(aq)$ and Ag(s) are also a conjugate redox pair. In this case, the oxidation number of silver decreases from +1 to 0, indicating that this is a reduction half-equation.

The relationship between changes in oxidation numbers and conjugate redox pairs can be seen by following the colour-coding in the equation in Figure 12.2.3. One conjugate redox pair is red and the other one is blue.

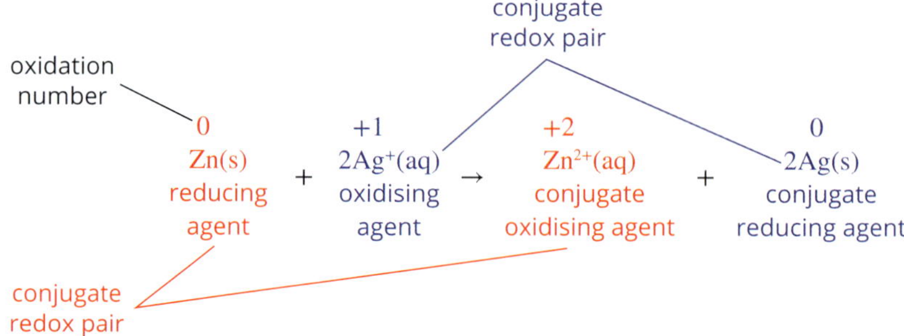

FIGURE 12.2.3 The change in oxidation numbers seen in conjugate redox pairs. For each redox reaction, there are two conjugate redox pairs.

12.2 Review

SUMMARY

- Oxidation numbers are calculated according to a set of rules.
  - Free elements have an oxidation number of 0.
  - In ionic compounds composed of simple ions, the oxidation number is equal to the charge on the ion.
  - Oxygen in a compound usually has an oxidation number of –2.
  - Hydrogen in a compound usually has an oxidation number of +1.
  - The sum of the oxidation numbers in a neutral compound is 0.
  - The sum of the oxidation numbers in a polyatomic ion is equal to the charge on the ion.
- Transition metals and some non-metals have variable oxidation numbers that can be calculated using the rules above.
- An increase in the oxidation number of an element in a reaction indicates oxidation has occurred.
- A decrease in the oxidation number of an element in a reaction indicates reduction has occurred.
- For oxidation to occur, there must be a corresponding reduction.
- If there is no change in the oxidation number of all elements in the equation for a reaction, then the reaction is not a redox reaction.
- A conjugate redox pair consists of an oxidising agent (a reactant) and the reducing agent (a product) that is formed when the oxidising agent gains electrons. In this case, the oxidation number of the oxidising agent decreases.
- The other conjugate redox pair in a redox reaction is made up of a reducing agent (a reactant) and the oxidising agent (a product) that is formed when the reducing agent loses electrons. In this case, the oxidation number of the reducing agent increases.

- The figure to the right summarises the redox terms that you need to understand from this section. This information is built up from the basic principles of redox that were introduced in Section 12.1.

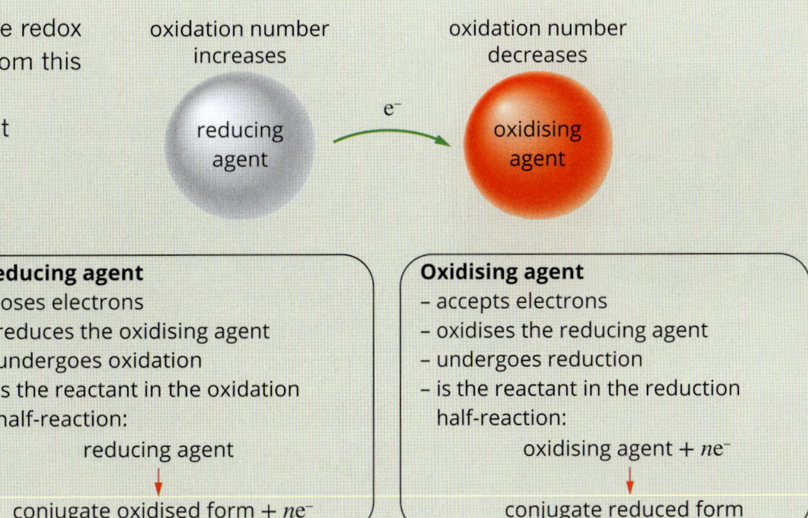

KEY QUESTIONS

1 State the oxidation number of carbon in:
 a CO
 b CO_2
 c CH_4
 d C (graphite)
 e HCO_3^-

2 Which one or more of the following substances contain manganese in the +6 oxidation state: $MnCl_2$, $MnCl_3$, MnO_2, K_2MnO_4, $KMnO_4$?

3 Find the oxidation numbers of each element in the following compounds or ions. Hint: For ionic compounds, use the charge on each ion to help you.
 a CaO
 b $CaCl_2$
 c HSO_4^-
 d MnO_4^-
 e F_2
 f SO_3^{2-}
 g $NaNO_3$
 h $K_2Cr_2O_7$

4 Assign oxidation numbers to each element in these equations, and hence identify the oxidising agents and reducing agents.
 a $Mg(s) + Cl_2(g) \rightarrow MgCl_2(s)$
 b $2SO_2(g) + O_2(g) \rightarrow 2SO_3(g)$
 c $Fe_2O_3(s) + 3CO(g) \rightarrow 2Fe(s) + 3CO_2(g)$
 d $2Fe^{2+}(aq) + H_2O_2(aq) + 2H^+(aq) \rightarrow 2Fe^{3+}(aq) + 2H_2O(l)$

5 For each of the following redox reactions, complete the table to show the conjugate redox pairs.

Redox reaction	Conjugate redox pair (oxidation process)	Conjugate redox pair (reduction process)
$Na(s) + Ag^+(aq) \rightarrow Na^+(aq) + Ag(s)$		
$Zn(s) + Cu^{2+}(aq) \rightarrow Zn^{2+}(aq) + Cu(s)$		
$2K(s) + Cl_2(g) \rightarrow 2K^+(s) + 2Cl^-(s)$		

12.3 Galvanic cells

Electronic devices such as mobile phones, notebook computers, cameras and hearing aids all depend on small portable sources of electricity: cells and batteries (Figure 12.3.1). Portable energy in the form of cells and batteries enables you to operate electrical equipment without the restrictions of a power cord.

The demand for electronic devices has stimulated the production of a variety of cells, from tiny button cells for watches and calculators, to the huge batteries used to operate lighthouses. The energy provided by cells and batteries may be more expensive than energy from other sources, such as fossil fuels, but this cost is offset by their convenience.

In this section, you will find out how galvanic cells are constructed, and how they can provide you with a source of electrical energy.

FIGURE 12.3.1 Your way of life depends on cells and batteries.

INTRODUCING GALVANIC CELLS: THE DANIELL CELL

A **galvanic cell** is a type of electrochemical cell in which chemical energy is converted into electrical energy. The cells in your mobile phone and laptop are galvanic cells.

If several cells are connected in series to obtain a higher potential difference or 'voltage', the combination of cells is called a **battery**. The term 'battery' strictly only applies to a combination of cells, but it is in everyday use to describe single cells as well.

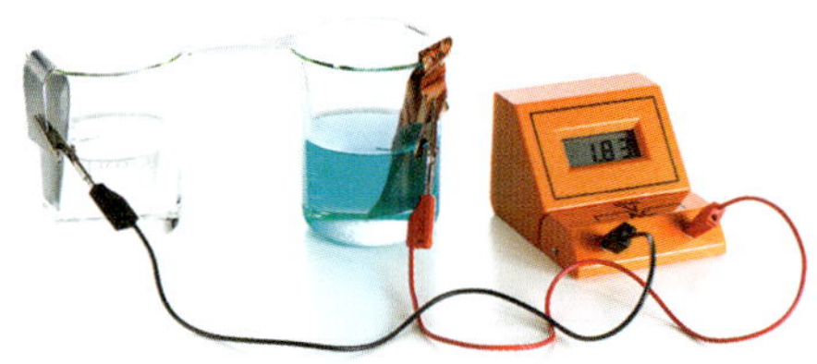

FIGURE 12.3.2 Construction of a galvanic cell from simple laboratory equipment.

Figure 12.3.2 shows how you can produce a galvanic cell from simple laboratory equipment.

In Figure 12.3.3, you can see a diagram of a cell called the Daniell cell, named after the scientist who invented it in 1836, John Daniell. The cell produces an electric current that flows through the wire and light globe. This part of the cell is called the **external circuit**. The globe converts the electrical energy of the current into light and heat.

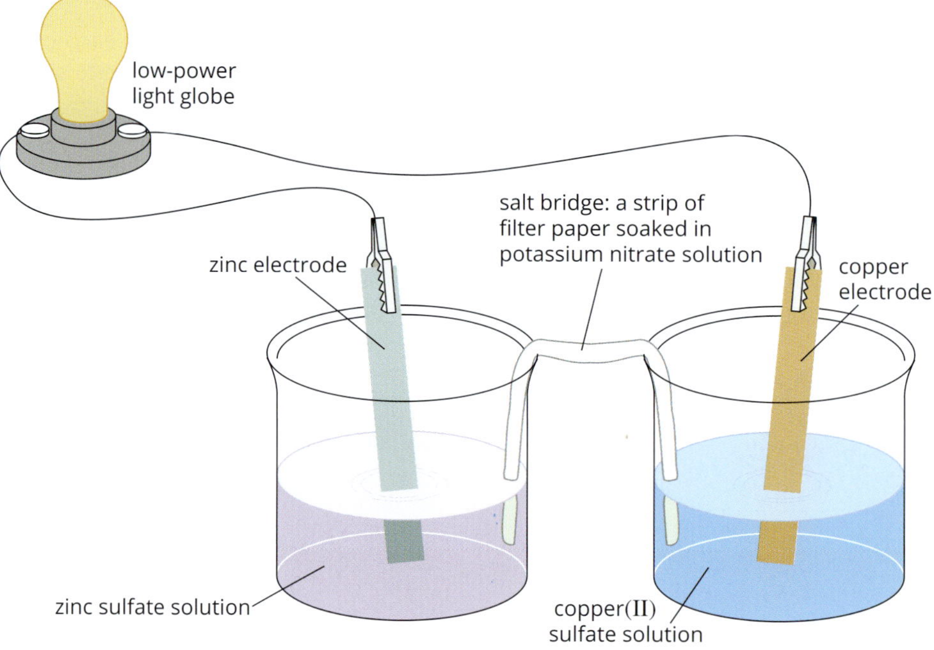

FIGURE 12.3.3 The Daniell cell is a type of galvanic cell.

The current flows because a chemical reaction is taking place in the cell. If you leave the cell with a light globe connected for several hours, you will see evidence of this reaction occurring: the zinc metal corrodes, the copper metal becomes covered with a furry dark brown deposit and the blue copper(II) sulfate solution loses some of its colour.

If you replace the light globe with a **galvanometer** (an instrument for detecting electric current), the galvanometer will indicate that electrons flow from the zinc **electrode** through the wire to the copper electrode. Current flows only if the two halves of the cell are connected by a **salt bridge**. A salt bridge is often made from filter paper soaked in a relatively unreactive **electrolyte**, such as a solution of potassium nitrate.

These observations suggest the following explanations about what is occurring in a galvanic cell:

- The reaction in the cell is a redox reaction because electrons are being transferred.
- The zinc electrode corrodes because the zinc metal forms zinc ions in solution:

$$Zn(s) \rightarrow Zn^{2+}(aq) + 2e^-$$

- The oxidation of the zinc metal releases electrons, which flow through the wire to the copper electrode.
- Electrons are accepted by copper(II) ions in the solution when the ions collide with the copper electrode:

$$Cu^{2+}(aq) + 2e^- \rightarrow Cu(s)$$

- The copper metal that is formed deposits on the electrode as a dark brown coating.

In this cell, there is a transformation of chemical energy to electrical energy. Figure 12.3.4 shows the processes that occur during the operation of the cell. The equation for the overall reaction is found by adding the two half-equations:

$$Cu^{2+}(aq) + Zn(s) \rightarrow Cu(s) + Zn^{2+}(aq)$$

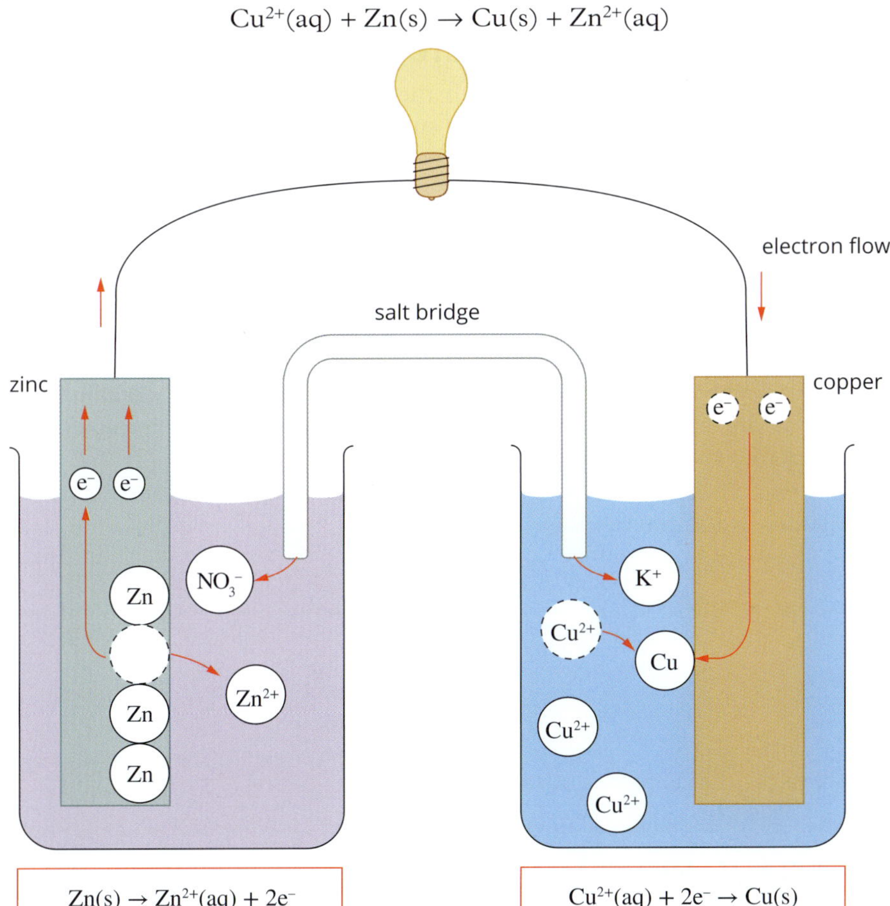

FIGURE 12.3.4 A representation of the electrode reactions in a Daniell cell. The salt bridge contains potassium nitrate solution.

This redox reaction is described as a **spontaneous reaction** because it occurs naturally. You will learn more about spontaneous reactions in Chapter 16.

GO TO ➤ Section 16.2 page 509

Copper(II) ions act as the oxidising agent and zinc metal acts as the reducing agent.

CHEMFILE S

A new rechargeable battery that improves environmental outcomes

In 1899, the first nickel–cadmium battery was invented. This battery was one of the first commercially available, rechargeable batteries. Most of these batteries ended up in landfill, where the cadmium inside the batteries could leak out over time and enter the local environment. The other main rechargeable battery through the 20th century was the lead–acid battery, shown in Figure 12.3.5; this battery has most commonly been used to start car engines. The lead–acid battery had similar issues to the nickel–cadmium battery, as it leaked another dangerous metal, lead, into the local environment when it was dumped in landfill. Both lead and cadmium have negative effects on humans, animals and plants, such as lowering bone strength and contributing to lung cancer.

In the 1970s, the first lithium-ion battery, now found in most tablets and smart phones, was proposed and constructed. By the 1990s, lithium-ion batteries were becoming more widely available and now make up around 70% of the market for rechargeable batteries. The metals present in the lithium-ion battery are much less dangerous for the environment and for people.

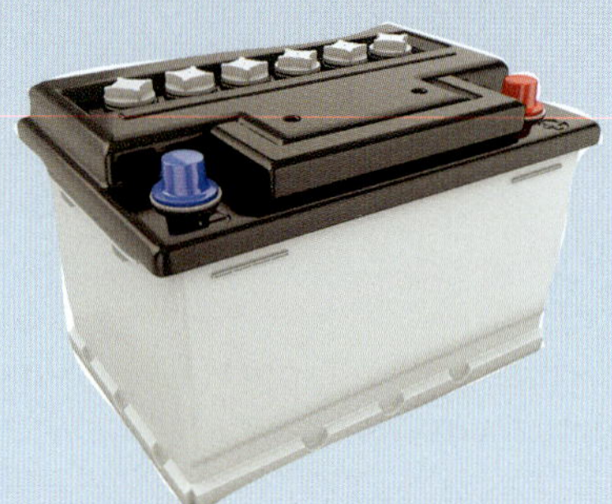

FIGURE 12.3.5 Lead–acid batteries are used to start cars, trucks and motorcycles.

FIGURE 12.3.6 Cu^{2+}(aq) ions in blue-coloured copper(II) sulfate solution reacting directly with a strip of zinc metal.

Energy transformations in direct reactions

You may have seen a similar reaction to the one that occurs in a Daniell cell if you have copper-plated a piece of metal, such as zinc. When zinc is immersed in an aqueous solution containing Cu^{2+}(aq) ions, the metal becomes coated in dark brown copper (Figure 12.3.6). At the same time, thermal energy is produced, which escapes into the surrounding environment as heat.

The overall equation for this metal displacement reaction is:

$$Cu^{2+}(aq) + Zn(s) \rightarrow Cu(s) + Zn^{2+}(aq)$$

If the reactants, Cu^{2+}(aq) and Zn(s), are allowed to come into direct contact with each other, their chemical energy is transformed directly to thermal energy. However, in a galvanic cell, the half-reactions occur in separate containers, and the electrons are transferred by the external circuit so that chemical energy is transformed into electrical energy.

The energy changes that occur in galvanic cells and when reactants undergo direct reaction are summarised in Figure 12.3.7.

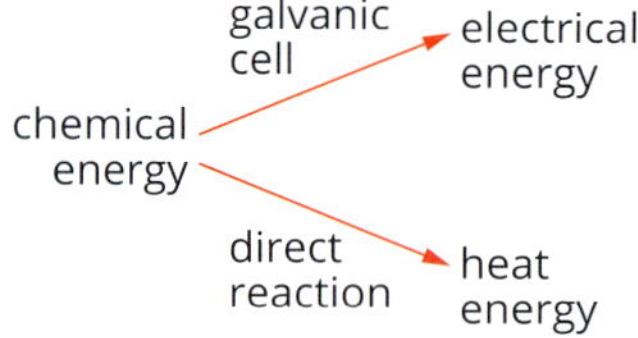

FIGURE 12.3.7 Energy transformations that occur when reactants are separated in a galvanic cell or when the reactants undergo direct reaction.

HOW A GALVANIC CELL OPERATES

A galvanic cell is designed so that half-reactions occur in two separate compartments of the cell. Because the oxidising agent and reducing agent do not come into direct contact with each other, electrons can only be transferred through an external circuit connecting the negative and positive electrodes.

This flow of electrons is an electric current. Therefore, the chemical energy of the reactants is transformed into electrical energy.

Half-cells

A galvanic cell can be regarded as consisting of two **half-cells**. Each half-cell contains an electrode in contact with a solution (Figure 12.3.8). In the Daniell cell, one half-cell contains Cu(s) and Cu^{2+}(aq); the other contains Zn(s) and Zn^{2+}(aq). The species present in each half-cell forms a conjugate redox pair (an oxidising agent and its corresponding reduced form).

If one member of the conjugate pair in a half-cell is a metal, it is usually used as the electrode. Some redox pairs, such as Br_2(aq)/Br^-(aq) and Fe^{3+}(aq)/Fe^{2+}(aq), do not involve solid metals. If no metal is present, an inert (unreactive) electrode, such as platinum or graphite, is used, as shown in Figure 12.3.8.

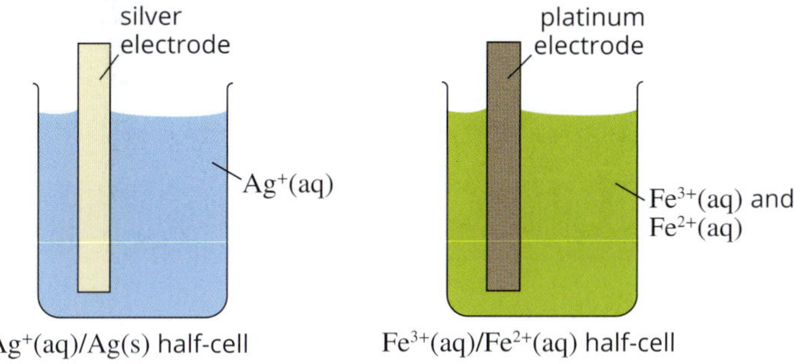

FIGURE 12.3.8 Typical half-cells. The Fe^{3+}(aq)/Fe^{2+}(aq) half-cell on the right does not contain a solid metal to act as an electrode, so an inert platinum electrode must be added.

In some half-cells, one of the conjugate pairs may be a gas. In such cases, a special 'gas electrode' like the one shown in Figure 12.3.9 for the H^+(aq)/H_2(g) half-cell is used. Note that half-cells usually contain other species not involved in the reaction, such as spectator ions and the solvent.

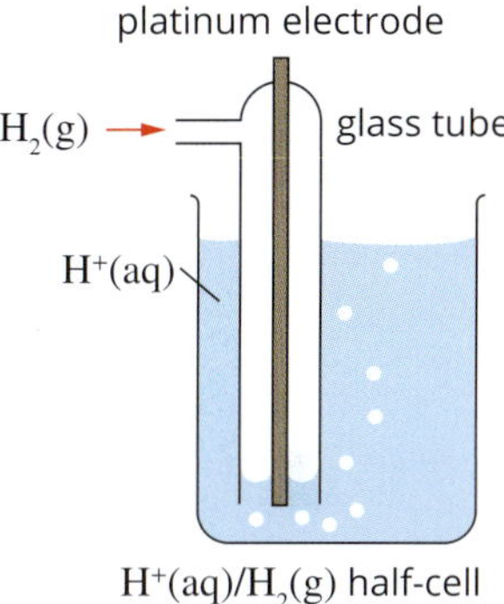

FIGURE 12.3.9 A H^+(aq)/H_2(g) half-cell consists of a platinum rod in a solution of H^+(aq) with H_2 gas bubbling through the solution.

The electrode at which oxidation occurs is called the **anode.** In a galvanic cell, the anode, where electrons are released, is described as the negative terminal. The electrode at which reduction occurs is called the **cathode**. The cathode, where electrons are gained, is the positive terminal in a galvanic cell.

The purpose of the salt bridge

The salt bridge contains ions that are free to move so that they can balance charges formed in the two compartments. *Cat*ions move towards the *cat*hode and *an*ions move towards the *an*ode.

Without a salt bridge, the solution in one compartment in the galvanic cell would accumulate negative charge and the solution in the other compartment would accumulate positive charge as the reaction proceeded. Such accumulation of charge would stop the reaction very quickly and, hence, prevent further reaction.

The salt bridge is also called the internal circuit.

> Cations in the salt bridge move towards the cathode and anions in the salt bridge move towards the anode.

WRITING HALF-CELL EQUATIONS

If a conjugate redox pair consists of an element and its corresponding ion, then the half-equation is relatively easy to write. For instance, knowing that a reduction reaction involves the conjugate redox pair of Zn^{2+} ions and Zn, you can quickly write the half-cell equation as:

$$Zn^{2+}(aq) + 2e^- \rightarrow Zn(s)$$

Half-cell equations involving polyatomic ions may be more complex to write. The equation for the reduction reaction in a half-cell containing the dichromate ion ($Cr_2O_7^{2-}$) and Cr^{3+} ion redox pair is:

$$Cr_2O_7^{2-}(aq) + 14H^+(aq) + 6e^- \rightarrow 2Cr^{3+}(aq) + 7H_2O(l)$$

You learnt how to write half-equations such as these in Section 12.1.

> In balanced half- and overall equations:
> - the number of atoms of each element is equal on both sides
> - the total charge on each side is equal.

The number of electrons lost in the oxidation reaction must equal the number of electrons gained in the reduction reaction.

Writing an overall equation for a cell reaction

The half-equations for the oxidation and reduction reactions that occur in a cell can be added together to obtain an overall (full) equation. An overall equation does not show any electrons; all the electrons lost in the oxidation reaction are gained in the reduction reaction. You may need to multiply one or both half-equations by a factor to ensure that the electrons balance and can be cancelled out in the overall equation.

DRAWING AND LABELLING A DIAGRAM OF A GALVANIC CELL

If you know the reaction occurring in a galvanic cell, then you can draw a diagram of the cell identifying key features, such as the anode, cathode, electrode polarity, direction of electron flow and direction of the flow of ions.

For example, consider a cell with the cell reaction:

$$Cu(s) + Cl_2(g) \rightarrow Cu^{2+}(aq) + 2Cl^-(aq)$$

From this reaction, you can see that:

- copper metal is at the anode (because copper is oxidised and oxidation occurs at the anode)
- chlorine gas is present at the cathode (because chlorine is reduced and reduction occurs at the cathode).

As in all galvanic cells:

- electrons flow through the external circuit from the anode (negative) to the cathode (positive)
- anions flow in the internal circuit to the anode and cations flow towards the cathode.

This information is used to draw and label the diagram of the cell as shown in Figure 12.3.10.

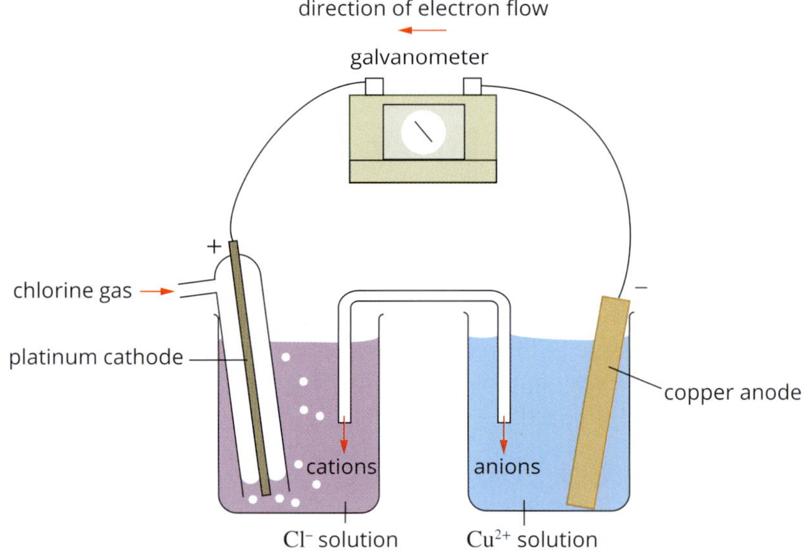

$Cl_2(g) + 2e^- \rightarrow 2Cl^-(aq)$ $Cu(s) \rightarrow Cu^{2+}(aq) + 2e^-$

FIGURE 12.3.10 A diagram of the cell containing $Cl_2(g)/Cl^-(aq)$ and $Cu^{2+}(aq)/Cu(s)$ half-cells.

12.3 Review

SUMMARY

- The reactions that occur in galvanic cells are spontaneous.
- In a galvanic cell, chemical energy is converted directly into electrical energy in a redox reaction.
- A cell is made from two half-cells.
- Each half-cell contains a conjugate redox pair.
- An oxidation reaction occurs in one half-cell and a reduction reaction occurs in the other half-cell.
- The electrode in the half-cell in which oxidation occurs is called the anode; the electrode in the half-cell in which reduction occurs is called the cathode.
- In galvanic cells, the anode is negative and the cathode is positive.
- Electrons flow through the external circuit from the anode to the cathode.
- A salt bridge allows a cell to produce electricity by preventing the accumulation of charge. Cations in the salt bridge move towards the cathode and anions move towards the anode.
- If the reactants in a galvanic cell reaction are allowed to come into direct contact, chemical energy is converted into heat energy rather than electrical energy.
- These key concepts about a galvanic cell are summarised in the figure below.

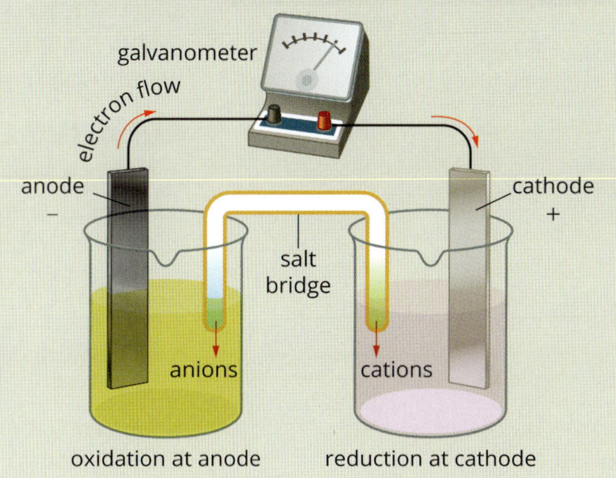

KEY QUESTIONS

1 Which one of the following statements about a galvanic cell is correct?

A The electrode where oxidation occurs is the cathode.

B Electrons flow towards the electrode where oxidation occurs.

C Anions flow into the half-cell containing the electrode where reduction occurs.

D The electrode where oxidation occurs has a negative polarity.

2 Draw labelled diagrams of the following half-cells.

a $Ni^{2+}(aq)/Ni(s)$

b $Sn^{4+}(aq)/Sn^{2+}(aq)$

c $H^{+}(aq)/H_2(g)$

3 The overall equation for the reaction that occurs in a cell made up of $Al^{3+}(aq)/Al(s)$ and $Sn^{2+}(aq)/Sn(s)$ half-cells is:

$$2Al(s) + 3Sn^{2+}(aq) \rightarrow 2Al^{3+}(aq) + 3Sn(s)$$

Write half-equations for the reaction occurring at the:

a cathode

b anode.

4 Draw a diagram of the cell from Question **3** and label the:

- half-equation for the reaction occurring in each half-cell
- anode and cathode
- direction of electron flow
- electrode polarities (which electrode is positive and which is negative)
- direction of anion and cation flow from the salt bridge.

5 In the spaces provided, label the galvanic cell in the figure below with the:

- anode
- cathode
- positive electrode
- negative electrode
- reduction
- oxidation
- salt bridge.

$$Zn(s) + Cu^{2+}(aq) \longrightarrow Cu(s) + Zn^{2+}(aq)$$

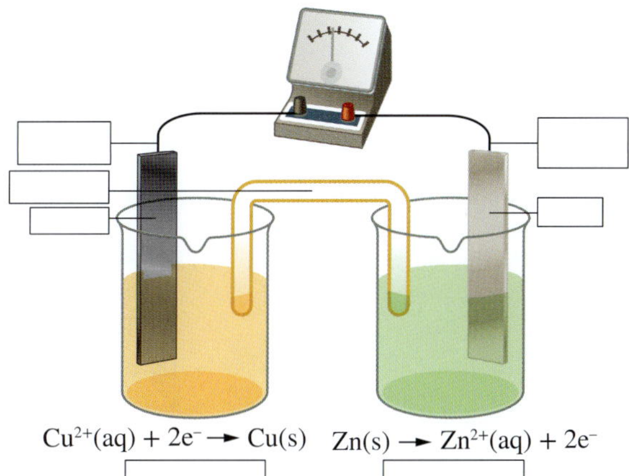

$Cu^{2+}(aq) + 2e^- \rightarrow Cu(s)$ $Zn(s) \rightarrow Zn^{2+}(aq) + 2e^-$

12.4 The table of standard reduction potentials

FIGURE 12.4.1 Water dropped onto sodium metal. Sodium is highly reactive; the products are sodium hydroxide and hydrogen gas.

As you saw in Chapter 11, metals vary in their reactivity. Platinum and gold are unreactive and widely used for jewellery. Other metals are very reactive. For example, sodium reacts so readily with oxygen and water (Figure 12.4.1) that it must be stored under paraffin oil.

Galvanic cells can help you compare the relative reactivity of metals. Galvanic cells can be constructed from various combinations of half-cells. The experimental data collected from these combinations allows chemists to determine the oxidising and reducing strengths of many different substances. This information is very useful, because it allows you to predict the products of various reactions, calculate the voltages of cells and develop more powerful and longer-lasting batteries.

RELATIVE OXIDISING AND REDUCING STRENGTHS

A half-cell contains a conjugate redox pair. The reactions that can occur in a half-cell can be written as reversible reactions, showing the relationship between the two chemicals in the redox pair. For example, the reaction in:

- a half-cell containing the $H^+(aq)/H_2(g)$ redox pair may be written as:

$$2H^+(aq) + 2e^- \rightleftharpoons H_2(g)$$

- a half-cell containing the $Zn^{2+}(aq)/Zn(s)$ redox pair may be written as:

$$Zn^{2+}(aq) + 2e^- \rightleftharpoons Zn(s)$$

Figure 12.4.2 shows a diagram of a cell constructed from $H^+(aq)/H_2(g)$ and $Zn^{2+}(aq)/Zn(s)$ half-cells.

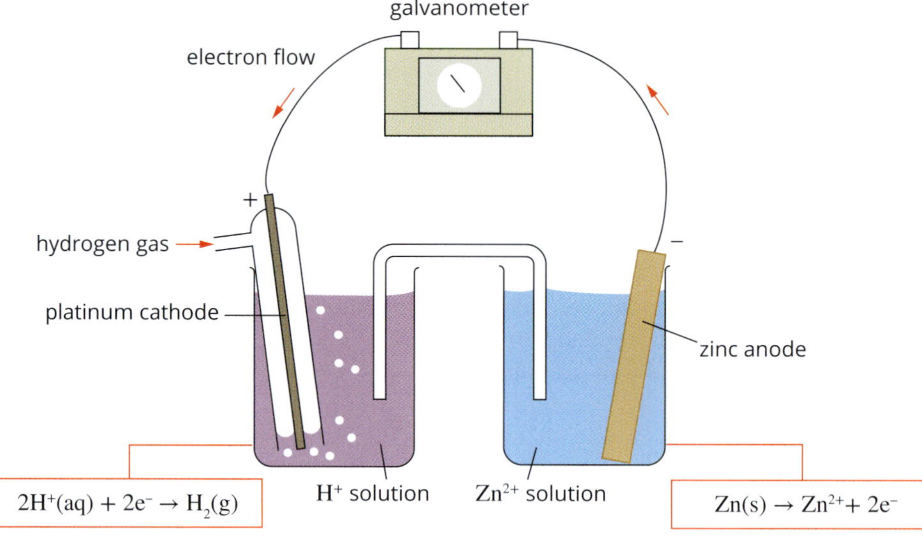

FIGURE 12.4.2 A diagram of a cell constructed from $H^+(aq)/H_2(g)$ and $Zn^{2+}(aq)/Zn(s)$ half-cells.

Notice that in this cell the zinc electrode is negative. The reactions that are occurring are:

$$2H^+(aq) + 2e^- \rightarrow H_2(g)$$

$$Zn(s) \rightarrow Zn^{2+}(aq) + 2e^-$$

Zn is oxidised to Zn^{2+}, so Zn is the reducing agent; H^+ is reduced to H_2, so H^+ is the oxidising agent.

Because electrons flow from the Zn^{2+}/Zn half-cell to the H^+/H_2 half-cell, it can be inferred that:

- zinc is a stronger reducing agent than H_2
- H^+ is a stronger oxidising agent than Zn^{2+} ions.

In a galvanic cell, the stronger reducing agent is in the half-cell with the negative electrode (anode). The stronger oxidising agent is in the half-cell with the positive electrode (cathode).

POTENTIAL DIFFERENCE

A current flows in a galvanic cell because one half-cell has a greater tendency to push electrons into the external circuit than the other half-cell. Chemists say that a **potential difference** exists between the two half-cells. The potential difference of a cell is sometimes also called the **electromotive force**, or emf, and it is commonly referred to as the **voltage**.

The potential difference of a cell, symbol E, has the unit of a **volt** (V) and is measured with a **voltmeter**.

Potential differences of cells are usually measured under the **standard conditions** of:

- 1 bar (100 kPa) of pressure
- 1 mol L^{-1} concentration of solutions.

The potential difference of a cell under standard conditions is given the symbol $E°$. Potential differences are usually measured at 25°C.

Standard electrode potentials

It is impossible to measure the potential difference of an isolated half-cell because both oxidation and reduction must take place for a potential difference to exist. However, you can assign a standard half-cell potential ($E°$) to each half-cell by connecting the cells to a standard reference half-cell and measuring the voltage produced.

A hydrogen half-cell, $H^+(aq)/H_2(g)$, under standard conditions is used for this purpose, and its $E°$ value is arbitrarily assigned as zero. This half-cell is known as the **standard hydrogen half-cell** or standard hydrogen electrode.

The **standard electrode potential** of other cells may then be measured by connecting them to the standard hydrogen half-cell, as shown in Figure 12.4.3 for the $Fe^{2+}(aq)/Fe(s)$ half-cell.

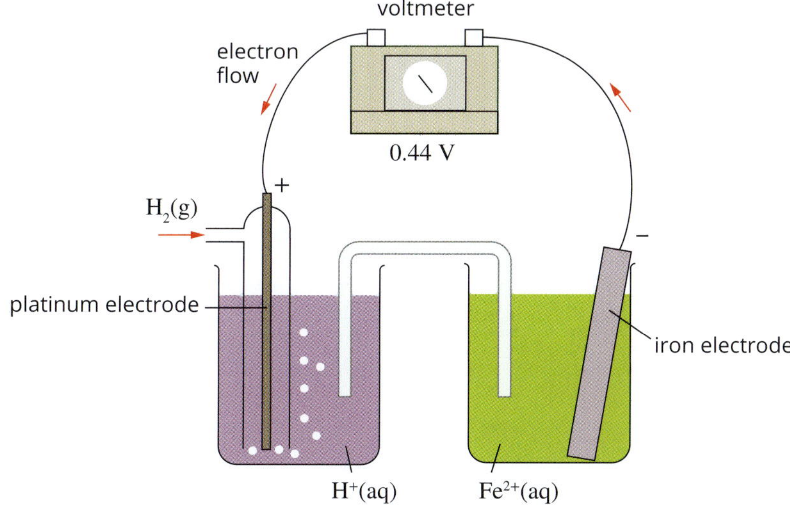

FIGURE 12.4.3 Measuring the standard half-cell electrode potential of a $Fe^{2+}(aq)/Fe(s)$ half-cell. From the voltmeter reading and negative polarity of the iron electrode, the $E°$ of this half-cell is −0.44 V.

We can summarise the information obtained from the measurement shown in Figure 12.4.3 as follows:

$$Fe^{2+}(aq) + 2e^- \rightleftharpoons Fe(s) \quad E° = -0.44 V$$

The negative sign indicates that the electrode in the half-cell was negative when connected to the hydrogen half-cell. Oxidation is occurring in the $Fe^{2+}(aq)/Fe(s)$ half-cell, and the electrons that are produced by the iron electrode move towards the hydrogen half-cell. The value of −0.44 V is known as both the standard electrode potential and the **standard reduction potential**. The standard reduction potential gives a numerical measure of the tendency of a half-cell reaction to occur as a reduction reaction.

TABLE OF STANDARD REDUCTION POTENTIALS

By connecting the standard hydrogen electrode to different half-cells and measuring their standard reduction potentials, chemists have developed a table called the **table of standard reduction potentials** or electrochemical series (Table 12.4.1).

TABLE 12.4.1 The table of standard reduction potentials. The strongest oxidising agents are at the bottom left-hand side of the table and the strongest reducing agents are at the top right-hand side.

Oxidising agents	⇌	Reducing agents	Standard reduction potential
$K^+(aq) + e^-$	⇌	$K(s)$	–2.94 V
$Ba^{2+}(aq) + 2e^-$	⇌	$Ba(s)$	–2.91 V
$Ca^{2+}(aq) + 2e^-$	⇌	$Ca(s)$	–2.87 V
$Na^+(aq) + e^-$	⇌	$Na(s)$	–2.71 V
$Mg^{2+}(aq) + 2e^-$	⇌	$Mg(s)$	–2.36 V
$Al^{3+}(aq) + 3e^-$	⇌	$Al(s)$	–1.68 V
$Mn^{2+}(aq) + 2e^-$	⇌	$Mn(s)$	–1.18 V
$H_2O(l) + e^-$	⇌	$\frac{1}{2}H_2(g) + OH^-(aq)$	–0.83 V
$Zn^{2+}(aq) + 2e^-$	⇌	$Zn(s)$	–0.76 V
$Fe^{2+}(aq) + 2e^-$	⇌	$Fe(s)$	–0.44 V
$Ni^{2+} + 2e^-$	⇌	$Ni(s)$	–0.24 V
$Sn^{2+}(aq) + 2e^-$	⇌	$Sn(s)$	–0.14 V
$Pb^{2+}(aq) + 2e^-$	⇌	$Pb(s)$	–0.13 V
$H^+(aq) + e^-$	⇌	$\frac{1}{2}H_2(g)$	0.00 V
$SO_4^{2-}(aq) + 4H^+(aq) + 2e^-$	⇌	$SO_2(aq) + 2H_2O(l)$	0.16 V
$Cu^{2+}(aq) + 2e^-$	⇌	$Cu(s)$	0.34 V
$\frac{1}{2}O_2(g) + H_2O(l) + 2e^-$	⇌	$2OH^-(aq)$	0.40 V
$Cu^+(aq) + e^-$	⇌	$Cu(s)$	0.52 V
$\frac{1}{2}I_2(s) + e^-$	⇌	$I^-(aq)$	0.54 V
$\frac{1}{2}I_2(aq) + e^-$	⇌	$I^-(aq)$	0.62 V
$Fe^{3+}(aq) + e^-$	⇌	$Fe^{2+}(aq)$	0.77 V
$Ag^+(aq) + e^-$	⇌	$Ag(s)$	0.80 V
$\frac{1}{2}Br_2(l) + e^-$	⇌	$Br^-(aq)$	1.08 V
$\frac{1}{2}Br_2(aq) + e^-$	⇌	$Br^-(aq)$	1.10 V
$\frac{1}{2}O_2(g) + 2H^+(aq) + 2e^-$	⇌	$H_2O(l)$	1.23 V
$\frac{1}{2}Cl_2(g) + e^-$	⇌	$Cl^-(aq)$	1.36 V
$\frac{1}{2}Cr_2O_7^{2-}(aq) + 7H^+(aq) + 3e^-$	⇌	$Cr^{3+}(aq) + \frac{7}{2}H_2O$	1.36 V
$\frac{1}{2}Cl_2(aq) + e^-$	⇌	$Cl^-(aq)$	1.40 V
$MnO_4^-(aq) + 8H^+(aq) + 5e^-$	⇌	$Mn^{2+}(aq) + 4H_2O(l)$	1.51 V
$\frac{1}{2}F_2(g) + e^-$	⇌	$F^-(aq)$	2.89 V

Under non-standard conditions, the order of the half-cells may change.

Notice the value of 0.00 V given for the $H^+(aq)/H_2(g)$ half-equation. All other E° values are relative to this arbitrary standard. The strongest oxidising agent, F_2, is at the bottom left-hand side of the table, and the strongest reducing agent, K, is at the top right-hand side of the table.

In a galvanic cell, the stronger reducing agent is oxidised so it is in the half-cell with the negative electrode (anode). The stronger oxidising agent is reduced so it is in the half-cell with the positive electrode (cathode).

 Strong reducing agents donate electrons more readily than weak ones.
Strong oxidising agents accept electrons more readily than weak ones.
Strong reducing agents have weak conjugate oxidising agents.
Strong oxidising agents have weak conjugate reducing agents.

USING THE TABLE OF STANDARD REDUCTION POTENTIALS

Predicting cell reactions

You can use the table of standard reduction potentials to predict what will happen when two specific half-cells are combined to form a cell. The strongest oxidising agent in the cell will react with the strongest reducing agent.

Another way to predict the electrode reactions is to remember that the half-reaction that is lower in the table of standard reduction potentials goes forward and the higher one is reversed. Therefore:

- a reduction reaction will occur in the half-cell with the most positive E° value, whereas an oxidation reaction will occur in the half-cell with the most negative E° value
- the positive electrode will be in the half-cell with the most positive E° value, whereas the negative electrode will be in the other half-cell.

The equation for the overall cell reaction is found by adding the two half-equations. Worked example 12.4.1 shows you how to use the table of standard reduction potentials to predict a cell reaction.

Worked example 12.4.1

PREDICTING THE OPERATION OF A GALVANIC CELL

A cell is made from $Ag^+(aq)/Ag(s)$ and $Fe^{2+}(aq)/Fe(s)$ half-cells under standard conditions and 25°C. Use the table of standard reduction potentials to predict the overall cell reaction, identify the anode and cathode, and determine the direction of electron flow.

Thinking	Working
Identify the two relevant half-equations in the table of standard reduction potentials.	$Fe^{2+}(aq) + 2e^- \rightleftharpoons Fe(s)$ $E^\circ = -0.44$ V $Ag^+(aq) + e^- \rightleftharpoons Ag(s)$ $E^\circ = 0.80$ V
Identify the strongest oxidising agent (the species on the left side of the table with the most positive E° value) and the strongest reducing agent (top right).	Because Ag^+ is lower on the left side of the table than Fe^{2+}, it is the stronger oxidising agent. Fe, being higher on the right side of the table than Ag, is a stronger reducing agent.
Write the two half-equations that will occur. The strongest oxidising agent will react with the strongest reducing agent. (Hint: The reduction equation has the most positive E° value, and the oxidation equation has the most negative E° value.)	Oxidation: $Fe(s) \rightarrow Fe^{2+}(aq) + 2e^-$ Reduction: $Ag^+(aq) + e^- \rightarrow Ag(s)$ Because this is the oxidation reaction, the equation should be written in reverse.

Write the overall cell equation.	Multiply the Ag^+/Ag half-cell equation by two so that the number of electrons in each half-equation is equal, and then add the two equations together: $Fe(s) \rightarrow Fe^{2+}(aq) + 2e^-$ $[Ag^+(aq) + e^- \rightarrow Ag(s)] \times 2$ $Fe(s) + 2Ag^+(aq) \rightarrow Fe^{2+}(aq) + 2Ag(s)$
Identify the anode and the cathode in this cell. The anode is the electrode at which oxidation occurs. The cathode is the electrode at which reduction occurs.	The silver electrode will be the cathode and the iron electrode will be the anode.
Determine the polarities of the electrodes and the direction of electron flow in the cell. The anode is negative; the cathode is positive.	Electrons flow from the negative electrode (anode) to the positive electrode (cathode) as shown in the figure below. voltmeter; electron flow; −; +; iron anode; silver cathode; $Fe^{2+}(aq)$; $Ag^+(aq)$

Worked example: Try yourself 12.4.1

PREDICTING THE OPERATION OF A GALVANIC CELL

A cell is made from $Sn^{2+}(aq)/Sn(s)$ and $Ni^{2+}(aq)/Ni(s)$ half-cells under standard conditions and at 25°C. Use the table of standard reduction potentials to predict the overall cell reaction, identify the anode and cathode, and determine the direction of electron flow.

Calculating the voltage of a cell

The maximum potential difference of a cell under standard conditions is the difference between the E° values of its two half-cells. It is defined as follows:

cell potential difference = E° of half-cell containing the oxidising agent − E° of half-cell containing the reducing agent

An easy way to remember this for galvanic cells is:

cell potential difference = most positive half-cell E° − most negative half-cell E°

For example, the maximum cell voltage of a cell constructed from $Ag^+(aq)/Ag(s)$ and $Fe^{2+}(aq)/Fe(s)$ half-cells under standard conditions can be calculated as follows:

cell potential difference = most positive half-cell E° − most negative half-cell E°

$$= E^\circ(Ag^+(aq)/Ag(s)) - E^\circ(Fe^{2+}(aq)/Fe(s))$$
$$= 0.80 - (-0.44)$$
$$= 1.24\text{ V}$$

As a galvanic cell discharges, the concentration or amount of reactants changes and the cell voltage eventually drops to zero—the cell is then referred to as 'flat'.

The easiest way to calculate the cell potential difference is to remember:
cell potential difference = most positive half-cell E° – most negative half-cell E°

PREDICTING DIRECT REDOX REACTIONS

The table of standard reduction potentials provides you with a ranking of the relative strengths of oxidising and reducing agents.

You can apply your understanding of the table of standard reduction potentials to predict the likelihood of a redox reaction taking place when different chemicals are combined.

Predicting reactions

If the contents of the half-cells of a galvanic cell were mixed, the reactants would react directly. Energy would be released as heat rather than as electrical energy. Reactions that occur in galvanic cells or when chemicals are directly mixed are described as naturally occurring reactions, or spontaneous reactions.

Earlier in this section, you saw that in a galvanic cell the strongest oxidising agent in the cell reacts with the strongest reducing agent. In other words, the lower half-reaction (the one with the most positive E° value) in the table of standard reduction potentials occurs in the forward direction (as reduction) and the higher half-reaction (the one with the most negative E° value) occurs in the reverse direction (as oxidation).

This principle applies equally to redox reactions that occur when reactants are mixed directly.

As shown in Figure 12.4.4, for a spontaneous reaction to occur, an oxidising agent (on the left-hand side of the table of standard reduction potentials) must react with a reducing agent (on the right-hand side) that is higher in the table of standard reduction potentials.

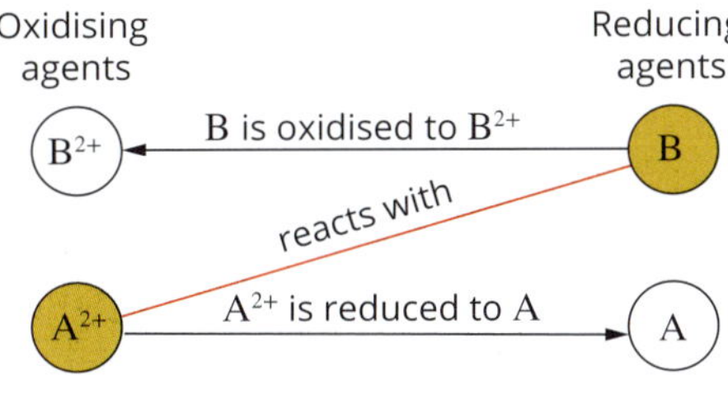

FIGURE 12.4.4 Oxidising agents only react significantly with reducing agents that are higher in the table of standard reduction potentials.

With half-equations arranged in order of increasing reducing agent strength, you can predict that a spontaneous reaction will occur by looking for substances that are arranged in a bottom-left/top-right position.

CHEMFILE 

The Nernst equation

The reduction potential of an equation is dependent on conditions such as temperature. Walther Nernst, a German physical chemist, first proposed a mathematical formula that could be used to predict the reduction potential of an equation under varying conditions. The Nernst equation is:

$$E = E^\circ - \frac{RT}{nF}\ln(Q)$$

This mathematical formula allows for reduction potential, temperature (in K), number of moles of electrons, the standard reduction potential and reaction quotient (Q) to be related using two different constants (the gas constant, R, and Faraday's constant, F). This formula plays an important role in selecting the conditions for many electrochemical reactions in industry and inside electrochemical cells such as batteries.

Worked example 12.4.2

PREDICTING DIRECT REDOX REACTIONS

Consider the following equations that appear in the order shown in the table of standard reduction potentials.

$$Mg^{2+}(aq) + 2e^- \rightleftharpoons Mg(s) \quad E^\circ = -2.36V$$
$$Ni^{2+}(aq) + 2e^- \rightleftharpoons Ni(s) \quad E^\circ = -0.24V$$
$$Ag^+(aq) + e^- \rightleftharpoons Ag(s) \quad E^\circ = +0.80V$$

Use the table of standard reduction potentials to predict the effect of mixing:

a $Ag^+(aq)$ and $Mg^{2+}(aq)$
b $Mg^{2+}(aq)$ and Ni(s)
c $Ni^{2+}(aq)$ and Mg(s).

Thinking	Working
Identify the two relevant half-equations in the table of standard reduction potentials.	**a** $Mg^{2+}(aq) + 2e^- \rightleftharpoons Mg(s)$ $E^\circ = -2.36\,V$ $Ag^+(aq) + e^- \rightleftharpoons Ag(s)$ $E^\circ = +0.80\,V$ **b** $Mg^{2+}(aq) + 2e^- \rightleftharpoons Mg(s)$ $E^\circ = -2.36\,V$ $Ni^{2+}(aq) + 2e^- \rightleftharpoons Ni(s)$ $E^\circ = -0.24\,V$ **c** $Mg^{2+}(aq) + 2e^- \rightleftharpoons Mg(s)$ $E^\circ = -2.36\,V$ $Ni^{2+}(aq) + 2e^- \rightleftharpoons Ni(s)$ $E^\circ = -0.24\,V$
Predict whether a reaction will occur. A chemical species on the left-hand side of the table of standard reduction potentials (an oxidising agent) reacts with a chemical species on the right-hand side that is higher in the series (a reducing agent).	**a** $Mg^{2+}(aq) + 2e^- \rightleftharpoons Mg(s)$ $E^\circ = -2.36\,V$ $Ag^+(aq) + e^- \rightleftharpoons Ag(s)$ $E^\circ = +0.80\,V$ No reaction occurs because both $Ag^+(aq)$ and $Mg^{2+}(aq)$ are oxidising agents. **b** $Mg^{2+}(aq) + 2e^- \rightleftharpoons Mg(s)$ $E^\circ = -2.36\,V$ $Ni^{2+}(aq) + 2e^- \rightleftharpoons Ni(s)$ $E^\circ = -0.24\,V$ No reaction occurs because the oxidising agent, Mg^{2+}, is above the reducing agent, Ni, in the table of standard reduction potentials. **c** $Mg^{2+}(aq) + 2e^- \rightleftharpoons Mg(s)$ $E^\circ = -2.36\,V$ $Ni^{2+}(aq) + 2e^- \rightleftharpoons Ni(s)$ $E^\circ = -0.24\,V$ A reaction occurs because the oxidising agent, Ni^{2+}, is below the reducing agent, Mg, in the table of standard reduction potentials. The lower half-equation occurs in the forward direction: $Ni^{2+}(aq) + 2e^- \rightarrow Ni(s)$ The higher half-equation occurs in the reverse direction: $Mg(s) \rightarrow Mg^{2+}(aq) + 2e^-$
Write the overall equation.	The overall equation is found by adding the half-equations: $Ni^{2+}(aq) + Mg(s) \rightarrow Ni(s) + Mg^{2+}(aq)$

Worked example: Try yourself 12.4.2

PREDICTING DIRECT REDOX REACTIONS

Consider the following equations that appear in the order shown in the table of standard reduction potentials.

$$Mn^{2+}(aq) + 2e^- \rightleftharpoons Mn(s) \quad E^\circ = -1.18\,V$$
$$Sn^{2+}(aq) + 2e^- \rightleftharpoons Sn(s) \quad E^\circ = -0.14\,V$$
$$Cu^+(aq) + e^- \rightleftharpoons Cu(s) \quad E^\circ = +0.52\,V$$

Use the table of standard reduction potentials to predict the effect of mixing:

a Mn(s) and $Cu^+(aq)$

b $Mn^{2+}(aq)$ and $Sn^{2+}(s)$

c Cu(s) and $Sn^{2+}(s)$.

CHEMISTRY IN ACTION 

The cost of corroding iron

Iron metal is the main component of steel, an alloy. Steel is widely used in building structures, vehicles and many other areas. Unfortunately, the iron in steel undergoes an oxidation reaction naturally, by reacting with oxygen in the air and with water. For example, 'wet corrosion' involves several steps:

- Step 1: Iron is oxidised to form Fe^{2+} ions at one region on the iron surface:

  $$Fe(s) \rightarrow Fe^{2+}(aq) + 2e^-$$

  At the same time, at another region on the surface and using the electrons produced by the oxidation process, oxygen is reduced in the presence of water to hydroxide ions:

  $$O_2(aq) + 2H_2O(l) + 4e^- \rightarrow 4OH^-(aq)$$

  The overall equation for step 1 is:

  $$2Fe(s) + O_2(aq) + 2H_2O(l) \rightarrow 2Fe^{2+}(aq) + 4OH^-(aq)$$

- Step 2: A precipitate of iron(II) hydroxide forms:

  $$Fe^{2+}(aq) + 2OH^-(aq) \rightarrow Fe(OH)_2(s)$$

  Steps 1 and 2 of the wet corrosion process are summarised in Figure 12.4.5.

This process costs the Australian industry large amounts of money, as measures have to be taken to prevent corrosion, and corroded iron structures have to be replaced. In Australia, corrosion costs approximately $100 billion each year, which makes up about 7% of gross domestic product (GPD). GDP is the amount of money exchanged in Australia over a year.

While there are many methods for preventing corrosion, one of the more common ways, especially on ships, is to attach a block of a stronger reducing agent than iron, which will react instead of the iron. Typically, zinc is used for this, because it is a slightly stronger reducing agent than iron but is not so strong that it will spontaneously react with water. This can be seen in the table of standard reduction potentials on page 398.

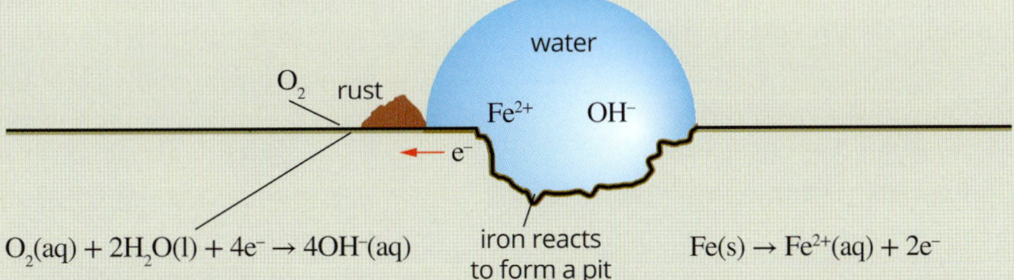

FIGURE 12.4.5 During wet corrosion, electrons are transferred through the iron from the area where oxidation occurs to the area where reduction occurs. Ions flow through the water droplets.

Limitations of predictions

The standard half-cell potentials given in the table of standard reduction potentials are measured under standard conditions. As you would expect, half-cell potentials can be different under other conditions.

When conditions are very different from standard conditions, the order of the half-reactions in the table of standard reduction potentials may also be different, and predictions of reactions based on the standard half-cell potentials may not be reliable.

It is also important to remember that the table of standard reduction potentials gives no information about the rate at which reactions occur.

12.4 Review

SUMMARY

- The standard hydrogen half-cell is used as the standard reference half-cell; its value is arbitrarily assigned as zero.
- The standard electrode potential or standard reduction potential (E°) of a half-cell is measured by connecting the half-cell to a standard hydrogen half-cell and measuring the voltage produced.
- The standard reduction potential gives a numerical measure of the tendency of a half-cell reaction to occur as a reduction reaction.
- Standard reduction potentials are used as the basis of the table of standard reduction potentials.
- In the table of standard reduction potentials, half-reactions are listed in order so that the strongest oxidising agent is at the bottom left-hand side of the series (with the most positive E° value) and the strongest reducing agent is at the top right-hand side (with the most negative E° value).
- The table of standard reduction potentials is valid for standard conditions; that is, gas pressures of 1 bar and solution concentrations of 1 mol L^{-1}. Standard reduction potentials are usually measured at 25°C.
- The relative strengths of oxidising and reducing agents can be determined by comparing their standard reduction potentials, and these can be used to predict half-cell and overall cell reactions.
- The maximum potential difference of a cell under standard conditions can be calculated from standard reduction potentials:

  cell potential difference = E° of half-cell containing the oxidising agent – E° of half-cell containing the reducing agent.

  A simple way of remembering this is:

  cell potential difference = most positive half-cell E° – most negative half-cell E°.
- The relative strength of oxidising and reducing agents can be determined by comparing standard reduction potentials, and these can be used to predict the tendency of a reaction to occur, both in galvanic cells and in direct reactions.
- For a spontaneous reaction to occur, an oxidising agent (on the left-hand of the table of standard reduction potentials) must react with a reducing agent (on the right-hand side) that is higher in the series.
- When reactants react in a galvanic cell, chemical energy is transformed into electrical energy. However, when they react directly, their chemical energy is transformed into heat energy.
- The standard reduction potentials in the table of standard reduction potentials are measured under standard conditions. Under other conditions, the order of the half-reactions may be different, and predictions based on the table of standard reduction potentials may not be reliable.

KEY QUESTIONS

1 A galvanic cell was constructed from Al^{3+}(aq)/Al(s) and Pb^{2+}(aq)/Pb(s) half-cells. Use the table of standard reduction potentials to:
  - a predict the oxidation and reduction half-equations
  - b predict the overall cell reaction
  - c identify the anode and cathode.

2 Draw a labelled diagram of a cell formed from Ag^{+}(aq)/Ag(s) and Sn^{2+}(aq)/Sn(s) half-cells. Use the table of standard reduction potentials to indicate the:
  - a half-cell reactions
  - b anode and cathode
  - c direction of electron flow
  - d electrode polarities (which electrode is positive and which is negative)
  - e directions of flow of the anions and cations in the salt bridge
  - f overall reaction.

3 Calculate the cell potential difference for the cell in Question **2**.

4 Using the table of standard reduction potentials, predict whether a reaction will occur in the following situations. If a reaction does occur, write the overall equation for the reaction.
  - a Zinc metal is placed in a solution containing copper(II) ions.
  - b Magnesium metal is placed into a solution containing calcium ions.
  - c A solution containing lead(II) ions is mixed with a solution containing iron(II) ions.

5 Iron nails are placed into 1 mol L^{-1} solutions of $CuSO_4$, $MgCl_2$, $Pb(NO_3)_2$ and $ZnCl_2$. Use the table of standard reduction potentials to identify the solution(s) in which you would expect a coating of a metal other than iron to appear on the nail.

Chapter review

12

KEY TERMS

anode
battery
cathode
conjugate redox pair
electrode
electrolyte
electromotive force
external circuit
galvanic cell
galvanometer
half-cell
half-equation
oxidant
oxidation
oxidation number
oxidised
oxidising agent
potential difference
redox reaction
reduced
reducing agent
reductant
reduction
salt bridge
spontaneous reaction
standard conditions
standard electrode potential
standard hydrogen half-cell
standard reduction potential
table of standard reduction potentials
transition element
volt
voltage
voltmeter

REVIEW QUESTIONS

1 Consider the following half-equations and the overall equation for the reaction between sodium metal and a solution of silver ions.

$$Na(s) \rightarrow Na^{+}(aq) + e^{-}$$
$$Ag^{+}(aq) + e^{-} \rightarrow Ag(s)$$
$$\overline{Na(s) + Ag^{+}(aq) \rightarrow Na^{+}(aq) + Ag(s)}$$

Which one of the following statements correctly describes this redox reaction?

A Na(s) is the oxidising agent, $Ag^{+}(aq)$ is the reducing agent, sodium metal is reduced.
B Na(s) is the reducing agent, $Ag^{+}(aq)$ is the oxidising agent, sodium metal is reduced.
C $Na^{+}(aq)$ is the oxidising agent, Ag(s) is the reducing agent, sodium metal is oxidised.
D Na(s) is the reducing agent, $Ag^{+}(aq)$ is the oxidising agent, sodium metal is oxidised.

2 Magnesium reacts with nickel(II) ions according to the following equation:

$$Mg(s) + Ni^{2+}(aq) \rightarrow Mg^{2+}(aq) + Ni(s)$$

Which one of the following is the correct set of conjugate redox pairs for this reaction?

A $Mg^{2+}(aq)/Mg(s)$ and $Ni^{2+}(aq)/Ni(s)$
B $Ni(s)/Mg^{2+}(aq)$ and $Ni^{2+}(aq)/Mg(s)$
C $Mg(s)/Ni(s)$ and $Ni^{2+}(aq)/Mg^{2+}(aq)$
D $Mg(s)/Ni^{2+}(aq)$ and $Ni(s)/Mg^{2+}(aq)$

3 Which one of the following alternatives describes what happens when magnesium and oxygen react?

A Each magnesium atom gains two electrons.
B Each magnesium atom loses two electrons.
C Each oxygen atom gains one electron.
D Each oxygen atom loses two electrons.

4 Complete the following sentences, which describe oxidation and reduction.

Oxidation and reduction occur together. Oxidation occurs when an atom ___________ electrons to form a ___________ ion, such as happens when a calcium atom, with an electronic configuration of 2,8,8,2 ___________ electrons to form a Ca^{2+} ion. Reduction occurs when an atom ___________ electrons to form a ___________ ion, or a cation ___________ electrons to become less positively charged. An example is when a bromine atom, with ___________ electrons in its valence shell, ___________ an electron to form a ___________ ion.

5 Lead metal is oxidised to form Pb^{2+} ions by reaction with silver ions in solution. Write half-equations for the reaction and then write the balanced overall equation.

6 Classify each of the following half-equations as either oxidation or reduction half-equations:

a $Mg(s) \rightarrow Mg^{2+}(aq) + 2e^{-}$
b $2Br^{-}(aq) \rightarrow Br_2(aq) + 2e^{-}$
c $Cu^{2+}(aq) + 2e^{-} \rightarrow Cu(s)$
d $K(s) \rightarrow K^{+}(aq) + e^{-}$
e $Fe^{3+}(aq) + e^{-} \rightarrow Fe^{2+}(aq)$
f $I_2(aq) + 2e^{-} \rightarrow 2I^{-}(aq)$

7 All of the following half-equations have a mistake in them. For each half-equation, state what the error is and then write the correct half-equation.

a $Ag(s) + e^{-} \rightarrow Ag^{+}(aq)$
b $Cu(s) + e^{-} \rightarrow Cu^{2+}(aq) + 3e^{-}$
c $Zn(aq) \rightarrow Zn^{2+}(s) + 2e^{-}$
d $I_2(aq) + e^{-} \rightarrow I^{-}(aq)$
e $Na^{+}(aq) - e^{-} \rightarrow Na(s)$

8 Complete the summary in the table as you balance the half-equation for the reduction of NO_3^- to NO_2 in acidic solution.

$$NO_3^-(aq) \rightarrow NO_2(g)$$

Step	Task	How it's done	Half-equation
1	Balance nitrogens.		$NO_3^-(aq) \rightarrow NO_2(g)$
2	Balance oxygens by adding ______.	Add ____ H_2O molecule(s) to the right-hand side of the equation.	
3	Balance hydrogens by adding _____.	Add _____ H^+ ion(s) to the ________ side of the equation.	
4	Balance charge by adding ______.	Charge on left-hand side = _____ Charge on right-hand side = _____ Add _____ e^- to the ______ side of the equation.	
5	Add state symbols to give the final half-equation.	Give the appropriate states for each reactant and product in the equation.	

9 The unbalanced half-equation for the reduction of the iodate ion (IO_3^-) is:

$$2IO_3^-(aq) + xH^+(aq) + ye^- \rightarrow I_2(aq) + zH_2O(l)$$

Complete the equation by inserting the correct coefficients for *x*, *y* and *z*.

10 In dry cells, commonly used in torches, an electric current is produced from the reaction of zinc metal with solid MnO_2. During this reaction, Zn^{2+} ions and solid Mn_2O_3 are formed. Write half-equations, and hence an overall equation, for the reaction.

11 Answer the following questions about the reaction of $H_3AsO_4(aq)$ with $I^-(aq)$. The unbalanced equation for the reaction is:

$$H_3AsO_4(aq) + I^-(aq) \rightarrow As_2O_3(s) + IO_3^-(aq) + H_2O(l)$$

a Write the two half-equations for the reaction.

b Write a balanced overall equation using your two half-equations from part **a**.

12 What is the oxidation number of sulfur in each of the following compounds?

a SO_2

b H_2S

c H_2SO_4

d SO_3

e Na_2SO_3

f $Na_2S_2O_3$

13 Complete the following table.

Compound	Element	Oxidation number
$CaCO_3$	Ca	
HNO_3	O	
H_2O_2		−1
HCO_3^-		+4
HNO_3	N	
$KMnO_4$	Mn	
H_2S	S	
Cr_2O_3	Cr	
N_2O_4	N	

14 Place the following substances in order of increasing oxidation number of nitrogen.

NO, K_3N, N_2O_4, N_2O, $Ca(NO_3)_2$, N_2O_3, N_2

15 Which of the following reactions are redox reactions? Give reasons for each of your answers.

a $BaCl_2 + H_2SO_4 \rightarrow BaSO_4 + 2HCl$

b $2Ag + Cl_2 \rightarrow 2AgCl$

c $2FeCl_3 + SnCl_2 \rightarrow 2FeCl_2 + SnCl_4$

d $ZnCO_3 \rightarrow ZnO + CO_2$

e $HPO_3^{2-} + I_2 + OH^- \rightarrow H_2PO_4^- + 2I^-$

f $2Cu^+ \rightarrow Cu^{2+} + Cu$

g $CaF_2 \rightarrow Ca^{2+} + 2F^-$

h $P_4 + 6H_2 \rightarrow 4PH_3$

16 Which one of the following statements best describes the role of the salt bridge in a galvanic cell?

A It allows positive charges to accumulate in one half-cell and negative charges to accumulate in the other.

B It provides a pathway for electrons to move between the half-cells.

C It allows reactants from one half-cell to mix with reactants from the other half-cell.

D It allows movement of ions to balance charges formed at the electrodes.

17 Which one of the following materials would be least suitable for use as an electrode in a $Cl_2(g)/Cl^-(aq)$ half-cell?

A iron

B platinum

C graphite

D gold

18 Explain the difference between an:

a oxidising agent and a reducing agent

b anode and a cathode

c external circuit and an internal circuit.

19 The overall reaction for a galvanic cell constructed from the $Cl_2(g)/Cl^-(aq)$ and $Pb^{2+}(aq)/Pb(s)$ half-cells is:

$$Cl_2(g) + Pb(s) \rightarrow 2Cl^-(g) + Pb^{2+}(aq)$$

Draw a diagram of the galvanic cell and on your diagram show:

a the direction of electron flow in the external circuit

b a half-equation for the reaction at each electrode

c which electrode is the anode

d which electrode is positive

e which way cations flow in the salt bridge.

20 Two half-cells are set up. One contains a solution of magnesium nitrate with a strip of magnesium as the electrode. The other contains lead(II) nitrate with a strip of lead as the electrode. The solutions in the two half-cells are connected by a piece of filter paper soaked in potassium nitrate solution. When the electrodes are connected by wires to a galvanometer, the magnesium electrode is shown to be negatively charged.

a Sketch the galvanic cell described. Label the positive and negative electrodes. Mark the direction of the electron flow.

b Write the half-equations for the reactions that occur in each half-cell and an equation for the overall reaction.

c Label the anode and cathode.

d Indicate the direction in which ions in the salt bridge migrate.

21 The two galvanic cells shown in the diagram were constructed under standard conditions.

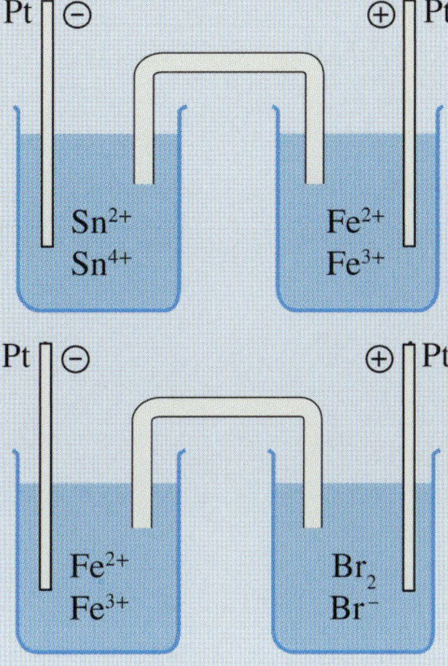

On the basis of the electrode polarities, determine the order of reducing agent strength from strongest to weakest.

22 Four half-cells $A^{2+}(aq)/A(s)$, $B^{2+}(aq)/B(s)$, $C^{2+}(aq)/C(s)$ and $D^{2+}(aq)/D(s)$ are used to make the cells shown in the diagram.

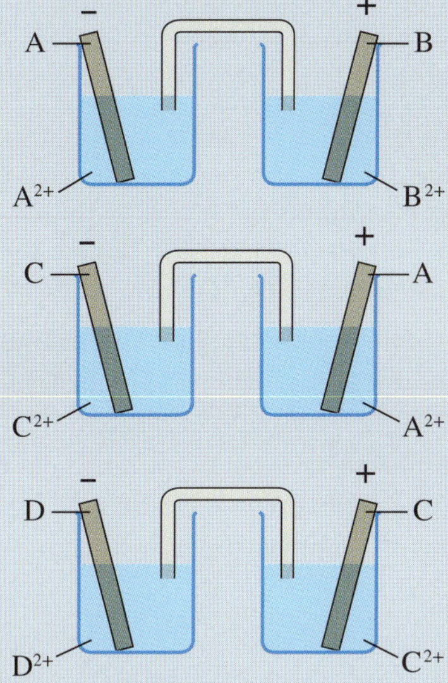

Rank the half-cells in order of their reduction half-cell potentials, from highest to lowest.

23 Use the table of standard reduction potentials to predict whether a reaction will occur in each of the following situations. Write an equation for each reaction that you predict will occur.

a Copper(II) sulfate solution is stored in an aluminium container.

b Sodium chloride solution is stored in a copper container.

c Silver nitrate solution is stored in a zinc container.

24 The following equations form part of the table of standard reduction potentials. They are ranked in the order shown.

$$Mg^{2+}(aq) + 2e^- \rightleftharpoons Mg(s)$$
$$Zn^{2+}(aq) + 2e^- \rightleftharpoons Zn(s)$$
$$Fe^{2+}(aq) + 2e^- \rightleftharpoons Fe(s)$$
$$Pb^{2+}(aq) + 2e^- \rightleftharpoons Pb(s)$$
$$Ag^{+}(aq) + e^- \rightleftharpoons Ag(s)$$

a Which species is the strongest oxidising agent and which species is the weakest oxidising agent?

b Which species is the strongest reducing agent and which species is the weakest reducing agent?

c Lead rods are placed in solutions of silver nitrate, iron(II) sulfate and magnesium chloride. In which solutions would you expect to see a coating of another metal form on the lead rod? Explain.

d Which of the metals silver, zinc or magnesium might be coated with lead when immersed in a solution of lead(II) nitrate?

25 As a result of a traffic accident, residents in a Wollongong suburb had to be evacuated when toxic fumes leaked from a container of sodium dithionite ($Na_2S_2O_4$). The dithionite ion reacts with water according to the equation:

$$2S_2O_4^{2-}(aq) + H_2O(l) \rightarrow S_2O_3^{2-}(aq) + 2HSO_3^{-}(aq)$$

a State the oxidation number of the sulfur in the following ions:

i $S_2O_4^{2-}$

ii $S_2O_3^{2-}$

iii HSO_3^{-}

b Write half-equations for the oxidation and reduction reactions that occur when sodium dithionite is mixed with water.

26 Many of the 'alkaline cells' on the market contain zinc electrodes in contact with an electrolyte containing hydroxide ions. The half-cell reaction can be represented as:

$$Zn(s) + 4OH^{-}(aq) \rightarrow Zn(OH)_4^{2-}(aq) + 2e^{-}$$

To investigate whether the standard reduction potential ($E°$) of these half-cells is the same as that of a half-cell reaction using a zinc electrode in contact with a zinc nitrate electrolyte (for which the electrode reaction would be $Zn(s) \rightarrow Zn^{2+}(aq) + 2e^{-}$), a student was provided with the two half-cells shown in the figure below, and a $Cu^{2+}(aq)/Cu(s)$ half-cell.

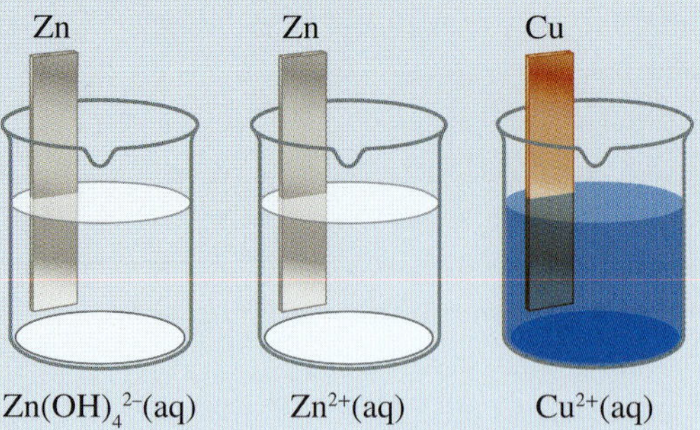

a Write an overall equation for the reaction in a galvanic cell in which the 'alkaline zinc half-cell' is connected to the Cu^{2+}/Cu half-cell.

b Carefully explain how the student could use the half-cells that were provided to determine whether the two different half-cells containing zinc had the same $E°$ value. Include fully labelled diagrams with your answer and explain how the results could be interpreted.

27 Reflect on the Inquiry activity on page 372.

a What type of reaction did you observe?

b How could you prevent this type of reaction from happening to the steel wool?

CHAPTER 13

Rates of reactions

Chemical reactions occur at many different rates. The explosion of gunpowder and the combustion of petrol in a car's engine occur very quickly. On the other hand, the ripening of fruit and the rusting of iron occur quite slowly.

It is important to appreciate that collisions between reactant particles do not always result in a chemical reaction. For example, while a car's fuel tank is being filled with petrol, the hydrocarbon molecules in the fuel are colliding with oxygen molecules in the air without a reaction occurring.

In this chapter, you will learn how rates of chemical reactions can be measured, and how varying the conditions of chemical reactions can affect the rate of a reaction. Using collision theory, you will learn to predict the effects of concentration of solutions, gas pressure, surface area and temperature on the rate of chemical reactions, and to explain these effects.

Content

INQUIRY QUESTION

What affects the rate of a chemical reaction?

By the end of this chapter, you will be able to:

- conduct a practical investigation, using appropriate tools (including digital technologies), to collect data, analyse and report on how the rate of a chemical reaction can be affected by a range of factors, including but not limited to: ICT N
  - temperature
  - surface area of reactant(s)
  - concentration of reactant(s)
- investigate the role of activation energy, collisions and molecular orientation in collision theory
- explain a change in reaction rate using collision theory (ACSCH003, ACSCH046) CCT

13.1 Collision theory

CHEMISTRY INQUIRY 

Observing the rate of a chemical reaction

Can you change the rate of a chemical reaction?

COLLECT THIS ...

- white vinegar
- 2 pieces of chalk of the same colour and size
- snaplock bag
- chopping board
- rolling pin
- 2 plastic containers of the same size
- measuring jug (250 mL – 1 L capacity)
- mobile phone or digital camera with video-recording capabilities

DO THIS ...

1. Take the equipment outside.
2. Place one piece of chalk into a plastic bag. Seal the bag and crush the chalk into a fine powder.
3. Pour the powdered chalk into one of the plastic containers.
4. Place the intact chalk in the other plastic container.
5. Draw up the table below and record your predictions for what will happen when you:
   a pour 200 mL of white vinegar over the crushed chalk
   b pour 200 mL of white vinegar over the intact piece of chalk.
6. Video your observations as you pour the white vinegar over the intact piece of chalk.
7. Video your observations as you pour the same amount of white vinegar over the crushed piece of chalk.

RECORD THIS ...

Draw up this table to record your predictions and observations.

Description of experiment	Prediction	Observations
a white vinegar and whole piece of chalk		
b white vinegar and crushed piece of chalk		

Test your predictions.

Record your observations.

REFLECT ON THIS ...

Explain the meaning of your observations.

What could you do next time to improve your experiment?

The chemical equation for a reaction indicates the nature of the reactants and products, but provides no information about the way in which the reaction proceeds.

Chemical reactions occur as a result of collisions between the reacting particles. This idea is part of the **collision theory** of reaction rates, which will be discussed in this section.

During chemical reactions, particles (atoms, molecules or ions) collide and are rearranged to produce new particles. Consider the decomposition reaction for hydrogen peroxide:

GO TO ➤ Section 10.3 page 319

$$2H_2O_2(l) \rightarrow 2H_2O(l) + O_2(g)$$

The collision that forms the first step of the reaction occurs between the two hydrogen peroxide molecules. If this collision is to result in the formation of molecules of water and oxygen, the collision must occur in such a way that the covalent bonds in the hydrogen peroxide break. To break bonds, energy is required.

The collision theory of reactions explains why some collisions result in reactions and others do not.

According to collision theory, for a reaction to occur, the reactant particles must:

- make contact (collide) with each other
- collide with sufficient energy to break the bonds within the reactants
- collide with the correct orientation to break the bonds within the reactants and so allow the formation of new products.

If a collision does not meet all of these requirements, then no reaction occurs. In fact, most collisions do not result in a chemical reaction. Collision theory explains why this is the case.

CHEMISTRY IN ACTION CC

Saved by a very fast chemical reaction

Imagine the scene. An 18-year-old borrows his parent's car to take his girlfriend for a drive to celebrate gaining his driver's licence. Roof down, enjoying the beautiful afternoon and the countryside, the driver rounds a corner to find the road wet. The car begins to slide on the wet surface. Due to his inexperience, the driver brakes sharply and the car starts to spin. Suddenly, the car is leaving the road and heading straight for a large tree. Then, bang!

Later, the car was estimated to have been travelling at 60 km h^{-1} when it hit the tree. The collision was a 'head-on', with the front and passenger side taking most of the impact. Yet the girl in the passenger seat suffered just a chipped tooth, and her boyfriend sustained only minor bruising.

This is the true story of a lucky escape, thanks to a very rapid chemical reaction. As the collision took place, airbags were inflating and then deflating as the travellers were slammed forward towards the windscreen. The driver described it as being 'all over in a flash' and had no clear recollection of the airbags going off.

Hidden in the car's steering wheel, dashboard and windscreen pillars, special nylon bags fill with gas within 30 ms of impact (Figure 13.1.1). As a consequence, the car occupants are prevented from smashing their heads against the steering wheel, dashboard, windscreen or front pillars, all within the blink of an eye. As the head and body strike the airbags, the cushion of gas is forced out of the bag through tiny vents, and within 100 ms the bag has completely deflated.

FIGURE 13.1.1 Airbags are deployed within 30 ms of an impact.

Air bags contain a mixture of crystalline solids—sodium azide (NaN_3), potassium nitrate (KNO_3) and silica (SiO_2)—stored in a canister. Sensors in the front of the car detect the difference between a bump and a life-threatening impact. When a response is required, an electronic impulse 'ignites' the sodium azide. Sodium metal and hot nitrogen gas are the products of this energy-releasing redox reaction:

$$2NaN_3(s) \rightarrow 2Na(l) + 3N_2(g)$$

The pulse of hot nitrogen gas released from this reaction starts to inflate the nylon bag. The molten sodium metal immediately reduces the potassium nitrate, generating more nitrogen gas, as well as sodium oxide and potassium oxide, which are white powdery solids.

The equation for this reaction is:

$$10Na(l) + 2KNO_3(s) \rightarrow K_2O(s) + 5Na_2O(s) + N_2(g)$$

A filtration system prevents any of the oxides leaving the nylon bag, while a third reaction 'captures' them, producing a harmless glassy solid.

In this reaction they combine with silica:

$$K_2O(s) + 5Na_2O(s) + SiO_2(s) \rightarrow \text{alkaline silicate ('glass')}$$

Chemical reactions do save lives!

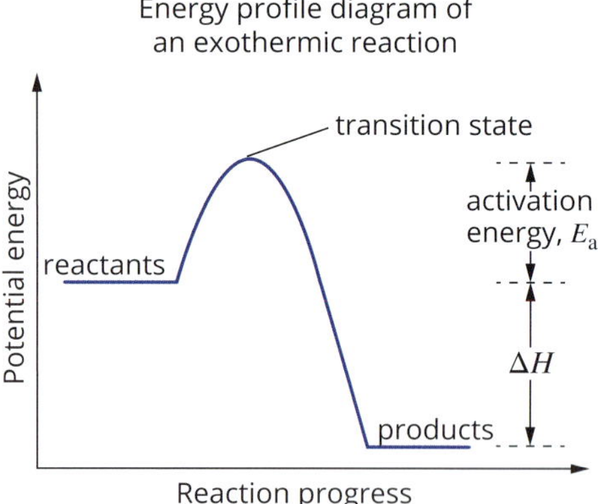

FIGURE 13.1.2 Energy profile diagram of an exothermic reaction such as burning natural gas.

GO TO ➤ Chapter 14 page 441

ACTIVATION ENERGY

For a reaction to occur between reactant molecules, the molecules must collide with a certain minimum amount of energy. Unless this minimum amount of energy is met or exceeded, the colliding molecules will rebound and simply move away from each other without reacting.

The minimum energy that a collision must possess for a reaction to occur is called the **activation energy**, E_a. When the energy of a collision is equal to or greater than the activation energy, a reaction can occur.

The potential energy changes associated with a chemical reaction can be represented as an **energy profile diagram**. An energy profile diagram for an exothermic reaction (one that releases heat energy to the surroundings) is shown in Figure 13.1.2.

On the energy profile diagram, the activation energy is measured from the energy of the reactants to the peak of the energy profile diagram.

Some reactions occur very readily because they have a very small activation energy, E_a. These reactions need only a small amount of energy to be absorbed for bonds in the reactants to be broken.

The difference in energy between the products and the reactants is called the enthalpy change of the reaction, ΔH. It can also be represented on an energy profile diagram. You will learn more about this aspect of a chemical reaction in the next chapter.

Reactant particles must have energy equal to or greater than the activation energy before a reaction can occur.

When the activation energy is absorbed, a new arrangement of the atoms known as the **transition state** forms. The transition state occurs at the stage of maximum potential energy in the reaction, when reactants have achieved the activation energy (Figure 13.1.3). Bond breaking and bond forming are both occurring at this stage, and the arrangement of the atoms is unstable. The atoms in the transition state rearrange to form the products as the reaction progresses.

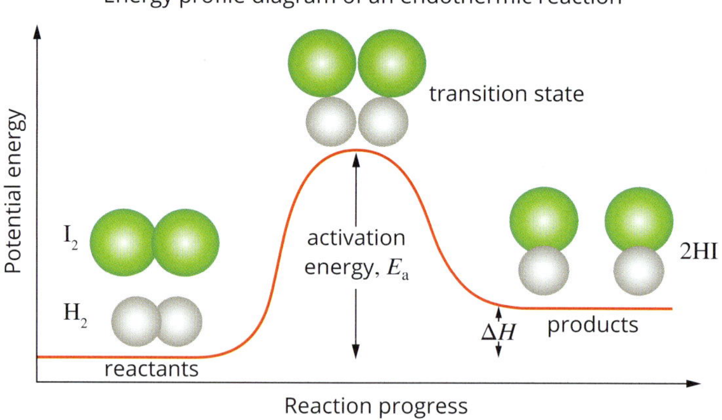

FIGURE 13.1.3 Energy profile diagram for the endothermic reaction $H_2(g) + I_2(g) \rightarrow 2HI(g)$.

CHEMFILE EU

A little too reactive!

In 1846, the Italian chemist Ascanio Sobrero reacted glycerol with a mixture of sulfuric and nitric acids to make the explosive liquid nitroglycerine. Nitroglycerine is so unstable that even a small bump can cause it to explode. It decomposes according to the equation:

$$4C_3H_5(NO_3)_3(l) \rightarrow 12CO_2(g) + 10H_2O(g) + 6N_2(g) + O_2(g)$$

Despite being many times more powerful than conventional gunpowder, it was far too dangerous to be practical. The reason for nitroglycerine's instability is the very small activation energy for its decomposition reaction (Figure 13.1.4).

Some years later, the Swedish scientist Alfred Nobel learnt how to manage nitroglycerine more safely through his invention of dynamite. Nobel was a talented chemist and inventor of chemical technology. He was also a successful businessman who died a wealthy man. While he spent a considerable amount of his lifetime developing explosives technology, linking him to warfare, Nobel was in fact inclined towards pacifism. In death he is remembered for his wider contributions to technological developments and for having left the bulk of his fortune for establishing annual prizes in chemistry, physics, medicine, literature and peace.

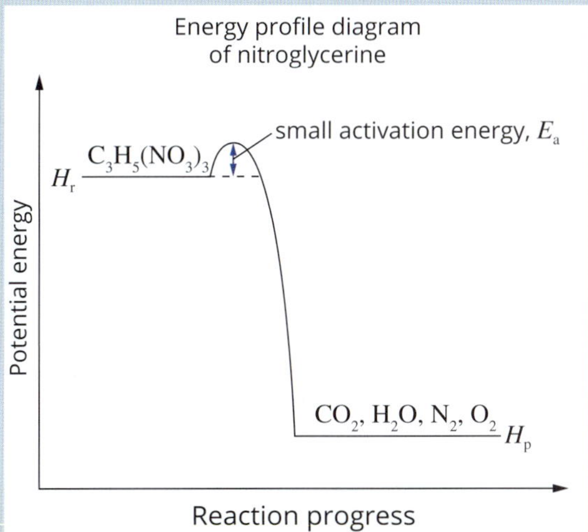

FIGURE 13.1.4 Nitroglycerine has a very low activation energy, making its rate of reaction very fast.

ORIENTATION OF COLLIDING PARTICLES

For a reaction to occur, reactants need to collide with enough energy to provide the activation energy. Reacting molecules must also collide with each other in the correct orientation so that particular bonds in the reactants are broken and new bonds are formed in the products.

Figure 13.1.5 shows the importance of collision orientation. In the decomposition of hydrogen iodide gas into hydrogen gas and iodine gas, two hydrogen iodide molecules must collide with the hydrogen atoms orientated towards one other (and the iodine atoms similarly orientated). If the collision orientation is incorrect, the particles simply bounce off each other, and no reaction occurs.

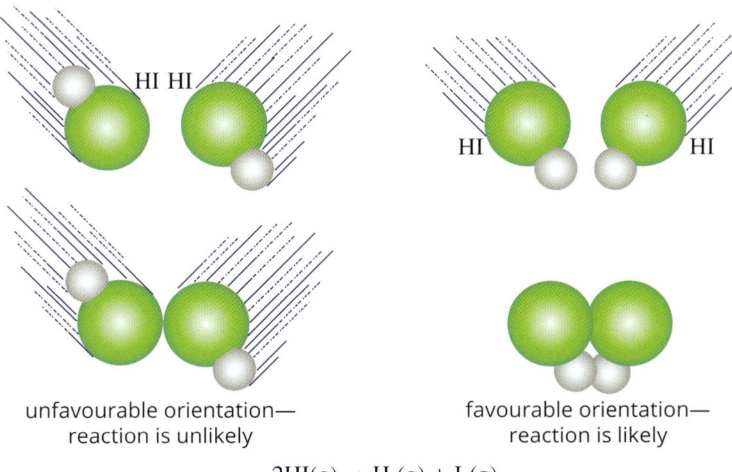

FIGURE 13.1.5 A reaction between colliding molecules is more likely to occur if the orientation of the collision is favourable.

Experimental investigations have shown that five main factors can change the rate of a chemical reaction:

- surface area of solid reactants
- concentration of reactants in a solution
- pressure of any gaseous reactant
- temperature of the reaction mixture
- presence of catalysts (examined further in Chapter 14).

GO TO ➤ Section 14.6 page 468

You can probably think of some examples of situations where one or more of these conditions is changed and a reaction becomes noticeably faster or slower. The later sections in this chapter will show how collision theory can be used to explain why each of these factors influences reaction rates.

13.1 Review

SUMMARY

- Collision theory is a theoretical model that accounts for the rates of chemical reactions in terms of collisions between particles during a chemical reaction.
- The activation energy (E_a) of a reaction is the minimum amount of energy that a collision between reactant particles must possess for a reaction to occur.
- The transition state is an arrangement of atoms that occurs when the activation energy is absorbed. The transition state is an unstable state in which bonds in the reactants are being broken and bonds in the products are starting to form.
- An energy profile diagram shows the activation energy as the highest potential energy between the energy of the reactants and the energy of the products.
- According to collision theory, for a reaction to occur, the reactant particles must:
  - collide with each other
  - collide with sufficient energy to break the bonds within the reactants
  - collide with the correct orientation to break the bonds within the reactants and so allow the formation of new products.

KEY QUESTIONS

1. According to the collision theory, which one of the following is not essential for a reaction to occur?
   - **A** The reactant particles should collide with double the energy of the activation energy.
   - **B** The reactant particles should collide with the correct orientation.
   - **C** The reactant particles should collide with enough energy to overcome the activation energy barrier.
   - **D** Molecules must collide to react.
2. **a** Write a balanced chemical equation for the complete combustion of methane gas (CH_4).
   b Using the energy profile diagram for this reaction, shown to the right, determine the activation energy for this reaction.
3. The decomposition of hydrogen iodide into its elements is represented by the equation:
   $$2HI(g) \rightarrow H_2(g) + I_2(g)$$
   This reaction has an activation energy of +139 kJ mol^{-1} and the heat of reaction, ΔH, is −28 kJ mol^{-1}. What is the activation energy for the reverse reaction, the formation of 2 mol of hydrogen iodide?
4. Reactions used in chemical explosives are often very fast. What characteristic of their activation energy do they often share?
5. Describe two features that are required of colliding gas molecules if they are to successfully react and produce new chemical products.

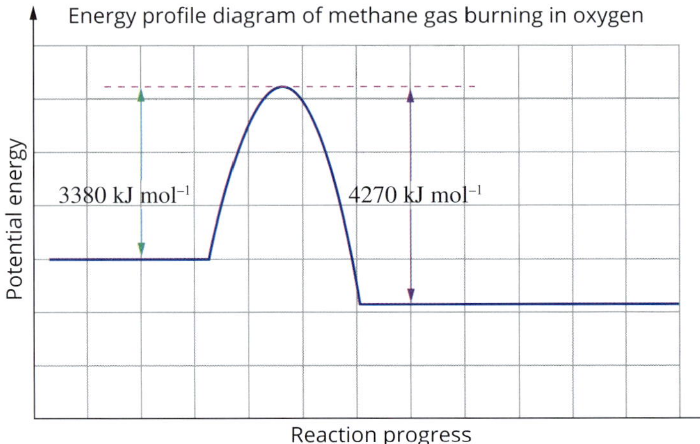

13.2 Measuring reaction rate

The time it takes for a batch of chocolate brownies to cook in the oven (Figure 13.2.1) and the time taken for a fibreglass patch on a surfboard to set depend on the rates of chemical reactions.

FIGURE 13.2.1 How quickly do the chemical reactions involved in baking occur?

Some chemical reactions are over in a flash. In a car accident, when a car's airbag needs to be inflated, the chemical reactions producing the gas that expands the airbag need to happen extremely quickly. On the other hand, it is fortunate that if the car's painted surface is scratched to expose the metal beneath, the rusting reactions take place at a very slow rate.

In this section, you will learn how chemists measure the rate at which a chemical reaction occurs. The **rate of reaction** is defined as the change in concentration of a reactant or product per unit time. The usual unit for rate of reaction is mole per litre per second ($mol\,L^{-1}\,s^{-1}$).

> The rate of reaction is defined as the change in concentration of a reactant or product per unit of time:
>
> $$\text{rate} = \frac{\text{change in concentration}}{\text{time}}$$

To experimentally determine the rate of reaction, either directly or indirectly, you need to measure how much of a reactant is being used up or how much of a product is being formed in a given time period. When a reaction involves gaseous products, this might involve measuring changes in gas volume or mass over time.

VOLUME OF GAS PRODUCED

The graph shown in Figure 13.2.2b was obtained by measuring the volume of carbon dioxide produced in the reaction between marble chips and hydrochloric acid. The reaction mixture is in a conical flask that is connected to a gas syringe, as shown in Figure 13.2.2a, enabling the gas volume to be measured at fixed time intervals. The experiment was performed twice, first with large marble chips, then with the same mass of small marble chips.

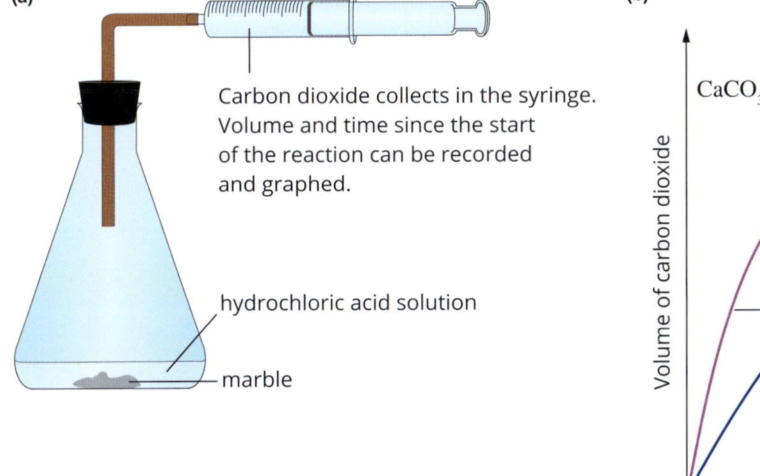

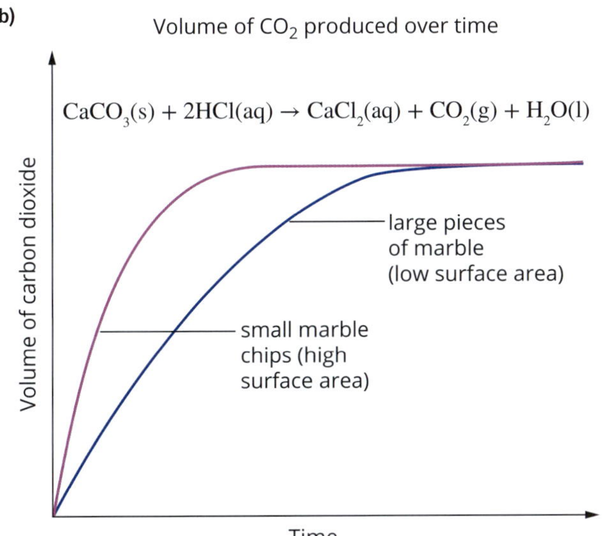

FIGURE 13.2.2 (a) Apparatus for monitoring the volume of gas produced over a period of time. (b) As carbon dioxide is produced from the reaction between marble chips and hydrochloric acid, it is collected in a gas syringe and its volume is recorded and graphed.

The steeper initial gradient of the graph with small marble chips indicates that the initial rate of production of carbon dioxide gas is faster with the smaller marble chips, which have a higher surface area. Because the same amount of marble chips and acid reacted in both these experiments, the total volume of carbon dioxide produced was the same.

This method of monitoring reaction rate could also be used for reactions between reactive metals and acids, where hydrogen gas is produced.

> To compare the rate of reaction in different conditions, for example at different temperatures, the volume of gas produced can be measured at fixed time intervals; from graphs of the data, the initial gradient of each curve gives an indication of the effect of different temperatures on the rate of the reaction.

MASS LOSS

To monitor mass changes during a reaction, the reaction mixture might be placed on a balance as shown in Figure 13.2.3. The cotton wool plug in the top of the flask allows carbon dioxide gas to escape, but prevents any drops of solution splashing out.

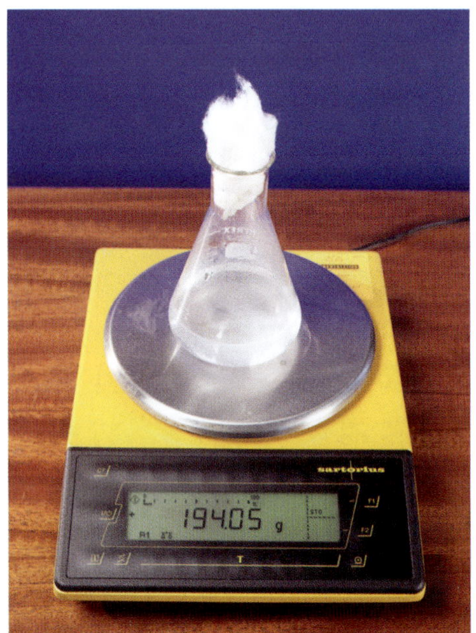

FIGURE 13.2.3 Apparatus to monitor the mass of carbon dioxide gas released during a reaction between marble chips and an acid.

If the mass of the apparatus is recorded at regular time intervals, the mass of gas released can be deduced and graphed against time, as shown in Figure 13.2.4. By drawing tangents to the curve, a measure of the rate of reaction can be calculated—the greater the gradient of the tangent, the faster the reaction rate. In this graph, the initial rate of reaction has the maximum gradient and is therefore the maximum rate.

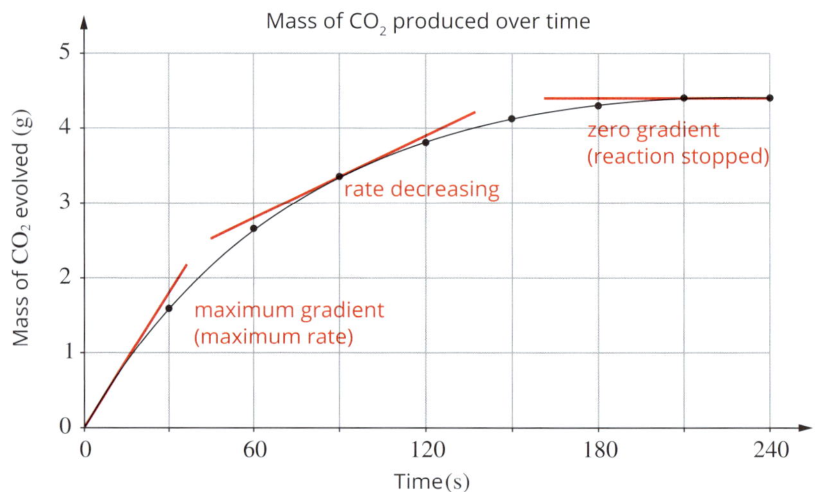

FIGURE 13.2.4 The rate of production of carbon dioxide gas is fastest initially and progressively decreases with time as the concentration of acid in the reaction mixture decreases.

> By drawing gradients of graphs of the change in mass or volume of a reacting mixture over time, changes in the rate of reaction can be deduced.

This method of monitoring reaction rate is not suited for reactions that produce gaseous products with a low molar mass, such as hydrogen, as the mass loss is extremely small.

CHANGE IN TEMPERATURE

An exothermic reaction occurring in solution will produce an increase in the temperature of the solution. The faster the reaction, the more rapidly the temperature of the solution will rise. If the same reaction is repeated using an excess of a reactant that varies in concentration, at higher concentrations the reaction will be faster and the temperature will increase more rapidly. The same total temperature increase would be expected in each reaction because the amount of the limiting reactant is constant.

In a similar way, temperature decreases in solutions in which endothermic reactions occur can be used to measure the reaction rate.

When the rate of temperature change is being used as a measure of reaction rate, the best way to compare the effects of a variable is to measure the initial rate of change, because temperature itself can influence the rate of the reaction.

Instead of using a thermometer and manually reading and recording the temperature of solutions, a digital temperature probe may be used, especially if the temperature rise is rapid.

CHANGE IN COLOUR

For reactions that involve a change in colour, either the time required for a particular colour change to occur can be measured, or the change in intensity of a specific colour can be monitored over time, using colorimetry. The colour intensity of the solution is directly related to the concentration of the coloured species.

One example you may be familiar with is the reaction between sodium thiosulfate and hydrochloric acid, two colourless solutions. They react to produce a suspension that contains fine solid particles of yellow sulfur. By placing the reacting mixture in a conical flask with a cross marked below it and observing the time required for enough sulfur to be produced to mask the cross (Figure 13.2.5), the rate of this reaction can be monitored.

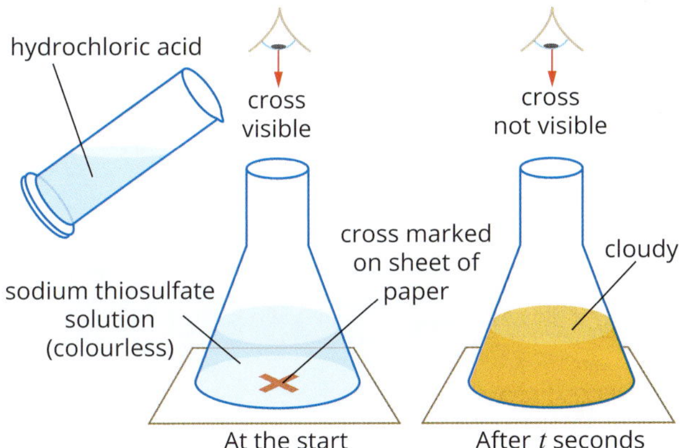

FIGURE 13.2.5 Observing the time taken for the cross to be masked by the suspension of sulfur formed during the reaction: $Na_2S_2O_3(aq) + 2HCl(aq) \rightarrow 2NaCl(aq) + SO_2(g) + S(s) + H_2O(l)$

When the time taken for a reaction to proceed to a specific point is measured, a lower time required indicates a faster reaction. The data in Table 13.2.1 show that the reaction occurred at a faster rate as temperature increased, because the required amount of sulfur was produced in a shorter time. In fact, it can be shown that the rate of this reaction is proportional to $\frac{1}{\text{time taken}}$.

TABLE 13.2.1 Data from colorimetric analysis of the rate of reaction between $Na_2S_2O_3(aq)$ and HCl(aq)

Experiment number	$[Na_2S_2O_3(aq)]$ (mol L^{-1})	Temperature (°C)	Time taken for the cross to be masked (s)
1	0.10	15	56
2	0.10	20	40
3	0.10	25	28

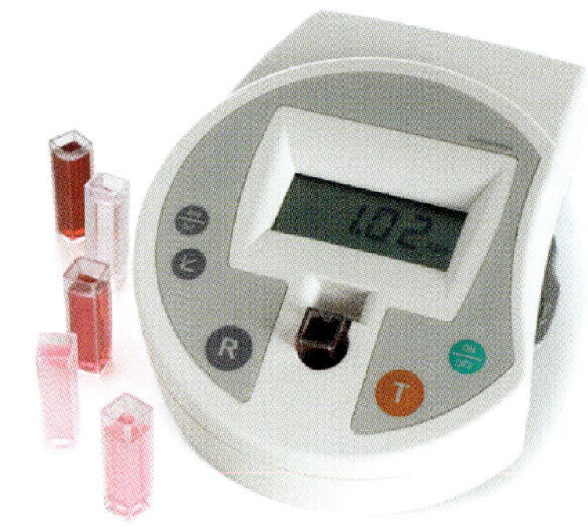

FIGURE 13.2.6 Colorimeter and cuvettes containing solutions with different concentrations of potassium permanganate.

In the same way that using a temperature probe assists an experimenter to record precise measurements, using a colorimeter (such as the one shown in Figure 13.2.6) can precisely indicate the amount of light absorbed by a coloured solution. When such an instrument is used, the subjectivity of estimating by eye is eliminated.

The student shown in Figure 13.2.7 is using a colorimeter that measures the amount of light absorbed by a purple solution containing potassium permanganate. If the solution contains potassium permanganate that is reacting with oxalic acid, the amount of light absorbed will decrease as the concentration of potassium permanganate in the solution decreases. The student can therefore time how long it takes for the absorbance to reach a particular value for different experimental conditions.

FIGURE 13.2.7 A student using a colorimeter to monitor the rate of reaction between potassium permanganate and oxalic acid: $2MnO_4^-(aq) + 5H_2C_2O_4(aq) + 6H^+(aq) \rightarrow 2Mn^{2+}(aq) + 10CO_2(g) + 8H_2O(l)$

A colorimeter measures the amount of visible light of a particular range of wavelengths absorbed by a solution. The higher the concentration of the absorbing species, the greater the amount of light it absorbs.

In addition, other methods of determining rate can also be applied to suitable reactions. For example:

- pH change with time may be monitored for reactions that involve acidic or basic reactants or products
- the change in electrical conductivity of solutions may be monitored for reactions involving a distinct difference in conductivity between reactants and products.

By using probes (measuring and recording temperature, gas pressure, absorbance of light, pH or conductivity) linked to a computer, many measurements can be recorded in a very short time, and often a graph can be automatically generated. This can be especially useful when a reaction occurs rapidly.

CHEMISTRY IN ACTION 

Monitoring cheesemaking reactions by pH measurements

At Bega Cheese Ltd, chemists use traditional cheesemaking methods to produce the cheese you may have included in your lunch. Milk is first pasteurised by heating, and then cooled to an optimum temperature for growth of the microorganisms of the 'starter culture'. The microbial starter culture and a coagulant are added to the milk. The coagulant immediately forms a gel. The consistency of the gel is critical for successful cheesemaking: it must be firm enough to be cut into cubes (Figure 13.2.8).

The next step in the process agitates the slurry of cubes in the whey. In these conditions, a variety of chemical reactions are occurring, some of which involve production of hydrogen ions. Consequently, the reactions are accompanied by a drop in pH as milk sugar is converted into lactic acid. At this stage of cheesemaking, the aim is for acid development to promote moisture removal from the curds. To monitor the rate of these reactions, the pH of the whey is constantly measured using pH probes (Figure 13.2.9).

Achieving the ideal pH development during cheesemaking provides a foundation that will result in the desired cheese texture, flavour and composition after maturation. The amount of starter added can be adjusted to give the desired acidification rate in subsequent batches.

FIGURE 13.2.8 Once a coagulant has been added to the milk, it is able to be cut into cubes.

FIGURE 13.2.9 A pH tester in cheese.

13.2 Review

SUMMARY

- The rate of a reaction is the change in concentration of reactants or products over time, with units of $mol\,L^{-1}\,s^{-1}$.
- The initial rate of reaction is an indication of the rate of reaction at time 0.
- A range of experimental methods can be used to measure the rate of a reaction, including measuring the following during specific time intervals:
  - volume of gas produced
  - mass loss (if a gaseous product is formed)
  - temperature changes
  - colour intensity changes (when a coloured species is involved in the reaction)
  - pH changes (when an acidic or basic species is involved in the reaction)
  - changes in electrical conductivity of a solution.
- A colorimeter allows changes in concentration of a coloured species to be monitored.
- Instruments such as a colorimeter and digital probes assist in making precise measurements.
- By comparing the gradients of tangents to a graph of mass loss versus time, gas volume versus time, solution temperature versus time or light absorbance versus time, faster reaction rates can be identified by steeper gradients.

KEY QUESTIONS

1 Hydrogen peroxide, a colourless solution, decomposes to produce oxygen gas and water. The reaction is also exothermic. Name three ways in which the rate of this reaction could be monitored.

2 Describe two methods that could be used to measure the initial rate of a reaction between magnesium ribbon and hydrochloric acid.

3 This graph shows the mass of carbon dioxide gas produced during a 4 min period from a reaction between marble chips (calcium carbonate) and $1.0\,mol\,L^{-1}$ nitric acid.

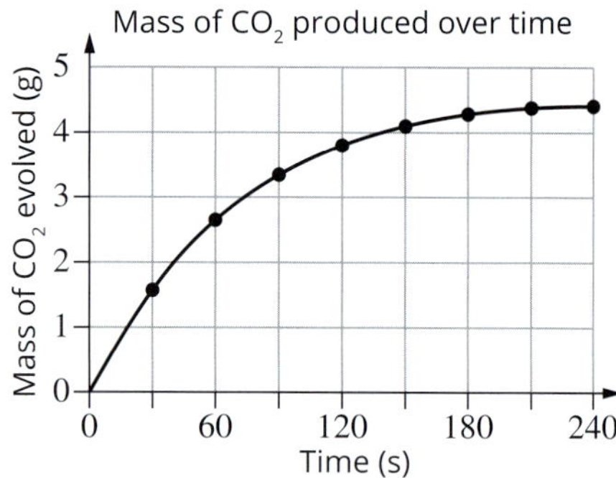

a Write a full chemical equation for this reaction.

b Explain whether the rate of this chemical reaction is increasing or decreasing over time.

4 The rate of decomposition of an aqueous solution of hydrogen peroxide is tracked by recording the volume of gas produced at fixed time intervals.

a Sketch a diagram of the apparatus that would be used.

b The following data was collected. Sketch a graph of this data and use the graph to determine the initial rate of the reaction.

Time (s)	Volume of gas (mL)
0	0
30	22
60	44
90	56
120	67
150	72
180	76
210	78
240	80
270	80
300	80

5 Two samples of copper(II) sulfate solution with concentrations of $0.080\,mol\,L^{-1}$ and $0.30\,mol\,L^{-1}$ were analysed in a colorimeter. Which solution would absorb the most light?

6 Hydrogen peroxide reacts with an acidified solution of iodide ions to produce iodine according to the following equation:

$$H_2O_2(aq) + 2I^-(aq) + 2H^+(aq) \rightarrow I_2(aq) + 2H_2O(l)$$

Suggest two methods that could be used to measure the rate of this reaction.

13.3 Effect of surface area, concentration and pressure on reaction rate

In any given reaction mixture, only a certain proportion of the collisions are successful in forming new products. To increase a reaction rate, you can increase the proportion of successful collisions by:

- increasing the frequency of successful collisions by increasing the number of collisions that can occur in a given time
- increasing the proportion of collisions that have energy that is equal to or greater than the activation energy (i.e. can overcome the activation energy barrier).

In this section, you will consider various changes in conditions that affect only the frequency of collisions between reactants which lead to increased reaction rates.

SURFACE AREA

When a solid is involved in a reaction, only the particles at the surface of the solid are involved in the reaction. The number of particles at the surface depends on the **surface area** of the substance. As you can see from Figure 13.3.1, breaking a solid into smaller parts means that more particles are present at the surface and available to react. The surface area has increased.

As a consequence of the greater number of exposed particles, the frequency of collisions between these particles and the other reactant particles increases, so the reaction occurs more rapidly.

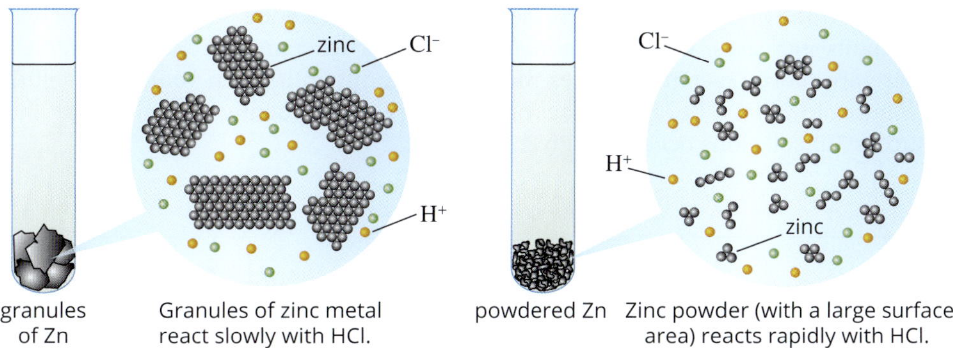

FIGURE 13.3.1 The reaction of hydrochloric acid and zinc. As the surface area of zinc increases, the rate of reaction with hydrochloric acid increases.

The effect of increasing surface area can be seen when setting up an open fire at home or on a camp. It is best to first light a pile of kindling, rather than trying to directly light large logs. The kindling has a larger surface area than the logs, so it catches fire more easily and will then burn rapidly, providing enough sustained heat energy to make the logs catch fire as well.

Manufacturers of fireworks modify the surface area of solid reactants to control the rate at which fuels in the fireworks burn and create different effects. For example, very small pieces of aluminium confined in a shell explode violently. If larger pieces of aluminium are used, the reaction is slower and pieces of burning metal are ejected, seen as sparks.

CHEMFILE A S

Fireworks and nanoparticles

Studies have shown that the fireworks associated with celebrations such as New Year's Eve (Figure 13.3.2) and the Lantern Festival in China can significantly increase air pollution levels. Sulfur dioxide and particles of metals are released when fireworks explode. These can cause breathing problems and lung disease.

Recent research aimed at lowering the environmental impact of fireworks has focused on reducing the particle size of the chemicals used as fuels in the fireworks. By using smaller nanoparticles, reaction rates are increased and smaller amounts of chemicals are needed for the same performance. Thus, fewer pollutants are released into the atmosphere.

However, this new approach carries increased risks because fireworks made of such small particles could be even more explosive.

FIGURE 13.3.2 Looking across the Huangpu River to a fireworks display in Shanghai at New Year.

CONCENTRATION

FIGURE 13.3.3 This limestone statue has become pitted in recent years. Limestone (calcium carbonate) reacts more rapidly with the increased acid concentration in rainwater.

The concentration of solutes dissolved in a solution can influence the rate of their reactions: higher concentrations lead to increased reaction rates.

Pollutants such as sulfur dioxide and nitrogen dioxide are released by cars and many industrial processes. When these compounds react with rainwater, acids such as sulfurous acid and nitric acid are formed. This produces what is known as **acid rain**. The increasing acidity of rain over the past 200 years has caused many famous marble buildings and statues to deteriorate much more rapidly due to the reaction between the marble and the acids (Figure 13.3.3).

The rate of a reaction increases when the frequency of collisions between reactants increases. When the concentration of a reactant in a solution increases, there are more reactant particles moving randomly in a given volume of solution (Figure 13.3.4). The frequency of collisions consequently increases, so more successful collisions occur in a given time.

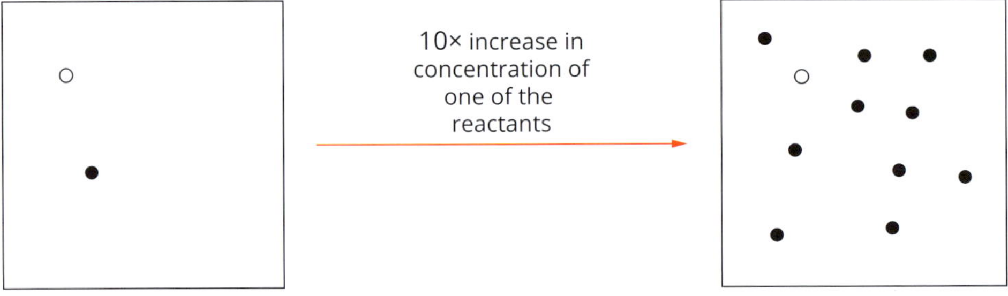

FIGURE 13.3.4 On the left, the concentration of both reactants is low. On the right, the concentration of one of the reactants has been increased 10-fold, resulting in an increase in collision frequency.

PRESSURE

In reactions involving gases, increasing the pressure on the gas increases the rate at which the reaction takes place. Increasing the pressure at a constant temperature will result (on average) in the reactant particles coming closer together. This will increase the chance of the gas molecules colliding, and therefore increase the rate of reaction.

For this reason, engineers often employ high gas pressures in their design of chemical processes that use gas-phase reactions. An example is the production of ammonia gas by reacting hydrogen gas and nitrogen gas. Increasing the pressure ensures a faster rate of reaction.

For a reaction in the gas phase, the pressure of the gases can be increased either by adding more gas to a fixed-volume container or by decreasing the volume of a container with a variable volume, such as a gas syringe. Increasing the pressure increases the concentration of the gas molecules, causing more frequent collisions and increasing the number of successful collisions in a given time.

When using collision theory to explain the effects of concentration, pressure and surface area on the rate of a chemical reaction, you must discuss the effect on either collision frequency or on the number of successful collisions per unit time.

Worked example 13.3.1

USING COLLISION THEORY TO EXPLAIN CHANGES IN RATES OF REACTIONS

Iron anchors recovered from shipwrecks at considerable depths can show little corrosion after years in the sea. Explain this observation in terms of collision theory.

Thinking	Working
Consider the states of the reactants.	The iron anchor is a solid. Oxygen is a gas dissolved in the sea water.
Relate the states of the reactants to the factor that affects the reaction rate and explain in terms of collision theory.	The concentration of oxygen at great depths in the ocean is low. The surface area of the iron anchor is relatively small. The frequency of collisions with reacting particles that could cause corrosion would be low, so the rate of corrosion would also be low.
Answer the question.	The concentration of oxygen at great depths is low, so the frequency of collisions with reacting particles would be low. The surface area of the iron anchor is relatively small, so the frequency of collisions is further reduced. Therefore, the rate of corrosion is low.

Worked example: Try yourself 13.3.1

USING COLLISION THEORY TO EXPLAIN CHANGES IN RATES OF REACTIONS

There have been many explosions in underground coal mines due to the presence of coal dust. Explain this observation in terms of collision theory.

13.3 Review

SUMMARY

- The rate of a reaction may be increased by:
  - increasing the surface area
  - increasing the concentration of a reactant in solution
  - increasing the pressure of a gaseous reactant.
- Collision theory can be used to explain increase in rate of reaction in terms of an increase in the concentration of a reactant in solution or of the pressure of a gaseous reactant: there are more reactant particles moving randomly in a given volume, so the frequency of collisions increases and more successful collisions occur in a given time.
- Collision theory can be used to explain the increase in rate of reaction caused by an increase in the surface area of a solid reactant: since there are a greater number of exposed reactant particles in smaller solid pieces, the frequency of collisions between these particles and the other reactant particles increases, so the reaction occurs much faster.
- Increase in concentration, pressure or surface area results in an increase in the:
  - frequency of collisions between reactant particles
  - number of successful collisions in a given time.

KEY QUESTIONS

1 Which one of the following changes would decrease the rate of the reaction between zinc metal and dilute hydrochloric acid solution?

A decreasing the pressure of the hydrochloric acid
B decreasing the size of the pieces of zinc
C decreasing the concentration of the hydrochloric acid
D decreasing the volume of hydrochloric acid used

2 Select the correct response in each of these statements about various ways in which reaction rates can be increased.

a increasing/decreasing the surface area of solid reactants
b increasing/decreasing the concentration of a reactant in solution
c increasing/decreasing the pressure of gaseous reactants
d increasing/decreasing the particle size of solid reactants

3 A number of experiments involving the reaction between 100 mL of hydrochloric acid and 5 g of calcium carbonate were carried out. Rearrange experiments **A–D** to place them in increasing order of rate of reaction (slowest to fastest).

A Powdered $CaCO_3$ and 1 mol L^{-1} HCl are mixed at 25°C.
B Large pieces of $CaCO_3$ and 0.5 mol L^{-1} HCl are mixed at 25°C.
C Powdered $CaCO_3$ and 0.5 mol L^{-1} HCl are mixed at 25°C.
D Small pieces of $CaCO_3$ and 0.5 mol L^{-1} HCl are mixed at 25°C.

4 Account for the following observation with reference to the collision model of particle behaviour: a bottle of fine aluminium powder has a caution sticker warning 'Highly flammable, dust explosion possible'.

5 Describe one way of increasing the rate of each of the following reactions:

a wood burning on a camp fire
b removing excess mortar from between bricks using brick-cleaner solution.

13.4 Effect of temperature on reaction rate

As you have seen, a reaction can be made to occur more rapidly by increasing the concentration of the reactants, or by increasing the surface area of solid reactants. A change in temperature can also have a major effect on the rate of a reaction.

Every cook knows that the temperature of an oven affects the rate of the chemical reactions occurring during baking. The higher the temperature, the more rapidly the reactions occur.

On the other hand, in hot weather it is wise to store fruit and vegetables in the refrigerator so that the chemical reactions that cause them to overripen and spoil will be slowed down at the lower temperatures (Figure 13.4.1).

An increase in temperature not only increases the frequency of collisions, it also increases the **kinetic energy (KE)** of the particles and hence the energy of their collisions.

FIGURE 13.4.1 Food is stored at low temperatures in a refrigerator to slow down the rate of reactions that cause food spoilage.

CHEMFILE IU

Ötzi the Iceman

In September 1991, Erika and Helmut Simon were walking in the Ötztal Alps near the border between Austria and Italy when they discovered the body of what they thought was a dead mountaineer (Figure 13.4.2). It was known that, in this region, bodies decompose very slowly because they are enclosed in ice. Closer examination of the body, and the Bronze Age items with it, eventually established that he had died approximately 5300 years ago.

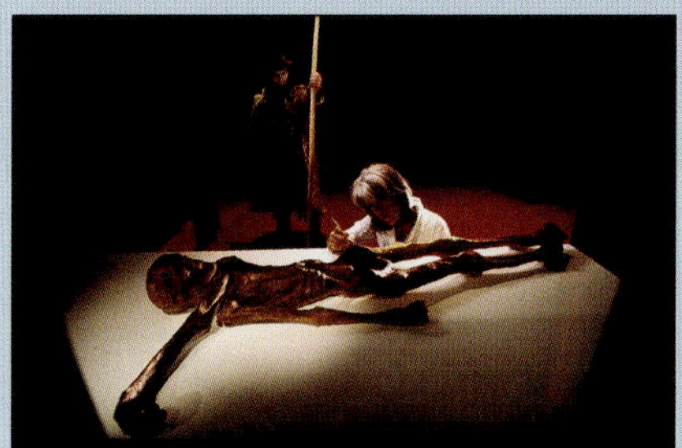

FIGURE 13.4.2 Ötzi the Iceman was so well preserved for 5300 years in the ice that his stomach contents could be identified, and pollen from a spring-flowering plant was found on his clothes.

MAXWELL–BOLTZMANN DISTRIBUTION

At any particular temperature, the particles in a substance have a range of kinetic energies. Although most of the particles have similar kinetic energies, there are always some particles with a higher energy or a lower energy. This range of energies is shown on a graph called a **Maxwell–Boltzmann distribution curve**, also known as a **kinetic energy distribution diagram**. Figure 13.4.3 shows how the distribution of energies is represented in a Maxwell–Boltzmann distribution curve.

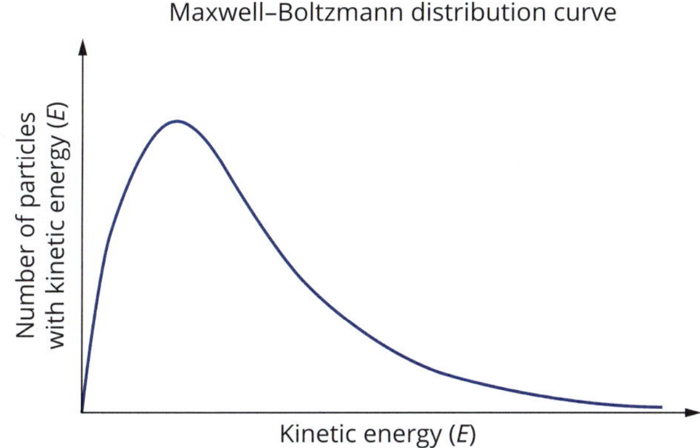

FIGURE 13.4.3 This Maxwell–Boltzmann curve shows the distribution of energies of particles in a sample at a particular temperature. The peak of the graph corresponds to the energy of the greatest number of particles.

During a reaction at a given temperature, only a small proportion of the reactant particles have kinetic energy that is equal to or greater than the activation energy and are able to react. You can see this in Figure 13.4.4 from the small shaded area to the right of the activation energy, E_a.

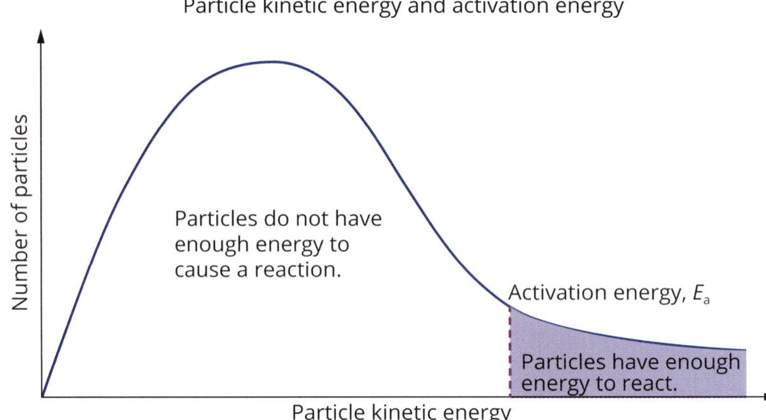

FIGURE 13.4.4 Only a small number of higher-energy particles (represented by the shaded area) have sufficient energy to overcome the activation energy barrier.

The Maxwell–Boltzmann curve shows the range of kinetic energies of particles at a particular temperature.

EFFECT OF TEMPERATURE ON RATE OF REACTION

The relationship between kinetic energy (KE), mass (m) and velocity (v) is given by the formula $KE = \frac{1}{2}mv^2$. As the temperature of a reaction system increases, the average kinetic energy of the particles increases. The average speed of the particles in the system increases as well.

This is illustrated in Figure 13.4.5, in which the range of kinetic energies for a gas at three different temperatures is shown. Note that the area under the curve, which is equal to the total number of particles in the sample, stays constant when the temperature is changed. As the temperature increases, the increasing average kinetic energy of the particles can be seen by the movement to the right of the peak in the Maxwell–Boltzmann curve.

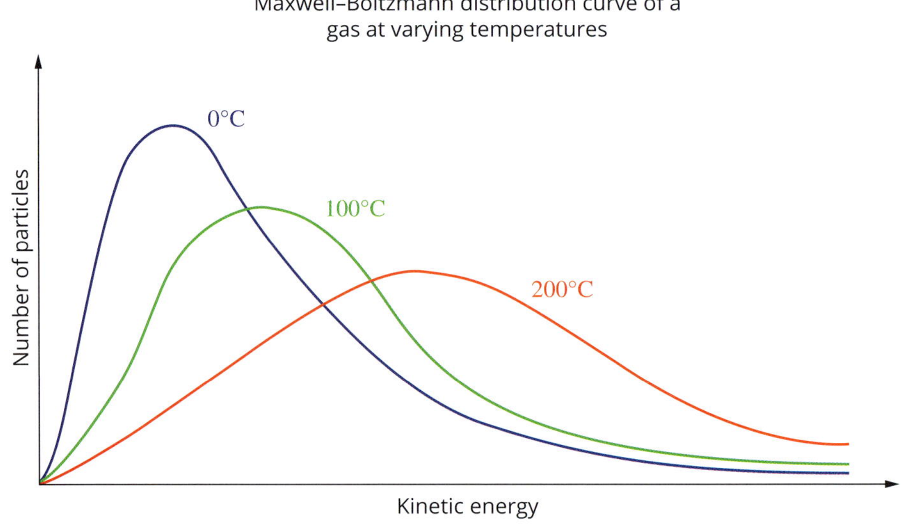

FIGURE 13.4.5 Kinetic energy distribution for a gas at a range of temperatures.

As the temperature of a reaction increases, the increased speed of the particles causes them to collide more often, increasing the frequency of collisions and the rate of reaction.

The increased kinetic energy of the particles also means that collisions occurring at higher temperatures have greater energy than those occurring at lower temperatures. More colliding particles will have energies that are greater than or equal to the activation energy, so the proportion of successful collisions also increases.

When the temperature increases, the increase in reaction rate due to the increased energy of the collisions is much greater than the increase in reaction rate due to the increased frequency of collisions.

A temperature increase of just 10°C causes the rate of many reactions to double (an increase of 100%), but it can be shown that the increased frequency of collisions is hardly significant. The frequency of collisions only increases by about 3% when the temperature increases by 10°C. The main reason why the reaction rate increases is that many more of the colliding particles have sufficient energy to overcome the activation energy barrier of the reaction.

The effect of increasing temperature on the rate of reaction is mainly through increasing the proportion of reacting particles that have energies equal to or greater than the activation energy for the reaction. This increases the proportion of successful collisions.

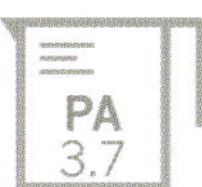

CHEMISTRY IN ACTION A

Activation energy and pollution in large cities

The size of the activation energy determines how easy it is for a reaction to occur and therefore what proportion of collisions would result in a successful reaction. For this reason, the reaction rate is dependent upon the average kinetic energy of the colliding molecules.

The existence of an activation energy for a reaction means that collisions between reactants do not always result in a chemical change. For example, nitrogen (N_2) and oxygen (O_2) molecules collide frequently in the air around us at room temperature. However, it is only when energy is provided by a spark, such as in car engines, that the energy of the collisions is increased enough to overcome the activation energy barrier. This allows nitrogen monoxide to be produced. The nitrogen monoxide formed in this reaction can then react to form brown nitrogen dioxide (NO_2), a poisonous gas that is a major contributor to the **photochemical smog** in busy cities such as Beijing (Figure 13.4.6), where there are large numbers of cars on the roads.

FIGURE 13.4.6 Photochemical smog, such as seen here over Beijing, is caused by NO_2, a poisonous gas.

13.4 Review

SUMMARY

- As temperature increases, the kinetic energy of particles increases.
- Kinetic energy, KE, is related to particle velocity, v, by the formula: $KE = \frac{1}{2} mv^2$.
- A Maxwell–Boltzmann curve is a graph that shows the distribution of energies of particles in a sample at a particular temperature.
- Increase in temperature results in an increase:
  - in the frequency of collisions between reactants
  - in the energy of collisions, so an increased proportion of collisions has an energy larger than the activation energy for the reaction
  - the number of successful collisions in a given time.

KEY QUESTIONS

1. Which one of the following makes the most significant contribution to the increase in reaction rate that occurs when temperature is increased?
   A The number of collisions between reactant particles increases.
   B The frequency of collisions between reactant particles increases.
   C More of the collisions between reactant particles have greater energy.
   D The activation energy of the reaction increases.
2. Which one or more of the following may be true if a reaction is observed to proceed very slowly?
   A The activation energy may be very large.
   B The temperature may be low.
   C Few collisions may be occurring with the correct orientation.
3. Account for the following observations with reference to the collision model of particle behaviour.
   a Surfboard manufacturers find that fibreglass plastics set within hours in summer, but may remain tacky for days in winter.
   b A potato cooks much more slowly in a pot of boiling water on a trekking holiday in Nepal than a potato boiled in a similar way in the Australian bush.
4. This graph shows the kinetic energy distribution of particles at 100°C.
   a What is the name given to this type of curve?

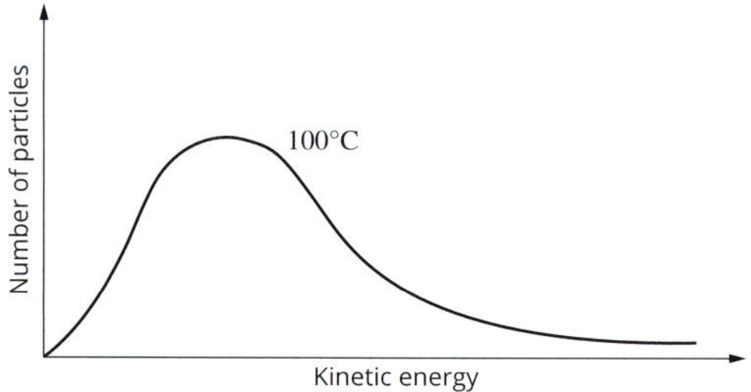

   b Copy this graph and add two labelled curves to show the change that occurs if:
   i the temperature is increased to 200°C.
   ii the temperature is decreased to 30°C.
   c Use the information shown by the graphs to explain why reactions can occur at a faster rate at 200°C, compared with at 100°C.

Chapter review

KEY TERMS

acid rain
activation energy
collision theory
energy profile diagram
kinetic energy
kinetic energy distribution diagram
Maxwell–Boltzmann distribution curve
photochemical smog
rate of reaction
surface area
transition state

REVIEW QUESTIONS

1 According to collision theory, what must happen for a reaction to occur?

2 This diagram shows the reaction between chloromethane and a hydroxide ion. (Also refer to the diagram in question 3.)
Using collision theory, explain why collision 1 might be successful, while collisions 2–4 will not be successful.

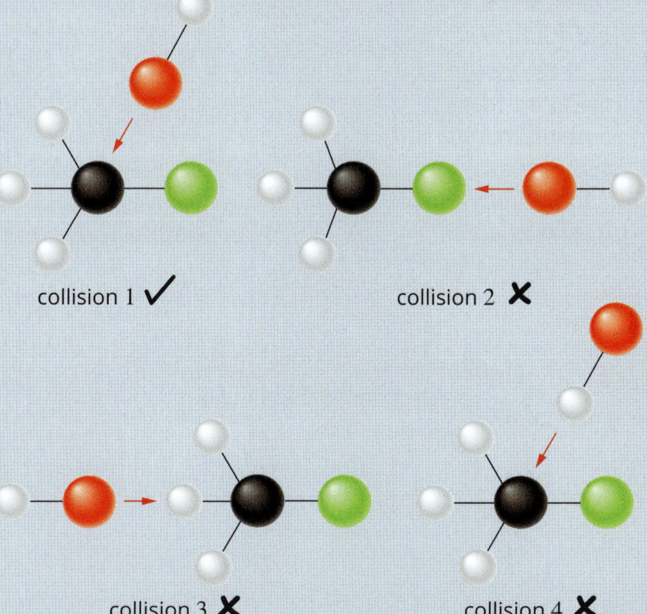

3 This graph is an energy profile diagram for the reaction between chloromethane and hydroxide ion.

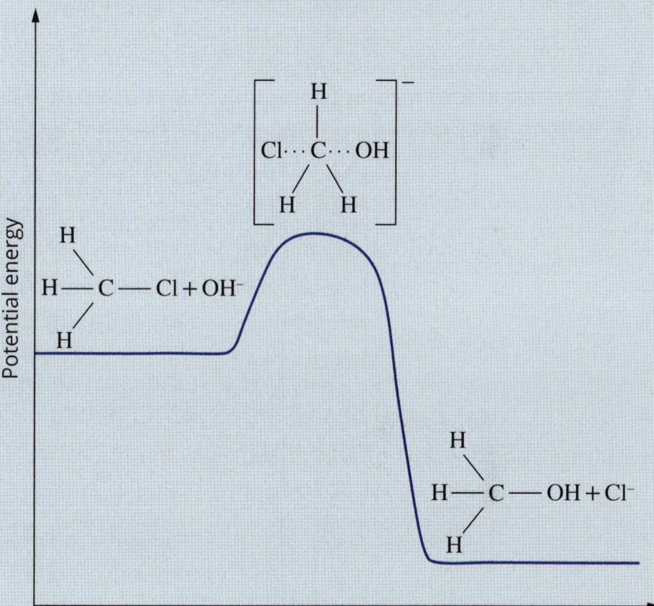

a Copy this diagram and label ΔH and activation energy E_a.
b Explain what is meant by the term 'activation energy'.
c Label the transition state of this reaction on your diagram.
d What bonds are beginning to be broken and formed to produce the transition state?

4 Hydrogen reacts explosively with oxygen to form water.
a What chemical bonds are broken in the reaction?
b What chemical bonds are formed?
c Why is there no reaction until the reaction mixture is ignited?

5 Which one of the following is the correct definition of rate of reaction?
   A the time it takes for all of a reactant to be used up
   B the change in concentration of reactants or products over time
   C how much a reaction is bubbling
   D how fast a reaction is going at the end of 1 minute

6 The following changes are made to a reaction mixture. Which change will lead to a decrease in reaction rate?
   A Smaller solid particles are used.
   B The temperature is decreased.
   C Reactant gas pressure is increased.
   D The concentration of an aqueous reactant is increased.

7 According to the definition of rate of reaction, what unit is used to measure rate of reaction?

8 A student was investigating the difference in the rate of the reaction between acidified potassium permanganate (a purple solution containing the permanganate ion) and oxalic acid ($H_2C_2O_4$) at two different temperatures. The graphs that were plotted are shown here, but do not have the vertical axis labelled. The equation for the reaction involved is:

$$2MnO_4^-(aq) + 5H_2C_2O_4(aq) + 6H^+(aq) \rightarrow 2Mn^{2+}(aq) + 10CO_2(g) + 8H_2O(l)$$

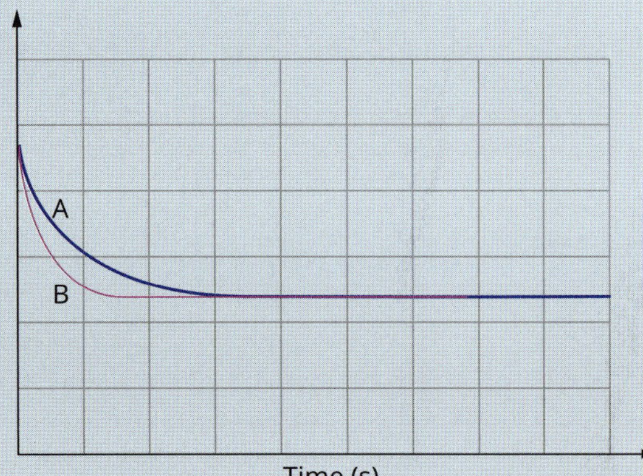

   a The student could have used the following techniques to monitor the rate of this reaction: colorimetry, or measuring the volume of carbon dioxide gas produced over time. Identify and explain which technique was chosen to monitor the rate of this reaction.
   b Identify whether graph A or B is likely to represent the higher-temperature reaction.

9 Which one of the following alternatives correctly explains why a sample of magnesium reacts more rapidly with 1 mol L^{-1} HCl than with 0.1 mol L^{-1} HCl?
   A The energy of collisions between reactant particles is greater for the reaction containing 1 mol L^{-1} HCl.
   B The rate of collisions between reactant particles is greater for the reaction containing 0.1 mol L^{-1} HCl.
   C There are fewer collisions between the magnesium and 1 mol L^{-1} hydrochloric acid.
   D The frequency of collisions between reactant particles is greater with 1 mol L^{-1} HCl than for the reaction containing 0.1 mol L^{-1} HCl.

10 **a** Write an equation for the reaction between marble chips ($CaCO_3$) and hydrochloric acid (HCl).
   b Select the combination of reactants that would produce the greatest initial rate of reaction from the following list.
   0.1 mol L^{-1} HCl, 1 mol L^{-1} HCl, 2 mol L^{-1} HCl, powdered $CaCO_3$, small marble chips, large marble chips

11 Account for the following observations with reference to the collision model of particle behaviour.
   a Refrigeration slows down the browning of sliced apples.
   b Hydrogen gas burns in air to produce water vapour. Using pure oxygen gas instead of air increases the rate of this reaction.

12 Lumps of limestone (calcium carbonate) react readily with dilute hydrochloric acid. Four large lumps of limestone (total mass 10.0 g) were reacted with 100 mL 0.100 mol L^{-1} acid.
   a Write a balanced equation to describe the reaction.
   b Which reactant is in excess? Use a calculation to support your answer.
   c Describe a technique that you could use in a school laboratory to measure the rate of the reaction.
   d 10.0 g of small lumps of limestone will react at a different rate from four large lumps. Will the rate of reaction with the smaller lumps be faster or slower? Explain your answer in terms of collision theory.
   e List two other ways in which the rate of this reaction can be altered. Explain your answer in terms of collision theory.

13 A 5.00 g piece of copper was dissolved in a beaker containing 500 mL of 2.00 mol L^{-1} nitric acid. The equation for the reaction that occurred is:

$$3Cu(s) + 8HNO_3(aq) \rightarrow 3Cu(NO_3)_2(aq) + 2NO(g) + 4H_2O(l)$$

The changing mass of the mixture was observed for a period of time, and this graph was obtained.

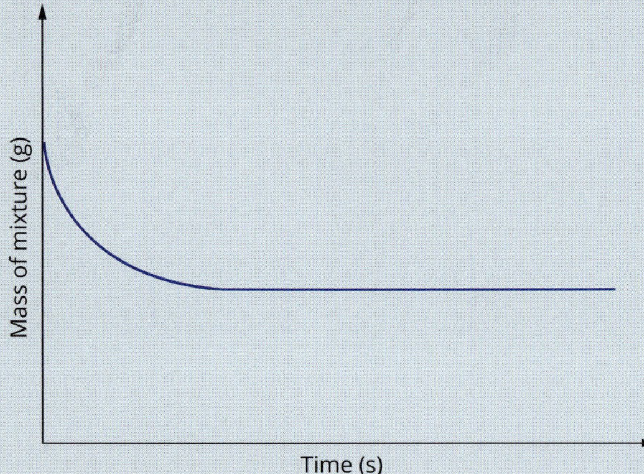

a Explain why the mass of the mixture initially decreases with increasing time.
b Based on the information provided, determine which reactant is limiting.
c Redraw the graph, then sketch the expected curve that would result if 500 mL of 1.00 mol L^{-1} nitric acid was used instead. Label your new graph line. Explain the difference in shape.
d Redraw the graph, then sketch the expected curve that would result if 5.00 g of powdered copper was used instead. Label this new graph line. Explain the difference in shape.

14 Which one of the following statements correctly describes what must occur when reactant particles collide and react?
A Colliding particles must have an equal amount of kinetic energy.
B Colliding particles must have different amounts of kinetic energy.
C Colliding particles must have kinetic energy equal to or greater than the average kinetic energy.
D Colliding particles must have kinetic energy equal to or greater than the activation energy of the reaction.

15 Chemical reactions in the body normally take place at 37°C. Explain how the rate of chemical reactions in the body can account for the following facts.
a The body often responds to illness by an increase in temperature, accompanied by a higher pulse rate and faster breathing.
b People rescued from drowning after 20–30 minutes in freezing water can sometimes survive and recover with no brain damage.

16 A number of experiments involving a reaction between 100 mL of hydrochloric acid and 5 g of calcium carbonate were carried out. Rearrange experiments I–VI to place them in increasing order of rate of reaction (slowest to fastest).
I Powdered $CaCO_3$ and 1 mol L^{-1} HCl were mixed at 40°C.
II Small pieces of $CaCO_3$ and 1 mol L^{-1} HCl were mixed at 15°C.
III Powdered $CaCO_3$ and 1 mol L^{-1} HCl were mixed at 25°C.
IV Large pieces of $CaCO_3$ and 0.5 mol L^{-1} HCl were mixed at 15°C.
V Powdered $CaCO_3$ and 1 mol L^{-1} HCl were mixed at 15°C.
VI Small pieces of $CaCO_3$ and 0.5 mol L^{-1} HCl were mixed at 15°C.

17 a What are the four factors (besides the presence of a catalyst) that influence the rate of a reaction?
b Classify the four factors from part (a) according to whether they increase the proportion of successful collisions by:
i increasing collision frequency
ii increasing the proportion of collisions that have energy equal to or greater than the activation energy.

18 The first step in most toffee recipes is to dissolve about 3 cups of sugar in 1 cup of water. Although sugar is quite soluble in water, this step can be time-consuming. Use your knowledge of reaction rates to suggest at least three things you could do to increase the rate of dissolution without ruining the toffee.

19 Read the article and answer the questions that follow.

Exploding iron

In 1996, while the Turkish ship MV *B. Onal* was riding at anchor in Delaware Bay, near Philadelphia in the USA, a two tonne hatch cover suddenly blew off. As the ship was carrying a cargo of iron, the surprised crew asked themselves, 'Can iron explode?'

As you may be aware, traditionally iron oxide (Fe_2O_3) is reduced to molten iron at 1800°C in a blast furnace.

A new process that uses less energy has been developed. Iron oxide is converted directly to solid iron without having to heat the reactants to the melting point of iron. Iron oxide is heated to 550°C in the presence of carbon monoxide and hydrogen gas. The iron oxide is reduced to iron by both gases, with the formation of carbon dioxide or water.

$$Fe_2O_3(s) + 3CO(g) \rightarrow 2Fe(s) + 3CO_2(g) \quad (1)$$

$$Fe_2O_3(s) + 3H_2(g) \rightarrow 2Fe(s) + 3H_2O\ (g) \quad (2)$$

The pellets of pure iron that are formed are extremely porous and full of many tiny holes, in contrast to the solid formed when the molten iron from a blast furnace cools. Under the right conditions, the iron pellets can be oxidised back to iron oxide.

In most cases, iron is oxidised slowly by oxygen back to iron oxide, and the resulting heat can readily escape. If the pellets are more than 1 metre deep, as in the hold of a ship, the heat cannot escape quickly enough and the temperature rises. This speeds up the reaction rate. If the temperature increases sufficiently and water is present, another reaction occurs and the oxidation rate is speeded up 100-fold, with the release of more heat:

$$Fe(s) + H_2O(g) \rightarrow FeO(s) + H_2(g) \quad (3)$$

Any spark or fire will set off an explosion of hydrogen gas, and that is what happened on the MV *B. Onal*.

a What is the main reason the new reduction process uses less energy than the old process?
b Write equations showing the oxidation of iron by oxygen to form iron(II) oxide and iron(III) oxide.

c List the factors that increased the rate of reaction in equation 3.

d Firefighters were not able to use water to put out the fire in the cargo hold. Why not? Suggest how they could put out the fire.

20 a By discussing the energy and frequency of collisions, explain why the rate of reaction between 1 $mol\,L^{-1}$ $CuSO_4$ and powdered zinc is greater than that with an equal amount of large zinc pieces at the same temperature.

b Write half-equations for the oxidation and reduction reactions in this redox reaction.

21 Reflect on the Inquiry activity on page 410. Why did the crushed chalk bubble more rapidly compared with the whole piece of chalk?

MODULE 3 • REVIEW

REVIEW QUESTIONS

Reactive chemistry

WS 3.9 MR 3

Multiple choice

1 Which of the following is not an example of a chemical change?

A Respiration produces CO_2 gas and water from glucose.

B Combustion of natural gas will produce CO_2 gas and water.

C When CO_2 gas passes through limewater, the limewater turns milky.

D When dry ice sublimes, solid CO_2 changes to a gas.

2 The equation for the decomposition of chalk ($CaCO_3$) is:

$CaCO_3(s) \rightarrow CaO(s) + CO_2(g)$

In this reaction, the mass of:

A CaO formed will equal the mass of CO_2 formed.

B CaO formed will be half of the mass of $CaCO_3$ reacting.

C CO_2 formed will be equal to the mass of CaO formed subtracted from the mass of $CaCO_3$.

D CaO formed will equal the mass of $CaCO_3$ reacting, as mass is conserved in the reaction.

3 Which of the following is a neutralisation reaction?

A $CH_3COOH + NaOH(aq) \rightarrow NaCH_3COO(aq) + H_2O(l)$

B $2HCl(aq) + Zn(s) \rightarrow ZnCl_2(aq) + H_2(g)$

C $CuCO_3(s) \rightarrow CuO(s) + CO_2(g)$

D $HCl(aq) + H_2O(l) \rightarrow H_3O^+(aq) + Cl^-(aq)$

4 Select the option that contains only compounds soluble in water.

A $NaNO_3$, AgCl, $BaSO_4$

B KNO_3, NH_4Cl, Na_2SO_4

C $CuCO_3$, $Pb(OH)_2$, $BaSO_4$

D $LiNO_3$, $ZnCO_3$, $Cu(OH)_2$

5 Which of the following represents a correctly balanced incomplete combustion reaction?

A $2C_2H_6(g) + 7O_2(g) \rightarrow 4CO_2(g) + 6H_2O(g)$

B $C_2H_6(g) + 3O_2(g) \rightarrow 2CO(g) + 3H_2O(g)$

C $2C_2H_6(g) + 5O_2(g) \rightarrow 4CO(g) + 6H_2O(g)$

D $C_3H_8(g) + 7O_2(g) \rightarrow 3CO(g) + 4H_2O(g)$

6 Which one of the following equations correctly describes the reaction that occurs when aluminium metal reacts with oxygen?

A $Al(s) + O_2(g) \rightarrow AlO_2(s)$

B $2Al(s) + O_2(g) \rightarrow 2AlO(s)$

C $4Al(s) + 3O_2(g) \rightarrow 2Al_2O_3(s)$

D $6Al(s) + 2O_2(g) \rightarrow 2Al_3O_2(s)$

7 Consider the reactions of the metals listed with the following reagents:

Reaction I: lithium + water

Reaction II: zinc + steam

Reaction III: copper + dilute hydrochloric acid

Which of these reactions will produce hydrogen gas?

A reaction I only

B reactions I and II only

C reactions II and III only

D reactions I, II and III

8 A reaction between which one of the following pairs is not classified as a displacement reaction?

A zinc and oxygen

B sodium and water

C iron and steam

D magnesium and copper(II) sulfate solution

9 Consider the information provided in the following table:

Property	Metal			
	I	II	III	IV
atomic radius (picometres)	174	215	122	130
first ionisation energy ($kJ\,mol^{-1}$)	590	403	745	520
electronegativity	1.0	0.8	1.9	1.0

On the basis of the data provided, which metal would you expect to be least reactive?

A I

B II

C III

D IV

10 Which one of the following is not a redox reaction?

A $2Fe^{3+}(aq) + Sn^{2+}(aq) \rightarrow 2Fe^{2+}(aq) + Sn^{4+}(aq)$

B $Mg(s) + O_2(g) \rightarrow 2MgO(s)$

C $2Na(s) + 2H_2O(l) \rightarrow H_2(g) + 2NaOH(aq)$

D $Ag^+(aq) + Cl^-(aq) \rightarrow AgCl(s)$

11 Which of the following species would not be expected to react with each other to a great extent?

A Cu(s) and $Mg^{2+}(aq)$

B Fe(s) and $Pb^{2+}(aq)$

C Mg(s) and $Fe^{2+}(aq)$

D Cu(s) and $Ag^+(aq)$

12 Consider the following equation:

$$Mg(s) + H_2SO_4(aq) \rightarrow MgSO_4(aq) + H_2(g)$$

Which one of the following statements is correct about this reaction?

A This is a neutralisation reaction.
B Mg has been reduced.
C H^+ is acting as an oxidising agent.
D Mg is acting as an oxidising agent.

13 Metal X was added to a solution of lead(II) nitrate ($Pb(NO_3)_2$). A reaction occurred and a precipitate of lead was produced. On the basis of this result, which one of the following can be deduced?

A Lead metal must be more reactive than metal X.
B Metal X must be able to react with hydrochloric acid to release hydrogen gas.
C Metal X must be able to react with a solution of $Mg(NO_3)_2$ to produce a precipitate of Mg.
D Lead metal must be able to react with a solution of $X(NO_3)_2$ to produce a precipitate of X.

14 Which of the following statements about a reducing agent is/are correct?

I A reducing agent is reduced in a reaction.
II A reducing agent causes another substance to be reduced.
III A reducing agent takes electrons away from another substance.

A I only
B II only
C I and III only
D II and III only

15 The following equations involve ions of the transition metal vanadium. The half-cell potential E° (V) is shown for each equation.

$$V^{3+}(aq) + e^- \rightarrow V^{2+}(aq) \qquad E^\circ = -0.25\,V$$
$$VO^{2+}(aq) + 2H^+(aq) + e^- \rightarrow V^{3+}(aq) + H_2O(l) \qquad E^\circ = +0.36\,V$$
$$VO_2^+(aq) + 2H^+(aq) + e^- \rightarrow VO^{2+}(aq) + H_2O(l) \qquad E^\circ = +1.00\,V$$

Of the ions listed, which is the strongest oxidising agent and which is the strongest reducing agent?

	Strongest oxidising agent	Strongest reducing agent
A	VO_2^+	V^{2+}
B	VO_2^+	VO^{2+}
C	V^{3+}	VO^{2+}
D	V^{2+}	VO_2^+

16 Which of the following best describes the features of an anode in a galvanic cell?

	Polarity	Electrode reaction
A	positive	oxidation
B	positive	reduction
C	negative	oxidation
D	negative	reduction

17 An electrochemical cell was made by dipping a copper rod into a solution of 1 mol L^{-1} $CuSO_4$ in one beaker and dipping a nickel rod into a solution of 1 mol L^{-1} $NiSO_4$ in another beaker. The metal electrodes were connected with wire, and the two solutions were connected by a piece of filter paper that had been soaked in a potassium nitrate solution. The cell is shown in the following diagram.

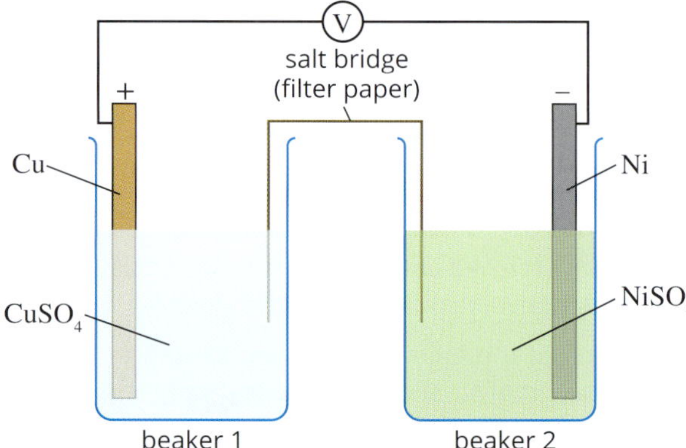

The solution in beaker 1 was initially coloured blue, owing to the presence of Cu^{2+} ions. The solution in beaker 2 was initially coloured green because of the presence of Ni^{2+} ions. Which of the following changes might it be possible to detect after the galvanic cell has been discharging for a period of time?

A The green colour in beaker 2 faded and the mass of the copper electrode increased.
B The blue colour in beaker 1 faded and the mass of the nickel electrode increased.
C The green colour in beaker 2 faded and the mass of the copper electrode decreased.
D The blue colour in beaker 1 faded and the mass of the nickel electrode decreased.

18 The diagram shows the typical Maxwell–Boltzmann distribution of kinetic energies of the particles in a reaction mixture.

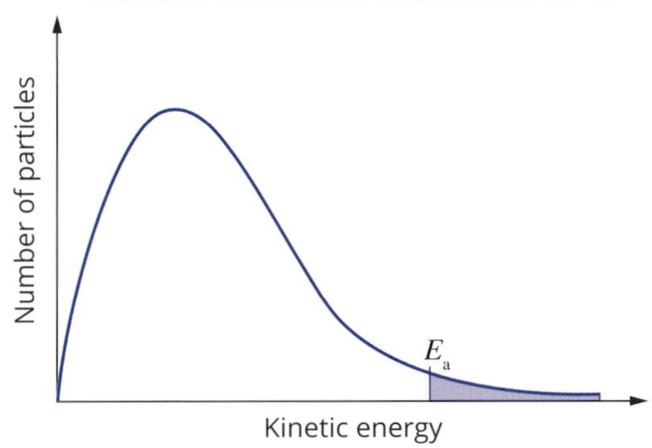

If E_a represents the activation energy for the reaction, what does the shaded area represent?

A the total amount of energy that will be released when the reaction is complete

B the proportion of reactant particles that have enough energy at a given instant for a collision to result in the formation of products

C the energy that must be added to the reactant mixture to initiate the reaction

D the proportion of reactant particles that will not be converted to products when the reaction is complete

19 Two changes are made to the reaction mixture from Question **18**, affecting both the distribution curve and the activation energy, as shown in red in the diagram.

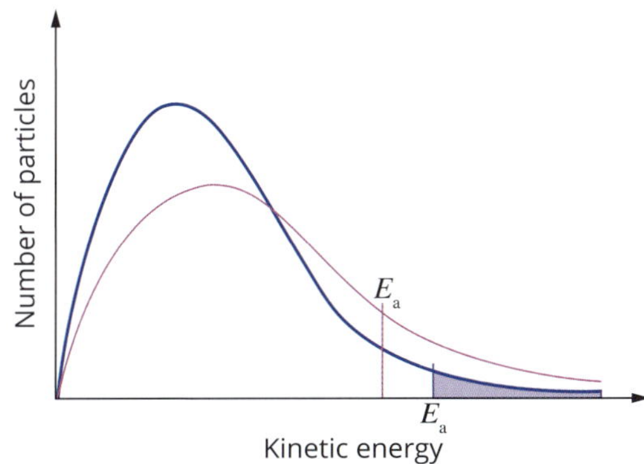

Which of the following gives changes in conditions that are consistent with these effects?

	Change in distribution curve	Change in activation energy
A	higher temperature	catalyst added
B	higher temperature	lower pressure
C	higher pressure	lower temperature
D	higher pressure	catalyst added

20 Which of the following statements about a catalyst is not true?

A A catalyst is not consumed in the course of a reaction.

B A catalyst alters the reaction pathway between reactants and products.

C A catalyst reduces the energy released or absorbed by a reaction.

D The proportion of molecules with sufficient energy to react is increased by a catalyst.

Short answer

1 Classify each of the following reactions as one of synthesis, decomposition, combustion, precipitation or acid base.

a $CuSO_4(aq) + 2NaOH(aq) \rightarrow Cu(OH)_2(s) + Na_2SO_4(aq)$

b $H_2SO_4(aq) + Mg(OH)_2(aq) \rightarrow MgSO_4(aq) + 2H_2O(l)$

c $C_2H_4(g) + H_2(g) \rightarrow C_2H_6(g)$

d $C_2H_4(g) + 2O_2(g) \rightarrow 2CO(g) + 2H_2O(g)$

e $Na_2CO_3(s) \rightarrow Na_2O(s) + CO_2(g)$

2 Predict the products of the following reactions to complete and balance the equations.

a $HCl(aq) + Na_2CO_3(s) \rightarrow$

b $C_4H_{10}(g) + O_2(g) \rightarrow$

c $Ba(s) + H_2SO_4(aq) \rightarrow$

d $Al(s) + Cl_2(g) \rightarrow$

e $HNO_3(aq) + KOH(aq) \rightarrow$

3 The chemical formula of ammonia is NH_3. Ammonia can be manufactured from a reaction between nitrogen and hydrogen gases. Once formed, ammonia can be reacted with oxygen gas to form nitrogen monoxide (NO) and water.

a Write a balanced chemical equation for the formation of ammonia from nitrogen and hydrogen gases.

b Which category of reaction is this?

c Write a balanced chemical equation for the formation of nitrogen monoxide gas and water vapour from ammonia and oxygen gases.

4 Write ionic equations for the reaction between:

a solutions of barium nitrate and potassium sulfate

b sodium hydroxide solution and dilute nitric acid

c dilute hydrochloric acid and magnesium carbonate solution

d zinc metal and copper(II) sulfate solution.

5 Write ionic equations for the reactions that occur when:

a an iron nail is placed in a test tube containing dilute hydrochloric acid

b a small piece of rubidium is dropped into a beaker of water

c a strip of zinc metal is added to a silver nitrate solution.

6 **a** Two descriptions of the relative reactivity of magnesium in dilute hydrochloric acid are provided.

I A strip of magnesium was placed into a test tube containing dilute hydrochloric acid. The magnesium initially reacted very vigorously, producing a large number of gas bubbles in a short period of time.

II 2.5 mL of hydrogen gas was collected in 60.0 s when a strip of magnesium weighing 0.5 g was placed into a test tube containing 10.0 mL of hydrochloric acid that had a concentration of 3.64 $g\,L^{-1}$.

i Which of these descriptions uses qualitative data? Explain your answer.

ii List the quantitative data provided in the quantitative description.

b The activity series and the table of standard reduction potentials both rank metals in order of their reactivity. The activity series is based on qualitative data, and the table of standard reduction potentials is based on quantitative data. List one similarity and one difference, other than the qualitative and quantitative nature of data, between the activity series and the table of standard reduction potentials.

7 Iron is an important structural material in our society, and the prevention of its corrosion is expensive.

The half-equations for the reaction resulting in iron corrosion are:

$$Fe(s) \rightarrow Fe^{2+}(aq) + 2e^-$$

$$O_2(g) + 2H_2O(l) + 4e^- \rightarrow 4OH^-(aq)$$

a Write an overall equation for this reaction.

b Identify the oxidising agent and the reducing agent.

c A moist iron nail corrodes readily.

i If a small piece of zinc metal is wrapped around only one part of the nail, corrosion is prevented. Explain why.

ii If tin is used instead of zinc, the nail needs to be completely coated with the tin to prevent corrosion. Explain why.

8 For each of the following unbalanced redox reactions (all occurring in acidic aqueous solution):

i write separate oxidation and reduction half-equations

ii write a balanced ionic equation for the overall reaction.

a $Al(s) + Br_2(aq) \rightarrow Al^{3+}(aq) + Br^-(aq)$

b $ClO^-(aq) + S_2O_3^{2-}(aq) \rightarrow SO_4^{2-}(aq) + Cl^-(aq)$

c $H_2O_2(aq) + MnO_4^-(aq) \rightarrow O_2(aq) + MnO_2(s)$

9 Referring to a table of standard reduction potentials and for each of the half-cell combinations listed, predict:

i the maximum cell voltage expected at standard conditions

ii which of the half-cells will contain the negative electrode

iii the ionic equation for the overall reaction occurring when the cell is discharging.

a $Sn^{2+}(aq)/Sn(s)$ and $Fe^{2+}(aq)/Fe(s)$

b $Fe^{3+}(aq)/Fe^{2+}(aq)$ and $Al^{3+}(aq)/Al(s)$

c $H^+(aq)/H_2(g)$ and $I_2(aq)/I^-(aq)$

10 The activation energy for the decomposition of hydrogen peroxide to oxygen was measured under two different conditions.

$$2H_2O_2(aq) \rightarrow 2H_2O(l) + O_2(g)$$

I The activation energy for the decomposition reaction when an enzyme was added was found to be +36.4 $kJ\,mol^{-1}$. The temperature of the reaction mixture increased.

II When platinum was added to another sample of the hydrogen peroxide solution, the activation energy was +49.0 $kJ\,mol^{-1}$. The temperature of the reaction mixture increased.

a What is the function of the enzyme and platinum in each of these reactions?

b Sketch, on the same set of axes, the energy profiles for the decomposition of hydrogen peroxide with the enzyme and the decomposition using the platinum.

c Which reaction system, I or II, would be faster? Explain your answer.

Extended response

1 Some potassium chloride solution is added to a solution containing 3.15 g of lead(II) nitrate. A white precipitate forms.

a Write a full chemical equation for the formation of the precipitate.

b Write an ionic equation for the formation of the precipitate.

c For this reaction, give the name of the:

i precipitate

ii spectator ions.

d Assume there was sufficient potassium chloride to react with all of the lead(II) nitrate.

i Calculate the amount, in mol, of lead(II) nitrate that reacted.

ii Calculate the mass of precipitate that would form.

2 An activity series of metals can be experimentally determined using displacement reactions between metals and metal ions in solution.

Imagine you have been provided with the following metals and 1.0 mol L^{-1} solutions of metal ions:

Metal	Metal ion solution	Metal	Metal ion solution
Cu	Cu^{2+}	Fe	Fe^{2+}
Pb	Pb^{2+}	Mg	Mg^{2+}
Sn	Sn^{2+}	Zn	Zn^{2+}

a Describe an experimental procedure that would enable you to rank metals in order of reactivity.

b Rank the metals in the table in order from most reactive to least reactive.

c i Identify one metal ion that will be displaced from solution by metallic zinc. Justify your selection in terms of the activity series.

ii Write a balanced ionic equation for the reaction that occurs between zinc and the ion you selected.

d i Identify one metal ion that will displace zinc from a solution containing Zn^{2+} ions. Justify your selection in terms of the activity series.

ii Write a balanced ionic equation for this reaction.

---

3 Describe the trends in metal activity in the periodic table. Use the information provided in the table below to explain why these trends correlate with:

i ionisation energy

ii atomic radius

iii electronegativity.

Property	Sodium	Magnesium	Potassium	Calcium
electronegativity	0.9	1.3	0.8	1.0
atomic radius (picometres)	160	140	200	174
first ionisation energy (kJ mol^{-1})	496	738	419	590
ionisation equation	$Na \rightarrow Na^+ + e^-$	$Mg \rightarrow Mg^{2+} + 2e^-$	$K \rightarrow K^+ + e^-$	$Ca \rightarrow Ca^{2+} + 2e^-$
energy required to form ion (kJ mol^{-1})	496	2200	419	1748

---

4 The diagram shows a version of an alkaline aluminium–air fuel cell being used to power a load, such as an electric motor.

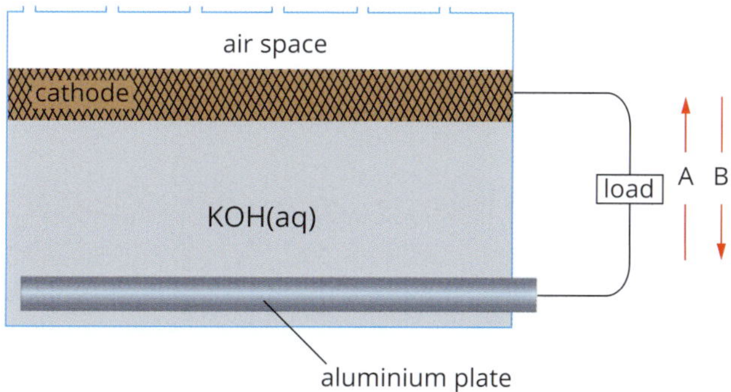

The half-reactions occurring are:

$Al(s) \rightarrow Al^{3+}(aq) + 3e^-$

$O_2(g) + 2H_2O(l) + 4e^- \rightarrow 4OH^-(aq)$

a Which arrow, A or B, shows the direction of electron flow through the load?

b What is the oxidising agent in this cell?

c Write a balanced equation for the overall cell reaction.

d Explain why it is important for the cathode to be porous.

Using a zinc (Zn) plate in place of the aluminium plate gave a cell that still functioned. However, when a silver (Ag) plate was used instead, the cell did not function.

e Account for this difference by referring to the relative strengths of the oxidising agents and reducing agents involved.

f Would the cell potential (voltage) produced by the cell with the zinc plate be larger or smaller than that with the original aluminium plate? Explain your answer.

g When the potassium hydroxide electrolyte was replaced with a sulfuric acid solution, the cell with the silver plate began to deliver current. Explain why changing the electrolyte made this difference.

5 A student investigated the factors affecting the rate of reaction between a solution of sodium thiosulfate and hydrochloric acid.

$$Na_2S_2O_3(aq) + 2HCl(aq) \rightarrow 2NaCl(aq) + SO_2(g) + S(s) + H_2O(l)$$

The reaction was carried out in a conical flask placed on top of a piece of white paper with a dark cross marked on it. The rate of reaction was determined by measuring the time taken for the cross to be masked by the suspension of sulfur formed during the reaction, as shown.

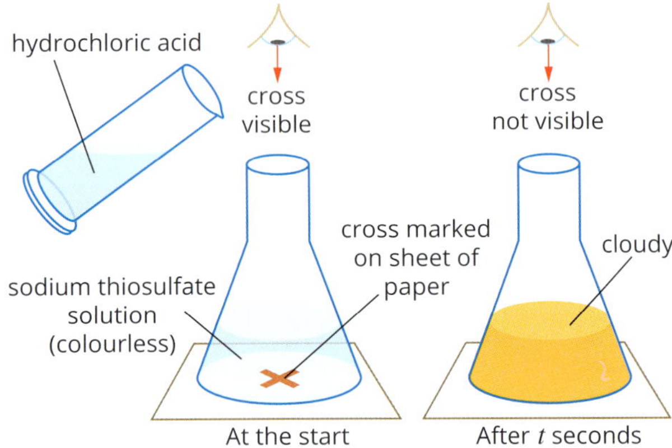

The rate was determined for different concentrations of sodium thiosulfate and for different temperatures. The volume of each solution and the concentration of hydrochloric acid was kept constant. The results are summarised in the table.

Experiment number	$[Na_2S_2O_3(aq)]$ ($mol\ L^{-1}$)	Temperature (°C)	Time taken for the cross to be masked (s)
1	0.1	20	36
2	0.2	20	20
3	0.1	25	28

a Explain, in terms of collision theory, why the rate in experiment 2 is higher than the rate in experiment 1.

b Explain, in terms of collision theory, why the rate in experiment 3 is higher than the rate in experiment 1.

c What factors, other than the two investigated in this experiment, may affect the rate of a reaction?

MODULE 4

Drivers of reactions

In this module, you will investigate the factors that initiate and drive a reaction. You will examine the relationship between enthalpy and entropy in calculating Gibbs free energy, and the roles that enthalpy and entropy play in the spontaneity of reactions. You will develop an understanding that all chemical reactions involve the creation of new substances and are accompanied by an associated energy transformation. This is observable as a change in temperature of the surrounding environment and/or emission of light.

You will learn how to measure the energy changes that occur in chemical reactions, to describe reactions using terms such as 'endothermic' and 'exothermic', and to explain reactions in terms of the law of conservation of energy. You will use Hess's law to calculate enthalpy changes involved in the breaking and making of bonds.

Outcomes

By the end of this module, you will be able to:

- develop and evaluate questions and hypotheses for scientific investigation (CH11-1)
- analyse and evaluate primary and secondary data and information (CH11-5)
- solve scientific problems using primary and secondary data, critical thinking skills and scientific processes (CH11-6)
- communicate scientific understanding using suitable language and terminology for a specific audience or purpose (CH11-7)
- analyse the energy considerations in the driving force for chemical reactions (CH11-11)

CHAPTER

14 Energy changes in chemical reactions

Energy, like matter, is all around us. In fact, Albert Einstein's most famous equation, $E = mc^2$, gives the relationship between energy and matter. In all chemical reactions, including those in the human body, there is a change in energy as bonds of different chemical species are broken and reformed. An average adult consumes approximately 8 MJ of energy per day from food. Historically, this was all the energy required by ancient hunter-gatherers. The phones, cars and computers of modern life require far more energy. Across 2014–2015, the total energy consumption for New South Wales was 1400 PJ (1.4×10^{18} J, where 1 PJ = 1 quadrillion joules). Dividing that among the 7.5 million citizens results in an average of 430 MJ used per person per day. Most of this energy is provided by burning fossil fuels.

By the end of this chapter, you will have a greater understanding of the way in which energy changes occur within chemical reactions and be able to write thermochemical equations, to represent chemical equations by drawing energy profile diagrams, and to determine the heat of combustion and dissolution experimentally by simple calorimetry.

The role of catalysts in changing the rate of chemical reactions will be considered and modelled in relation to the energy changes that occur during reactions.

Content

INQUIRY QUESTION

What energy changes occur in chemical reactions?

By the end of this chapter, you will be able to:

- conduct a practical investigation using appropriate tools (including digital technologies) to collect data, analyse and report on how the rate of a chemical reaction can be affected by a range of factors, including but not limited to: ICT N
  - catalysts (ACSCH042)
- conduct practical investigations to measure temperature changes in examples of endothermic and exothermic reactions, including: CCT ICT L
  - combustion
  - dissociation of ionic substances in aqueous solution (ACSCH018, ACSCH037) N
- investigate enthalpy changes in reactions using calorimetry and $q = mc\Delta T$ (heat capacity formula) to calculate, analyse and compare experimental results with reliable secondary-sourced data, and to explain any differences CCT ICT N
- construct energy profile diagrams to represent and analyse the enthalpy changes and activation energy associated with a chemical reaction (ACSCH072) ICT
- model and analyse the role of catalysts in reactions (ACSCH073) ICT

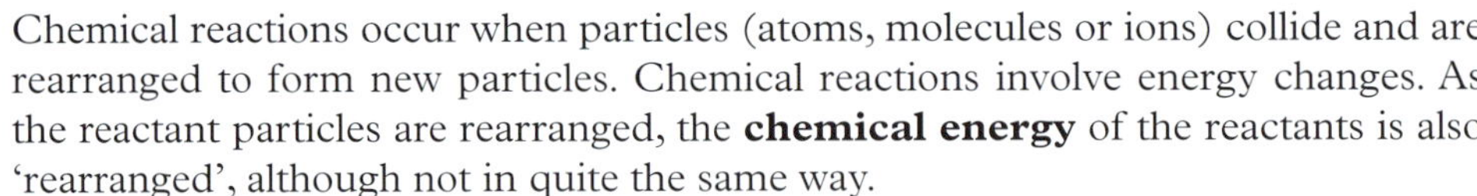

14.1 Exothermic and endothermic reactions

Chemical reactions occur when particles (atoms, molecules or ions) collide and are rearranged to form new particles. Chemical reactions involve energy changes. As the reactant particles are rearranged, the **chemical energy** of the reactants is also 'rearranged', although not in quite the same way.

In some chemical reactions, including the **combustion** of fuels, the rearrangement of atoms causes energy to be released to the surroundings. For example, the amount of energy released in the combustion of wood is large enough to be easily seen or felt (Figure 14.1.1).

FIGURE 14.1.1 The energy released by the combustion of wood in a fire is easily seen and felt. Exothermic reactions feel hot because they are releasing mainly thermal energy to the surrounding environment.

In other chemical reactions, energy is absorbed from the surroundings as the chemical reaction takes place. An example of this can be seen in instant ice packs (Figure 14.1.7 on page 446).

The energy change in some reactions is very small and can only be determined with specialised equipment.

In this section, you will learn about the energy changes that occur during chemical reactions. You will also learn how to classify chemical reactions based on their energy changes.

CHEMICAL ENERGY

There are many different forms of energy. You will be more familiar with some than others. For example, you are in contact with forms of heat (thermal energy), light (radiant energy), sound energy and electrical energy every day. You can probably see, hear or feel some of them as you are reading these pages.

All substances have a form of energy called chemical energy. It is stored in the bonds between atoms and molecules. This energy results from:

- attractions between electrons and protons in atoms
- repulsions between nuclei
- repulsions between electrons
- movement of electrons
- vibrations and rotations around bonds.

When you eat a meal, the bonds within the food molecules have stored energy that you can access to provide energy for other chemical and physical activities that take place in your body (Figure 14.1.2).

FIGURE 14.1.2 When you eat food, you access the chemical energy stored in the food. This energy powers all of the chemical reactions and physical activities that take place in your body.

The SI unit for energy is the **joule** (J). It gets its name from English physicist James Prescott Joule (1818–1889), who conducted ground-breaking studies into electricity, energy and mechanics. As 1 J is a relatively small amount of energy (approximately the amount required to lift a 100 g chocolate block up 1 metre against the gravity of Earth), it is also common to see larger units used for quantifying energy, such as kilojoules (kJ) and megajoules (MJ). Making mistakes while converting units is common, so it is important that you are familiar with these conversions and the use of your calculator.

GO TO ➤ SkillBuilder page 127

The relationship between joules, kilojoules and megajoules is:

$$1\,\text{J} = 10^{-3}\,\text{kJ} = 10^{-6}\,\text{MJ}$$

This can also be expressed as:

$$1\,\text{MJ} = 1000\,\text{kJ} = 1\,000\,000\,\text{J} = 1.0 \times 10^{6}\,\text{J}$$

> The SI unit for energy is the joule (J). In chemistry, the larger units kilojoule (kJ) and megajoule (MJ) are often used.

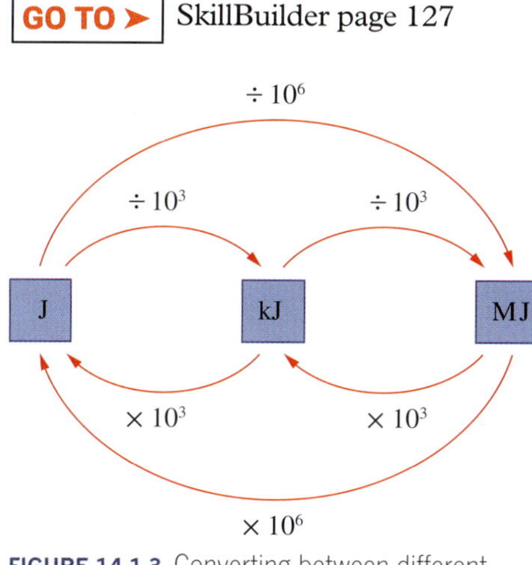

FIGURE 14.1.3 Converting between different energy units.

Figure 14.1.3 shows how you can convert between different units of energy.

ENERGY CONSERVATION

Much like the conservation of mass discussed in Chapter 6, there is also the **law of conservation of energy**: energy cannot be created or destroyed. However, energy can change forms. When energy is transformed from one form to another, the total amount of energy remains the same.

GO TO ➤ Section 6.1 page 206

Systems and surroundings

When we talk about energy changes in chemical reactions, we often refer to a **system** and its **surroundings**.

In chemistry, the system is usually the chemical reaction. When we say that energy is released or absorbed by a system, we are referring to energy changes that occur as bonds are broken and formed between the atoms of the elements involved in the reaction.

The surroundings are usually regarded as everything else. For example, the walls of a container in which a reaction takes place in the gas phase, or the water in a solution in which a reaction takes place in the aqueous phase, can be regarded as the surroundings for the reaction. Energy leaves the system (the reaction) and enters the surroundings, or leaves the surroundings and enters the system.

Energy changes during chemical reactions

The reactants in a chemical reaction have a certain amount of chemical energy stored in their bonds. The products that form as a result of the rearrangement of particles during the chemical reaction have different bonds and so have a different amount of chemical energy. Energy will be released or absorbed during the reaction, depending on the relative energies of the bonds within the reactants and products.

Often the energy released to or absorbed from the surroundings is in the form of heat (thermal energy). However, chemical energy can be converted into other types of energy, including light and electricity.

CHEMISTRY IN ACTION S

Glow-in-the-dark light sticks

You might have seen glow-in-the-dark hoops, necklaces and bracelets similar to those shown in Figure 14.1.4 at festivals or concerts, especially those held at night.

Glow-in-the-dark bracelets contain chemicals held in separate containers. When these bracelets are bent, the containers break and the chemicals combine. Light is produced through a process called **chemiluminescence**.

The chemistry of a glow stick is fairly straightforward. The aqueous reactants are hydrogen peroxide in one compartment and diphenyl oxalate in another compartment. When they mix, energy is released from the reaction that occurs. This reaction is shown in Figure 14.1.5. It is important that the plastic casing remains intact, as while the contents may be non-toxic, it is not advised to consume them or expose your skin or eyes to the mixture due to its irritating nature.

Instead of the energy from this reaction being released to the surroundings solely as heat, a carrier molecule transfers the energy to a chemiluminescent dye in the glow stick. The electrons in the dye are excited to higher energy levels. Light is emitted as these electrons return to their original lower-energy levels. The light from the glow stick is simply the emission spectrum of the dye molecule. However, they are one-use-only devices with a limited lifespan and are not easily recycled, so they contribute to landfill. Future development into such devices may make them safer, longer lasting and more sustainable. Existing alternatives include coloured LED bands, which are made with recyclable materials and function for a period limited only by the batteries that operate them.

FIGURE 14.1.4 Glow-in-the-dark bracelets give off light that is the result of chemiluminescence.

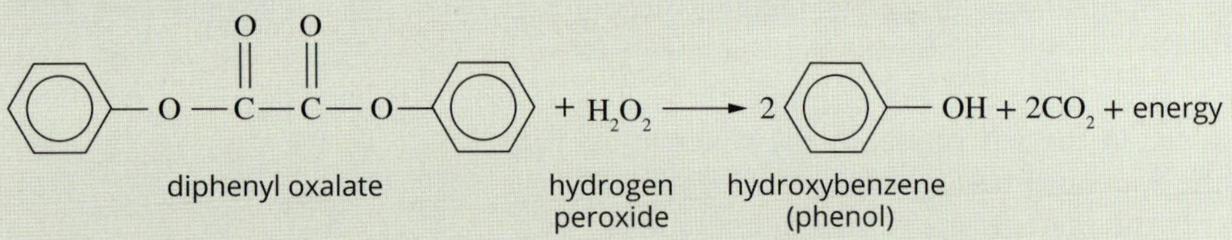

$$C_6H_5-O-\overset{\displaystyle O}{\overset{\|}{C}}-\overset{\displaystyle O}{\overset{\|}{C}}-O-C_6H_5 + H_2O_2 \longrightarrow 2\,C_6H_5-OH + 2CO_2 + \text{energy}$$

diphenyl oxalate | hydrogen peroxide | hydroxybenzene (phenol)

FIGURE 14.1.5 This reaction occurs in a glow stick. The hexagon with a circle in the middle represents a phenyl group, which has the formula C_6H_5.

GO TO ➤ Section 12.1 page 372

CHEMFILE CCT

Glow-worms

Glow-worms (Figure 14.1.6) apply similar strategies to chemiluminescence for their glow-in-the-dark **bioluminescence**. Three chemicals within the worm combine. However, they require oxygen to produce light. When the worm breathes, oxygen acts as the oxidising agent in the chemical reaction between the three reactants producing the bioluminescence. Worms are able to control the amount of 'glow' by breathing in more or less oxygen. Greater understanding of these biochemical processes may lead to future lighting technologies.

FIGURE 14.1.6 A female glow-worm. The luminescent abdominal organs are visible.

EXOTHERMIC AND ENDOTHERMIC SYSTEMS

When the total chemical energy of the products of a chemical reaction is less than the total chemical energy of the reactants, the excess energy is released to the surroundings. Energy 'exits' the reaction system and the chemical reaction is called an **exothermic** reaction.

The released energy can be shown in a chemical equation by writing 'energy' on the product side of the arrow.

For example, the production of water from the reaction between hydrogen and oxygen gas is an exothermic reaction. This can be represented by the equation:

$$2H_2(g) + O_2(g) \rightarrow 2H_2O(l) + \text{energy}$$

Another example of an exothermic reaction is the combustion of methane gas:

$$CH_4(g) + 2O_2(g) \rightarrow CO_2(g) + 2H_2O(g) + \text{energy}$$

Heat is given off to the surroundings. All combustion reactions give off heat (energy) to the surroundings and are therefore exothermic reactions.

When the total chemical energy of the products of a chemical reaction is greater than the total chemical energy of the reactants, energy is absorbed from the surrounding environment. Energy 'enters' the reaction system and the chemical reaction is called an **endothermic** reaction.

If an endothermic reaction takes place in a container, the container may feel cold to the touch. This is because the reaction system is absorbing heat from the surroundings, leaving the environment cooler.

In the chemical equation of an endothermic reaction, the energy that is required can be written on the reactant side of the equation arrow.

For example, the **decomposition** of calcium carbonate is an endothermic process. This can be represented by the equation:

$$CaCO_3(s) + \text{energy} \rightarrow CaO(s) + CO_2(g)$$

Endothermic reactions require the constant input of energy.

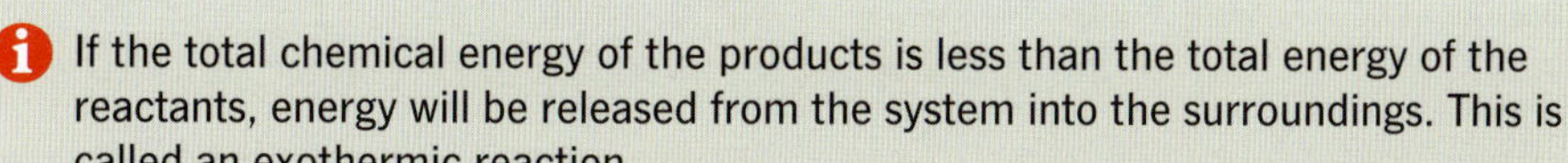

If the total chemical energy of the products is less than the total energy of the reactants, energy will be released from the system into the surroundings. This is called an exothermic reaction.

If the total chemical energy of the products is greater than the total energy of the reactants, energy will be absorbed from the surroundings. This is called an endothermic reaction.

Changes of state

Changes of state, such as a solid melting to form a liquid, are a type of physical change, rather than a chemical change. These changes also involve energy being absorbed or released.

For example, the melting of ice into water requires the absorption of energy, making it an endothermic process. The boiling of water to produce water vapour is also endothermic.

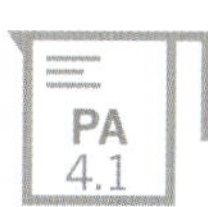

Conversely, condensing a gas to a liquid and freezing a liquid to form a solid both release heat to the surroundings and are thus exothermic processes.

CHEMFILE S

Instant cold packs

Instant cold packs are often carried in first aid kits at sporting events (Figure 14.1.7). One type of cold pack contains a sealed bag of water surrounded by solid ammonium nitrate. When the cold pack is squeezed, the water bag is broken and ammonium nitrate dissolves in the water. The process is endothermic, absorbing heat from the surroundings and quickly lowering the pack's temperature. However useful they may be, the environmental implications of these instant, single-use devices must be considered, particularly as re-usable icepacks, while less convenient, are available.

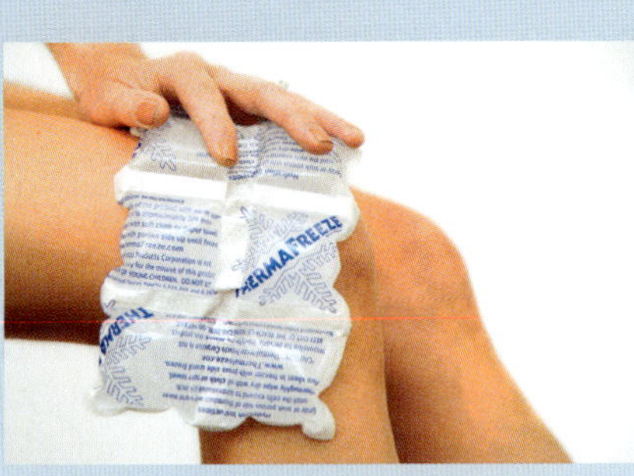

FIGURE 14.1.7 An endothermic reaction produces an instant cold pack. Endothermic reactions feel cold because they are absorbing energy from the surrounding environment.

14.1 Review

SUMMARY

- Energy is measured in J, kJ or MJ:
  $1\,J = 10^{-3}\,kJ = 10^{-6}\,MJ$
- Chemical energy is stored in the bonds between atoms.
- Energy is conserved during a chemical reaction; energy cannot be created or destroyed.
- Chemical reactions and changes of state involve energy changes.
- A chemical reaction that releases energy to the surroundings is called an exothermic reaction.
- A chemical reaction in which energy is absorbed from the surroundings is called an endothermic reaction.
- All combustion reactions are exothermic reactions.

KEY QUESTIONS

1. Which of the following statements about combustion reactions is correct?
   A Combustion reactions are usually exothermic, meaning energy is absorbed during the reaction.
   B Combustion reactions are always exothermic, meaning energy is absorbed during the reaction.
   C Combustion reactions are usually endothermic, meaning energy is absorbed by the system.
   D Combustion reactions are always exothermic, meaning energy is released during the reaction.
2. Convert the following energy values to kJ.
   a 0.180 MJ
   b 1.5×10^{6} J
   c 10.0 J
   d 2.0×10^{-3} J
3. Explain the difference between the terms 'system' and 'surroundings' in relation to a chemical reaction.
4. Explain the term 'endothermic' in relation to the total amount of chemical energy of the reactants and products.

14.2 Thermochemical equations and energy profile diagrams

In combustion reactions, energy is released in the form of heat. As you saw in the previous section, the word 'energy' can be included in a combustion equation to show that energy is released. However, it is generally more useful to be precise about the magnitude (size) of the energy change that takes place.

Thermochemical equations achieve this by including a sign and numerical value for the energy change that occurs in the reaction.

It is also useful to be able to show the energy changes that occur as a reaction proceeds. All reactions absorb some energy before they can proceed, even if energy is released overall. An **energy profile diagram** represents the energy changes that occur during the course of a reaction.

In this section, you will learn how to write and interpret thermochemical equations and draw energy profile diagrams.

REPRESENTING ENERGY CHANGE IN A CHEMICAL EQUATION

Enthalpy change

The chemical energy of a substance is sometimes called its **heat content** or **enthalpy**. It is given the symbol H. The enthalpy of the reactants in a chemical reaction is given the symbol H_r. The enthalpy of the products is given the symbol H_p.

Most chemical processes take place in open systems under a constant pressure (usually atmospheric pressure). The exchange of heat energy between the system and its surroundings under constant pressure is referred to as the **enthalpy change**, or **heat of reaction**, and it is given the symbol ΔH. The capital delta symbol (Δ) is commonly used in chemistry to represent 'change in'. For example, ΔT is the symbol for change in temperature.

For the general reaction:

$$\text{reactants} \rightarrow \text{products}$$

the enthalpy change (ΔH) is calculated by:

$$\Delta H = H_p - H_r$$

Enthalpy change is a measure of the amount of energy absorbed or released during chemical reactions. It is given the symbol ΔH and is determined by subtracting the enthalpy of the reactants (H_r) from the enthalpy of the products (H_p).

Enthalpy change in exothermic reactions

When H_p is less than H_r, energy is released from the system into the surroundings, so the reaction is exothermic. The system has lost energy, so ΔH has a negative value.

Therefore, for combustion reactions (which are exothermic reactions), $\Delta H < 0$ (Figure 14.2.1).

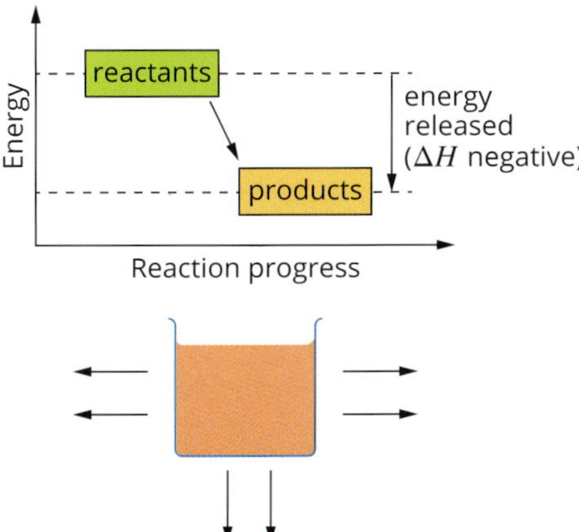

FIGURE 14.2.1 In the combustion of a fuel, the enthalpy of the reactants is greater than the enthalpy of the products, so energy is released to the surroundings during the reaction.

Enthalpy change in endothermic reactions

When H_p is greater than H_r, energy must be absorbed from the surroundings, so the reaction is endothermic. The system has gained energy, so ΔH has a positive value, i.e. $\Delta H > 0$ (Figure 14.2.2).

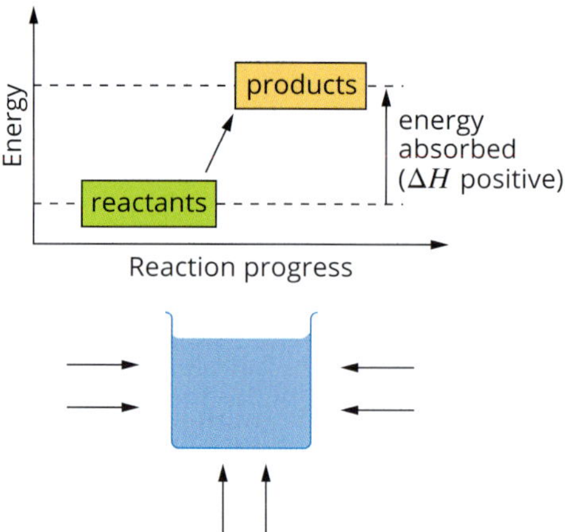

FIGURE 14.2.2 For an endothermic reaction, the enthalpy of the reactants is less than the enthalpy of the products, so energy is absorbed from the surroundings during the reaction.

Thermochemical equations

The enthalpy change can be shown by writing the ΔH value to the right of the chemical equation. Such an equation is called a **thermochemical equation**. The ΔH value in a thermochemical equation usually has the units kJ mol^{-1}. This means that the amount of energy (in kJ) signified by the ΔH value corresponds to the mole amounts specified by the coefficients in the equation.

For example, respiration in most living things can be considered as a type of combustion reaction of glucose. Therefore, respiration is an exothermic reaction:

$$C_6H_{12}O_6(aq) + 6O_2(g) \rightarrow 6CO_2(g) + 6H_2O(l) \qquad \Delta H = -2803\,\text{kJ mol}^{-1}$$

This thermochemical equation tells you that when 1 mol of glucose reacts with 6 mol of oxygen to produce 6 mol of carbon dioxide and 6 mol of water, 2803 kJ of energy is released.

Enthalpy changes also occur during physical changes, so thermochemical equations can be written for physical changes. Melting ice is an example of a physical change. It is an endothermic process because heat must be absorbed by solid ice in order to convert it into liquid water. The thermochemical equation for this reaction is:

$$H_2O(s) \rightarrow H_2O(l) \qquad \Delta H = +6.01\,\text{kJ mol}^{-1}$$

The ΔH value is positive because this is an endothermic reaction.

CHEMISTRY IN ACTION N

The chemistry behind an explosion

When chemical explosives (such as ammonium nitrate, trinitrotoluene (TNT) and nitroglycerine) decompose, they release large amounts of energy and gaseous products very quickly.

This is the thermochemical equation for the decomposition of nitroglycerine:

$$4C_3H_5N_3O_9(l) \rightarrow 12CO_2(g) + 10H_2O(g) + 6N_2(g) + O_2(g) \qquad \Delta H = -1456\,\text{kJ mol}^{-1}$$

Notice that 29 mol of gas (the total number of mol of all products) are produced from 4 mol of nitroglycerine. The negative ΔH value indicates that this is an exothermic reaction. At atmospheric pressure, the reactant products would expand to fill a volume more than 10000 times larger than the volume of the nitroglycerine! During a blast, this gas is usually produced within a small cavity into which the explosive has been placed, creating huge pressures that shatter the surrounding rock or structure.

CHEMISTRY IN ACTION EU

Explosives—a blast of chemical energy

Humans have been using chemicals to make explosions since 919 BCE. The Chinese were the first to mix saltpetre (potassium nitrate), sulfur and charcoal with explosive results. They quickly realised that there were many uses for this mixture, which later became known as gunpowder. It was put to military use and eventually led to the development of bombs, cannons and guns.

Today, explosives are an essential tool in mining and in other engineering works, such as road construction, tunnelling, building and demolition (Figure 14.2.3). However, explosives can be used for warfare. Alfred Nobel (1833–1896) was well known as the inventor of dynamite, a composition of a stabilised form of nitroglycerine, a highly explosive compound. After his death he left much of his estate to the creation of the Nobel prizes, including the Nobel Peace Prize, thought to be an attempt to compensate for the destructive forces of his creation.

Explosives have the potential to transform chemical energy into large quantities of thermal energy very quickly. Although thermal energy is also released when fuels such as petrol and natural gas burn, the rate of combustion in these reactions is limited by the availability of oxygen gas to the fuel. In contrast, the compounds making up an explosive contain sufficient oxygen for a complete (or almost complete) reaction to occur very quickly.

FIGURE 14.2.3 An old bridge is demolished with the help of explosives.

THERMOCHEMICAL EQUATIONS AND MOLE RATIOS

The ΔH value in a thermochemical equation corresponds to the mole amounts specified by the equation. If the coefficients in the equation are changed, the ΔH value will also change.

For example, the thermochemical equation for the combustion of methanol can be written as:

$$CH_3OH(l) + \tfrac{3}{2}O_2(g) \rightarrow CO_2(g) + 2H_2O(l) \qquad \Delta H = -726\,kJ\,mol^{-1}$$

This means that 726 kJ of energy is released when 1 mol of methanol reacts with 1.5 mol of oxygen gas, to produce 1 mol of carbon dioxide and 2 mol of water.

If twice as much methanol were to react, then twice as much energy would be released. So, if the coefficients of the equation are doubled, the ΔH value is also doubled:

$$2CH_3OH(l) + 3O_2(g) \rightarrow 2CO_2(g) + 4H_2O(l) \qquad \Delta H = -1452\,kJ\,mol^{-1}$$

If the mole amounts are tripled, the ΔH value is also tripled:

$$3CH_3OH(l) + \tfrac{9}{2}O_2(g) \rightarrow 3CO_2(g) + 6H_2O(l) \qquad \Delta H = -2178\,kJ\,mol^{-1}$$

The importance of states

It is very important to always include state symbols in thermochemical equations. Physical changes involve an enthalpy change, so the state of the species in a chemical reaction affects the enthalpy change of the reaction.

For example, both of the following equations represent physical changes involving water. They have different ΔH values because the states are different.

$$H_2O(s) \rightarrow H_2O(l) \qquad \Delta H = +6.01\ kJ\ mol^{-1}$$

$$H_2O(l) \rightarrow H_2O(g) \qquad \Delta H = +40.7\ kJ\ mol^{-1}$$

You can see that it requires more energy to boil water than it does to melt ice.

Effect on ΔH of reversing a chemical reaction

Reversing a chemical equation changes the sign but not the magnitude of ΔH.

For example, methane (CH_4) reacts with oxygen gas to produce carbon dioxide gas and water in an exothermic reaction:

$$CH_4(g) + 2O_2(g) \rightarrow CO_2(g) + 2H_2O(l) \qquad \Delta H = -890\ kJ\ mol^{-1}$$

If this reaction is reversed, the magnitude of ΔH remains the same because the enthalpies of the individual chemicals have not changed, but the sign changes to indicate that energy must be absorbed for this reaction to proceed.

$$CO_2(g) + 2H_2O(l) \rightarrow CH_4(g) + 2O_2(g) \qquad \Delta H = +890\ kJ\ mol^{-1}$$

Worked example 14.2.1

CALCULATING ΔH FOR A RELATED EQUATION

Iron reacts with oxygen according to the equation:

$$3Fe(s) + 2O_2(g) \rightarrow Fe_3O_4(s) \qquad \Delta H = -1121\ kJ\ mol^{-1}$$

Calculate ΔH for the reaction represented by the equation:

$$2Fe_3O_4(s) \rightarrow 6Fe(s) + 4O_2(g)$$

Thinking	Working
The reaction has been reversed in the second equation, so the sign for ΔH is changed to the opposite sign.	ΔH for the second equation is positive.
Identify how the mole amounts in the equation have changed.	The mole amount of Fe_3O_4 has changed from 1 to 2, of O_2 has changed from 2 to 4 and of Fe has changed from 3 to 6. They have all doubled.
Identify how the magnitude of ΔH will have changed for the second equation.	The mole amounts of the chemicals have all doubled, so ΔH will also have doubled.
Calculate the new magnitude of ΔH. (You will write the sign of ΔH in the next step.)	$2 \times 1121 = 2242$
Write ΔH for the second equation, including the sign.	$\Delta H = +2242\ kJ\ mol^{-1}$

Worked example: Try yourself 14.2.1

CALCULATING ΔH FOR A RELATED EQUATION

Carbon reacts with hydrogen according to the equation:

$$6C(s) + 3H_2(g) \rightarrow C_6H_6(g) \qquad \Delta H = +49\ kJ\ mol^{-1}$$

Calculate ΔH for the reaction represented by the equation:

$$3C_6H_6(g) \rightarrow 18C(s) + 9H_2(g)$$

Activation energy

Activation energy is the minimum amount of energy colliding reactant particles need in order for successful collisions to occur that lead to a reaction. The activation energy is an energy barrier that must be overcome before a reaction can get started. (The concept of activation energy was discussed in more detail back in Chapter 13.)

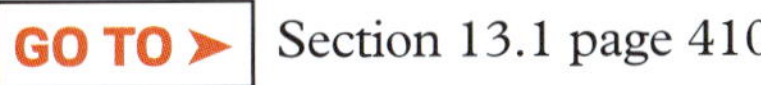
GO TO ➤ Section 13.1 page 410

A reaction cannot proceed unless the reactant particles collide with energy equal to or greater than the reaction's activation energy.

An activation energy barrier exists for both exothermic and endothermic reactions. If the activation energy for a reaction is very low, the chemical reaction can be initiated as soon as the reactants come into contact, because the reactants already have sufficient energy for a reaction to take place. Special conditions are not always required for reactions to occur. An example of this can be seen in the reaction between zinc and hydrochloric acid in Figure 14.2.4.

The reaction between zinc and hydrochloric acid produces hydrogen gas:

$$Zn(s) + 2HCl(aq) \rightarrow ZnCl_2(aq) + H_2(g)$$

As you can see in Figure 14.2.4, bubbles of hydrogen gas are vigorously produced as soon as zinc is added to the acid.

FIGURE 14.2.4 When zinc comes into contact with hydrochloric acid, it reacts almost immediately. The reactants have sufficient energy to overcome the activation energy barrier.

ENERGY PROFILE DIAGRAMS

You may recall from Section 13.1 that the energy changes occurring during the course of a chemical reaction can be shown on an energy profile diagram.

The energy profile diagram for an exothermic combustion reaction like the one shown in Figure 14.2.5 indicates that the enthalpy of the products is always less than the enthalpy of the reactants. Overall, energy is released and so the ΔH value is negative. The energy profile also shows that, even in exothermic reactions, the activation energy must first be absorbed before the reaction can start.

The energy profile diagram for an endothermic reaction (Figure 14.2.6) shows that the enthalpy of the products is greater than the enthalpy of the reactants. Overall, energy is absorbed and so the ΔH value is positive. The energy profile also shows the absorption of the activation energy before the release of energy as products form.

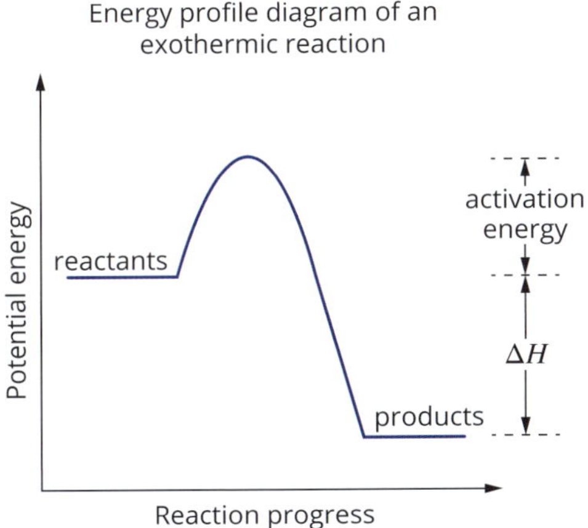

FIGURE 14.2.5 The characteristic shape of an energy profile diagram for an exothermic reaction.

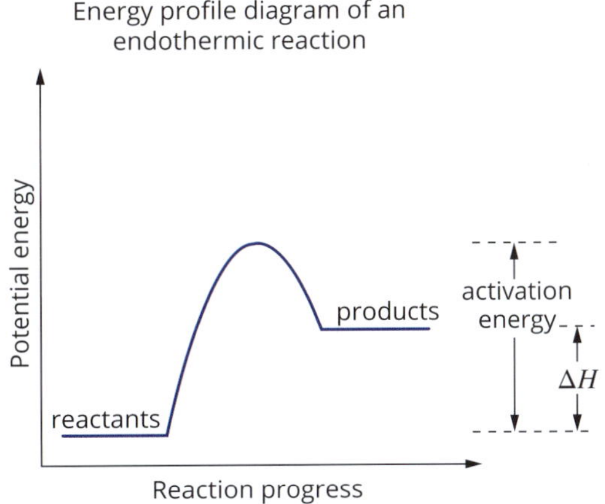

FIGURE 14.2.6 The characteristic shape of an energy profile diagram for an endothermic reaction.

14.2 Review

SUMMARY

- Thermochemical equations include a ΔH value for the chemical reaction. The unit of ΔH is usually $kJ\,mol^{-1}$.
- The value of ΔH indicates the magnitude of the energy change and whether the energy is absorbed (a positive value) or released (a negative value).
- Reversing an equation causes the sign of ΔH to change, as the reaction changes from exothermic to endothermic, or vice versa.
- Doubling the coefficients in a chemical reaction causes the ΔH value to also double, as twice as many reactants react to produce or absorb twice as much energy.
- States of matter must be included in thermochemical equations because changes of state also involve enthalpy changes.
- Activation energy is the minimum amount of energy colliding reactant particles need in order for successful collisions to occur that lead to a reaction. Both endothermic and exothermic reactions require activation energy.
- Energy profile diagrams show energy changes over the course of a reaction. They show the relative enthalpies of reactants and products, and the activation energy.
- Combustion reactions are always exothermic and so always have a negative ΔH value.

KEY QUESTIONS

1 Explain what a negative ΔH value indicates about a chemical reaction, in terms of the relative enthalpies of the reactants and products.

2 When 1 mol of methane gas undergoes combustion in oxygen to produce carbon dioxide and water, 890 kJ of energy is released. Write a balanced thermochemical equation for this reaction.

3 The combustion of octane to form carbon dioxide and liquid water can be written as:

$$C_8H_{18}(g) + \frac{25}{2}O_2(g) \rightarrow 8CO_2(g) + 9H_2O(l) \quad \Delta H = -5450\,kJ\,mol^{-1}$$

The combustion of octane to form carbon dioxide and water vapour can be written as:

$$C_8H_{18}(g) + \frac{25}{2}O_2(g) \rightarrow 8CO_2(g) + 9H_2O(g)$$

How would the energy released by the combustion of 1 mol of octane to form water vapour compare with the energy released by 1 mol of octane to form liquid water?

4 The energy profile diagram shows the energy changes during the reaction of hydrogen and iodine to form hydrogen iodide:

$$H_2(g) + I_2(g) \rightarrow 2HI(g)$$

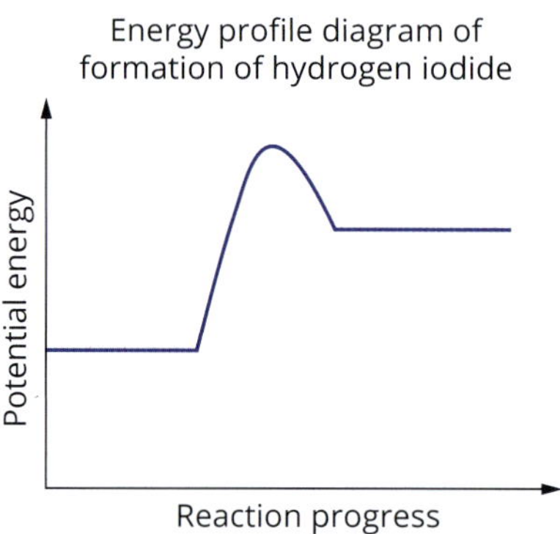

a Is the reaction exothermic or endothermic?

b Describe the relative enthalpies of the reactants and products.

c Comment on the size of the activation energy compared with the magnitude of ΔH.

5 The reaction for photosynthesis is the opposite of the reaction for respiration (page 448, under Thermochemical equations). Write a thermochemical equation for photosynthesis.

14.3 Heat of combustion

The **heat of combustion** is defined as the enthalpy change that occurs when a specified amount (e.g. 1 mol) of a substance burns completely in oxygen. It is usually measured at conditions of 298 K and 100 kPa, which means that the water produced should be shown in the liquid state. The heat of combustion (symbol ΔH_c and measured in kJ mol^{-1}) can be given for a **pure substance**, which is a single substance by itself, not a mixture.

The heats of combustion for some common elements and compounds are listed in Table 14.3.1. Heat energy is released during combustion, so ΔH_c always has a negative value. When hydrogen gas (H_2) is combusted, the only product formed is water, as no carbon is present. The equation is:

$$H_2(g) + \tfrac{1}{2}O_2(g) \rightarrow H_2O(l)$$

When metals combust, the oxide of the metal is the product. For example, the equation for the combustion of magnesium is:

$$2Mg(s) + O_2(g) \rightarrow 2MgO(s)$$

When materials are combusted, not all of the energy is released as heat energy, and not all of the heat energy liberated can be harnessed for a particular purpose, such as powering a vehicle or producing electricity. However, even if the energy transformation process is not 100% efficient, you can still use these values to compare the relative amounts of energy released from the combustion of different compounds.

As Worked example 14.3.1 shows, you can use the data in Table 14.3.1 to calculate the energy released during combustion of a specified mass of one of the substances.

The heat of combustion is usually measured at conditions of 298 K and 100 kPa. This means that the water produced should be shown in the liquid state.

TABLE 14.3.1 Heats of combustion for some common elements and compounds

Substance	Heat of combustion, ΔH_c (kJ mol^{-1})
methane	–890
ethane	–1560
propane	–2220
butane	–2886
octane	–5450
methanol	–725
ethanol	–1367
hydrogen	–286
carbon (graphite)	–394

The relationship discussed in Chapter 7 will be used throughout this chapter, relating the amount of a substance (n) (in mol) to its molar mass (M) (in g mol^{-1}) and the given mass of the substance (m) (in g):

$$n = \frac{m}{M}$$

Where needed, this relationship can be rearranged to:

$$m = n \times M \text{ or } M = \frac{m}{n}$$

GO TO ➤ Section 7.2 page 229

The energy released when n mol of a fuel burns is given by the equation:

$$\text{energy} = n \times \Delta H_c$$

Worked example 14.3.1

CALCULATING ENERGY RELEASED BY A SPECIFIED MASS OF A PURE SUBSTANCE

Calculate the amount of energy released when 3.60 kg of butane (C_4H_{10}) is burnt in an unlimited supply of oxygen.

Thinking	Working
Calculate the number of mol of the compound using: $n = \frac{m\text{(in grams)}}{M}$	$n(C_4H_{10}) = \frac{m}{M}$ $= \frac{3.60 \times 10^3}{58.12}$ $= 61.9$ mol
Multiply the number of mol by the heat of combustion.	Energy $= n \times \Delta H_c$ $= 61.9 \times 2886$ $= 1.79 \times 10^5$ kJ or 179 MJ

Worked example: Try yourself 14.3.1

CALCULATING ENERGY RELEASED BY A SPECIFIED MASS OF A PURE SUBSTANCE

Calculate the amount of energy released when 5.40 kg of propane (C_3H_8) is burnt in an unlimited supply of oxygen.

14.3 Review

SUMMARY

- Heat of combustion, ΔH_c, indicates the maximum amount of energy that can be released when a specified amount of a pure substance undergoes complete combustion.
- The relative amounts of energy released by different substances can be compared by referring to heats of combustion.
- For a pure substance with a heat of combustion, ΔH_c, measured in kJ mol^{-1}, the energy released when *n* mol of the fuel burns is given by the equation:

$$\text{Energy} = n \times \Delta H_c$$

KEY QUESTIONS

1 Write a balanced equation for the complete combustion of liquid benzene (C_6H_6).

2 Information about two hydrocarbon fuels, propane and octane, is given in the table below.

Characteristic	Propane (C_3H_8)	Octane (C_8H_{18})
Heat of combustion (kJ mol^{-1})	–2220	–5450
Molar mass (g mol^{-1})	44.09	114.22

Calculate the amount of energy released during the combustion of 1.000 g of each fuel, and use your answer to state which fuel produces more energy per unit of mass.

3 Write a balanced equation for the incomplete combustion of ethanol (C_2H_5OH), when carbon monoxide is formed.

4 Using the information in Table 14.3.1 (page 453), calculate the amount of energy released when the following amounts of each fuel undergo complete combustion:

a 250 g of methane

b 9.64 kg of propane

c 403 kg of ethanol

14.4 Determining the heat of combustion

Knowing the energy released by similar quantities of different substances helps you to compare them and determine their suitability for specific purposes. For example, the fuel used to power an aeroplane (Figure 14.4.1) is different from the fuel used to power a car or bus.

FIGURE 14.4.1 F14 jets being refuelled by a Boeing 707. The type of fuel suitable for use in aircraft is different from the type of fuel suitable for use in cars or buses.

In previous sections, you learnt ways of representing the energy change for fuels involved in combustion reactions.

- The heat of combustion (ΔH_c) of a fuel gives the amount of energy change when a specified amount of the fuel burns in oxygen.
- A thermochemical equation for a combustion reaction includes a ΔH value, which shows the amount of heat released when a fuel undergoes combustion. The value of ΔH is based on the stoichiometric ratios in the equation.

In this section, you will learn how knowledge of the specific heat capacity of water can be used to obtain an experimental estimate of the energy released in the combustion of a fuel.

SPECIFIC HEAT CAPACITY OF WATER

The **specific heat capacity** of a substance is a measure of the amount of energy (usually in joules) needed to increase the temperature of a specific quantity of that substance (usually 1 g) by 1°C.

Specific heat capacity is given the symbol *c* and is usually expressed in joules per gram per kelvin ($J\,g^{-1}\,K^{-1}$). It can also be expressed in joules per gram per degree Celsius ($J\,g^{-1}\,°C^{-1}$) (an increase of 1 K is the same as an increase of 1°C).

The specific heat capacities of some common substances are listed in Table 14.4.1. You can see that the value for water is relatively high.

TABLE 14.4.1 Specific heat capacities of common substances

Substance	Specific heat capacity ($J\,g^{-1}\,K^{-1}$)
water	4.18
glycerine	2.43
ethanol	2.46
sand	0.48
copper	0.39
lead	0.16

The specific heat capacity of a substance is the amount of energy required to increase the temperature of 1 g of the substance by 1 K . Specific heat capacity is frequently written using the units $J\,g^{-1}\,K^{-1}$.

Comparison of temperature verses heat energy supplied when heating 1 kg of water or glycerine

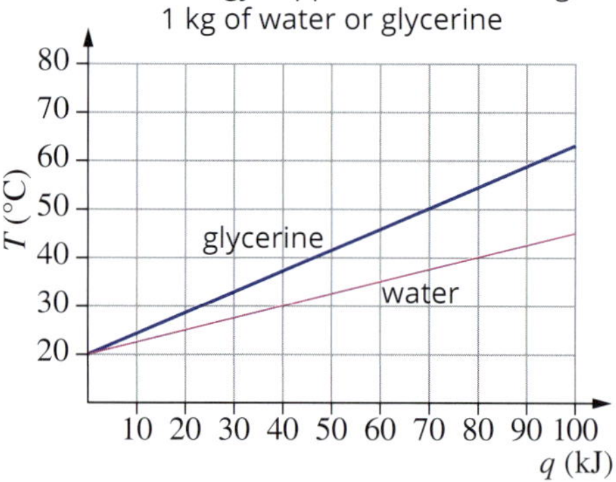

FIGURE 14.4.2 A comparison of the effect of the different specific heat capacities of water and glycerine on their increase in temperature (T). Water has a very high heat capacity so it requires more heat energy (q) to increase its temperature by 1 K.

Water is distinguished by its very high heat capacity, which is a consequence of the hydrogen bonding between its molecules.

The specific heat capacity of a substance is a reflection of the types of bonds holding the molecules, ions or atoms together in the substance.

Water has a specific heat capacity of $4.18\,J\,g^{-1}\,K^{-1}$. This means that 4.18 J of heat energy are needed to increase the temperature of 1 g of water by 1 K. This relatively high value is due to the hydrogen bonds between the water molecules. The higher the specific heat capacity, the more effectively a material stores heat energy.

When a substance is being heated, its temperature rises. The temperature of 1 g of water increases by 1 K when it is supplied with 4.18 J of heat energy. In comparison, 2.43 J of heat energy is required to increase the temperature of 1 g of glycerine by 1 K. The effect of their different specific heat capacities of on their temperatures when heated can be seen in Figure 14.4.2.

Calculations using specific heat capacity

The specific heat capacity of water can be used to calculate the heat energy in joules needed to increase the temperature of a given mass of water by a particular amount. Heat energy is given the symbol q. Specific heat capacity is given the symbol c.

A useful formula can be written:

Heat energy = mass × specific heat capacity × temperature change

Using symbols, the formula can be written as:

$$q = m \times c \times \Delta T$$

The heat energy required to increase the temperature of a given mass of water by a particular amount can be calculated using the formula:

$$q = m \times c \times \Delta T$$

where

q is the amount of heat energy (in J)

c is the specific heat capacity (in $J\,g^{-1}\,K^{-1}$)

m is the mass in g

ΔT is the temperature change (in K).

Worked example 14.4.1

CALCULATING THE AMOUNT OF ENERGY REQUIRED TO HEAT A SPECIFIED MASS OF WATER (USING SPECIFIC HEAT CAPACITY)

Calculate the heat energy, in kJ, needed to increase the temperature of 500 mL of water by 15°C.

Thinking	Working
Change the volume of water, in mL, to mass of water, in g. Remember that 1 mL of water has a mass of 1 g.	1 mL of water has a mass of 1 g, so 500 mL of water has a mass of 500 g.
Find the specific heat capacity of water from the data in Table 14.4.1 (page 455).	The specific heat capacity of water is $4.18\,J\,g^{-1}\,K^{-1}$.
To calculate the quantity of heat energy in joules, use the formula: $q = m \times c \times \Delta T$ Remember, a change of 1°C is the same as a change of 1 K.	$q = 500 \times 4.18 \times 15$ $= 3.14 \times 10^4$ J
Express the quantity of energy in kJ. Remember that to convert from J to kJ, you multiply by 10^{-3}.	$q = 3.14 \times 10^4 \times 10^{-3}$ $= 31.4$ kJ

Worked example: Try yourself 14.4.1

CALCULATING THE AMOUNT OF ENERGY REQUIRED TO HEAT A SPECIFIED MASS OF WATER (USING SPECIFIC HEAT CAPACITY)

Calculate the heat energy, in kJ, needed to increase the temperature of 375 mL of water by 45°C.

EXPERIMENTAL DETERMINATION OF HEAT OF COMBUSTION BY SIMPLE CALORIMETRY

When a combustion reaction takes place, chemical energy is converted to thermal energy. You can use the thermal energy released by a specific quantity of fuel as it undergoes combustion to heat a measured volume of water. If you measure the temperature change of the water, it can be used to determine the approximate amount of energy released by the fuel.

An experimental arrangement for estimating the heat of combustion of a liquid fuel, such as ethanol, is shown in Figure 14.4.3.

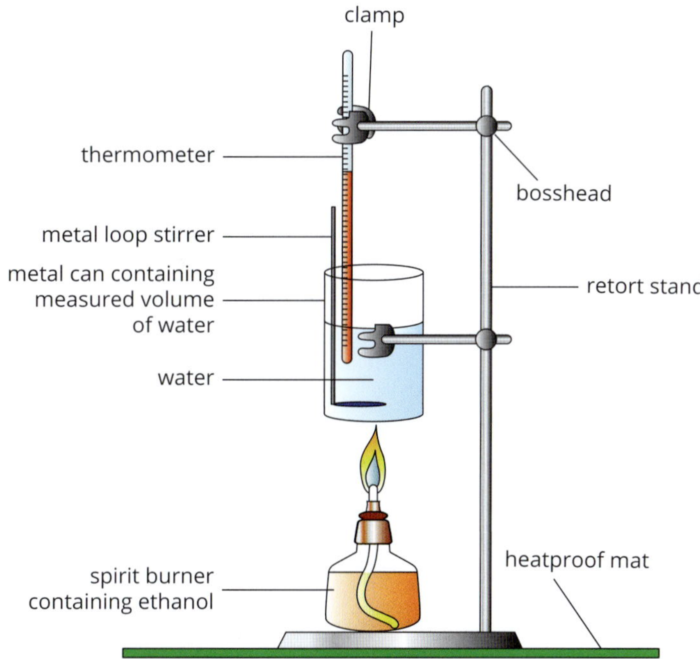

FIGURE 14.4.3 Apparatus for measuring heat of combustion of a fuel (for example, ethanol). A metal tin containing a measured volume of water is held above the wick of a spirit burner.

Figure 14.4.4 summarises the steps followed in this experiment.

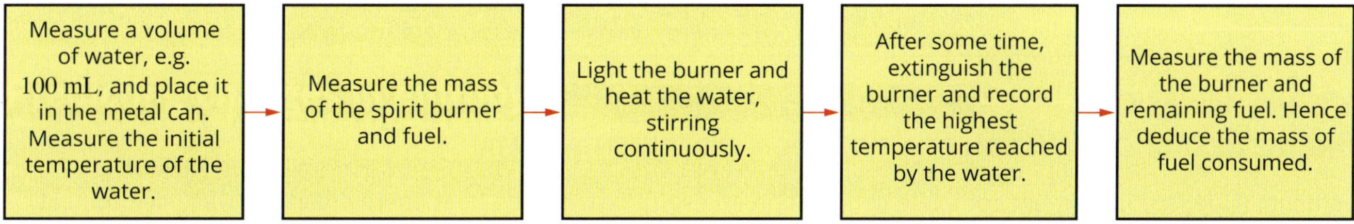

FIGURE 14.4.4 Flow chart of the steps followed when using the specific heat capacity of water to determine the heat of combustion of a fuel.

Three key pieces of information collected from this procedure are the:

- mass of water (because the density of water is 1.00 g mL^{-1}, the volume of water, measured in mL, is equal to its mass, in g)
- change in temperature of the water, ΔT
- mass of fuel consumed, m.

This data can be used to determine the heat of combustion of the fuel, as shown in Worked example 14.4.2.

Note that when performing these calculations, it is assumed that all of the energy released by the combustion of the fuel is used to heat the water. In reality, some of the energy heats the metal tin and is lost to the surroundings.

For this reason, measurements of heats of combustion from these experiments only give approximate values. More accurate measurements are obtained using bomb calorimetry. A **calorimeter** is a device that measures the energy changes occurring during chemical and physical reactions.

Worked example 14.4.2

CALCULATING THE HEAT OF COMBUSTION OF A FUEL FROM EXPERIMENTAL DATA

0.355 g of methanol (CH_3OH) underwent complete combustion in a spirit burner. The heat energy released was used to heat 100 mL of water. The temperature of the water rose from 20.24°C to 37.65°C. Calculate the heat of combustion of methanol in kJ mol^{-1}.

Thinking	Working
Calculate the temperature change of the water. Remember, a change of 1°C is the same as a change of 1 K.	$\Delta T = 37.65 - 20.24$ $= 17.41$ K
Use the specific heat capacity of water to determine the energy used to heat the water. Use the formula: $q = m \times c \times \Delta T$ (m in this formula is the mass of water.)	$q = 100 \times 4.18 \times 17.41$ $= 7277$ J
Express the quantity of energy in kJ. Remember that to convert from J to kJ, you multiply by 10^{-3}.	$q = 7277 \times 10^{-3}$ $= 7.277$ kJ
Calculate the amount, in mol, of methanol, using the formula: $n = \frac{m}{M}$	$n = \frac{0.355}{32.04}$ $= 0.0111$ mol
Determine the heat of combustion of methanol, in kJ mol^{-1}. Heat of combustion $= \frac{\text{heat energy released by sample}}{\text{amount of sample (in mol)}}$	heat of combustion $= \frac{-7.277}{0.0111}$ $= -656$ kJ mol^{-1} (Note: The negative sign indicates that the reaction released energy, causing the temperature to rise.)

Worked example: Try yourself 14.4.2

CALCULATING THE HEAT OF COMBUSTION OF A FUEL FROM EXPERIMENTAL DATA

0.295 g of ethanol (C_2H_5OH) underwent complete combustion in a spirit burner. The heat energy released was used to heat 100 mL of water. The temperature of the water rose from 19.56°C to 38.85°C. Calculate the heat of combustion of ethanol in kJ mol^{-1}.

+ ADDITIONAL

Bomb calorimetry

Figure 14.4.5 shows the components of a bomb calorimeter, which is used for reactions that involve gaseous reactants or products. The reaction vessel in a bomb calorimeter is designed to withstand the high pressures that may build up during reactions.

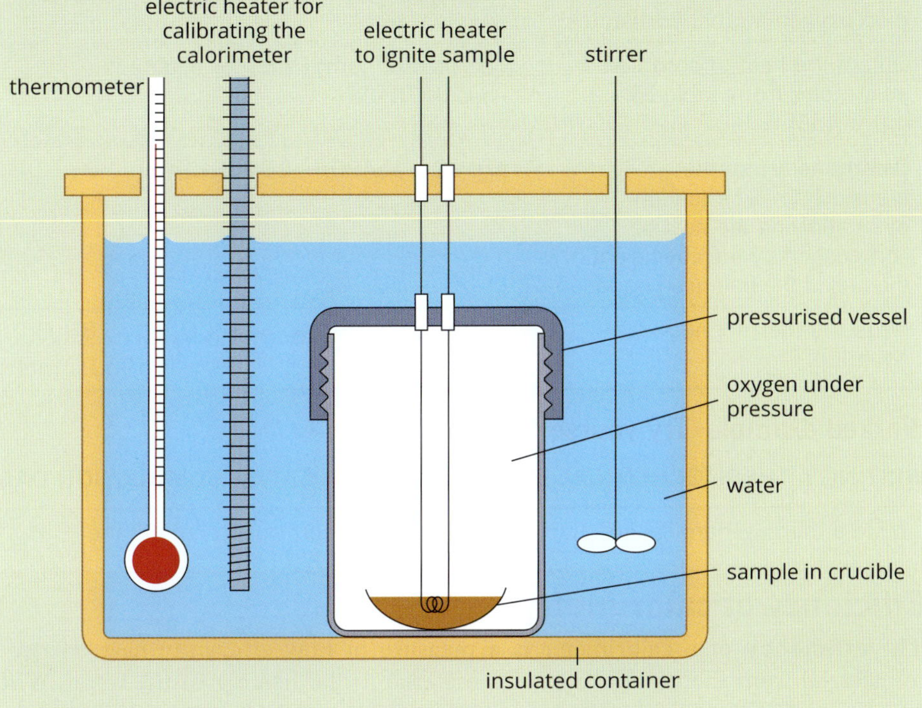

FIGURE 14.4.5 A bomb calorimeter is used for reactions that involve gases, such as the combustion of foods.

Determining the ΔH value for thermochemical equations

Experimentally determined heats of combustion for different elements and compounds can be listed in a data table, such as that shown in Table 14.4.2.

This data can be used to write thermochemical equations for the combustion of these substances. For example, the equation for the complete combustion of propane is:

$$C_3H_8(g) + 5O_2(g) \rightarrow 3CO_2(g) + 4H_2O(l) \qquad \Delta H = -2220\,kJ\,mol^{-1}$$

In this equation, the coefficient of propane is 1, so the ΔH value is the same as the heat of combustion.

If the coefficient of the substance in the combustion equation is 2 or more, the heat of combustion must be multiplied by this number to determine ΔH. For example, the thermochemical equation for the complete combustion of butane is written as:

$$2C_4H_{10}(g) + 13O_2(g) \rightarrow 8CO_2(g) + 10H_2O(l) \qquad \Delta H = -5772\,kJ\,mol^{-1}$$

ΔH in this case is twice the heat of combustion, $2 \times -2886 = -5772\,kJ\,mol^{-1}$.

Combustion reactions are always exothermic, so the enthalpy change always has a negative sign.

TABLE 14.4.2 Experimentally determined heats of combustion of various fuels

Fuel	Heat of combustion, ΔH_c ($kJ\,mol^{-1}$)
octane	–5450
butane	–2886
propane	–2220
ethane	–1560
ethanol	–1367
methane	–890
methanol	–725
carbon (graphite)	–394
hydrogen	–286

Worked example 14.4.3

WRITING A THERMOCHEMICAL EQUATION USING HEAT OF COMBUSTION DATA

Write a thermochemical equation for the complete combustion of octane (C_8H_{18}).

Thinking	Working
Add oxygen as a reactant and carbon dioxide and water as the products in the equation.	$C_8H_{18} + O_2 \rightarrow CO_2 + H_2O$
Balance the carbon, hydrogen, then oxygen atoms. Add states.	$2C_8H_{18}(g) + 25O_2(g) \rightarrow 16CO_2(g) + 18H_2O(l)$
Obtain the heat of combustion, in kJ mol^{-1}, from Table 14.4.2 (page 459).	The heat of combustion of octane is -5450 kJ mol^{-1}.
Determine ΔH for the thermochemical equation by multiplying the heat of combustion by the coefficient of the fuel in the balanced equation.	$\Delta H = 2 \times -5450 = -10\,900$ kJ mol^{-1}
Write the thermochemical equation.	$2C_8H_{18}(g) + 25O_2(g) \rightarrow 16CO_2(g) + 18H_2O(l)$ $\Delta H = -10\,900$ kJ mol^{-1}

Worked example: Try yourself 14.4.3

WRITING A THERMOCHEMICAL EQUATION USING HEAT OF COMBUSTION DATA

Write a thermochemical equation for the complete combustion of ethane (C_2H_6).

Efficiency of calorimetry

The **efficiency** of a calorimeter is a measure of how efficiently heat energy is transferred from the combustion of a material to the heating of the water. When energy is transferred from a combusted material across an open space, heat is lost to the surroundings, such as the air around the material. Similarly, if there is no lid on a container of water, heat will be lost from the surface of the water. This heat loss is illustrated in Figure 14.4.6.

When some heat energy from the combusted material is transferred to the surrounding air, the temperature of water does not increase as much as it would if all the energy was used to heat the water. A lower change in temperature (ΔT) of the water results in a lower energy value (q).

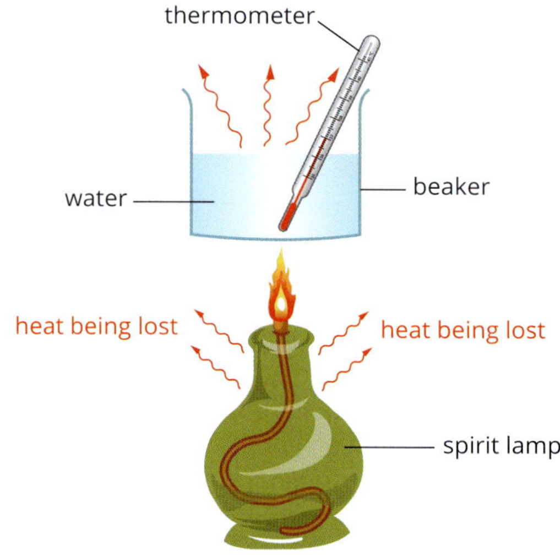

FIGURE 14.4.6 Heat is lost to the surroundings when fuel is burned to heat some water.

There are a number of ways to reduce heat loss during the calorimetry experiment shown in Figure 14.4.6, including:

- putting a lid on the container
- insulating the beaker of water (with a flame-proof material such as aluminium)
- placing insulating material around the combusting material, although sufficient oxygen must reach the material for combustion to be complete.

The greater the heat lost to the surrounding environment, the lower the efficiency of the calorimeter. Using theoretical or expected values of the material being combusted, the efficiency of the calorimeter can be calculated using the following relationship:

$$\%\text{ efficiency} = \frac{\text{expected energy absorbed}}{\text{actual energy supplied}} \times 100\%$$

 When determining the **% efficiency** of a calorimeter, ensure that the expected and experimental values are given in the same units.

Worked example 14.4.4

DETERMINING % EFFICIENCY FROM EXPERIMENTAL DATA

A student adds 125 kJ of heat energy to a calorimeter containing 200.0 g of water. The temperature of the water rises from 25.0°C to 95.0°C. What is the percentage efficiency of the calorimeter?

Thinking	Working
Identify the experimental value of energy.	125 kJ is the experimental value of energy.
Calculate the temperature change of the water. Remember, a change of 1°C is the same as a change of 1 K.	$\Delta T = 95.0 - 25.0$ $= 70.0\text{ K}$
Use the specific heat capacity of water to determine the expected energy required to heat the water. Use the formula: $q = m \times c \times \Delta T$ (*m* in this formula is the mass of water.)	$q = m \times c \times \Delta T$ $= 200.0 \times 4.18 \times 70.0$ $= 58\,520\text{ J}$ or 58.52 kJ
Use the % efficiency relationship to determine the efficiency of the calorimeter. $\%\text{ efficiency} = \frac{\text{expected energy absorbed}}{\text{actual energy supplied}} \times 100\%$ Remember, expected and experimental values must be in the same units.	$\%\text{ efficiency} = \frac{\text{expected energy absorbed}}{\text{actual energy supplied}} \times 100\%$ $= \frac{58.52}{125} \times 100\%$ $= 46.8\%$
Answer the question.	46.8% of the energy supplied was absorbed by the calorimeter, and the rest was lost to the surrounding environment.

Worked example: Try yourself 14.4.4

DETERMINING % EFFICIENCY FROM EXPERIMENTAL DATA

A student adds 26.0 kJ of heat energy to a calorimeter containing 100.0 g of water. The temperature of the water rises from 23.5°C to 58.1°C. What is the percentage efficiency of the calorimeter?

14.4 Review

SUMMARY

- The specific heat capacity of a substance measures the quantity of energy (usually in joules) needed to increase the temperature of a specified quantity of that substance (usually 1 g) by 1 K.
- The specific heat capacity of water is $4.18\,J\,g^{-1}\,K^{-1}$.
- The heat energy required to increase the temperature of a given mass of substance by a particular amount can be calculated using the formula:

$$q = m \times c \times \Delta T$$

- Heat of combustion is the energy change when a specified quantity of a pure substance (usually 1 mol) is burnt.
- The specific heat capacity of water can be used to experimentally determine the approximate amount of heat energy released in the combustion of a fuel.
- Heat of combustion data can be used to determine the enthalpy change (ΔH) in a thermochemical equation.
- Calorimetry is the experimental method by which the heat energy released or absorbed in a chemical reaction or physical process is measured.
- If the insulation around a calorimeter is insufficient, or missing, or if there is no lid, heat will be lost from the water to the surroundings, and the value measured for ΔT will be lower than the calculated value.
- The efficiency of a calorimeter is a measure of how much heat energy has been transferred from the combustion of a material to the heating of the water.
- % efficiency can be calculated using the relationship:

$$\%\ \text{efficiency} = \frac{\text{expected energy absorbed}}{\text{actual energy supplied}} \times 100\%$$

KEY QUESTIONS

1 Calculate the heat energy (in kJ) needed to increase the temperature of 1.00 kg of water by 25.0°C.

2 What assumptions are made when making calculations (relating to the heat of combustion of fuels) using experimental data obtained from the apparatus shown in Figure 14.4.3 (page 457)?

3 A temperature rise of 1.78°C was observed when 1.00×10^{-3} mol of propane gas (C_3H_8) was burnt and used to heat 300 mL of water. Calculate the heat of combustion for propane in $kJ\,mol^{-1}$, assuming all the heat released was used to heat the water.

4 A temperature rise of 11.5°C was observed when 0.500 g of butane gas (C_4H_{10}) was burnt and used to heat 500 mL of water. Calculate the heat of combustion, in $kJ\,mol^{-1}$, for butane, assuming all the heat released was used to heat the water.

5 The heat of combustion of methane (CH_4) is $-890\,kJ\,mol^{-1}$.

a Write a thermochemical equation for the complete combustion of methane.

b Determine the mass of methane (in g) that has to be burnt in order to heat 500 mL of water from 20.0°C to boiling. Assume all the heat released was used to heat the water.

6 A student adds 73.2 kJ of heat energy to a calorimeter containing 150.0 g of water. The temperature of the water rises from 16.0°C to 97.0°C. What is the percentage efficiency of the calorimeter?

14.5 Enthalpy of dissolution

CHEMISTRY INQUIRY 

No-freezer ice-cream

What energy changes occur in chemical reactions?

COLLECT THIS...

- 1 cup milk
- 2 tablespoons sugar
- $\frac{1}{4}$ teaspoon vanilla extract or $\frac{1}{2}$ teaspoon of vanilla essence
- 6–8 cups of ice cubes
- 50 mL of water
- $\frac{1}{2}$ cup of normal table salt
- 1 small (approx. 18 cm × 15 cm) snaplock bag
- 1 large (approx. 25 cm × 40 cm) snaplock bag

DO THIS...

1. Mix the milk, sugar and vanilla together and pour into the small plastic bag. Seal it and ensure there are no leaks. If necessary, double-bag it.
2. Place the sealed small snaplock bag into the large snaplock bag. Add the ice and water, and sprinkle the salt over the ice. Seal the large snaplock bag.
3. Vigorously shake the two bags together for about 10 minutes. Do not do this in a lab; you may want to do this outside or over a sink.
4. Pull the small snaplock bag out of the large bag, rinse off the salt water, open it and enjoy your ice cream.

RECORD THIS...

Describe what is happening to the ice cubes and ice-cream in terms of heat energy transfers throughout the process.

Present your observations.

REFLECT ON THIS...

What is the purpose of shaking the bags? What would occur if you did not shake the bag?

Why is salt added to the water?

When it is very cold, why is salt used on roads and roofs to stop them from freezing?

Many ionic substances are soluble in water. In this section, you will look at how ionic substances dissolve and learn how to experimentally determine the enthalpy change associated with **dissolution**.

DISSOLUTION OF AN IONIC LATTICE IN WATER

Many ionic compounds dissolve readily in water. Sodium chloride is a typical ionic compound that exists as a solid at room temperature. In Figure 14.5.1, you can see the arrangement of sodium cations (Na^+) and chloride anions (Cl^-) in a three-dimensional ionic lattice. The ions are held together by strong electrostatic forces between the positive and negative charges of the ions.

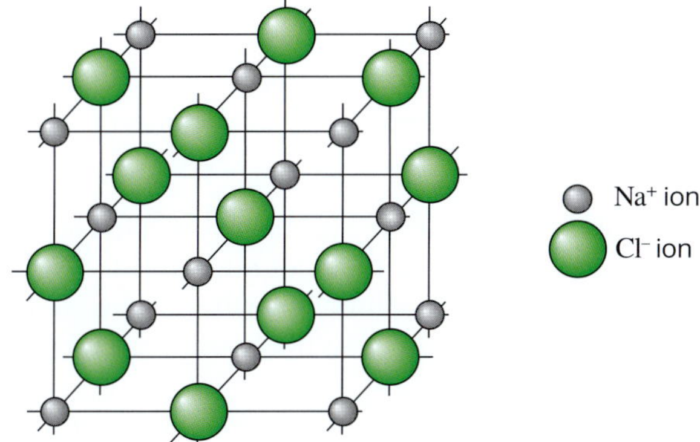

FIGURE 14.5.1 A representation of the crystal lattice of the ionic compound sodium chloride.

When an ionic compound such as sodium chloride is added to water, the positive ends of the water molecules are attracted to the negatively charged chloride ions and the negative ends of the water molecules are attracted to the positively charged sodium ions (Figure 14.5.2). The attraction between an ion and a polar molecule such as water is described as an **ion–dipole attraction**.

The positive end of a polar molecule is indicated by δ+. The negative end of a polar molecular is indicated by δ–.

---- represents ion–dipole attraction between ions in lattice and water dipoles

FIGURE 14.5.2 Electrostatic attraction occurs between the negative ions in a NaCl lattice and the hydrogen atoms in polar water molecules. Electrostatic attraction also occurs between the positive ions in the lattice and the oxygen atoms in water molecules.

Water molecules are in a continuous state of random motion. If the ion–dipole attractions between the ions and the water molecules are strong enough, the water molecules can pull the sodium and chloride ions on the outer part of the crystal out of the lattice and into the surrounding solution.

Sodium ions and chloride ions pulled out of the lattice become surrounded by water molecules. These ions are said to be **hydrated**. Water molecules are arranged around the ions as shown in Figure 14.5.3. Note the different arrangements of the water molecules around the positive and negative ions. The hydrogen atoms in the water molecule are more positive, so they are orientated towards the negative chloride ion. The positive sodium ion is surrounded by the more negative oxygen atoms of the water molecules.

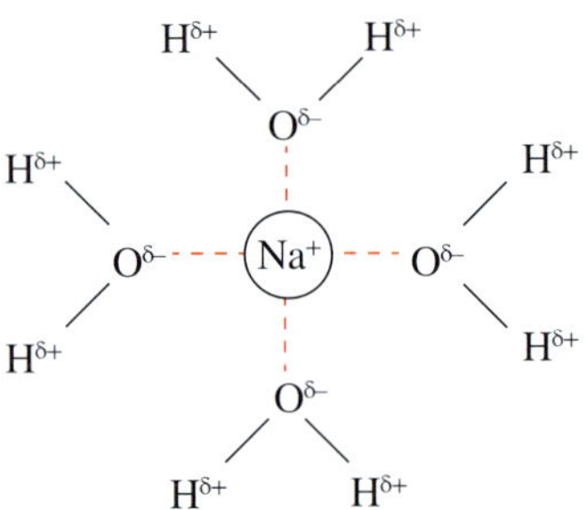

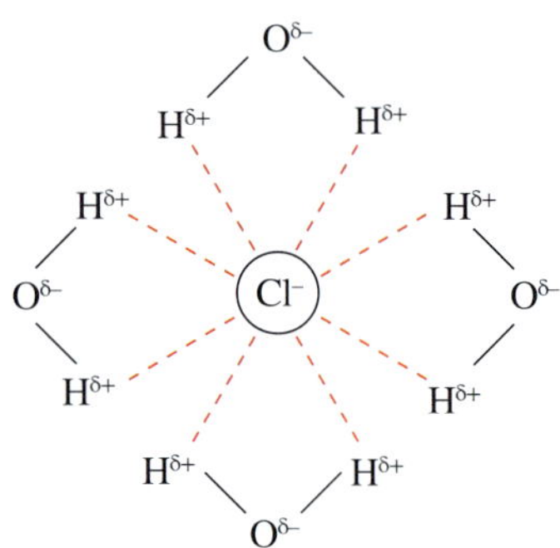

---- represents ion–dipole interaction

FIGURE 14.5.3 Ion–dipole attraction between the ions and adjacent water molecules to form hydrated sodium and chloride ions.

The process of separating positive and negative ions from a solid ionic compound to form hydrated ions when an ionic compound dissolves in water is called **dissociation**.

Although the ionic bonds within the lattice are strong, the ions can be pulled away from the lattice by the interactions of many water molecules.

In summary, when sodium chloride dissolves in water:

- ionic bonds within the sodium chloride lattice are broken
- hydrogen bonds between water molecules are broken
- ion–dipole attractions form between ions and polar water molecules.

An equation can be written to represent the dissociation process:

$$NaCl(s) \xrightarrow{H_2O(l)} Na^+(aq) + Cl^-(aq)$$

Note that the formula of water sits above the arrow. This is because there is no direct reaction between the water and the sodium chloride. No chemical change occurs; only the state symbol for sodium chloride is altered from (s) to (aq), indicating it is now dissolved in water. (You may omit the H_2O from this equation if you wish.)

It is important to note that dissociation of ionic compounds is simply freeing ions from the lattice so that they can move freely throughout the solution.

Ionic substances dissolve by dissociation. Ion–dipole bonds are formed between the ions and water molecules.

EXPERIMENTAL DETERMINATION OF ENTHALPY OF DISSOLUTION

A **solution calorimeter** is a type of calorimeter that is designed for measuring the energy changes in physical and chemical changes that take place in solution. This may be as simple as a polystyrene foam coffee cup with a lid, as shown in Figure 14.5.4.

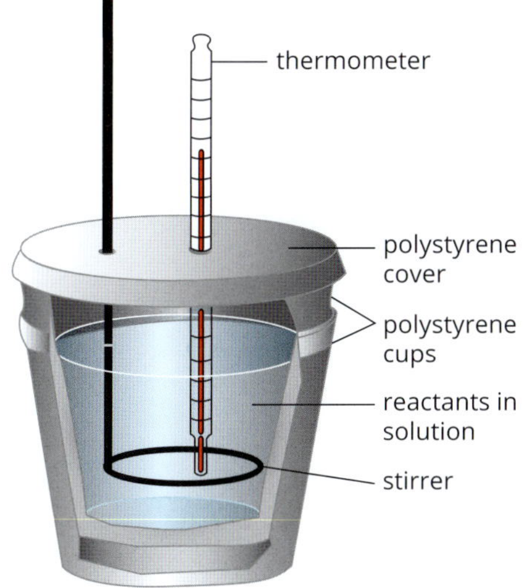

FIGURE 14.5.4 A simple 'coffee cup' calorimeter.

The insulation provided by the polystyrene foam prevents the transfer of heat to or from the surroundings of the calorimeter. The reaction is carried out in the calorimeter with an accurately known volume of water. The initial and final temperatures are measured and recorded, as are the amounts of the reactants used.

If the temperature of the water in the calorimeter increases, the reaction occurring is an exothermic reaction. The reaction has released heat energy, the water in the calorimeter has absorbed that energy, and the temperature of the water has increased. Experimentally, in exothermic reactions at some stage the temperature stops rising or starts to decrease, and the maximum temperature is the final temperature to be recorded.

Similarly, if the temperature of the water decreases, the reaction occurring in the calorimeter has absorbed energy from the water and is an endothermic reaction. Experimentally, in endothermic reactions at some stage the temperature stops falling or starts to rise again, and the minimum temperature is the final temperature to be recorded.

A coffee cup calorimeter has some limitations. The polystyrene container absorbs some of the heat, so the temperature change is smaller than it would otherwise be, and the calculated value for the heat released or absorbed by the reaction will be smaller than it should be.

Solution calorimeters can be used to determine the energy change when sodium chloride or other ionic solids dissolve in water. Once the energy change is determined, this can be used to determine the **enthalpy of dissolution** (given the symbol ΔH_{soln}). This is calculated by dividing the energy change by the amount (in mol) of the substance being dissolved.

Enthalpy of dissolution is calculated by the following expression:

$$\Delta H_{soln} = \frac{q}{n(\text{solute})}$$

where

ΔH_{soln} is the enthalpy of dissolution (in kJ mol^{-1})

q is the amount of energy released or absorbed by the reaction (in kJ)

n (solute) is the amount of substance (in mol) of the solute

The construction of a laboratory solution calorimeter is shown in Figure 14.5.5. The stirrer is used to ensure that the distribution of the solute and the temperature of the water is constant throughout the solution.

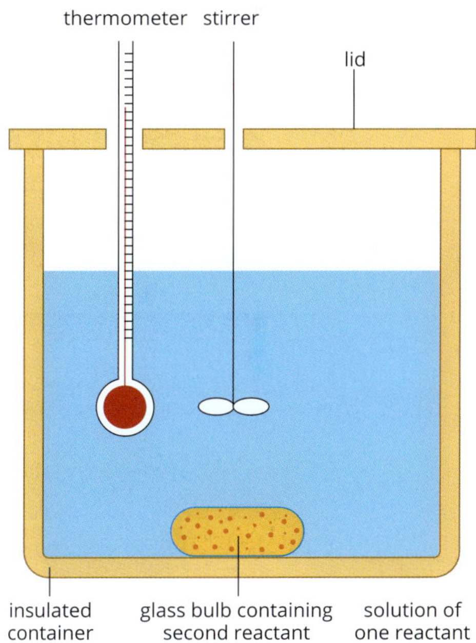

FIGURE 14.5.5 A solution calorimeter: breaking the glass bulb starts the reaction.

Worked example 14.5.1

DETERMINING THE ENTHALPY OF A DISSOCIATION PROCESS

When 5.00 g of magnesium sulfate ($MgSO_4$) (M = 120.38 g mol^{-1}) is added to 100.0 g of water in a solution calorimeter, it dissolves according to the equation:

$$MgSO_4(s) \xrightarrow{H_2O(l)} Mg^{2+}(aq) + SO_4^{2-}(aq)$$

The temperature of the water increases from 23.00°C to a maximum temperature of 32.05°C. Determine the enthalpy of dissolution.

Thinking	Working
Determine the temperature change (ΔT). Remember, a change of 1°C is the same as a change of 1 K.	$\Delta T = 32.05 - 23.00$ $= 9.05\ K$
Determine the amount of energy absorbed/released by the water, resulting in the temperature change, using the expression: $q = m \times c \times \Delta T$ where m is the mass of the water.	$q = m \times c \times \Delta T$ $= 100.0 \times 4.18 \times 9.05$ $= 3783\ J$
Convert energy to kJ.	$3783 \times 10^{-3} = 3.78\ kJ$
Determine the amount of solute, in mol, using the formula: $n = \frac{m}{M}$	$n = \frac{m}{M}$ $= \frac{5.00}{120.38}$ $= 0.041535$ mol
Determine ΔH_{soln} for the dissolution reaction by dividing the energy by the amount of substance according to: $\Delta H_{soln} = \frac{q}{n(solute)}$ As the temperature increased, the reaction is exothermic and hence ΔH_{soln} is negative.	$\Delta H_{soln} = \frac{q}{n(solute)}$ $= \frac{-3.78}{0.04154}$ $= -91.0\ kJ\ mol^{-1}$

Worked example: Try yourself 14.5.1

DETERMINING THE ENTHALPY OF A DISSOCIATION PROCESS

When 5.00 g of ammonium nitrate (NH_4NO_3) (M = 80.05 g mol^{-1}) is added to 100.0 g of water in a solution calorimeter, it dissolves according to the equation:

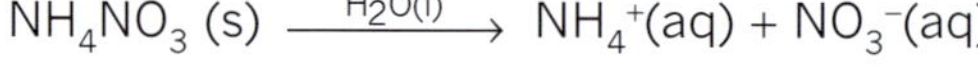

$$NH_4NO_3(s) \xrightarrow{H_2O(l)} NH_4^+(aq) + NO_3^-(aq)$$

The temperature of the water decreases from 23.0°C to a minimum temperature of 19.1°C. Determine the enthalpy of dissolution.

14.5 Review

SUMMARY

- Soluble ionic compounds dissociate in water to form hydrated ions.
- In a hydrated ion, water molecules are attracted to the central ion by ion–dipole attractions.
- A solution calorimeter is an insulated container that holds a known volume of water, and in it the enthalpy change associated with the dissociation of an ionic solid in aqueous solution can be measured.
- During calorimetry, energy is transferred to a fixed volume of water. The measured temperature change can be used to calculate the energy transferred using the formula:

$$q = m \times c \times \Delta T$$

- The enthalpy of dissolution (ΔH_{soln}) can be calculated using the expression:

$$\Delta H_{soln} = \frac{q}{n(\text{solute})}$$

KEY QUESTIONS

1. Sodium nitrate ($NaNO_3$) and calcium hydroxide ($Ca(OH)_2$) will both dissociate when they dissolve in water. Write chemical equations to represent the dissolving process for each of these compounds.
2. Redraw the representations of Na^+ and Cl^- ions below, then show the correct orientation of the water molecules around hydrated sodium and chloride ions. Use a V-shape to represent the shape of the water molecules.

3. Potassium nitrate (KNO_3) is soluble in water. Explain the steps involved when solid potassium nitrate dissolves to form an aqueous potassium nitrate solution. What is the name given to this process?
4. Use the following terms to complete the statements about calorimeters: absorbed, decreases, endothermic, exothermic, increases, insulating, lid, lost required, stays the same, thermometer, using.
   ___________ a calorimeter improves the accuracy of measurement of the energy changes occurring during a chemical reaction. Heat energy can be ___________ from a calorimeter, so a(an) ___________ is a useful form of insulation.
   If the reaction occurring in a calorimeter is ___________, the temperature of the water ___________. If the reaction occurring is ___________, the temperature of the water in a calorimeter ___________.
5. When 6.00 g of calcium chloride ($CaCl_2$) (M = 110.98 g mol^{-1}) is added to 50.00 g of water in a solution calorimeter, it dissolves according to the equation:

$$CaCl_2(s) \xrightarrow{H_2O(l)} Ca^{2+}(aq) + 2Cl^-(aq)$$

   The temperature of the water increases from 17.90°C to a maximum temperature of 38.90°C. Determine the enthalpy of dissolution.
6. 2.00 g of KNO_3 is dissolved in 50.00 mL of water with an initial temperature of 24.0°C. The thermochemical equation for the dissociation of KNO_3 in water is:

$$KNO_3(s) \rightarrow K^+(aq) + NO_3^-(aq) \qquad \Delta H_{soln} = +34.89 \text{ kJ mol}^{-1}$$

   Determine the final temperature of the water.

14.6 Catalysts

Some chemical reactions occur much more rapidly if another substance is added to the reaction mixture. Such substances are called **catalysts**. Catalysts allow the reaction to follow a more energetically favourable pathway.

For example, if you chew a piece of dry biscuit or bread for several minutes, you may notice it tasting much sweeter. This happens because there is a catalyst present in your saliva that speeds up the breakdown of starch into sweet-tasting sugars.

The change in reaction rate when a catalyst is present is often very substantial. The action of a catalyst can be understood using collision theory and knowledge of the changes in energy that occur during a chemical reaction.

In this section, you will learn how catalysts increase the rate of reaction and how they play an important role in industrial chemistry, in limiting air pollution and in controlling biochemical processes (Figure 14.6.1).

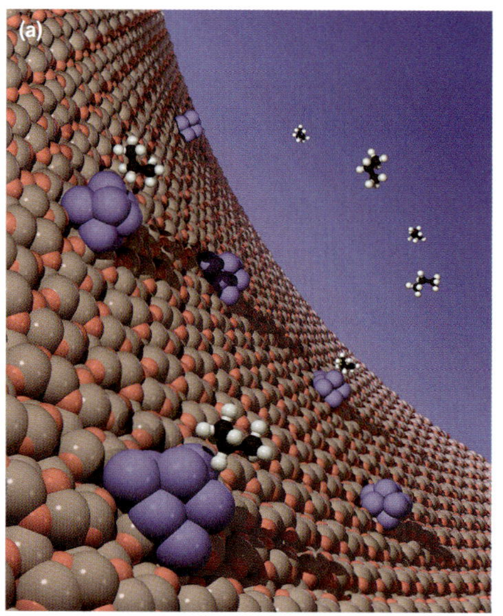

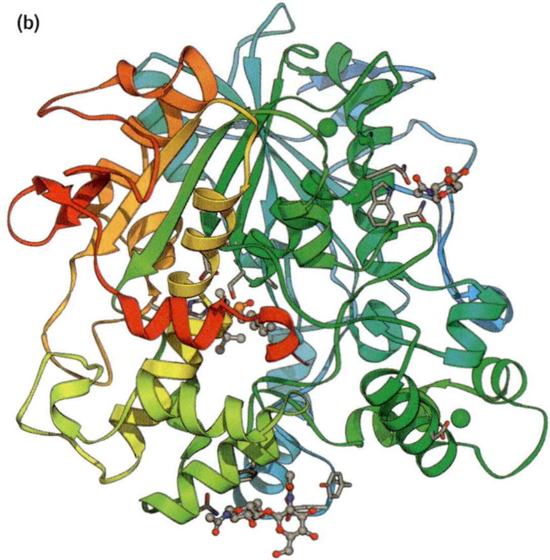

FIGURE 14.6.1 (a) A representation of the structure of a synthetic zeolite catalyst that is widely used in petroleum refineries to break down large hydrocarbon molecules into smaller, more useful molecules. Zeolite is a silica–alumina mineral. (b) A model of a lipase enzyme. Lipase is a catalyst that breaks down fats in the digestive system of the human body.

CHEMISTRY IN ACTION S

Catalytic converters in cars

All new cars sold in Australia have a catalyst fitted between the engine and the exhaust pipe (Figure 14.6.2). The purpose of the catalyst is to clean the exhaust gases from the engine and to reduce the air pollution that could be caused if these gases entered the atmosphere.

Catalysts in cars convert carbon monoxide and nitrogen oxide, formed in the engine, to the non-toxic gases carbon dioxide and nitrogen. Several reactions are involved in this process, including:

$$2NO(g) \rightarrow N_2(g) + O_2(g)$$

$$2CO(g) + O_2(g) \rightarrow 2CO_2(g)$$

Unburnt hydrocarbons are also converted by the catalyst to carbon dioxide and water:

$$2C_8H_{18}(g) + 25O_2(g) \rightarrow 16CO_2(g) + 18H_2O(g)$$

The catalyst is usually a mixture of platinum, palladium and rhodium metals, and aluminium oxide, and it is mounted on a honeycomb-shaped support. Millions of tiny pores in the metals provide a large surface area.

Exhaust gases enter the catalyst chamber, pass quickly over the metals, and leave the exhaust pipe in a purified condition. The catalyst is unchanged by the reaction and can be used without replacement for many years.

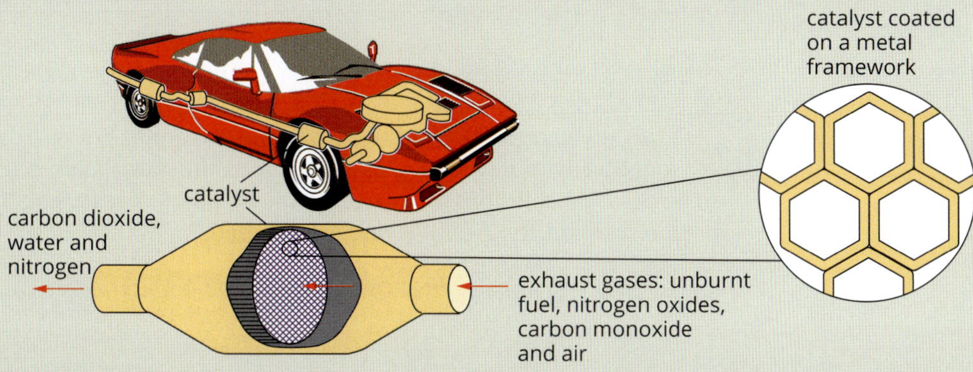

FIGURE 14.6.2 A catalyst fitted in the exhaust system of a car reduces air pollution.

CATALYSTS AND ACTIVATION ENERGY

You have already learnt that the potential energy changes associated with a reaction can be represented by an energy profile diagram of the reaction.

The activation energy (E_a) is the minimum amount of energy required for a reaction to take place. On the energy profile diagram, the activation energy is measured from the energy of the reactants to the peak of the energy profile diagram.

GO TO ➤ Figure 14.2.5 page 451

GO TO ➤ Figure 14.2.6 page 451

Some reactions occur very readily because they have very small activation energies. In these reactions, a large proportion of the reactant particles colliding have energy equal to or greater than the reaction's activation energy.

Many reactions can occur much more rapidly in the presence of a particular element or compound, known as a catalyst. Catalysts are not consumed during the reactions they speed up, and so do not appear as either reactants or products in reaction equations. They do not alter the surface area, or the concentration or energy of the reactants, and therefore do not change the frequency of collisions between particles nor the energy of the colliding particles.

Catalysts are able to increase the rate of a reaction because they provide a new **reaction pathway**, which causes the activation energy barrier of the overall reaction to be dramatically reduced, as seen in the energy profile diagram in Figure 14.6.3.

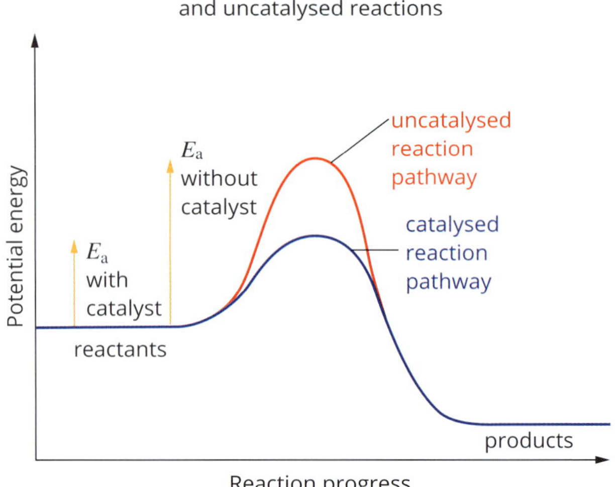

FIGURE 14.6.3 Energy profile diagram of a catalysed reaction and an uncatalysed reaction.

With the catalyst present, and a lower activation energy, the colliding particles are more likely to have energies that exceed this lower barrier. As a result, a greater frequency of collisions are successful, in other words lead to the formation of products. Thus, the reaction rate is increased.

Catalysts only lower the activation energy for a reaction. There is no change to ΔH for the reaction.

You may recall from Chapter 13 that the Maxwell–Boltzmann curve represents the distribution of energies of particles in a sample at a particular temperature. In the presence of a catalyst, the shape of the Maxwell–Boltzmann curve is not altered. However, Figure 14.6.4 shows that a smaller number of particles have energies that exceed the activation energy in an uncatalysed reaction (shaded in red) than in a catalysed reaction (regions shaded in red and blue).

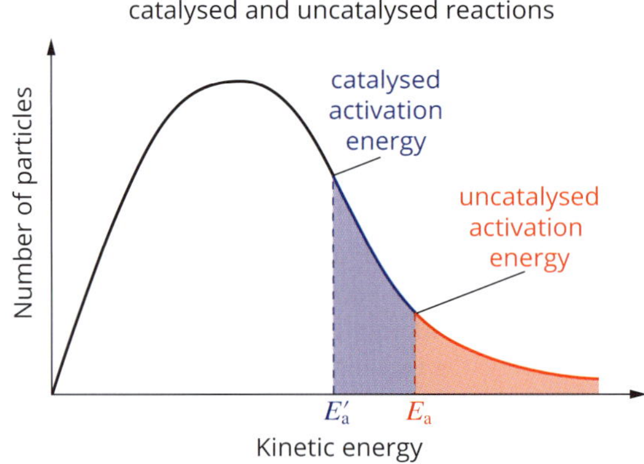

FIGURE 14.6.4 A catalyst provides a reaction pathway with a lower activation energy, increasing the frequency of collisions that exceed the activation energy and lead to a reaction.

Catalysts lower the activation energy by providing an alternative reaction pathway for the reaction. This increases the frequency of reactant particles with energies equal to or greater than the activation energy, which increases the frequency of successful collisions.

Figure 14.6.5 shows an analogy for the way that a catalyst works: two groups of tourists travelling from Sydney to Canberra by two different routes. The trip via the Blue Mountains will take considerably longer—this slower, scenic journey can be likened to the progress of an uncatalysed reaction. The faster trip via the foothills can be likened to the progress of a catalysed reaction.

FIGURE 14.6.5 An analogy for the role of a catalyst in a chemical reaction.

TYPES OF CATALYSTS

Depending on the physical states of the catalyst and the chemicals involved in the reaction, catalysts can be divided into two groups.

- **Homogeneous catalysts** are in the same physical state as the reactants and products of the reaction.
- **Heterogeneous catalysts** are in a different physical state from the reactants and products of the reaction.

An example of homogeneous catalysis occurs in the upper atmosphere and has contributed to the depletion of the ozone layer. Chlorine atoms in the gaseous state act as catalysts in the decomposition of ozone gas into oxygen gas. The chlorine atoms may have come from chlorofluorocarbons (CFCs) released into the atmosphere from refrigerators or air conditioners.

On the other hand, you may be familiar with the catalysed decomposition of a hydrogen peroxide solution using the black powder manganese(IV) oxide (MnO_2). This is an example of the use of a heterogeneous catalyst.

Another example of a heterogeneous catalyst is shown in Figure 14.6.1 (page 468). The solid zeolite catalyst is widely used in petroleum refineries to break down large gaseous hydrocarbon molecules into smaller, more useful molecules.

CATALYSTS IN INDUSTRY

The chemical industry uses catalysts extensively. Chemists prefer to use heterogeneous catalysts for industrial processes because they are:

- more easily separated from the products of a reaction
- much easier to reuse
- able to be used at high temperatures.

Particles at the surface of some solids of high surface area tend to adsorb (form a bond with) gas molecules that strike the surface. **Adsorption** distorts bonds in the gas molecules or may even break them completely. This can allow a reaction to proceed more easily than it would if the solid were absent.

These solid surfaces provide a new way for the reaction to occur (a new reaction pathway) that has a significantly lower activation energy.

A powdered or sponge-like form of a solid catalyst is often used in order to provide the greatest possible surface area. With a larger surface area, more reactant molecules can be adsorbed, and the reaction can be even faster.

+ ADDITIONAL

The Haber process

At Kooragang Island, near Newcastle, Orica uses the Haber process to produce ammonia gas. Ammonia is a very important inorganic chemical that is used to make fertilisers, nylon, explosives and some pharmaceuticals. In the Haber process, hydrogen and nitrogen gas are converted to ammonia (NH_3), using a catalyst of powdered iron.

The reaction is represented by the equation:

$$N_2(g) + 3H_2(g) \rightarrow 2NH_3(g) \qquad \Delta H = -91\ kJ\ mol^{-1}$$

Hydrogen and nitrogen molecules adsorb onto the iron surface (Figure 14.6.6a). As they attach themselves to the surface, the covalent bonds within their molecules break (Figure 14.6.6b).

The hydrogen and nitrogen atoms now readily combine to form ammonia molecules and move away from the iron surface (Figure 14.6.6c). The catalyst remains unaltered by the reaction.

Without a catalyst, temperatures over 3000°C are needed for a significant reaction to occur. The catalyst allows the manufacture of ammonia to proceed at an economical rate, using a temperature of about 500°C.

The iron catalyst provides an alternative reaction pathway that dramatically reduces the activation energy barrier—the energy needed to break the covalent bonds in the nitrogen and hydrogen molecules. Even though collisions are less frequent at 500°C than at 3000°C, a greater proportion of colliding particles have sufficient energy to successfully react.

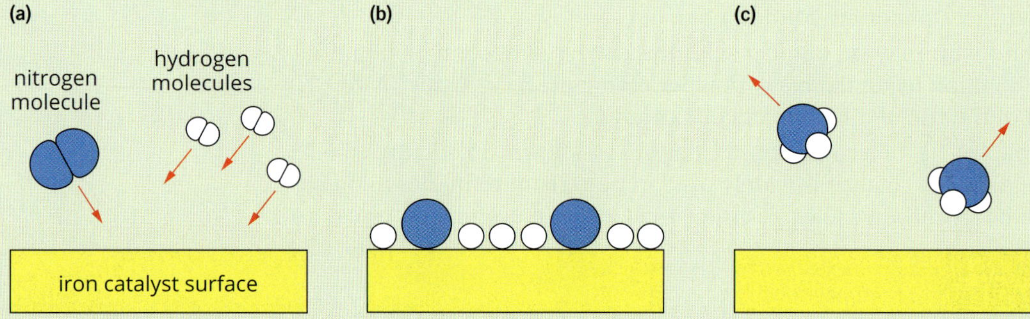

FIGURE 14.6.6 Behaviour of a catalyst in ammonia production. (a) Nitrogen and hydrogen molecules approach the iron catalyst surface. (b) The nitrogen and hydrogen molecules adsorb on the surface of the catalyst and their covalent bonds are broken. (c) The separated nitrogen and hydrogen atoms readily combine to form ammonia molecules and move away from the iron surface.

14.6 Review

SUMMARY

- The rate of a reaction can be increased by using a catalyst.
- A catalyst provides an alternative reaction pathway that has a lower activation energy.
- Energy profile diagrams (like the one to the right), which can represent catalysed and uncatalysed pathways, may be used to indicate the enthalpy change and activation energies associated with a chemical reaction.
- A catalyst provides a new reaction pathway, and it is not used up in the reaction.
- When a catalyst is present, a greater proportion of the collisions between particles exceed the activation energy barrier of the reaction. This leads to an increased rate of reaction.

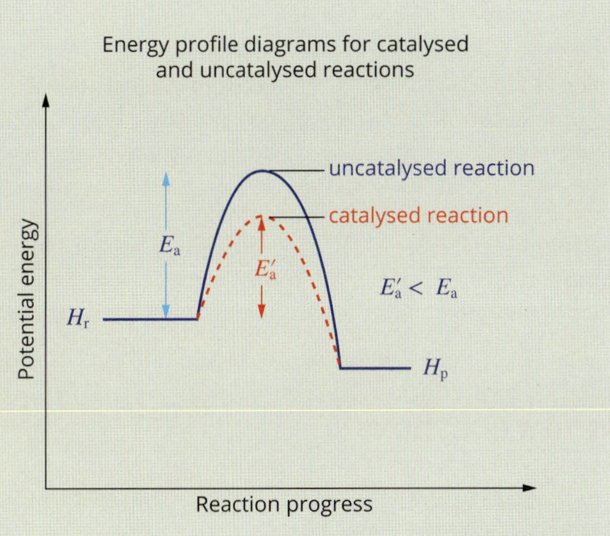

KEY QUESTIONS

1 Consider the reaction between solutions V and W that produces X and Z according to the equation:

$$V(aq) + W(aq) \rightarrow X(aq) + Z(aq)$$

The energy profile diagram for this process is shown below.

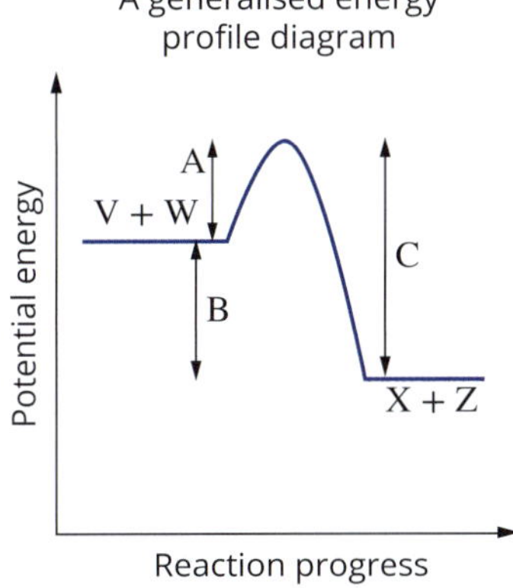

Which one of the following alternatives describes the change that a catalyst produces to increase the reaction rate?

A B only is decreased.
B A only is decreased.
C A, B and C are decreased.
D A and C only are decreased.

2 Explain the meaning of:

a catalyst
b activation energy.

3 If a sugar cube is held in the flame of a candle, the sugar melts and browns, but does not burn. However, the cube burns if salt is first rubbed into it, even though the salt does not react. Explain the effect of the salt on the activation energy of this combustion reaction.

4 **a** Explain why surface properties are important to the operation of catalysts.

b Many industrial catalysts are made into porous pellets. What is the reason for this?

5 The graph below shows the energy profile diagram for the reaction between hydrogen and iodine to form hydrogen iodide:

$$H_2(g) + I_2(s) \rightarrow 2HI(g)$$

Energy profile diagram for hydrogen iodide formation

Potential energy

Reaction progress

a Copy the diagram and label the following: $H_2(g)$ and $I_2(s)$; $2HI(g)$; ΔH; activation energy.

b Is the reaction endothermic or exothermic?

c On the diagram, draw the energy profile that would result if a catalyst was used in the reaction.

Chapter review

14

KEY TERMS

activation energy
adsorption
bioluminescence
calorimeter
catalyst
chemical energy
chemiluminescence
combustion
decomposition
dissociation
dissolution
efficiency (calorimetry)
endothermic
energy profile diagram
enthalpy
enthalpy change
enthalpy of dissolution
exothermic
heat content
heat of combustion
heat of reaction
heterogeneous catalyst
homogeneous catalyst
hydrated
ion–dipole attraction
joule
law of conservation of energy
pure substance
reaction pathway
solution calorimeter
specific heat capacity
surroundings
system
thermochemical equation
% efficiency (calorimetry)

REVIEW QUESTIONS

1 Convert the following energy values to the units given (to 3 significant figures).
 a 2205 J to kJ
 b 0.152 kJ to J
 c 1 890 000 J to MJ
 d 0.0125 MJ to kJ

2 Decide whether the following processes are exothermic or endothermic. Give reasons for your answers.
 a burning of wood
 b melting of ice
 c recharging of a car battery
 d decomposition of plants in a compost heap

3 Which one of the following statements about the energy profile diagrams of both endothermic and exothermic reactions is correct?
 A There is always less energy absorbed than released.
 B The enthalpy of the products is always less than the enthalpy of the reactants.
 C Some energy is always absorbed to break bonds in the reactants.
 D The ΔH value is the difference between the enthalpy of the reactants and the highest energy point reached on the energy profile diagram.

4 State whether each of the following statements related to activation energy is true or false.
 a Activation energy is the minimum amount of energy reactant particles need to have for successful collisions.
 b Reactions that start immediately do not have an activation energy.
 c Reactions that release energy overall do not need to absorb activation energy.
 d The match used to light a fire is providing activation energy.

5 The combustion reaction of ethyne gas that occurs in a welding machine can be represented by the thermochemical equation:

$$2C_2H_2(g) + 5O_2(g) \rightarrow 4CO_2(g) + 2H_2O(l) \quad \Delta H = -2619\ kJ\ mol^{-1}$$

 a Is this reaction endothermic or exothermic?
 b What would be the new value of ΔH if the equation was written as follows?

$$4C_2H_2(g) + 10O_2(g) \rightarrow 8CO_2(g) + 4H_2O(l)$$

6 Explain why reversing a chemical reaction reverses the sign of ΔH.

7 The combustion of butane gas in portable stoves can be represented by the thermochemical equation:

$$2C_4H_{10}(g) + 13O_2(g) \rightarrow 8CO_2(g) + 10H_2O(l) \quad \Delta H = -5772\ kJ\ mol^{-1}$$

 a How does the overall energy of the bonds in the reactants compare with the overall energy of the bonds in the products?
 b Draw an energy profile diagram for the reaction, labelling ΔH and the activation energy.

8 Write a balanced equation for the complete combustion of liquid butanol (C_4H_9OH).

9 Write a balanced equation for the incomplete combustion of butane gas (C_4H_{10}), where carbon monoxide is formed.

10 Calculate the energy released when 5.00 kg of methane (CH_4) is burnt in an unlimited supply of oxygen. The heat of combustion of methane is $-890\ kJ\ mol^{-1}$. (Give your answer in megajoules, MJ.)

11 Calculate the energy needed to heat:
 a 100 mL of water from 20.0°C to 80.0°C
 b 250 mL of water from 25.0°C to 100.0°C
 c 1.5 kg of water from 20.0°C to 30.0°C
 d 2300 g of water from 18.0°C to 100.0°C
 e 300 g of cooking oil from 18.0°C to 100.0°C (c(water) $= 4.18\ J\ g^{-1}\ K^{-1}$. ($c$(cooking oil) $= 2.2\ J\ g^{-1}\ K^{-1}$).

12 A 200 mL beaker of water at a temperature of 21.0°C is heated with 10.0 kJ of energy. Calculate the temperature reached by the beaker of water.

13 The heat of combustion of hydrogen is –286 kJ mol^{-1}. Write a thermochemical equation for the complete combustion of hydrogen.

14 Write the formulae for the ions produced when these compounds dissolve in water:

a sodium carbonate

b calcium nitrate

c potassium bromide

d iron(III) sulfate

e copper(II) chloride

15 What is the name given to the process that soluble ionic solids undergo when dissolving in water?

16 Write equations to show the dissolution of the following compounds when they are added to water.

a magnesium sulfate

b sodium sulfide

c potassium hydroxide

d copper(II) acetate

e lithium sulfate

17 Briefly describe what happens to the forces between solute and solvent substances when an ionic substance (such as potassium bromide) dissolves in water.

18 When 6.00 g of sodium chloride (NaCl) (M = 58.44 g mol^{-1}) is added to 50.0 g of water in a solution calorimeter, it dissolves according to the equation:

$$NaCl(s) \xrightarrow{H_2O(l)} Na^+(aq) + Cl^-(aq)$$

The temperature of the water decreases from 22.0°C to a minimum temperature of 20.4°C. Determine the enthalpy of dissolution.

19 When 7.50 g of potassium hydroxide (KOH) (M = 56.11 g mol^{-1}) is added to 300.0 g of water in a solution calorimeter, it dissolves according to the equation:

$$KOH(s) \xrightarrow{H_2O(l)} K^+(aq) + OH^-(aq)$$

The temperature of the water increases from 18.0°C to a maximum temperature of 24.1°C. Determine the enthalpy of dissolution.

20 1.50 g of NaOH is dissolved in 100.0 mL of water. The solution that is formed heats up to a final temperature of 28.0°C. The enthalpy of dissolution for NaOH is ΔH_{soln} = –44.51 kJ mol^{-1}. Determine the initial temperature of the water.

21 Which one of the following factors is not likely to increase the rate of decomposition of hydrogen peroxide?

$$2H_2O_2(aq) \rightarrow 2H_2O(l) + O_2(g)$$

A increasing the pressure of oxygen gas

B increasing the concentration of hydrogen peroxide

C increasing the temperature of hydrogen peroxide

D adding a potassium iodide catalyst

22 The Haber process involves the reaction of nitrogen gas with hydrogen gas to make ammonia gas. Describe two ways in which the rate of this reaction could be increased at constant temperature. Using collision theory, explain why the rate would be increased.

23 Some car manufacturers have plans for hydrogen-powered cars. In the fuel cells of these cars, hydrogen reacts with oxygen from the air to produce water according to the equation:

$$2H_2(g) + O_2(g) \rightarrow 2H_2O(l)$$

The energy profile diagram for the reaction is shown below.

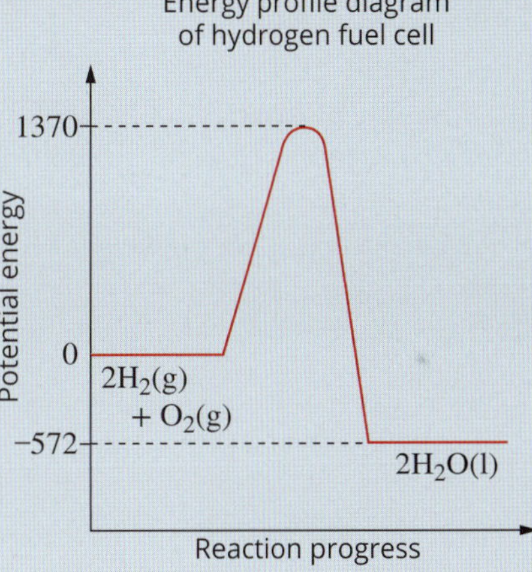

a What is the magnitude of the activation energy of this reaction?

b What is the ΔH for the reaction between hydrogen gas and oxygen gas to produce water?

c Several groups of scientists have claimed to have split water into hydrogen and oxygen using a molybdenum catalyst:

$$2H_2O(l) \xrightarrow{Mo} 2H_2(g) + O_2(g)$$

Sketch energy profile diagrams for this reaction with and without the presence of a catalyst.

d What is the value of ΔH for this water-splitting equation?

24 By comparing both the collision frequency and the size of the activation energy, explain how each of the following changes cause an increase in reaction rate.

a addition of a catalyst

b an increase in the concentration of a reactant

25 In a steelworks, carbon monoxide present in the exhaust gases of the blast furnace can be used as a fuel elsewhere in the plant. It reacts according to the equation:

$$CO(g) + \frac{1}{2}O_2(g) \rightarrow CO_2(g) \quad \Delta H = -283\,kJ\,mol^{-1}$$

a Which has the greater total enthalpy: 1 mol of $CO(g)$ and 0.5 mol of $O_2(g)$, or 1 mol of $CO_2(g)$?

b Write the value of ΔH for the following equations:

i $2CO(g) + O_2(g) \rightarrow 2CO_2(g)$

ii $2CO_2(g) \rightarrow 2CO(g) + O_2(g)$

26 A temperature rise of 5.52°C was observed when 0.0450 g of ethane gas (C_2H_6) was combusted and used to heat 100.0 mL of water. Use this information to write a balanced thermochemical equation for the complete combustion of ethane gas.

27 When 3.00 g of potassium hydroxide (KOH) ($M = 56.11\,g\,mol^{-1}$) is added to 100.0 g of water in a solution calorimeter, it dissolves according to the equation:

$$KOH\,(s) \xrightarrow{H_2O(l)} K^+(aq) + OH^-(aq)$$

The temperature of the water increases from 25.0°C to a maximum temperature of 31.0°C.

a Determine the enthalpy of dissolution.

b The expected enthalpy of dissolution for potassium hydroxide is $-57.61\,kJ\,mol^{-1}$. Use this to determine the % efficiency of the calorimeter.

c Suggest one way to improve the efficiency of the calorimeter.

28 a The graph shows the distribution of the energies of the particles in a substance at two different temperatures, 40°C and 60°C. Indicate which temperature is represented by each of curves A and B in this figure.

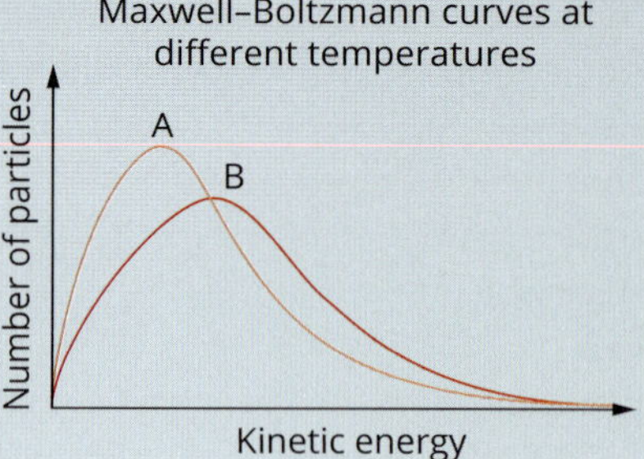

b Copy this diagram for temperature B, and use the diagram to show the effect of a catalyst on a reaction.

c Use the diagram you have drawn in part **b** to explain in terms of collision theory how a catalyst increases the rate of a reaction.

29 Reflect on the Inquiry activity on page 463. What is the importance of the salt?

CHAPTER 15

Enthalpy and Hess's law

In this chapter, you will learn about Hess's law. Hess's law is an expression of the law of conservation of energy. You will learn to apply Hess's law to determine the enthalpy change for a reaction, using average bond energies and enthalpies of formation.

Content

INQUIRY QUESTION

How much energy does it take to break bonds, and how much is released when bonds are formed?

By the end of this chapter, you will be able to:

- explain the enthalpy changes in a reaction in terms of breaking and reforming bonds, and relate this to: L
  - the law of conservation of energy
- investigate Hess's law in quantifying the enthalpy change for a stepped reaction using standard enthalpy change data and bond energy data, for example: (ACSCH037) CCT ICT N
  - carbon reacting with oxygen to form carbon dioxide via carbon monoxide
- apply Hess's law to simple energy cycles and solve problems to quantify enthalpy changes within reactions, including but not limited to: ICT N
  - heat of combustion
  - enthalpy changes involved in photosynthesis
  - enthalpy changes involved in respiration (ACSCH037)

15.1 Latent heat

CHEMISTRY INQUIRY CCT

The latent heat of fusion of ice

What happens to the temperature of a solid substance as it melts?

This activity will investigate what happens when the physical bonds that hold water molecules in a more or less fixed position in the solid state are weakened so that the water transforms into the liquid state. This will be observed by looking at the temperature of ice as it melts.

COLLECT THIS ...

- melting ice (so that it is at 0°C)
- a polystyrene cup
- 110°C thermometer or data-logger temperature probe

DO THIS ...

1 Fill a polystyrene cup with ice that is just beginning to melt.
2 Every minute, record the temperature of the ice until it has just melted.
3 To speed up the melting of the ice, place the cup into a sink filled with warm water.

RECORD THIS ...

Describe what has happened to the ice.

Present your results in a table:

Initial temperature of melting ice =
Final temperature of just melted ice =
Change in temperature of the ice =

REFLECT ON THIS ...

What happened to the temperature as the ice was melting?

Where did the energy come from to melt the ice?

Why does water always boil at 100°C at sea level? No matter how long the water is heated, its temperature never rises above 100°C. Why is this?

This section explores the energy associated with the making and breaking of intermolecular bonds during changes of state for molecular substances.

ENTHALPY OF VAPORISATION

In your earlier science studies, you may have heated iced water over a Bunsen burner, as shown in Figure 15.1.1, and produced a temperature versus time graph similar to the one in Figure 15.1.2. Characteristically, the graph has two plateaus: the first at 0°C, while ice remains in the beaker, and a second while the liquid water continues to boil. No matter how long the water is heated, its temperature never rises above 100°C.

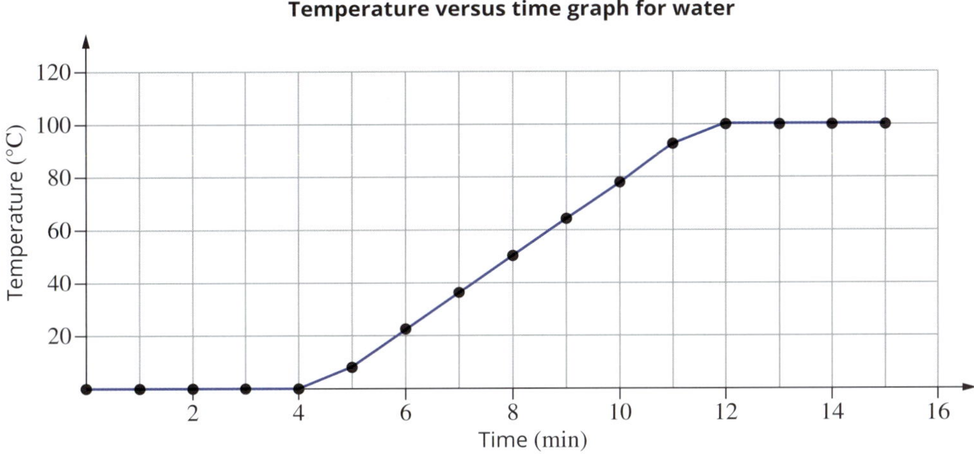

FIGURE 15.1.2 A temperature versus time graph for water, showing the plateaus at 0°C and 100°C.

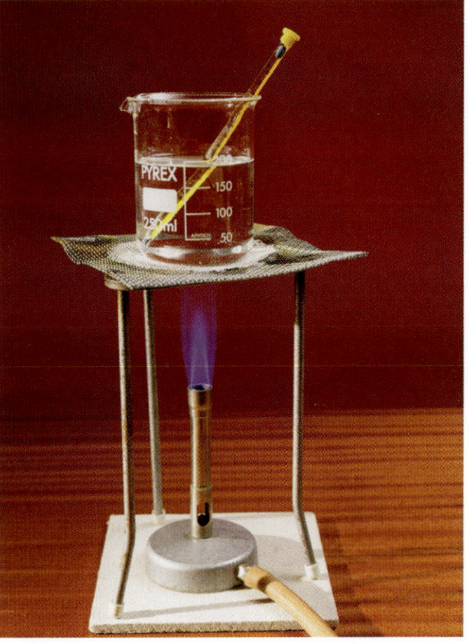

FIGURE 15.1.1 Heating water in a beaker over a Bunsen burner. Pure water boils at a temperature of 100°C at an atmospheric pressure of 101.3 kPa.

The boiling temperature is constant for all pure substances at a given pressure. This is because the heat energy being supplied is used to change the state of the substance. You will recall from Chapter 2 that a change of state in water occurs when the heat energy supplied increases the kinetic energy of the water molecules. Some molecules are then able to move beyond the influence of the intermolecular hydrogen bonds that bind them together in the liquid state. The molecules move into the gaseous state, where they move more randomly and freely with respect to one another.

GO TO ➤ Section 2.2 page 54

The term **enthalpy of vaporisation** (symbol ΔH_{vap}), also known as latent heat of vaporisation (or just heat of vaporisation), is the amount of energy required to transform 1 mol of a substance from a liquid into a gas. The enthalpy of vaporisation is typically measured at one atmosphere pressure (1 atm).

> The term **enthalpy** is a measure of the energy in a thermodynamic system. In some chemistry literature, the term enthalpy is used interchangeably with the term 'heat'.

A total of 40.7 kilojoules (kJ) of heat energy are required to vaporise 1 mol of liquid water at 100°C to 1 mol of steam. That is, the enthalpy of vaporisation of water is 40.7 kJ mol^{-1}. This process is endothermic (i.e. heat is absorbed by the water).

The thermochemical equation for this process is:

$$H_2O(l) \rightarrow H_2O(g) \quad \Delta H = +40.7\,kJ\,mol^{-1}$$

Conversely, the enthalpy change involved in the condensation of 1 mol of steam to 1 mol of water can be written as:

$$H_2O(g) \rightarrow H_2O(l) \quad \Delta H = -40.7\,kJ\,mol^{-1}$$

This is an exothermic process, as indicated by the negative sign (i.e. heat is released during condensation).

Vaporisation of a substance requires energy and therefore is an endothermic process. Conversely, condensation of the same substance releases exactly the same amount of energy and is therefore an exothermic process.

Worked example 15.1.1

CALCULATING THE ENTHALPY CHANGE OF VAPORISATION

Calculate the amount of heat energy required to transform 2 mol of liquid water at its boiling point to steam.

Thinking	Working
State the amount of energy required to transform 1 mol of liquid water to steam.	$H_2O(l) \rightarrow H_2O(g)$ $\Delta H = +40.7\ \text{kJ mol}^{-1}$
2 mol of water will require twice the amount of energy to transform it from liquid to steam as 1 mol.	$2H_2O(l) \rightarrow 2H_2O(g)$ Energy $= 2 \times +40.7$ $= +81.4\ \text{kJ}$
Answer the question.	81.4 kJ of heat energy is required to transform 2 mol of liquid water at its boiling point to steam.

Worked example: Try yourself 15.1.1

CALCULATING THE ENTHALPY CHANGE OF VAPORISATION

Calculate the amount of heat energy required to transform 5 mol of liquid water at its boiling point to steam.

CHEMFILE IU

Joseph Black and latent heat

Joseph Black was a Scottish physician and chemist (Figure 15.1.3). He attended medical school in Glasgow and Edinburgh. He is known for several scientific discoveries, including the fact that carbon dioxide can kill you.

Black was engaged by frugal local distillers of whisky to determine the minimum amount of fuel required to make their product. From his experiments, he was able to show the distillers the minimum amount of wood they needed to burn in order to evaporate a given quantity of whisky mash, and then how much cold water it would take to condense the whisky out of the steam.

While saving the whisky distillers money, Black found that changes of state take place at a constant temperature, regardless of heat input. He applied the term **latent heat** to describe the 'hidden' heat that was released during the process of condensation and defined the property 'specific heat of substances'.

His friend, the Scottish inventor James Watt, used Black's theory of latent heat to make important improvements to the steam engine.

FIGURE 15.1.3 The Scottish chemist Joseph Black (1728–1799) lecturing on latent heat at Glasgow University.

ENTHALPY OF FUSION

The **enthalpy of fusion** (ΔH_{fus}), also known as the latent heat of fusion, is the energy associated with melting 1 mol of a solid to its liquid form. The enthalpy of fusion is typically measured at 1 atm.

The enthalpy of fusion for water at its melting point (0°C) is 6.01 kJ mol^{-1}.

This can be written as:

$$H_2O(s) \rightarrow H_2O(l) \qquad \Delta H = +6.01\ kJ\ mol^{-1}$$

For water, this change represents a change in intermolecular bonding. In solid ice, the hydrogen bonds are fixed. In liquid water, however, the molecules have more energy and the hydrogen bonds are continually being broken and formed. The liquid water molecules remain within one another's sphere of influence, but can move relative to one another.

The values for the enthalpies of fusion and vaporisation are experimentally determined, so there may be slight variations in values from different sources. Table 15.1.1 shows that the enthalpy of fusion per mol of a substance at its melting point is usually much smaller than the enthalpy of vaporisation per mol of a substance at its boiling point.

Melting requires energy and is therefore an endothermic process. Conversely, solidification releases energy and is therefore an exothermic process.

TABLE 15.1.1 Enthalpy of fusion and vaporisation for some common substances

Name	Chemical formula	ΔH_{fus} (kJ mol^{-1})	ΔH_{vap} (kJ mol^{-1})
ammonia	NH_3	+5.97	+23.4
benzene	C_6H_6	+9.95	+30.8
ethanol	C_2H_5OH	+5.02	+38.6
helium	He	+0.02	+0.10
mercury	Hg	+2.33	+59.2
water	H_2O	+6.01	+40.7

Worked example 15.1.2

CALCULATING THE ENTHALPY CHANGE OF MELTING

Calculate the amount of heat energy required to transform 34 g of solid ammonia at its melting point into a liquid.

Thinking	Working
Convert grams to moles.	$n = \frac{m}{M}$ $= \frac{34}{17.03}$ $= 2.0$ mol
Write the thermochemical equation for fusion for 1 mol of the substance. (Use the data in Table 15.1.1.)	$NH_3(s) \rightarrow NH_3(l) \qquad \Delta H = +5.97\ kJ\ mol^{-1}$
Multiply this by the number of mol of ammonia.	$2NH_3(s) \rightarrow 2NH_3(l) \qquad$ Energy $= 2.0 \times +5.97$ $= +12$ kJ
Answer the question.	12 kJ of heat energy is required to transform 34 g of solid ammonia at its melting point into a liquid.

Worked example: Try yourself 15.1.2

CALCULATING THE ENTHALPY CHANGE OF MELTING

Calculate the amount of heat energy required to transform 23 g of solid ethanol at its melting point into a liquid.

15.1 Review

SUMMARY

- The enthalpy of vaporisation is the energy required to transform a liquid substance at its boiling point into a gaseous substance.
- The enthalpy of fusion is the energy required to melt a solid to its liquid form at its melting point.
- Melting and vaporisation are endothermic processes. Conversely, solidification and condensation are exothermic processes.
- The sign of the enthalpy value indicates whether the process is endothermic (+) or exothermic (–).
- Standard enthalpies of fusion and vaporisation are given in kilojoules per mole (kJ mol^{-1}). They are experimentally determined.

KEY QUESTIONS

1 Rank the following systems in order of increasing amounts of average kinetic energy of their molecules:
   A ice at 0°C
   B steam at 170°C
   C liquid water at 45°C
   D ice at –34°C
   E liquid water at 98°C

2 Use data from Table 15.1.1 (page 481) to compare the intermolecular bond strength of water with that of ethanol.

3 The ΔH_{vap} of ammonia is +23.4 kJ mol^{-1}. Is this process endothermic or exothermic?

4 How much heat energy would be to require to vaporise 12 mol of liquid ethanol? Use the data provided in Table 15.1.1.

5 What is the enthalpy change for the condensation of 1 mol of benzene? Use the data provided in Table 15.1.1.

6 Calculate the amount of heat energy required to melt 100 g of ice at 0°C.

7 How much energy is released when 1 kg of steam condenses to water?

15.2 Bond energy

In a chemical reaction, energy is required to break the bonds within the reactant molecules. The term **bond energy** refers to the amount of energy required to break 1 mol of a stated bond into its constituent gaseous atoms under standard state conditions (25°C, 1 bar).

For example, 436 kJ of energy is required to break the bonds in 1 mol of hydrogen molecules, forming 2 mol of hydrogen atoms. This process can be written as:

$$H_2(g) \rightarrow 2H(g) \qquad \Delta H = +436\,\text{kJ mol}^{-1}$$

In Chapter 13, you learnt that bond breaking is an endothermic process because it takes energy to break the chemical bonds within molecules.

The bond energies of diatomic molecules can be accurately measured. However, in larger molecules it is often difficult to break bonds one at a time to measure the bond energies. In these larger molecules, it is usual to quote the **average bond energy**.

Consider the complete dissociation of methane:

$$CH_4(g) \rightarrow C(g) + 4H(g) \qquad \Delta H = +1659\,\text{kJ mol}^{-1}$$

This is made up of the four processes:

$$CH_4(g) \rightarrow CH_3(g) + H(g) \qquad \Delta H = +435\,\text{kJ mol}^{-1}$$
$$CH_3(g) \rightarrow CH_2(g) + H(g) \qquad \Delta H = +444\,\text{kJ mol}^{-1}$$
$$CH_2(g) \rightarrow CH(g) + H(g) \qquad \Delta H = +444\,\text{kJ mol}^{-1}$$
$$CH(g) \rightarrow C(g) + H(g) \qquad \Delta H = +339\,\text{kJ mol}^{-1}$$

All of these equations represent the breaking of a C–H bond. However, they are not all identical, because the bond is being broken from a different reactant particle each time.

The average bond enthalpy of the C–H bond can be determined by averaging the above values:

$$\text{average bond energy for C–H bond} = \frac{435 + 444 + 444 + 339}{4} = \frac{1662}{4} = 416\,\text{kJ mol}^{-1}$$

The bond energy for a particular bond may not be exactly the same for all the different molecules in which that bond occurs. Table 15.2.1 shows the average bond energies for various single and multiple bonds. Since the values are derived from experimental observation, there may be slight variation between the values in different publications.

TABLE 15.2.1 Average bond energies in kJ mol^{-1}

Single bonds								Multiple bonds	
C–H	414	Si–O	466	H–I	298	Cl–F	255	C=C	614
C–C	346	N–H	391	O–H	463	Cl–Cl	242	C≡C	839
C–N	286	N–C	286	O–O	144	Br–F	249	C=N	615
C–O	358	N–N	158	O–F	191	Br–Cl	219	C≡N	890
C–F	492	N–O	214	O–Cl	206	Br–Br	193	C=O	804
C–Cl	324	N–F	278	O–I	201	I–Cl	211	N=N	470
C–Br	285	N–Cl	192	S–H	364	I–Br	178	N≡N	945
C–I	228	N–Br	243	S–F	327	I–I	151	O=O	498
C–S	289	H–H	436	S–Cl	271			S=O	523
Si–H	323	H–F	567	S–Br	218			S=S	429
Si–Si	226	H–Cl	431	S–S	266				
Si–C	307	H–Br	366	F–F	159				

> ℹ The enthalpy change for a reaction is the overall difference in energy between the amount of energy absorbed to break bonds in the reactants and the amount of energy released when bonds are formed in making the products.

Enthalpy changes associated with a chemical reaction can be understood in terms of the energy associated with the making and breaking of the chemical bonds during the reaction.

Consider the formation of hydrogen chloride:

$$H_2(g) + Cl_2(g) \rightarrow 2HCl(g)$$

In this reaction, the covalent bonds in hydrogen (H–H) and chlorine (Cl–Cl) are broken, and two new covalent bonds (H–Cl and H–Cl) are formed.

Using data from Table 15.2.1 (page 483), the enthalpy change for this reaction can be calculated. Note that bond breaking requires energy; the process is endothermic, so the sign attached to the bond energy value is positive. Conversely, bond formation is an exothermic process, so the sign attached to the bond energy will be negative.

		Average bond energy ($kJ\,mol^{-1}$)
bonds broken	H–H	+436
	Cl–Cl	+242
bonds formed	2 × H–Cl	2 × −431
enthalpy change		−184

Therefore, the thermochemical equation for this reaction can be written as:

$$H_2(g) + Cl_2(g) \rightarrow 2HCl(g) \qquad \Delta H = -184\,kJ\,mol^{-1}$$

Worked example 15.2.1

USING AVERAGE BOND ENERGIES TO CALCULATE THE ENTHALPY CHANGE FOR A REACTION

Use data from Table 15.2.1 (page 483) to estimate the enthalpy change for the following reaction:

$$N_2(g) + 3H_2(g) \rightarrow 2NH_3(g)$$

Thinking	Working
Identify the number and type of bonds broken and made.	N≡N + H—H, H—H, H—H → H—N(—H)—H, H—N(—H)—H bonds broken: 1 × N≡N, 3 × H–H; bonds formed: 6 × N–H
Sum the values for the bond energies of the bonds broken and bonds formed. Pay special attention to the sign.	see table below
State the answer.	$N_2(g) + 3H_2(g) \rightarrow 2NH_3(g)$ $\quad \Delta H = -93\,kJ\,mol^{-1}$

bonds broken	bonds formed
1 × N≡N	6 × N–H
3 × H–H	

		Average bond energy ($kJ\,mol^{-1}$)
bonds broken	1 × N≡N	+945
	3 × H–H	3 × +436
bonds formed	6 × N–H	6 × −391
enthalpy change		−93

Worked example: Try yourself 15.2.1

USING AVERAGE BOND ENERGIES TO CALCULATE THE ENTHALPY CHANGE FOR A REACTION

Use data from Table 15.2.1 (page 483) to estimate the enthalpy change for the following reaction:

$$H_2(g) + F_2(g) \rightarrow 2HF(g)$$

15.2 Review

SUMMARY

- Bond energy is the amount of energy required to break 1 mol of a stated bond into its constituent gaseous atoms. Data tables of average bond energies can be used to estimate the enthalpy change for a reaction.
- The enthalpy change for a reaction is the overall difference in energy between the amount of energy absorbed to break bonds in the reactants and the amount of energy released when bonds are formed in making the products.

KEY QUESTIONS

1 Refer to the following changes:

i $Br_2(g) \rightarrow Br_2(l)$

ii $Br_2(g) \rightarrow 2Br(g)$

iii $\frac{1}{2}Br_2(g) \rightarrow Br(g)$

a Which change is associated with an energy change equal to the bond energy of bromine?

b Which change is associated with the energy change that occurs during the condensation of bromine?

c Which change is exothermic?

2 List the numbers and types of bonds in the following molecules.

Name	Formula
hydrogen	H_2
water	H_2O
methane	CH_4
oxygen	O_2
ammonia	NH_3
carbon dioxide	CO_2

3 The bond energy of chlorine (Cl–Cl) is 1.7 times that of iodine (I–I). Suggest a reason for this.

4 Use the average bond energies in Table 15.2.1 (page 483) to estimate the enthalpy changes for the following reactions:

a $2HF(g) \rightarrow H_2(g) + F_2(g)$

b $2H_2(g) + O_2(g) \rightarrow 2H_2O(g)$

c $C(g) + 2H_2(g) \rightarrow CH_4(g)$

d $CH_4(g) + 2O_2(g) \rightarrow CO_2(g) + 2H_2O(g)$

15.3 Hess's law

In 1840, the Swiss–Russian chemist Germain Henri Hess (1802–1850) made an important discovery, which has become known as Hess's law. Hess found that the amount of heat energy released or absorbed in a chemical reaction is constant, irrespective of the number of steps or the kind of steps in which the reaction is carried out, provided that the same reactants and products are involved.

Hess's law is consistent with the law of conservation of energy.

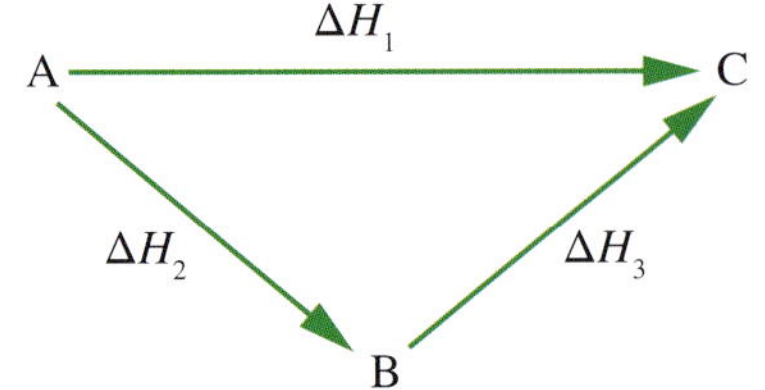

FIGURE 15.3.1 An energy cycle is a diagram showing how three reactions are related. Energy cycles can be constructed to determine unknown enthalpy changes (by adding up the enthalpy changes for steps from an alternative pathway leading to the same final products).

ENERGY CYCLES

The **energy cycle** in Figure 15.3.1 shows that the consequence of Hess's law is that no matter what the pathway of a chemical reaction is, the difference in enthalpy between two substances is independent of the route taken for the conversion. Since the total energy in a system is constant:

$$\Delta H_1 = \Delta H_2 + \Delta H_3$$

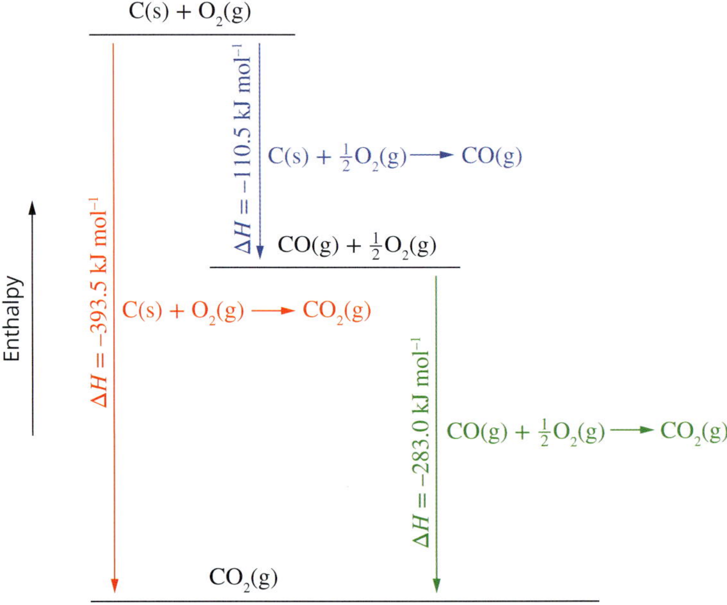

FIGURE 15.3.2 An energy level diagram for the formation of carbon dioxide from carbon and oxygen via carbon monoxide.

Figure 15.3.2 shows an energy level diagram (an energy cycle in a different format) for the formation of carbon dioxide from carbon and oxygen, via carbon monoxide. The diagram shows that, whichever reaction pathway is taken to form carbon dioxide, the enthalpy change involved is the same.

$$C(s) + \tfrac{1}{2}O_2(g) \rightarrow CO(g) \qquad \Delta H = -110.5\,kJ\,mol^{-1}$$

$$CO(g) + \tfrac{1}{2}O_2(g) \rightarrow CO_2(g) \qquad \Delta H = -283.0\,kJ\,mol^{-1}$$

$$C(s) + O_2(g) \rightarrow CO_2(g) \qquad \Delta H = -110.5 + (-283.0) = -393.5\,kJ\,mol^{-1}$$

USING HESS'S LAW TO DETERMINE THE ENTHALPY CHANGE FOR A REACTION

Many reactions do not occur under standard state conditions, so the enthalpy change cannot be directly measured. Hess's law is invaluable in chemistry, as it enables the calculation of enthalpy changes for a multitude of chemical reactions (including oxidation reactions and combustion reactions) from limited enthalpy change data. Such calculations are based on three key properties of chemical reactions:

1. Chemical reactions can be considered to be reversible if the appropriate conditions are available.
2. Energy is always released or absorbed in a chemical reaction. The products of an exothermic reaction contain less energy than the starting reactants: ΔH for an exothermic reaction is negative. Conversely, ΔH for the reverse endothermic reaction is positive.

   For example, the combustion of carbon to form carbon dioxide can be reversed:

   $$C(s) + O_2(g) \rightarrow CO_2(g) \qquad \Delta H = -393.5\,kJ\,mol^{-1}$$
   $$CO_2(g) \rightarrow C(s) + O_2(g) \qquad \Delta H = +393.5\,kJ\,mol^{-1}$$
3. Equations for chemical reactions can be added and subtracted in the manner of ordinary algebraic equations.

For example, calculate the enthalpy change for the reaction of carbon and hydrogen to form ethyne:

$$2C(s) + H_2(g) \rightarrow C_2H_2(g) \qquad \Delta H_4 = ?$$

given the following enthalpies of reaction:

$$C_2H_2(g) + \tfrac{5}{2}O_2(g) \rightarrow 2CO_2(g) + H_2O(l) \qquad \Delta H_1 = -1299.5\,kJ\,mol^{-1} \quad (1)$$

$$C(s) + O_2(g) \rightarrow CO_2(g) \qquad \Delta H_2 = -393.5\,kJ\,mol^{-1} \quad (2)$$

$$H_2(g) + \tfrac{1}{2}O_2(g) \rightarrow H_2O(l) \qquad \Delta H_3 = -285.8\,kJ\,mol^{-1} \quad (3)$$

The solution to finding the enthalpy change for the reaction, ΔH_4, can be found using the algebraic method demonstrated below.

- Step 1. Reverse equation (1) to make $C_2H_2(g)$ a product. Also reverse the sign of the ΔH_1 value.

  $$2CO_2(g) + H_2O(l) \rightarrow C_2H_2(g) + \tfrac{5}{2}O_2(g) \qquad -\Delta H_1 = +1299.5\,kJ\,mol^{-1}$$
- Step 2. Rewrite equations (2) and (3) so $CO_2(g)$ and $H_2O(l)$ are products.

  $$C(s) + O_2(g) \rightarrow CO_2(g) \qquad \Delta H_2 = -393.5\,kJ\,mol^{-1}$$
  $$H_2(g) + \tfrac{1}{2}O_2(g) \rightarrow H_2O(l) \qquad \Delta H_3 = -285.8\,kJ\,mol^{-1}$$
- Step 3. Apply a multiplier to each equation if necessary.

  $$2C(s) + 2O_2(g) \rightarrow 2CO_2(g) \qquad 2\Delta H_2 = 2 \times -393.5 = -787.0\,kJ\,mol^{-1}$$
- Step 4. Add the equations and their ΔH values. Cancel reactants and products that occur on both sides of the equation:

~~$2CO_2(g) + H_2O(l)$~~	$\rightarrow$	$C_2H_2(g)$ + ~~$\tfrac{5}{2}O_2(g)$~~	$-\Delta H_1 = +1299.5\,kJ\,mol^{-1}$
$2C(s)$ + ~~$2O_2(g)$~~	$\rightarrow$	~~$2CO_2(g)$~~	$2\Delta H_2 = -787.0\,kJ\,mol^{-1}$
$H_2(g)$ + ~~$\tfrac{1}{2}O_2(g)$~~	$\rightarrow$	~~$H_2O(l)$~~	$\Delta H_3 = -285.8\,kJ\,mol^{-1}$
$2C(s) + H_2(g)$	$\rightarrow$	$C_2H_2(g)$	$\Delta H_4 = +226.7\,kJ\,mol^{-1}$

The reaction of carbon and hydrogen to form ethyne has an enthalpy change of $+226.7\,kJ\,mol^{-1}$.

The energy cycle shown in Figure 15.3.3 shows two pathways to ethyne from carbon and hydrogen. According to Hess's law, the enthalpy change for the pathway via carbon dioxide and water, $2\Delta H_2 + \Delta H_3 + (-\Delta H_1)$, is equal to the enthalpy change via the direct pathway, ΔH_4.

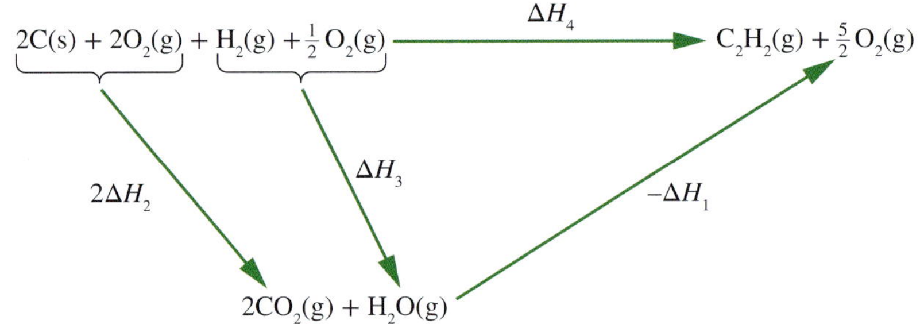

FIGURE 15.3.3 An energy cycle for the formation of ethyne (C_2H_2) from carbon and hydrogen.

Worked example 15.3.1

CALCULATING THE ENTHALPY CHANGE FOR A REACTION

Calculate the enthalpy change for the following reaction:

$CH_3OH(g) \rightarrow H_2CO(g) + H_2(g)$

given the following enthalpies of combustion:

$CH_3OH(g) + \frac{3}{2}O_2(g) \rightarrow CO_2(g) + 2H_2O(g) \quad \Delta H = -676.5\ kJ\ mol^{-1}$ (1)

$H_2CO(g) + O_2(g) \rightarrow CO_2(g) + H_2O(g) \quad \Delta H = -520\ kJ\ mol^{-1}$ (2)

$H_2(g) + \frac{1}{2}O_2(g) \rightarrow H_2O(g) \quad \Delta H = -241.8\ kJ\ mol^{-1}$ (3)

Thinking	Working
Rewrite equations to make the products for the reaction. If necessary, reverse the sign of the ΔH values.	Reverse (2): $CO_2(g) + H_2O(g) \rightarrow H_2CO(g) + O_2(g)$ $\Delta H = +520\ kJ\ mol^{-1}$ Reverse (3): $H_2O(g) \rightarrow H_2(g) + \frac{1}{2}O_2(g)$ $\Delta H = +241.8\ kJ\ mol^{-1}$
Rewrite equations to make reactants for the reaction. If necessary, reverse the sign of the ΔH values.	Equation (1) $CH_3OH(g) + \frac{3}{2}O_2(g) \rightarrow CO_2(g) + 2H_2O(g)$ $\Delta H = -676.5\ kJ\ mol^{-1}$
Make sure the molar quantities are correct. Apply a multiplier to each equation if necessary.	No multiplier is required for any of the equations.
Add the equations and their ΔH values. Cancel reactants and products that occur on both sides of the equation.	$\cancel{CO_2(g)} + \cancel{H_2O(g)} \rightarrow H_2CO(g) + \cancel{O_2(g)}$ $\Delta H = +520\ kJ\ mol^{-1}$ $\cancel{H_2O(g)} \rightarrow H_2(g) + \cancel{\frac{1}{2}O_2(g)}$ $\Delta H = +241.8\ kJ\ mol^{-1}$ $CH_3OH(g) + \cancel{\frac{3}{2}O_2(g)} \rightarrow \cancel{CO_2(g)} + \cancel{2H_2O(g)}$ $\Delta H = -676.5\ kJ\ mol^{-1}$ $CH_3OH(g) \rightarrow H_2CO(g) + H_2(g)$ $\Delta H = +85\ kJ\ mol^{-1}$
State the answer.	The enthalpy change for $CH_3OH(g) \rightarrow H_2CO(g) + H_2(g)$ is $+85\ kJ\ mol^{-1}$.

Worked example: Try yourself 15.3.1

CALCULATING THE ENTHALPY CHANGE FOR A REACTION

Calculate the enthalpy change for the following reaction:

$CH_3COOH(l) + H_2(g) \rightarrow CH_3CH_2OH(l) + \frac{1}{2}O_2(g)$

given the following enthalpies of combustion:

$CH_3CH_2OH(l) + 3O_2(g) \rightarrow 2CO_2(g) + 3H_2O(g)$ $\Delta H = -1360\ kJ\ mol^{-1}$ (1)

$CH_3COOH(l) + 2O_2(g) \rightarrow 2CO_2(g) + 2H_2O(l)$ $\Delta H = -876\ kJ\ mol^{-1}$ (2)

$H_2(g) + \frac{1}{2}O_2(g) \rightarrow H_2O(l)$ $\Delta H = -285.8\ kJ\ mol^{-1}$ (3)

STANDARD ENTHALPY OF FORMATION

The **standard enthalpy of formation** (ΔH°_f) is the change in enthalpy when 1 mol of a compound is formed in its **standard state** from its constituent elements in their standard state. The standard state of a chemical substance is its phase (solid, liquid, gas) at 25.0°C and 1 bar. A superscript circle symbol is added to the ΔH (ΔH°) to indicate standard state conditions.

For example, the thermochemical equation for the formation of ammonia from nitrogen and hydrogen in their standard states is:

$$\frac{1}{2}N_2(g) + \frac{3}{2}H_2(g) \rightarrow NH_3(g) \qquad \Delta H^\circ_f = -45.9\ kJ\ mol^{-1}$$

Table 15.3.1 gives data for the enthalpies of formation for selected substances. Note that the enthalpy of formation for elements in their standard state is defined as zero. These values are experimentally determined, so there may be minor differences in the chemistry literature.

TABLE 15.3.1 Selected standard enthalpies of formation

Substance	Formula	ΔH°_f (kJ mol^{-1})
ammonia	$NH_3(g)$	–45.9
butane	$C_4H_{10}(g)$	–126
carbon dioxide	$CO_2(g)$	–393.5
carbon monoxide	$CO(g)$	–110.5
ethane	$C_2H_6(g)$	–84.0
ethanol	$C_2H_5OH(l)$	–277.7
ethene	$C_2H_4(g)$	+52.0
ethyne (acetylene)	$C_2H_2(g)$	+228
glucose	$C_6H_{12}O_6(s)$	–1271
hydrogen	$H_2(g)$	0
hydrogen bromide	$HBr(g)$	–36.3
hydrogen chloride	$HCl(g)$	–92.3
hydrogen fluoride	$HF(g)$	–273.3
hydrogen iodide	$HI(g)$	+26.5
methane	$CH_4(g)$	–74.8
nitrogen	$N_2(g)$	0
nitrogen dioxide	$NO_2(g)$	+33.1
nitrogen monoxide	$NO(g)$	+90.2
octane	$C_8H_{18}(l)$	–250.3
oxygen	$O_2(g)$	0
propane	$C_3H_8(g)$	–104.6
water(g)	$H_2O(g)$	–241.8
water(l)	$H_2O(l)$	–285.8

Using standard enthalpies of formation

Consider a general reaction represented by the equation:

elements → products

The sum of the standard energies of formation of the products can be represented mathematically by $\Sigma\Delta H^\circ_f$(products), where the symbol Σ means 'the sum of'.

Similarly, the sum of the standard energies of formation of the reactants can be represented mathematically by $\Sigma\Delta H^\circ_f$(reactants).

Therefore, according to Hess's law, the enthalpy change for any reaction is:

$\Delta H^\circ_{reaction}$ = (sum of enthalpies of formation of products) − (sum of enthalpies of formation of reactants)

$= \Sigma\Delta H^\circ_f(\text{products}) - \Sigma\Delta H^\circ_f(\text{reactants})$

The energy cycle in Figure 15.3.4 shows this diagramatically.

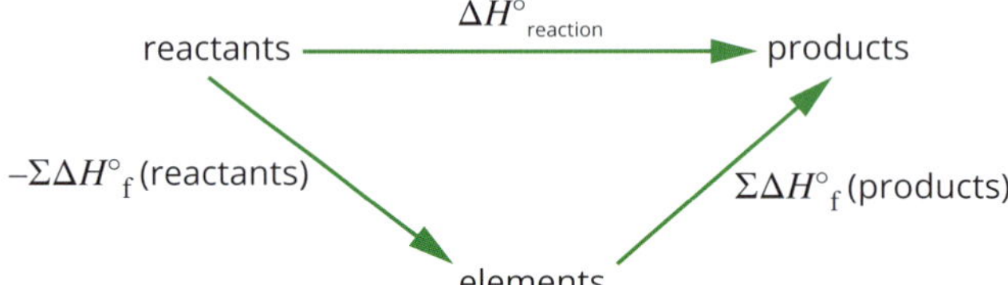

FIGURE 15.3.4 An energy cycle showing how the enthalpy change of a reaction can be determined from the enthalpies of formation of the reactants and products.

The enthalpy change for a chemical reaction at standard state conditions can be calculated from its standard enthalpies of formation using the following expression:

$\Delta H^\circ_{reaction} = \Sigma\Delta H^\circ_f(\text{products}) - \Sigma\Delta H^\circ_f(\text{reactants})$

Worked example 15.3.2

USING STANDARD ENTHALPY OF FORMATION DATA TO CALCULATE THE ENTHALPY CHANGE FOR A REACTION

Calculate the enthalpy change for the following reaction:

$NH_3(g) + HCl(g) \rightarrow NH_4Cl(s)$

given the following standard enthalpies of formation:

$\frac{1}{2}N_2(g) + \frac{3}{2}H_2(g) \rightarrow NH_3(g)$ $\quad \Delta H^\circ_f = -45.9\ \text{kJ mol}^{-1}$

$\frac{1}{2}H_2(g) + \frac{1}{2}Cl_2(g) \rightarrow HCl(g)$ $\quad \Delta H^\circ_f = -92.3\ \text{kJ mol}^{-1}$

$\frac{1}{2}N_2(g) + 2H_2(g) + \frac{1}{2}Cl_2(g) \rightarrow NH_4Cl(s)$ $\quad \Delta H^\circ_f = -314.9\ \text{kJ mol}^{-1}$

Thinking	Working
Calculate the sum of the enthalpies of formation of products, $\Sigma\Delta H^\circ_f$(products). As ΔH°_f is given per mol, multiply the ΔH°_f of each product by its coefficient in the overall equation.	$\Sigma\Delta H^\circ_f(\text{products}) = 1 \times \Delta H^\circ_f(NH_4Cl)$ $= 1 \times -314.9$ $= -314.9\ \text{kJ mol}^{-1}$
Calculate the sum of the enthalpies of formation of the reactants, $\Sigma\Delta H^\circ_f$(reactants). As ΔH°_f is given per mol, multiply the ΔH°_f of each reactant by its coefficient in the overall equation.	$\Sigma\Delta H^\circ_f(\text{reactants}) = 1 \times \Delta H^\circ_f(NH_3) + 1 \times \Delta H^\circ_f(HCl)$ $= (1 \times -45.9) + (1 \times -92.3)$ $= -138.2\ \text{kJ mol}^{-1}$
Calculate $\Delta H^\circ_{reaction}$ using the expression: $\Delta H^\circ_{reaction} = \Sigma\Delta H^\circ_f(\text{products}) - \Sigma\Delta H^\circ_f(\text{reactants})$	$\Delta H^\circ_{reaction} = \Sigma\Delta H^\circ_f(\text{products}) - \Sigma\Delta H^\circ_f(\text{reactants})$ $= -314.9 - (-138.2)$ $= -176.7\ \text{kJ mol}^{-1}$

Worked example: Try yourself 15.3.2

USING STANDARD ENTHALPY OF FORMATION DATA TO CALCULATE THE ENTHALPY CHANGE FOR A REACTION

Calculate the enthalpy change for the following reaction:

$Cl_2(g) + C_3H_8(g) \rightarrow C_3H_7Cl(g) + HCl(g)$

given the following standard enthalpies of formation:

$3C(s) + 4H_2(g) \rightarrow C_3H_8(g)$	$\Delta H^\circ_f = -104.6\ kJ\ mol^{-1}$
$\frac{1}{2}H_2(g) + \frac{1}{2}Cl_2(g) \rightarrow HCl(g)$	$\Delta H^\circ_f = -92.3\ kJ\ mol^{-1}$
$3C(s) + \frac{7}{2}H_2(g) + \frac{1}{2}Cl_2(g) \rightarrow C_3H_7Cl(g)$	$\Delta H^\circ_f = -132.5\ kJ\ mol^{-1}$

Combustion: calculation of enthalpy of reaction

Spirit burners, such as the one illustrated in Figure 15.3.5, are commonplace in school laboratories. They are used during practical activities to empirically determine the enthalpy of combustion of various alcohols and other fuels.

However, the enthalpy of combustion of any fuel can also be calculated using enthalpy of formation data.

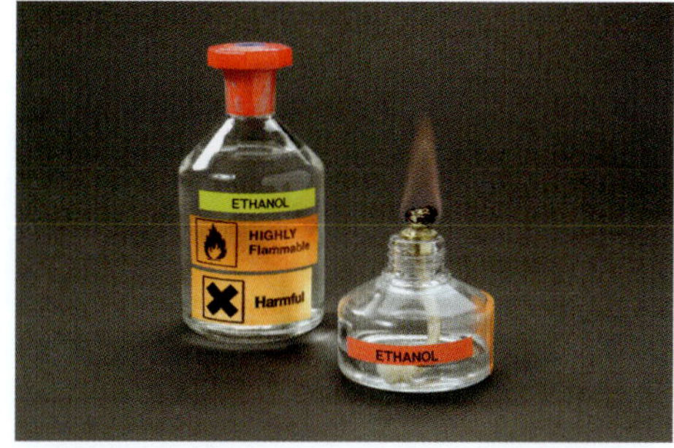

FIGURE 15.3.5 Spirit burner burning ethanol as a fuel.

Worked example 15.3.3

USING STANDARD ENTHALPY OF FORMATION DATA TO CALCULATE ENTHALPY OF COMBUSTION

Calculate the enthalpy change for the complete combustion of ethanol:

$C_2H_5OH(l) + 3O_2(g) \rightarrow 2CO_2(g) + 3H_2O(l)$

given the following standard enthalpies of formation.

$C(s) + O_2(g) \rightarrow CO_2(g)$	$\Delta H^\circ_f = -393.5\ kJ\ mol^{-1}$
$H_2(g) + \frac{1}{2}O_2(g) \rightarrow H_2O(l)$	$\Delta H^\circ_f = -285.8\ kJ\ mol^{-1}$
$2C(s) + \frac{1}{2}O_2(g) + 3H_2(g) \rightarrow C_2H_5OH(l)$	$\Delta H^\circ_f = -277.7\ kJ\ mol^{-1}$

Thinking	Working
Calculate the sum of the enthalpies of formation of products, $\Sigma\Delta H^\circ_f$(products). As ΔH°_f is given per mol, multiply the ΔH°_f of each product by its coefficient in the overall equation.	$\Sigma\Delta H^\circ_f(\text{products}) = 2 \times \Delta H^\circ_f(CO_2) + 3 \times \Delta H^\circ_f(H_2O)$ $= (2 \times -393.5) + (3 \times -285.8)$ $= -1644.4\ kJ\ mol^{-1}$
Calculate the sum of the enthalpies of formation of reactants, $\Sigma\Delta H^\circ_f$(reactants). As ΔH°_f is given per mol, multiply the ΔH°_f of each reactant by its coefficient in the overall equation.	$\Sigma\Delta H^\circ_f(\text{reactants}) = 1 \times \Delta H^\circ_f(C_2H_5OH) + 3 \times \Delta H^\circ_f(O_2)$ $= (1 \times -277.7) + (3 \times 0)$ $= -277.7\ kJ\ mol^{-1}$
Calculate $\Delta H^\circ_{reaction}$ using the expression: $\Delta H^\circ_{reaction} = \Sigma\Delta H^\circ_f(\text{products}) - \Sigma\Delta H^\circ_f(\text{reactants})$	$\Delta H^\circ_{reaction} = \Sigma\Delta H^\circ_f(\text{products}) - \Sigma\Delta H^\circ_f(\text{reactants})$ $= -1644.4 - (-277.7)$ $= -1366.7\ kJ\ mol^{-1}$

Worked example: Try yourself 15.3.3

USING STANDARD ENTHALPY OF FORMATION DATA TO CALCULATE ENTHALPY OF COMBUSTION

Calculate the enthalpy change for the complete combustion of butan-1-ol:

$C_4H_9OH(l) + 6O_2(g) \rightarrow 4CO_2(g) + 5H_2O(l)$

given the following standard enthalpies of formation:

$$C(s) + O_2(g) \rightarrow CO_2(g) \qquad \Delta H^\circ_f = -393.5\,kJ\,mol^{-1}$$

$$H_2(g) + \tfrac{1}{2}O_2(g) \rightarrow H_2O(l) \qquad \Delta H^\circ_f = -285.8\,kJ\,mol^{-1}$$

$$4C(s) + \tfrac{1}{2}O_2(g) + 5H_2(g) \rightarrow C_4H_9OH(l) \qquad \Delta H^\circ_f = -325.8\,kJ\,mol^{-1}$$

Photosynthesis: calculation of enthalpy of reaction

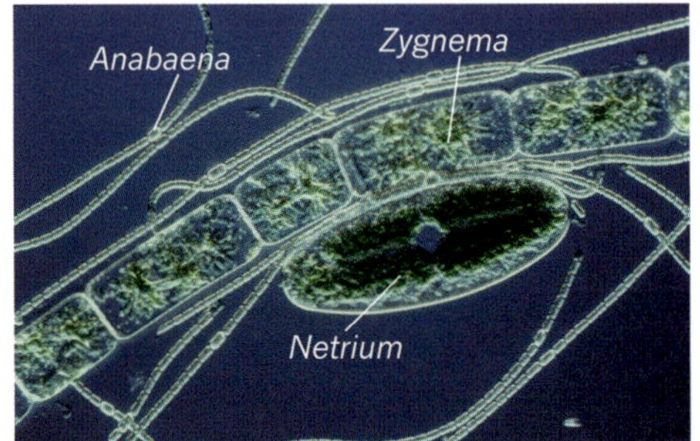

FIGURE 15.3.6 Photosynthesis occurs in cyanobacteria (such as *Anabaena*, shown here as strings of tiny bead-like cells) and in the chloroplasts of these green algae (*Netrium* and *Zygnema*).

Photosynthesis is the process used by plants and certain bacteria (Figure 15.3.6) to make sugars and other essential molecules, using carbon dioxide and water in the presence of light.

While the photosynthesis reaction pathway is quite complex, the process can be summarised as:

$$\text{carbon dioxide} + \text{water} \xrightarrow{\text{sunlight}} \text{glucose} + \text{oxygen}$$

Figure 15.3.7 sets out the raw materials for and the products of photosynthesis.

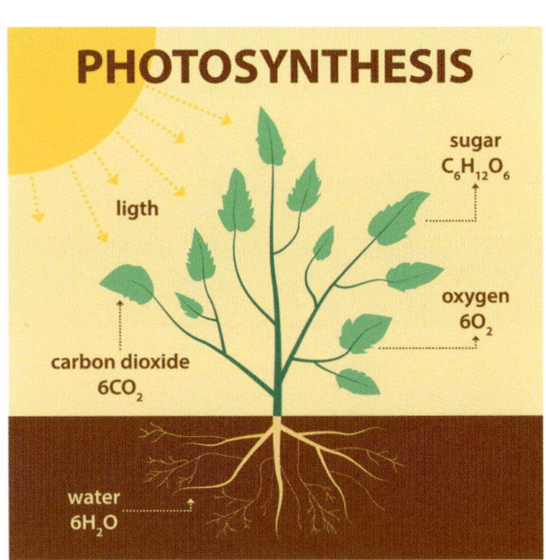

FIGURE 15.3.7 Photosynthesis is the process used by plants and certain bacteria to make sugars (such as glucose) and other essential molecules, using carbon dioxide and water in the presence of light.

The overall chemical equation for photosynthesis is:

$$6H_2O(l) + 6CO_2(g) \xrightarrow{\text{sunlight}} C_6H_{12}O_6(s) + 6O_2(g)$$

The enthalpy change involved in the photosynthesis reaction can be determined using Hess's law and data from a standard table of enthalpies of formation.

The following enthalpies of formation are obtained from Table 15.3.1 (page 489):

$$C(s) + O_2(g) \rightarrow CO_2(g) \qquad \Delta H^\circ_f = -393.5\,kJ\,mol^{-1}$$

$$H_2(g) + \tfrac{1}{2}O_2(g) \rightarrow H_2O(l) \qquad \Delta H^\circ_f = -285.8\,kJ\,mol^{-1}$$

$$6C(s) + 6H_2(g) + 3O_2(g) \rightarrow C_6H_{12}O_6(s) \qquad \Delta H^\circ_f = -1271\,kJ\,mol^{-1}$$

- Step 1. Calculate the sum of the enthalpies of formation of the products, $\Sigma\Delta H^\circ_f$(products). As ΔH°_f is given per mol, multiply the ΔH°_f of each product by its coefficient in the overall equation.

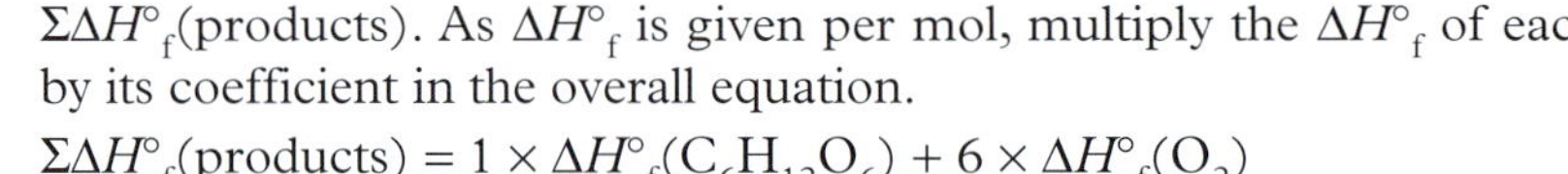

$$\Sigma\Delta H^\circ_f(\text{products}) = 1 \times \Delta H^\circ_f(C_6H_{12}O_6) + 6 \times \Delta H^\circ_f(O_2)$$
$$= (1 \times -1271) + (6 \times 0)$$
$$= -1271\,kJ\,mol^{-1}$$

- Step 2. Calculate the sum of the enthalpies of formation of the reactants, $\Sigma\Delta H^\circ_f$(reactants). As ΔH°_f is given per mol, multiply the ΔH°_f of each reactant by its coefficient in the overall equation.

$$\Sigma\Delta H^\circ_f(\text{reactants}) = 6 \times \Delta H^\circ_f(H_2O) + 6 \times \Delta H^\circ_f(CO_2)$$
$$= (6 \times -285.8) + (6 \times -393.5)$$
$$= -4075.8\,kJ\,mol^{-1}$$

- Step 3. Calculate $\Delta H^\circ_{reaction}$ using the expression:

$$\Delta H^\circ_{reaction} = \Sigma\Delta H^\circ_f(\text{products}) - \Sigma\Delta H^\circ_f(\text{reactants})$$

$$\Delta H^\circ_{reaction} = -1271 - (-4075.8)$$
$$= +2805\,kJ\,mol^{-1}$$

The enthalpy change for photosynthesis is $+2805\,kJ\,mol^{-1}$.

Respiration: calculation of enthalpy of reaction

Cellular respiration is the process by which cells obtain energy to do work. Aerobic cellular respiration occurs within mitochondria (Figure 15.3.8), organelles found inside the cytoplasm of some cells. It is a complex oxidation process; however, it may be summarised as:

glucose + oxygen → carbon dioxide + water + energy

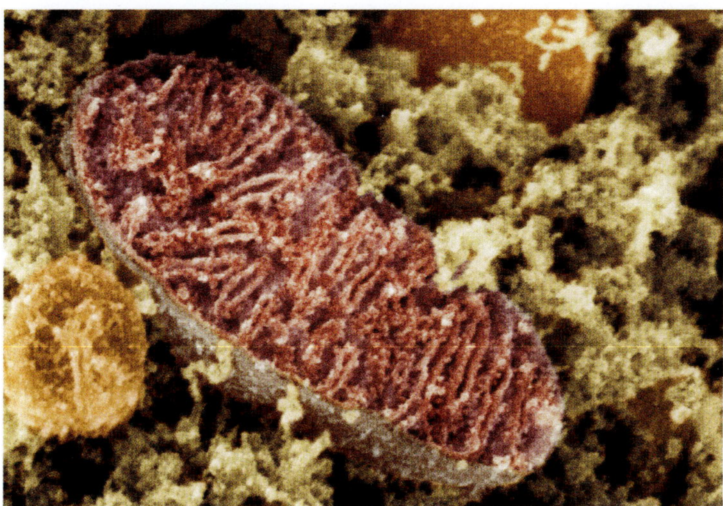

FIGURE 15.3.8 Aerobic cellular respiration occurs in the mitochondria. This scanning electron micrograph of a mitochondrion shows the internal structures on which respiration takes place.

The overall chemical equation for aerobic cellular respiration is the reverse of the equation for photosynthesis:

$$C_6H_{12}O_6(s) + 6O_2(g) \rightarrow 6H_2O(l) + 6CO_2(g)$$

The enthalpy change involved in aerobic cellular respiration can be determined from the overall summary equation using Hess's law and data from a standard table of enthalpies of formation.

The following enthalpies of formation are obtained from Table 15.3.1 (page 489):

$$C(s) + O_2(g) \rightarrow CO_2(g) \qquad \Delta H^\circ_f = -393.5\,kJ\,mol^{-1}$$
$$H_2(g) + O_2(g) \rightarrow H_2O(l) \qquad \Delta H^\circ_f = -285.8\,kJ\,mol^{-1}$$
$$6C(s) + 6H_2(g) + 3O_2(g) \rightarrow C_6H_{12}O_6(s) \qquad \Delta H^\circ_f = -1271\,kJ\,mol^{-1}$$

- Step 1. Calculate the sum of the enthalpies of formation of the products, $\Sigma\Delta H^\circ_f$(products). As ΔH°_f is given per mol, multiply the ΔH°_f of each product by its coefficient in the overall equation.

  $\Sigma\Delta H^\circ_f(\text{products}) = 6 \times \Delta H^\circ_f(H_2O) + 6 \times \Delta H^\circ_f(CO_2)$
  $= (6 \times -285.8) + (6 \times -393.5)$
  $= -4075.8\,kJ\,mol^{-1}$

- Step 2. Calculate the sum of the enthalpies of formation of the reactants, $\Sigma\Delta H^\circ_f$(reactants). As ΔH°_f is given per mol, multiply the ΔH°_f of each reactant by its coefficient in the overall equation.

  $\Sigma\Delta H^\circ_f(\text{reactants}) = 1 \times \Delta H^\circ_f(C_6H_{12}O_6) + 6 \times \Delta H^\circ_f(O_2)$
  $= (1 \times -1271) + (6 \times 0)$
  $= -1271\,kJ\,mol^{-1}$

- Step 3. Calculate $\Delta H^\circ_{reaction}$ using the expression:

  $\Delta H^\circ_{reaction} = \Sigma\Delta H^\circ_f(\text{products}) - \Sigma\Delta H^\circ_f(\text{reactants})$
  $= -4075.8 - (-1271)$
  $= -2805\,kJ\,mol^{-1}$

The overall change for aerobic cellular respiration is $-2805\,kJ\,mol^{-1}$. Note that the energy change for this reaction has the same magnitude as the energy change for photosynthesis, but is opposite in sign. The reactions are the reverse of each other.

Worked example 15.3.4

CALCULATING THE ENTHALPY CHANGE FOR ANAEROBIC RESPIRATION

Anaerobic respiration is a type of respiration not requiring oxygen, in which glucose is converted into lactic acid.

In humans, anaerobic respiration occurs in our muscle cells when they are depleted of oxygen (e.g. during strenuous exercise).

Calculate the enthalpy change associated with the anaerobic respiration of glucose:

$$C_6H_{12}O_6(s) \rightarrow 2C_3H_6O_3(s)$$

given the following standard enthalpies of formation:

$$6C(s) + 6H_2(g) + 3O_2(g) \rightarrow C_6H_{12}O_6(s) \qquad \Delta H^\circ_f = -1271\ kJ\ mol^{-1}$$

$$3C(s) + O_2(g) + 3H_2(g) \rightarrow C_3H_6O_3(s) \qquad \Delta H^\circ_f = -694.1\ kJ\ mol^{-1}$$

Thinking	Working
Calculate the sum of the enthalpies of formation of the products, $\Sigma\Delta H^\circ f$. As ΔH°_f is given per mol, multiply the ΔH°_f of each product by its coefficient in the overall equation.	$\Sigma\Delta H^\circ_f(\text{products}) = 2 \times \Delta H^\circ_f(C_3H_6O_3)$ $= 2 \times -694.1$ $= -1388.2\ kJ\ mol^{-1}$
Calculate the sum of the enthalpies of formation of the reactants, $\Sigma\Delta H^\circ_f(\text{reactants})$. As ΔH°_f is given per mol, multiply the ΔH°_f of each reactant by its coefficient in the overall equation.	$\Sigma\Delta H^\circ_f(\text{reactants}) = 1 \times \Delta H^\circ_f(C_6H_{12}O_6)$ $= 1 \times -1271$ $= -1271\ kJ\ mol^{-1}$
Calculate $\Delta H^\circ_{reaction}$ using the expression: $\Delta H^\circ_{reaction} = \Sigma\Delta H^\circ_f(\text{products}) - \Sigma\Delta H^\circ_f(\text{reactants})$	$\Delta H^\circ_{reaction} = \Sigma\Delta H^\circ_f(\text{products}) - \Sigma\Delta H^\circ_f(\text{reactants})$ $= -1388.2 - (-1271)$ $= -117\ kJ\ mol^{-1}$

Worked example: Try yourself 15.3.4

CALCULATING THE ENTHALPY CHANGE FOR ANAEROBIC RESPIRATION

Fermentation is a form of anaerobic respiration performed by yeasts. Fermentation of glucose to alcohol and carbon dioxide by yeasts is crucially important in the bread making and alcohol industries.

Calculate the enthalpy change associated with the fermentation of glucose:

$C_6H_{12}O_6(s) \rightarrow 2C_2H_5OH(l) + 2CO_2(g)$

given the following standard enthalpies of formation:

$$6C(s) + 6H_2(g) + 3O_2(g) \rightarrow C_6H_{12}O_6(s) \qquad \Delta H^\circ_f = -1271\ kJ\ mol^{-1}$$

$$C(s) + O_2(g) \rightarrow CO_2(g) \qquad \Delta H^\circ_f = -393.5\ kJ\ mol^{-1}$$

$$2C(s) + \tfrac{1}{2}O_2(g) + 3H_2(g) \rightarrow C_2H_5OH(l) \qquad \Delta H^\circ_f = -277.7\ kJ\ mol^{-1}$$

CHEMFILE CCT

The chemistry of hand warmers

Hand warmers (Figure 15.3.9) are popular with snowboarders, skiers, alpine hikers and anyone who must work outside in cold weather.

Hand warmers are small disposable packets that come in various shapes and sizes, but all work in about the same way. You just rip open the cellophane bag, exposing the warmer to air, producing instant warmth lasting for hours. You can put one in each pocket to keep your fingers from getting numb. There are even warmers to fit in your boots to keep your toes toasty. They are invaluable when teamed with insulated clothing for warding off frostbite in extreme conditions, as well as for soothing muscular or joint aches.

FIGURE 15.3.9 Air-activated hand warmers produce heat from the exothermic oxidation of iron when exposed to air.

The air-activated hand warmers contain cellulose, iron, water, activated carbon (which evenly distributes heat), vermiculite (a water reservoir) and salt (the catalyst). The heat is generated by an oxidation reaction of iron powder with the help of the activated carbon and salty water. The reaction is represented by the equation:

$$Fe(s) + \tfrac{3}{4}O_2(g) + \tfrac{3}{2}H_2O(l) \rightarrow Fe(OH)_3(s)$$

Heat is produced from the exothermic oxidation of iron when exposed to air. Energy is produced for up to 10h, although this rapidly diminishes after 1–2h. Typically, the hand warmers reach a temperature of between 55°C and 65°C.

The enthalpy change for this reaction can be calculated using Hess's law and the standard enthalpies of formation.

$$\tfrac{3}{2}H_2O(l) \rightarrow \tfrac{3}{2}H_2(g) + \tfrac{3}{4}O_2(g) \qquad \Delta H^\circ = +428.7\,kJ\,mol^{-1}$$

$$Fe(s) + \tfrac{3}{2}O_2(g) + \tfrac{3}{2}H_2(g) \rightarrow Fe(OH)_3(s) \qquad \Delta H^\circ = -823.0\,kJ\,mol^{-1}$$

$$Fe(s) + \tfrac{3}{4}O_2(g) + \tfrac{3}{2}H_2O(l) \rightarrow Fe(OH)_3(s) \qquad \Delta H^\circ = +428.7 + (-823.0) = -394.3\,kJ\,mol^{-1}$$

15.3 Review

SUMMARY

- Hess's law states that, regardless of the multiple stages or steps of a reaction, the total enthalpy change for the reaction is equal to the sum of the enthalpy changes for each of the steps.
- Hess's law is consistent with the law of conservation of energy.
- Energy cycles represent the enthalpy changes within the stages of a chemical reaction. They are a graphical representation of Hess's law.
- Enthalpy of formation data can be used to calculate the enthalpy change associated with a chemical reaction.

KEY QUESTIONS

1 Find the enthalpy change for the following reaction:
$2NOCl(g) \rightarrow 2NO(g) + Cl_2(g)$
given the following data:
$2NO(g) \rightarrow N_2(g) + O_2(g)$ $\Delta H° = -180.6\ kJ\ mol^{-1}$ (1)
$N_2(g) + O_2(g) + Cl_2(g) \rightarrow 2NOCl(g)$
$\Delta H° = +103.4\ kJ\ mol^{-1}$ (2)

2 Find the enthalpy change for the following reaction:
$H_2SO_4(l) \rightarrow SO_3(g) + H_2O(g)$
given the following data:
$H_2S(g) + 2O_2(g) \rightarrow H_2SO_4(l)$ $\Delta H° = -235.5\ kJ\ mol^{-1}$ (1)
$H_2S(g) + 2O_2(g) \rightarrow SO_3(g) + H_2O(l)$
$\Delta H° = -207\ kJ\ mol^{-1}$ (2)
$H_2O(l) \rightarrow H_2O(g)$ $\Delta H° = +44\ kJ\ mol^{-1}$ (3)

3 Find the enthalpy change for the reaction:
$CaCO_3(s) + 2HCl(g) \rightarrow CaCl_2(s) + H_2O(l) + CO_2(g)$
given the following information:
$CaCO_3(s) \rightarrow CaO(s) + CO_2(g)$ $\Delta H° = +175\ kJ\ mol^{-1}$ (1)
$Ca(OH)_2(s) \rightarrow H_2O(l) + CaO(s)$ $\Delta H° = +67\ kJ\ mol^{-1}$ (2)
$Ca(OH)_2(s) + 2HCl(g) \rightarrow CaCl_2(s) + 2H_2O(l)$
$\Delta H° = -198\ kJ\ mol^{-1}$ (3)

4 Using the standard enthalpy of formation data in Table 15.3.1 (page 489), calculate the enthalpy of combustion of methane:
$$CH_4(g) + 2O_2(g) \rightarrow CO_2(g) + 2H_2O(l)$$

5 Liquid hydrazine (N_2H_4) is a rocket fuel. The enthalpy of formation of liquid hydrazine is $+50.6\ kJ\ mol^{-1}$. Using this value and the enthalpy of formation data in Table 15.3.1 (page 489), calculate the enthalpy of combustion of liquid hydrazine:
$$N_2H_4(l) + O_2(g) \rightarrow N_2(g) + 2H_2O(g)$$

6 Calculate the enthalpy change for the following reaction:
$C_2H_4(g) + 4F_2(g) \rightarrow C_2F_4(g) + 4HF(g)$
given the following standard enthalpies of formation:
$\Delta H°_f(C_2H_4(g)) = +52.0\ kJ\ mol^{-1}$
$\Delta H°_f(C_2F_4(g)) = -675.0\ kJ\ mol^{-1}$
$\Delta H°_f(HF(g)) = -273.3\ kJ\ mol^{-1}$

7 Calculate the enthalpy change for the following reaction:
$CH_4(g) + NH_3(g) \rightarrow HCN(g) + 3H_2(g)$
given:
$N_2(g) + 3H_2(g) \rightarrow 2NH_3(g)$ $\Delta H° = -91.8\ kJ\ mol^{-1}$
$C(s) + 2H_2(g) \rightarrow CH_4(g)$ $\Delta H° = -74.8\ kJ\ mol^{-1}$
$H_2(g) + 2C(s) + N_2(g) \rightarrow 2HCN(g)$
$\Delta H° = +270.3\ kJ\ mol^{-1}$

8 **a** Write a balanced chemical equation for the complete combustion of gaseous ethyne (C_2H_2). Assume that liquid water is formed.
b Using the enthalpy of formation data in Table 15.3.1 (page 489), calculate the enthalpy of combustion of ethyne.
c Calculate the amount of energy released by the complete combustion of 100.0 g of ethyne.

Chapter review

KEY TERMS

average bond energy
bond energy
energy cycle
enthalpy
enthalpy of fusion
enthalpy of vaporisation
latent heat
standard enthalpy of formation
standard state

REVIEW QUESTIONS

1 Consider the following thermochemical equations:

- **i** $H_2(g) + I_2(g) \rightarrow 2HI(g)$ $\Delta H^\circ = +52\,kJ\,mol^{-1}$
- **ii** $H_2O(l) \rightarrow H_2O(g)$ $\Delta H^\circ = +40.7\,kJ\,mol^{-1}$
- **iii** $2Na(s) + Cl_2(g) \rightarrow 2NaCl(s)$ $\Delta H^\circ = +822\,kJ\,mol^{-1}$
- **iv** $C(s) + O_2(g) \rightarrow CO_2(g)$ $\Delta H^\circ = -393.5\,kJ\,mol^{-1}$

a Which one of the equations represents an exothermic process?

b Which one of the reactions is not a chemical change?

2 Given the information:

$$H_2O(l) \rightarrow H_2O(g) \quad \Delta H_{vap} = +40.7\,kJ\,mol^{-1}$$

calculate the amount of heat energy required for the vaporisation of:

a 10 mol of liquid water

b 1 kg of liquid water.

3 Consider the following data for the enthalpies of vaporisation for water and mercury:

$H_2O(l) \rightarrow H_2O(g)$ $\Delta H_{vap} = +40.7\,kJ\,mol^{-1}$
$Hg(l) \rightarrow Hg(g)$ $\Delta H_{vap} = +56.9\,kJ\,mol^{-1}$

a What is the significance of the positive value of the enthalpy of vaporisation?

b Identify the type of bonds being broken when water and mercury are vaporised.

c Explain the difference in the enthalpies of vaporisation between water and mercury.

4 The ΔH_{vap} for a liquid is $+30.8\,kJ\,mol^{-1}$.

a Is the process endothermic or exothermic?

b How much heat energy would be required to vaporise 3 mol of this liquid?

5 The enthalpy of fusion of sodium chloride at its melting point is $+28.6\,kJ\,mol^{-1}$.
The enthalpy of fusion for water at its melting point is $+6.01\,kJ\,mol^{-1}$.
Explain the difference in the enthalpies of fusion in terms of the bonding within each of the solids.

6 Explain why steam at 100°C has a greater energy content than water at 100°C.

7 Use the average bond energies from Table 15.2.1 (page 483) to calculate the enthalpy change for the following reactions:

a $C(g) + O_2(g) \rightarrow CO_2(g)$

b $H_2(g) + F_2(g) \rightarrow 2HF(g)$

c $N_2(g) + 3H_2(g) \rightarrow 2NH_3(g)$

8 Domestic oil heaters use long-chain hydrocarbons with an average length of 15 carbons ($C_{15}H_{32}$).

a Write the equation for the complete combustion of $C_{15}H_{32}(l)$.

b Use Table 15.2.1 (page 483) to estimate the energy released by the complete combustion of 1 mol of domestic heating oil.

9 **a** Write a balanced chemical equation for the complete combustion of ethane, assuming all species are gaseous.

b Use Table 15.2.1 (page 483) to calculate the enthalpy of combustion of ethane.

10 Using the average bond energies in Table 15.2.1 (page 483), calculate the enthalpy change for the following reaction:

$$CH_4(g) + Cl_2(g) \rightarrow CH_3Cl(g) + HCl(g)$$

11 Urea ($(NH_2)_2CO$) is used as an ingredient in most fertilisers. It was originally obtained from the guano deposits of birds. However, it has been produced industrially since World War I from ammonia and carbon dioxide. The structural formulae of the reacting species are as shown here:

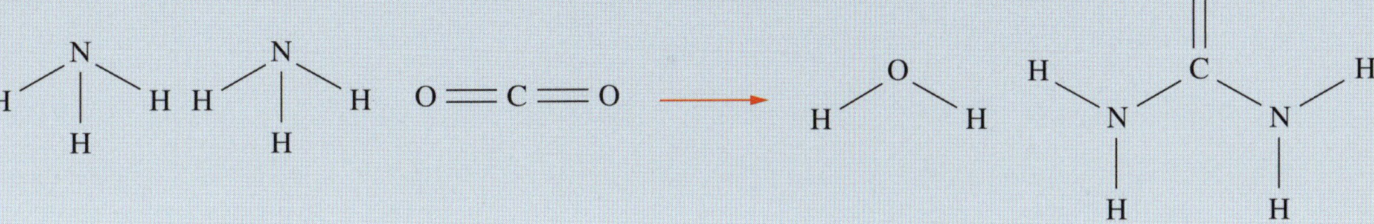

Use the values for average bond energies in Table 15.2.1 (page 483) to calculate the enthalpy change for the reaction:

$$2NH_3(g) + CO_2(g) \rightarrow H_2O(g) + NH_2CONH_2(g)$$

12 The bond energy of H_2 is about 2.8 times stronger than that of F_2. Suggest a reason for this.

13 This graph shows the trend in the enthalpy of combustion for successive alcohols.
Explain the trend in terms of bond energies.

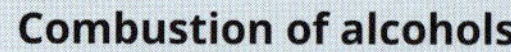

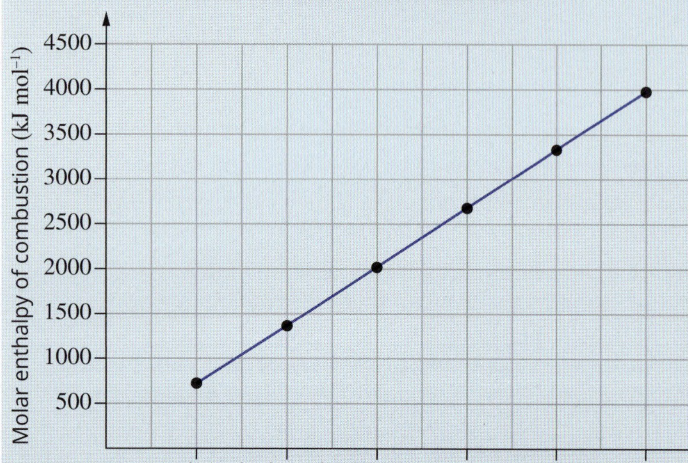

14 The thermite reaction between powdered aluminium and iron(III) oxide produces so much heat that it has been used to weld railway tracks together.
Calculate $\Delta H°$ for the reaction:

$Fe_2O_3(s) + 2Al(s) \rightarrow Al_2O_3(s) + 2Fe(s)$

given the following enthalpies of reaction:

$2Al(s) + \frac{3}{2}O_2(g) \rightarrow Al_2O_3(s)$ $\quad \Delta H° = -1676\ \text{kJ mol}^{-1}$ (1)

$2Fe(s) + \frac{3}{2}O_2(g) \rightarrow Fe_2O_3(s)$ $\quad \Delta H° = -824\ \text{kJ mol}^{-1}$ (2)

15 Calculate the enthalpy of reaction for the oxidation of carbon monoxide.

$CO(g) + \frac{1}{2}O_2(g) \rightarrow CO_2(g)$

given the following enthalpies of reaction:

$C(s) + O_2(g) \rightarrow CO_2(g)$ $\quad \Delta H° = -394\ \text{kJ mol}^{-1}$ (1)

$C(s) + \frac{1}{2}O_2(g) \rightarrow CO(g)$ $\quad \Delta H° = -111\ \text{kJ mol}^{-1}$ (2)

16 Hydrogen can be produced industrially by the following reaction:

$$C(s) + 2H_2O(g) \rightarrow CO_2(g) + 2H_2(g)$$

Find $\Delta H°$ for this reaction, given the following data:

$C(s) + H_2O(g) \rightarrow CO(g) + H_2(g)$ $\quad \Delta H° = +131\ \text{kJ mol}^{-1}$ (1)

$CO_2(g) + H_2(g) \rightarrow CO(g) + H_2O(g)$ $\quad \Delta H° = +40\ \text{kJ mol}^{-1}$ (2)

17 Use this energy cycle to determine ΔH_3.

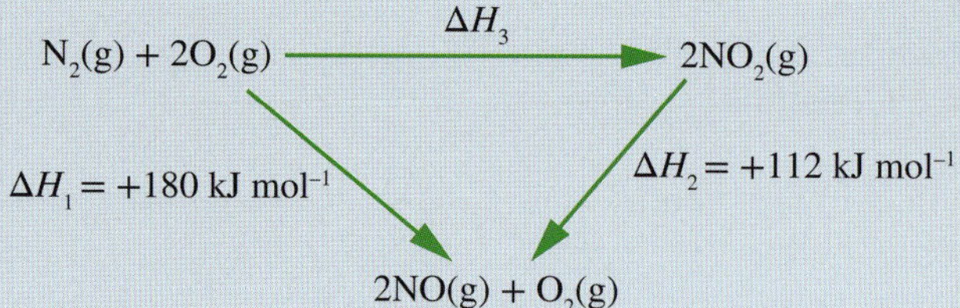

18 Carbon disulfide $CS_2(l)$ is a colourless, volatile liquid often used as an industrial solvent. Calculate the enthalpy change of the following reaction:

$CS_2(l) + 3O_2(g) \rightarrow CO_2(g) + 2SO_2(g)$

given the following standard enthalpies of formation:

$C(s) + O_2(g) \rightarrow CO_2(g)$ $\quad \Delta H°_f = -393.5\ \text{kJ mol}^{-1}$

$S(s) + O_2(g) \rightarrow SO_2(g)$ $\quad \Delta H°_f = -296.8\ \text{kJ mol}^{-1}$

$C(s) + 2S(s) \rightarrow CS_2(l)$ $\quad \Delta H°_f = +87.9\ \text{kJ mol}^{-1}$

19 Why do helium and neon have enthalpies of formation of zero?

20 Calculate the enthalpy of formation of $SO_2(g)$ from the following data:

$S(s) + \frac{3}{2}O_2(g) \rightarrow SO_3(g)$ $\quad \Delta H°_f = -295\ \text{kJ mol}^{-1}$

$SO_2(g) + \frac{1}{2}O_2(g) \rightarrow SO_3(g)$ $\quad \Delta H°_f = +98\ \text{kJ mol}^{-1}$

Use the enthalpy of formation data in Table 15.3.1 (page 489) to answer Questions **21–25**.

21 a Write a balanced chemical equation for the formation of gaseous chloroethene (C_2H_3Cl) and gaseous hydrogen chloride (HCl) from ethene (C_2H_4) and chlorine (Cl_2).

b Determine the enthalpy change for the reaction that produces chloroethene from ethene and chlorine.

$$\Delta H°_f(C_2H_3Cl(g)) = +22.0\ \text{kJ mol}^{-1}$$

22 Calculate the enthalpy change involved in the formation of ethanol from the catalytic hydration of ethene using steam. The equation for the reaction is:

$$C_2H_4(g) + H_2O(g) \rightarrow C_2H_5OH(l)$$

23 Methane (natural gas) can be used to make hydrogen and carbon monoxide through a catalytic reaction with steam:

$$CH_4(g) + H_2O(g) \rightarrow CO(g) + 3H_2(g)$$

Calculate the enthalpy change for this reaction.

24 Methane can burn with insufficient oxygen to form carbon (soot, C(s)) and liquid water.

a Write the equation for this incomplete combustion of methane.

b Calculate the enthalpy change for this reaction.

25 It is possible to reduce iron(III) oxide to iron using hydrogen gas:

$$Fe_2O_3(s) + 3H_2(g) \rightarrow 2Fe(s) + 3H_2O(g)$$

Calculate the enthalpy change for this reduction reaction of iron(III) oxide.

$$\Delta H^\circ_f(Fe_2O_3(s)) = -824.2\ \text{kJ mol}^{-1}$$

26 ΔH_1, ΔH_2 and ΔH_3 are enthalpy change values for steps in this energy cycle:

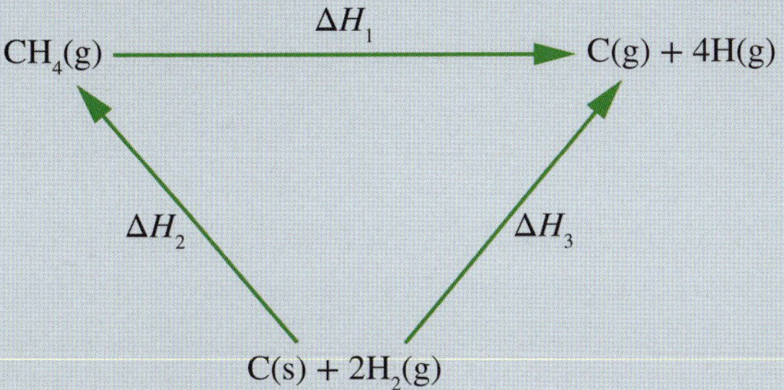

Given the following data:

$\Delta H^\circ_f(CH_4) = -74.8\ \text{kJ mol}^{-1}$

$C(s) \rightarrow C(g) \quad \Delta H^\circ = +717\ \text{kJ mol}^{-1}$

H–H bond energy = $+436\ \text{kJ mol}^{-1}$

a which enthalpy is represented by ΔH_2?

b Calculate ΔH_3.

c Calculate ΔH_1.

27 Use this energy level diagram to calculate ΔH_3.

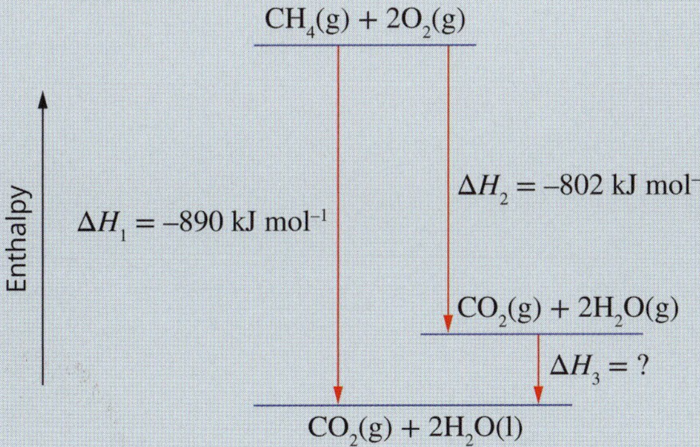

28 Butane (C_4H_{10}) is a hydrocarbon used in gas lighters.

a Write a balanced chemical equation for the complete combustion of butane.

b Use the following standard enthalpy of formation data to calculate the enthalpy of combustion of butane.

$\Delta H^\circ_f(CO_2(g)) = -393.5\ \text{kJ mol}^{-1}$

$\Delta H^\circ_f(H_2O(l)) = -285.8\ \text{kJ mol}^{-1}$

$\Delta H^\circ_f(C_4H_{10}(g)) = -126\ \text{kJ mol}^{-1}$

29 Reflect on the Inquiry activity on page 478. Using concepts from this chapter, explain your observation of the ice's temperature as it was melting.

CHAPTER

16 Entropy and Gibbs free energy

In previous chapters, you have explored important questions about chemical reactions, such as how fast they proceed and what energy changes occur as reactants combine. It is now time to ask perhaps the most profound question of all: why do reactions occur? This chapter will allow you to answer this most fundamental of questions.

In this chapter, you will develop an understanding of the laws of thermodynamics, and what determines whether a reaction will take place. Building on the concept of enthalpy, this chapter introduces you to a new quantity called entropy. You will learn how the measurements of enthalpy and entropy can be combined to determine whether a chemical process occurs naturally.

Outcomes

INQUIRY QUESTION

How can enthalpy and entropy be used to explain reaction spontaneity?

By the end of this chapter, you will be able to:

- analyse the differences between entropy and enthalpy L
- use modelling to illustrate entropy changes in reactions CCT ICT
- predict entropy changes from balanced chemical reactions to classify as increasing or decreasing entropy CCT
- explain reaction spontaneity using terminology, including: (ACSCH072) L
  - Gibbs free energy
  - enthalpy
  - entropy
- solve problems using standard references and $\Delta G^\circ = \Delta H^\circ - T\Delta S^\circ$ (Gibbs free energy formula) to classify reactions as spontaneous or nonspontaneous ICT N
- predict the effect of temperature changes on spontaneity (ACSCH070) CCT

16.1 Energy and entropy changes in chemical reactions

CHEMISTRY INQUIRY 

Spontaneous reactions

How can enthalpy and entropy be used to explain reaction spontaneity?

COLLECT THIS ...

- 4 drinking straws
- scissors
- wide shallow pan
- water
- stirring rod

DO THIS ...

1. Cut the straws into short 1.5–2 cm lengths.
2. Fill the pan with water to a depth of about 5 cm.
3. Drop the straws into the pan.
4. Push all straws underwater. They will float to the surface, but this ensures that there is no air trapped within the straw and that all surfaces are wet.
5. Give the water a quick stir. The straws will become distributed over the surface of the water.
6. Start the timer.
7. After a few minutes some of the straws will have assembled into conglomerates (Figure 16.1.1).
8. Give the water a gentle shake or stir to introduce some energy into the system. The straws will move and the self-assembly will continue.

RECORD THIS ...

Record how the straws assemble themselves, taking note of how many straws are in each unit and how the straws are arranged.

Record how long it takes for units of three or more assembled straws to appear.

Does the gentle mix aid the assembly of conglomerates?

Do conglomerates reform quickly?

REFLECT ON THIS ...

What would happen if the water was warm or hot? (Be careful not to submerge the straws.)

Can you identify other simple materials that spontaneously self-assemble?

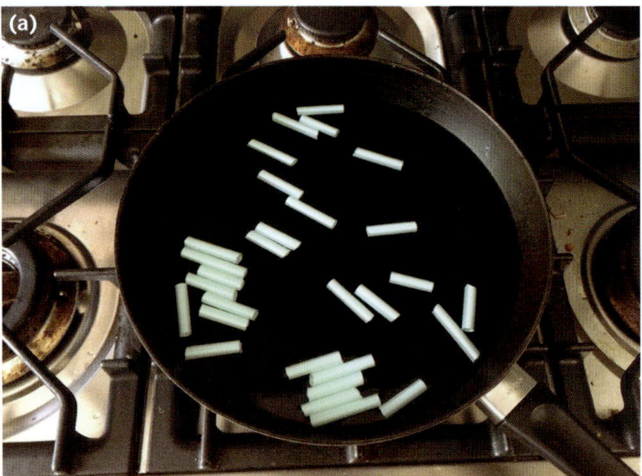

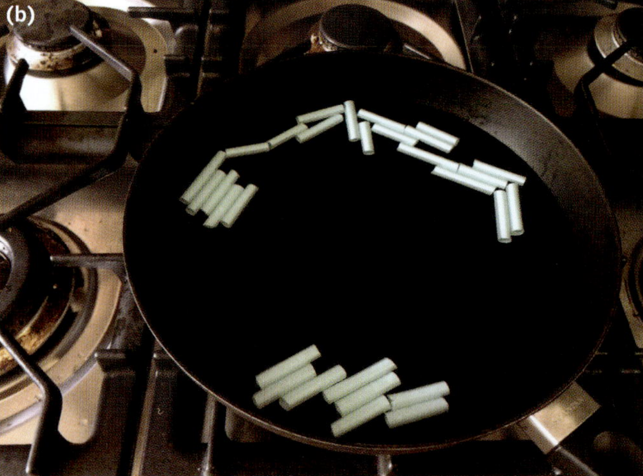

FIGURE 16.1.1 Short pieces of straw (a) in a random pattern and (b) self-assembling into conglomerates on the surface of the water

Thermodynamics is the science of heat, temperature and work. In this section, you will learn about some of the basic ideas of thermodynamics. While you will be applying the principles of thermodynamics to chemical applications in this course, the same principles can be applied much more widely. An astronomer might apply thermodynamics to predict the result of two galaxies colliding, a biologist to studies of the functioning of a human cell, and an ecologist to an investigation of the effects of diet on animal populations.

This section introduces you to a quantity from thermodynamics called entropy. Later in this chapter, you will use entropy to determine whether reactions occur naturally.

SYSTEMS AND SURROUNDINGS

As you learnt in Chapter 14, it is often useful to define a **system** and its **surroundings** (Figure 16.1.2). What is included in the system can be chosen freely. Anything outside the system is called the surroundings, and the system and surroundings are separated by a boundary. The boundary needs to be defined clearly so you know what is in the system. If matter can cross the boundary, the system is **open**. If matter cannot cross the boundary, the system is **closed**. In either case, energy can still cross the boundary, and be exchanged between the system and the surroundings. If neither matter nor energy can cross the boundary, the system is said to be **isolated**.

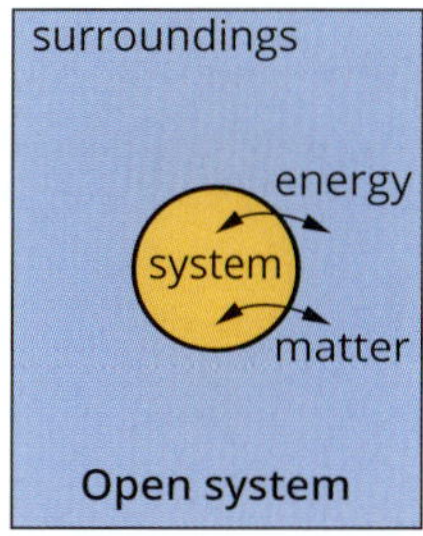

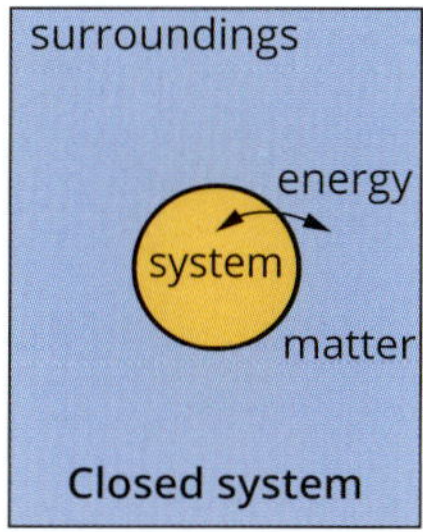

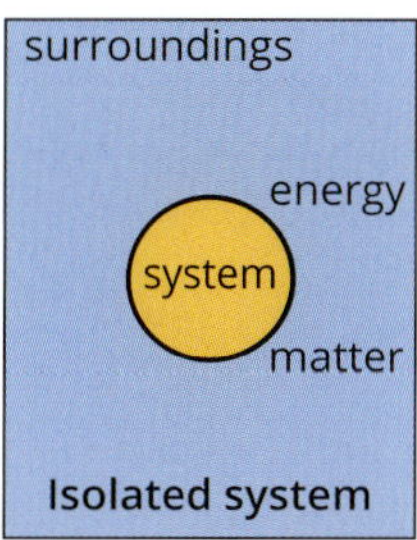

FIGURE 16.1.2 Open systems exchange energy and matter with the surroundings. Closed systems only exchange energy with the surroundings. Isolated systems do not exchange matter or energy with the surroundings.

An unsealed flask of hot water is an open system because heat and water vapour can be lost to the air. The same flask with a tightly fitted stopper approximates a closed system. Placing the stoppered flask in an insulated box approximates an isolated system. A truly isolated system is hard to achieve.

Energy, temperature and heat

All systems have energy. This energy is divided into the contributions of kinetic energy and potential energy. Kinetic energy is associated with motions and vibrations in the material. Potential energy involves other energies in the system, such as energy in the chemical bonds.

In common language, temperature and heat can mean almost the same thing; in thermodynamics, temperature and heat are different. Temperature is a measure of the average kinetic energy of the atoms, ions or molecules in the system. Temperature does not change, no matter how much of the substance is present. Heat, on the other hand, is the energy transferred between a system and its surroundings. When, for example, a gas absorbs heat, the average kinetic energy of its molecules will increase, and its temperature will also increase.

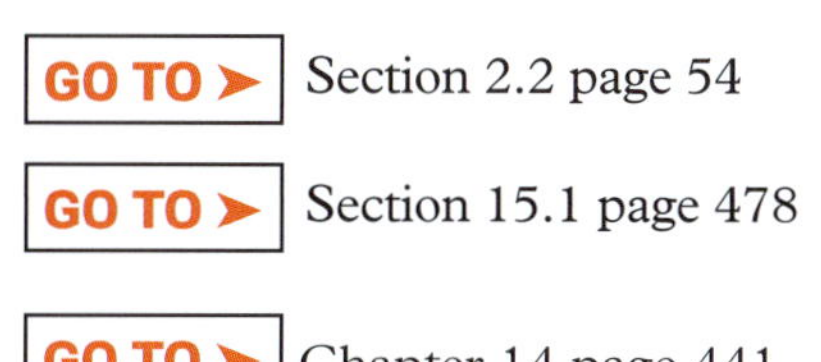

THE FIRST LAW OF THERMODYNAMICS

Famously, thermodynamics has three laws. The first law of thermodynamics states that energy cannot be created or destroyed. This is often called the law of conservation of energy.

The first law of thermodynamics is a version of the law of conservation of energy. It states that the total energy of the universe is constant; energy can be transformed from one form to another, but cannot be created or destroyed.

For example, a light bulb converts electrical energy into light energy. The bulb also feels warmer, so some fraction of the electrical energy becomes thermal energy. Similarly, in a chemical reaction the total energy before and after the reaction is unchanged. At constant pressure, the change in heat involved in a reaction is equal to the change in **enthalpy**, the energy used in breaking and creating chemical bonds.

ENTROPY

In this section, you will learn about the second and third laws of thermodynamics. These two laws involve a quantity called **entropy**. Entropy is often regarded as being a measure of the degree of disorder or randomness in the system being studied. While a thermometer measures temperature and a barometer measures pressure, there is no 'entropy meter'. However, entropy is a physical quantity, and changes in entropy can be measured experimentally.

While entropy can be thought of as a measure of the disorder of a system, there is a more technical definition. For any collection of objects, entropy is a measure of how many possible arrangements there can be with the same overall energy. A system with more possible arrangements but the same energy has more entropy. Consider the matches in the matchbox shown in Figure 16.1.3.

FIGURE 16.1.3 Two systems of matchsticks in a matchbox. If matches are required to fit neatly into a matchbox (the system shown on the left), fewer arrangements are possible than if the matches can be disordered. The system on the left has lower entropy than the system on the right.

The two matchboxes have different rules for their arrangements of matches. If the matches in the matchbox on the left must fit neatly into the matchbox, then clearly there are fewer possible arrangements for this system than for the messy matchbox on the right. The messy matchbox system therefore has more entropy than a system that only allows neatly ordered matchsticks.

Now consider the small sports stadium with 10 seats shown in Figure 16.1.4. As spectators enter the stadium, Spectator A has 10 choices of places to sit, so there are 10 possible arrangements in this system. Spectator B has 9 choices, so now there are 90 (10×9) possible arrangements for seating the two spectators. Spectator C has 8 choices and now the system has 720 ($10 \times 9 \times 8$) possible arrangements for seating the three spectators. As more spectators come into the stadium, more arrangements are possible and the amount of entropy increases.

At constant temperature, the entropy of a gas increases as the volume of the gas increases.

If we expand our system to a sports stadium with 20 seats, each spectator would have more possible seating arrangements. In this case, there would be 6840 ($20 \times 19 \times 18$) possible arrangements. A larger system permits more possible arrangements, and the measured entropy is greater than for a smaller system.

In a similar way, the entropy of a sample of gas increases as the volume of the gas increases. With an increase in volume, there are more arrangements of gas particles that are possible.

A hotter system allows the components more freedom to move than does a colder system, and thus has more possible arrangements. A hotter system will have greater entropy.

Back in the small stadium with 10 seats, if we make the game exciting (think of this as increasing the 'temperature' of the system) each spectator could also be sitting, standing or jumping. Spectator A has 10 seat choices and three position choices, meaning there are 30 (10×3) possible arrangements. In addition, spectator B has 9 seat choices and three position choices, giving a combined $((10 \times 3) \times (9 \times 3)) = 810$ possible arrangements. Adding spectator C gives 19440 possible arrangements for our simple model $((10 \times 3) \times (9 \times 3) \times (8 \times 3))$. A rise in temperature increases the number of possible arrangements and therefore the entropy.

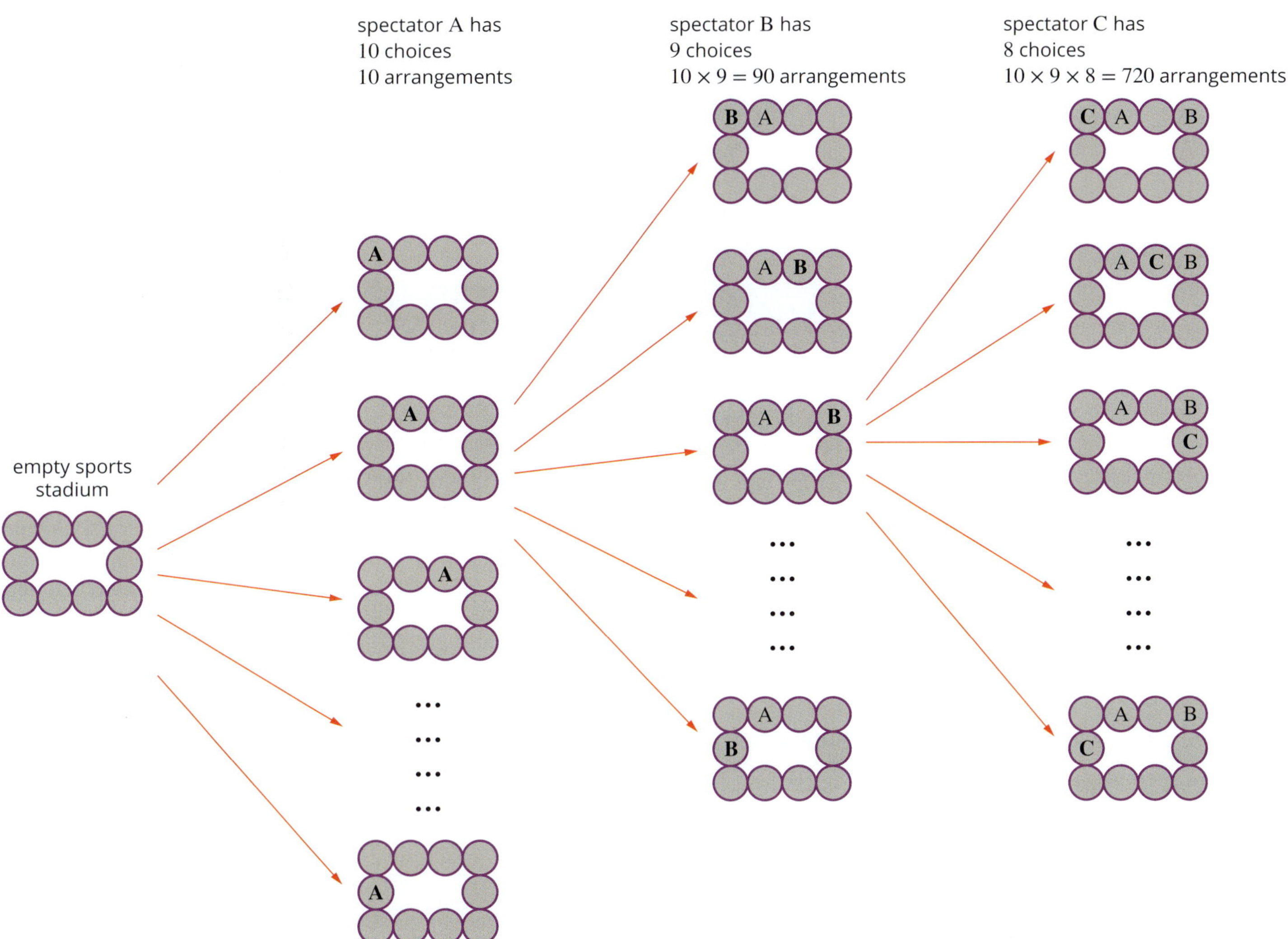

FIGURE 16.1.4 This represents a small sports stadium with 10 seats. As more spectators are introduced, there are more ways to arrange the spectators, so the entropy increases.

Entropy is represented by the symbol S. Like enthalpy, the change in entropy of a system, ΔS, is the difference between the entropy of the final and initial states of the system. For a chemical reaction, ΔS is the difference between the entropy of the products and the reactants:

$$\Delta S = S(\text{products}) - S(\text{reactants})$$

GO TO ➤ SkillBuilder page 280

As for temperature, there is a zero value for entropy, which means it is possible to calculate or experimentally determine the entropy of a substance at a specific temperature. However, like enthalpy, it is the change in entropy that occurs during a chemical or physical change that is usually of most interest to chemists.

Predicting changes in entropy

The following generalisations can be used to predict the change in entropy for a system.

- As the temperature of a substance increases, its entropy increases. The higher kinetic energy of particles in a hotter system allows more possible arrangements of particles. For example, the entropy of a block of iron at 200°C is greater than its entropy at 25°C.
- A substance in its liquid state has more entropy than the same amount of the substance in the solid state (Figure 16.1.5 on page 506). Solids have a regular structure, whereas liquids have no fixed arrangement. Particles in a liquid can be arranged in more ways than in the same amount of a solid, and so are more disordered.

  For example, as ice melts the entropy of the system increases.

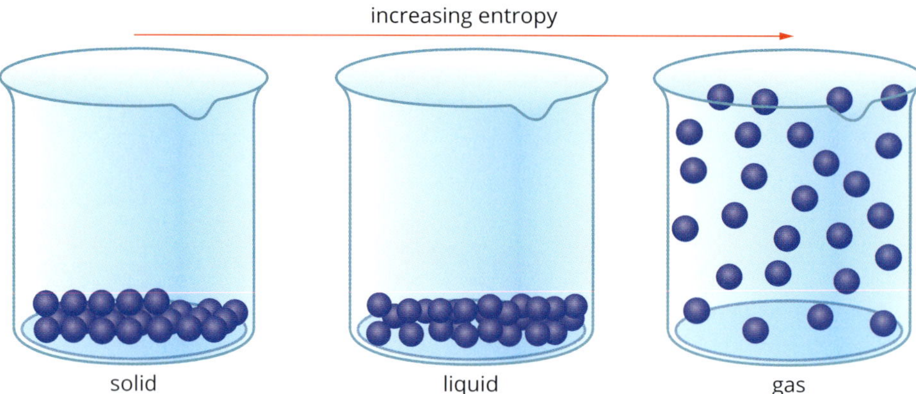

FIGURE 16.1.5 The solid has a relatively low entropy because its particles are fixed in position. The liquid has a higher entropy because its particles can move more freely. Since the gas particles are randomly distributed throughout the container, the gas has the highest entropy.

- A substance in its gaseous state has more entropy than the same amount of the substance in its liquid state (Figure 16.1.5). Gas particles move continually in random directions and occupy their entire container. Particles in a gas can be arranged in more ways than in a liquid, and so are more disordered. For example, as water boils the entropy of the system increases.

  Table 16.1.1 summarises the changes in entropy for various changes in state.

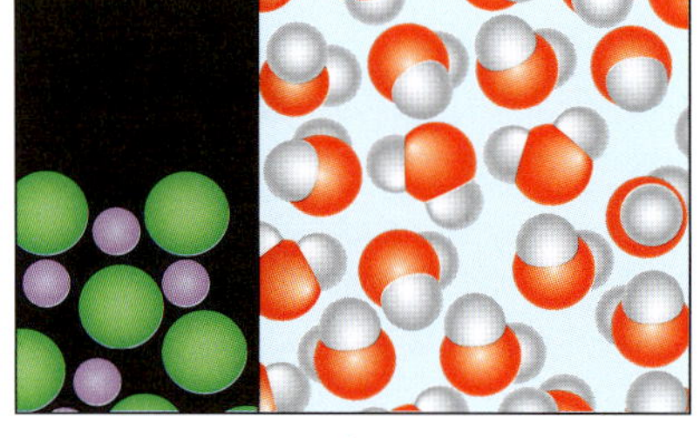

FIGURE 16.1.7 When salt dissolves in water, the ions separate and the entropy increases.

TABLE 16.1.1 Changes in entropy for various changes in state

Change in state	ΔS
solid → liquid	increases, ΔS positive
liquid → gas	increases, ΔS positive
gas → liquid	decreases, ΔS negative
liquid → solid	decreases, ΔS negative

- Adding more particles to a system creates more possible arrangements, and therefore increases the amount of disorder and entropy. For example, if more nitrogen is added to a 1 L container of nitrogen gas at 25°C, the entropy of the system in the container increases (Figure 16.1.6).

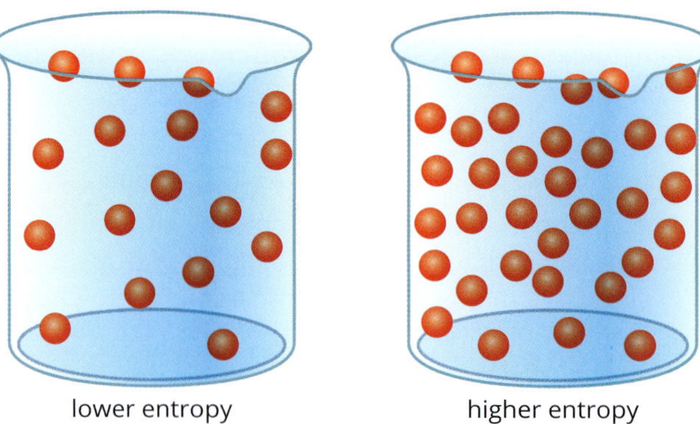

FIGURE 16.1.6 Entropy increases as the number of particles in a system increases.

- When a solute dissolves in a solvent, the solute particles in solution can be arranged in more ways than when they were in the solid, so the amount of disorder and entropy usually increases. For example, when salt dissolves in water, its entropy increases (Figure 16.1.7).

The following Worked example shows you how to apply the generalisations listed above to predict the change in entropy for a particular chemical reaction.

Worked example 16.1.1

DETERMINING THE ENTROPY CHANGE OF A CHEMICAL REACTION

Determine the change in entropy of the following chemical reaction: $CaO(s) + H_2O(l) \rightarrow Ca(OH)_2(s)$

Thinking	Working
Determine whether there is a change in the number of particles in the system and, if so, the effect on entropy.	There are 2 mol of reactant particles and 1 mol of product particles. The system becomes less disordered, so ΔS is negative.
Determine whether there is a change of state in the system and, if so, the effect on entropy.	There is a decrease in the number of mol of liquid and an increase in the number of moles of solid during the reaction. The system becomes less disordered, so ΔS is negative.
Determine the overall change in entropy for the system.	Overall, the system becomes less disordered, so ΔS is negative.

Worked example: Try yourself 16.1.1

DETERMINING THE ENTROPY CHANGE OF A CHEMICAL REACTION

Determine the change in entropy of the following chemical reaction: $(NH_4)_2CO_3(s) \rightarrow 2NH_3(g) + CO_2(g) + H_2O(l)$

CHEMFILE ICT

Claude Shannon: information and entropy

Clear communication requires that the message sent is the same as the message received, and nowhere is this more important than in electronic communication. A transmitter encodes a message, the channel transfers the message, and the receiver seeks to recreate the original message. How can the receiver be certain that the message is correct, and how efficiently can the transmitted message be encoded?

Claude Shannon (1916–2001; Figure 16.1.8) used the idea of entropy, but in reverse. The more uncertain or random the message (higher information entropy), the more information there is in the event.

FIGURE 16.1.8 Claude Shannon developed the ideas of a 'bit' for digitised information, maximum transmission rates ('bandwidth'), and error correction for digital transmission.

Think of spelling out an unfamiliar word. If each letter is randomly distributed, the letters transmitted so far give no indication of what the next letter will be. In a real language, letters are not distributed evenly. On hearing a 'T' then an 'H', you would most likely expect 'A', 'E', 'I', 'O', 'U' or 'R'. On hearing a 'Q' you would likely expect a 'U'. On hearing a 'Z' you would not expect to hear a 'Q' or 'X'. This predictability gives a lower information entropy than a random, unpredictable string.

So, a more efficient way to encode a word in English would be in 'chunks' such as 'THA', 'THE', 'THI', 'THO', 'THU' and 'THR'. These new chunks more efficiently transfer the message and reduce the uncertainty – the information entropy – that the receipt of a 'T' generates.

Shannon's approach to entropy showed how efficiently a message could be encoded ('compressed'), and he laid down a theory that led to the processes used when you download an image or movie from the internet (Figure 16.1.9).

FIGURE 16.1.9 Thanks to Claude Shannon's ideas, you can now quickly download images or movies from the internet.

16.1 Review

SUMMARY

- The universe can be divided into systems (each consisting of the part of the universe being studied, its boundary and its surroundings).
- Temperature is a measure of the average kinetic energy of the particles in an object or a system; heat is a transfer of energy.
- Entropy is a measure of the number of possible arrangements of a system. It can be regarded as the degree of disorder or randomness in the system.
- As the number of particles in a system increases, its entropy increases.
- As the temperature of a substance increases, its entropy increases.
- At constant temperature, the entropy of a gas increases as the volume of the gas increases.
- When a substance changes state, its solid state has less entropy than its liquid state, which in turn has less entropy than its gaseous state.

KEY QUESTIONS

1 A student added 3 mL of a cold dilute acid to 10 mL of a cold dilute base in a flask on a laboratory bench, then stoppered the flask. What is the system and boundary, and is the system open, closed or isolated?

2 You place three buttons in a cup, then shake it. When you look into the cup afterwards, does the cup have more entropy compared with before it was shaken?

3 A cylinder contains some argon gas. A piston compresses the gas in the cylinder to half its volume. If the temperature remains constant, has the entropy of the system increased, decreased or remained the same?

4 Predict and explain the changes in entropy for the following reactions:

a $N_2(g) + 3H_2(g) \rightarrow 2NH_3(g)$

b $2Mg(s) + O_2(g) \rightarrow 2MgO(s)$

c $CaCO_3(s) \rightarrow CaO(s) + CO_2(g)$

d $H_2O(l) \rightarrow H_2O(g)$

5 Arrange the following in order from least to most entropy.

I liquid water, 25°C

II steam, 100°C

III ice, −25°C

IV liquid water, 0°C

V steam, 125°C

VI ice, 0°C

16.2 Entropy and spontaneous processes

Our everyday experience tells us that some changes have a natural direction. A padlock made of iron rusts, but over time a rusty padlock does not become shiny metal. Petrol burns in oxygen to produce carbon dioxide and water vapour, providing energy, yet we never see carbon dioxide and water forming petrol and oxygen. The changes are not reversed. More broadly, consider what happens when a glass is dropped, shattering into many pieces, or to the human body as it ages. These changes also have a natural direction.

In the previous section, you learnt about the quantity called entropy. In this section, you will learn about the key role that entropy plays in determining the direction in which chemical and physical changes occur.

THE SECOND LAW OF THERMODYNAMICS

The second law of thermodynamics states that, over time, the entropy of the universe is increasing. Since a system will be located within the universe, and any system plus its surroundings is effectively the universe, the total entropy of the system and surroundings always increases.

The second law can be expressed mathematically as:

$$\Delta S_{system} + \Delta S_{surroundings} = \Delta S_{universe} > 0$$

The second law of thermodynamics states that the entropy of the universe is increasing.

Spontaneity

A **spontaneous** process is one that occurs of its own accord, without any ongoing addition of external energy. Such a process is often described as one that 'occurs naturally'. When a spontaneous process occurs, there is always an increase in the entropy of the universe ($\Delta S_{universe} > 0$).

A ball falling from your hand is an example of a spontaneous process. No energy is required to make this happen. The same ball does not jump from the ground spontaneously back into your hand, as energy is needed to lift the ball. For chemical reactions, this means that exothermic reactions tend to be spontaneous.

For example, zinc will spontaneously react with a solution of copper(II) sulfate to produce zinc sulfate and copper. However, zinc sulfate does not react with copper to form zinc and copper(II) sulfate.

$$Zn(s) + CuSO_4(aq) \rightarrow ZnSO_4(aq) + Cu(s) \quad \Delta H = -218\,kJ\,mol^{-1} \quad \text{(spontaneous)}$$

$$ZnSO_4(aq) + Cu(s) \not\rightarrow Zn(s) + CuSO_4(aq) \quad \Delta H = +218\,kJ\,mol^{-1} \quad \text{(non-spontaneous)}$$

You are familiar with numerous other spontaneous chemical and physical changes that are exothermic. The burning of natural gas, rusting of iron, condensing of steam at 100°C and respiration in the body are just some examples. In each case, energy is released, which increases the entropy of the surroundings ($\Delta S_{surroundings} > 0$).

Entropy changes and spontaneity

It might be tempting to expect that the only spontaneous chemical processes are exothermic. This, however, is not the case. Consider, for example, the melting of ice at temperatures greater than 0°C (Figure 16.2.1):

$$H_2O(s) \rightarrow H_2O(l) \quad \Delta H = +6.01\,kJ\,mol^{-1}$$

FIGURE 16.2.1 Ice melts spontaneously at temperatures above 0°C. Although the process is endothermic, there is an overall increase in entropy, which makes it spontaneous.

A process is spontaneous if it leads to an increase in the entropy of the universe.

FIGURE 16.2.2 Solid barium hydroxide and solid ammonium chloride have been mixed in a flask that is sitting on a damp piece of wood. Although the reaction is endothermic (its temperature drops to well below 0°C and freezes the flask to the wood), it occurs spontaneously due to an increase in entropy of the system.

As you know, even though this process is endothermic, ice melts spontaneously above 0°C. As the ice melts, the water molecules become more disordered and the entropy of the system increases.

You can see from the previous examples (zinc reacting with copper(II) sulfate, and ice melting) that whether or not a process is spontaneous is determined by two factors: the change in enthalpy, ΔH, and the change in entropy, ΔS, of the system. When ice melts, the endothermic nature of the change does not favour spontaneity, because energy is absorbed by the water molecules, which lowers the entropy of the surroundings. However, this is offset by the large increase in entropy that occurs within the melting ice. Overall, the entropy of the universe increases and the process of melting is spontaneous.

Figure 16.2.2 shows another example of an endothermic process that is spontaneous because it involves a large increase in entropy.

Non-spontaneous reactions

It is important to understand that not all of the reactions that occur in the world are spontaneous reactions. Non-spontaneous reactions can be forced to occur if they are combined, or coupled, with a spontaneous process, provided that the second law of thermodynamics is obeyed overall.

For example, recharging the battery in a mobile phone involves a non-spontaneous reaction that is driven by the flow of electric current. Once the current is turned off the process stops. The source of the electric current is a spontaneous process that occurs at the power station. Even though the reaction that occurs in the phone during recharging decreases the entropy of the universe, if the entropy changes involved in supplying the electric current are taken into account there is an overall increase in entropy.

Similarly, the process of photosynthesis is non-spontaneous. Photosynthesis is a chemical process that occurs in the cells of plants, algae and some bacteria. In this process, carbon dioxide and water are converted into glucose and oxygen. Biological systems use spontaneous reactions to drive non-spontaneous reactions like photosynthesis; the non-spontaneous reactions are coupled to spontaneous reactions so that the overall process is spontaneous.

CHEMFILE S

How is life possible?

Life needs order. Enzymes need a precise structure in order to function. Cells have a distinct level of organisation. An entire organism requires all of its components to be working in harmony. Organisms at the base of the food chain transform simple inorganic and organic materials into more complex substances such as carbohydrates, proteins and DNA. Each step seems to lead to a decrease in entropy. How could life occur spontaneously?

The second law of thermodynamics cannot be broken. Life is permissible because organisms absorb energy from sunlight and consume energy-rich, low-entropy compounds, and release heat and high-entropy compounds, such as water and carbon dioxide, into their surroundings (Figure 16.2.3). The entropy changes of the surroundings increase greatly, so the large increase in the entropy of the surroundings $\Delta S_{surroundings}$ makes up for the decrease in the entropy of the cell ΔS_{system}. Overall, the entropy of the universe increases over time and the processes are spontaneous.

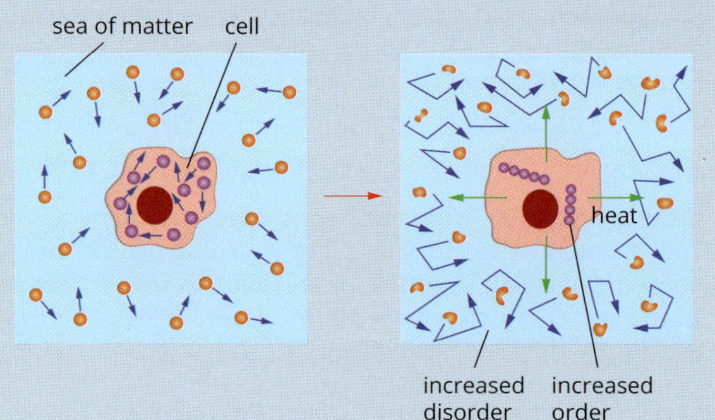

FIGURE 16.2.3 A cell is permitted to become more complex and ordered through making its environment more disordered.

THE THIRD LAW OF THERMODYNAMICS

The third law of thermodynamics states that the entropy of a perfect crystal is zero at absolute zero.

Consider a diatomic gas like hydrogen chloride (HCl). The gas particles move randomly, but when cooled to a solid the particles freeze into a lattice (Figure 16.2.4). As the temperature decreases, the particle vibrations reduce, so the number of possible arrangements reduces too. Therefore, the entropy of the system is reduced as the temperature decreases. Eventually, at absolute zero, there will be no vibrations, and each particle will be locked in a set position. Within their positions, the HCl molecules will retain variations in their orientation. As there are many possible arrangements of the HCl molecules in the lattice, the system retains some degree of disorder, so some minimum entropy will be retained.

For a perfect crystal at absolute zero, there will be one single arrangement and no vibrations from kinetic energy. For this special, theoretical case, the entropy is defined as zero.

According to the third law of thermodynamics, the entropy of a perfect crystal is zero at absolute zero.

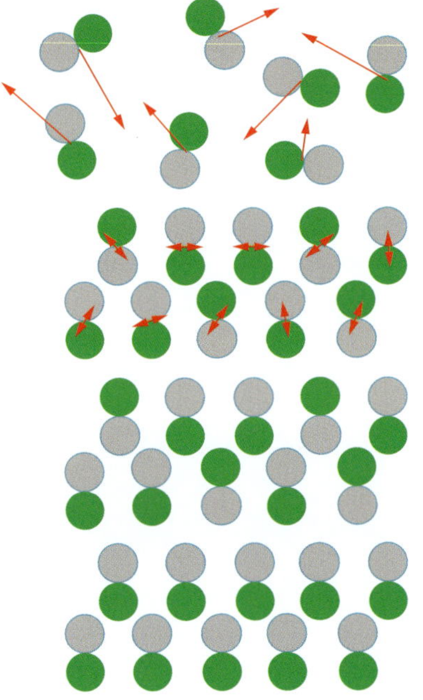

FIGURE 16.2.4 On cooling, entropy decreases. A solid has less entropy than a gas. For a perfect crystal at absolute zero, the entropy is defined as zero.

STANDARD ENTROPY

By defining a zero value of entropy, the absolute entropy of a substance can be determined. Absolute entropy is calculated at standard state conditions (25°C and 1 bar pressure) and is called **standard entropy**. Its symbol is S°. (You learnt about standard state conditions in Chapter 15.)

GO TO ➤ Section 15.3 page 486

Table 16.2.1 lists the standard entropy of sulfur trioxide (SO_3) in its different states. In keeping with what you have learnt in the previous section, the standard entropy increases from a solid to a liquid to a gas.

TABLE 16.2.1 Standard entropies of sulfur trioxide

State	S° (J mol^{-1} K^{-1})
solid	71
liquid	114
gas	257

Table 16.2.2 gives data for the standard entropies for selected substances. Unlike the standard enthalpies of formation discussed in Chapter 15, all standard entropies are positive.

TABLE 16.2.2 Selected standard entropies

Substance	Formula	S° (J mol^{-1} K^{-1})
ammonia	$NH_3(g)$	+192.8
butane	$C_4H_{10}(g)$	+310
carbon dioxide	$CO_2(g)$	+213.8
carbon monoxide	$CO(g)$	+197.7
ethane	$C_2H_6(g)$	+230
ethanol	$C_2H_5OH(l)$	+161
ethene	$C_2H_4(g)$	+220
ethyne (acetylene)	$C_2H_2(g)$	+201
glucose	$C_6H_{12}O_6(s)$	+209.2
hydrogen	$H_2(g)$	+130.7
hydrogen bromide	$HBr(g)$	+198.7
hydrogen chloride	$HCl(g)$	+186.9
hydrogen fluoride	$HF(g)$	+173.8
hydrogen iodide	$HI(g)$	+206.6
methane	$CH_4(g)$	+186
nitrogen	$N_2(g)$	+191.6
nitrogen dioxide	$NO_2(g)$	+240.0
nitrogen monoxide	$NO(g)$	+210.8
octane	$C_8H_{18}(l)$	+359.8
oxygen	$O_2(g)$	+205
propane	$C_3H_8(g)$	+270
water(g)	$H_2O(g)$	+188.8
water(l)	$H_2O(l)$	+70.0

Using standard entropies

The change in standard entropy for a chemical reaction (ΔS°) can be calculated by determining the difference between the total entropy of the products and the total entropy of the reactants. This may be expressed mathematically as:

$$\Delta S^\circ = \Sigma S^\circ(\text{products}) - \Sigma S^\circ(\text{reactants})$$

The following Worked example shows you how to calculate the entropy change for a particular chemical reaction.

Worked example 16.2.1

DETERMINING THE STANDARD ENTROPY CHANGE OF A CHEMICAL REACTION FROM STANDARD ENTROPIES

Determine the entropy change of the following reaction:

$$C(s) + O_2(g) \rightarrow CO_2(g)$$

given the following standard entropies:

$$S^\circ(C) = +158.2\,J\,mol^{-1}\,K^{-1}$$
$$S^\circ(O_2) = +205.0\,J\,mol^{-1}\,K^{-1}$$
$$S^\circ(CO_2) = +213.8\,J\,mol^{-1}\,K^{-1}$$

Thinking	Working
Calculate the total entropy of the products, ΣS°(products). As S° is given per mol, multiply the S° of each product by its coefficient in the overall equation.	$\Sigma S^\circ(\text{products}) = 1 \times S^\circ(CO_2)$ $= 1 \times +213.8$ $= +213.8\,J\,mol^{-1}\,K^{-1}$
Calculate the total entropy of the reactants, ΣS°(reactants). As S° is given per mol, multiply the S° of each reactant by its coefficient in the overall equation.	$\Sigma S^\circ(\text{reactants}) = 1 \times S^\circ(C) + 1 \times S^\circ(O_2)$ $= (1 \times +158.2) + (1 \times +205.0)$ $= +363.2\,J\,mol^{-1}\,K^{-1}$
Calculate the entropy change of the reaction, ΔS°, using the expression: $\Delta S^\circ = \Sigma S^\circ(\text{products}) - \Sigma S^\circ(\text{reactants})$	$\Delta S^\circ = \Sigma S^\circ(\text{products}) - \Sigma S^\circ(\text{reactants})$ $= +213.8 - (+363.2)$ $= -149.4\,J\,mol^{-1}\,K^{-1}$

Worked example: Try yourself 16.2.1

DETERMINING THE STANDARD ENTROPY CHANGE OF A CHEMICAL REACTION FROM STANDARD ENTROPIES

Determine the entropy change of the following reaction:

$$H_2(g) + I_2(s) \rightarrow 2HI(g)$$

given the following standard entropies:

$$S^\circ(H_2) = +130.6\,J\,mol^{-1}\,K^{-1}$$
$$S^\circ(I_2) = +116.7\,J\,mol^{-1}\,K^{-1}$$
$$S^\circ(HI) = +206.3\,J\,mol^{-1}\,K^{-1}$$

16.2 Review

SUMMARY

- The second law of thermodynamics states that the overall entropy of the universe is increasing.
- During a spontaneous process, the overall entropy of the universe increases.
- The entropy change for a chemical reaction (ΔS°) can be determined from the following expression:
  $\Delta S^\circ = \Sigma S^\circ(\text{products}) - \Sigma S^\circ(\text{reactants})$
- Non-spontaneous reactions can be forced to occur if they are combined, or coupled, with a spontaneous process, provided that the second law of thermodynamics is obeyed overall.

KEY QUESTIONS

1 Calculate the standard entropy change (ΔS°) for the following reaction at 298 K:
$S(s) + O_2(g) \rightarrow SO_2(g)$
given the following standard entropies:

	S° ($J\,mol^{-1}\,K^{-1}$)
S	+32
O_2	+205
SO_2	+257

2 For the following reactions, explain why the reaction is spontaneous:

a $Fe_2O_3(s) + 2Al(s) \rightarrow 2Fe(s) + Al_2O_3(s)$
$\Delta H = -815.5\,kJ\,mol^{-1}$ (the 'thermite reaction')

b $2H_2(g) + O_2(g) \rightarrow 2H_2O(g)$
$\Delta H = -483.6\,kJ\,mol^{-1}$

c $NH_4NO_3(s) \rightarrow NH_4^+(aq) + NO_3^-(aq)$
$\Delta H = +25.7\,kJ\,mol^{-1}$

3 **a** Calculate the entropy change for the following reaction:
$Mg(s) + 2HCl(g) \rightarrow MgCl_2(s) + H_2(g)$
given the following standard entropies:

	S° ($J\,mol^{-1}\,K^{-1}$)
Mg(s)	+32.7
HCl(g)	+186.9
$MgCl_2(s)$	+89.6
$H_2(g)$	+130.6

b Based on the chemical equation above, would you predict an increase or decrease in entropy for this reaction? Does your prediction match the results of your calculation in part **a**?

4 Nitrogen (N_2) reacts with oxygen (O_2) to form nitric oxide (NO) according to the following equation:
$N_2(g) + O_2(g) \rightarrow 2NO(g)$
The standard entropy change (ΔS°) for this reaction is $+25\,J\,mol^{-1}\,K^{-1}$.
Calculate the standard entropy of NO, S°(NO), given the following standard entropies:

	S° ($J\,mol^{-1}\,K^{-1}$)
$N_2(g)$	+191.6
$O_2(g)$	+205

16.3 Gibbs free energy

In the previous section, you learnt that for a reaction to be spontaneous, the entropy of the universe must increase. While most spontaneous reactions are exothermic, there are also some endothermic reactions that are spontaneous. The change in enthalpy and the change in entropy both play a role in determining whether a reaction proceeds under natural conditions.

In this section, you will be introduced to a quantity called Gibbs free energy, which is used to determine whether a particular reaction will be spontaneous. Temperature can also affect the spontaneity of a reaction, and you will learn how to predict the effect of temperature on individual reactions.

GIBBS FREE ENERGY AND CHEMICAL REACTIONS

Scientists have defined a quantity called **Gibbs free energy** (also known simply as 'free energy'), ΔG, which can be used to determine whether a process (such as a chemical reaction) is spontaneous. The change in Gibbs free energy is defined as:

$$\Delta G = \Delta H - T\Delta S$$

For a reaction to be spontaneous, ΔG must be negative.

If the enthalpy and entropy changes are measured under **standard state conditions** of 25°C and 1 bar pressure, the Gibbs free energy change formula can be written as:

$$\Delta G^\circ = \Delta H^\circ - T\Delta S^\circ$$

Very simply, the change in Gibbs free energy is a measure of the effects of the enthalpy change of a reaction, ΔH, and the effects of temperature and entropy change, $T\Delta S$, on the spontaneity of a reaction. ΔG can have a value that is positive, zero or negative.

- If $\Delta G < 0$, the reaction is spontaneous.
- If $\Delta G = 0$, the reaction is neither spontaneous nor **non-spontaneous**. (The reaction is described as being in equilibrium. Equilibrium is discussed in Module 5.)
- If $\Delta G > 0$, the reaction is non-spontaneous.

The units of ΔG and ΔH are usually given in kJ mol^{-1}, whereas the units of ΔS are usually given in J mol^{-1} K^{-1}. When doing calculations involving $\Delta G = \Delta H - T\Delta S$, you must divide the values of ΔS by 1000 to convert to kJ mol^{-1} K^{-1}. Importantly, temperature, T, is measured in kelvin (K), not degrees Celsius (°C). To convert from °C to K, add 273.

DETERMINING WHETHER A REACTION IS SPONTANEOUS OR NON-SPONTANEOUS

The change in Gibbs free energy of a chemical reaction depends on the terms ΔH and $T\Delta S$. The enthalpy change can be positive or negative. Since the entropy change can be positive or negative, and temperature is always positive, the $T\Delta S$ term in the Gibbs free energy formula will have the sign of the entropy change. The spontaneity of a reaction will depend on the sign and magnitude of the two terms. Table 16.3.1 lists the conditions under which a reaction is spontaneous.

TABLE 16.3.1 Conditions under which a reaction is spontaneous

Spontaneous ($\Delta G < 0$)?	when $\Delta H < 0$	when $\Delta H > 0$
when $\Delta S > 0$	**spontaneous at all T** As $\Delta H < 0$ and $T\Delta S > 0$ so $\Delta H - T\Delta S < 0$ always	**spontaneous at high T** As $\Delta H < T\Delta S$ so $\Delta H - T\Delta S < 0$ if T large
when $\Delta S < 0$	**spontaneous at low T** As $\Delta H > T\Delta S$ so $\Delta H - T\Delta S < 0$ if T small	**not spontaneous at any T** As $\Delta H > 0$ and $T\Delta S < 0$ so $\Delta H - T\Delta S > 0$ always

Note that exothermic reactions that increase entropy are always spontaneous, whereas endothermic reactions that decrease entropy are never spontaneous.

Worked example 16.3.1 shows you how to determine the spontaneity of a particular reaction at a given temperature. In these calculations, it is assumed that ΔH° and ΔS° do not change with temperature. In fact, they are slightly temperature dependent, but the answer obtained from a calculation with this assumption is a reasonable approximation.

Worked example 16.3.1

CALCULATING THE SPONTANEITY OF A REACTION

Ice melting is a change of state in which solid water melts to liquid water.

The reaction is:

$H_2O(s) \rightarrow H_2O(l)$

For this reaction, $\Delta H^\circ = +6.01\ kJ\ mol^{-1}$ and $\Delta S^\circ = +22.1\ J\ mol^{-1}\ K^{-1}$. Is this reaction spontaneous at 40°C?

Thinking	Working
State the known values.	$\Delta H^\circ = +6.01\ kJ\ mol^{-1}$ $\Delta S^\circ = +22.1\ J\ mol^{-1}\ K^{-1}$ $T = 40°C$
Convert ΔS° to $kJ\ mol^{-1}\ K^{-1}$ by dividing by 1000.	$\Delta S^\circ = +22.1\ J\ mol^{-1}\ K^{-1}$ $= +0.0221\ kJ\ mol^{-1}\ K^{-1}$
Convert T to kelvin by adding 273.	$T = 40°C$ $= 40 + 273$ $= 313\ K$
Substitute the values into the Gibbs free energy change formula.	$\Delta G^\circ = \Delta H^\circ - T\Delta S^\circ$ $= +6.01 - (313 \times +0.0221)$ $= 6.01 - 6.92$ $= -0.91\ kJ\ mol^{-1}$
Answer the question.	As $\Delta G^\circ = -0.91\ kJ\ mol^{-1}$, ice will spontaneously melt at 40°C.
Check your answer using the values of ΔH° and ΔS° and Table 16.3.1 (page 515).	As both ΔH° and ΔS° are positive, Table 16.3.1 states that this reaction is spontaneous at relatively high T. This agrees with the calculated answer.

Worked example: Try yourself 16.3.1

CALCULATING THE SPONTANEITY OF A REACTION

Ice melting is a change of state in which solid water melts to liquid water.

The reaction is

$$H_2O(s) \rightarrow H_2O(l)$$

For this reaction, $\Delta H^\circ = +6.01\ kJ\ mol^{-1}$ and $\Delta S^\circ = +22.1\ J\ mol^{-1}\ K^{-1}$. Is this reaction spontaneous at −40°C?

It is important to note that even though the sign of ΔG indicates whether a reaction is spontaneous or not, it does not give any indication of how fast a reaction will be. Some spontaneous reactions are extremely slow. For example, the reaction between methane and oxygen at room temperature has a large negative value of ΔG and the reaction is therefore highly spontaneous. Nevertheless, the rate of the reaction is almost immeasurably slow at that temperature.

CHEMFILE N

Diamonds and graphite

Diamonds and graphite are both forms of carbon (Figure 16.3.1). It is often said that diamonds are forever, but is this true?

Diamond can be converted to another allotrope of carbon, graphite. The reaction may be represented as:

$$C_{diamond}(s) \rightarrow C_{graphite}(s)$$

In this process, the carbon remains solid, but the structure of the covalent lattice changes. The standard enthalpy change of the reaction is $-1.9\,kJ\,mol^{-1}$, and the entropy change is $+3.4\,J\,mol^{-1}\,K^{-1}$. From the formula $\Delta G^\circ = \Delta H^\circ - T\Delta S^\circ$, you can see that, since $\Delta H^\circ < 0$ and $\Delta S^\circ > 0$, the reaction is spontaneous, and it would be expected that all diamonds will eventually convert to graphite. More precisely, the Gibbs free energy change under standard state conditions (ΔG°) is $-2.9\,kJ\,mol^{-1}$.

So why are diamonds not seen to spontaneously change into graphite? Although diamond is slightly less stable than graphite, there is a very large activation energy barrier to be overcome. At standard state conditions, 1 g of diamond changes to 1 g of graphite in about 10^{80} s, or 3.0×10^{72} years. Our Sun has existed for about 4.6×10^9 years, and the universe for about 1.4×10^{10} years. Under standard state conditions, diamonds effectively last forever.

FIGURE 16.3.1 Diamond, the well-known allotrope of carbon, is not as stable as graphite. Thermodynamics predicts that diamonds will eventually become graphite, but the rate of conversion is negligible under standard state conditions.

Heating diamonds increases the rate of the conversion of diamond into graphite. At 1200°C—a very hot temperature, but achievable—diamond is converted into graphite within a few days. The first scientist to try this, proving that diamond is made of carbon, was Antoine Lavoisier (1743–1794), who also discovered the role of oxygen in combustion and helped develop the metric system. During the French Revolution, Lavoisier was beheaded following charges of aristocratic extravagance, which cited his experiment of burning a diamond!

Calculating the temperature at which a reaction becomes spontaneous

For some reactions, if both ΔS° and ΔH° have the same sign, temperature determines whether the reaction will be spontaneous. The Gibbs free energy change formula, $\Delta G^\circ = \Delta H^\circ - T\Delta S^\circ$, can be used to determine the minimum temperature at which these reactions switch from being non-spontaneous to spontaneous.

For example, the Gibbs free energy change formula can be used to determine the temperature at which ice will spontaneously melt. For the process $H_2O(s) \rightarrow H_2O(l)$, $\Delta H^\circ = +6.01\,kJ\,mol^{-1}$ and $\Delta S^\circ = +22.1\,J\,mol^{-1}\,K^{-1}$. Substituting $\Delta G^\circ = 0$ (when the reaction is neither spontaneous nor non-spontaneous) in the Gibbs free energy change formula gives:

$$\Delta G^\circ = \Delta H^\circ - T\Delta S^\circ$$

$$0 = +6.01 - T \times (+0.0221)$$

Solving for T gives:

$$T = \frac{6.01}{0.0221}$$

$$= 273\,K$$

A temperature of 273 K is 0°C and, unsurprisingly, this is the melting point of water. Above 273 K, ΔG° will be negative and the process (solid water becoming liquid water) will be spontaneous (Figure 16.3.2). Below 273 K, ΔG° will be positive and the reverse process (liquid water becoming solid water) will be spontaneous.

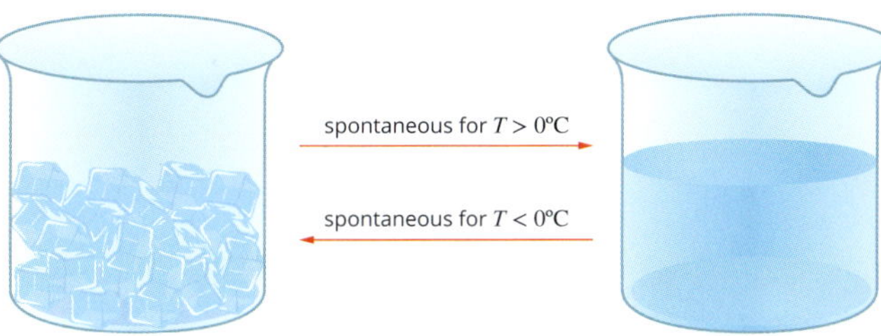

FIGURE 16.3.2 Ice melts spontaneously at temperatures above 0°C. Water freezes spontaneously below 0°C. At 0°C, ice and water are described as being in equilibrium, and no spontaneous process occurs.

Worked example 16.3.2

CALCULATING THE TEMPERATURE AT WHICH A REACTION BECOMES SPONTANEOUS

The Haber process is used to manufacture ammonia. The equation for the reaction is:

$$N_2(g) + 3H_2(g) \rightarrow 2NH_3(g)$$

For this reaction, $\Delta H^\circ = -91.8\,kJ\,mol^{-1}$ and $\Delta S^\circ = -198\,J\,mol^{-1}\,K^{-1}$. Calculate at what temperature this reaction becomes spontaneous.

Thinking	Working
State the known variables.	$\Delta H^\circ = -91.8\,kJ\,mol^{-1}$ $\Delta S^\circ = -198\,J\,mol^{-1}\,K^{-1}$ $\Delta G^\circ = 0$ $T = ?$
Convert ΔS° to $kJ\,mol^{-1}\,K^{-1}$ by dividing by 1000.	$\Delta S = -198\,J\,mol^{-1}\,K^{-1}$ $= -0.198\,kJ\,mol^{-1}\,K^{-1}$
Substitute the values into the Gibbs free energy change formula.	$\Delta G^\circ = \Delta H^\circ - T\Delta S^\circ$ $0 = -91.8 - T \times (-0.198)$ Solving for T: $T = \frac{-91.8}{-0.198}$ $= 464\,K$
Answer the question.	The Haber process will become spontaneous at 464 K or 191°C.
Check your answer using the values of ΔH° and ΔS° and Table 16.3.1 (page 515).	As both the ΔH° and ΔS° are negative, Table 16.3.1 states that this reaction is temperature dependent and favoured at a relatively low T.

Worked example: Try yourself 16.3.2

CALCULATING THE TEMPERATURE AT WHICH A REACTION BECOMES SPONTANEOUS

Ammonia reacts with hydrogen chloride according to the equation:

$$NH_3(g) + HCl(g) \rightarrow NH_4Cl(s)$$

For this reaction, $\Delta H = -176.7\,kJ\,mol^{-1}$ and $\Delta S = -284.8\,J\,mol^{-1}\,K^{-1}$. Calculate the highest temperature at which the reaction is spontaneous.

16.3 Review

SUMMARY

- The change in Gibbs free energy is defined as:
  $$\Delta G = \Delta H - T\Delta S$$
  At standard state conditions, this is written as:
  $$\Delta G^\circ = \Delta H^\circ - T\Delta S^\circ$$
- The change in Gibbs free energy of a reaction indicates whether the reaction is spontaneous ($\Delta G^\circ < 0$) or non-spontaneous ($\Delta G^\circ > 0$).
- The value of the Gibbs free energy change of a reaction is dependent on temperature, and changes in temperature may determine whether a reaction is spontaneous or not.
- Exothermic reactions that increase entropy are always spontaneous; endothermic reactions that decrease entropy are never spontaneous.
- The temperature at which a spontaneous reaction becomes non-spontaneous, or vice versa, can be calculated from the Gibbs free energy change formula.

KEY QUESTIONS

1 A spontaneous chemical reaction is best described as:
   A a reaction that proceeds at a fast rate
   B a reaction in which ΔG is negative
   C a reaction with an increase in entropy
   D a reaction with a decrease in entropy

2 What values of ΔH and ΔS for a reaction result in a process that is always spontaneous?

3 The biologically important ATP–ADP reaction, which occurs within the mitochondria of plant and animal cells, has a Gibbs free energy change at standard state conditions of –31 kJ mol^{-1}. In a human cell, however, the Gibbs free energy change is nearly –57 kJ mol^{-1}. Why are these values different for the same reaction?

4 For the following reaction:

$$Mg(s) + \tfrac{1}{2}O_2(g) \rightarrow MgO(s)$$

determine the change in entropy in the reaction from the following standard entropies:

	S° (J mol^{-1} K^{-1})
Mg(s)	+32.7
O_2(g)	+205.5
MgO(s)	26.9

5 For the following reaction:

$$Mg(s) + 2HCl(g) \rightarrow MgCl_2(s) + H_2(g)$$

determine whether it is spontaneous under standard state conditions. Use the following thermodynamic data.

	ΔH°_f (kJ mol^{-1})	S° (J mol^{-1} K^{-1})
Mg(s)	0	+32.7
HCl(g)	–92.3	+186.9
$MgCl_2$(s)	–641.6	+89.6
H_2(g)	0	+130.7

Chapter review

16

KEY TERMS

closed system
enthalpy
entropy
Gibbs free energy
isolated system
open system
non-spontaneous
spontaneous reaction
standard state conditions
standard entropy
surroundings
system
thermodynamics

REVIEW QUESTIONS

1 How does a closed system differ from an open system?

2 In which of the following scenarios will there be an increase in entropy?
- **A** A molecule breaks into two smaller molecules.
- **B** A reaction results in an increase in the number of molecules of gas.
- **C** A liquid becomes a solid.
- **D** A solid becomes a gas.
- **E** A gas becomes a solid.

3 Predict whether there will be an increase or decrease in entropy for each of these systems:
- **a** $H_2O(g) \rightarrow H_2O(l)$
- **b** $N_2(g) + 3H_2(g) \rightarrow 2NH_3(g)$
- **c** $Na_2CO_3(s) \rightarrow Na_2O(s) + CO_2(g)$

4 Calculate the standard entropy change for the following reactions involving phosphine (PH_3), using the following standard entropies:
$S^\circ(P(l)) = +43.0\,J\,mol^{-1}\,K^{-1}$
$S^\circ(P_2O_5(l)) = +114.4\,J\,mol^{-1}\,K^{-1}$
$S^\circ(H_2(g)) = +130.7\,J\,mol^{-1}\,K^{-1}$
$S^\circ(H_2O(g)) = +188.8\,J\,mol^{-1}\,K^{-1}$
$S^\circ(O_2(g)) = +205.0\,J\,mol^{-1}\,K^{-1}$
$S^\circ(PH_3(g)) = +210.0\,J\,mol^{-1}\,K^{-1}$
- **a** $P(l) + \frac{3}{2}H_2(g) \rightarrow PH_3(g)$
- **b** $2PH_3(g) + 4O_2(g) \rightarrow P_2O_5(l) + 3H_2O(g)$

5 Why is the decomposition of hydrogen peroxide spontaneous?
$2H_2O_2(l) \rightarrow 2H_2O(l) + O_2(g)\ \Delta H = -98.0\,kJ\,mol^{-1}$

6 In the presence of a flame, 1 mol of liquid petrol burns to produce 8 mol of carbon dioxide and 9 mol of water. Is the reverse reaction spontaneous?

7 If a chemical reaction is spontaneous, what will happen to the total entropy of the system and its surroundings (i.e. to the total entropy of the universe) if:
- **a** the reaction is exothermic
- **b** the reaction is endothermic.

8 An exothermic reaction has a ΔS_{system} that is positive.
- **a** Do the reactants or products have a greater degree of disorder?
- **b** Explain whether the reaction is spontaneous or not.

9 A solid melts and becomes a liquid when heated.
- **a** Is the melting process endothermic or exothermic?
- **b** What would be the sign of the change in entropy for this system?
- **c** Explain whether or not this is a spontaneous process.

10
- **a** Write an equation for the complete combustion of liquid pentane (C_5H_{12}) to produce carbon dioxide and liquid water.
- **b** Determine the standard entropy of liquid pentane. Use the following data:
$\Delta S^\circ = -414.5\,J\,mol^{-1}\,K^{-1}$
$S^\circ(CO_2(g)) = +213.8\,J\,mol^{-1}\,K^{-1}$
$S^\circ(H_2O(l)) = +70.0\,J\,mol^{-1}\,K^{-1}$
$S^\circ(O_2(g)) = +205\,J\,mol^{-1}\,K^{-1}$

11 For the reaction
$A + B \rightarrow C + D$
$\Delta H^\circ = +30\,kJ\,mol^{-1}$ and $\Delta S^\circ = +50\,J\,mol^{-1}\,K^{-1}$
which one of the following statements is true?
- **A** The reaction is spontaneous at all temperatures.
- **B** The reaction is spontaneous at temperatures above 600 K.
- **C** The reaction is spontaneous at temperatures below 600 K.
- **D** The reaction is non-spontaneous at all temperatures.

12 Suppose a reaction has $\Delta H^\circ = -25\,kJ\,mol^{-1}$ and $\Delta S^\circ = -58\,J\,mol^{-1}\,K^{-1}$. At what temperature will it change from being non-spontaneous to spontaneous?

13 Estimate the boiling point of bromine in °C. The process can be represented by the equation:
$Br_2(l) \rightarrow Br_2(g)$
Use the following data:
$\Delta H^\circ = +30.9\,kJ\,mol^{-1}$
$\Delta S^\circ = +93.0\,J\,mol^{-1}\,K^{-1}$

14 For the reaction:

$$SBr_4(g) \rightarrow S(g) + 2Br_2(l)$$

calculate the change in standard Gibbs free energy.
Use the following data:
$\Delta H^\circ = +115\,kJ\,mol^{-1}$
$\Delta S^\circ = +125\,J\,mol^{-1}\,K^{-1}$

15 **a** Calculate ΔG° for the formation of 1 mol of liquid carbon tetrachloride (CCl_4) from elemental carbon and chlorine. The standard enthalpy and entropy changes of the reaction are $-128\,kJ\,mol^{-1}$ and $+205\,J\,mol^{-1}\,K^{-1}$, respectively.

b Calculate ΔG° for the formation of 1 mol of gaseous carbon tetrachloride (CCl_4) from elemental carbon and chlorine. The standard enthalpy and entropy changes of the reaction are $-100\,kJ\,mol^{-1}$ and $+309\,J\,mol^{-1}\,K^{-1}$, respectively.

c Is the formation of liquid or gaseous carbon tetrachloride spontaneous? Which is the preferred reaction?

16 For the preparation of methanol from its component elements, $\Delta H^\circ = -238\,kJ\,mol^{-1}$ and $\Delta S^\circ = +127\,J\,mol^{-1}\,K^{-1}$. Assuming these values are temperature independent, is the formation of methanol spontaneous at 10°C, 25°C or 70°C?

17 Calculate ΔG° for the reaction:

$$H_2(g) + O_2(g) \rightarrow H_2O_2(l)$$

Is the reaction spontaneous?
Use the following data:
$\Delta H^\circ = -187.8\,kJ\,mol^{-1}$
$S^\circ(H_2O_2(l)) = +109.6\,J\,mol^{-1}\,K^{-1}$
$S^\circ(O_2(g)) = +205.2\,J\,mol^{-1}\,K^{-1}$
$S^\circ(H_2(g)) = +130.7\,J\,mol^{-1}\,K^{-1}$

18 Calculate ΔG° for the reaction at standard conditions:

$$2H_2(g) + N_2(g) \rightarrow N_2H_4(l)$$

Is the reaction spontaneous?
Use the following data:
$\Delta H^\circ = +50.6\,kJ\,mol^{-1}$
$S^\circ(N_2H_4(l)) = +121.2\,J\,mol^{-1}\,K^{-1}$
$S^\circ(N_2(g)) = +191.6\,J\,mol^{-1}\,K^{-1}$
$S^\circ(H_2(g)) = +130.7\,J\,mol^{-1}\,K^{-1}$

19 Buta-1,3-diene (C_4H_6) can be hydrogenated to form butane (C_4H_{10}) according to the following equation:

$$C_4H_6(g) + 2H_2(g) \rightarrow C_4H_{10}(g)$$

Determine the entropy change for this reaction.
Use the following data:
$\Delta G^\circ = -168\,kJ\,mol^{-1}$
ΔH°_f of $(C_4H_6(g)) = +110\,kJ\,mol^{-1}$
ΔH°_f of $(C_4H_{10}(g)) = -126\,kJ\,mol^{-1}$

20 Rubber is made of long, kinked molecular strands. In repeatedly stretching a rubber band, the rubber band feels warmer. Does this reflect an increase or decrease in entropy in the rubber band?
(Hints: Stretching is a non-spontaneous process; refer to the formula $\Delta G = \Delta H - T\Delta S$.)

21 Reflect on the Inquiry activity on page 502. Using concepts from this chapter, answer the following questions.

a Is the self-assembly of drinking straws a spontaneous or non-spontaneous process? What is the sign of ΔG for the process?

b Does the entropy increase or decrease? What is the sign of ΔS for the process?

c Is ΔH for the process positive or negative? Is the process endothermic or exothermic?

d Use the Gibbs free energy change formula to predict the effect of raising the temperature of the water on the self-assembly of the straws.

e Compare the drinking straw activity with the process of crystallisation. What can you say about the spontaneity and energy changes that occur as crystals form in a solution?

MODULE 4 • REVIEW

REVIEW QUESTIONS

Drivers of reactions

Multiple choice

1 What is 3.7×10^{-2} J in MJ?

A 37 MJ
B 3.7×10^{-2} MJ
C 3.7×10^{-5} MJ
D 3.7×10^{-8} MJ

2 An endothermic reaction is one that:

A has a positive ΔH and heat is released to the surrounding environment
B has a positive ΔH and heat is absorbed from the surrounding environment
C has a negative ΔH and heat is released to the surrounding environment
D has a negative ΔH and heat is absorbed from the surrounding environment

3 The heat of combustion of propan-1-ol (C_3H_7OH) ($M = 60\,g\,mol^{-1}$) is $-2021\,kJ\,mol^{-1}$. How much heat is released when 16.0 g of propan-1-ol is combusted in excess oxygen?

A 539 kJ
B 7578 kJ
C 3.23×10^4 kJ
D 1.94×10^6 kJ

4 How much heat energy, in kJ, is required to heat 250.0 g of water from 23.0°C to 80.0°C?

A 3.41
B 83 600
C 59.6
D 83.6

The following information relates to questions 5–7.

As a response to rising petrol prices, the Australian Government in 2006 offered car owners a $2000 incentive to convert their cars from petrol to LPG. Propane is one of the gases in the mixture sold as LPG. The thermochemical equation for the complete combustion of propane is:

$C_3H_8(g) + 5O_2(g) \rightarrow 3CO_2(g) + 4H_2O(l)\ \ \Delta H = -2220\,kJ\,mol^{-1}$

5 Which energy profile diagram best represents the energy changes that take place during this reaction?

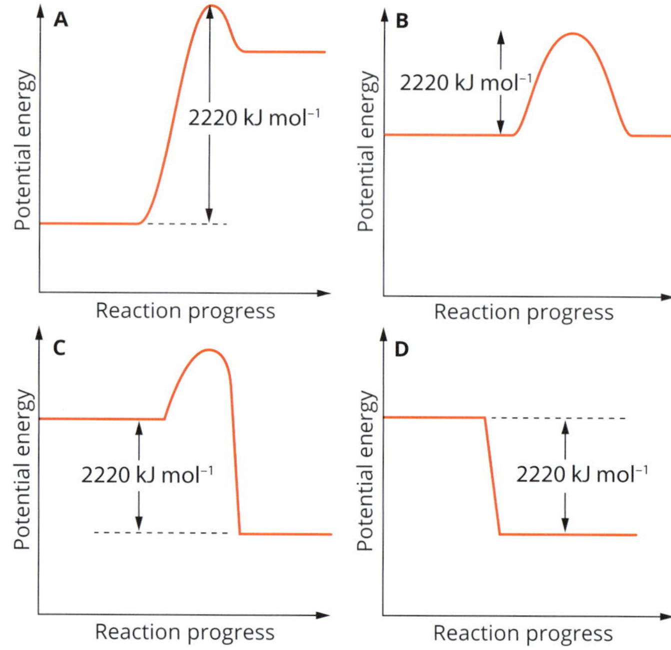

6 The mass of 1 mol of propane is (44.09 g). When 100 g of propane undergoes complete combustion:

A 5.04×10^3 kJ of energy is absorbed
B 5.04×10^3 kJ of energy is released
C 3.92×10^1 kJ of energy is absorbed
D 3.92×10^1 kJ of energy is released

7 When 1 MJ of energy is released in this reaction, the volume of $CO_2(g)$ produced, at standard laboratory conditions (SLC), is:

A 3.38×10^4 L
B 33.8 L
C 1.12×10^4 L
D 11.2 L

8 Which of these equations represents the reaction for the enthalpy of formation (ΔH°_f) of methanol ($CH_3OH(l)$)?

A $C(g) + 4H(g) + O(g) \rightarrow CH_3OH(l)$
B $2H_2(g) + C(s) + \frac{1}{2}O_2(g) \rightarrow CH_3OH(l)$
C $CH_3OH(l) + \frac{3}{2}O_2(g) \rightarrow CO_2(g) + 2H_2O(l)$
D $CH_3Cl(g) + KOH(aq) \rightarrow CH_3OH(l) + KCl(aq)$

9 Which of these reactions could be used to calculate the average bond energy of C–H in methane?

A $CH_4(g) \rightarrow CH_4(l)$
B $CH_4(g) \rightarrow C(g) + 4H(g)$
C $CH_4(g) \rightarrow 2H_2(g) + C(s)$
D $CH_4(g) + O_2(g) \rightarrow CO_2(g) + H_2O(l)$

10 Consider the following bond energy data:

Bond	Bond energy ($kJ\,mol^{-1}$)
H–H	436
O=O	498
O–H	463

Using the bond energy data, calculate the enthalpy change, in $kJ\,mol^{-1}$, for this reaction.

$$O_2(g) + 2H_2(g) \rightarrow 2H_2O(g)$$

A −444
B +444
C −482
D +482

11 Use the following energy cycle to determine the value for the enthalpy of formation (ΔH°_f) for methane.

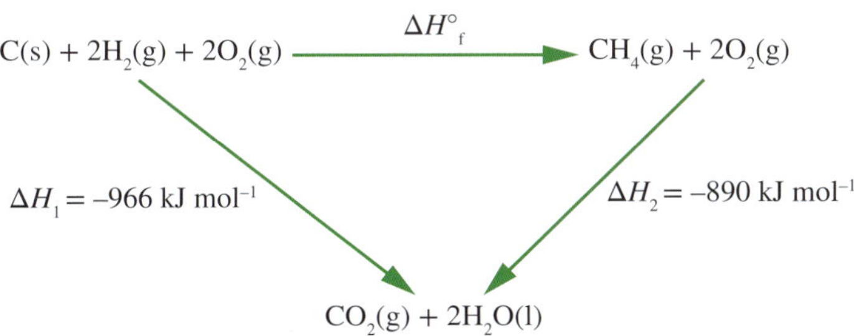

A $-76\,kJ\,mol^{-1}$
B $-890\,kJ\,mol^{-1}$
C $-966\,kJ\,mol^{-1}$
D $-1856\,kJ\,mol^{-1}$

12 Using the following energy level diagram, calculate the enthalpy change for the reaction:

$$C_2H_6(g) \rightarrow 3C(s) + 3H_2(g)$$

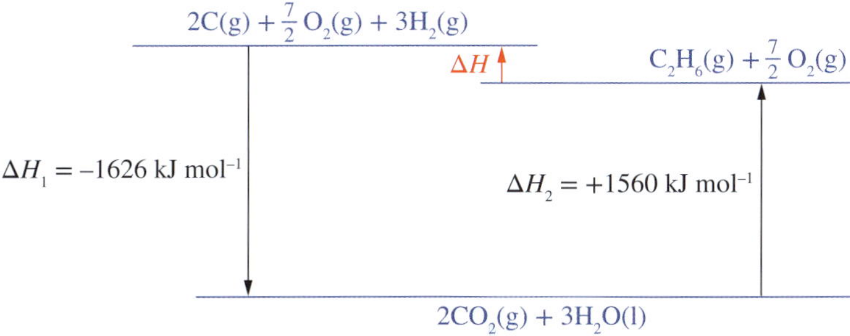

A $-86\,kJ\,mol^{-1}$
B $+86\,kJ\,mol^{-1}$
C $+1560\,kJ\,mol^{-1}$
D $-1626\,kJ\,mol^{-1}$

13 Calculate the enthalpy change for the following reaction:

$NO_2(g) + \frac{7}{2}H_2(g) \rightarrow 2H_2O(l) + NH_3(g)$

given:

$NH_3(g) \rightarrow \frac{1}{2}N_2(g) + \frac{3}{2}H_2(g)$ $\quad \Delta H^\circ = +46\,kJ\,mol^{-1}$

$\frac{1}{2}N_2(g) + 2H_2O(l) \rightarrow NO_2(g) + 2H_2(g)$ $\quad \Delta H^\circ = +170\,kJ\,mol^{-1}$

A $+124\,kJ\,mol^{-1}$
B $-124\,kJ\,mol^{-1}$
C $+216\,kJ\,mol^{-1}$
D $-216\,kJ\,mol^{-1}$

14 Hydrogen gas can be made through the nickel-catalysed process in which methane reacts with steam according to the following equation:

$CH_4(g) + H_2O(g) \rightarrow CO(g) + 3H_2(g)$ $\quad \Delta H^\circ = +205.3\,kJ\,mol^{-1}$

Calculate the standard enthalpy of formation of $CH_4(g)$, given:

$\Delta H^\circ_f(CO(g)) = -110.5\,kJ\,mol^{-1}$

$\Delta H^\circ_f(H_2O(g)) = -241.8\,kJ\,mol^{-1}$

A −205.3 − 110.5 + 241.8
B −205.3 + 110.5 − 241.8
C +205.3 − 110.5 + 241.8
D +205.3 − 110.5 + 241.8

15 Matter in a/an ________ state has a ________ entropy than matter in a/an ________ state.

A gaseous; lower; solid
B liquid; greater; gaseous
C solid: greater; aqueous
D aqueous; lower; gaseous

16 Which of the following reactions is associated with a decrease in entropy?

A $H_2O(s) \rightarrow H_2O(l)$
B $2C_6H_{14}(l) + 19O_2(g) \rightarrow 12CO_2(g) + 14H_2O(l)$
C $CO(g) \rightarrow CO(s)$
D $CuCO_3(s) \rightarrow CuO(s) + CO_2(g)$

17 What are the signs for the entropy changes associated with this process?

$$CO_2(s) \rightarrow CO_2(g)$$

	$\Delta S_{surroundings}$	ΔS_{system}
A	+	−
B	+	+
C	−	−
D	−	+

18 For a process in which $\Delta S_{system} = -50.1\ J\ mol^{-1}\ K^{-1}$ and $\Delta H_{system} = +23.9\ kJ\ mol^{-1}$, what is the temperature at which it is spontaneous?

A 50°C
B 150°C
C 250°C
D at no temperature

19 What is the change in entropy, $\Delta S°$, for the reaction:

$NO_2(g) + \frac{7}{2}H_2(g) \rightarrow 2H_2O(l) + NH_3(g)$

given these standard entropies?

$S°(NO_2) = 240.0\ J\ mol^{-1}\ K^{-1}$
$S°(H_2) = 130.6\ J\ mol^{-1}\ K^{-1}$
$S°(H_2O) = 70.0\ J\ mol^{-1}\ K^{-1}$
$S°(NH_3) = 192.3\ J\ mol^{-}\ K^{-1}$

A $-364.8\ J\ mol^{-1}\ K^{-1}$
B $-108.3\ J\ mol^{-1}\ K^{-1}$
C $332.3\ J\ mol^{-1}\ K^{-1}$
D $364.8\ J\ mol^{-1}\ K^{-1}$

20 Consider the following reaction:

$$\frac{1}{2}A_2 + 2B_2 + C \rightarrow CAB_4$$

The change in enthalpy is $-104.0\ kJ\ mol^{-1}$ and the change in entropy is $-60.8\ J\ mol^{-1}\ K^{-1}$. What is the Gibbs free energy change of the reaction at 30°C? Comment on whether the reaction is spontaneous or not at this temperature.

A $-85.6\ kJ\ mol^{-1}$, spontaneous
B $-18.3\ kJ\ mol^{-1}$, not spontaneous
C $18.3\ kJ\ mol^{-1}$, spontaneous
D $85.6\ kJ\ mol^{-1}$, not spontaneous

Short answer

1 The reaction that occurs in one type of fire extinguisher can be best represented by the following thermochemical equation:

$$Al_2(SO_4)_3(s) + 6NaHCO_3(s) \rightarrow 2Al(OH)_3(s) + 3Na_2SO_4(s) + 6CO_2(s) \qquad \Delta H = +129.2\ kJ\ mol^{-1}$$

a Is this reaction endothermic or exothermic?

b What would be the new value of ΔH if the equation was now written as follows?

$$3Al_2(SO_4)_3(s) + 18NaHCO_3(s) \rightarrow 6Al(OH)_3(s) + 9Na_2SO_4(s) + 18CO_2(s)$$

c How much energy would be released/absorbed if 30.0 g of $Al_2(SO_4)_3$ ($M = 342.17\ g\ mol^{-1}$) reacted to completion?

d Explain why changing states in the chemical equation will affect the ΔH of the reaction.

2 Calculate the energy needed to heat:

a 50.0 mL of water from 15.0°C to 70.0°C
b 700 mL of water from 32.0°C to 98.0°C
c 2.6 kg of water from 8.0°C to 10.0°C
d 1.00 kg of nitrogen gas from 175°C to 200°C ($c(N_2(g)) = 1.039\ J\ g^{-1}\ K^{-1}$)
e 270 g of a pure copper rod from 26.0°C to 54.0°C (c(copper) $= 0.385\ J\ g^{-1}\ K^{-1}$)

3 Consider the following thermochemical equations involving the formation of nitrogen oxides:

$N_2(g) + O_2(g) \rightarrow 2NO(g) \qquad \Delta H = +181\ kJ\ mol^{-1}$
$2NO(g) + O_2(g) \rightarrow 2NO_2(g) \qquad \Delta H = -114\ kJ\ mol^{-1}$

Use this information to calculate ΔH for each of the following equations:

a $2N_2(g) + 2O_2(g) \rightarrow 4NO(g)$
b $2NO(g) \rightarrow N_2(g) + O_2(g)$
c $NO_2(g) \rightarrow NO(g) + \frac{1}{2}O_2(g)$
d $N_2(g) + 2O_2(g) \rightarrow 2NO_2(g)$

4 Calculate the enthalpy of formation of acetone ($C_3H_6O(l)$), given:

$C_3H_6O(l) + O_2(g) \rightarrow 3CO_2(g) + 3H_2O(l)$
$\Delta H_c = -1790\ kJ\ mol^{-1}$
$\Delta H°_f(CO_2(g)) = -393.5\ kJ\ mol^{-1}$
$\Delta H°_f(H_2O(l)) = -285.8\ kJ\ mol^{-1}$

5 Calculate the enthalpy change for the following reaction:

$2N_2O_3(g) \rightarrow 2NO(g) + N_2O_4(g)$

given:

$NO(g) + NO_2(g) \rightarrow N_2O_3(g) \qquad \Delta H = -40\ kJ\ mol^{-1}$
$N_2O_4(g) \rightarrow 2NO_2(g) \qquad \Delta H = +58\ kJ\ mol^{-1}$

6 Calculate the enthalpy of formation for potassium chlorate, given:

$K(s) + \frac{1}{2}Cl_2(g) + \frac{3}{2}O_2(g) \rightarrow KClO_3(s)$

$KClO_3(s) + 3Mg(s) \rightarrow KCl(s) + 3MgO(s) \qquad \Delta H = -1852\ kJ\ mol^{-1}$
$K(s) + \frac{1}{2}Cl_2(g) \rightarrow KCl(s) \qquad \Delta H = -437\ kJ\ mol^{-1}$
$Mg(s) + \frac{1}{2}O_2(g) \rightarrow MgO(s) \qquad \Delta H = -602\ kJ\ mol^{-1}$

7 About 10% of natural gas used in the home and in industry is ethane.

a Using appropriate bond energies from the table given, calculate enthalpy change for the following reaction:

$$C_2H_6(g) + \frac{7}{2}O_2(g) \rightarrow 2CO_2(g) + 3H_2O(g)$$

Bond energies ($kJ\,mol^{-1}$)			
C–H	414	O–O	144
C–C	346	O=O	498
C–O	358	O–H	463
C=O	804		

b Given that ΔH°_{vap} for water is $+44\,kJ\,mol^{-1}$, calculate the heat of combustion of ethane using the table data.

8 For the following endothermic reaction, the change in enthalpy is $-44.1\,kJ\,mol^{-1}$ and change in entropy is $-90.8\,J\,mol^{-1}\,K^{-1}$:

$$A + 2B_2 \rightarrow AB_4$$

a At what temperature does the reaction become non-spontaneous?

b Does the reaction remain non-spontaneous above or below this temperature?

9 Consider the reaction and data shown.

$$C_2H_2(g) + 2H_2(g) \rightarrow C_2H_6(g)$$

Substance	S° ($J\,mol^{-1}\,K^{-1}$)	ΔH°_f ($kJ\,mol^{-1}$)
$C_2H_2(g)$	+200.9	+226.7
$H_2(g)$	+130.7	0
$C_2H_6(g)$		−84.7

The standard entropy change, ΔS°, for the reaction is $-232.7\,J\,mol^{-1}\,K^{-1}$.

a Calculate the standard entropy, S°, of C_2H_6 gas.

b Calculate the value of the standard Gibbs free energy change, ΔG°, for the reaction.

c Is the reaction spontaneous?

10 For the reaction shown, it is observed that greater amounts of PCl_3 and Cl_2 are produced as the temperature is increased.

$$PCl_5(g) \rightarrow PCl_3(g) + Cl_2(g)$$

What is the sign of ΔS° for the reaction, and what change, if any, will occur in ΔG° for the reaction as the temperature is increased?

Extended response

1 Despite its unpleasant aroma, pentan-1-ol ($C_5H_{11}OH$) ($M = 88.15\,g\,mol^{-1}$), which is liquid at room temperature, can be used as an alternative fuel source because it can be derived from fractional distillation of fusel oil. Fusel oil is a mixture of alcohols produced as a by-product of alcohol fermentation.

a Write a balanced chemical equation for the complete combustion of pentan-1-ol.

b A simple calorimeter was set up. A mass of 0.40 g of pentan-1-ol was combusted in plenty of oxygen and caused the temperature of 100.0 g of water to increase by 30.0°C. Calculate the heat of combustion for pentan-1-ol.

c Pentan-1-ol has a known molar heat of combustion of $-3330\,kJ\,mol^{-1}$. Determine the percentage efficiency of the calorimeter.

d If a car engine is only 45% efficient at converting the chemical energy released by the combustion of pentan-1-ol to mechanical energy, how much mechanical energy would be produced if 42.5 kg of pentan-1-ol were to be combusted in excess oxygen?

2 Sodium acetate (also known as sodium ethanoate) ($NaCH_3COO$) ($M = 82.03\,g\,mol^{-1}$) is often used as an additive to potato chips to give them their salt and vinegar flavour.

a Write an equation to show the dissociation of sodium ethanoate when it is added to water.

b With reference to various electrostatic forces of attraction, describe what happens when sodium ethanoate dissolves in water.

c When 3.50 g of sodium acetate is added to 75.0 g of water in a solution calorimeter, it dissolves and the temperature of the water increases from 25.0°C to a maximum temperature of 26.8°C. Determine the enthalpy of dissolution.

d The expected enthalpy of dissolution for this reaction is $-17.3\,kJ\,mol^{-1}$. Use your answer to part **c** to determine the percentage efficiency of the calorimeter.

3 Lighter fluid is mainly butane (molar enthalpy of combustion $-2876\,kJ\,mol^{-1}$), which undergoes combustion according to the following equation:

$$2C_4H_{10}(g) + 13O_2(g) \rightarrow 8CO_2(g) + 10H_2O(l)$$

a What is the value of ΔH for the equation?

b Sketch and label an energy-level diagram (showing the relative enthalpies of reactants and products) illustrating the energy changes that occur during this reaction.

c Use your diagram from part **b** to explain why butane does not spontaneously ignite when exposed to air; that is, why a spark is necessary to begin combustion.

d Calculate the heat of combustion of butane in $kJ\,g^{-1}$.

e An inexperienced hiker wishes to use his lighter, which contains 3.00 g of butane, to heat the water in his mug. If his mug contains 150 mL of water initially at 20.0°C, and 70.0% of the heat generated by the lighter is lost to the surroundings, what will be the temperature of the water when the fuel in the lighter is exhausted?

4 Explain why it is possible to have an absolute entropy but not an absolute enthalpy.

5 The equation for the reaction between hydrazine (N_2H_4) and oxygen is shown:

$$N_2H_4(g) + O_2(g) \rightarrow N_2(g) + 2H_2O(g)$$

a Draw the structural formula for hydrazine.

b Using the bond energy data shown, calculate the standard enthalpy change for this reaction.

Bond	Bond energy ($kJ\,mol^{-1}$)
N–H	391
N–N	158
O=O	498
N≡N	945
O–H	463

c Using the standard enthalpy of formation data shown, calculate the standard enthalpy change for this reaction.

Substance	ΔH°_f ($kJ\,mol^{-1}$)
$N_2H_4(g)$	+95.4
$H_2O(g)$	−241.8

d Compare your standard enthalpy change values calculated from the two methods from parts **b** and **c**. Explain the difference in the two values.

e Using the standard entropy data shown, calculate the standard entropy change for this reaction.

Substance	S° ($J\,mol^{-1}\,K^{-1}$)
$N_2H_4(g)$	+238.7
$O_2(g)$	+243.6
$N_2(g)$	+191.6
$H_2O(g)$	+188.8

f Using your answers from parts **c** and **e**, calculate the standard Gibbs free energy for this reaction. Comment on whether this reaction is spontaneous or not.

APPENDIX 1 Symbols, units and fundamental constants

TABLE 1 Units and symbols based on the SI system*

Quantity	Symbol for physical quantity	Corresponding SI unit	Symbol for SI unit	Definition of SI unit
Mechanics				
length	l	metre	m	fundamental unit
area	A	square metre	m^2	
volume	V	cubic metre	m^3	
mass	m	kilogram	kg	fundamental unit
density	d	–	$kg\,m^{-3}$	
time	t	second	s	fundamental unit
force	F	newton	N	$kg\,m\,s^{-2}$
pressure	P	pascal	Pa	$N\,m^{-2}$
energy	E	joule	J	N m
Electricity				
electric current	I	ampere	A	fundamental unit
electric charge	Q	coulomb	C	A s
electric potential difference	V	volt	V	$J\,A^{-1}\,s^{-1}$
Nuclear and chemical quantities				
atomic number	Z	–	–	–
neutron number	N	–	–	–
mass number	A	–	–	$Z + N$
amount of substance	n	mole	mol	fundamental unit
relative atomic mass	A_r	–	–	–
relative molecular mass	M_r	–	–	–
molar mass	M	–	–	$kg\,mol^{-1}$
molar volume	V_m	–	–	$m^3\,mol^{-1}$
concentration	c	–	–	$mol\,m^{-3}$
Thermal quantities				
temperature	T	kelvin	K	fundamental unit
specific heat capacity	c	–	–	$J\,kg^{-1}\,K^{-1}$

*Units listed in red are the arbitrarily defined fundamental units of the SI system.

TABLE 2 SI prefixes, their symbols and values

SI prefix	Symbol	Value
pico	p	10^{-12}
nano	n	10^{-9}
micro	μ	10^{-6}
milli	m	10^{-3}
centi	c	10^{-2}
deci	d	10^{-1}
kilo	k	10^{3}
mega	M	10^{6}
giga	G	10^{9}
tera	T	10^{12}

TABLE 3 Some physical constants

Description	Symbol	Value
Avogadro's constant	N_A	$6.022 \times 10^{23}\,mol^{-1}$
charge of electron	e	$-1.60 \times 10^{-19}\,C$
mass of electron	m_e	$9.109 \times 10^{-31}\,kg$
mass of proton	m_p	$1.673 \times 10^{-27}\,kg$
mass of neutron	m_n	$1.675 \times 10^{-27}\,kg$
gas constant	R	$8.314\,J\,mol^{-1}\,K^{-1}$
ionic product for water	K_w	$1.0 \times 10^{-14}\,mol^2\,L^{-2}$ at 298 K
molar volume of an ideal gas	V_m	
at 273 K, 100 kPa		$22.71\,L\,mol^{-1}$
at 298 K, 100 kPa		$24.79\,L\,mol^{-1}$
specific heat capacity of water	c	$4.18\,J\,g^{-1}\,K^{-1}$
density of water	d	$1.00\,g\,mL^{-1}$ at 298 K

APPENDIX 2 Significant figures and standard form

SIGNIFICANT FIGURES

The number of significant figures of a piece of data indicates the precision of a measurement. For example, compare the following sets of data:

- A jogger takes 20 minutes to cover 4 kilometres.
- A sprinter takes 10.21 seconds to cover 100.0 metres.

The sprinter's data has been measured more precisely than that of the jogger. This is indicated by the greater number of significant figures in the second set of data.

Which figures are significant?

A significant figure is a non-zero integer or a zero that follows a non-zero integer.

In the data above:

- the distance '4 kilometres' has one significant figure
- the time '20 minutes' has two significant figures (the zero follows the integer 2)
- the 10.21 seconds and 100.0 metres each have four significant figures.

A zero that comes before any integers, however, is not significant. For example:

- the value 0.0004 has only one significant figure, whereas the value 0.0400 has three significant figures. The zeroes that come before the integer 4 are not significant, whereas those that follow the integer are significant.

Using significant figures

In chemistry you will often need to calculate a value from a set of data. It is important to remember that the final value you calculate is only as precise as your least precise piece of data.

Addition and subtraction

When adding or subtracting values, the answer should have no more digits to the right of the decimal place than the value with the least number of digits to the right of the decimal place.

Example

12.78 mL of water was added to 10.0 mL of water. What is the total volume of water?

12.78 mL + 10.0 mL = 22.78 = 22.8 mL

Because one of the values (10.0 mL) has only one digit to the right of the decimal place, the answer will need to be adjusted so that it too has only one digit to the right of the decimal place.

Multiplication and division

When multiplying and dividing values, the answer should have no more significant figures than the value with the least number of significant figures.

Example

An athlete takes 3.5 minutes to complete four laps of an oval. What is the average time taken for one lap?

Average time = $\frac{3.5}{4}$ = 0.875 = 0.88 minutes

Because the data (3.5 minutes) has only two significant figures, the answer will need to be adjusted to two significant figures so that it has the same degree of precision as the data. (Note: The 'four' is taken to indicate a precise number of laps and so is considered to have as many significant figures as the calculation requires. This applies to values that describe quantities rather than measurements.)

Rounding off

When adjusting the number of significant figures, if the integer after the last significant figure is equal to or greater than 5, then the last significant integer is rounded up. Otherwise, it 'stays the same'.

Statistics

If a calculated statistic such as mean or uncertainty has more significant figures than the data, it is usual practice to round the value to one more significant figure than there is in the data. It is important to note that this is only common practice and not a rule. If in doubt, always abide by the rules detailed above.

For example, for the data set [4, 5, 6, 6, 7, 7] the mean is 5.833, so this would conventionally be rounded to 5.8.

The uncertainty is 5.833 – 4 = 1.833, so this would be rounded to 1.8.

If you do round multiple values, as in this example, make sure you round to a consistent and appropriate number of significant figures.

STANDARD FORM

A value written in standard form is expressed as a number equal to or greater than 1 and less than 10 multiplied by 10^x, where x is an integer. For example, when written in standard form:

- 360.0 becomes 3.600×10^2
- 0.360 becomes 3.60×10^{-1}
- 0.000456 becomes 4.56×10^{-4}.

Sometimes you will need to use standard form to indicate the precision of a value.

APPENDIX 3 Table of relative atomic masses

TABLE 1 Table of relative atomic masses*

Element name	Symbol	Atomic number	Relative atomic mass	Element name	Symbol	Atomic number	Relative atomic mass	Element name	Symbol	Atomic number	Relative atomic mass
actinium	Ac	89	–	hafnium	Hf	72	178.49	praseodymium	Pr	59	140.9077
aluminium	Al	13	26.9815	hassium	Hs	108	–	promethium	Pm	61	–
americium	Am	95	–	helium	He	2	4.00260	protactinium	Pa	91	231.0359
antimony	Sb	51	121.76	holmium	Ho	67	164.9303	radium	Ra	88	–
argon	Ar	18	39.948	hydrogen	H	1	1.0080	radon	Rn	86	–
arsenic	As	33	74.9216	indium	In	49	114.82	rhenium	Re	75	186.21
astatine	At	85	–	iodine	I	53	126.9045	rhodium	Rh	45	102.9055
barium	Ba	56	137.33	iridium	Ir	77	192.22	roentgenium	Rg	111	–
berkelium	Bk	97	–	iron	Fe	26	55.845	rubidium	Rb	37	85.468
beryllium	Be	4	9.01218	krypton	Kr	36	83.80	ruthenium	Ru	44	101.07
bismuth	Bi	83	208.9804	lanthanum	La	57	138.9055	rutherfordium	Rf	104	–
bohrium	Bh	107	–	lawrencium	Lr	103	–	samarium	Sm	62	150.4
boron	B	5	10.81	lead	Pb	82	207.2	scandium	Sc	21	44.9559
bromine	Br	35	79.904	lithium	Li	3	6.94	seaborgium	Sg	106	–
cadmium	Cd	48	112.41	livermorium	Lv	116	–	selenium	Se	34	78.97
caesium	Cs	55	132.9055	lutetium	Lu	71	174.967	silicon	Si	14	28.086
calcium	Ca	20	40.08	magnesium	Mg	12	24.305	silver	Ag	47	107.868
californium	Cf	98	–	manganese	Mn	25	54.9380	sodium	Na	11	22.9898
carbon	C	6	12.011	meitnerium	Mt	109	–	strontium	Sr	38	87.62
cerium	Ce	58	140.12	mendelevium	Md	101	–	sulfur	S	16	32.06
chlorine	Cl	17	35.453	mercury	Hg	80	200.59	tantalum	Ta	73	180.9479
chromium	Cr	24	51.996	molybdenum	Mo	42	95.95	technetium	Tc	43	–
cobalt	Co	27	58.9332	moscovium	Mc	115	–	tellurium	Te	52	127.60
copernicium	Cn	112	–	neodymium	Nd	60	144.24	tennessine	Ts	117	–
copper	Cu	29	63.55	neon	Ne	10	20.180	terbium	Tb	65	158.9254
curium	Cm	96	–	neptunium	Np	93	–	thallium	Tl	81	204.384
darmstadtium	Ds	110	–	nickel	Ni	28	58.693	thorium	Th	90	232.038
dubnium	Db	105	–	nihonium	Nh	113	–	thulium	Tm	69	168.9342
dysprosium	Dy	66	162.50	niobium	Nb	41	92.9064	tin	Sn	50	118.71
einsteinium	Es	99	–	nitrogen	N	7	14.0067	titanium	Ti	22	47.87
erbium	Er	68	167.26	nobelium	No	102	–	tungsten	W	74	183.84
europium	Eu	63	151.96	oganesson	Og	118	–	uranium	U	92	238.0289
fermium	Fm	100	–	osmium	Os	76	190.2	vanadium	V	23	50.942
flerovium	Fl	114	–	oxygen	O	8	15.9994	xenon	Xe	54	131.29
fluorine	F	9	18.9984	palladium	Pd	46	106.4	ytterbium	Yb	70	173.05
francium	Fr	87	–	phosphorus	P	15	30.9738	yttrium	Y	39	88.9058
gadolinium	Gd	64	157.25	platinum	Pt	78	195.08	zinc	Zn	30	65.38
gallium	Ga	31	69.72	plutonium	Pu	94	–	zirconium	Zr	40	91.22
germanium	Ge	32	72.63	polonium	Po	84	–				
gold	Au	79	196.9666	potassium	K	19	39.098				

*Based on the atomic mass of $^{12}C = 12$.
The values for relative atomic masses given in the table apply to elements as they exist in nature (without artificial alteration of their isotopic composition) and to natural mixtures that do not include isotopes of radiogenic origin.

APPENDIX 4 Atomic radii and boiling temperatures of selected elements

Atomic radius ($\times10^{-12}$ m)
Symbol
Boiling temperature (K)

32 **H** 20																	37 **He** 4
130 **Li** 1615	99 **Be** 2741											84 **B** 4723	75 **C** 5100	71 **N** 77	64 **O** 90	60 **F** 85	62 **Ne** 27
160 **Na** 1156	140 **Mg** 1363											124 **Al** 2792	114 **Si** 3538	109 **P** 554	104 **S** 718	100 **Cl** 239	101 **Ar** 87
200 **K** 1032	174 **Ca** 1757	159 **Sc** 3109	148 **Ti** 3560	144 **V** 3680	130 **Cr** 2944	129 **Mn** 2334	124 **Fe** 3134	118 **Co** 3200	117 **Ni** 3186	122 **Cu** 2833	120 **Zn** 1180	123 **Ga** 2502	120 **Ge** 3106	120 **As** 886	118 **Se** 958	117 **Br** 332	116 **Kr** 120
215 **Rb** 961	190 **Sr** 1650	176 **Y** 3618	164 **Zr** 4679	156 **Nb** 5014	146 **Mo** 4912	138 **Tc** 4535	136 **Ru** 4420	134 **Rh** 3968	130 **Pd** 3236	136 **Ag** 2435	140 **Cd** 1040	142 **In** 2345	140 **Sn** 2875	140 **Sb** 1860	137 **Te** 1261	136 **I** 457	136 **Xe** 165
238 **Cs** 944	206 **Ba** 2118	194 **La** 3737	164 **Hf** 4873	158 **Ta** 5728	150 **W** 5828	141 **Re** 5869	136 **Os** 5281	132 **Ir** 4701	130 **Pt** 4098	130 **Au** 3109	132 **Hg** 630	144 **Tl** 1746	145 **Pb** 2022	150 **Bi** 1837	142 **Po** 1235	148 **At** 640	146 **Rn** 211
242 **Fr** 950	211 **Ra** 1413	201 **Ac** 3473															

APPENDIX 5 Common ions and solubilities of ionic compounds

TABLE 1 Names and formulae of some common positive and negative ions

Positive ions (cations)						Negative ions (anions)			
+1		+2		+3		−1		−2	
caesium	Cs^{+}	barium	Ba^{2+}	aluminium	Al^{3+}	acetate (ethanoate)	CH_3COO^{-}	carbonate	CO_3^{2-}
copper(I)	Cu^{+}	cadmium(II)	Cd^{2+}	chromium(III)	Cr^{3+}	bromide	Br^{-}	chromate	CrO_4^{2-}
gold(I)	Au^{+}	calcium	Ca^{2+}	gold(III)	Au^{3+}	chloride	Cl^{-}	dichromate	$Cr_2O_7^{2-}$
lithium	Li^{+}	cobalt(II)	Co^{2+}	iron(III)	Fe^{3+}	cyanide	CN^{-}	hydrogen phosphate	HPO_4^{2-}
potassium	K^{+}	copper(II)	Cu^{2+}			dihydrogen phosphate	$H_2PO_4^{-}$	oxalate	$C_2O_4^{2-}$
rubidium	Rb^{+}	iron(II)	Fe^{2+}	+4		fluoride	F^{-}	oxide	O^{2-}
silver	Ag^{+}	lead(II)	Pb^{2+}	lead(IV)	Pb^{4+}	hydrogen carbonate	HCO_3^{-}	sulfate	SO_4^{2-}
sodium	Na^{+}	magnesium	Mg^{2+}	tin(IV)	Sn^{4+}	hydrogen sulfate	HSO_4^{-}	sulfide	S^{2-}
		manganese(II)	Mn^{2+}			hydrogen sulfide	HS^{-}	sulfite	SO_3^{2-}
		mercury(II)	Hg^{2+}			hydrogen sulfite	HSO_3^{-}		
		nickel	Ni^{2+}			hydroxide	OH^{-}	−3	
		strontium	Sr^{2+}			iodide	I^{-}	nitride	N^{3-}
		tin(II)	Sn^{2+}			nitrate	NO_3^{-}	phosphate	PO_4^{3-}
		zinc	Zn^{2+}			nitrite	NO_2^{-}	phosphide	P^{3-}
						permanganate	MnO_4^{-}		

TABLE 2 Solubility of common ionic compounds in water

Soluble ionic compounds		
Soluble in water (>0.1 mol dissolves per L at 25°C)	**Exceptions: insoluble (<0.01 mol dissolves per L at 25°C)**	**Exceptions: slightly soluble (0.01–0.1 mol dissolves per L at 25°C)**
most chlorides (Cl^{-}), bromides (Br^{-}) and iodides (I^{-})	AgCl, AgBr, AgI, PbI_2	$PbCl_2$, $PbBr_2$
all nitrates (NO_3^{-})	no exceptions	no exceptions
all ammonium (NH_4^{+}) salts	no exceptions	no exceptions
all sodium (Na^{+}) and potassium (K^{+}) salts	no exceptions	no exceptions
all ethanoates (CH_3COO^{-})	no exceptions	no exceptions
most sulfates (SO_4^{2-})	$SrSO_4$, $BaSO_4$, $PbSO_4$	$CaSO_4$, Ag_2SO_4
Insoluble ionic compounds		
Insoluble in water	**Exceptions: soluble**	**Exceptions: slightly soluble**
most hydroxides (OH^{-})	NaOH, KOH, $Ba(OH)_2$, NH_4OH*, AgOH**	$Ca(OH)_2$, $Sr(OH)_2$
most carbonates (CO_3^{2-})	Na_2CO_3,K_2CO_3, $(NH_4)_2CO_3$	no exceptions
most phosphates (PO_4^{3-})	Na_3PO_4, K_3PO_4, $(NH_4)_3PO_4$	no exceptions
most sulfides (S^{2-})	Na_2S, K_2S, $(NH_4)_2S$	no exceptions

*NH_4OH does not exist in significant amounts in an ammonia solution. Ammonium and hydroxide ions readily combine to form ammonia and water.
**AgOH readily decomposes to form a precipitate of silver oxide and water.

APPENDIX 6 Selected thermodynamic data

TABLE 1 Bond energies and average bond energies

Single bonds (kJ mol^{-1})							
C–H	414	N–H	391	O–H	463	F–F	159
C–C	346	N–C	286	O–O	144		
C–N	286	N–N	158	O–F	191	Cl–F	255
C–O	358	N–O	214	O–Cl	206	Cl–Cl	242
C–F	492	N–F	278	O–I	201		
C–Cl	324	N–Cl	192			Br–F	249
C–Br	285	N–Br	243	S–H	364	Br–Cl	219
C–I	228			S–F	327	Br–Br	193
C–S	289	H–H	436	S–Cl	271		
		H–F	567	S–Br	218	I–Cl	211
Si–H	323	H–Cl	431	S–S	266	I–Br	178
Si–Si	226	H–Br	366			I–I	151
Si–C	307	H–I	298				
Si–O	466						
Multiple bonds (kJ mol^{-1})							
C=C	614	N=N	470	O=O	498		
C≡C	839	N≡N	945				
C=N	615			S=O	523		
C≡N	890			S=S	429		
C=O	804						

Selected thermodynamic data *(continued)*

TABLE 2 Standard enthalpies of formation and standard entropies

Substance	Formula	ΔH°_{f} (kJ mol^{-1})	S° (J mol^{-1} K^{-1})
ammonia	$NH_3(g)$	–45.9	+192.8
butane	$C_4H_{10}(g)$	–126	+310
carbon dioxide	$CO_2(g)$	–393.5	+213.8
carbon monoxide	$CO(g)$	–110.5	+197.7
ethane	$C_2H_6(g)$	–84.0	+230
ethanol	$C_2H_5OH(l)$	–277.7	+161
ethene	$C_2H_4(g)$	+52.0	+220
ethyne (acetylene)	$C_2H_2(g)$	+228	+201
glucose	$C_6H_{12}O_6(s)$	–1271	+209.2
hydrogen	$H_2(g)$	0	+130.7
hydrogen bromide	$HBr(g)$	–36.3	+198.7
hydrogen chloride	$HCl(g)$	–92.3	+186.9
hydrogen fluoride	$HF(g)$	–273.3	+173.8
hydrogen iodide	$HI(g)$	+26.5	+206.6
methane	$CH_4(g)$	–74.8	+186
nitrogen	$N_2(g)$	0	+191.6
nitrogen dioxide	$NO_2(g)$	+33.1	+240.0
nitrogen monoxide	$NO(g)$	+90.2	+210.8
octane	$C_8H_{18}(l)$	–250.3	+359.8
oxygen	$O_2(g)$	0	+205
propane	$C_3H_8(g)$	–104.6	+270
water(g)	$H_2O(g)$	–241.8	+188.8
water(l)	$H_2O(l)$	–285.8	+70.0

TABLE 3 Enthalpies of combustion*

Substance	Formula	ΔH°_{c} (kJ mol^{-1})
butane	$C_4H_{10}(g)$	–2886
carbon (graphite)	$C(s)$	–394
ethane	$C_2H_6(g)$	–1560
ethene	$C_2H_4(g)$	–1411
ethanol	C_2H_5OH	–1367
glucose	$C_6H_{12}O_6(s)$	–2803
hydrogen	$H_2(g)$	–286
methane	$CH_4(g)$	–890
methanol	$CH_3OH(l)$	–725
octane	$C_8H_{18}(l)$	–5450
propane	$C_3H_8(g)$	–2220

*Values are measured at standard state conditions (298 K, 1 bar).

Answers

Chapter 1 Working scientifically

1.1 Questioning and predicting

1 A hypothesis is a statement that can be tested. This involves making a prediction based on previous knowledge and evidence or observations.
2 A
3 a electrical conductivity
b concentration of lead
c electrical conductivity
d pH
4 qualitative observation
5 C
6 C

1.2 Planning investigations

1 B
2 a In a controlled experiment, two groups of subjects are tested; the groups, or the tests performed on them, are identical except for a single factor (the independent variable).
b The dependent variable is the variable that is measured to determine the effect of changes in the independent variable. The independent (experimental) variable is the variable that is changed in an experiment. For example, in an experiment testing the effect of soil pH on flower colour, the independent variable would be soil pH and the dependent variable would be flower colour.
3 a type of soft drink
b pH
c Student answers will vary, e.g. temperature of solutions, type of equipment used, storage and preparation of solutions.
4 a Litmus paper and universal indicator give qualitative information about pH (e.g. acidic, basic, or neutral) through colour, and therefore is not an accurate method for determining pH.
b A calibrated pH meter will give quantitative information and is more accurate than using litmus paper or universal indicator.
5 a valid
b reliable
c accurate

1.3 Conducting investigations

1 Data set A: mistake, because there is one unexpected value (1.5) in the data.
Data set B: systematic error, because the error is not obvious and may be due to a consistent equipment or operator error.
2 a systematic errors
b random errors
3 17.34 mL, 17.38 mL and 17.44 mL
4 a systematic
b mistake
c random
5 There could be many reasons why the same experimental results cannot be obtained. The experimental design may be poor because of a lack of objectivity, clear and simple instructions and appropriate equipment; or a failure to control variables. Other problems not specifically related to the experiment could be a poor hypothesis that could not be tested objectively, conclusions that do not agree with the results and interpretations that are subjective.

1.4 Processing data and information

TY 1.4.1 ±3°C
1 a 23 b 19 c 23
2 Add a trend line or line of best fit.
3 a

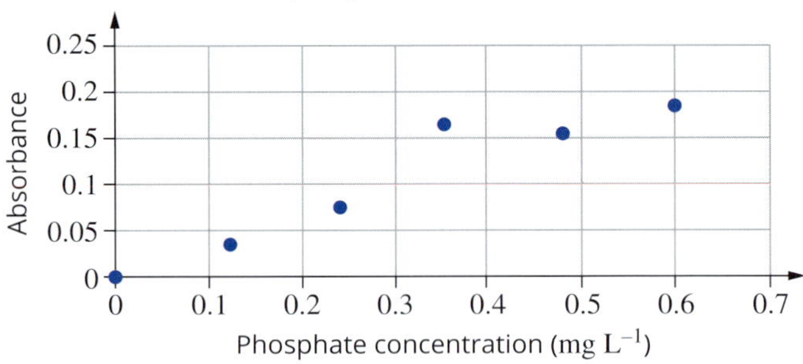

b Data point 4 (0.36, 0.159) is an outlier.
c An outlier is a point in the data that does not fit the trend and may be the result of error.
4 mean = 23; uncertainty = 5

1.5 Analysing data and information

1 A sloping linear graph shows a directly proportional relationship between two variables.
2 inversely proportional relationship
3 directly proportional
4 time restraints and limited resources
5 An increase in the amount of limescale on the heating element of the kettle from 10% to 20% produced a 14% decrease in the efficiency of the kettle.

1.6 Problem solving

1 B
2 Implications. Generalisations apply facts to a broader context that they have not been specifically tested for. Implications use facts and accepted knowledge to logically make a connection between two things, such as the results of a chemical experiment and a possible commercial use of that chemical.
3 The statement 'Many repeats of the procedure were conducted' is unquantified. 'Thirty repeats of the procedure were conducted' is better because the number of trials is quantified.
4 A

1.7 Communicating

1 B 2 D
3 a $g\,mol^{-1}$ b $J\,g^{-1}\,K^{-1}$ c $mol\,L^{-1}$
4 Divide the value in g by 1000.
5 This might need to be done to reflect how we use quantities of varying units of measurement in the laboratory or to make the numbers easier to comprehend.

Chapter 1 review

1 a concentration of lead—dependent variable
b analytical technique, temperature of water sample, type of sampling container—controlled variables
c source and location of water—independent variable
2 a reflect b create c analyse d investigate e apply
f identify g describe
3 independent variable: source of the water; dependent variable: phosphate concentration; controlled variables: temperature, time of testing, method of testing
4 a bar graph b line graph
c scatter graph (with line of best fit) d pie chart
5 a It can cause severe burns, and dissolve or eat away at substances, including tissues such as your skin or lungs.
b It is poisonous if inhaled or ingested.
c It is a highly combustible liquid that could catch fire.

6 Accuracy refers to the ability of the method to allow the measurement to be close to a true or accepted value. Validity refers to whether an experiment or investigation enables testing of the set hypothesis and achievement of the purpose.
7 The uncertainty of a set of data is the maximum variance from the mean of the data. In this example, the mean is 6.63. The maximum variance is 7.20 – 6.63 = 0.57 = the uncertainty.
8 Mean. The mode is the most common value in a set of data. The median is the middle value. The mean is the average of the values in a data set. The inclusion of an outlier will have the most effect on the mean value; therefore, outliers are excluded from calculating the mean value.
9 **a** mistake **b** random error **c** systematic error
10 a non-linear relationship, e.g. exponential, inverse
11 **a** reliability **b** validity **c** accuracy **d** precision
12 Evaluate the method; identify issues that could affect validity, accuracy, precision and reliability of data; state systematic sources of error and uncertainty; and recommend improvements to the investigation if it is to be repeated.
13 A trend is a pattern or relationship that can be seen between the dependent and independent variables. It may be linear, in which case the variables change in direct proportion to each other to produce a straight trend line. The relationship may be in indirect proportion to each other and non-linear, giving a curved trend line. The relationship may also be inverse, when one variable decreases in response to the other variable increasing. This relationship can also be linear or non-linear.
14 Limitations are issues that could have affected the validity, accuracy, precision or reliability of the data, plus any sources of error or uncertainty.
15 Bias is a form of systematic error resulting from a researcher's personal preferences or motivations.
16 The purpose of referencing and acknowledgements is to ensure creators and sources are properly credited for their work, and to demonstrate that prior research has been undertaken to support the validity of the hypothesis.
17 C 18 A
19 0.030 00 L or 3.000×10^{-2} L
20 **a** Purpose: to determine the effect of increasing water temperature on the electrical conductivity of water.
b independent variable: water temperature; dependent variable: electrical conductivity of water; controlled variables: pH, water source, type of sampling container
c The data collected would be electrical conductivity, measured using a probe, and therefore it would be quantitative.
d Raw data is data collected in the field and recorded as measurements are taken. Processed data is tabulated in a form in which the reader can clearly see the temperature of the water and the conductivity at each separate temperature value. This can then be processed and graphed with the independent variable on the **x**-axis (temperature) and dependent variable on the **y**-axis (conductivity). If the hypothesis is correct, the graph for this experiment could look like this:

Water temperature and electrical conductivity

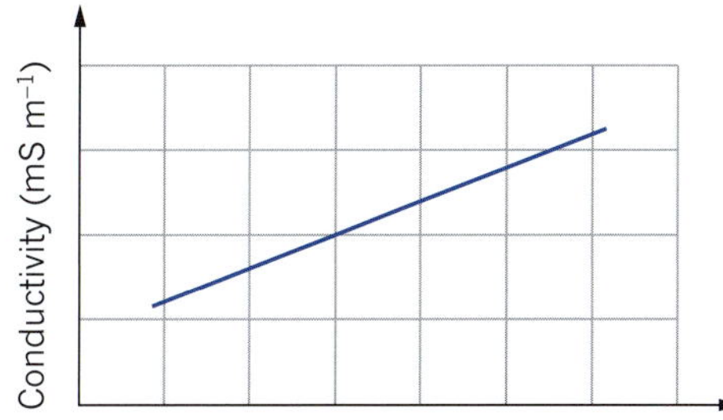

A non-linear increase could also fit the hypothesis, because a directly proportional relationship between conductivity and temperature has not been specified.

Chapter 2 Properties of matter

2.1 Types of matter

1 An element is made up of just one type of atom. A compound is a substance that contains at least two different types of atoms, bonded together in a definite proportion.
2 **a** element **b** element **c** compound **d** compound **e** element **f** compound **g** element **h** compound
3 **a** solvent **b** solution **c** solute **d** solute
4 They all have water as the solvent.
5 **a** The batter has different compositions depending on where it is sampled. The batter is a heterogeneous mixture.
b The vigorous mixing would make the batter a homogeneous mixture. A sample of batter taken from the middle, left or right of the bowl should now taste the same.
6 The gases are combining physically to form a mixture. The component gases retain their individual character. To form water, hydrogen needs to react with oxygen and combine chemically, not just mix physically.
7 A compound has two or more different types of atoms in a definite proportion, and any sample of a homogeneous mixture (wherever it is sampled from) will have the same combination of materials in the same proportion. In this case, you cannot tell if it is a chemically combined compound or a physically combined mixture. You can only tell that it is neither a heterogeneous mixture nor an element.

2.2 Physical properties and changes of state

1 The particles are (1) hard, (2) indivisible, (3) attracted to each other and (4) constantly in motion.
2 liquid and gas
3 solid
4 1 kg of steam. A change of state does not create or destroy mass.
5 The particle model assumes particles are attracted to one another. The particles in solids and liquids are in contact and there is limited space between the particles to allow them to come closer together. The particles in a gas are not in contact, and there is space that allows the particles to come closer together.
6 The hot day means the temperature is higher, and the particles of water in the liquid move more rapidly and have enough energy to overcome the forces of attraction between particles to form water vapour. As the wind blows, the particles in the water vapour are blown away from the puddle, increasing the rate of evaporation. Eventually the puddle will dry out.

2.3 Separating mixtures

1 A
2 by (fractional) distillation
3 A magnet. Neither iron nor sand is a volatile material.
4 Heating the pot containing the solution will evaporate the water. Eventually the magnesium sulfate will precipitate from solution. When cool, the magnesium sulfate crystals can be filtered from the solution.
5 The sodium phosphate is soluble and remains dissolved in the unevaporated water. Recovering all the material from solution is difficult.

2.4 Calculating percentage composition

TY 2.4.1 14 g nitrogen, 48 g oxygen, 23 g sodium
TY 2.4.2 2.1% carbon, 10.8% hydrogen, 87.1% oxygen
1 27% carbon, 73% oxygen
2 26% nitrogen, 67% chlorine, 7% hydrogen
3 No, because the total returned mass is 63 g, more than the original mass of the sample. There is an error: either the sample was contaminated during analysis or the original mass is incorrect.
4 77% carbon dioxide, 23% oxygen
5 15 g gold, 4 g copper, 1 g silver
6 **a** 7 g oxygen **b** 60% oxygen, 13% sulfur
7 29% carbon

2.5 Elements and the periodic table

1 **a** iron **b** potassium **c** tungsten **d** lead

2 Malleable. The property of being ductile is similar, but refers to being able to be drawn into a wire.

3 Oxygen. Oxygen is a non-metal. Rhodium and calcium are metals, so are good conductors of heat and electricity. Silicon is a metalloid, so will also conduct heat and electricity.

4 The unreactivity of copper with water is a chemical property. The colour of the element and the melting point are physical properties. An alloy is a physical mixture and not a chemical property.

5 The properties of being a good thermal conductor and having a high melting point are useful. Cookware needs to transfer heat from the stove or oven to the food, but not melt itself. Materials that are a good conductor of electricity and are ductile are useful for making wires.

6 Sulfur then silicon then scandium. Sulfur is a non-metal; silicon, a metalloid; and scandium, a metal. A metal is a good conductor of electricity, so it is a poor insulator; a non-metal is a poor conductor of electricity, so it is a good insulator.

Chapter 2 review

1 Compounds are made of two or more different types of atom in fixed ratios.

2 Atoms are hard, indivisible structures.

3 Atoms are the basic building blocks of matter. Molecules are composed of a number of atoms chemically bonded together.

4 A discrete molecule has a definite number of atoms chemically bonded together. A network has an indefinite number of atoms chemically bonded together.

5 monatomic: helium, neon, krypton; molecules: sulfur, oxygen, nitrogen; large network: copper, gold, carbon, tin

6 A compound consists of two or more elements chemically bonded together in a fixed ratio. A mixture has two or more pure substances physically combined in no fixed ratio.

7 The milk with cream is a mixture that differs depending on whether it is sampled from the top or with the bottom. It is a heterogeneous mixture.

8 In general, no. Dalton defined matter as being composed of indivisible particles. Particle theory views matter as particles in motion, but does not assume that the particles are atoms. In Dalton's atomic theory, an atom of sulfur is fundamental. In particle theory, the ring of eight sulfur atoms is the fundamental unit.

9 More dense. The balloon when blown up is taut and squeezes the air inside slightly. The air inside is slightly more pressured, and has a greater mass of air in the volume of the balloon compared with the volume of unpressurised air.

10 This is a physical change. The particles in the metal are rearranged and ordered differently, but the particles themselves have not changed. If the particles had changed, this would be a chemical change.

11 **a** The temperature stays the same during the transition.
b The attractive forces change—they are being formed as the temperature decreases (condensing, freezing), and being broken as the temperature increases (melting, vaporising).

12 Toothpaste keeps its shape like a solid, but when squeezed it flows like a liquid.

13 No. The ink solution is a homogeneous mixture, so filtration is not the best choice.

14 Most filters are too coarse and the blood cells would remain in the filtrate. A very fine filter would retain the blood cells, but it would be a slow process. Techniques such as centrifugation are simpler.

15 The puddle will evaporate more quickly. Evaporation takes place on the surface of the water. The cup of water has a much smaller surface exposed to air than the puddle.

16 The liquid with the lower boiling point is more volatile than the other liquid. Therefore, the vapour will contain a greater proportion of the liquid with the lower boiling point.

17 Na 31% + O 21% means the balance, 48%, must be Cl.

18 6.3 g

19 74% Fe, 18% Cr, 8% Ni

20 2.4 g

21 39% carbon

22 **a** O **b** C **c** He **d** Ni **e** K **f** Au

23 **a** nitrogen **b** calcium **c** chlorine **d** silver **e** mercury

24 **a** A metalloid. It is shiny but a poor conductor of heat, so it is not a metal.
b the element that conducts electricity

25 Iron is a metal and is malleable. A fine, sharp edge can be formed by flattening the metal. Sulfur is a non-metal and not malleable. When hammered, the sulfur becomes a powder.

26 They are all non-metals, so they are all likely to be poor conductors. The other properties are properties of metals.

27 Properties **a**, **b** and **d** are physical properties, because the diamond, sugar and aluminium have not changed. Properties **c** and **e** are chemical properties, because the carbon in the wood burns and changes into carbon dioxide, and the lithium reacts with water and changes into lithium hydroxide.

28 **a** Lithium and sodium are positioned in the same group, so their chemical properties are likely to be similar. Magnesium is in a different group so its chemical properties will differ. All of these elements are metals, so their physical properties are likely to be similar, in that all are malleable, ductile, good conductors etc.
b These elements are, respectively, a metal, a metalloid and a non-metal. You would not expect their physical properties to be similar.
c From group 1, potassium, to group 13, gallium, the elements are metals and likely to be good conductors of heat. In group 14, germanium, and group 15, arsenic, the elements are metalloids and moderate conductors of heat. From group 16 onwards the non-metals are expected to be poor conductors of heat.

29 The crust is a heterogeneous mixture of minerals and rocks, and the rocks are in turn mixtures of minerals. The first problem was in separating the mixtures to generate a sufficient quantity of pure aluminium-bearing minerals from the soil and other minerals.
The main problem in isolating metallic aluminium from its ores was that it is a very reactive metal and therefore hard to separate chemically from the other elements in its compounds.
For modern industry, bauxite (a mix of $Al(OH)_3$ and $AlO(OH)$) is the most common aluminium-bearing mineral. Through the Bayer process, the bauxite is separated from other minerals and purified to alumina (Al_2O_3). Molten alumina is converted to elemental aluminium by electrolysis. This means the large-scale production of aluminium was only possible from the late 18th century onwards.

30 The mixture is a gas. Filtration is inappropriate as there are no solids. Evaporation is inappropriate as there are no liquids. However, on cooling, the gases will become liquids. Oxygen boils at −183°C (90 K), argon at −186°C (87 K), and nitrogen at −196°C (77 K). Cooling the mixture to less than the boiling point of the least volatile compound (nitrogen) means that all of the components will become liquids. Distillation is now possible. Allowing temperatures to increase slightly, first nitrogen will evaporate, then argon, leaving behind liquid oxygen.
Argon freezes at −189°C (84 K), nitrogen at −210°C (63 K) and oxygen at −219°C (54 K). In principle, the frozen solids could be collected and filtered, but with much more difficulty than for simple distillation.
In industrial practice, this fractional distillation process is actually used, but the precise boiling points of the gas mixtures shift a little when the liquid mixture is rich in oxygen.

31 Malleability. This property supports aluminium being a metal, because metals can be beaten into different shapes without breaking or shattering.

Chapter 3 Atomic structure and atomic mass

3.1 Inside atoms

1 10000–100000 times larger

2 protons and neutrons, found in the nucleusprotons and neutrons, found in the nucleus

3 The electrons are held within the cloud surrounding the nucleus by the electrostatic attraction of the protons; the negative electrons are attracted to the positive protons and pulled towards them.

4 **a** −1 **b** +1 **c** 0

3.2 Classifying atoms

TY 3.2.1 92 protons; 143 neutrons; 92 electrons

1 mass number

2 15 protons; 16 neutrons; 15 electrons

3 nitrogen

4 Isotopes of the same element have the same atomic number and therefore the same number of protons and electrons. The number of neutrons differs between isotopes of the same element; therefore, they have different mass numbers.

5 2

6 Alpha particles are made up of two protons and two neutrons (a helium nucleus). Beta particles are high-energy electrons. Gamma radiation is a type of high-energy electromagnetic radiation.

7 **a** ${}^{234}_{90}\text{Th} \rightarrow {}^{234}_{91}\text{Pa} + {}^{0}_{-1}\text{e}$

b ${}^{234}_{91}\text{Pa} \rightarrow {}^{230}_{89}\text{Ac} + {}^{4}_{2}\text{He}$

c ${}^{220}_{86}\text{Rn} \rightarrow {}^{216}_{84}\text{Po} + {}^{4}_{2}\text{He}$

d ${}^{216}_{84}\text{Po} \rightarrow {}^{216}_{85}\text{At} + {}^{0}_{-1}\text{e}$

3.3 Masses of particles

TY 3.3.1 10.81

TY 3.3.2 % abundance 206Pb = 28%; % abundance 207Pb = 22%; % abundance 208Pb = 50%

TY 3.3.3 relative abundance of Cu with isotopic mass of 62.95 is 70.50%; relative abundance of Cu with isotopic mass of 64.95 is 29.50%

TY 3.3.4 63.02 **TY 3.3.5** 187.57

1 B

2 **a** 15.999 **b** 107.9 **c** 1.008

3 8%

4 **a** % abundance 90Zr = 51%; % abundance 91Zr = 11%; % abundance 92Zr = 17%; % abundance 94Zr = 17%; % abundance 96Zr = 4%
b 91

4 % abundance 90Zr = 51%; % abundance 91Zr = 11%; % abundance 92Zr = 17%; % abundance 94Zr = 17%; % abundance 96Zr = 4%

5 **a** 98.09 **b** 17.03 **c** 30.07

6 **a** 74.55 **b** 105.99 **c** 342.17

3.4 Electronic structure of atoms

1 electrons; higher; emit; lower

2 An emission spectrum is the spectrum of the light emitted by an element when it is heated. Atoms are heated so that electrons move to higher energy levels, before returning to lower levels and emitting light.

3 Each line in an emission spectrum corresponds to a specific amount of energy. This energy is emitted when electrons from higher-energy electron shells transition to a lower-energy shell. The different spectral lines indicate that there are energy differences between shells. This is evidence that electrons are found in shells with discrete energy levels.

4 Electrons revolve around the nucleus in fixed, circular orbits. Electrons' orbits correspond to specific energy levels in the atom. Electrons can only occupy fixed energy levels and cannot exist between two energy levels. Orbits of larger radii correspond to higher energy levels.

5 light energy

3.5 Electronic configuration and the shell model

TY 3.5.1 2,8,18,6

TY 3.5.2 6

1 **a** 2,3 **b** 2,8,2 **c** 2,8,8,2 **d** 2,8,18,7

2 **a** 2,2 **b** 2,8,6 **c** 2,8,8 **d** 2,8,2 **e** 2,8

3

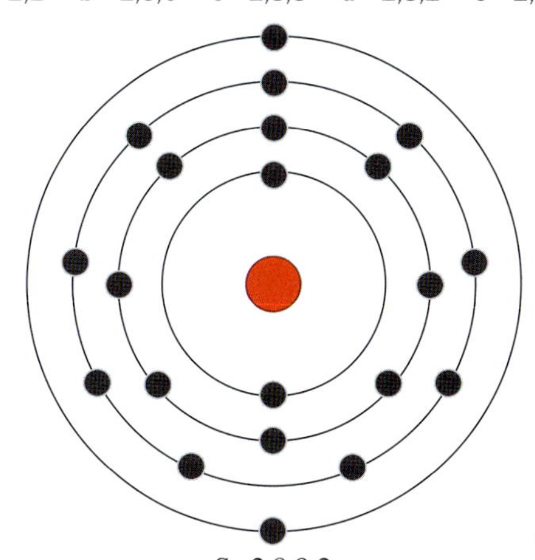

4 **a** helium, He **b** fluorine, F **c** aluminium, Al
d nitrogen, N **e** chlorine, Cl

5 7

3.6 The Schrödinger model of the atom

TY 3.6.1 $1s^2 2s^2 2p^6 3s^2 3p^6 3d^2 4s^2$

1 B

2

Element (atomic number)	Electronic configuration using the shell model	Electronic configuration using the subshell model
boron (5)	2,3	$1s^2 2s^2 2p^1$
lithium (3)	2,1	$1s^2 2s^1$
chlorine (17)	2,8,7	$1s^2 2s^2 2p^6 3s^2 3p^5$
sodium (11)	2,8,1	$1s^2 2s^2 2p^6 3s^1$
neon (10)	2,8	$1s^2 2s^2 2p^6$
potassium (19)	2,8,8,1	$1s^2 2s^2 2p^6 3s^2 3p^6 4s^1$
scandium (21)	2,8,9,2	$1s^2 2p^2 2p^6 3s^2 3p^6 3d^1 4s^2$
iron (26)	2,8,14,2	$1s^2 2s^2 2p^6 3s^2 3p^6 3d^6 4s^2$
bromine (35)	2,8,18,7	$1s^2 2s^2 2p^6 3s^2 3p^6 3d^{10} 4s^2 4p^5$

3 The subshell model is a refinement of the shell model. The shell model proposed that all electrons in the one shell were of equal energy. Evidence from emission spectra indicated that there were different electronic energy levels (called subshells) within a shell.

4 **a** [↑↓] 1s [↑↓] 2s

b

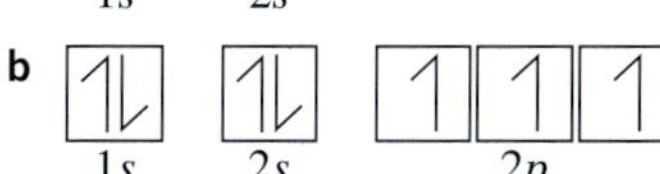

c [↑↓] 1s [↑↓] 2s [↑↓][↑↓][↑↓] 2p [↑↓] 3s [↑][↑][ ] 3p

d [↑↓][↑↓][↑↓][↑][↑] 3d [↑↓] 4s

Chapter 3 review

1 in the nucleus

2 The protons and neutrons form the nucleus. The electrons are grouped in shells and occupy the space around the nucleus.

3 The mass of a proton is approximately equal to the mass of a neutron and is about 1800 times the mass of an electron. The proton and electron have equal but opposite charges, and the neutron has no charge.

4 **a** 1.20×10^{-10} m **b** 2.16×10^{-10} m **c** 3.48×10^{-10} m

5 **a** atomic number is 24; mass number is 52
b 24 electrons, 24 protons, 28 neutrons

6 No. Isotopes have the same number of protons (atomic number) but different numbers of neutrons (and therefore different mass numbers). These atoms have different atomic numbers and different mass numbers.

7 Atoms are electrically neutral. The positive charge on one proton balances the negative charge on one electron. For electrical neutrality, there must be an equal number of protons and electrons.

8 Most elements have more than one isotope, so they will have more than one mass number. Bromine atoms are unique in having 35 protons in their nuclei (i.e. an atomic number of 35). Isotopes of bromine, however, differ in their mass numbers, so mass number is not fixed for an element (except for those elements such as sodium, which have only one naturally occurring isotope). In addition, an isotope of one element may have the same mass number as an isotope of another element.

9 **a** 18.02 **b** 123.88 **c** 28.01

10 **a** 225.18 **b** 171.3 **c** 291.73

11 **a** ${}^{231}_{87}\text{Fr} \rightarrow {}^{231}_{88}\text{Ra} + {}^{0}_{-1}\text{e}$

b ${}^{239}_{94}\text{Pu} \rightarrow {}^{235}_{92}\text{U} + {}^{4}_{2}\text{He}$

c ${}^{231}_{91}\text{Pa} \rightarrow {}^{227}_{89}\text{Ac} + {}^{4}_{2}\text{He}$

d ${}^{214}_{84}\text{Po} \rightarrow {}^{210}_{82}\text{Pb} + {}^{4}_{2}\text{He}$

e ${}^{149}_{62}\text{Sm} \rightarrow {}^{145}_{61}\text{Pm} + {}^{4}_{2}\text{He} + {}^{0}_{-1}\text{e}$

12 The relative atomic mass of carbon is the weighted average of the isotopic masses of all carbon isotopes (i.e. ^{12}C, ^{13}C and ^{14}C). Small amounts of ^{13}C and ^{14}C make this average slightly greater than 12, the relative isotopic mass of the ^{12}C isotope.

13 106.4

14 **a** A_r(Ar) = 39.96 and A_r(K) = 39.11
b Although potassium atoms have one more proton than argon atoms, the most abundant isotope of argon has 22 neutrons, giving it a relative atomic mass close to 40. The most abundant isotope of potassium has only 20 neutrons, giving it a relative atomic mass close to 39.

15 **a** Peak heights: ^{50}Cr = 0.3 units, ^{52}Cr = 12 units, ^{53}Cr = 1 unit, ^{54}Cr = 0.2 units; total height = 13.5 units
Percentages: ^{50}Cr = 2.2%, ^{52}Cr = 88.9%, ^{53}Cr = 7.4%, ^{54}Cr = 1.5%
b 52

16 48.0% and 52.0%

17 **a** 61.5% **b** 20.2%

18 $n = 1$

19 $n = 3$. The first 2 electrons fill the first shell, the next 8 the second shell, the next 8 the third shell. The next 2 electrons go into the fourth shell and the remaining 10 electrons go into the third shell.

20 magnesium

21 **a** $1s^2$ **b** $1s^22s^22p^2$ **c** $1s^22s^22p^5$ **d** $1s^22s^22p^63s^23p^1$
e $1s^22s^22p^63s^23p^6$ **f** $1s^22s^22p^63s^23p^63d^84s^2$
g $1s^22s^22p^63s^23p^63d^{10}4s^24p^5$

22 A fluorine atom contains nine electrons. The electrons are arranged in energy levels called shells; two electrons are in the first shell and seven electrons are in the second shell, which has higher energy. The electron arrangement in the shells can be written as 2,7. Shells are regarded as being made up of energy levels called subshells. The first shell contains an *s*-type subshell, which is labelled '1*s*'. The second shell contains both *s*- and *p*-type subshells, labelled '2*s*' and '2*p*', respectively. Within subshells, electrons occupy regions of space known as orbitals. An orbital can hold up to two electrons. Subshells of an *s*-type contain one orbital, whereas *p*-type subshells contain three orbitals. The electron arrangement in the subshells of a fluorine atom can be represented as $1s^22s^22p^5$.

23 $1s^22s^22p^63s^23p^4$

24 In the Schrödinger model of the atom, electron shells are divided into subshells, and each subshell can have a different energy level. According to the Schrödinger model, the 4*s*-subshell is lower in energy than the 3*d*-subshell. Therefore, the 4*s*-subshell begins filling after the 3*s*- and 3*p*-subshells, but before the 3*d*-subshell.

25 There is very little difference between the energy levels of the 3*d*- and 4*s*-subshells. As orbitals fill in sequence, one electron at a time, it is more stable for chromium to have all *d*-orbitals exactly half-filled than to have one empty *d*-orbital. Likewise, for copper it is more stable to have all *d*-orbitals completely filled than to have one half-filled and a full 4*s*-orbital.

26 **a** 1*s*: [↑↓] 2*s*: [↑↓] 2*p*: [↑↓][↑][↑]

b 1*s*: [↑↓] 2*s*: [↑↓] 2*p*: [↑↓][↑↓][↑↓] 3*s*: [↑↓]

c 3*d*: [↑][↑][↑][↑][↑] 4*s*: [↑↓]

d 3*d*: [↑↓][↑↓][↑↓][↑↓][↑↓] 4*s*: [↑↓] 4*p*: [↑][↑][↑]

27 nothing

28 **a** Atomic models before Bohr's accounted for a number of atomic properties, but were not able to account for the characteristic emission spectrum of each element. Previous models were also in conflict with the principles of classical physics, which suggested that electrons moving in circular orbits should continuously lose energy and spiral into the nucleus.
b The Bohr model of the atom did not adequately explain why electrons adopted some energy levels but not others. In addition, calculated frequencies for lines in the emission spectra of atoms with more than one electron gave poor agreement with measured values.

29 Student answers will vary. Possible answers include the lack of a magnetic field and that the balls representing the particles are not charged.

Chapter 4 Periodicity

4.1 The periodic table

1 row

2 groups 1, 2 and 13–18

3 **a** 1 **b** 5 **c** 7 **d** 2

4 The element is in period 3 and therefore has three occupied shells. As the element is in group 2, it will have two valence electrons. This gives an electron configuration of 2,8,2 or $1s^22s^22p^63s^2$.

5 **a** **i** 13 **ii** 17 **iii** 1 **iv** 18 **v** 14 **vi** 14
b **i** 4 **ii** 2 **iii** 1 **iv** 1 **v** 7 **vi** 3
c **i** silicon, Si, 2,8,4 or $1s^22s^22p^63s^23p^2$
ii beryllium, Be, 2,2 or $1s^22s^2$
iii argon, Ar, 2,8,8 or $1s^22s^22p^63s^23p^6$

6 Atoms are listed in the periodic table in order of atomic number because the atomic number determines the number of protons and thus electrons, and this in turn sets the properties that make each element unique. Relative atomic mass does not necessarily increase as atomic number increases.

4.2 Trends in the periodic table: Part 1

TY 4.2.1 +7

1 +4

2 As core charge increases, electrons are more strongly attracted to the nucleus of an atom, so electronegativity increases.

3 **a** **i** fluorine, F **ii** francium, Fr
b **i** group 17 **ii** group 1
c Elements in group 18, the noble gases, have a very stable electronic configuration and so are unreactive.

4 F, O, P, Mg, Ca

5 hydrogen, helium, nitrogen, oxygen, fluorine, neon, chlorine, argon, krypton, xenon and radon

4.3 Trends in the periodic table: Part 2

TY 4.3.1 C

1 **a** First ionisation energy is the amount of energy needed to remove an electron from an atom in the gas phase.
b The factors that affect first ionisation energy across a period are the size of the atom (i.e. the distance of the outermost (highest energy) electron from the nucleus) and the core charge.

2 A metalloid is an element that exhibits both metallic and non-metallic properties.

3 Across a period, the number of occupied shells in the atoms remains constant, but core charge increases. The valence electrons become more strongly attracted to the nucleus, so more energy is required to remove an electron from an atom. Therefore, the first ionisation energy increases across a period.

4 The valence electrons in strontium are less tightly held than those of beryllium. They are easily lost in reactions, such as with water.

5 **a** fluorine
b Aluminium and magnesium are metals. Fluorine has the highest core charge (it is located farthest right on the periodic table), while also containing the least number of electron shells. This means that fluorine attracts an electron more strongly than other elements and is therefore the most reactive non-metal.

Chapter 4 review

1 The atomic number is determined by the number of protons in the nucleus of an atom.

2 **a** period 1, *s*-block
b period 2, *p*-block
c period 3, *p*-block
d period 4, *d*-block
e period 7, *f*-block

3 **a** period 2, group 2
b period 3, group 14
c period 4, group 13
d period 1, group 18

4 **a** Elements in the *s*-block are filling an *s*-subshell. As an *s*-subshell accommodates a maximum of two electrons, there are only two groups of elements in the *s*-block. They have outer-shell configurations of s^1 and s^2.
b Elements in the *p*-block are filling a *p*-subshell. As a *p*-subshell accommodates a maximum of six electrons, there are six groups of elements in the *p*-block. They have outer-shell configurations of s^2p^1 to s^2p^6.
c Elements in the *d*-block are filling a *d*-subshell. As a *d*-subshell accommodates a maximum of 10 electrons, there are 10 elements in each transition series.
d The lanthanoids and actinoids are filling an *f*-subshell. As an *f*-subshell accommodates a maximum of 14 electrons, there are 14 elements in each of these series.

5 **a** silicon **b** potassium, caesium
c bromine **d** nitrogen, arsenic

6 Melting points increase across periods from groups 1 to 14 and then are drastically lower in groups 15–18.

7 bromine and mercury

8 As you move from left to right across groups 1, 2 and 13–17, the charge on the nucleus increases. Each time the atomic number increases by one, the electrons are attracted to an increasingly positive nucleus. Within a period, the outer electrons are in the same shell—that is, they have the same number of inner-shell electrons shielding them from the nucleus. Therefore, the additional nuclear charge attracts the electrons more strongly, drawing them closer to the nucleus and so decreasing the size of the atom.

9 **a** Magnesium and phosphorus, with outer electrons in the third shell, are in the same period. Magnesium has a nuclear charge of +12, but with completed inner shells of $1s^22s^22p^6$, the outer electrons experience the attraction of a core charge of +2. The outer-shell electrons of phosphorus, which has a nuclear charge of +15 and the same number of inner shells as magnesium, are attracted by a core charge of +5. The stronger attraction of the phosphorus electrons to the core means that more energy is required to remove an electron from a phosphorus atom than from a magnesium atom.
b Both fluorine and iodine are in group 17, so the outer electrons of each atom experience the attraction of the same core charge. Because the outer-shell electrons of a fluorine atom are closer to the nucleus than those of an iodine atom, they are attracted more strongly and so more energy is needed to remove one.

10 **a** $1s^22s^22p^3$ **b** period 2 and group 15
c 5 **d** +5

11 As the atomic number increases across a period of the periodic table, the number of protons in the nucleus increases, pulling the electrons added to the outermost shell towards the nucleus more strongly. This results in a decreased atomic radius. Once each shell is filled, a new period is started and additional electrons are added to a new shell, making the radius increase.

12 60 pm

13 **a** The radius of the atoms decreases as the core charge increases.
b There is a trend from metals (lithium, beryllium) to metalloids (boron) to non-metals (carbon, nitrogen, oxygen, fluorine and neon).
c Electronegativity increases as the core charge increases and size of the atoms decreases.

14 **a** lithium, sodium, potassium, rubidium
b The reactivity of metals increases down a group. This is because the number of electron shells increases down a group, so the valence electrons are farther from the nucleus and more easily lost. Since metals lose electrons in their reactions, those that lose electrons most easily will be most reactive.

15 **a** nitrogen **b** chlorine **c** chlorine

16 The reactivity decreases then increases. Metal atoms lose electrons in chemical reactions, and it is easier for atoms with a low core charge to do this. These atoms are found on the left-hand side of the periodic table. Moving across the period from left to right, core charge increases and reactivity decreases. However, in the non-metals, which are located on the right-hand side of the period, electrons are gained or shared by their atoms in chemical reactions, so as the core charge increases from left to right across the periodic table, the reactivity of non-metals increases.

17 Metals conduct electricity, are usually solid at room temperature and are located on the left side of the periodic table. They often contain one, two or three valence electrons.

18 The first ionisation energy decreases from Mg to Al because of their electronic configurations. Mg ($1s^22s^22p^63s^2$) has a completely filled 3*s*-subshell. In Al ($1s^22s^22p^63s^23p^1$), the valence electron is in the 3*p*-subshell and slightly farther from the nucleus, and it experiences a little bit more shielding from the nuclear charge due to the electrons in the 3*s*-subshell. Therefore, the valence electron in Al are not as strongly attracted to the nucleus, so Al has a lower first ionisation energy.

19 Student answers will vary. Possible answer: Periodicity refers to the periodic, or regularly recurring, pattern of properties of the elements when they are listed in order of atomic number.

20 Student answers will vary. Possible answer: The first periodic table, Benfey's spiral periodic table, illustrates the sizes of different periods (two periods of eight elements, two periods of 18 elements, two periods of 32 elements, etc.). The second, the Janet form of the periodic table, arranges the elements in order of the filling of electron subshells. Student answers will vary for additional forms of the periodic table.

21 Student answers will vary. Ensure the assigned labels are internally consistent and accurately reflect atomic radii.

Chapter 5 Bonding

5.1 Metallic bonding

TY 5.1.1 Mg has an atomic number of 12. This means that a neutral atom of Mg has 12 electrons. The electronic configuration is $1s^2 2s^2 2p^6 3s^2$. Mg therefore has two electrons in its outer shell (the $3s^2$ electrons). Mg atoms will tend to lose these two valence electrons to form a cation with a charge of +2. The outer-shell electrons become delocalised and form the sea of delocalised electrons within the metal lattice. An electric current occurs when there are free-moving charged particles. If the Mg is part of an electric circuit, the delocalised electrons can move through the lattice towards a positively charged electrode.

1 a Li atoms have three electrons. The electronic structure is $1s^2 2s^1$. There is one electron in the outer shell. The charge of the cation will therefore be +1.

b Mg atoms have 12 electrons. The electronic structure is $1s^2 2s^2 2p^6 3s^2$. There are two electrons in the outer shell. The charge of the cation will therefore be +2.

c Ga atoms have 31 electrons. The electronic structure is $1s^2 2s^2 2p^6 3s^2 3p^6 3d^{10} 4s^2 4p^1$. There are three electrons in the outer shell. The charge of the cation will therefore be +3.

d Ba atoms have 56 electrons. The electronic structure is $1s^2 2s^2 2p^6 3s^2 3p^6 3d^{10} 4s^2 4p^6 4d^{10} 5s^2 5p^6 6s^2$. There are two electrons in the outer shell. The charge of the cation will therefore be +2. Another way to approach this is to notice that barium is in the same group as magnesium. All alkaline earth metals form +2 cations by losing their two outershell electrons.

2 a Both potassium and gold have good thermal and electrical conductivity. However, gold has a higher density and higher melting and boiling points than potassium.

b sodium **c** silver

d Sodium and potassium are in group 1. Gold and silver are in the transition metals.

3 a silver, copper, gold, aluminium

b Availability and cost need to be considered. Other properties might include malleability and ductility.

4 Sodium belongs to the alkali metals which have relatively low melting and boiling points, relatively low density and are relatively soft. Iron is a transition metal. Transition metals have relatively high melting and boiling points, relatively high density and are relatively hard.

5 tensile strength, cost, availability

6 a

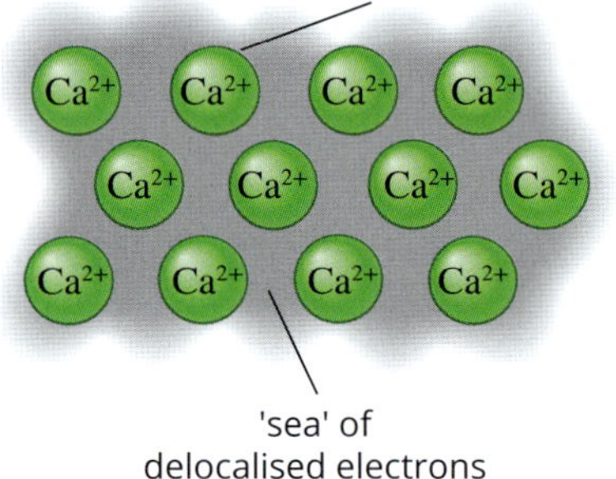

b strong electrostatic forces of attraction between Ca^{2+} ions and the delocalised valence electrons

7 Barium has a high melting point because there are strong attractive forces between the positive ions and the delocalised electrons. Barium conducts electricity because the delocalised electrons in the outer shell are free to move.

8 a Both graphite and metals are lustrous and conduct heat and electricity.

b These properties are explained by the presence of free-moving electrons. Both graphite and metals must contain delocalised electrons.

5.2 Ionic bonding

TY 5.2.1 BaF_2

1 B

2 a The diagram shows that in solid sodium chloride, the sodium and chloride ions are held in fixed positions in the crystal lattice and are not free to move and conduct electricity.

b Molten sodium chloride contains sodium and chloride ions that are free to move and, therefore, it can conduct electricity.

3 The electrostatic forces of attraction between the positive and negative ions holding the lattice together are very strong, and a lot of energy is required to break them apart.

4 a

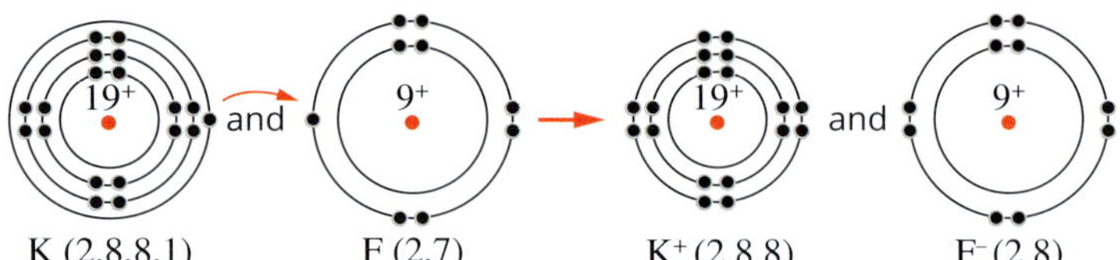

b

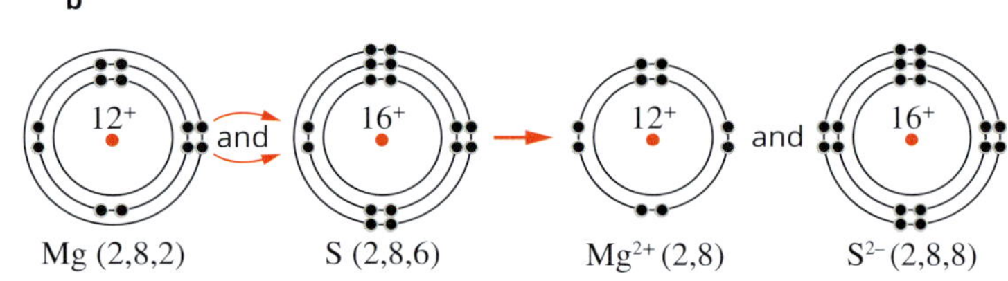

c

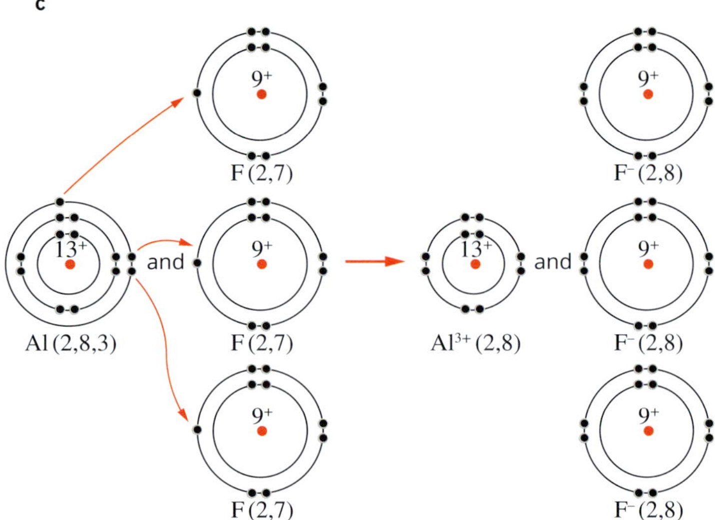

d

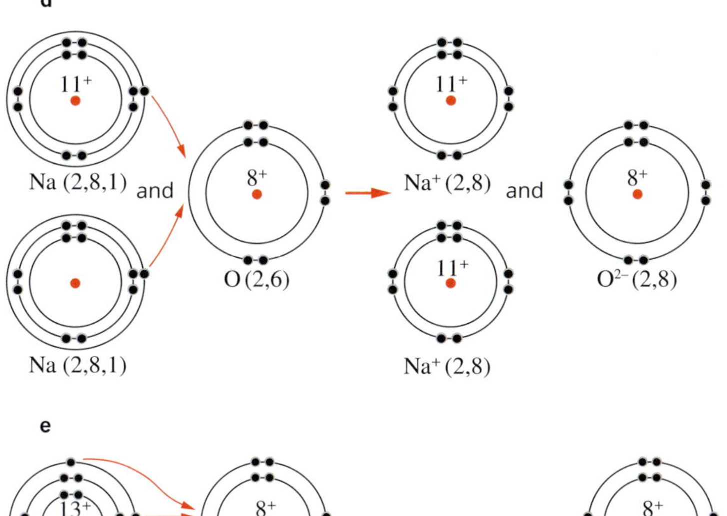

e

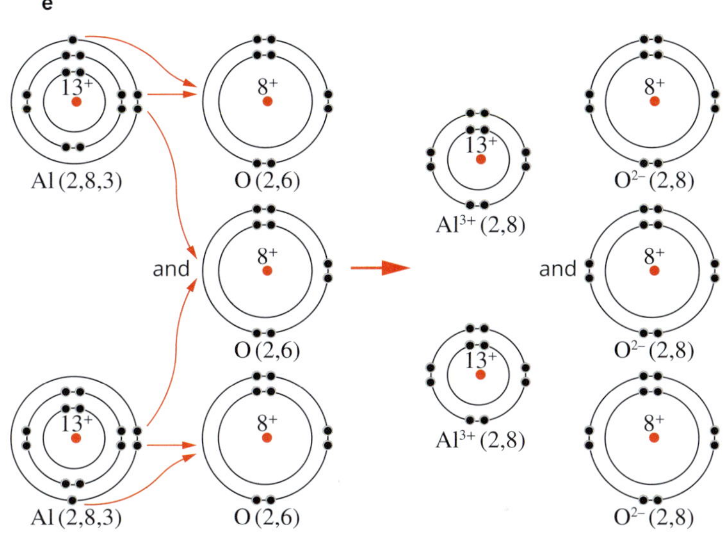

5 The electronic configuration of an atom of potassium is $1s^2 2s^2 2p^6 3s^2 3p^6 4s^1$ and that of an atom of chlorine is $1s^2 2s^2 2p^6 3s^2 3p^5$. Because an atom of K has one more electron than an atom of a noble gas and an atom of Cl has one less, in a reaction one K atom can donate one electron to one Cl atom to give K^+ ($1s^2 2s^2 2p^6 3s^2 3p^6$) and Cl^- ($1s^2 2s^2 2p^6 3s^2 3p^6$), both of which have the electronic configuration of the noble gas argon. The formula of this compound is, therefore, KCl. Calcium, however, has the electronic configuration of $1s^2 2s^2 2p^6 3s^2 3p^6 4s^2$. A Ca atom will lose two electrons to gain a noble gas configuration. Because each Cl atom will gain only one electron, there needs to be two chlorine atoms for each Ca atom. This reaction will therefore produce Ca^{2+} ($1s^2 2s^2 2p^6 3s^2 3p^6$) ions and Cl^- ($1s^2 2s^2 2p^6 3s^2 3p^6$) ions. The formula of the compound is, therefore, $CaCl_2$.

6 **a** NaCl **b** KBr **c** $ZnCl_2$ **d** K_2O **e** $BaBr_2$
f Al_2I_3 **g** AgBr **h** ZnO **i** BaO **j** Al_2S_3

7 **a** potassium chloride **b** calcium oxide **c** magnesium sulfide
d potassium oxide **e** sodium fluoride

8 **a** Na_2CO_3 **b** $Ba(NO_3)_2$ **c** $Al(NO_3)_3$ **d** $Ca(OH)_2$
e $Zn(SO_4)_2$ **f** KOH **g** KNO_3 **h** $ZnCO_3$
i K_2SO_4 **j** $Ba(OH)_2$

5.3 Covalent bonding

TY 5.3.1

H
H : N :
H

1 **a** 1 **b** 3 **c** 2 **d** 1

2 fluorine (F_2)

: F : F :

hydrogen fluoride (HF)

H : F :

water (H_2O)

H : O : H

tetrachloromethane (CCl_4)

Cl
Cl : C : Cl
Cl

phosphine (PH_3)

H : P : H
H

butane (C_4H_{10})

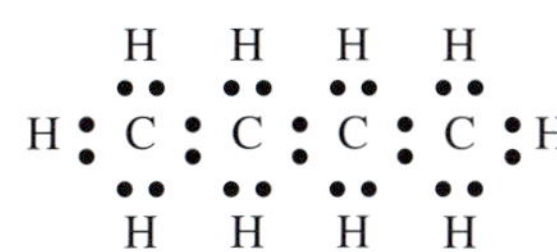

carbon dioxide (CO_2)

O :: C :: O

3 **a** 1 **b** 2 **c** 3 **d** 4 **e** 1 **f** 0

4 To complete its outer shell, the oxygen atom uses two of its outer-shell electrons to form two single bonds or a double bond with suitable non-metal atoms. The remaining four electrons in the outer shell are not required for bonding, as the outer shell is now complete, and they arrange themselves as two lone pairs around the oxygen atom.

5 **a** CCl_4, carbon tetrachloride
b NBr_3, nitrogen tribromide
c SiO_2, silicon dioxide
d HF, hydrogen fluoride
e PF_3, phosphorus trifluoride

6 A conductor must contain free-moving charged particles, such as ion or electrons. Covalent compounds do not contain ions or delocalised electrons and therefore do not conduct an electric current.

7 **a** Non-metallic atoms bond together by sharing one or more valence electrons so that there are eight electrons in the outer shell of each atom.
b The hydrogen atom has only a single electron. This electron is in the first shell, which can hold a maximum of two electrons. The first shell is filled when hydrogen shares one electron from a non-metal atom.

8 **a** Allotropes are different forms of the same element in the same physical state. Allotropes differ in the way atoms are bonded within their structures.
b **i** oxygen and ozone
ii red, white and black phosphorus

5.4 Intermolecular forces

TY 5.4.1 The sulfur and hydrogen atoms are arranged in a bent or V-shape.

TY 5.4.2 HCl is more polar than NO.

1 **a** V-shaped **b** linear **c** tetrahedral **d** pyramidal **e** linear

2 **a** O **b** C **c** N **d** N **e** F **f** F

3

4 **a** polar **b** polar **c** polar **d** polar **e** non-polar

5 **a** ionic **b** polar covalent **c** non-polar covalent

6 B and E **7** C

8 dipole–dipole forces: b, c, d, e, f; hydrogen bonds: a, g, h; neither: i (dispersion forces between H_2 molecules)

5.5 Covalent network structures

1 **a** Diamond is hard because it has strong covalent bonds throughout the network, with all atoms being held in fixed positions. Graphite is soft because there are weak dispersion forces between the layers in graphite, so layers can be made to slide over each other easily.
b Diamond is a non-conductor of electricity because all of its electrons are localised in covalent bonds and are not free to move. Graphite can conduct electricity because it has delocalised electrons between its layers of carbon atoms.

2 **a** Graphite is used as a dry lubricant because the dispersion forces between the layers in graphite enable the layers to slide over each other easily and to reduce the friction between moving parts.
b The strong covalent bonding throughout the network means that the carbon atoms are fixed in place. This makes the diamond very hard and suitable as a material for cutting other less hard materials.

3 Fullerenes are similar to graphite in that they are allotropes of carbon in which each carbon atom has bonds to three other carbon atoms. Fullerenes conduct electricity and heat, similar to graphite. Fullerenes differ from graphite in that they are nanomaterials, which give them different physical properties.

4 Each carbon atom in a buckyball is covalently bonded to three other carbon atoms. Each carbon atom has one free electron, which is shared between the other carbon atoms in the sphere. The structure consists of hexagonal and pentagonal rings of atoms.

5 Each carbon atom in a graphene sheet is covalently bonded to three other carbon atoms. Each carbon atom has one free electron that is shared with the other carbon atoms in the sheet. The structure consists of hexagonal rings of atoms. The sheet is one atom thick but may be of any length and width.

6 The structure of silicon dioxide is a covalent network structure. Each silicon atom sits at the centre of a tetrahedron and is covalently bonded to four oxygen atoms. Each oxygen atom is bonded to two silicon atoms.

Chapter 5 review

1 Electrical conductivity in the solid state.

2 **a** When a current is applied to copper wire, the free-moving, delocalised electrons move from one end to the other, so the copper wire conducts electricity.
b The delocalised electrons in the metal spoon obtain energy from the boiling mixture and move more quickly. These electrons move freely throughout the spoon, colliding with other electrons and metal ions, transferring energy, so the spoon becomes warmer and eventually too hot to hold.
c A lot of energy is required to overcome the strong forces of attraction between the iron ions and the delocalised electrons in the metal lattice.
d Because of the strong forces of attraction between them, the lead ions and the delocalised electrons form a closely packed three-dimensional structure. Also, the lead atom itself has a higher mass-to-volume ratio than the sulfur atom. This means that the density—the mass per volume—is high.
e As the copper is drawn out, the copper ions are forced apart. The delocalised electrons rearrange themselves around these ions and re-establish strong forces of attraction.

3 **a** **i** valence electrons that are not restricted to a region between two atoms
ii a regular three-dimensional arrangement of a very large number of positive ions (cations)
iii the electrostatic attraction between a lattice of cations and delocalised electrons
b valence electrons

4 **a** Assemble equipment to test conductivity. Add a globe to the circuit. When the electrodes are touching the solid magnesium chloride, the globe will not light up.
b Using the same equipment with molten sodium chloride, the globe will glow. Care is needed, as sodium chloride melts at 801°C.
c If a crystal of sodium chloride was hit firmly with a hammer, it would shatter. Again, care is needed—safety glasses must be worn.

5 **a** The electrostatic forces of attraction between the positive and negative ions are strong and will be overcome only at high temperatures.
b The strong electrostatic forces of attraction between the ions mean that a strong force is needed to break up the lattice, giving the ionic crystals the property of hardness. However, the crystal lattice will shatter when a strong force is applied, suddenly causing ions of like charge to become adjacent to one another and be repelled.
c In the solid state, the ions are not free to move. However, when the solid melts or dissolves in water, the ions are free to move and conduct electricity.

6 **a**

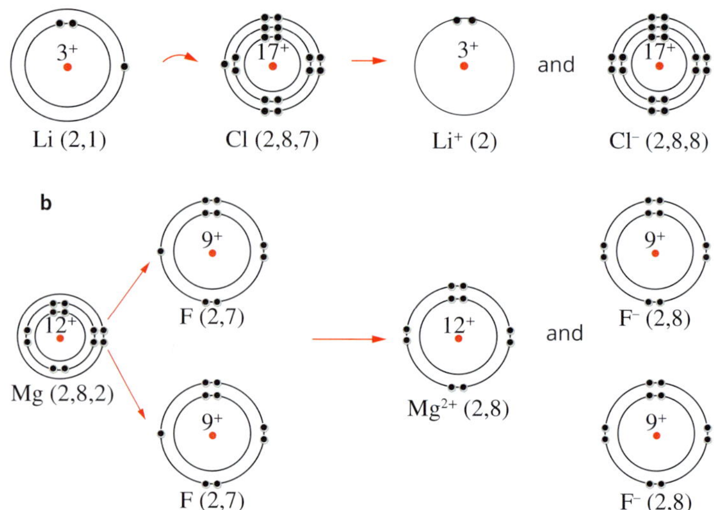

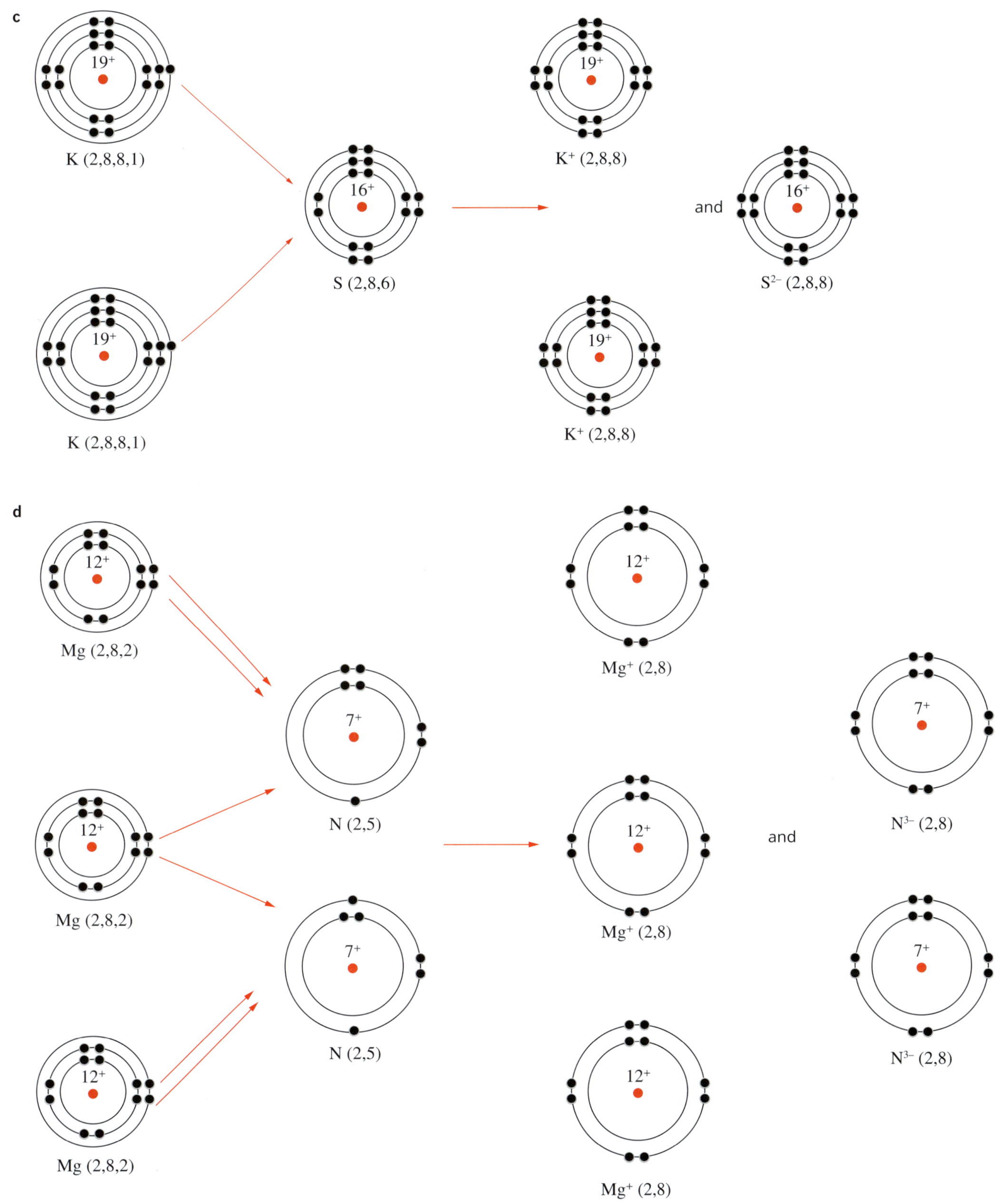

7 **a** KBr **b** MgI_2 **c** CaO **d** AlF_3 **e** Ca_3N_2

8 **b** CD_3 **c** EF **d** G_3H **e** KL

9 **a** CuCl **b** Ag_2O **c** Li_3N **d** KI **e** $Cu(NO_3)_2$ **f** CrF_2 **g** K_2CO_3 **h** $Mg(HCO_3)_2$ **i** $Ni_3(PO_4)_2$

10 **a** ammonium carbonate
b copper(II) nitrate
c chromium(III) bromide

11 The strength of the intermolecular bonds in pure hydrogen chloride must be relatively weak. Since pure hydrochloride exists as a gas at room temperature, it must have a low boiling temperature, indicating that not much energy is required to break the intermolecular bonds between molecules.

12 D

13 **a** 6 **b** 3 **c** 4

14 Atom X requires 2 electrons to form a stable outer shell. Each atom of element Y can share 1 electron. Therefore, the molecular formula will be XY_2.

15 The diagrams of each molecule below show the number of electrons.

a 2 bonding electrons, 6 non-bonding electrons

H : Br :

b 6 bonding electrons, 8 non-bonding electrons

H : O : O : H

c 8 bonding electrons, 24 non-bonding electrons

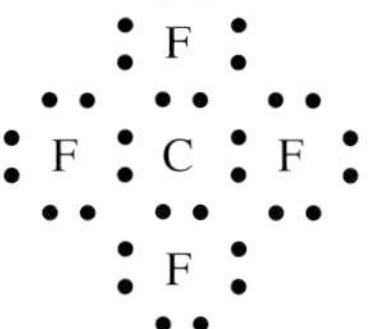

d 14 bonding electrons, 0 non-bonding electrons

H H
H : C : C : H
H H

e 6 bonding electrons, 20 non-bonding electrons

: F : P : F :
: F :

f 4 bonding electrons, 16 non-bonding electrons

: Cl : O : Cl :

g 8 bonding electrons

H
H : C : H
H

h 4 bonding electrons, 4 non-bonding electrons

: S : H
H

16 **a** The bonds are similar in that they all involve the sharing of electron pairs between two atoms; that is, they are covalent bonds.

b They differ in the number of electron pairs shared: one pair (fluorine), two pairs (oxygen) and three pairs (nitrogen).

17 PCl_3—pyramidal, HOCl—V-shaped, $CHCl_3$—tetrahedral, HF—linear

18 **a** non-polar

S═C═S

b polar

O
Cl Cl

c non-polar

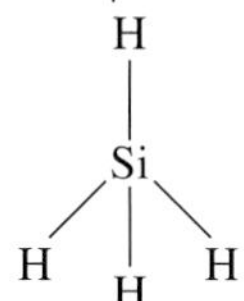

d polar

H
C
H H Cl

e non-polar

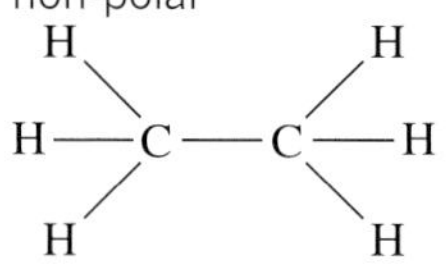

19 Si–O, H–Br, N–O, O–Cl, F–F

20

a	SO_3	**i** non-polar	**ii** dispersion forces
b	$SiCl_4$	**i** non-polar	**ii** dispersion forces
c	CF_4	**i** non-polar	**ii** dispersion forces
d	NF_3	**i** polar	**ii** dipole–dipole attraction
e	CH_3NH_2	**i** polar	**ii** hydrogen bonding

21 **a** CCl_4

b CH_4 and CCl_4 are both non-polar, so their molecules are held together only by dispersion forces. CCl_4 is the larger of these two molecules and has more electrons, so the dispersion forces between CCl_4 molecules will be greater than those between CH_4 molecules. As there are stronger dispersion forces between molecules of CCl_4 than between molecules of CH_4, it takes more energy to vaporise CCl_4.

22 When sugar turns to a liquid, it is melting; the intermolecular bonds are being broken. When the liquid turns black and a gas is produced, the intramolecular bonds are being broken, allowing new substances to be produced.

23 Refer to Figure 5.5.5 (page 189). Each carbon atom bonds covalently to three other carbon atoms in graphite to form layers. These layers consist of hexagonal rings of atoms, connected to one another. The fourth electron in each carbon atom is delocalised which explains its electrical conductivity. There are weak dispersion forces between the layers of graphite, allowing the layers to slide over each other, thus enabling it to act as a lubricant.

24 **a** Methane is an example of a molecular substance. It has strong, covalent intramolecular bonds and very weak intermolecular bonds. Diamond is an example of a covalent network structure. It has strong covalent bonds throughout its structure.

b Due to the weak dispersion forces between its molecules, methane will have extremely low boiling and melting points. If it were a solid it would be crystalline, brittle and soft. Due to its extended covalent network, diamond is extremely hard, does not exist as a liquid and has a very high sublimation point.

25 Diamond and silicon dioxide have a three-dimensional covalent network structure with strong covalent bonding. Each carbon and silicon atom is at the centre of a tetrahedron. In diamond, each carbon atom is bonded to four other carbon atoms. In silicon dioxide, each silicon atom is bonded to four oxygen atoms by covalent bonds. Each oxygen atom is bonded to two silicon atoms. In both substances, the strong covalent bonds extend throughout the network and account for its hardness and high melting points. All four valence electrons in each carbon and silicon atom are involved in the formation of covalent bonds, and there are no free electrons to carry an electric current.

26 Metallic. Ag and Cu. Ag and Cu are both transition metals. Since metals have low electronegativities and ionisation energies, they easily lose their valence electrons. A metal lattice consists of metal cations surrounded by a sea of delocalised valence electrons. Ionic. $CuCl_2$ and CaS. Ionic compound form by a transfer of electrons between metals and non-metals. Metals have low electronegativities and non-metals have high electronegativities. Metals are mainly found in groups 1 and 2 and in the transition block in the periodic table. Some metals are also found at the lower ends of the other groups.

Non-metals are found in groups 15, 16 and 17 of the periodic table.
Polar covalent. HCl, H_2O and NH_3 have polar covalent bonds between their constituent atoms. Oxygen and nitrogen are more electronegative than hydrogen. The bonding electrons are not shared equally because they are attracted to the more electronegative atom. Polar covalent bonds are formed between the atoms of different non-metals found in groups 15, 16, and 17 of the periodic table. The asymmetrical nature of both molecules means they have molecular dipoles making them polar molecules.
Non-polar covalent. Cl_2. Since both atoms in Cl_2 are identical, there is no electronegativity difference between them and the bonding electrons are shared equally. Non-polar covalent bonds are formed between identical atoms of non-metals found in groups 15, 16 and 17 in the periodic table.

27 **a** both metallic and ionic lattices
b both metallic and ionic lattices
c both metallic and ionic lattices
d ionic lattices only
e both metallic and ionic lattices

28 **a** **i** N_2
:N≡N:
ii O_2
:O=O:
iii CO_2
:O=C=O:
b **i** NH_3, HCl, H_2O, $CHCl_3$
ii N_2, Cl_2, O_2, CH_4, CO_2, CCl_4
iii NH_3, H_2O

29 The intermolecular bonds that hold molecules together in covalent molecular substances are much weaker (100 times weaker) than the chemical bonds holding the atoms together in ionic, metallic and covalent network substances. As a result, it takes much less heat energy to break the intermolecular bonds holding covalent molecular solids and liquids together, and these substances have relatively low melting and boiling points.

30 **a**

Name	Nitrogen	Hydrogen	Ammonia	Iron	Diamond	Potassium fluoride
formula	N_2	HBr	NH_3	Fe	C	KF
melting point (°C)	–210	-87	–78	1535	>3550 (sublimes)	858
bond type	dispersion forces	dipole-dipole forces (and dispersion forces)	hydrogen bonding (and dispersion forces)	metallic bonding	covalent bonding	ionic bonding

b Nitrogen (N_2). In solid nitrogen, weak dispersion forces exist between nitrogen molecules. These dispersion forces are a result of electrostatic attraction between the positive and negative ends of instantaneous dipoles in neighbouring molecules. The instantaneous dipoles are a result of random fluctuations in electron density around the two nitrogen atoms.
Hydrogen bromide (HBr). Dipole–dipole forces are the main cause of attraction between molecules when HBr is in the solid state. HBr molecules are polar because there is an electronegativity difference between the hydrogen and bromine atoms. Dipole–dipole attractions occur between polar HBr molecules: the partially positive charged hydrogen end of one HBr molecule is attracted to the partially negatively charged bromine end of another HBr molecule.
Ammonia (NH_3). Ammonia molecules are polar because nitrogen is more electronegative than hydrogen. There is an electrostatic attraction between the partially negative nitrogen atom on one ammonia molecule and the partially positive hydrogen atom on a nearby ammonia molecule. This special form of dipole-dipole bonding is called hydrogen bonding when hydrogen is bonded to fluorine, oxygen or nitrogen.
Iron (Fe). Metallic bonding is an electrostatic attraction between positively charged cations in the metal lattice and negatively charged delocalised electrons.
Diamond (C). The bonding in covalent network structures such as diamond is covalent. In diamond each carbon atom shares one valence electron with each of its four neighbouring carbon atoms. There is an electrostatic attraction between the shared electrons and the positive nuclei of the two carbon atoms involved in the formation of each bond.
Potassium fluoride (KF). An ionic bond is formed by the electrostatic attraction between the positively charged potassium cations and the negatively charged fluoride anions.

31 Student answers will vary, but will need to discuss how the salt and bicarbonate of soda are ionic substances that dissolve in water due to ions breaking away from the ionic lattice.

Module 1 Properties and structure of matter

Multiple choice

1 A 2 B 3 C 4 B 5 D 6 C
7 C 8 B 9 B 10 A 11 B 12 C
13 D 14 D 15 B 16 A 17 C
18 C 19 A 20 A

Short answer

1 31.0% aluminium, 22.0% potassium, 20.0% chlorine, 27.0% oxygen

2 **a** mass spectrometer

b $A_r = \frac{(37.30 \times 190.97) + (62.70 \times 192.97)}{100} = 192.2$

3 **a** Sodium has 11 electrons, which are distributed in three main shells or energy levels: two electrons in the first shell closest to the nucleus, eight electrons in the second shell and one in the third shell.

Shells are divided into subshells labelled *s*, *p*, *d* and *f*. Two of sodium's second-shell electrons are in the *s*-subshell and the remaining six are in the *p*-subshell.

s-, *p*-, *d*- and *f*-subshells are further divided into orbitals, which are regions of space in which electrons may be found. The *s*-subshell consists of only one orbital, the *p*-subshell has three. The six electrons of sodium in the 2*p*-subshell are distributed in pairs in the three *p*-orbitals.

b It represents an excited atom. A ground state electronic configuration would contain an electron in the 4*s*-subshell, because the 4*s*-subshell has a lower energy than the 3*d*-subshell.

4 **a** Chlorine is on the right side of the periodic table and sodium is on the left. Atomic radius decreases across a period because the increasing core charge pulls the outer-shell electrons closer to the nucleus, causing the volume of the atom to decrease.

b Fluorine is further to the right on the periodic table than lithium, and core charge increases from left to right across the periodic table. As core charge increases, the electrons are held closer to the nucleus, and more energy is required to remove the first one.

c Ba and Be are in the same group, with Be higher than Ba. Going down a group, atomic size increases, meaning the outer-shell electrons are further from the nucleus. Therefore, the outer electrons of Be are held more tightly and are less readily lost.

d The *s*-block elements have an *s*-subshell as their outer occupied electron subshell. The *s*-subshell can take one or two electrons, so the block only has two groups.

5 **a** K_3PO_4 **b** Al_2O_3 **c** $NaNO_3$ **d** Fe_2S_3

6 **a** Student answers will vary, e.g. NaCl preserves food, $NaHCO_3$ is baking soda, NaF is in toothpaste to harden tooth enamel.

b **i** Although charged particles are present, they are held in fixed positions in the lattice and so cannot move to carry a current.

ii If a strong force is applied to a crystal of the compound, the layers of ions will move relative to one another, causing ions of like charge to be adjacent and hence repel. The crystal thus shatters.

iii A large amount of energy is needed to overcome the electrostatic attraction between oppositely charged ions.

7 **a** V-shaped/bent and polar

b triangular pyramid and polar

c tetrahedral and non-polar

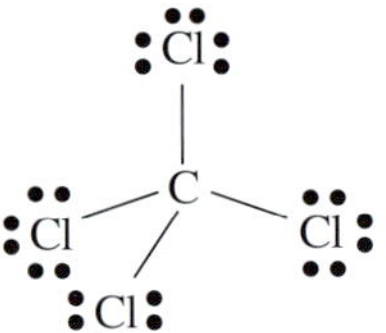

d linear and non-polar

e tetrahedral and non-polar

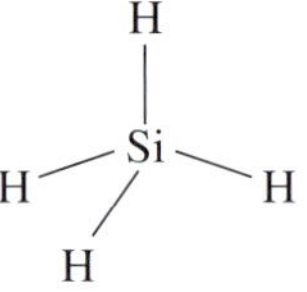

8 **a** The bonds between the molecules of ice are hydrogen bonds. The bonds between the atoms within the water molecules are covalent bonds. Covalent bonds are much stronger than hydrogen bonds and so require much more energy, and thus a higher temperature, to break.

b In ice, each water molecule forms hydrogen bonds with four other water molecules. This arrangement is very open, causing ice to have a lower density than liquid water.

c Ethyne has the structure CHCH. The C atoms have a triple bond between them, each using three of their four valence electrons to form the triple bond. The fourth valence electron of each carbon atom forms a covalent bond with a hydrogen atom. There are no lone pairs and only two bonding regions on the carbon atoms. So these adopt a linear arrangement. Hydrogen peroxide has the following structure:

Each oxygen has six valence electrons. Two are involved in bonding, leaving two pairs of non-bonding electrons. So there are four regions where electrons are present. These assume a tetrahedral arrangement to minimise repulsion, so the molecule is not linear.

9 **a** HCl, HF, H_2O, H_2S **b** HCl, H_2S **c** HCl, HF, F_2, H_2, O_2
d H_2O, HF **e** O_2
f H_2, as it is the smallest non-polar molecule in the list

Extended response

1 The fertiliser is a mixture of insoluble (rock, coal) and soluble (urea, potash) materials, so the first step is to filter out the insolubles. Mixing the fertiliser with water and passing it through a filter will separate the coal and phosphorus rock. Coal is less dense than water, so mixing the insolubles with water will make the coal float and the rock settle, which will allow the floating coal to be decanted. The coal and rock are now separated but are still mixed with water. A final filtration step isolates these from the water. Of the solubles, the urea derivatives are less soluble than potash. By evaporating the water, the urea derivatives will precipitate first, and can be separated by decanting or filtration. The final component, potash, can be separated from water by evaporating to dryness.

2 **a** $1s^22s^22p^63s^2$

b **i** 0.140 nm

ii The Na atom would have a larger radius because there are fewer protons in the nucleus, so the attraction for the outer-shell electron would be weaker, pulling it less strongly towards the nucleus.

c **i** It is a lattice of positively charged magnesium ions surrounded by a 'sea' of valence electrons. The lattice is held together by the electrostatic attraction between electrons and cations.

ii The electrons are not localised, but are free to move, so metals can conduct an electric current.

iii The model cannot explain the differences in melting points/densities/electrical conductivities/magnetism between metals.

d any suitable example such as K, Na, Ca

e Transition elements have an incomplete *d*-subshell of electrons.

3 a 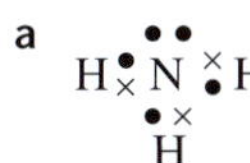

b The four electron pairs form a tetrahedral arrangement around the atom due to the repulsion of the electron pairs. The result is a pyramid-shaped molecule.

covalent bond

hydrogen bond and dispersion forces

c i N≡N

ii O=C=O

d i N_2 has no polar bonds, so the intermolecular bonds are weak dispersion forces. Although the bonds within CO_2 are polar, the molecule overall is symmetrical, so the bond dipoles cancel and the overall molecule is non-polar. Thus, the only intermolecular forces are dispersion forces.

ii Nitrogen is a highly electronegative element, and consequently the bonds between the atoms of nitrogen and hydrogen are highly polarised. The ammonia molecule is a dipole because its shape is not symmetrical. There is an electrostatic attraction between the nitrogen atom of one ammonia molecule and the hydrogen atom of a nearby ammonia molecule. The attractions between these dipoles are known as hydrogen bonds.

4 a Allotropes are different physical forms of the same element.

b Both consist of carbon atoms covalently bonded to other carbon atoms.

c Diamond is a three-dimensional lattice in which each carbon atom is covalently bonded to four other carbon atoms in a tetrahedral configuration. Thus, strong bonding extends throughout the lattice. Graphite consists of layers of carbon atoms in which each atom is covalently bonded to three other carbon atoms, making strong layers. There are weaker dispersion forces between the layers. The one electron not involved in bonding is delocalised.

d The delocalised electrons in graphite are free to move and serve as current, so graphite is an excellent conductor. In diamond, each carbon atom is bonded with four other carbon atoms, so there are no free electrons.

e Because of the weak bonding between the layers of graphite, the layers can slide over one another and thus can slide onto a page.

5 a H : P : H
H

b PH_3 is a polar molecule. It is asymmetrical, so it has an overall molecular dipole. (The P–H bond is not particularly polar, but the overall asymmetry of the molecule results in an overall molecular dipole.)

c In a PH_3 molecule, there are four pairs of electrons around the central P atom. These electron pairs adopt a tetrahedral geometry. Since there is one lone pair, the molecular geometry is trigonal pyramidal.

d $M_r(PH_3) = 33.99$. $M_r(NH_3) = 17.03$.

e Since H is bonded to N in ammonia, NH_3 molecules are able to form hydrogen bonds between molecules. Between PH_3 molecules there are dipole–dipole forces. Hydrogen bonding is stronger than dipole–dipole forces, so NH_3 has the higher melting point.

f $^{32}_{15}P$

g $^{32}_{15}P \rightarrow {}^{32}_{16}S + {}^{0}_{-1}e$

h 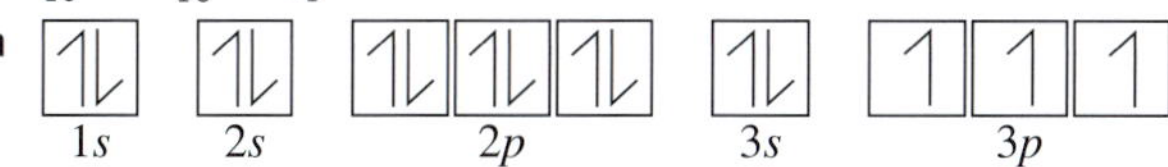

Chapter 6 Chemical reactions and stoichiometry

6.1 Writing chemical equations

TY 6.1.1 $3H_2(g) + N_2(g) \rightarrow 2NH_3(g)$

TY 6.1.2 $3AgNO_3(aq) + K_3PO_4(aq) \rightarrow Ag_3PO_4(s) + 3KNO_3(aq)$

TY 6.1.3 $Ca_3(PO_4)_2(s) + 3SiO_2(s) + 5C(s) \rightarrow 3CaSiO_3(s) + 5CO(g) + 2P(s)$

1 C

2 a $2NO(g) + 5H_2(g) \rightarrow 2NH_3(g) + 2H_2O(l)$

b $2C(s) + O_2(g) \rightarrow 2CO(g)$

c $5CuO(s) + 2NH_3(g) \rightarrow 5Cu(s) + 2NO(g) + 3H_2O(l)$

3 a $2KClO_3(s) \rightarrow 2KCl(s) + 3O_2(g)$

b $2C_2H_2(g) + 5O_2(g) \rightarrow 4CO_2(g) + 2H_2O(g)$

c $2Na(s) + 2H_2O(l) \rightarrow 2NaOH(aq) + H_2(g)$

4 a Ag: 16; S: 8

b Mg: 6; P: 4: O: 16

c Na: 3; Cl: 3; O: 9

5 a Zinc metal is added to copper(II) sulphate solution. The zinc becomes coated with copper and colourless zinc sulphate solution is produced.

b Magnesium ribbon reacts with carbon dioxide gas. Black carbon (soot) and some magnesium oxide solid is produced.

c A potassium hydroxide solution reacts with a solution of carbonic acid (dihydrogen carbonate) to form liquid water and a potassium carbonate solution.

6.2 Problems involving conservation of mass

TY 6.2.1 15.3 g

TY 6.2.2 3.50 g

TY 6.2.3 68.0 g hydrogen peroxide produced; 64.0 g oxygen consumed

TY 6.2.4 175 g

1 a true

b False. Density is a measure of the mass of a substance divided by the volume that it occupies.

c true

d False. Volume is not always conserved in a chemical reaction.

2 a 8.29 g **b** 16.0 g **c** 18.30 g

3 a Reactants: 267.6 g. Products: 267.6 g. As the mass of the reactants and products is equal, this satisfies the law of conservation of mass.

b 32.4 g

c 35.0 g

d Following the law of conservation of mass, if all reactant quantities are doubled and all reactants are consumed in the chemical reaction, then the total mass of the products will also double.

4 a 4.28 g **b** 2.36 L

Chapter 6 review

1 D **2** D **3** B

4 a $4Li(s) + O_2(g) \rightarrow 2Li_2O(s)$

b $2Al(OH)_3(s) \rightarrow Al_2O_3(s) + 3H_2O(l)$

c $10Rb(s) + 2RbNO_3(aq) \rightarrow 6Rb_2O(s) + N_2(g)$

d $4Fe(OH)_2(aq) + O_2(g) \rightarrow 2Fe_2O_3(s) + 4H_2O(l)$

e $2NaOCl(aq) \rightarrow 2NaCl(aq) + O_2(g)$

f $4H_2SiCl_2(s) + 4H_2O(l) \rightarrow H_8Si_4O_4(aq) + 8HCl(aq)$

5 a $2HNO_3(aq) + Na_2CO_3(aq) \rightarrow 2NaNO_3(aq) + H_2O(l) + CO_2(g)$

b $H_2SO_4(aq) + 2KOH(aq) \rightarrow K_2SO_4(aq) + 2H_2O(l)$

c $Ba(NO_3)_2(aq) + H_2CO_3(aq) \rightarrow BaCO_3(s) + 2HNO_3(aq)$

6 a $3Cu(s) + 8HNO_3(aq) \rightarrow 3Cu(NO_3)_2(aq) + 2NO(g) + 4H_2O(l)$
b $8CO(g) + 17H_2(g) \rightarrow C_8H_{18}(l) + 8H_2O(l)$
c $(NH_4)_2Cr_2O_7(s) \rightarrow N_2(g) + 4H_2O(g) + Cr_2O_3(s)$
d $10P_2I_4(s) + 13P_4(s) + 128H_2O(l) \rightarrow 40PH_4I(aq) + 32H_3PO_4(aq)$

7 a 30.6 g $H_2O(l)$ b 41.0 g $CO_2(g)$ c 1.7 g $O_2(g)$

8 a 0.4 g b 0.34 g c 0.022 g

9 a A chemical change is said to have occurred when a new substance has been formed. The evidence to suggest this includes a colour change; heat given off or absorbed; light emitted; a smell, odour or gas produced; or a precipitate formed. In the case of vinegar and milk, a precipitate is formed.
b The atoms from the two substances are rearranged to form the new products. The number and type of atoms in a chemical reaction remains the same.
c 55.0 g

10 a $2Ca(s) + O_2(g) \rightarrow 2CaO(s)$ b 14.0 g c 6.0 g

11 0.16 g

12 a 31.5 g b 9.52 mL c 2100 g or 2.10 kg
Hint: You need to convert L to mL, as density is given in $g\,mL^{-1}$.

13 a 58.1 L b 12.9 g c 8.72×10^3 L

14 a As temperature increases, particles have more kinetic energy and spread further apart, thus reducing the density of that substance.
b As pressure increases, particles are forced closer together, thus the density increases.

15 a $C_2H_5OH(l) + C_3H_7COOH(l) \rightarrow C_6H_{12}O_2(l) + H_2O(l)$
b 7.89 g c 15.1 g d 19.9 g e 22.6 mL

16 a $2NH_4NO_3(s) \rightarrow 2N_2(g) + O_2(g) + 4H_2O(g)$
Note: As the question stipulates water vapour, it must be listed in gaseous state.
b 18.0 g H_2O c 11.2 L d 5.6 L e 41.5 L
f As the volume goes from 23.1 mL to 41.5 L, it can be seen that the volume increases by more than 1000 times. The rapid expansion and high kinetic energy of the particles during the reaction accounts for the explosive properties of ammonium nitrate.

17 a Though the total number of particles has decreased in the reaction, the total mass of the products must equal the total mass of the reactants according to the law of conservation of mass.
b Gases do have mass, but can be difficult to measure.
c Density is a measure of how close particles are, whereas viscosity is a measure of how easily a substance flows or how 'thick' it appears.

18 $2HgO(s) \rightarrow 2Hg(l) + O_2(g)$
a 2.16 g b 4.23 g

a 1s [↑↓] 2s [↑↓] 2p [↑↓][↑][↑]

b 1s [↑↓] 2s [↑↓] 2p [↑↓][↑↓][↑↓] 3s [↑↓]

c 3d [↑][↑][↑][↑][↑] 4s [↑↓]

d 3d [↑↓][↑↓][↑↓][↑↓][↑↓] 4s [↑↓] 4p [↑][↑][↑]

19 a $4Na(s) + O_2(g) \rightarrow 2Na_2O(s)$
b The mass of the product (sodium oxide) produced is equal to the sum of the mass of the sodium and the mass of the oxygen reacting.

20 a reactant: copper(II) nitrate; products: oxygen gas, copper(II) oxide, nitrogen dioxide gas
b $2Cu(NO_3)_2(s) \rightarrow 2CuO(s) + 4NO_2(g) + O_2(g)$

21 It is not a closed system and it is likely that some of the carbon dioxide gas evolved from the reaction has escaped from the beaker, resulting in the apparent mass loss in the reaction.

Chapter 7 The mole concept

7.1 Introducing the mole

TY 7.1.1 9.6×10^{23}
TY 7.1.2 8.4×10^{23} atoms
TY 7.1.3 0.0013 mol
TY 7.1.4 1.5 mol

1 Number of particles = amount (mol) $\times N_A$
a 1.2×10^{24} atoms b 6.0×10^{22} molecules
c 1.20×10^{25} atoms d 2.5×10^{24} molecules
e 6.0×10^{21} atoms f 2.78×10^{19} molecules

2 a 0.50 mol b 0.25 mol c 70 mol d 70 mol

3 a 1.7×10^{-4} mol b 1.7×10^{-4} mol c 1.7×10^{-4} mol

4 a 0.8 mol b 4.8 mol c 0.72 mol d 6.0 mol

7.2 Molar mass

TY 7.2.1 496 g
TY 7.2.2 7.4×10^{21}

1 a 28.02 $g\,mol^{-1}$ b 17.03 $g\,mol^{-1}$ c 98.09 $g\,mol^{-1}$
d 241.88 $g\,mol^{-1}$ e 60.05 $g\,mol^{-1}$ f 32.07 $g\,mol^{-1}$
g 176.12 $g\,mol^{-1}$ h 249.72 $g\,mol^{-1}$

2 a 23 g b 64 g c 1.6 g d 25 g

3 a 2.5 mol b 5.0 mol c 0.10 mol d 0.025 mol
e 0.0031 mol f 0.0063 mol g 9.7×10^{-6} mol
h 3.9×10^{-5} mol

4 a 6.0×10^{23} atoms b 6.0×10^{22} atoms
c 6.02×10^{21} atoms d 3.06×10^{22} atoms

5 a i 3.0×10^{23} molecules
ii 6.0×10^{22} molecules
b 6.0×10^{22} atoms.
c 4.08×10^{25} atoms.

7.3 Percentage composition and empirical formula

TY 7.3.1 35.00%
TY 7.3.2 $MgCl_2$
TY 7.3.3 C_4H_{10}

1 a 69.94% b 84.80% c 26.19% d 51.18%

2 a HCl b CO c MgO d CH_4

3 a C_6H_6 b H_2O_2 c $C_3H_6O_3$ d NO_2 e $C_{11}H_{22}$

4 a CH_2 b C_5H_{10}

5 a CH_2O b $C_6H_{12}O_6$

7.4 Calculations based on the amount of a reactant or product

TY 7.4.1 0.013 mol
TY 7.4.2 1.486 g

1 81.9 g

2 a 0.020 mol b $\frac{2}{1}$ c 8.52 g

3 276.92 g

4 3.38 g

7.5 Calculations based on the amounts of two reactants

TY 7.5.1 a O_2 b 7.08×10^8 g

1 B, D, A, E, C

2

Nitrogen molecules available	Hydrogen molecules available	Ammonia molecules produced	Nitrogen molecules in excess	Hydrogen molecules in excess
2	10	4	0	4
879	477	318	720	0
9 mol	6 mol	4 mol	7 mol	0 mol

3 a $2Na(s) + Cl_2 \rightarrow 2NaCl(s)$ b 63.8 g

4 a $Pb(NO_3)_2$ is in excess by 0.50 mol.
b $Pb(NO_3)_2$ is in excess by 1.8 mol
c KI will be in excess. $m(PbI_2) = 1.39$ g

Chapter 7 review

1 **a** 0.75 mol **b** 15.0 mol **c** 3.8×10^4 mol **d** 1.7×10^{-24} mol

2 **a** i 8.73×10^{23} molecules
ii 3.49×10^{24} atoms
iii 3 significant figures
b i 3.47×10^{23} molecules
ii 1.04×10^{24} atoms
iii 3 significant figures
c i 9.21×10^{21} molecules
ii 4.61×10^{22} atoms
iii 3 significant figures
d i 1.5×10^{24} molecules
ii 6.8×10^{25} atoms
iii 2 significant figures

3 The molar mass, M, has the same numerical value as the relative molecular mass, M_r, which is the sum of the relative atomic masses, A_r, of the elements in the compound. The molar mass, M, is the actual mass of 1 mole and so has the unit $g\,mol^{-1}$.

4 **a** $55.85\,g\,mol^{-1}$ **b** $98.09\,g\,mol^{-1}$ **c** $61.98\,g\,mol^{-1}$
d $189.40\,g\,mol^{-1}$ **e** $75.07\,g\,mol^{-1}$ **f** $342.17\,g\,mol^{-1}$
g $270.30\,g\,mol^{-1}$

5 **a** 6.656×10^{-23} g **b** 2.992×10^{-23} g **c** 7.310×10^{-23} g

6 **a** i 0.032 mol
ii 1.9×10^{22} molecules
iii 7.8×10^{22} atoms
b i 0.292 mol
ii 1.76×10^{23} molecules
iii 1.41×10^{24} atoms
c i 0.0088 mol
ii 5.3×10^{21} molecules
iii 1.1×10^{22} atoms
d i 1.22×10^{-4} mol
ii 7.3×10^{19} molecules
iii 1.8×10^{21} atoms

7 **a** i $n(NaCl) = 0.100$ mol
ii $n(Na^+) = 0.100$ mol
$n(Cl^-) = 0.100$ mol
b i $n(CaCl_2) = 0.405$ mol
ii $n(Ca^{2+}) = 0.405$ mol
$n(Cl^-) = 0.810$ mol
c i $n(Fe_2(SO_4)_3) = 0.00420$ mol
ii $n(Fe^{3+}) = 0.00840$ mol
$n(SO_4^{2-}) = 0.0126$ mol

8 **a** $40\,g\,mol^{-1}$ **b** $98\,g\,mol^{-1}$ **c** $44\,g\,mol^{-1}$ **d** $106\,g\,mol^{-1}$

9 B

10 **a** $1.25 \times 10^4\,g\,mol^{-1}$ **b** 1.6×10^{-7} mol **c** 9.6×10^{16} molecules

11 **a** Al 52.92%; O 47.07%
b Cu 65.1%; O 32.8%; H 2.1%
c Mg 12.0%; Cl 34.9%; H = 5.9%; O 47.2%
d Fe 27.9%; S 24.1%; O 48.0%
e H 1.0%; Cl 35.3%; O 63.7%

12 $C_2H_5NO_2$

13 **a** C_2H_6O **b** C_2H_6O

14 C_3H_4 **15** $58.9\,g\,mol^{-1}$ **16** $183.9\,g\,mol^{-1}$

17 **a** C_4H_{10} **b** P_4O_{10} **c** $C_6H_{12}O_6$ **d** $H_2S_2O_7$

18 **a** The relative isotopic mass of an isotope is the mass of an atom of that isotope relative to the mass of an atom of ^{12}C, taken as 12 units exactly. For example, the relative isotopic mass of the lighter of the two chlorine isotopes is 34.969.
b The relative atomic mass of an element is the weighted average of the relative masses of the isotopes of the element on the ^{12}C scale; e.g. the relative atomic mass of boron is 10.81.
c The relative molecular mass (M_r) of a compound is the mass of one molecule of that substance relative to the mass of a ^{12}C atom, which is 12 exactly; e.g. the relative molecular mass of carbon dioxide is 44.01.
d Relative formula mass is calculated by taking the sum of the relative atomic masses of the elements in the formula. Relative formula mass (rather than relative molecular mass) is the appropriate term to use for ionic compounds, as they do not contain molecules. For example, the relative formula mass of sodium chloride is 58.44.
e The molar mass of an element is the mass of 1 mol of the element. It is equal to the relative atomic mass of the element expressed in grams. For example, the molar mass of magnesium is $24.31\,g\,mol^{-1}$. Note that relative atomic mass and molar mass of an element are numerically equal. However, relative atomic mass has no units because it is the mass of one atom of the element compared with the mass of one atom of the carbon-12 isotope. The molar mass of a compound is the mass of 1 mol of the compound. It is equal to the relative molecular or relative formula mass of the compound expressed in grams. For example, the molar mass of sodium chloride is $58.44\,g\,mol^{-1}$.

19 **a** $C_4H_5N_2O$ **b** $194\,g\,mol^{-1}$ **c** $C_8H_{10}N_4O_2$ **d** 5.15×10^{-3} mol
e 3.10×10^{21} molecules of caffeine **f** 7.44×10^{22} atoms

20 **a** D, F, E, A, C, B
b

	Metal	Oxygen
mass (g)	0.542	0.216
relative atomic mass	40.08	16.00
moles	0.0135 mol	0.0135 mol
ratio	1	1

c calcium

21

$Ca(NO_3)_2$	Na_3PO_4	$Ca_3(PO_4)_2$	$NaNO_3$
27 mol	18	9.0	54
0.72	0.48 mol	0.24	0.44
0.54	0.36	0.18 mol	1.08
1.2	0.8	0.4	2.4 mol

22 $n(KOH) = \frac{2}{1} \times n(Fe(OH)_2)$

$n(FeSO_4) = \frac{2}{1} \times n(KOH)$

$n(KOH) = \frac{2}{1} \times n(K_2SO_4)$

$n(Fe(OH)_2) = \frac{1}{1} \times n(FeSO_4)$

23 150 g **24** 50.0 g

25

Carbon atoms available	Oxygen molecules available	Carbon dioxide molecules produced	Carbon atoms in excess	Oxygen molecules in excess
8	20	8	0	12
1000	3000	1000	0	2000
9 mol	6 mol	6 mol	3 mol	0 mol

26 **a** oxygen **b** 1.5 mol

27 **a** 22.6 g **b** 4.93 g

28 **a** $m(P_4O_6)$ in excess = 2.77 g **b** 3.37 g **c** 1.26 g
d Total mass of products = 4.63 g. 4.63 g of products formed plus 2.78 g unreacted P_4O_6 = 7.40 g, which is consistent with the total mass of reactants used initially.

29 Student answers will vary. Possible answers include: a large molar mass, like that of sucrose, gives a smaller number of mol; there is no connection between the type of bonding and the number of mol in half a cup of a substance; low density and high molar mass substances have a low number of mol in half a cup of the substance.

Chapter 8 Concentration and molarity

8.1 Concentration of solutions

TY 8.1.1 20.0 g L^{-1}
TY 8.1.2 215 ppm
1 D
2 **a** 1.11%(w/v) **b** 11.1 g L^{-1}
3 **a** 32.5%(v/v) **b** 17 mL
4 **a** 5.0 ppm **b** 625 ppm **c** 27 ppm
5 sugar: 14.0%(w/v); fat: 3.0%(w/v)
6 2.1×10^2 ppb
7 0.24%(w/w) Therefore, the fluoride concentration is above the legal limit.

8.2 Molar concentration

TY 8.2.1 0.48 mol L^{-1}
TY 8.2.2 0.00250 mol
TY 8.2.3 0.666 mol L^{-1}
1 B
2 **a** 0.080 mol L^{-1} **b** 0.30 mol L^{-1} **c** 0.19 mol L^{-1}
3 **a** 0.25 mol L^{-1} **b** 0.50 mol L^{-1} **c** 0.438 mol L^{-1}
4 1.3 mol L^{-1}
5 **a** 2.2×10^{-2} mol **b** 6.4×10^{-3} mol
c 2.34×10^{-4} mol **d** 7.8×10^{-4} mol
6 0.120 mol L^{-1}
7 7.11×10^{-3} mol L^{-1}
8 0.0498 mol L^{-1}

8.3 Dilution

TY 8.3.1 0.0250 mol L^{-1}
TY 8.3.2 400 ppm
1 0.40 mol L^{-1}
2 **a** 0.40 mol L^{-1} **b** 0.075 mol L^{-1} **c** 0.025 mol L^{-1}
3 D
4 8.0×10^{-2} mol L^{-1}
5 4.25 mol L^{-1}
6 Calculate the volume of concentrated acid required using $c_1V_1 = c_2V_2$. $V_1 = 0.011$ L = 11 mL. Add about a litre of water to a 2.0 L volumetric flask. Carefully add small amounts of concentrated sulfuric acid until you have added 11 mL. Make the volume up to the mark of 2.0 L.

8.4 Standard solutions

TY 8.4.1 4.004 mol L^{-1}
1 C
2 Weigh the solid primary standard on an electronic balance. Transfer the solid into the volumetric flask using a clean, dry funnel. Rinse any remaining solid particles into the flask using deionised water. Half fill the flask with deionised water and swirl vigorously to dissolve the solid. Fill the flask with deionised water to just below the calibration mark. Add deionised water drop by drop up to the calibration line on the flask until the bottom of the meniscus touches the line. Stopper and shake the solution to ensure an even concentration throughout.
3 0.2000 mol L^{-1}
4 13.2 g **5** 12.6 g

Chapter 8 review

1 B
2 **a** 2.0 ppm **b** 2.0×10^{-4}%(w/w)
3 5.0%(w/w)
4 4.0×10^2 mg L^{-1} = 4.0×10^{-2}%(w/v)
5 **a** 19.5 g **b** 1281 g
6 **a** 80 mg L^{-1} **b** 8.0%(w/v)
7 2.38 mol L^{-1}
8 0.38 mol L^{-1}
9 0.300 mol L^{-1}
10 **a** 2.6×10^{-3} mol **b** 3.75×10^{-3} mol **c** 2.3×10^{-2} mol
11 **a** 2.04 g **b** 1.7 g
12 **a** 91.2 g **b** 3.65%(w/v)
13 D **14** 1.00 mol L^{-1} **15** 38 mL **16** 0.532 ppm
17 9 **18** C **19** 0.7998 mol L^{-1} **20** 21.2 g
21 A
22 **a** The chemical formula of hydrated sodium carbonate ($Na_2CO_3 \cdot 10H_2O$) changes over time. It can lose water, changing the ratio of sodium carbonate to water.
b Sodium hydroxide absorbs moisture from the air, and it also reacts readily with carbon dioxide in the air. This means the sodium hydroxide is relatively unstable, and therefore is not a completely pure substance.
c Sodium carbonate is heated to ensure it is dry so that the mass of solid is accurate.
d 5.30 g
23 **a** Product B **b** Product A
24

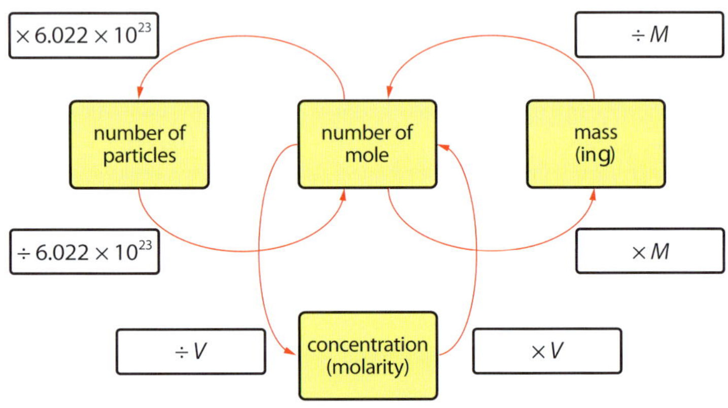

25 0.80 mol L^{-1}
26 **a** 1.65 kg **b** 297 kg **c** 66.0 g L^{-1} **d** 0.53 mol L^{-1}
27 **a** CO_2
b acetic acid + sodium hydrogen carbonate → sodium acetate + carbon dioxide + water
c Baking soda will release carbon dioxide gas during cooking. The gas is trapped in the cake, making the cake rise and the texture light.

Chapter 9 Gas laws

9.1 Introducing gases

TY 9.1.1 **a** 700 cm^3
b 0.700 L
c 7.00×10^{-4} m^3
TY 9.1.2 **a** 95.9 kPa
b 0.947 atm
c 95.9 kPa
d 0.959 bar
1 **a** Molecules of gases are in constant, rapid, random motion, and the forces between molecules are negligible. They continue to move outwards until stopped by the walls of the container, filling all the space available.
b Most of the volume occupied by a gas is space, so compression can be achieved by reducing the space between the particles.
c The molecules in a gas are spread much further apart than those of a liquid. A given mass of gas would occupy a much greater volume than the same mass of the liquid phase. Therefore, the density of the gas is less.
d Gases mix easily together because of the large amount of space between the molecules.
e The pressure exerted by a gas depends on the number of collisions of gas particles with the walls of the container. The pressure is independent of the type of gas involved. The total pressure exerted by a mixture of gases will depend on the total number of collisions each gas has with the container walls.

2 a Tyres have a recommended maximum pressure for giving a comfortable ride as well as good traction on the road. If the pressure in a tyre is too high, the gas inside cannot be compressed as easily and passengers will be more aware of bumps on the road.
  b During a long journey on a hot day, the air in a tyre warms up. This means the air molecules have increased kinetic energy, and collisions with the walls of the tyres will increase in frequency, exerting more force. This will cause pressure to increase.
  c Particles from the cooking food escape the pot and move randomly through the house. If the food has an odour, and if there are enough particles in the air, you will detect the odour as you enter the house.
  d As air is pumped into a balloon, air molecules collide with the rubber of the balloon, forcing it to expand. If too much air is pumped in, the balloon reaches a stage where it cannot stretch any further. If the number of collisions with molecules per given surface area is increased still further, the rubber will break.

3 a As temperature increases, the average kinetic energy of gas molecules in the can will increase. This will lead to an increase in the frequency and force of collisions of gas molecules with the inside wall of the aerosol can. This will cause an increase in pressure.
  b As the syringe is compressed, the inside surface area of the syringe will decrease. The number of collisions of molecules per unit area per second with the inside walls of the syringe will increase. This will cause a pressure increase.

4 a 1.40×10^5 Pa
  b 92 kPa
  c 3.22×10^3 mmHg; 4.30×10^5 Pa
  d 900 mmHg; 1.18 atm; 1.20 bar
  e 1.84 atm; 1.87×10^5 Pa; 1.87 bar
  f 0.790 atm; 600 mmHg; 0.800 bar

5 a 2×10^3 mL b 4.5×10^{-3} m^3 c 2.250 L d 0.120 L
  e 5.6×10^{-3} L f 3.7×10^3 L g 2.85×10^{-4} m^3
  h 4.70 L; 4.70×10^3 cm^3

6 a 97.5 kPa b 21.1%

9.2 The gas laws

TY 9.2.1 15 mL
TY 9.2.2 373 K
TY 9.2.3 514 L
TY 9.2.4 204°C
TY 9.2.5 345 kPa

1 a 273 K b 298 K c 373 K d 448 K e 128 K
2 a 2 L b 49.5 mL c 0.23 mL d 6.7 L e 0.50 L
3 a 300 mL b 338 mL c 19.2 L d 750 K e 374°C
4 a 15 atm b 514 mmHg c 407°C d 500 K e 433 atm
5 a 13.9 L b 0.29 atm c 8.14 bar d 596 K e 14.5 L
6 0.37 atm
7 34 kPa
8 a 1.1×10^4 m^3
  b i 0.887 bar
    ii 1100 m

9.3 The ideal gas law

TY 9.3.1 87 L
TY 9.3.2 150 L

1 a 35 L b 25 mL c 1.2 L
2 a 2.28 g b 64.5 g c 0.249 g
3 1.2×10^2 kPa
4 a 4.5 L b 86.8 L c 6.22 L
5 16.8 g
6 304 K = 31°C
7 $n(N_2) = 0.16$ mol; $n(\text{He}) = 0.12$ mol
  ∴ There is a greater amount of nitrogen.

9.4 Stoichiometric calculations involving gases

TY 9.4.1 506 L
TY 9.4.2 2.24×10^3 L
TY 9.4.3 50 mL
TY 9.4.4 a C_4H_{10}
  b 111 L

1 a i 62 L ii 37 L
  b i 46.4 L ii 27.8 L
  c i 9.56×10^3 L
    ii 5.74×10^3 L
2 24.4 g 3 4.07 L 4 0.5 L 5 75 mL
6 2.27×10^3 kPa 7 5.0 L

Chapter 9 review

1 B 2 A
3 smaller, straight-line, weak, elastic, directly, K
4 a As volume is reduced, there is an increase in the frequency of gas particle collisions per unit wall area. This is measured as an increase in pressure.
  b When the temperature of a gas is lowered, the average kinetic energy of the particles decreases. The rate of collisions between particles and the walls of the container decreases and particles collide with less force. As pressure is a measure of the force of particle collisions per unit wall area of the container, pressure is found to decrease.
  c In a mixture of gases, the particles of each gas are moving and colliding with the walls of the container, independently of one another. Each gas therefore exerts a pressure. As the gases behave independently of one another, total pressure is simply the sum of the individual gas (or partial) pressures.
  d When more gas is added to a container, the total number of particles in the container increases. Provided that the volume of the container and the temperature do not change, the collisions of these additional particles means that the total pressure in the container will increase.
5 a The pressure inside the container is reduced when some of the gas escapes.
  b There are fewer gas particles to collide with one another and with the walls of the container. Pressure is the force exerted by the particles over a defined area, so this will decrease.
6 a A graph of the data reveals a straight-line relationship between P and T. That is, over this range the propane is acting as an ideal gas.
  b Student answers will vary. Real gases will deviate from ideal gas behaviour because real gas particles have volume, which becomes significant at high pressures, when the gas volume is significantly compressed.
  Alternate response: Real gases deviate from ideal gas behaviour at low temperatures because real gas particles have intermolecular forces of attraction between the particles, which become significant at low temperatures.
7 a V = kT. For an ideal gas, there is a positive linear relationship between the volume and absolute temperature of the gas.
  b The gas particles have negligible volume, there are no interactions between the particles.
  c At 90 K, the oxygen is so compressed it forms a liquid. The particles are so close they begin to interact (dispersion forces become significant), causing oxygen to liquefy.
  d Helium gas is monotomic. There are weak dispersion forces between helium atoms, so its behaviour is very close to that of an ideal gas. Oxygen molecules are larger than helium atoms. Oxygen molecules have more electrons and therefore form stronger instantaneous dipoles, resulting in stronger dispersion forces. This accounts for its deviation from ideal gas behaviour at extremely low temperatures.
8 1.67 L 9 2.1 bar 10 55°C 11 282 mL
12 64.6 g

13 a 6.2 L b 19 g
14 B 15 5.56 L 16 13.8 g 17 770 K = 497°C
18 a 44.0 g b 24.79 L. c 1.78 g L^{-1} d greater
19 a 0.22 mol b 46 g mol^{-1}
20 container A
21 a 8.94×10^3 mol b 257 kg
22 24.4 g 23 35.0 L
24 a 5.0 L b 10 L c 4.6 g
25 a 100 mL of oxygen in excess
b $V(CO_2)$ = 240 mL
$V(H_2O)$ = 320 mL
c There was an increase in volume of 80 mL.
26 a 25 L. b 44.7 g
27 35.7 L 28 7.59×10^{12} L
29 a They have equal numbers of molecules. With pressure, volume and temperature the same, *n* will be the same.
b carbon dioxide
c carbon dioxide
30 a $C_2H_5OH(l) + 3O_2(g) \rightarrow 2CO_2(g) + 3H_2O(g)$
b i 1.91 kg ii 1.08×10^3 L
c 39.3 kg
31 Charles' law applies because the Inquiry activity investigated volume and temperature.

Module 2 Introduction to quantitative chemistry

Multiple choice

1 C 2 B 3 D 4 B 5 D 6 A
7 A 8 B 9 C 10 D 11 C 12 A
13 B 14 A 15 C 16 D 17 D 18 B
19 D 20 C

Short answer

1 a $2C_3H_6(g) + 9O_2(g) \rightarrow 6CO_2(g) + 6H_2O(g)$
b $Al(OH)_3(aq) + 3HBr(aq) \rightarrow AlBr_3(aq) + 3H_2O(l)$
c $MgCl_2(aq) + 2AgNO_3(aq) \rightarrow 2AgCl(s) + Mg(NO_3)_2(aq)$
d $2HNO_3(aq) + 6FeSO_4(aq) + 3H_2SO_4(aq) \rightarrow 3Fe_2(SO_4)_3(aq) + 4H_2O(l) + 2NO(g)$
2 a $3CO(g) + Fe_2O_3(s) \rightarrow 2Fe(s) + 3CO_2(g)$
b $2Na(s) + 2H_2O(l) \rightarrow 2NaOH(aq) + H_2(g)$
c $CaBr_2(l) \rightarrow Ca(l) + Br_2(g)$
3 a 17.85 g b 44.0 g c 42.06 g
4 a 0.143 mol b 0.430 mol
c 1.12×10^{24} N atoms d 67.6%
5 WS_2
6 a CuO
b If the sample was a mixture of CuO and Cu_2O, the ratio of Cu to O would be greater and so the mass of Cu extracted would be greater than 1.704 g.
7 a 39.1 g mol^{-1} b potassium
8 42.5 g
9 a 0.67% b 6.7×10^3 ppm
10 a i 163 atm
ii 1.24×10^5 mmHg
b 690 g c 4.27×10^3 L d 617

Extended response

1 a $Ba(ClO_3)_2(s) \rightarrow BaCl_2(s) + 3O_2(g)$
b 2.29 g of oxygen
c 2.25 g more of barium chlorate required to produce 4.96 g of barium chloride.
d If you tripled the amount of barium chlorate, the total mass of the products would also triple, assuming the reaction goes to completion.
e 1.81 g of barium chlorate did not decompose.
2 a $MgCO_3(s) + 2HCl(aq) \rightarrow MgCl_2(aq) + H_2O(l) + CO_2(g)$
b 0.105 mol c 69.1% d 0.42 L
e i $CO_2(g) + Ca(OH)_2(aq) \rightarrow CaCO_3(s) + H_2O(l)$
ii 10.5 g
3 a 0.592 mol L^{-1} b 0.370 mol L^{-1}
c i $Pb(NO_3)_2(aq) + 2NaI(aq) \rightarrow PbI_2(s) + 2NaNO_3(aq)$
ii 0.0325 mol iii NaI is limiting. iv 7.49 g
4 a The volume decreases non-linearly as the pressure increases.
b PV = k or $P \propto \frac{1}{V}$
c Choose a convenient initial condition, say P_1 = 70 kPa, V_1 = 62 mL, P_2 = 160 kPa, V_2 = ?
$P_1V_1 = P_2V_2$
$V_2 = \frac{P_1V_1}{P_2}$
$V_2 = \frac{70 \times 62}{160}$
V_2 = 27 mL (This can be verified by extrapolation of the graph.)
d The relationship $P \propto \frac{1}{V}$ holds for an ideal gas of negligibly sized particles that have no interactions. At high pressures, the gas particle size does become significant, as do the interactions between particles. Consequently, the inverse relationship will no longer hold.
e $n = 1.8 \times 10^{-3}$ mol
f If CO_2 was replaced by H_2 in this particular experiment, close to identical results should be obtained. At these conditions of low pressure and moderate temperatures, all gases approximate the behaviour of an ideal gas. That is, the particle size is negligible compared with the volume of the gas, and the interactions between particles are insignificant. At high pressures, H_2 will behave more like an ideal gas due to its weaker dispersion forces and smaller molecular size compared with CO_2.
5 a $P_4(s) + 3KOH(aq) + 3H_2O(l) \rightarrow PH_3(g) + 3KH_2PO_2(aq)$
b 0.04801 mol c KOH was limiting.
d $n(P_4)$excess = 0.02128 mol e 0.3968 L
f 0.3208 mol L^{-1}

Chapter 10 Chemical reactions

10.1 Chemical change

TY 10.1.1 $2Al(s) + 6HCl(aq) \rightarrow 2AlCl_3(aq) + 3H_2(g)$
1 C
2 a chemical b physical c physical d chemical
e physical f chemical
3 a reactants: carbon dioxide and water; products: glucose and oxygen
b reactant: water; products: hydrogen gas and oxygen gas
c reactant: calcium carbonate; products: calcium oxide and carbon dioxide
d reactants: carbon and oxygen; product: carbon dioxide
4 a sulfur + oxygen gas → sulfur dioxide gas
$S(s) + O_2(g) \rightarrow SO_2(g)$
b silver nitrate + potassium iodide → silver iodide + potassium nitrate
$AgNO_3(aq) + KI(aq) \rightarrow AgI(s) + KNO_3(aq)$
5 a Stage one is a physical change. The solid turns to a liquid, but the bonds in the sugar molecules have not changed.
b Stage two is a chemical change. The bonds in the sugar molecules have been broken and new substances have been formed.
6 96 g
7 a The mass of the bag and contents will slowly decrease because a gas is being evolved from the bag. Gases have mass.
b The mass of the bag and contents will not change because the gas has been trapped in the bag. No products are lost from the bag.
8 a Student answers will vary, e.g. unpleasant odour, change of colour of flesh, bloating of corpse that indicates a gas has been produced
b light from candle or mass loss
c surface of the trophy has a film on it, colour change on the surface of the trophy

10.2 Synthesis reactions

1 A
2 Synthesis reactions might use elements as reactants, but reactions that produce elements are not synthesis reactions.
3 a hydrogen + chlorine → hydrogen chloride
 b $H_2(g) + Cl_2(g) \rightarrow 2HCl(g)$
 c 40 molecules: 20 molecules of hydrogen gas are required and 20 molecules of chlorine gas.
 d The mass of chlorine gas required will be much greater than the mass of hydrogen required. The number of molecules is the same, but the mass of chlorine molecules is greater than that of hydrogen molecules.
4 a sulfur dioxide + oxygen → sulfur trioxide
 b $2SO_2(g) + O_2(g) \rightarrow 2SO_3(g)$
 c The mass of SO_3 will equal the combined mass of SO_2 and O_2 that reacted.
 d 200 g of SO_3 will not be not formed, because the reactants are not used in the mole ratio of 2:1. Some reactant will remain after the reaction. (The reactant left over will be oxygen gas.)

10.3 Decomposition reactions

1 C
2 $2KClO_3(s) \rightarrow 2KCl(s) + 2O_2(g)$
3 No new products are formed when water boils. There is only a change of state, which is a physical change. The bonds between water molecules can reform easily. The water has not chemically changed.
4 $MgCl_2(l) \rightarrow Mg(l) + Cl_2(g)$
5 a $H_2CO_3(aq) \rightarrow H_2O(l) + CO_2(g)$
 b Once the container is opened, the carbonic acid starts to react. As its concentration drops and carbon dioxide gas escapes the container, the soft drink becomes less gaseous.
6 a $MgCO_3(s) \rightarrow MgO(s) + CO_2(g)$
 b metal carbonate → metal oxide + carbon dioxide

10.4 Combustion reactions

TY 10.4.1 $2C_6H_{14}(l) + 19O_2(g) \rightarrow 12CO_2(g) + 14H_2O(l)$
TY 10.4.2 $2CH_3OH(l) + 3O_2(g) \rightarrow 2CO_2(g) + 4H_2O(l)$
TY 10.4.3 $CH_3OH(l) + O_2(g) \rightarrow CO(g) + 2H_2O(l)$
1 D
2 $2C_6H_6(l) + 15O_2(g) \rightarrow 12CO_2(g) + 6H_2O(l)$
3 $C_2H_5OH(l) + 2O_2(g) \rightarrow 2CO(g) + 3H_2O(l)$
4 a A bellows is used to increase the supply of oxygen to the coals, ensuring rapid and complete combustion is occurring.
 b Air flow into a closed caravan is often low. Opening the door increases the supply of oxygen, often leading to increased combustion.
 c A kerosene lamp often operates with a low oxygen supply. The incomplete combustion leads to carbon as one of the products.
 d The temperature inside a haystack can be very high, and dry grass is a good fuel. However, the lack of oxygen penetrating to the heat source limits the degree of combustion.
5 a If air is limited, incomplete combustion occurs and the toxic gas carbon monoxide forms.
 b reactants: diesel and oxygen; products: carbon dioxide and water
 c Diesel is reacting with oxygen to form carbon dioxide and water. The mass of diesel and oxygen consumed will equal the mass of carbon dioxide and water produced.

10.5 Precipitation reactions

TY 10.5.1 copper(II) sulfide
TY 10.5.2 $CuSO_4(aq) + 2NaOH(aq) \rightarrow Cu(OH)_2(s) + Na_2SO_4(aq)$
spectator ions: $Na^+(aq)$ and $SO_4^{2-}(aq)$
TY 10.5.3 soluble
1 B
2 A, B, D, E, H
3 A, C, D, F, H
4 a silver carbonate
 b lead(II) hydroxide
 c magnesium sulfide
 d no precipitate
5 a i magnesium sulfide
  ii silver chloride
  iii aluminium hydroxide
  iv magnesium hydroxide
 b i $K_2S(aq) + MgCl_2(aq) \rightarrow MgS(s) + 2KCl(aq)$
  ii $CuCl_2(aq) + 2AgNO_3(aq) \rightarrow 2AgCl(s) + Cu(NO_3)_2(aq)$
  iii $AlCl_3(aq) + 3KOH(aq) \rightarrow Al(OH)_3(s) + 3KCl(aq)$
  iv $MgSO_4(aq) + 2NaOH(aq) \rightarrow Mg(OH)_2(s) + Na_2SO_4(aq)$
6 a i $AgNO_3(aq) + NaCl(aq) \rightarrow AgCl(s) + NaNO_3(aq)$
  ii $Ag^+(aq) + Cl^-(aq) \rightarrow AgCl(s)$
 b i $CuSO_4(aq) + Na_2CO_3(aq) \rightarrow CuCO_3(s) + Na_2SO_4(aq)$
  ii $Cu^{2+}(aq) + CO_3^{2-}(aq) \rightarrow CuCO_3(s)$
 c i $(NH_4)_2SO_4(aq) + BaCl_2(aq) \rightarrow BaSO_4(s) + 2NH_4Cl(aq)$
  ii $SO_4^{2-}(aq) + Ba^{2+}(aq) \rightarrow BaSO_4(s)$
 d i $K_2S(aq) + Pb(NO_3)_2(aq) \rightarrow PbS(s) + 2KNO_3(aq)$
  ii $S^{2-}(aq) + Pb^{2+}(aq) \rightarrow PbS(s)$
 e i $3CaCl_2(aq) + 2Na_3PO_4(aq) \rightarrow 6NaCl(aq) + Ca_3(PO_4)_2(s)$
  ii $3Ca^{2+}(aq) + 2PO_4^{3-}(aq) \rightarrow Ca_3(PO_4)_2(s)$
 f i $2NaOH(aq) + Pb(NO_3)_2(aq) \rightarrow 2NaNO_3(aq) + Pb(OH)_2(s)$
  ii $2OH^-(aq) + Pb^{2+}(aq) \rightarrow Pb(OH)_2(s)$
7 a Na^+/NO_3^- b Na^+/SO_4^{2-} c NH_4^+/Cl^-
 d K^+/NO_3^- e Na^+/Cl^- f Na^+/NO_3^-

10.6 Reactions of acids and bases

TY 10.6.1 $6H^+(aq) + 2Al(s) \rightarrow 2Al^{3+}(aq) + 3H_2(g)$
TY 10.6.2 $H^+(aq) + OH^-(aq) \rightarrow H_2O(l)$
TY 10.6.3 products: sodium chloride in solution, water and carbon dioxide gas
$2H^+(aq) + CO_3^{2-}(aq) \rightarrow H_2O(l) + CO_2(g)$
1 C
2 a $Mg(s) + H_2SO_4(aq) \rightarrow MgSO_4(aq) + H_2(g)$
 $Mg(s) + 2H^+(aq) \rightarrow Mg^{2+}(aq) + H_2(g)$
 b $Ca(s) + 2HCl(aq) \rightarrow CaCl_2(aq) + H_2(g)$
 $Ca(s) + 2H^+(aq) \rightarrow Ca^{2+}(aq) + H_2(g)$
 c $Zn(s) + 2CH_3COOH(aq) \rightarrow Zn(CH_3COO)_2 + H_2(g)$
 $Zn(s) + 2H^+(aq) \rightarrow Zn^{2+}(aq) + H_2(g)$
 d $2Al(s) + 6HCl(aq) \rightarrow 2AlCl_3(aq) + 3H_2(g)$
 $2Al(s) + 6H^+(aq) \rightarrow 2Al^{3+}(aq) + 3H_2(g)$
3 a magnesium sulfate
 b calcium chloride
 c zinc acetate
 d aluminium chloride
4 a i $ZnO(s) + H_2SO_4(aq) \rightarrow ZnSO_4(aq) + H_2O(l)$
  ii $ZnO(s) + 2H^+(aq) \rightarrow Zn^{2+}(aq) + H_2O(l)$
 b i $Ca(s) + 2HNO_3(aq) \rightarrow Ca(NO_3)_2(aq) + H_2(g)$
  ii $Ca(s) + 2H^+(aq) \rightarrow Ca^{2+}(aq) + H_2(g)$
 c i $Cu(OH)_2(s) + 2HNO_3(aq) \rightarrow Cu(NO_3)_2(aq) + 2H_2O(l)$
  ii $Cu(OH)_2(s) + 2H^+(aq) \rightarrow Cu^{2+}(aq) + 2H_2O(l)$
 d i $Mg(HCO_3)_2(s) + 2HCl(aq) \rightarrow MgCl_2(s) + 2H_2O(l) + 2CO_2(g)$
  ii $Mg(HCO_3)_2(s) + 2H^+(aq) \rightarrow Mg^{2+}(aq) + 2H_2O(l) + 2CO_2(g)$
 e i $SnCO_3(s) + H_2SO_4(aq) \rightarrow SnSO_4(s) + H_2O(l) + CO_2(g)$
  ii $SnCO_3(s) + 2H^+(aq) \rightarrow Sn^{2+}(s) + H_2O(l) + CO_2(g)$
5 a $2KOH(aq) + H_2SO_4(aq) \rightarrow K_2SO_4(aq) + 2H_2O(l)$
 $OH^-(aq) + H^+(aq) \rightarrow H_2O(l)$
 b $NaOH(aq) + HNO_3(aq) \rightarrow NaNO_3(aq) + H_2O(l)$
 $OH^-(aq) + H^+(aq) \rightarrow H_2O(l)$
 c $Mg(OH)_2(s) + 2HCl(aq) \rightarrow MgCl_2(aq) + 2H_2O(l)$
 $Mg(OH)_2(s) + 2H^+(aq) \rightarrow Mg^{2+}(aq) + 2H_2O(l)$
 d $CuCO_3(s) + H_2SO_4(aq) \rightarrow CuSO_4(aq) + H_2O(l) + CO_2(g)$
 $CuCO_3(s) + 2H^+(aq) \rightarrow Cu^{2+}(aq) + H_2O(l) + CO_2(g)$
 e $KHCO_3(aq) + HF(aq) \rightarrow KF(aq) + H_2O(l) + CO_2(g)$
 $HCO_3^-(aq) + H^+(aq) \rightarrow H_2O(l) + CO_2(g)$
 f $Zn(s) + 2HNO_3(aq) \rightarrow Zn(NO_3)_2(aq) + H_2(g)$
 $Zn(s) + 2H^+(aq) \rightarrow Zn^{2+}(aq) + H_2(g)$

g $CaCO_3(s) + 2HCl(aq) \rightarrow CaCl_2(aq) + H_2O(l) + CO_2(g)$
$CaCO_3(s) + 2H^+(aq) \rightarrow Ca^{2+}(aq) + H_2O(l) + CO_2(g)$

h $NaHCO_3(s) + CH_3COOH(aq) \rightarrow NaCH_3COO(aq) + H_2O(l) + CO_2(g)$
$NaHCO_3(s) + H^+(aq) \rightarrow Na^+(aq) + H_2O(l) + CO_2(g)$

6 a reactants: zinc metal and hydrochloric acid
b $Zn(s) + 2HCl(aq) \rightarrow ZnCl_2(aq) + H_2(g)$

7 a reactants: sulfuric acid and lithium hydroxide
b $H_2SO_4(aq) + 2LiOH(aq) \rightarrow Li_2SO_4(aq) + 2H_2O(l)$

10.7 Removing toxins from food

1 C

2 a Grinding the food will increase the surface area. It will be easier to remove each toxin particle if it is not protected by other parts of the food.
b Some of the toxins will be relatively soluble in water. Sitting the ground-up food in a water supply will allow time for soluble toxins to wash from the food.
c Cooking the food will lead to the decomposition reaction of some of the toxins into less harmful substances.

3 Fire leads to more effective seeding of cycads. The eventual number of cycad bushes in a particular area will be greater if the bushes are regularly exposed to fire.

4 Coffee beans can be placed in a solvent. The caffeine dissolves in the solvent. The solvent and caffeine are then washed from the coffee beans.

Chapter 10 review

1 a change of colour or odour
b unpleasant odour, curdled appearance or discolouration
c change in colour of egg white from colourless to white
d exhaust fumes or soot or sound emitted
e rapid volume change or temperature change

2 Gold – physical change. The gold melts to a liquid, then solidifies to form a gold bar. The substance stays as gold.
Mercury(II) oxide – chemical change. The mercury(II) oxide decomposes to mercury and oxygen gas.
Aluminium – chemical change. When the aluminium burns, it reacts with oxygen to form aluminium oxide.

3 a combustion of petrol, reaction of nitrogen in air with oxygen, or reactions that occur in the battery.
b action of baking powder to create a gas, cooking of an egg, or meat turning brown when cooked

4 Aluminium metal can react with oxygen gas to form solid aluminium oxide. The ratio of aluminium atoms to oxygen molecules required is 4:3.

5 a magnesium + sulfur → magnesium sulfide
$Mg(s) + S(s) \rightarrow MgS(s)$
b zinc + copper(II) sulfate → copper + zinc sulfate
$Zn(s) + CuSO_4(aq) \rightarrow ZnSO_4(aq) + Cu(s)$
c magnesium carbonate → magnesium oxide + carbon dioxide
$MgCO_3(s) \rightarrow MgO(s) + CO_2(g)$

6 A

7 a nitrogen + oxygen → nitrogen monoxide
$N_2(g) + O_2(g) \rightarrow 2NO(g)$
b nitrogen monoxide + oxygen → nitrogen dioxide
$2NO(g) + O_2(g) \rightarrow 2NO_2(g)$

8 a thermal decomposition
b $2KClO_3(s) \rightarrow 2KCl(s) + 3O_2(g)$

9 a electrolysis
b $2Al_2O_3(l) \rightarrow 4Al(l) + 3O_2(g)$

10 a photolysis
b $2AgCl(s) \rightarrow 2Ag(s) + Cl_2(g)$

11 C

12 a methane + oxygen → carbon dioxide and water
b $CH_4(g) + 2O_2(g) \rightarrow CO_2(g) + 2H_2O(g)$
c $2CH_4(g) + 3O_2(g) \rightarrow 2CO(g) + 4H_2O(g)$
d The Bunsen burner has an air valve. When the valve is open, complete combustion occurs, producing a hot blue flame. When the valve is closed, the flame is orange and produces less heat.

13 a Heaters such as this have an air intake valve or adjustment. When the valve is open, oxygen is readily available and combustion is relatively complete.
b If combustion is limited by air flow, less energy is released, but the fuel lasts longer. More soot is produced.
c Carbon dioxide, carbon monoxide and water are also produced in the reaction and escape through the chimney flue, as this is not a closed system. Once all the products are taken into account, the mass of the products is found to equal that of the reactants.

14 a true b true c false d false e true f false

15

	NaOH	KBr	NaI	$MgSO_4$	$BaCl_2$
$Pb(NO_3)_2$	$Pb(OH)_2$	$PbBr_2$*	PbI_2	$PbSO_4$	$PbCl_2$*
KI	–	–	–	–	–
$CaCl_2$	$Ca(OH)_2$*	–	–	CaSO4*	–
Na_2CO_3	–	–	–	$MgCO_3$	$BaCO_3$
Na_2S	–	–	–	MgS	BaS

*These compounds are only slightly soluble.

16 PO_4^{3-} and S^{2-}

17 a barium sulfate b none c lead(II) sulfate d none

18 a $NH_4Cl(aq) + AgNO_3(aq) \rightarrow NH_4NO_3(aq) + AgCl(s)$
$Ag^+(aq) + Cl^-(aq) \rightarrow AgCl(s)$
b $FeCl_2(aq) + Na_2S(aq) \rightarrow FeS(s) + 2NaCl(aq)$
$Fe^{2+}(aq) + S^{2-}(aq) \rightarrow FeS(s)$
c $Fe(NO_3)_3(aq) + 3KOH(aq) \rightarrow 3KNO_3(aq) + Fe(OH)_3(s)$
$Fe^{3+}(aq) + 3OH^-(aq) \rightarrow Fe(OH)_3(s)$
d $CuSO_4(aq) + 2NaOH(aq) \rightarrow Cu(OH)_2(s) + Na_2SO_4(aq)$
$Cu^{2+}(aq) + 2OH^-(aq) \rightarrow Cu(OH)_2(s)$
e $Ba(NO_3)_2(aq) + Na_2SO_4(aq) \rightarrow BaSO_4(s) + 2NaNO_3(aq)$
$Ba^{2+}(aq) + SO_4^{2-}(aq) \rightarrow BaSO_4(s)$

19 a $CuSO_4(aq) + Na_2CO_3(aq) \rightarrow CuCO_3(s) + Na_2SO_4(aq)$
$Cu^{2+}(aq) + CO_3^{2-}(aq) \rightarrow CuCO_3(s)$
spectator ions: Na+, SO_4^{2-}
b $AgNO_3(aq) + KCl(aq) \rightarrow AgCl(s) + KNO_3(aq)$
$Ag^+(aq) + Cl^-(aq) \rightarrow AgCl(s)$
spectator ions: K^+, NO_3^-
c $Na_2S(aq) + Pb(NO_3)_2(aq) \rightarrow PbS(s) + 2NaNO_3(aq)$
$Pb^{2+}(aq) + S^{2-}(aq) \rightarrow PbS(s)$
spectator ions: Na^+, NO_3^-
d $FeCl_3(aq) + 3NaOH(aq) \rightarrow Fe(OH)_3(s) + 3NaCl(aq)$
$Fe^{3+}(aq) + 3OH^-(aq) \rightarrow Fe(OH)_3(s)$
spectator ions: Na+, Cl^-
e $Fe_2(SO_4)_3(aq) + 6KOH(aq) \rightarrow 2Fe(OH)_3(s) + 3K_2SO_4(aq)$
$Fe^{3+}(aq) + 3OH^-(aq) \rightarrow Fe(OH)_3(s)$
spectator ions: K+, SO_4^{2-}

20 a $HNO_3(aq) + KOH(aq) \rightarrow KNO_3(aq) + H_2O(l)$
b $H_2SO_4(aq) + K_2CO_3(aq) \rightarrow K_2SO_4(aq) + H_2O(l) + CO_2(g)$
c $2H_3PO_4(aq) + 3Ca(HCO_3)_2(s) \rightarrow Ca_3(PO_4)_2(s) + 6H_2O(l) + 6CO_2(g)$
d $2HF(aq) + Zn(OH)_2(s) \rightarrow ZnF_2(aq) + 2H_2O(l)$

21 E

22 $2H^+(aq) + CO_3^{2-}(aq) \rightarrow H_2O(l) + CO_2(g)$

23 a metal carbonate or metal hydrogen carbonate

24 a $2Al(s) + 3H_2SO_4(aq) \rightarrow Al_2(SO_4)_3(aq) + 3H_2(g)$
b $2Al(s) + 6H^+(aq) \rightarrow 2Al^{3+}(aq) + 3H_2(g)$

25 a decomposition b synthesis c acid base
d precipitation e combustion

26 a combustion b precipitation c synthesis
d decomposition e acid base

27 a sodium + chlorine → sodium chloride
b $2Na(s) + Cl_2(g) \rightarrow 2NaCl(s)$ c synthesis

28 a $HCl(aq) + KOH(aq) \rightarrow KCl(aq) + H_2O(l)$
b $2HCl(aq) + Ca(s) \rightarrow CaCl_2(aq) + H_2(g)$
c $2HNO_3(aq) + CaCO_3(s) \rightarrow Ca(NO_3)_2(aq) + H_2O(l) + CO_2(g)$
d $2Ca(s) + O_2(g) \rightarrow 2CaO(s)$
e $2C_2H_6(g) + 7O_2(g) \rightarrow 4CO_2(g) + 6H_2O(l)$

29 acid and metal carbonate reaction

Chapter 11 Predicting reactions of metals

11.1 Reactions of metals

1 a sodium oxide b magnesium oxide
c aluminium oxide d tin(II) oxide

2 a potassium, calcium, zinc, iron and silver
b $4Na(s) + O_2(g) \rightarrow 2Na_2O(s)$
$2Ca(s) + O_2(g) \rightarrow 2CaO$
$2Zn(s) + O_2(g) \rightarrow 2ZnO$
$4Fe(s) + 3O_2(g) \rightarrow 2Fe_2O_3(s)$
$Ag(s) + O_2(g) \rightarrow$ no reaction

3 a i K, Na or Ca
ii Mg, Al, Zn or Fe
iii Sn, Pb, Cu or Ag
b i hydrogen gas, metal ions and hydroxide ions (or metal hydroxides)
ii hydrogen gas and metal oxides

4 a $Zn(s) + 2H^+(aq) \rightarrow Zn^{2+}(aq) + H_2(g)$
b $2Al(s) + 6H^+(aq) \rightarrow 2Al^{3+}(aq) + 3H_2(g)$
c $Sn(s) + 2H^+(aq) \rightarrow Sn^{2+}(aq) + H_2(g)$

5 The metal displaces hydrogen ions (H^+) from the acid to form hydrogen gas (H2).

6 a i A reaction occurs producing silver metal and zinc ions.
ii no reaction
b $2Ag^+(aq) + Zn(s) \rightarrow 2Ag(s) + Zn^{2+}(aq)$

11.2 The activity series of metals

TY 11.2.1 There will be a reaction: $Cu(s) \rightarrow Cu^{2+}(aq) + 2e^-$

1 a magnesium > manganese > copper
b magnesium > manganese > copper

2 $Ni(s) \rightarrow Ni^{2+}(aq) + 2e^-$
$Cu^{2+}(aq) + 2e^- \rightarrow Cu(s)$
$Ni(s) + Cu^{2+}(aq) \rightarrow Ni^{2+}(aq) + Cu(s)$

3 For reactions to occur, the metal ion in the solution must be lower down the activity series than the metal.
a no b no c yes d yes e yes f yes g no

4 a $3Cu^{2+}(aq) + 2Al(s) \rightarrow 3Cu(s) + 2Al^{3+}(aq)$
b no reaction
c $2Ag^+(aq) + Zn(s) \rightarrow 2Ag(s) + Zn^{2+}(aq)$

5 Use iron and lead to test the solutions. The copper(II) nitrate will react with both iron and lead. The tin(II) nitrate will react only with the iron. The zinc nitrate will not react with either of the metals.

6 The activity series ranks the reactivity of metals in decreasing order from most reactive to least reactive. The table of standard reduction potentials ranks the reactivity of metals and non-metals in increasing order of standard reduction potential, from the lowest potential at the top of the table to the highest at the bottom.
In the activity series, the reactivity of metals is determined in conditions that are not specified or controlled. In the table of standard reduction potentials, the reactivity of metals is measured under standard, specified conditions.
The activity series of metals is based on observations of the relative vigour of the reactions of different metals. The table of standard reduction potentials ranks both metals and non-metals in order of reactivity on the basis of a numerical value, electrode potential, derived from precise measurements made under standardised conditions.

11.3 Metal activity and the periodic table

1 In metals, reactivity increases down a group and reactivity decreases across the period.

2 Calcium has one more electron shell than magnesium, and its atomic radius is consequently greater. Because calcium has a greater atomic radius, the electrostatic force of attraction between its valence electrons and the nucleus is weaker. As a result, the ionisation energy and electronegativity of calcium are less than that of magnesium. A calcium atom is more reactive, since it will lose the outer-shell or valence electrons more readily than a magnesium atom does.

3 The potassium atom has one more electron shell than the sodium atom. Because there is a greater distance between the valence electron and the nucleus, the electrostatic attraction is weaker, so it is easier for the potassium atom to lose its valence electron. The potassium atom has a core charge of +1, whereas the calcium atom has a core charge of +2. The greater core charge of the calcium atom means that the attraction between its valence electrons and nucleus is stronger than the attraction between the valence electron and nucleus of the potassium atom.

4 Metal II is the most reactive. The most reactive metal will most readily lose its valence electron. Metal II has the greatest atomic radius. The force of attraction between the nucleus and valence electrons will be less, since the distance between the nucleus and the outer electron shell is the greatest. Metal II also has the lowest ionisation energy, indicating that it requires less energy to lose an electron. Electronegativity is a measure of the ability of an atom to attract electrons. Metal II has the lowest electronegativity, so it has the least ability to attract electrons.

Chapter 11 review

1 a calcium oxide
b hydrogen and hydroxide ions (or potassium hydroxide solution)
c hydrogen and $Zn^{2+}(aq)$ ions

2 B 3 B 4 Cu 5 D

6 Add the mixture of metals to some dilute acid. The zinc metal will dissolve in the dilute acid, whereas copper metal will not dissolve.
$Zn(s) + 2H^+(aq) \rightarrow Zn^{2+}(aq) + H_2(g)$
The copper can be recovered from the solution by filtration.

7 C

8 a $Ag^+(aq) + e^- \rightarrow Ag(s)$
b $Cu(s) \rightarrow Cu^{2+}(aq) + 2e^-$
c $Zn(s) \rightarrow Zn^{2+}(aq) + 2e^-$

9 Na, Li, Ca , Mg, Al, Fe, Sn, Cu, Ag

10 The activity series of metals is based on experimental observations of the relative rate of reaction of metals with other substances such as oxygen, water and dilute acids. The rates of reaction are described using terms such as vigorous fast, slow etc. Qualitative data is based on observations, whereas quantitative data is based on measurements.

11 a $Zn(s) + 2Ag^+(aq) \rightarrow Zn^{2+}(aq) + 2Ag(s)$
b no reaction c no reaction
d $Mg(s) + Pb^{2+}(aq) \rightarrow Mg^{2+}(aq) + Pb(s)$
e no reaction f no reaction g no reaction
h $Pb(s) + 2Ag^{2+}(aq) \rightarrow Pb^{2+}(aq) + 2Ag(s)$

12 a no
b $Zn(s) + 2AgNO_3(aq) \rightarrow Zn(NO_3)_2(aq) + 2Ag(s)$
c $Zn(s) + SnCl_2(aq) \rightarrow ZnCl_2(aq) + Sn(s)$
d $Zn(s) + CuSO_4(aq) \rightarrow ZnSO_4(aq) + Cu(s)$

13 Place 10 mL of each solution into a clean test-tube. Add a small piece of magnesium ribbon to each solution. The magnesium should be coated by displaced metal in the silver nitrate and lead(II) nitrate solutions, but not in the sodium nitrate solution. The lead coating will be black. Over time, the silver coating will change from black to silver as more metal is deposited. To confirm the identity of the silver nitrate solution, take a fresh sample and add a small piece of copper. Copper will displace silver from the solution, giving a silver deposit and a blue solution. Copper will not displace lead from the solution.

14 Zinc. Zinc is more reactive than iron, but less reactive than aluminium, so will only displace iron from solution. Cobalt is less reactive than both iron and aluminium so will displace neither from solution.

15 Coatings of metals other than iron would be expected on the nails placed in $1\ mol\ L^{-1}$ solutions of $CuSO_4$ and $Pb(NO_3)_2$. Iron is more reactive than copper and lead, so iron will displace copper(II) and lead(II) ions from solution.

16 **a** Magnesium and sodium, with outer electrons in the third shell, are in the same period. Magnesium has a nuclear charge of +12, but with completed inner shells of $1s^2 2s^2 2p^6$, the outer electrons experience the attraction of a core charge of +2. The outer-shell electrons of sodium, which has a nuclear charge of +11 and the same number of inner shells as magnesium, are attracted by a core charge of +1. The stronger attraction of the magnesium electrons to the core means that more energy is required to remove an electron from a magnesium atom than from a sodium atom.

b Both lithium and sodium are in group 1, so the outer electron of each atom experiences the attraction of the same core charge. Because the outer-shell electrons of a lithium atom are closer to the nucleus than those of a sodium atom, they are attracted more strongly and more energy is needed to remove one.

17 **a** aluminium

b The aluminium atom has a core charge of +3 and a magnesium atom has a core charge of +2. Aluminium is the least reactive because the valence electrons in aluminium are more strongly attracted to the nucleus than are the valence electrons in magnesium. Sulfur, chlorine, fluorine and oxygen are non-metals.

18 **a** aluminium. The elements are all members of group 13 and have a core charge of +3. Aluminium has fewer electron shells than the other elements in the list. Since the valence electrons in aluminium are closer to the nucleus, the force of attraction between the valence electrons and the nucleus is stronger. Consequently, more energy is required to remove the first electron from the valence shell of an aluminium atom than from the valence shell of a gallium or indium atom.

b aluminium. The aluminium atom has a core charge of +3, magnesium atoms have a core charge of +2 and a sodium atom has a core charge of +1. The valence electrons in aluminium are more strongly attracted to the nucleus than the valence electrons in magnesium or sodium. Consequently, more energy is required to remove the first electron from the valence shell of an aluminium atom than from the valence shell of a magnesium or sodium atom.

19 **a** $Mg(s)$

b $Ag^+(aq)$

c A coating of silver will form on the lead when it is placed in silver nitrate solution.

d zinc and magnesium

20 Experiment 1 gives you evidence that T is a more reactive metal than R. Experiment 2 suggests that T is a more reactive metal than S, because S does not displace T from a solution of its salt, $T(NO_3)_2$. Experiment 3 indicates that R is a more reactive metal than S, because S does not displace R from a solution of its salt, RNO_3. This information tells you that S is less reactive than both T and R and that R is less reactive than T, so the correct order of decreasing reactivity is T > R > S.

21 List A contains metals, which tend to lose electrons to non-metals when they react. List B contains non-metals, which tend to gain electrons when they react.

22 Between Na and Ca: K, Na, X, Ca, Mg, Al, Zn, Cu, Ag.
The electronic structure shows that there are two electrons in the valence shell and that all the inner electron shells are filled, indicating that X is a group 2 metal. As the valence electrons are in the 5th shell, X is in the 5th period. This places X below calcium in group 2. The reactivity of group 2 metals in water increases down the group, so you would expect X to be more reactive than Ca. Metal X is placed after Na, since the reaction of group 1 metals in water is more vigorous than the reaction of group 2 metals in water.

23 More reactive metals have a larger atomic radius, lower ionisation energies and lower electronegativities than less reactive metals. The order of reactivity of the metals is II > IV > III > I.

24 The activity series of metals lists metals in decreasing order of reducing strength. Zinc is placed higher on the activity series of metals than iron, and therefore zinc metal is a stronger reducing agent than iron metal.

Chapter 12 Redox reactions and galvanic cells

12.1 Introducing redox reactions

TY 12.1.1 $Na(s) \rightarrow Na^+(s) + e^-$
$Cl_2(g) + 2e^- \rightarrow 2Cl^-(s)$

TY 12.1.2 $Ag^+(aq) + e^- \rightarrow Ag(s)$ (The $Ag^+(aq)$ is being reduced.)
$Cu(s) \rightarrow Cu^{2+}(aq) + 2e^-$ (The Cu(s) is being oxidised.)

TY 12.1.3 $O_2(g) + 4e^- \rightarrow 2O^{2-}(s)$
$4K(s) \rightarrow 4K^+(s) + 4e^-$
When the electrons have been cancelled, the overall equation is:
$4K(s) + O_2(g) \rightarrow 2K_2O(s)$

TY 12.1.4 $MnO_4^-(aq) + 4H^+(aq) + 3e^- \rightarrow MnO_2(s) + 2H_2O(l)$

TY 12.1.5
half-equations:
$SO_3^{2-}(aq) + 8H^+(aq) + 6e^- \rightarrow H_2S(g) + 3H_2O(l)$
$ClO^-(aq) + 2H_2O(l) \rightarrow ClO_3^-(aq) + 4H^+(aq) + 4e^-$
overall equation:
$2SO_3^{2-}(aq) + 4H^+(aq) + 3ClO^-(aq) \rightarrow 2H_2S(g) + 3ClO_3^-(aq)$

1 **a** oxidation **b** reduction **c** reduction **d** oxidation

2 **a** $Fe(s) \rightarrow Fe^{3+}(aq) + 3e^-$ (oxidation)

b $K(s) \rightarrow K^+(aq) + e^-$ (oxidation)

c $F_2(g) + 2e^- \rightarrow 2F^-(aq)$ (reduction)

d $O_2(g) + 4e^- \rightarrow 2O^{2-}(aq)$ (reduction)

3 **a** Fe(s) has been oxidised to $Fe^{2+}(aq)$.

b $Fe(s) \rightarrow Fe^{2+}(aq) + 2e^-$

c $H^+(aq)$

d $H^+(aq)$ has been reduced to $H_2(g)$.

e $2H^+(aq) + 2e^- \rightarrow H_2(g)$

f Fe(s)

g $Fe^{2+}(aq)/Fe(s)$ and $H^+(aq)/H_2(g)$

4 **a** Magnesium is oxidised, and copper ions are reduced.

b $Mg(s) \rightarrow Mg^{2+}(aq) + 2e^-$

c $Cu^{2+}(aq) + 2e^- \rightarrow Cu(s)$

d $Mg(s) + Cu^{2+}(aq) \rightarrow Mg^{2+}(aq) + Cu(s)$

e oxidizing agent Cu^{2+}; reducing agent Mg

f Copper(II) ions cause a solution to appear blue. The solution loses some of its blue colour due to the loss of $Cu^{2+}(aq)$, which is reduced to form Cu(s).

5 **a** $MnO_2(s) + 4H^+(aq) + 2e^- \rightarrow Mn^{2+}(aq) + 2H_2O(l)$

b $MnO_4^-(aq) + 4H^+(aq) + 3e^- \rightarrow MnO_2(s) + 2H_2O(l)$

c $SO_4^{2-}(aq) + 10H^+(aq) + 8e^- \rightarrow H_2S(g) + 4H_2O(l)$

d $SO_2(g) + 2H_2O(l) \rightarrow SO_4^{2-}(aq) + 4H^+(aq) + 2e^-$

e $H_2S(g) \rightarrow S(s) + 2H^+(aq) + 2e^-$

f $SO_3^{2-}(aq) + H_2O(l) \rightarrow SO_4^{2-}(aq) + 2H^+(aq) + 2e^-$

6 **a** CaO **b** Ca(s) **c** $Ca(s) \rightarrow Ca^{2+}(s) + 2e^-$

d $O_2(g)$ **e** $O_2(g) + 4e^- \rightarrow 2O_2^-(s)$ **f** $2Ca(s) + O_2(g) \rightarrow 2CaO(s)$

g Calcium has been *oxidised* by oxygen to calcium ions. The *calcium* has lost electrons to the *oxygen*.
The oxygen has been *reduced* by calcium to oxide ions. The oxygen has gained electrons from the *calcium*.

12.2 Oxidation numbers

TY 12.2.1 Oxidation numbers are written above the elements in the formula: $\overset{+1}{Na}\overset{+5}{N}\overset{-2}{O_3}$
(Note that the oxidation number of oxygen is written as −2 (not as −6), even though there are three oxygen atoms in the formula.)

TY 12.2.2 The oxidation number of copper has decreased from +2 to 0, so the copper in CuO has been reduced. The oxidation number of H has increased from 0 to +1, so H_2 has been oxidised. The oxidation number of O has not changed.

1 **a** +2 **b** +4 **c** −4 **d** 0 **e** +4

2 K_2MnO_4: the oxidation state of K is +1, O is −2
$(2 \times +1) + x + (4 \times -2) = 0$; hence, $x = +6$

3 **a** Ca: +2; O: −2 **b** Ca: +2; Cl: −1 **c** H: +1; S: +6; O: −2
d Mn: +7; O: −2 **e** F: 0 **f** S: +4; O: −2
g Na: +1; N: +5; O: −2 **h** K: +1; Cr: +6; O: −2

4 **a** $\overset{0}{Mg}(s) + \overset{0}{Cl_2} \rightarrow \overset{+2}{Mg}\overset{-1}{Cl_2}(s)$ $\therefore$ oxidant Cl_2; reductant Mg

b $2\overset{+4}{S}\overset{-2}{O_2}(g) + \overset{0}{O_2}(g) \rightarrow 2\overset{+6}{S}\overset{-2}{O_3}(g)$ $\therefore$ oxidant O_2; reductant SO_2

c $\overset{+3}{Fe_2}\overset{-2}{O_3}(s) + 3\overset{+2}{C}\overset{-2}{O}(g) \rightarrow 2\overset{0}{Fe}(s) + 3\overset{+4}{C}\overset{-2}{O_2}(g)$
$\therefore$ oxidant Fe_2O_3; reductant CO

d $2\overset{+2}{Fe^{2+}}(aq) + \overset{+1}{H_2}\overset{-1}{O_2}(aq) + 2\overset{+1}{H^+}(aq) \rightarrow 2\overset{+3}{Fe^{3+}}(aq) + 2\overset{+1}{H_2}\overset{-2}{O}(l)$
$\therefore$ oxidant H_2O_2; reductant Fe^{2+}

5

Redox reaction	Conjugate redox pair (oxidation process)	Conjugate redox pair (reduction process)
$Na(s) + Ag^+(aq) \rightarrow Na^+(aq) + Ag(s)$	$Na^+(aq)/Na(s)$	$Ag^+(aq)/Ag(s)$
$Zn(s) + Cu^{2+}(aq) \rightarrow Zn^{2+}(aq) + Cu(s)$	$Zn^{2+}(aq)/Zn(s)$	$Cu^{2+}(aq)/Cu(s)$
$2K(s) + Cl_2(g) \rightarrow 2K^+(s) + 2Cl^-(s)$	$K^+(s)/K(s)$	$Cl_2(g)/Cl^-(s)$

12.3 Galvanic cells

1 D

2 **a** $Ni^{2+}(aq)/Ni(s)$ half-cell

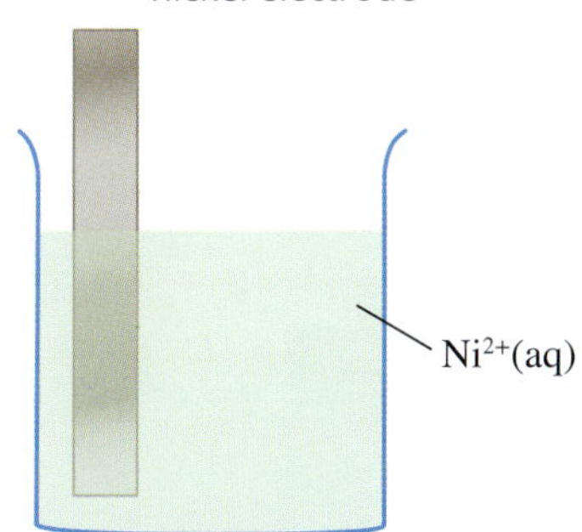

b $Sn^{4+}(aq)/Sn^{2+}(aq)$ half-cell

platinum electrode
Sn⁴⁺(aq)
and Sn²⁺(aq)

c $H^+(aq)/H_2(g)$ half-cell

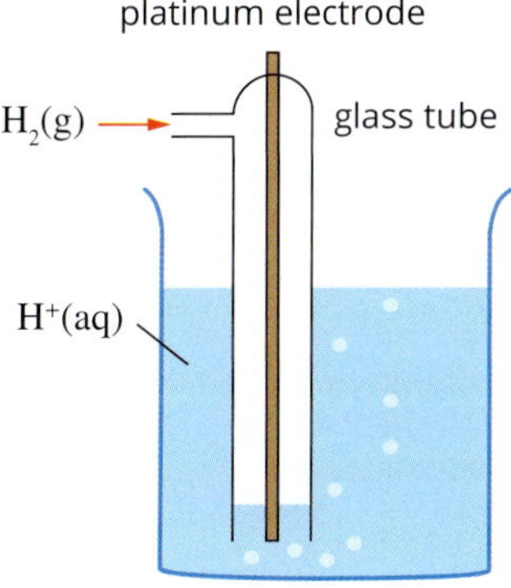

3 **a** Reduction occurs at the cathode: $Sn^{2+}(aq) + 2e^- \rightarrow Sn(s)$
b Oxidation occurs at the anode: $Al(s) \rightarrow Al^{3+}(aq) + 3e^-$

4

voltmeter
electron flow
aluminium anode
tin cathode
−
+
anions
cations
Sn²⁺(aq)+2e⁻ → Sn(s)
Al(s) → Al³⁺(aq)+3e⁻
Al³⁺(aq)
Sn²⁺(aq)

5 $Zn(s) + Cu^{2+}(aq) \rightarrow Cu(s) + Zn^{2+}(aq)$

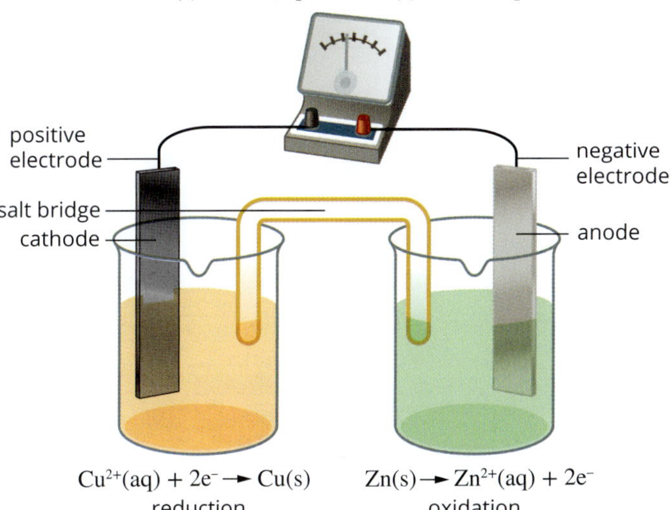

12.4 The table of standard reduction potentials

TY 12.4.1 $Sn^{2+}(aq) + Ni(s) \rightarrow Sn(s) + Ni^{2+}(aq)$
The nickel electrode will be the anode and the tin electrode will be the cathode. Electrons flow from the negative electrode (anode) to the positive electrode (cathode).

TY 12.4.2a The two relevant half-equations are:

$Mn^{2+}(aq) + 2e^- \rightarrow Mn(s)$ $E^\circ = -1.18$ V

$Cu^+(aq) + e^- \rightarrow Cu(s)$ $E^\circ = +0.52$ V

A reaction occurs because the oxidising agent, Cu^+, is below the reducing agent, Mn, in the table of standard reduction potentials.

The lower half-equation occurs in the forward direction:

$Cu^+(aq) + e^- \rightarrow Cu(s)$

The higher half-equation occurs in the reverse direction:

$Mn(s) \rightarrow Mn^{2+}(aq) + 2e^-$

The overall equation is found by adding the half-equations:

$2Cu^+(aq) + Mn(s) \rightarrow 2Cu(s) + Mn^{2+}(aq)$

b The two relevant half-equations are:

$Mn^{2+}(aq) + 2e^- \rightarrow Mn(s)$ $E^\circ = -1.18$ V

$Sn^{2+}(aq) + 2e^- \rightarrow Sn(s)$ $E^\circ = -0.14$ V

No reaction occurs because the Mn^{2+} and Sn^{2+} are both oxidising agents.

c The two relevant half-equations are:

$Sn^{2+}(aq) + 2e^- \rightarrow Sn(s)$ $E^\circ = -0.14$ V

$Cu^+(aq) + e^- \rightarrow Cu(s)$ $E^\circ = +0.52$ V

No reaction occurs because the oxidising agent, Sn^{2+}, is above the reducing agent, Cu, in the table of standard reduction potentials.

1 **a** reduction: $Pb^{2+}(aq) + 2e^- \rightarrow Pb(s)$
oxidation: $Al(s) \rightarrow Al^{3+}(aq) + 3e^-$

b overall cell reaction: $3Pb^{2+}(aq) + 2Al(s) \rightarrow 3Pb(s) + 2Al^{3+}(aq)$

c The lead electrode will be the cathode and the aluminium electrode will be the anode.

2

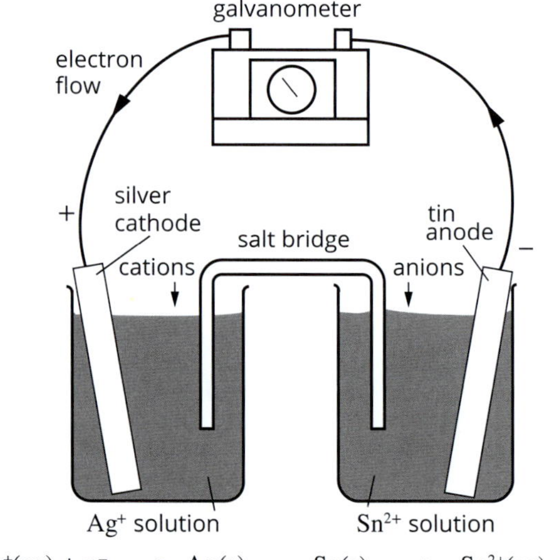

$Ag^+(ag) + e^- \longrightarrow Ag(s)$ $\quad Sn(s) \longrightarrow Sn^{2+}(aq) + 2e^-$

Overall reaction: $2Ag^+(aq) + Sn(s) \longrightarrow 2Ag(s) + Sn^{2+}(aq)$

3 0.94 V

4 **a** $Cu^{2+}(aq) + Zn(s) \rightarrow Zn^{2+}(aq) + Cu(s)$

b no reaction

c no reaction

5 For a reaction to occur, the reducing agent, Fe, must be above the oxidising agent in the table of standard reduction potentials.

$CuSO_4$: A reaction occurs because the reducing agent, Fe, is above the oxidising agent, Cu^{2+}. Cu^{2+} is reduced, forming a coating of copper on the nail.

$MgCl_2$: No reaction occurs because the reducing agent, Fe, is below the oxidising agent, Mg^{2+}. No reaction occurs between Fe and Cl^- because they are both reducing agents.

$Pb(NO_3)_2$: A reaction occurs because the reducing agent, Fe, is above the oxidising agent, Pb^{2+}. Pb^{2+} is reduced, forming a coating of lead on the nail.

$ZnCl_2$: No reaction occurs because the reducing agent, Fe, is below the oxidising agent, Zn^{2+}. No reaction occurs between Fe and Cl^- because they are both reducing agents.

Chapter 12 review

1 D **2** A **3** B

4 loses, positive, loses, gains, negative, gains, 7, gains, bromide

5 oxidation half-equation: $Pb(s) \rightarrow Pb^{2+}(aq) + 2e^-$
reduction half-equation: $Ag^+(aq) + e^- \rightarrow Ag(s)$
overall: $Pb(s) + 2Ag^+(aq) \rightarrow Pb^{2+}(aq) + 2Ag(s)$

6 **a** oxidation **b** oxidation **c** reduction
d oxidation **e** reduction **f** reduction

7 **a** Charges are not balanced. $Ag(s) \rightarrow Ag^+(aq) + e^-$

b Electrons appear on both sides of the equation.
$Cu(s) \rightarrow Cu^{2+}(aq) + 2e^-$

c State symbols are incorrect. $Zn(s) \rightarrow Zn^{2+}(aq) + 2e^-$

d Atoms are not balanced. $I_2(aq) + 2e^- \rightarrow 2I^-(aq)$

e Electrons are being subtracted. $Na^+(aq) + e^- \rightarrow Na(s)$

8

Step	Task	How it's done	Half-equation
1	Balance nitrogens.	already balanced	$NO_3^- \rightarrow NO_2$
2	Balance oxygens by adding *H_2O*.	Add *one* H_2O molecule(s) to the right-hand side of the equation.	*$NO_3^- \rightarrow NO_2 + H_2O$*
3	Balance hydrogens by adding *H^+*.	Add *$2H^+$* ion(s) to the *left-hand side* of the equation.	*$NO_3^- \rightarrow NO_2 + H_2O$*
4	Balance charge by adding *electrons*.	Charge on left-hand side = *−1 + 2 = +1* Charge on right-hand side = *0* Add *one* e– to the *left-hand side* of the equation.	*$NO_3^- + 2H^+ + e^- \rightarrow NO_2 + H_2O$*
5	Add state symbols to give the final half-equation.	Give the appropriate states for each reactant and product in the equation.	*$NO_3^-(aq) + 2H^+(aq) + e^- \rightarrow NO_2(g) + H_2O(l)$*

9 $2IO_3^-(aq) + 12H^+(aq) + 10e^- \rightarrow I_2(aq) + 6H_2O(l)$

10 half-equations:

$$Zn(s) \rightarrow Zn^{2+}(aq) + 2e^-$$
$$2MnO_2(s) + 2H^+(aq) + 2e^- \rightarrow Mn_2O_3(s) + H_2O(l)$$

overall equation:

$Zn(s) + 2MnO_2(s) + 2H^+(aq) \rightarrow Zn^{2+}(aq) + Mn_2O_3(s) + H_2O(l)$

11 **a** half-equations:

$$2H_3AsO_4(aq) + 4H^+(aq) + 4e^- \rightarrow As2O_3(s) + 5H_2O(l)$$
$$I^-(aq) + 3H_2O(l) \rightarrow IO_3^-(aq) + 6H^+(aq) + 6e^-$$

b overall equation:

$$6H_3AsO_4(aq) + 2I^-(aq) \rightarrow 3As_2O_3(s) + 2IO_3^-(aq) + 9H_2O(l)$$

12 **a** +4 **b** −2 **c** +6 **d** +6 **e** +4 **f** +2

13

Compound	Element	Oxidation number
$CaCO_3$	Ca	+2
HNO_3	O	−2
H_2O_2	O	−1
HCO_3^-	C	+4
HNO_3	N	+5
$KMnO_4$	Mn	+7
H_2S	S	−2
Cr_2O_3	Cr	+3
N_2O_4	N	+4

14 K_3N, N_2, N_2O, NO, N_2O_3, N_2O_4, $Ca(NO_3)_2$
Oxidation numbers for nitrogen are as follows:
K_3N: −3
N_2: 0
N_2O: +1
NO: +2
N_2O_3: +3
N_2O_4: +4
$Ca(NO_3)_2$: +5

15 **b**, **c**, **e**, **f** and **h** are redox reactions because the elements in the reactions undergo changes in oxidation number during the course of the reaction.
The changes in oxidation number which occur are:
- **b** Ag from 0 to +1; Cl from 0 to −1
- **c** Fe from +3 to +2; Sn from +2 to +4
- **e** P from +3 to +5; I from 0 to −1
- **f** Cu from +1 to +2; Cu from +1 to 0
- **h** P from 0 to −3; H from 0 to +1

16 D **17** A

18 **a** Oxidising agent: a substance that causes another substance to be oxidised and is reduced in the process.
Reducing agent: a substance that causes another substance to be reduced and is oxidised in the process.
b Anode: electrode at which oxidation occurs. Cathode: electrode at which reduction occurs.
c External circuit: section of a circuit where the electrons flow, e.g. through wires. Internal circuit: part of cell where the current is due to the movement of ions, e.g. in the salt bridge.

19

electron flow
polarity: + polarity: -
cathode anode
platinum electrode lead electrode
$Cl_2(g)$ cations
$Cl^-(aq)$ $Pb^{2+}(aq)$
half-equation: half-equation:
$Cl_2(g) + 2e^- \rightarrow 2Cl^-(aq)$ $Pb(s) \rightarrow Pb^{2+}(aq) + 2e^-$

20 **a**

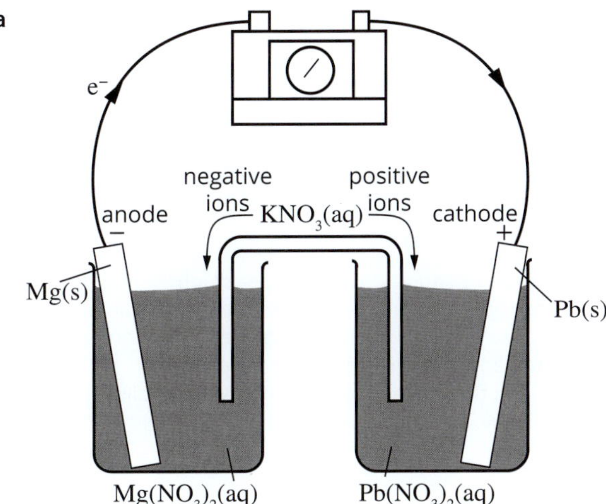

b
$$Mg(s) \rightarrow Mg^{2+}(aq) + 2e^-$$
$$Pb^{2+}(aq) + 2e^- \rightarrow Pb(s)$$
Overall: $Mg(s) + Pb^{2+}(aq) \rightarrow Mg^{2+}(aq) + Pb(s)$

c The lead electrode is the cathode; the magnesium electrode is the anode.
d Anions will migrate to the $Mg^{2+}(aq)/Mg(s)$ half-cell; cations to the $Pb^{2+}(aq)/Pb(s)$ half-cell.

21 $Sn^{2+} > Fe^{2+} > Br^-$

22 $B^{2+} > A^{2+} > C^{2+} > D^{2+}$

23 **a** $3Cu^{2+}(aq) + 2Al(s) \rightarrow 3Cu(s) + 2Al^{3+}(aq)$
b no reaction
c $2Ag^+(aq) + Zn(s) \rightarrow 2Ag(s) + Zn^{2+}(aq)$

24 **a** $Ag^+(aq)$; $Mg^{2+}(aq)$
b Mg(s); Ag(s)
c A coating of silver will form on the lead when it is placed in silver nitrate solution, because Ag^+ ions are stronger oxidants than Pb^{2+} ions.
d zinc and magnesium

25 **a** **i** +3
ii +2
iii +4
b oxidation: $S_2O_4^{2-}(aq) + 2H_2O(l) \rightarrow 2HSO_3^-(aq) + 2H^+(aq) + 2e^-$
reduction: $S_2O_4^{2-}(aq) + 2H^+(aq) + 2e^- \rightarrow S_2O_3^{2-}(aq) + H_2O(l)$

26 **a** $Zn(s) + 4OH^-(aq) + Cu^{2+}(aq) \rightarrow Zn(OH)_4^{2-}(aq) + Cu(s)$
b As shown in the diagrams, the student should make two electrochemical cells consisting of the:
- Cu^{2+}/Cu half-cell and the 'alkaline zinc half-cell'
- Cu^{2+}/Cu half-cell and the Zn^{2+}/Zn half-cell.

The cell voltages should be measured. And if they are identical, the two half-cells have the same *E*° values. (They would not be expected to have the same *E*° values.)

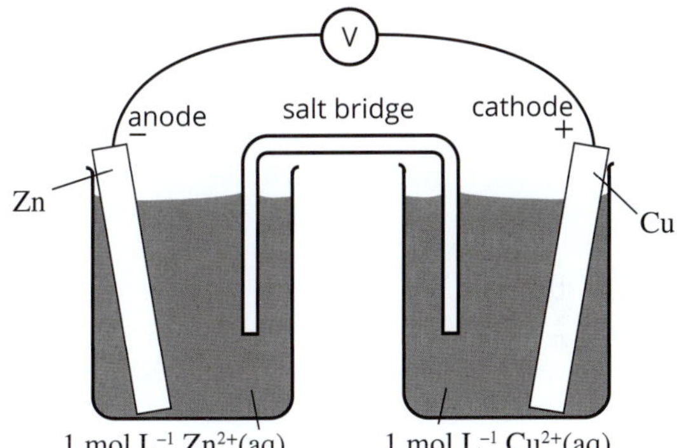

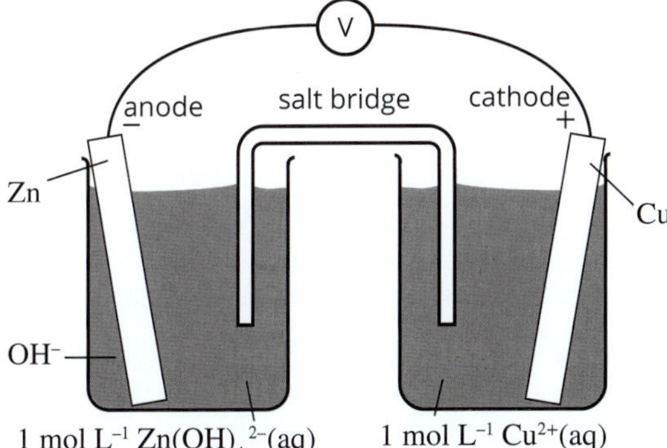

27 **a** The steel wool has undergone an oxidation reaction.
b Student answers will vary. Possible answers include: coating the steel with a stronger reducing agent and keeping the steel wool dry.

Chapter 13 Rates of reactions

13.1 Collision theory

1 A
2 a $CH_4(g) + 2O_2(g) \rightarrow CO_2(g) + 2H_2O(g)$
  b $+3380\ kJ\ mol^{-1}$
3 $+167\ kJ\ mol^{-1}$
4 a very small activation energy
5 They must collide with the correct orientation and also have sufficient energy to allow bond breaking to occur (greater than or equal to the activation energy).

13.2 Measuring reaction rate

1 mass loss over time, gas volume produced over time, temperature change over time
2 • React Mg and HCl in a conical flask that is connected to a gas syringe and measure the gas volume at fixed time intervals; graph the data and determine the initial gradient.
  • React Mg and HCl in a conical flask that has a calibrated pH probe in it; measure the pH at fixed time intervals; graph the data and determine the initial gradient.
3 a $CaCO_3(s) + 2HNO_3(aq) \rightarrow Ca(NO_3)_2(aq) + CO_2(g) + H_2O(l)$
  b The reaction rate is decreasing, because the gradient of the graph is decreasing.
4 a

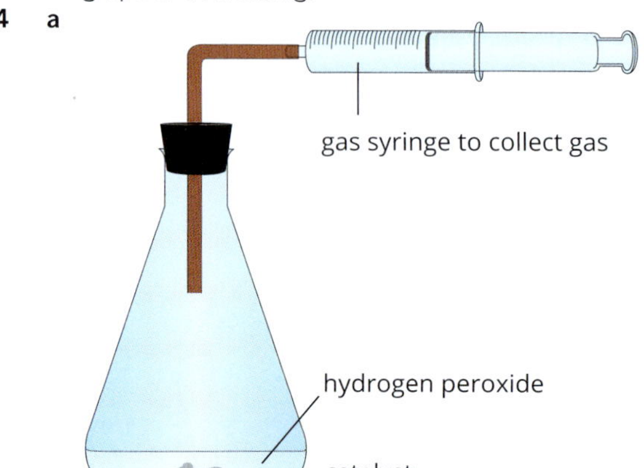

  b

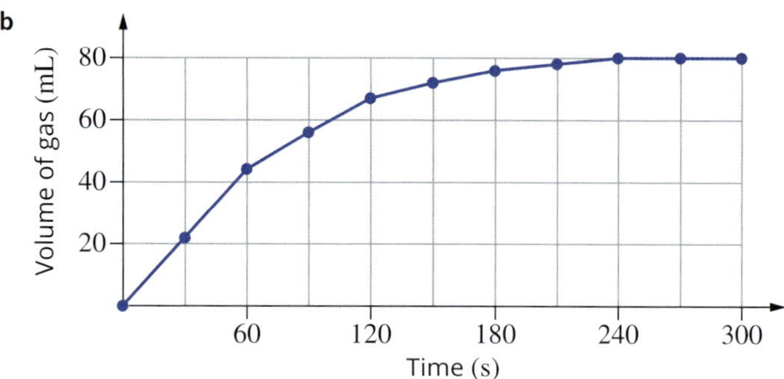

  initial rate = $0.73\ mL\ s^{-1}$
5 $0.30\ mol\ L^{-1}$ has a higher concentration of copper(II) ions, so would absorb more light.
6 At regular time intervals: measure pH; use a colourimeter to measure the amount of light of an appropriate wavelength absorbed (the I_2 produced will cause the solution to turn orange/brown); or measure the electrical conductivity of the solution.

13.3 Effect of surface area, concentration and pressure on reaction rate

TY 13.3.1 Coal is a solid. In the mine, there would be lumps of coal and also powdered coal present. The surface area of powdered coal is greater than that of solid coal. When the surface area increases, the frequency of collisions increases, so the rate of reaction increases. An explosion is a very fast reaction. The very large surface area of the coal dust allows for an increase in the frequency of collisions with reacting particles, which increases the reaction rate so much that explosions occur.

1 C
2 a increasing b increasing c increasing d decreasing
3 B, D, C, A
4 Fine particles have a large surface area, resulting in a high frequency of collisions between aluminium particles and gas molecules (such as oxygen) in the air and hence a rapid reaction rate. The aluminium can burn vigorously and release a large quantity of heat.
5 a using smaller pieces of wood with a larger surface area
  b using a brick cleaner with a higher concentration

13.4 Effect of temperature on reaction rate

1 C
2 A, B and C
3 a At higher temperatures, the molecules that react to form fibreglass plastics have greater energy. They collide more frequently and are more likely to have a total energy exceeding the activation energy of the reaction involved, increasing the rate of reaction.
  b At high altitude, such as in Nepal, air pressure is considerably lower than at any location in the Australian bush, so the water boils at a lower temperature in Nepal (up to 30°C lower). Thus, the average kinetic energy of the molecules in the potato will be lower, and they are less likely to have a total energy exceeding the activation energy of the reactions involved in cooking the potato. As a result, the potato will cook more slowly.
4 a a Maxwell–Boltzmann (distribution) curve or a kinetic energy distribution diagram
  b

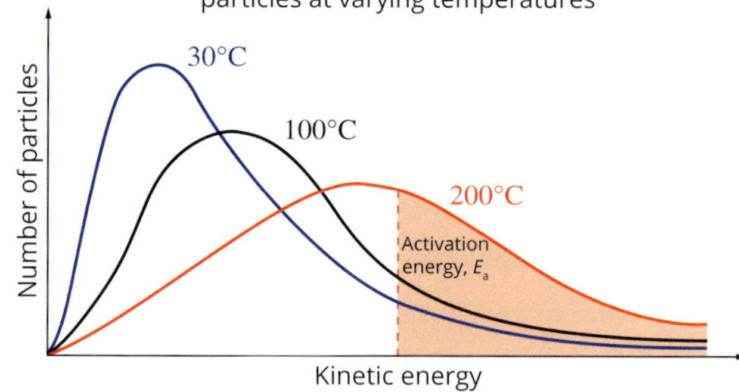

  c The number of particles that have energy in excess of the activation energy is represented by the area under the curve to the right of the dashed line. At 200°C, a greater proportion of particles have energy greater than or equal to the activation energy of the reaction, E_a, than at 100°C.

Chapter 13 review

1 Reactant particles must: collide with each other, collide with sufficient energy to break the bonds within the reactants, and collide with the correct orientation to break the bonds within the reactants to allow the formation of new products.
2 Collision 1 has the correct collision orientation, allowing bonds to break within the reactants and bonds to form within the products.

3 **a** Energy profile diagram for the reaction between CH_3Cl and OH^-

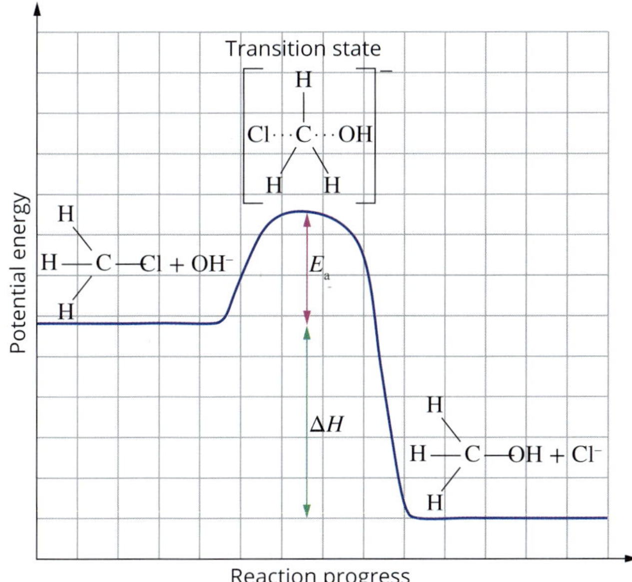

b Activation energy is the minimum amount of energy required by the reactants in order to form products in a reaction.

c

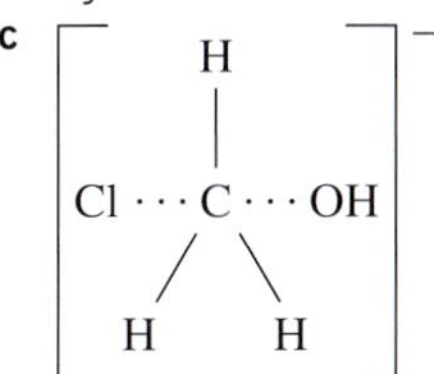

d The C–Cl bond is being broken in chloromethane. The C–O bond is beginning to form between chloromethane and the hydroxide ion.

4 **a** The single H–H bond in each hydrogen molecule and the double O=O bond in each oxygen molecule are broken in the course of this reaction.

b Two H–O bonds are formed in each new water molecule during the reaction.

c No reaction occurs until sufficient energy is supplied to give the reactant particles enough energy to overcome the activation energy barrier.

5 B **6** B

7 The rate of reaction is a measure of the change in concentration of a substance over time. Units that correspond to concentration per unit time, such as $mol\,L^{-1}\,s^{-1}$ or $mol\,L^{-1}\,min^{-1}$, would be suitable

8 **a** The student may have conducted a colourimetric investigation, as shown in Figures 13.2.6 and 13.2.7 (page 418). As the reaction proceeds, the concentration of permanganate ion decreases, so the intensity of the purple solution should decrease. The student did not measure the volume of carbon dioxide produced, because if it was collected during the experiment it would have increased with time.

b graph B

9 D

10 **a** $CaCO_3(s) + 2HCl(aq) \rightarrow CaCl_2(aq) + H_2O(l) + CO_2(g)$

b $2\,mol\,L^{-1}$ HCl and powdered $CaCO_3$

11 **a** At lower temperatures, the molecules that react to cause the apple to brown have less energy. They collide less frequently and are less likely to have a total energy exceeding the activation energy of the reaction involved, decreasing the rate of reaction.

b Using pure oxygen gas instead of air increases the concentration of oxygen. This results in a higher frequency of collisions between hydrogen and oxygen molecules, and hence an increased reaction rate.

12 **a** $CaCO_3(s) + 2HCl(aq) \rightarrow CaCl_2(aq) + CO_2(g) + H_2O(l)$

b $n(CaCO_3) = \frac{10.0}{100.09} = 0.0999\,mol$

$n(HCl) = 0.100\,mol\,L^{-1} \times 0.100\,L = 0.0100\,mol$

$\therefore$ $CaCO_3$ is in excess

c The rate of reaction could be measured by:
- the decrease in mass of the reaction mixture as $CO_2(g)$ escapes to the atmosphere
- the increase in pH measaured with a pH probe as acid is consumed.

d The rate of reaction with the smaller lumps will be faster. The smaller lumps have a larger surface area, so more collisions can occur per second.

e The rate of reaction could be increased by:
- increasing temperature: the average kinetic energy of the particles increases, causing increased collision frequency and also causing a greater proportion of the collisions between particles to have energy greater than the activation energy.
- increasing the concentration of the hydrochloric acid: there would be more HCl(aq) particles per unit volume, which would lead to more frequent collisions between reactants.

13 **a** A gas is produced, so mass is lost from the mixture.

b Cu

c

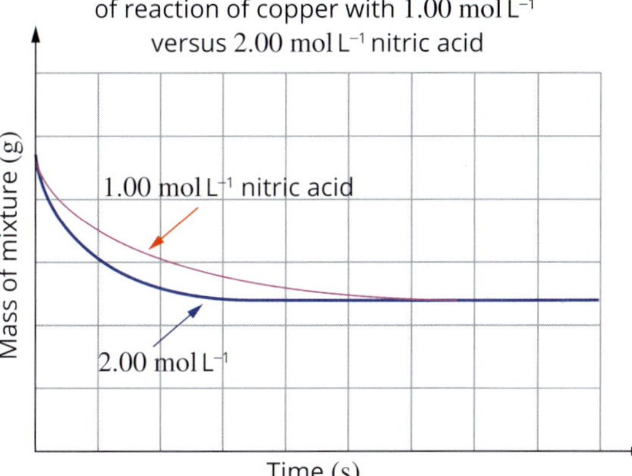

There would be a decreased rate of mass loss due to the lower nitric acid concentration.

($n(HNO_3) = cV = 1.00 \times 0.500 = 0.500\,mol$)

Copper would still be limiting, so the final mass would remain the same.

d

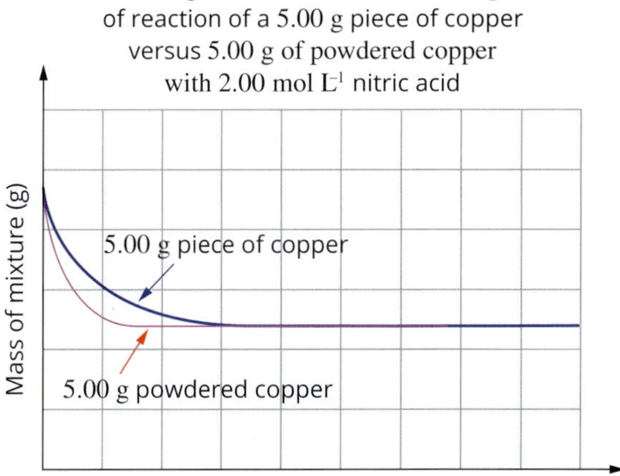

There would be an increased rate of mass loss due to the increased copper surface area. Copper would still be limiting, so the final mass would remain the same.

14 D

15 **a** Higher body temperature increases the rate of reactions. Increased pulse and breathing rate increases the concentration of reactants, which also increases the rate of reactions.

b Lower body temperature decreases the rate of the metabolic reactions in the body that could cause brain damage.

16 IV, VI, II, V, III, I

17 a surface area of a solid reactant, concentration of reactants in a solution, pressure of any gaseous reactants, temperature of the reaction

b i surface area of a solid reactant, concentration of reactants in a solution, pressure of any gaseous reactants

ii temperature of the reaction (also increases collision frequency)

18 (Student answers will vary.) You could:

- grind up the sugar crystals or use caster sugar
- use a cup of hot water to dissolve the sugar
- gently heat the sugar and water mixture while the sugar is dissolving
- stir the sugar and water mixture while the sugar is dissolving.

19 a The temperature of the new process (550°C) is much lower than the temperature of 1800°C in the blast furnace.

b $2Fe(s) + O_2(g) \rightarrow 2FeO(s)$

$4Fe(s) + 3O_2(g) \rightarrow 2Fe_2O_3(s)$

c high surface area of iron pellets; high temperature caused by trapped heat that was unable to escape rapidly

d Water would have caused the production of more hydrogen and increased the fire. The method used by the firefighters to extinguish the fire was to flood the hold with liquid nitrogen, which extinguished the surface fire but did not stop the deeper burning. A crane and clamshell bucket was then used to unload the iron into piles less than 1 m deep so the heat could escape.

20 a The energy of collisions is the same in both reactions because the temperature is the same. Because of an increase in surface area using powdered zinc compared with large zinc particles, the frequency of collisions between reactant particles increases.

b oxidation: $Zn(s) \rightarrow Zn^{2+}(aq) + 2e^-$;

reduction: $Cu^{2+}(aq) + 2e^- \rightarrow Cu(s)$

21 Crushed chalk has a greater surface area, which means the vinegar and chalk particles collide more frequently, increasing the rate of reaction.

Module 3 Reactive chemistry

Multiple choice

1	D	**2**	C	**3**	A	**4**	B	**5**	C
6	C	**7**	B	**8**	A	**9**	C	**10**	D
11	A	**12**	C	**13**	B	**14**	B	**15**	A
16	C	**17**	D	**18**	B	**19**	A	**20**	C

Short answer

1 a precipitation b acid base c synthesis

d combustion e decomposition

2 a $2HCl(aq) + Na_2CO_3(s) \rightarrow 2NaCl(aq) + H_2O(l) + CO_2(g)$

b $2C_4H_{10}(g) + 13O_2(g) \rightarrow 8CO_2(g) + 10H_2O(g)$

c $Ba(s) + H_2SO_4(aq) \rightarrow BaSO_4(s) + H_2(g)$

d $2Al(s) + 3Cl_2(g) \rightarrow 2AlCl_3(s)$

e $HNO_3(aq) + KOH(aq) \rightarrow KNO_3(aq) + H_2O(l)$

3 a $N_2(g) + 3H_2(g) \rightarrow 2NH_3(g)$

b synthesis

c $4NH_3(g) + 5O_2(g) \rightarrow 4NO(g) + 6H_2O(g)$

4 a $Ba^{2+}(aq) + SO_4^{2-}(aq) \rightarrow BaSO_4(s)$

b $H^+(aq) + OH^-(aq) \rightarrow H_2O(l)$

c $2H^+(aq) + CO_3^{2-}(aq) \rightarrow CO_2(g) + H_2O(l)$

d $Zn(s) + Cu^{2+}(aq) \rightarrow Zn^{2+}(aq) + Cu(s)$

5 a $Fe(s) + 2H^+(aq) \rightarrow Fe^{2+}(aq) + H_2(g)$

b $2Rb(s) + 2H_2O(l) \rightarrow 2Rb^+(aq) + 2OH^-(aq) + H_2(g)$

c $Zn(s) + 2Ag^+ \rightarrow Zn^{2+}(aq) + 2Ag(s)$

6 a i Statement I uses qualitative data. Qualitative data can be observed but not measured. Quantative data can be measured and expressed in numerical form.

ii quantitative data: 2.5 mL, 60.0 s, 0.5 g, 10.0 mL, 3.64 g L^{-1}

b Similarities: (Student answers will vary. They may include one of the following.)

- Both series rank metals in relative order of reactivity.
- With some minor differences, the ranking are in agreement.

Differences: (Student answers will vary. They may include one of the following.)

- The activity series only ranks metals; the table of standard reduction potentials also includes non-metals and polyatomic ions in its ranking.
- The ranking in the table of standard reduction potentials is determined under controlled conditions. These conditions are not controlled in the determination of the activity series.
- While both series can be used to predict the reactions of metals with ions of other metals, the table of standard reduction potentials can also be used to predict reactions involving non-metals.

7 a $O_2(g) + 2H_2O(l) + 2Fe(s) \rightarrow 4OH^-(aq) + 2Fe^{2+}(aq)$

b The oxidising agent is the reactant that causes the other reactant to be oxidised. Fe loses electrons so it is oxidised. Oxygen is therefore the oxidising agent and Fe is the reducing agent.

c i Zn(s) is more reactive than iron and so will lose electrons in preference to Fe. So when there is some Zn attached to the Fe, the Zn is oxidised and the Fe is protected.

ii Sn is less reactive than Fe, so if Fe were to be in contact with oxygen, it is the Fe that will preferentially corrode. If the Sn completely covers the Fe, the oxygen is prevented from reaching the Fe and the Fe is protected.

8 a i Oxidation: $Al(s) \rightarrow Al^{3+}(aq) + 3e^-$

Reduction: $Br_2(aq) + 2e^- \rightarrow 2Br^-(aq)$

ii Overall: $2Al(s) + 3Br_2(aq) \rightarrow 2Al^{3+}(aq) + 6Br^-(aq)$

b i Oxidation: $S_2O_3^{2-}(aq) + 5H_2O(l) \rightarrow 2SO_4^{2-}(aq) + 10H^+(aq) + 8e^-$

Reduction: $ClO^-(aq) + 2H^+(aq) + 2e^- \rightarrow Cl^-(aq) + H_2O(l)$

ii Overall: $4ClO^-(aq) + S_2O_3^{2-}(aq) + H_2O(l) \rightarrow 2SO_4^{2-}(aq) + 4Cl^-(aq) + 2H^+(aq)$

c i Oxidation: $H_2O_2(aq) \rightarrow O_2(aq) + 2H^+(aq) + 2e^-$

Reduction: $MnO_4^-(aq) + 4H^+(aq) + 3e^- \rightarrow MnO_2(s) + 2H_2O(l)$

ii Overall: $3H_2O_2(aq) + 2MnO_4^-(aq) + 2H^+(aq) \rightarrow 3O_2(aq) + 2MnO_2(s) + 4H_2O(l)$

9 a i 0.27 V

ii Fe^{2+}/Fe

iii $Sn^{2+}(aq) + Fe(s) \rightarrow Sn(s) + Fe^{2+}(aq)$

b i 2.48 V

ii Al^{3+}/Al

iii $3Fe^{3+}(aq) + Al(s) \rightarrow 3Fe^{2+}(s) + Al^{3+}(aq)$

c i 0.54 V

ii H^+/H_2

iii $I_2(aq) + H_2(g) \rightarrow 2I^-(aq) + 2H^+(aq)$

10 a The function is to act as a catalyst and lower the activation energy, giving a higher reaction rate.

b

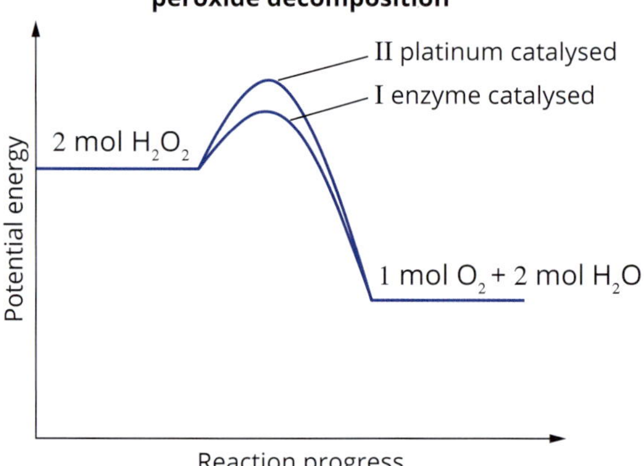

c System 1 would be faster. The lower activation energy requirement means that there is a greater probability of any given collision having sufficient energy for a successful reaction, so fruitful collisions will occur more frequently.

Extended response

1 a $Pb(NO_3)_2(aq) + 2KCl(aq) \rightarrow PbCl_2(s) + 2KNO_3(aq)$
  b $Pb^{2+}(aq) + 2Cl^-(aq) \rightarrow PbCl_2(s)$
  c i lead(II) chloride
   ii potassium ions and nitrate ions
  d i 0.00951 mol
   ii 2.64 g

2 a Place a sample of each metal into separate test-tubes containing a solution of Cu^{2+} ions. Observe and record which metals develop a copper coating. The copper coating indicates that the metal has displaced Cu^{2+} ions from solution. Repeat the experiment with each of the other metal ion solutions.
  b The activity series of the metals listed is Mg > Zn > Fe > Sn > Pb > Cu.
  c i Fe^{2+} or Pb^{2+} or Sn^{2+} or Cu^{2+}. A metal higher in the activity series will displace, from solution, the ion of a metal lower down the series. Zinc metal displaces Fe^{2+}, Pb^{2+}, Sn^{2+} and Cu^{2+} ions from solution.
   ii $Zn(s) + Fe^{2+}(aq) \rightarrow Zn^{2+}(aq) + Fe(s)$ or
   $Zn(s) + Pb^{2+}(aq) \rightarrow Zn^{2+}(aq) + Pb(s)$ or
   $Zn(s) + Sn^{2+}(aq) \rightarrow Zn^{2+}(aq) + Sn(s)$ or
   $Zn(s) + Cu^{2+}(aq) \rightarrow Zn^{2+}(aq) + Cu(s)$
  d i Magnesium. Zinc ions are displaced from solution by a metal that is more reactive than zinc metal, i.e. magnesium.
   ii $Mg(s) + Zn^{2+}(aq) \rightarrow Mg^{2+}(aq) + Zn(s)$

3 Metal reactivity decreases from left to right across a period and increases down a group. The reactivity of a metal is determined by the ease with which a metal atom loses its valence electrons.
Sodium and potassium are both group 1 metals. Sodium is the first element in the third period and potassium is the first element in the fourth period. Since potassium has one more electron shell than sodium its atomic radius is greater. The shielding effect of the inner electron shells is also greater in potassium. Consequently, the force of attraction between the potassium atom's valence electron and its nucleus is less. As a result, the ionisation energy and electronegativity of potassium are less than that of sodium. The potassium atom is more reactive, since it will lose the outer shell or valence electrons more readily than a sodium atom.
Sodium and magnesium are sequential elements in the third period of the periodic table and are found in groups 1 and 2, respectively.
The electronic configuration of sodium is 2,8,1, and magnesium has an electronic configuration of 2,8,2. The number of occupied or partially occupied shells stays constant, but the core charge increases with increasing atomic number. Thus, the strength of attraction between valence electrons and the nucleus increases from sodium to magnesium. As a consequence, the atomic radius decreases across the period from left to right. The first ionisation energy and electronegativity increase due to the combined effects of increased core charge and decreased atomic radius.
The energy required to form an ion also depends upon the number of electrons that are lost in formation of the metal cation. The sodium atom only loses a single electron when it forms the Na^+ ion. The formation of a magnesium ion involves the loss of two electrons. A greater amount of energy is required to remove the second valence electron than the first valence electron. More energy is required to form the Mg^{2+} ion than the Na^+ ion.
As a result of the factors described above, the ability of metals to form ions by losing electrons decreases across a period and increases down a group. This is reflected in the decreasing order of their reactivities: K > Na > Ca > Mg.

4 a arrow A
  b O_2 is the oxidising agent, and it enters the 'air space' from outside.
  c $4Al(s) + 3O_2(g) + 6H_2O(l) \rightarrow 4Al^{3+}(aq) + 12OH^-(aq)$
  d The cathode is the site of the reduction reaction involving oxygen/air, which must be in contact with the electrolyte. A porous electrode will allow a large surface area for this reaction and thereby maximise the rate of reaction and hence the current/power output of the cell.
  e From the table of standard reduction potentials:
  cathode reaction: $O_2(g) + 2H_2O(l) + 4e^- \rightarrow 4OH^-(aq)$ $E° = +0.40\,V$
  anode reactions: $Zn(s) \rightarrow Zn^{2+}(aq) + 2e^-$ $E° = -0.76\,V$
  $Ag(s) \rightarrow Ag^+(aq) + e^-$ $E° = +0.80\,V$
  Since $E°(Zn^{2+}/Zn) < E°(O_2,H_2O/OH^-)$, a spontaneous cell reaction will occur.
  Since $E°(Ag^+/Ag) > E°(O_2,H_2O/OH^-)$, a spontaneous cell reaction will not occur.
  f smaller, since $E°(Zn^{2+}/Zn) > E°(Al^{3+}/Al)$
  $E°(O_2,H_2O/OH^-) - E°(Zn^{2+}/Zn) = (+0.40) - (-0.76) = 1.16\,V$
  $E°(O_2,H_2O/OH^-) - E°(Al^{3+}/Al) = (+0.40) - (-1.71) = 2.10\,V$
  g With an acidic electrolyte, the cathode reaction becomes:
  cathode reaction: $O_2(g) + 4H^+(aq) + 4e^- - 2H_2O(l)$ $E° = +1.23\,V$
  anode reactions: $Zn(s) \rightarrow Zn^{2+}(aq) + 2e^-$ $E° = -0.76\,V$
  $Ag(s) \rightarrow Ag^+(aq) + e^-$ $E° = +0.80\,V$
  Since $E°(Zn^{2+}/Zn) < E°(O_2,H^+/H_2O)$, a spontaneous cell reaction will occur.
  Since $E°(Ag^+/Ag) < E°(O_2,H^+/H_2O)$, a spontaneous cell reaction will occur.
  Essentially, O_2 becomes a stronger oxidising agent in acidic conditions, and is capable of spontaneous reaction with the relatively weak reducing agent, Ag.

5 a The concentration of sodium thiosulfate is higher, so there will be more frequent collisions between reactant particles. While the probability of any particular collision being successful is unchanged, the higher frequency of collisions overall will increase the frequency of successful collisions and hence the rate of reaction.
  b The higher temperature means the particles in the mixture will be moving faster and hence colliding more frequently. Also, and more significantly, having faster-moving particles means there will be a higher proportion of collisions that are successful, since there is a greater chance of a collision having energy greater than the activation energy. These two factors combine to give a higher frequency of fruitful collisions and hence a higher rate of reaction.
  c Apart from changing concentration and temperature, reaction rates can be affected by changing the concentration of HCl.

Chapter 14 Energy changes in chemical reactions

14.1 Exothermic and endothermic reactions

1 D
2 a 180 kJ b 1.5×10^3 kJ
  c 0.0100 kJ or 1.00×10^{-2} kJ
  d 2.0×10^{-6} kJ
3 In chemistry, the system is usually the chemical reaction, whereas the surroundings refers to everything else (for example, the beaker or test-tube in which the reaction takes place).
4 In any reaction, the total amount of the chemical energy of the reactants is made up of the bonds between atoms within the reactants. If the total amount of chemical energy within the reactants is less than the total amount of chemical energy within the products, energy must be supplied to the system, and the reaction is said to be endothermic.

14.2 Thermochemical equations and energy profile diagrams

TY 14.2.1 $-147\,kJ\,mol^{-1}$
1 A negative ΔH value indicates that a reaction is exothermic. This is because the enthalpy of the reactants is greater than the enthalpy of the products. Energy is being released *to* the surroundings.
2 $CH_4(g) + 2O_2(g) \rightarrow CO_2(g) + 2H_2O(l)$ $\Delta H = -890\,kJ\,mol^{-1}$
3 It would be lower because the change of state of the H_2O from liquid to gas will require that energy be absorbed.

4 a endothermic
b The total enthalpy of the product (HI) is greater than that of the reactants (hydrogen gas and iodine gas).
c The activation energy is greater than the magnitude of the ΔH value.
5 $6CO_2(g) + 6H_2O(l) \rightarrow C_6H_{12}O_6(aq) + 6O_2(g)$ $\Delta H = +2803\ kJ\ mol^{-1}$

14.3 Heat of combustion

TY 14.3.1 2.72×10^5 kJ or 272 MJ

1 $2C_6H_6(l) + 15O_2(g) \rightarrow 12CO_2(g) + 6H_2O(l)$
2 propane: $E = 50.35$ kJ
octane: $E = 47.71$ kJ
Propane releases more energy per gram than octane does.
3 $C_2H_5OH(l) + 2O_2(g) \rightarrow 2CO(g) + 3H_2O(l)$
4 a 1.39×10^4 kJ or 13.9 MJ
b 4.85×10^5 kJ or 485 MJ
c 1.20×10^7 kJ or 1.20×10^4 MJ or 12.0 GJ

14.4 Determining the heat of combustion

TY 14.4.1 70.5 kJ
TY 14.4.2 $-1.26 \times 10^3\ kJ\ mol^{-1}$
TY 14.4.3 $2C_2H_6(g) + 7O_2(g) \rightarrow 4CO_2(g) + 6H_2O(l)$ $\Delta H = -3120\ kJ\ mol^{-1}$
TY 14.4.4 55.6%

1 1.05×10^2 kJ
2 It is assumed that all of the energy released by the combustion of the fuel is used to heat the water. In reality, some of the energy will be used to heat the metal can or be lost to the surroundings.
3 $-2.23 \times 10^3\ kJ\ mol^{-1}$ (The answer is negative because the reaction is exothermic.)
4 $2.79 \times 10^3\ kJ\ mol^{-1}$
5 a $CH_4(g) + 2O_2(g) \rightarrow CO_2(g) + 2H_2O(l)\Delta H = -890\ kJ\ mol^{-1}$
b 3.01 g
6 69.4%

14.5 Enthalpy of dissolution

TY 14.5.1 $+26\ kJ\ mol^{-1}$

1 $NaNO_3(s) \xrightarrow{H_2O(l)} Na^+(aq) + NO_3^-(aq)$
$Ca(OH)_2(s) \xrightarrow{H_2O(l)} Ca^{2+}(aq) + 2OH^-(aq)$

2 The positive sodium ion attracts the partial negative charges on the oxygen atoms in the water molecule. The negative chloride ion attracts the partial positive charges on the hydrogen atoms in the water molecule.

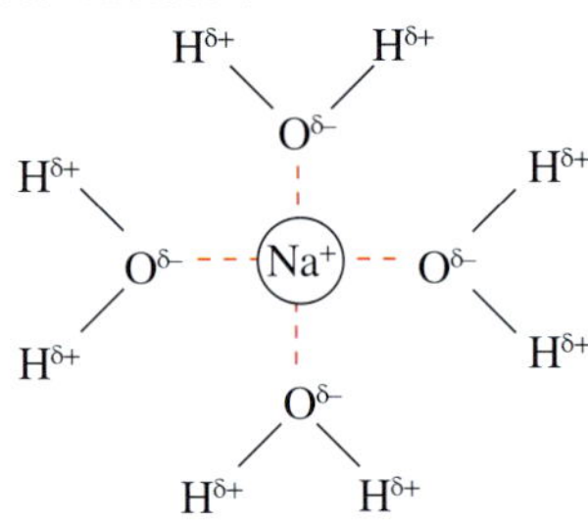

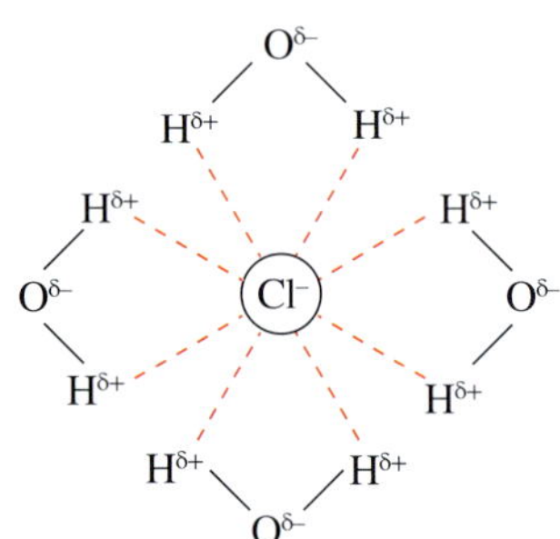

- - - represents ion–dipole interaction

3 When solid potassium nitrate dissolves in water, the ionic bonds between ions must be broken, and new ion–dipole bonds form between ions and water molecules. The ions are said to be hydrated and the process is known as dissociation.
4 insulating, lost, lid, exothermic, increases, endothermic.
5 $-81.2\ kJ\ mol^{-1}$
6 20.7°C

14.6 Catalysts

1 D
2 a A catalyst is a substance that increases the rate of a chemical reaction without itself undergoing permanent change.
b Activation energy is the minimum amount of energy that colliding reactant particles need in order for successful collisions to occur that lead to a reaction.
3 When salt is mixed with sugar, the salt acts as a catalyst and lowers the activation energy of the combustion reaction between sugar and oxygen.
4 a Reactions involving a heterogeneous catalyst take place at the surface of the catalyst. Reactants form bonds with the catalyst, lowering the activation energy of reactions and allowing them to proceed more rapidly.
b A porous pellet has a much larger surface area than a solid lump. More reactants may be in contact with the surface of a porous pellet at any instant, producing a faster rate of reaction.
5 a

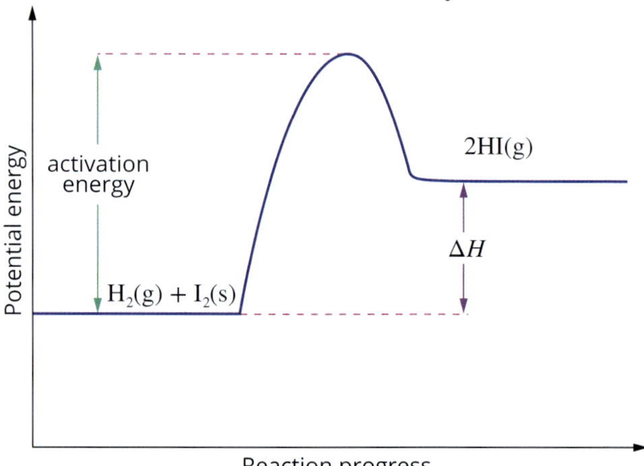

b endothermic
c

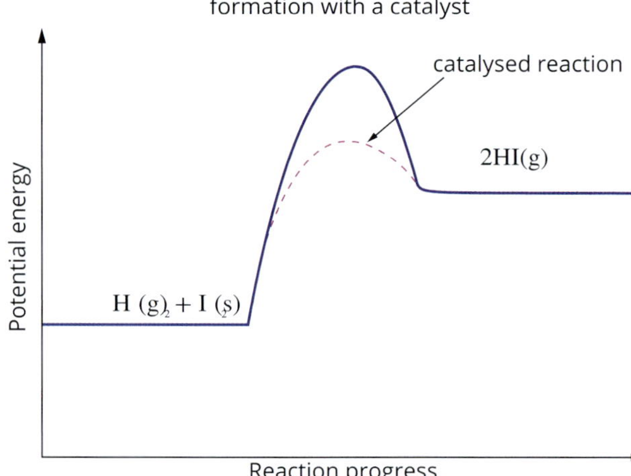

Chapter 14 review

1 a 2.21 kJ
b 152 J
c 1.89 MJ
d 12.5 kJ

2 a exothermic (Heat and light energy are released to the surrounding environment by the combustion of wood.)
b endothermic (Thermal energy is absorbed from the surrounding environment to melt the ice.)
c endothermic (Electrical energy is consumed from a power supply as the battery is recharged.)
d exothermic (Heat energy is released to the surrounding environment as organisms in the compost heap decompose the plant material. The temperature of the compost heap rises as a consequence.)

3 C

4 a true
b false
c false
d true

5 a exothermic
b $-5238\,\text{kJ mol}^{-1}$ (doubled)

6 If a chemical equation is written for an endothermic reaction, ΔH is positive, telling you that energy is absorbed as the reaction proceeds. The enthalpy of the products must be higher than the enthalpy of the reactants. If this reaction is reversed, the enthalpy of the reactants is now higher than the enthalpy of the products. The reaction releases energy as it proceeds, so is exothermic. ΔH becomes negative.

7 a higher energy in bonds of reactants
b Energy profile diagram of butane combustion

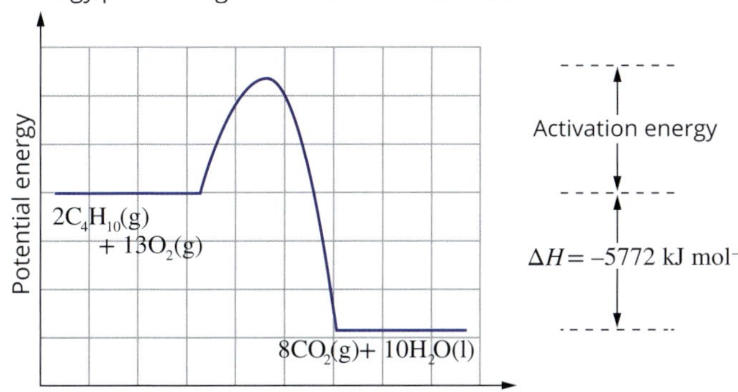

8 $C_4H_9OH(l) + 6O_2(g) \rightarrow 4CO_2(g) + 5H_2O(l)$

9 $2C_4H_{10}(g) + 9O_2(g) \rightarrow 8CO(g) + 10H_2O(l)$

10 277 MJ

11 a 25.1 kJ
b 78.4 kJ
c 63 kJ
d 788 kJ
e 54 kJ

12 33.0°C

13 $H_2(g) + \frac{1}{2}O_2(g) \rightarrow H_2O(l)$ $\Delta H = -286\,\text{kJ mol}^{-1}$
or
$2H_2(g) + O_2(g) \rightarrow 2H_2O(l)$ $\Delta H = -572\,\text{kJ mol}^{-1}$

14 a Na^+/CO_3^{2-}
b Ca^{2+}/NO_3^-
c K^+/Br^-
d Fe^{3+}/SO_4^{2-}
e Cu^{2+}/Cl^-

15 dissociation

16 a $MgSO4(s) \xrightarrow{H_2O(l)} Mg^{2+}(aq) + SO_4^{2-}(aq)$
b $Na_2S(s) \xrightarrow{H_2O(l)} 2Na^+(aq) + S^{2-}(aq)$
c $KOH(s) \xrightarrow{H_2O(l)} K^+(aq) + OH^-(aq)$
d $Cu(CH_3COO)_2(s) \xrightarrow{H_2O(l)} Cu^{2+}(aq) + 2CH_3COO^-(aq)$
e $Li_2SO_4(s) \xrightarrow{H_2O(l)} 2Li^+(aq) + SO_4^{2-}(aq)$

17 Potassium ions and bromide ions are held in the ionic lattice by ionic bonds that are based on electrostatic attraction. These ionic bonds in the solute break when it dissolves in water. Water, the solvent, has hydrogen bonds between the water molecules. When potassium bromide is added to water, the partially positive hydrogen atoms of the water molecules are attracted to the negative bromide ions, and the partially negative oxygen atoms of the water molecules are attracted to the positive potassium ions. Ion–dipole bonds form between the ions and water molecules, and the surface ions are pulled into solution. Gradually, the ionic lattice dissociates and a solution is formed.

18 $+3.3\,\text{kJ mol}^{-1}$

19 $-57\,\text{kJ mol}^{-1}$

20 24.0°C

21 A

22 i Increasing the pressure of the gases would cause an increase in the number of collisions between the reactant molecules in a given time, so more frequent collisions between molecules with the correct orientation for reacting and with energy greater than or equal to the activation energy would occur. As a result, the rate of reaction would increase.
ii Adding a catalyst would allow the reaction to occur by a different pathway with a lower activation energy. The proportion of collisions with energy greater than the activation energy would thus be increased. As a result, the rate of reaction would increase.

23 a $1370\,\text{kJ mol}^{-1}$
b $+572\,\text{kJ mol}^{-1}$
c

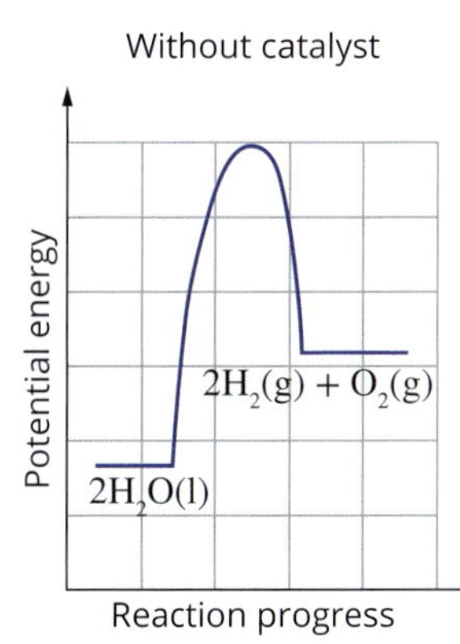

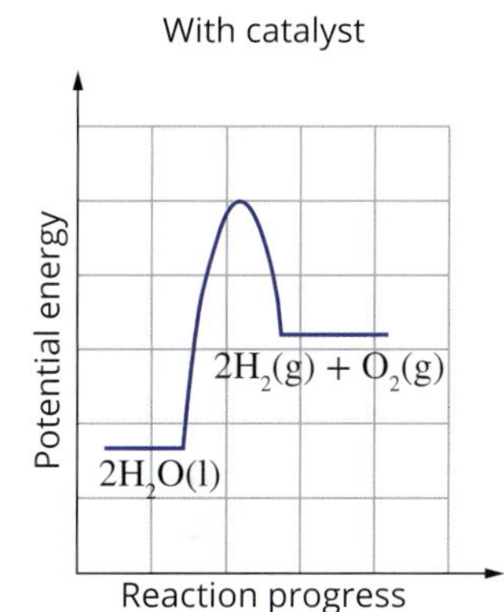

d $+572\,\text{kJ mol}^{-1}$

24 a The collision frequency is not changed when a catalyst is added. However, the activation energy for the reaction is lowered so that a greater proportion of the collisions occurring involve particles with energy greater than or equal to the activation energy.
b With more particles per unit volume, the frequency of collisions between reactants is increased. The activation energy is not changed.

25 a 1 mol of CO(g) and 0.5 mol of $O_2(g)$
b i $-566\,\text{kJ mol}^{-1}$
ii $+566\,\text{kJ mol}^{-1}$

26 $C_2H_6(g) + \frac{7}{2}O_2(g) \rightarrow 2CO_2(g) + 3H_2O(l)$ $\Delta H = -1.54 \times 10^3\,\text{kJ mol}^{-1}$
or
$2C_2H_6(g) + 7O_2(g) \rightarrow 4CO_2(g) + 6H_2O(l)$ $\Delta H = -3.08 \times 10^3\,\text{kJ mol}^{-1}$

27 a $-47\,\text{kJ mol}^{-1}$
b 81%
c Efficiency can be improved in any one of the following ways:
- putting a lid on the container (or if a lid is present, ensuring that the lid is tightly fastened, or the seal of the lid is improved)
- improving the insulation of the calorimeter
- ensuring that the solution is continuously stirred so that the heat is homogenous throughout the solution.

28 a

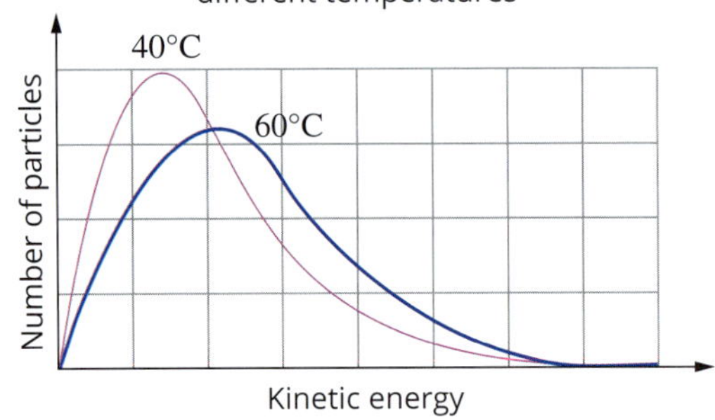

b

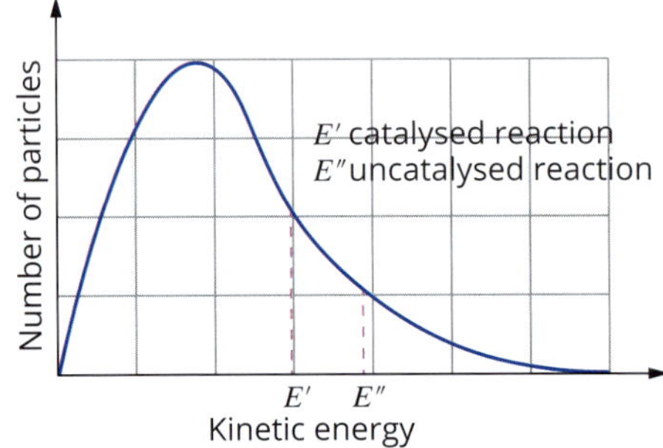

c When a catalyst is present, the reaction proceeds by an alternative reaction pathway with a lower activation energy than in the uncatalysed reaction. At a given temperature, presence of a catalyst means that there will be a greater proportion of the reacting particles that have a kinetic energy equal to or greater than the activation energy. As more reactants have sufficient energy to react, the rate of reaction increases.

29 The dissolution of sodium chloride in water is an endothermic physical reaction. As the salt dissolves into the water, heat energy is absorbed from the milk and sugar mixture.

Chapter 15 Enthalpy and Hess's law

15.1 Latent heat

TY 15.1.1 203.5 kJ
TY 15.1.2 2.5 kJ

1 D, A, C, E, B

2 The ΔH_{vap} is an indication of the intermolecular bond strength, because intermolecular bonds that hold the substance in the liquid state are broken when a substance vaporises.
The ΔH_{vap} of water is +40.7 kJ mol^{-1}, whereas that of ethanol is +38.6 kJ mol^{-1}. This suggests that the intermolecular bonds in water are stronger than those in ethanol.

3 endothermic

4 463 kJ

5 $C_6H_6(g) \rightarrow C_6H_6(l)$ $\Delta H = -30.8$ kJ mol^{-1}

6 33.4 kJ

7 2.26×10^{-3} kJ

15.2 Bond energy

TY 15.2.1 −543 kJ mol^{-1}

1 **a** (ii)
b (i)
c (i)

2

Name	Formula	Number and type of bonds
hydrogen	H_2	1 × H–H
water	H_2O	2 × O–H
methane	CH_4	4 × C–H
oxygen	O_2	1 × O=O
ammonia	NH_3	3 × N–H
carbon dioxide	CO_2	2 × C=O

3 Iodine is a larger atom than chlorine. The valence (bonding) electrons are a greater distance from the nucleus. Consequently, the electrons which form the covalent bond are less tightly held by the nuclei. Therefore, the bond energy of iodine is lower.

4 **a** +543 kJ mol^{-1} **b** −482 kJ mol^{-1}
c −784 kJ mol^{-1} **d** −808 kJ mol^{-1}

15.3 Hess's law

TY 15.3.1 +198 kJ mol^{-1}
TY 15.3.2 −120.2 kJ mol^{-1}
TY 15.3.3 −2677.2 kJ mol^{-1}
TY 15.3.4 −71 kJ mol^{-1}

1 $\Delta H^\circ = +77.2$ kJ mol^{-1}

2 $\Delta H^\circ = +73$ kJ mol^{-1}

3 $\Delta H^\circ = -90$ kJ mol^{-1}

4 $\Delta H^\circ_{reaction} = -890.3$ kJ mol^{-1}

5 $\Delta H^\circ_{reaction} = -534.2$ kJ mol^{-1}

6 $\Delta H^\circ_{reaction} = -1820.5$ kJ mol^{-1}

7 $\Delta H^\circ_{reaction} = +255.9$ kJ mol^{-1}

8 **a** $C_2H_2(g) + \frac{5}{2}O_2(g) \rightarrow 2CO_2(g) + H_2O(l)$
b $\Delta H^\circ_{reaction} = -1300.8$ kJ mol^{-1}
c 4995 kJ of energy.

Chapter 15 review

1 **a** iv
b ii

2 **a** $+4.1 \times 10^2$ kJ
b 2.26×10^3 kJ

3 **a** A positive value of ΔH_{vap} indicates an endothermic process, i.e. vaporisation requires energy.
b Intermolecular bonds are broken during vaporisation of water. Metallic bonds are broken during vaporisation of mercury.
c The metallic bonding in mercury is stronger than the hydrogen bonds between water molecules in water, so more heat energy is required to vaporise mercury.

4 **a** endothermic
b 92.4 kJ

5 The greater value for the ΔH_{fus} for NaCl indicates stronger bonding between the particles in solid NaCl than in solid water. In fact, the ions of NaCl(s) are held together by strong ionic bonds, whereas water molecules are held together by hydrogen bonding.

6 The particles of steam have more kinetic energy than those of water.

7 **a** −1110 kJ mol^{-1}
b −539 kJ mol^{-1}
c −93 kJ mol^{-1}

8 **a** $C_{15}H_{32}(l) + 23O_2(g) \rightarrow 15CO_2(g) + 16H_2O(g)$
b −9390 kJ mol^{-1}

9 **a** $C_2H_6(g) + \frac{7}{2}O_2(g) \rightarrow 2CO_2(g) + 3H_2O(g)$
b −1421 kJ mol^{-1}

10 −499 kJ mol^{-1}

11 +37 kJ mol^{-1}

12 One reason is that H–H is a much smaller molecule than F–F. Hydrogen's atomic radius is smaller, so the shared valence electrons in the covalent bond are more strongly attracted to the hydrogen nuclei. Second, the fluorine molecule has many more electron pairs, so there is a significant amount of coulombic (electrostatic) repulsion that reduces its bond strength.

13 A straight-line relationship is apparent in the graph because with each successive alcohol, an extra C–C bond, 2 C–H bonds and $\frac{3}{2}$O=O bonds are broken, while an extra 2 C=O bonds and 2 O–H bonds are formed.
14 $\Delta H^\circ = -852\,kJ\,mol^{-1}$
15 $\Delta H^\circ = -283\,kJ\,mol^{-1}$
16 $\Delta H^\circ = +91\,kJ\,mol^{-1}$
17 $\Delta H_3 = +68\,kJ$
18 $\Delta H^\circ_{reaction} = -1075.0\,kJ\,mol^{-1}$
19 Helium and neon are elements, so their enthalpies of formation are zero.
20 $\Delta H^\circ = -393\,kJ\,mol^{-1}$
21 a $C_2H_4(g) + Cl_2(g) \rightarrow C_2H_3Cl(g) + HCl$
b $\Delta H^\circ_{reaction} = -122.6\,kJ\,mol^{-1}$
22 $\Delta H^\circ_{reaction} = -88.2\,kJ\,mol^{-1}$
23 $\Delta H^\circ_{reaction} = +206.1\,kJ\,mol^{-1}$
24 a $CH_4(g) + O_2(g) \rightarrow C(s) + 2H_2O(l)$
b $\Delta H^\circ_{reaction} = -442.2\,kJ\,mol^{-1}$
25 $\Delta H^\circ_{reaction} = +98.8\,kJ\,mol^{-1}$
26 a $\Delta H^\circ_f(CH_4) = -74.8\,kJ\,mol^{-1}$
b $\Delta H_3 = +1589\,kJ\,mol^{-1}$
c $\Delta H1 = +1664\,kJ\,mol^{-1}$
27 $\Delta H_3 = -88\,kJ\,mol^{-1}$
28 a $C_4H_{10}(g) + \frac{13}{2}O_2(g) \rightarrow 4CO_2(g) + 5H_2O(l)$
b $\Delta H^\circ_{reaction} = -2877\,kJ\,mol^{-1}$
29 Melting requires energy—the enthalpy of fusion. For melting ice to liquid water, this is $6.01\,kJ\,mol^{-1}$. Heat energy is absorbed to cause a change of state instead of a change in temperature.

Chapter 16 Entropy and Gibbs free energy

16.1 Energy and entropy changes in chemical reactions

TY 16.1.1 There is an increase in the number of moles of liquid and gas during the reaction, so the system becomes more disordered and ΔS is positive.
1 While the system can be chosen freely, a sensible choice would be to choose the volume of the two liquids as the system. The boundary would be the flask, and the system is closed.
2 no change in entropy
3 decreased
4 a decreased b decreased
c increased d increased
5 III → VI → IV → I → II → V

16.2 Entropy and spontaneous processes

TY 16.2.1 $+165.3\,J\,mol^{-1}\,K^{-1}$
1 $+20\,J\,mol^{-1}\,K^{-1}$
2 a There is little change in ΔS_{system} because neither a gas nor a liquid is produced. The thermite reaction is exothermic ($\Delta H < 0$), so $\Delta S_{surrounding} > 0$. $\Delta S_{system} + \Delta S_{surrounding} > 0$, so the reaction is spontaneous.
b The exothermic reaction releases heat to the surroundings ($\Delta H < 0$), so $\Delta S_{surroundings} > 0$. The reaction has a loss of entropy ($\Delta S_{system} < 0$), because the reaction reduces 3 mol of gas to 2 mol of gas. The magnitude of $\Delta S_{surroundings}$ is larger than the magnitude of ΔS_{system}, so the sum of the two values is positive and the reaction is spontaneous.
c The process is endothermic ($\Delta H > 0$), so $\Delta S_{surroundings} < 0$. The increase in entropy when 1 mol of solid is dissolved to produce 2 mol of ions in solution is high enough ($\Delta S_{system} > \Delta S_{surroundings}$), so $\Delta S_{system} + \Delta S_{surroundings} > 0$; thus, the process is spontaneous.
3 a $-186.3\,J\,mol^{-1}\,K^{-1}$
b There is a net decrease in entropy (indicated by the net loss of 1 mol of gas). This matches the negative entropy change calculated in part **a**.
4 $+211\,J\,mol^{-1}\,K^{-1}$

16.3 Gibbs free energy

TY 16.3.1 As $\Delta G^\circ = +0.86\,kJ\,mol^{-1}$, ice will not spontaneously melt at −40°C.
TY 16.3.2 620.4 K or 347.4°C
1 B
2 $\Delta H^\circ < 0$ and $\Delta S^\circ > 0$
3 Standard state conditions are defined at a temperature of 25°C, but the human cell operates at around 37°C. The difference in temperature will give a different Gibbs free energy value.
4 entropy decreases: $-108.6\,J\,mol^{-1}\,K^{-1}$
5 spontaneous

Chapter 16 review

1 A closed system allows energy but not mass to pass through the boundary; an open system allows both energy and matter to be transferred between the system and the surroundings.
2 A, B, D
3 a decrease
b decrease
c increase
4 a $-29.1\,J\,mol^{-1}\,K^{-1}$
b $-559.2\,J\,mol^{-1}\,K^{-1}$
5 The reaction is exothermic, so $\Delta S_{surroundings} > 0$. The reaction generates 1 mol of gas plus 2 mol of liquid per 2 mol of liquid reactant, so there is a net increase in the entropy of the system, $\Delta S_{system} > 0$. As $\Delta S_{system} + \Delta S_{surroundings} > 0$, the reaction is spontaneous.
6 no
7 a The total entropy of the universe will increase.
b The total entropy of the universe will increase. (The total entropy of the universe will always increase if the reaction is spontaneous.)
8 a ΔS_{system} increases, so the products have greater degree of disorder.
b An exothermic reaction will result in $\Delta S_{surroundings} > 0$. This coupled with $\Delta S_{system} > 0$ results in $\Delta S_{universe} > 0$, so the reaction is spontaneous.
9 a endothermic
b A liquid has more entropy than a solid, so $\Delta S_{system} > 0$.
c An endothermic process will result in $\Delta S_{surroundings} < 0$. This fact, coupled with $\Delta S_{system} > 0$, results in $\Delta S_{universe} > 0$ only at relatively high temperatures. Therefore, the process is only spontaneous above the melting point.
10 a $C_5H_{12}(l) + 8O_2(g) \rightarrow 5CO_2(g) + 6H_2O(l)$
b $+264\,J\,mol^{-1}\,K^{-1}$
11 B
12 The reaction is spontaneous at $4.3 \times 10^2\,K$ and above.
13 59°C
14 $+78\,kJ\,mol^{-1}$
15 a $-189\,kJ\,mol^{-1}$
b $-192\,kJ\,mol^{-1}$
c The formation of both are spontaneous ($\Delta G^\circ < 0$). As $\Delta G^\circ_{gas} < \Delta G^\circ_{liquid}$, the formation of the gas is preferred (more spontaneous).
16 Since $\Delta H^\circ < 0$ and $\Delta S^\circ > 0$, the reaction will be spontaneous at all temperatures.
17 $\Delta S^\circ = -0.2263\,kJ\,mol^{-1}\,K^{-1}$ and $\Delta G^\circ = -120.4\,kJ\,mol^{-1}$ (The reaction is spontaneous.)
18 $\Delta G^\circ = +149.5\,kJ\,mol^{-1}$ (The reaction is not spontaneous.)
19 $-228\,J\,mol^{-1}\,K^{-1}$
20 In stretching a rubber band, the released warmth indicates that the processes are exothermic, so ΔH is negative. The stretching requires effort and is not spontaneous, so $\Delta G > 0$. From the Gibbs free energy change formula, the sign of ΔS must also be negative. Stretching a rubber band reduces the entropy in the rubber band system. The stretching straightens out the kinks in rubber, which reduces the disorder and is reflected in the decrease in entropy.

21 a The self-assembly of drinking straws is spontaneous, so ΔG is negative.

b The entropy decreases. ΔS is negative.

c Since $\Delta G = \Delta H - T\Delta S$ (the Gibbs free energy change formula), then $\Delta H = \Delta G + T\Delta S$. If ΔG and ΔS are both negative, then ΔH must be negative. The process is exothermic.

d Since $\Delta G = \Delta H - T\Delta S$ and ΔS is negative, the term $-T\Delta S$ is positive. Increasing the temperature therefore makes the process less spontaneous (ΔG become more positive) and the straws become less likely to self-assemble.

e The same process occurs during crystallisation. When crystals are forming in a solution, under those conditions the entropy of the system is decreasing and the process is spontaneous and exothermic.

Module 4 Drivers of reactions

Multiple choice

1 D
2 B
3 A
4 C
5 C
6 B
7 B
8 B
9 B
10 C
11 A
12 B
13 D
14 A
15 D
16 C
17 D
18 D
19 A
20 A

Short answer

1 a endothermic

b $387.6\ \text{kJ mol}^{-1}$

c 11.3 kJ absorbed

d Changing states changes the bond energies between the particles involved in the chemical reaction. By changing the bond energies, the energy changes, ΔH, in the reaction will also be affected.

2 a 11.5 kJ

b 193 kJ

c 21.8 kJ

d 26.0 kJ

e 2.9 kJ

3 a $+336\ \text{kJ mol}^{-1}$

b $-181\ \text{kJ mol}^{-1}$

c $+57\ \text{kJ mol}^{-1}$

d $+67\ \text{kJ mol}^{-1}$

4 $-247.9\ \text{kJ mol}^{-1}$

5 $+22\ \text{kJ mol}^{-1}$

6 $-391\ \text{kJ mol}^{-1}$

7 a $\Delta H^\circ = -1421\ \text{kJ mol}^{-1}$

b $\Delta H_c = -1553\ \text{kJ mol}^{-1}$

8 a 213°C.

b non-spontaneous at values of $T > 213$°C

9 a $S^\circ(C_2H_6(g)) = +229.6\ \text{J mol}^{-1}\ \text{K}^{-1}$

b $\Delta G^\circ = -242.1\ \text{kJ mol}^{-1}$

c spontaneous

10 DS° is positive. As T increases, the value of ΔG° decreases.

Extended response

1 a $2C_5H_{11}OH(l) + 15O_2(g) \rightarrow 10CO_2(g) + 12H_2O(g)$

b $\Delta H_c = -2.77 \times 10^3\ \text{kJ mol}^{-1}$

c 82.8%

d 7.2×10^2 MJ

2 a $NaCH_3COO(s) \xrightarrow{H_2O(l)} CH_3COO^-(aq) + Na^+(aq)$

b Sodium ions and ethanoate ions are held in the ionic lattice by ionic bonds that are based on electrostatic attraction. These ionic bonds in the solute break when it dissolves in water. Water, the solvent, has hydrogen bonds between water molecules. When sodium ethanoate is added to water, the hydrogen atoms of the water molecules are attracted to the negative ethanoate ions, and the oxygen atoms of the water molecules are attracted to the positive sodium ions. Ion–dipole bonds form between the ions and water molecules and the surface ions are pulled into solution. Gradually the ionic lattice dissociates and a solution is formed.

c $\Delta H_{soln} = -13.2\ \text{kJ mol}^{-1}$ **d** 76.5%

3 a $\Delta H = -5752\ \text{kJ mol}^{-1}$

b

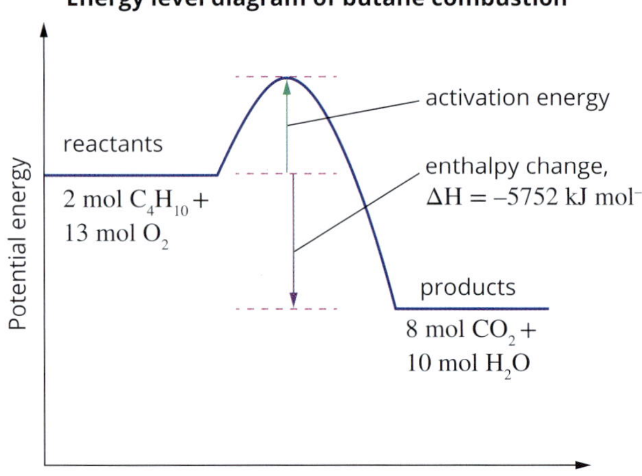

c Some initial input of energy is required to allow some reactants to overcome the activation energy barrier (associated with breaking bonds in reactants) to initiate the rapid reaction.

d heat of combustion $= -49.48\ \text{kJ g}^{-1}$

e 91.0°C

4 A perfect ordered substance at 0 K is defined to have a zero entropy. By measuring the entropy at any temperature, the change in entropy can be provided as an absolute value. There is no analogous zero point for enthalpy. The enthalpy of formation of an element is given as zero, but this is an arbitrary mark. There is no way of knowing the total enthalpy of a substance; only changes in enthalpy can be measured.

5 a

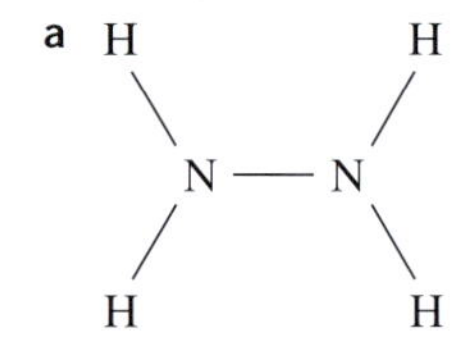

b $-577\ \text{kJ mol}^{-1}$

c $\Delta H^\circ = -579.4\ \text{kJ mol}^{-1}$

d There is a slight difference between the standard enthalpy change calculated using bond energies (–577) and using standard enthalpies of formation (–579.4). This difference is due to the use of average bond energies for the N–H, N–N and O–H bonds, which are approximations of the actual bond energies of these bonds within the reactants and products of this reaction.

e $\Delta S^\circ = 0.0869\ \text{kJ mol}^{-1}\ \text{K}^{-1}$

f $\Delta G^\circ = -553\ \text{kJ mol}^{-1}$
spontaneous

Glossary

A

absolute temperature scale A scale that starts at absolute zero, the temperature at which atoms and molecules have minimum kinetic energy. The absolute temperature scale is measured in kelvin.

absolute zero A temperature of −273°C or 0 K. Molecules and atoms have minimum kinetic energy at this temperature.

accurate When a set of measurements of a quantity is very close to the true or accepted value of the quantity.

acid rain Rainwater that has reacted with acidic emissions from industry and has a pH less than 5.5.

activation energy The minimum energy required by reactants for a reaction to occur; symbol E_a. This energy is needed to break the bonds between atoms in the reactants to allow products to form.

activity series of metals A ranking of metals in increasing order of their reactivity (ability to be oxidised), with the half-equations written as reduction equations of the corresponding ion. Least reactive metals are at the top and most reactive metals are at the bottom.

adsorption The attraction of one substance to the surface of another.

affiliation Connecting or associating with a person or organisation.

alkali metal A group 1 metal—Li, Na, K, Rb, Cs or Fr.

allotropes Different forms of the same element in which the atoms combine in different ways.

alloy A substance formed when other materials (e.g. carbon, other metals) are mixed with a metal.

alpha particle A positively charged particle formed from ionising radiation.

amount of substance A measure used by chemists for counting particles; the unit is the mole.

anhydrous Containing no water.

anion A negatively charged ion, e.g. a chloride ion, Cl^-.

anode An electrode at which an oxidation reaction occurs.

aqueous The solution resulting when a chemical species has been dissolved in water. This can be shown by writing '(aq)' after the name or symbol of the chemical.

asymmetrical molecule A molecule in which the polar bonds are unevenly (or asymmetrically) distributed. The bond dipoles do not cancel and an overall molecular dipole is created.

atom The basic building block of matter. It is made up of subatomic particles—protons, neutrons and electrons.

atomic number The number of protons in the nucleus of an atom; identical to the charge number of the nucleus; symbol Z.

atomic radius A measurement used for the size of atoms; the distance from the nucleus to the outermost electrons.

atomic theory of matter A theory proposed by John Dalton in 1802 that states that all matter is made of atoms. He said that atoms are indivisible, atoms of the same element are identical and compounds are made up of different types of atoms in fixed ratios.

Aufbau principle The lowest energy orbitals are always filled with electrons first.

average bond energy The measure of bond strength in a chemical bond. IUPAC defines bond energy as the average value of the gas-phase bond dissociation energies (usually at a temperature of 298 K) for all bonds of the same type within the same chemical species.

Avogadro's constant The number of particles in 1 mol; symbol $N_A = 6.022 \times 10^{23}\,mol^{-1}$.

B

balanced chemical equation A formula representation of a chemical reaction according to the conservation of elements.

balanced nuclear reaction A balanced reaction, represented by an equation where the sum of the mass numbers and the sum of the atomic numbers balance on either side of the equation.

band of stability The area on a graph of the number of neutrons versus the number of protons in nuclei where the nuclei are stable and do not emit alpha or beta particles.

bar A unit of pressure. 1 bar = 100 kPa. It is about equal to atmospheric pressure at sea level.

barometer Any instrument that measures atmospheric pressure.

battery A combination of cells connected in series.

beta particle An electron or positron ejected from the nucleus of a radioactive nuclide.

bias Measured values consistently in one direction from the actual value; they may be too high or too low.

biochar Charcoal produced from plant matter and stored in the soil.

biodegradation The breakdown of a substance by living organisms, such as bacteria and fungi.

bioluminescence The light that is emitted by a chemical reaction within a living organism.

block (periodic table) One of four main parts of the periodic table where elements have the same highest energy subshell filled, i.e. *s*, *p*, *d* or *f* subshell.

Bohr model A theory of the atom proposed by Niels Bohr that states that electrons in an atom occupy fixed, circular orbits that correspond to specific energy levels.

bond energy The amount of energy required to break a covalent bond.

brittle Shatters when given a sharp tap.

buckyball A ball-like polyhedral molecule consisting of carbon atoms of the type found in fullerenes.

C

calorimeter An instrument designed to measure energy changes in a reaction. It is made up of an insulated container of water in which the reaction occurs, with a stirrer and thermometer to measure the temperature change during the reaction. A lid is an important part of the insulation.

carbon-12 The isotope of carbon that has a mass number of 12. The isotope contains 6 protons and 6 neutrons. One atom of carbon-12 is taken as having a mass of exactly 12 units. This is the standard from which all other relative masses are calculated.

carborundum The common name for silicon carbide, a widely used abrasive.

catalyst A substance that increases the rate of a reaction but is not consumed in the reaction. The catalyst provides a new reaction pathway with a lower activation energy.

cathode An electrode at which a reduction reaction occurs.

cation A positively charged ion.

ceramic A material that is produced by the firing (heating followed by cooling) of clay.

chemical change A process where one or more substances are altered into one or more new and different substances. It is usually an irreversible change.

chemical energy The sum of the chemical potential energy and kinetic energy in a substance. Chemical energy is stored in the bonds between atoms and molecules. The energy results from things such as attractions between electrons and protons in atoms, repulsions between nuclei, repulsions between electrons, movement of electrons, and vibrations and rotations around bonds.

chemical equation A symbolic representation of a chemical reaction in the form of symbols and formulae. The reactants are shown on the left-hand side of the arrow and the products on the right-hand side.

chemical formula A representation of a substance using symbols for its constituent elements. It shows the ratio of atoms present in the substance.

chemical hazard codes A set of codes used in a system of labelling containers of hazardous chemicals or places in which such chemicals are used or stored, to alert people to the potential danger.

chemical reaction The interaction of two or more chemicals that produces one or more new chemical compounds.

chemical species Generally a type of element or compound that exhibits unique chemical and physical properties; it can be a reactant or product in a reaction.

chemical symbol A symbolic representation of an element, usually one or two letters, where the first letter is capitalised and the second letter is lower case, e.g. carbon's symbol is C and sodium's symbol is Na.

chemiluminescence The light that is emitted by a chemical reaction that does not produce significant amounts of heat.

closed system A system where matter does not enter or leave. Energy can still be exchanged between the system and its surroundings.

collision theory A theoretical model that accounts for the rates of chemical reactions in terms of collisions between particles occurring during a chemical reaction.

combustion A rapid reaction with oxygen accompanied by the release of large amounts of heat; also called burning.

complete combustion The complete reaction of a fuel because there is plentiful oxygen. A hydrocarbon undergoes complete combustion with oxygen at high temperatures when the only products are carbon dioxide and water.

component Each chemical in a mixture. The components can be separated by chromatography.

compound A pure substance made up of different types of atoms combined in a fixed ratio.

concentrated solution A solution that has a relatively high ratio of solute to solvent.

concentration A measure of how much solute is dissolved in a specified volume of solution.

conductor An object or type of material that permits the flow of electric charges, e.g. a wire is an electrical conductor that can carry electricity along its length.

conjugate redox pair An oxidising agent and its corresponding reduced form, e.g. Cu^{2+}/Cu.

conservation of elements The idea that in a chemical reaction the total numbers of each type of element in the reactants (left side of the chemical equation) must equal the total numbers of those elements in the products (right side of the chemical equation).

conservation of mass The idea that the total mass of products at the end of a reaction is equal to the total mass of the reactants at the beginning of the reaction.

controlled variable A variable that must be kept constant during an investigation.

core charge The effective nuclear charge experienced by the outer-shell electrons in an atom. It indicates the attractive force felt by the valence electrons towards the nucleus.

covalent bond The force of attraction formed when one or more pairs of electrons are shared between two nuclei.

covalent layer network An arrangement of atoms in a lattice where there are strong covalent bonds between the atoms that have formed a layer.

covalent network structure An arrangement of atoms in a network in which there are strong covalent bonds between the atoms in three dimensions.

credible Reliable and can be backed up with evidence; a credible source provides information that one can believe to be true.

crystal A solid made up of atoms or molecules arranged in a repeating three-dimensional pattern.

crystal lattice The symmetrical three-dimensional arrangement of atoms or ions inside a crystal.

crystallisation The process in which solid crystals are deposited when the concentration of a solute in a solution increases past the point of saturation.

D

decanting A separation technique in which a layer of liquid is removed by carefully pouring it off from a settled solid or more dense immiscible liquid.

decomposition reaction A reaction in which a compound is broken down into smaller parts.

delocalise Spread out.

delocalised electron An electron that is not restricted to the region between two atoms.

density A measure of the amount of mass per unit volume. It has the SI units of kg m^{-3}, but is commonly quoted in g cm^{-3}.

dependent variable The variable that is measured or observed in order to determine the effect of changes in the independent variable.

diamond A form of pure carbon that is the hardest naturally occurring substance.

diatomic molecule A molecule formed from two atoms only, e.g. Cl_2.

dilute solution A solution that has a relatively low ratio of solute to solvent.

dilution The addition of a solvent to a solution to reduce its concentration.

dipole The separation of positive and negative charges in a molecule.

dipole–dipole force A form of intermolecular force that occurs between polar molecules, where the partially positively charged end of one molecule is attracted to the partially negatively charged end of another molecule.

discrete molecule A definite number of atoms covalently bonded together with weak intermolecular forces.

dispersion force The force of attraction between molecules due to temporary dipoles induced in the molecules. The temporary dipoles are the result of random fluctuations in the electron density.

displace Cause the transfer of electrons from an element to a positive ion, which results in the ion leaving the solution as an element, e.g. when zinc is placed in a solution of copper(II) ions the displacement reaction is:

$Cu^{2+}(aq) + Zn(s) \rightarrow Zn^{2+}(aq) + Cu(s)$

dissociate Break up.

dissociation A process in which molecules or ionic compounds separate or split into smaller particles such as atoms or ions. Examples of dissociation reactions include the solution of NaCl solid in water, forming $Na^+(aq)$ and $Cl^-(aq)$ ions, and the reaction of HCl gas with water, forming $H^+(aq)$ and $Cl^-(aq)$.

dissolution The process by which a solute dissolves into a solvent to become a solution.

dissolved Incorporated a solid or gas into a liquid so as to form a solution.

distillation Separation of liquids based on selective evaporation.

double covalent bond A covalent bond in which four electrons (two electron pairs) are shared.

ductile Able to be drawn out into a wire.

E

efficiency (calorimetry) A measure of how much heat energy has been transferred from the combustion of a material to the heating of the water.

electric dipole The separation of positive and negative charges in a bond or a molecule.

electrode A solid conductor in a half-cell at which oxidation or reduction reactions occur.

electrolysis The production of a reaction by the passage of electrical energy from a power supply through a conducting liquid.

electrolyte A solution or molten substance that conducts electricity by means of the movement of ions, e.g. a solution of sodium chloride (table salt).

electromagnetic radiation A form of energy that moves through space. Visible light, radio waves and X-rays are forms of electromagnetic radiation.

electromotive force The 'electrical pressure' between two points in a circuit, such as the electrodes of an electrochemical cell; a measure of the energy given to electrons in a circuit.

electron A negatively charged, subatomic particle that occupies the region around the nucleus of an atom.

electron density The concentration of electrons; this usually refer to the regions around an atom or molecule.

electronegativity The ability of an atom to attract the electrons in a covalent bond towards itself.

electronic configuration In the shell model of an atom, the electronic configuration is a means of representing the number of electrons in each shell.

electron shell In the shell model of an atom, an electron shell is the fixed energy level that corresponds to a circular orbit of the electrons. In the Schrödinger model, a shell contains subshells and orbitals of equal or similar energy.

electron transfer diagram A diagram that shows how electrons move from a metal atom to a non-metal atom to form ions.

electrostatic attraction The force of attraction between a positively charged particle and a negatively charged particle.

electrovalency The charge on an ion.

element A substance made up of atoms with the same atomic number.

emission spectrum A spectrum produced when an element is excited by heat or radiation. It appears as distinct lines characteristic of the element.

empirical formula A formula that shows the simplest whole number ratio of the elements in a compound, e.g. CH_2 is the empirical formula of propene (C_3H_6).

endothermic A reaction that absorbs energy from the surroundings; ΔH is positive.

energy cycle A diagram showing how three or more reactions are related. Energy cycles can be constructed to find unknown energy changes, by adding up the energies of steps from a different route to the same final products.

energy level One of the different shells of an atom in which an electron can be found.

energy profile diagram A diagram that shows the potential energy changes during the course of a reaction.

enthalpy Heat content. The sum of the chemical potential and kinetic energies in a substance; symbol H.

enthalpy change The difference between the total enthalpy of the products and the total enthalpy of the reactants; symbol ΔH. Also known as heat of reaction.

enthalpy of dissolution The amount of heat per amount of substance (e.g. mol) which is absorbed or released as that substance is dissolved in water, given the symbol ΔH_{soln}. This is calculated by dividing the energy change by the amount of substance, in mol, of the substance being dissolved.

enthalpy of fusion Also known as (latent) heat of fusion. The change in enthalpy of 1 mol of a substance when it changes from a solid to a liquid. Conventionally, documented at the melting point of the substance at a pressure of 1 bar (100 kPa).

enthalpy of vaporisation Also known as (latent) heat of vaporisation. The change in enthalpy of 1 mol of a substance when it changes from a liquid to a gas. Conventionally, documented at the boiling point of the substance at a pressure of 1 bar (100 kPa).

entropy A measure of the number of possible arrangements in a thermodynamic system.

excess reactant A reactant that is not completely consumed in a chemical reaction.

excited state The state of an atom in which electrons occupy higher energy levels than their lowest possible energy levels.

exothermic A reaction that releases energy to the surroundings; ΔH is negative.

expertise Expert knowledge or skills in a field.

external circuit The section of an electrochemical cell in which electrons move. This section of the circuit will include the wires attached to the electrodes.

F

fermentation The breakdown of sugar solutions, by the action of enzymes in yeasts, into ethanol and carbon dioxide. The chemical equation for the fermentation of glucose is:
$C_6H_{12}O_6(aq) \rightarrow 2CH_3CH_2OH(aq) + 2CO_2(g)$

filtrate The liquid that has passed through a filter.

first ionisation energy The energy required to remove one electron from an atom of an element in the gas phase.

flame test A test used to detect the presence of or identify elements. The sample is burned and the emission spectrum analysed.

fluid A substance with no fixed shape; a liquid or a gas.

full equation A representation of a reaction that uses formulae.

fullerene A molecule composed entirely of carbon, in the form of a hollow sphere or tube. Other shapes are possible. Each carbon atom is bonded to three other carbon atoms.

G

galvanic cell A type of electrochemical cell also known as a voltaic cell; a device that converts chemical energy into electrical energy.

galvanometer An instrument for detecting electric current.

gamma radiation High-energy electromagnetic radiation ejected from the nucleus of a radioactive nuclide.

gas constant The constant, R, in the universal gas equation $PV = nRT$. $R = 8.314\,J\,mol^{-1}\,K^{-1}$ when pressure is measured in kPa, volume in L and temperature in K.

gemstone A precious stone that is usually ionic in structure.

giant molecule A type of very large molecule formed by some non-metals, e.g. a large network of covalently bonded atoms, such as diamond or graphite.

Gibbs free energy A measure of the effects of enthalpy and entropy on whether a reaction is spontaneous or not.

graphene A form of carbon consisting of planar sheets one atom thick, in which each carbon atom is bonded to three neighbouring carbon atoms.

graphite A form of carbon in which the carbon atoms are arranged in layers.

ground state A term used to describe an atom in which the electrons occupy the lowest possible energy levels.

group (periodic table) A vertical column of elements in the periodic table.

H

half-cell Half an electrochemical cell, which contains the oxidant and conjugate reductant. When two half-cells are combined, a galvanic cell is formed.

half-equation A balanced chemical equation that shows the loss or gain of electrons by a species during oxidation or reduction, e.g. the oxidation of magnesium is written as the half-equation $Mg(s) \rightarrow Mg^{2+}(aq) + 2e^-$.

hard solid, firm and rigid; not easily broken or shaped.

heat content At a simple level, the chemical energy of a substance; symbol H.

heat of combustion The energy released when a specified quantity (e.g. 1 mol) of a substance burns completely in oxygen.

heat of reaction The exchange of heat between a system and its surroundings during a chemical reaction under constant pressure; symbol ΔH. Also known as enthalpy change.

heterogeneous Diverse, different. A heterogeneous substance or solution possesses two or more different types of phases in the one sample, e.g. a suspension.

heterogeneous catalyst A catalyst that has a different physical state (phase) from the reactants and products.

homogeneous Uniform. The components of a homogeneous substance are uniformly distributed throughout the substance, e.g. a solution is homogeneous because the solute and the solvent cannot be distinguished from each one another.

homogeneous catalyst A catalyst that has the same physical state (phase) as the reactants and products.

Hund's rule The rule that every orbital in a subshell must first be filled with one electron with the same spin before an orbital is filled with a second electron.

hydrated An ion surrounded by water molecules. Hydrated ions can be found in aqueous solutions or crystalline solids.

hydrocarbon A compound that contains carbon and hydrogen only, e.g. the alkanes, alkenes and alkynes.

hydrogen bond A type of intermolecular, dipole–dipole force where a hydrogen atom is covalently bonded to a highly electronegative atom such as oxygen, nitrogen or fluorine. Due to the disparity in the electronegativity values between the atoms involved, the hydrogen develops a partial positive charge and bonds to lone pairs of electrons on neighbouring atoms of oxygen, nitrogen or fluorine.

I

ideal gas A gas that obeys the gas equations at all temperatures and pressures.

ideal gas law A law describing the behaviour of a hypothetical gas that obeys the gas equations at all temperatures and pressures.

incomplete combustion A combustion reaction that takes place when oxygen is limited. Incomplete combustion of hydrocarbons produces carbon and carbon monoxide as well as carbon dioxide.

independent variable A variable that is changed by the researcher.

instantaneous dipole A net dipole formed in a molecule due to temporary fluctuations in the electron density in the molecule.

intermolecular bond The net forces of attraction that act between neighbouring molecules.

intermolecular force An electrostatic force of attraction between molecules, in the form of dipole–dipole forces, hydrogen bonds or dispersion forces.

intramolecular bond A force that holds the atoms within a molecule together.

ion A positively or negatively charged atom or group of atoms.

ion–dipole attraction The attraction that forms between dissociated ions and polar water molecules when an ionic solid dissolves in water.

ionic bonding A type of chemical bonding that involves the electrostatic attraction between oppositely charged ions.

ionic compound A type of chemical compound that involves the electrostatic attraction between oppositely charged ions.

ionic equation An equation for a reaction that only includes the ions that are involved in the reaction, e.g.
$Cu^{2+}(aq) + 2OH^-(aq) \rightarrow Cu(OH)_2(s)$

ionisation (i) The removal of one or more electrons from an atom or ion; (ii) the reaction of a molecular substance with a solvent to form ions in solution.

ionisation energy The energy required to remove one electron from an atom of an element in the gas phase.

ionise The reaction of a molecular substance with a solvent to form ions in solution. When some polar molecules dissolve in water, they ionise to form a hydronium ion, e.g.
$HCl(g) + H_2O(l) \rightarrow H_3O^+(aq) + Cl^-(aq)$

isolated system A system where neither matter nor energy enter or leave.

isotope Each of two or more forms of the same element that contain equal numbers of protons but different numbers of neutrons in their nuclei, e.g. ^{12}C and ^{13}C are isotopes of carbon.

J

joule The standard unit of energy. One joule equates to one newton of force moving one metre in the direction of that force.

K

Kelvin scale The absolute temperature scale; measured in kelvin.

kinetic energy The energy that a particle or body has due to its motion.

kinetic energy distribution diagram A graph of kinetic energy against number of particles that shows the range of energies of the particles in a sample of a gas or a liquid at a given temperature. Also known as a Maxwell–Boltzmann distribution curve.

kinetic molecular theory A theory that aims to explain the behaviour of gases by assuming gases are composed of a large number of particles in random motion; these particles move in straight lines and have elastic collisions; the gas particles are very small and there is no attraction between the particles; the average kinetic energy of the gas particles is related to the temperature of the gas.

L

latent heat The heat energy required to change the state of a substance without changing the temperature.

latent heat of fusion The energy required to change a fixed amount of solid to liquid at its melting temperature. The latent heat of fusion of water is $6.0\,kJ\,mol^{-1}$, meaning 6.0 kJ of energy is needed to change 1 mol of water from a solid to a liquid at 0°C.

latent heat of vaporisation The energy required to change a fixed amount of liquid to a gas at its boiling temperature. The latent heat of vaporisation of water is $44.0\,kJ\,mol^{-1}$, meaning 44.0 kJ is needed to change 1 mol of water from a liquid to a gas at 100°C.

law of conservation of energy A scientific law stating that energy cannot be created or destroyed; it can only be transformed from one type of energy to another.

law of conservation of mass A scientific law stating that the total mass of the components of a chemical reaction in a closed system will remain constant before and after the reaction has taken place.

leaching The removal of soluble material from a substance, such as soil or rock, through the percolation of water.

Lewis dot diagram A representation of the electron arrangement in a molecule in which outer-shell electrons are represented by dots or crosses.

limewater test A test for carbon dioxide gas. The presence of carbon dioxide is detected by bubbling the gas through a calcium hydroxide solution ($Ca(OH)_2(aq)$). The limewater reacts with the carbon dioxide and turns milky.

limiting reactant A reactant that is completely consumed in a reaction and that determines the amount of products formed.

lone electron pair An outer-shell electron pair that does not form a bond with other atoms.

lone pair See *lone electron pair.*

lustrous Shiny; glossy.

M

main group element An element in groups 1, 2 or 13–18 in the periodic table.

malleable Able to be bent or beaten into sheets.

mass number The number of protons and neutrons in the nucleus of an atom.

mass spectrometer An instrument that measures the mass-to-charge ratio of particles.

mass spectrum A plot of the isotopic mass, relative to the mass of carbon-12 taken as 12 units exactly, against the relative abundance of each isotope present in a sample.

matter Anything that has mass and occupies space.

Maxwell–Boltzmann distribution curve A graph of kinetic energy against number of particles that shows the range of energies in a sample of a gas or a liquid at a given temperature. Also known as a kinetic energy distribution diagram.

mean The sum of all the values in a data set divided by the number of values in the data set. It is commonly known as the average of a set of numbers.

median The middle value of an ordered data set. To calculate the median, arrange the data set in ascending order, then count the number of data values. If the number of values is odd, select the middle value. If the number of values is even, select the two middle values, add them together and divide by 2.

metal Elements that are crystalline when solid and many of which are characterised by opacity, ductility, conductivity, and a peculiar lustre when freshly fractured. Metals are found on the left side and the centre of the periodic table.

metallic bonding The electrostatic attractive forces between delocalised valence electrons and positively charged metal ions.

metallic bonding model A description that explains the properties and behaviour of metals in terms of the particles in metals.

metalloid An element that displays both metallic and non-metallic properties, e.g. germanium, silicon, arsenic, tellurium.

mistake An avoidable error.

mixture A substance made from two or more other substances combined together.

mode The most frequent value in a data set. Arrange the data set in ascending order, then count the number of each data value.

model A description that scientists use to represent the important features of what they are trying to describe.

molarity The amount of solute, in mol, dissolved in 1 L of solution ($mol\,L^{-1}$).

molar mass The mass of 1 mol or equivalent of a compound.

molar volume The volume occupied by 1 mol of a gas at a specified set of conditions. At standard temperature (0°C or 273 K) and pressure (100 kPa), the molar volume of a gas is $22.7\,L\,mol^{-1}$.

mole The amount of substance that contains the same number of fundamental particles as there are atoms in 12 g of carbon-12; symbol *n*; unit abbreviation mol.

molecular formula A formula of a compound that gives the actual number and type of atoms present in a molecule. It may be the same as or different from the empirical formula.

molecule A group of two or more atoms covalently bonded together, and representing the smallest fundamental unit of a chemical compound.

mole ratio The ratio of species involved in a chemical reaction, based on the ratio of their coefficients in the reaction equation.

molten Normally found as a solid, but liquid (melted) due to elevated temperature.

monatomic Composed of single atoms, e.g. neon gas.

N

nanomaterial A material with nanoscale features.

nanotube An allotrope of carbon that consists of layers of carbon atoms formed into a long cylinder.

neutralisation reaction A reaction in which an acid reacts with a base in stoichiometric proportions to form a salt plus water.

neutralise To react an acid with a base in stoichiometric proportions to form a solution of a salt and water.

neutron An uncharged subatomic particle found in the nucleus of an atom.

noble gas An unreactive gaseous element in group 18 of the periodic table. With the exception of helium, noble gases have eight electrons in their outer shells.

nomenclature A system of names or terms.

non-bonding electron An outer-shell electron that is not shared between atoms.

non-metal An element that is not a metal.

non-polar Bonds or molecules that do not have a permanent dipole. They have an even distribution of charge.

non-polar solvent A liquid or solvent that is a compound of two or more elements whose electronegativities are almost the same, e.g. oil.

non-spontaneous A reaction that cannot proceed without a continuous supply of external energy. It has a positive Gibbs free energy change and is the reverse of a spontaneous reaction.

nucleon A particle that makes up the nucleus of an atom, i.e. a proton or a neutron.

nucleus The positively charged core at the centre of an atom, consisting of protons and neutrons.

O

octet rule A rule used as part of the explanation for electron configuration and bonding. The rule is that during a chemical reaction, atoms tend to lose, gain or share their valence electrons so that there are eight electrons in the outer shell.

open system A system in which energy and matter are exchanged between the system and its surroundings.

orbital In the Schrödinger model, an orbital is a component of a subshell. It is a region of space in which electrons move. Each orbital can hold two electrons.

orbital diagram A visual representation of electron placement in subshells.

outlier A value that lies outside most of the other values in a set of data.

oxidant See *oxidising agent*

oxidation The process by which a chemical species (such as a metal atom or a non-metal ion) loses electrons. An oxidation half-equation will show the electrons as products (on the right-hand side of the arrow).

oxidation number A number assigned to an atom in a compound or as the free element, which represents the charge that atom would have if it was an ion. A series of oxidation number rules are used to determine the oxidation number of an element in a compound. Oxidation numbers are used to identify redox reactions.

oxidised The state of having lost electrons or increased in oxidation number. When a substance is oxidised, the electrons are written on the right-hand side of the arrow in the half-equation.

oxidising agent A reactant that causes another reactant to lose electrons during a redox reaction. This reactant is, itself, reduced and gains electrons; for example, in the reaction between magnesium and oxygen, the oxygen is the oxidising agent, as it causes magnesium to lose electrons and form Mg^{2+}:
$2Mg(s) + O_2(g) \rightarrow 2MgO(s)$

P

parallax error The perceived shift in an object's position as it is viewed from different angles.

partial pressure The pressure exerted by one component of a mixture of gases. The total pressure of a mixture of gases is equal to the sum of the individual partial pressures of each component in the mixture.

pascal A unit of pressure equal to 1 newton per square metre ($1\,N\,m^{-2}$).

Pauli exclusion principle The principle stating that each orbital can contain a maximum of two electrons, with each electron having a different spin.

percentage composition The proportion by mass of the different elements in a compound.

percentage error The difference between an exact value and an approximate value, expressed as a percentage of the exact value.

period (periodic table) A horizontal row of elements in the periodic table. The start of a new period corresponds to the outer electron of that element beginning a new shell.

periodicity The periodic pattern of properties of the elements.

periodic law The way properties of elements vary periodically with their atomic number.

periodic table A table that organises the elements by grouping them according to their electronic configurations.

personal protective equipment (PPE) Equipment such as laboratory coats, gloves and goggles, designed to be worn by an individual for protection against hazards.

persuasion An attempt to convince a person to believe something or to do something.

phenomenon Something that occurs and can be observed (or felt, heard etc.) to have occurred.

photochemical smog Atmospheric pollution produced through the action of sunlight on nitrogen oxides and unburned hydrocarbons to form ozone and other pollutants. The nitrogen oxides are formed in high temperature reactions such as those that occur in car engines and lightning strikes.

photolysis The decomposition of substances through the action of light.

photovoltaic cell A device constructed from a specialised semiconductor that can produce a flow of electrons from light energy.

physical change A change affecting the form of a chemical substance but not its chemical composition.

polar Describing bonds or molecules with a permanent dipole. They have an uneven distribution of charge.

polarity The measure of how polar a molecule or bond is. The difference in charge between the positive and negative ends of an electric dipole. The difference in charge between the positive and negative ends of a polar molecule or covalent bond.

polyatomic ion An ion that is made up of more than one element, for example the carbonate ion (CO_3^{2-}).

polyatomic molecule A molecule that consists of more than two atoms, e.g. H_2O.

polymer A long-chain molecule that is formed by the reaction of large numbers of repeating units (monomers).

potential difference The electromotive force between two points in a circuit, such as the electrodes of an electrochemical cell.

ppb (parts per billion) A unit of concentration that states the number of grams of solute in one billion grams of solution. It is equivalent to the number of milligrams of solute per tonne of solution.

ppm (parts per million) A unit of concentration that states the number of grams of solute in one million grams of solution. It is equivalent to the number of milligrams of solute per kilogram of solution.

precipitate The solid formed during a reaction in which two or more solutions are mixed.

precipitation reaction A reaction between substances in solution in which one of the products is insoluble.

precise When repeated measurements of the same quantity give values that are in close agreement.

pressure The force exerted per unit area of a surface.

primary source A source that is a first-hand account.

primary standard A substance of known high purity that may be dissolved in a known volume of solvent.

products Chemical species that are produced in a chemical reaction.

proton A positively charged, subatomic particle usually bound to neutrons in the nucleus of an atom.

pure substance A chemical substance that contains only chemical species of a single type, not a mixture.

Q

qualitative Relating to quality and not measured values.

quantised In specific quantities or chunks.

quantitative Relating to measured values rather than quality.

quantum mechanics A branch of science that describes the behaviour of extremely small particles such as electrons.

R

radioactive Spontaneously undergoing nuclear decay to produce radiation such as beta particles, alpha particles and gamma rays.

radioisotope An isotope of an element that is radioactive.

random error An error that follows no regular pattern. (The effects of random errors can be reduced by taking the average of many observations.)

rate of reaction The change in concentration of a reactant or product over a period of time (usually one second).

raw data The information and results collected and recorded during an experiment.

reactants Chemical species that are used up in a chemical reaction.

reaction pathway A series of chemical reactions that converts a starting material into a product in a number of steps.

reactivity The ease with which a chemical can undergo reactions.

redox reaction A reaction in which electron transfer occurs from the reducing agent to the oxidising agent. In a redox reaction, both oxidation and reduction occur.

reduced The state of having gained electrons or decreased in oxidation number. When a substance is reduced, the electrons are written on the left-hand side of the arrow in the half-equation.

reducing agent A reactant that causes another reactant to gain electrons during a redox reaction. This reactant is oxidised and loses electrons; for example, in the reaction between magnesium and oxygen, the magnesium is the reducing agent, as it causes oxygen to gain electrons and form O^{2-} ions:
$2Mg(s) + O_2(g) \rightarrow 2MgO(s)$

reductant See *reducing agent.*

reduction The process by which a chemical species gains electrons. A reduction half-equation will show the electrons on the reactant side (left-hand side) of the equation.

relative atomic mass The weighted average of the relative isotopic masses of an element on the scale where ^{12}C is 12 units exactly; symbol A_r.

relative formula mass The mass of a formula unit relative to the mass of an atom of ^{12}C taken as 12 units exactly. It is numerically equal to the sum of the relative atomic masses of the atoms making up the formula. Substances that contain atoms or ions bonded in network structures have a relative formula mass. Such compounds include ionic compounds and covalent network substances.

relative isotopic abundance The percentage abundance of a particular isotope in a sample of an element.

relative isotopic mass The mass of an atom of the isotope relative to the mass of an atom of carbon-12 taken as 12 units exactly.

relative molecular mass The mass of a molecule relative to the mass of an atom of ^{12}C, taken as 12 units exactly.

reliability The ability to be repeated experimentally several times with consistent results.

reputation The perceived characteristics about someone or something.

rhetoric Speech or writing that has the aim of persuading.

S

safety data sheet (SDS) A summary of the risks of using a particular chemical, including measures to be followed to reduce risk.

salinity The presence of salt in water or soil that can damage plants or inhibit their growth.

salt A substance formed from a metal or ammonium cation and an anion. Salts are the products of reactions between acids and bases, metal oxides, carbonates and reactive metals.

salt bridge An electrical connection between the two half-cells in a galvanic cell; it is usually made from a material saturated in electrolyte solution.

Schrödinger model A model for the behaviour of electrons in atoms. It describes electrons as having wave-like properties.

secondary source A source derived from the original data or account.

significant figures The number of digits used in a measurement value to indicate the measurement accuracy.

single covalent bond A covalent bond in which two electrons are shared between two nuclei. It is depicted in a valence structure as a line between the two atoms involved.

SI units Physical units within an internationally accepted system, the International System of Units. For example, the SI unit for energy is the joule, J.

solubility table A reference table that can be used to predict the solubility of ionic compounds.

solute A substance that dissolves in a solvent; for example, sugar is the solute when it dissolves in water.

solution A homogeneous mixture of a solute dissolved in a solvent.

solution calorimeter An insulated container that holds a known volume of water and in which a reaction in solution, such as dissolution of a solid or a neutralisation reaction, can be carried out.

solvent A substance, usually a liquid, that is able to dissolve a solute to form a solution. Water is a very good solvent.

specific heat capacity The amount of energy required to raise the temperature of an amount of a substance, usually 1 g, by 1°C. The SI unit for specific heat capacity is $J\,kg^{-1}\,K^{-1}$. The specific heat capacity of water is $4.18\,J\,g^{-1}\,K^{-1}$.

spectator ion An ion that remains in solution and is unchanged in the course of a reaction. Spectator ions are not included in ionic equations.

spin A property of a nucleus with an odd number of protons or neutrons that causes it to interact with a magnetic field. The nuclear spin can be either with or against an external magnetic field.

spontaneous reaction A reaction that occurs naturally.

stakeholder A person with an interest or concern in the matter being discussed.

standard atmosphere The pressure required (1 atm) to support 760 millimetres of mercury (760 mmHg) in a mercury barometer at 25°C. One atmosphere pressure is approximately the pressure experienced at sea level.

standard state conditions Conditions of temperature and pressure relevant to the standard state of a substance, where temperature is 298 K (25°C) and the pressure is 1 bar.

standard electrode potential The electromotive force that is measured when a half-cell, at standard laboratory conditions (gas pressures 1 bar, solution concentrations of $1.0\,mol^{-1}\,L^{-1}$ and temperature of 25°C), is connected to a standard hydrogen half-cell; symbol $E°$. It gives a numerical measure of the tendency of a half-cell reaction to occur as a reduction reaction.

standard enthalpy of formation Also known as heat of formation. The change of enthalpy during the formation of 1 mol of the compound from its constituent elements. Conventionally, it is documented with all substances in their standard states, and at a pressure of 1 bar (100 kPa) and a temperature of 25°C (298.15 K).

standard entropy Absolute entropy that has been calculated at standard state conditions.

standard hydrogen half-cell A $H^+(aq)/H_2(g)$ half-cell; made from a platinum electrode placed in acid solution with hydrogen gas bubbled over it. The pressure of the gas is 1 bar, the H^+ concentration is 1 mol L^{-1} and the temperature is 25°C.

standard laboratory conditions (SLC) Conditions of temperature and pressure relevant to a gas, where the temperature is 298 K (25°C) and the pressure is 100 kPa.

standard reduction potential A numerical measure of the tendency of a half-cell reaction to occur as a reduction reaction. It is measured by connecting a half-cell, at standard laboratory conditions (gas pressures 1 bar, solution concentrations of 1.0 1 mol L^{-1} and temperature of 25°C), to a standard hydrogen half-cell.

standard solution A solution that has an accurately known concentration.

standard state The phase (solid, liquid or gas) in which a substance exists at 25°C and at 1 bar; used in calculations of enthalpy, entropy and Gibbs free energy.

standard temperature and pressure (STP) Conditions of temperature and pressure relevant to a gas, where the temperature is 0°C (273 K) and pressure is 100 kPa.

stoichiometric calculation The calculation of relative amounts of reactants and products in a chemical reaction.

stoichiometry The calculation of relative amounts of reactants and products in a chemical reaction. Chemical equations give the ratios of the amounts (moles) of the reactants and products.

subatomic particle One of the particles that make up an atom—protons, neutrons and electrons.

sublimation The process by which a substance changes directly from the solid phase to the gaseous phase, without passing through a liquid phase.

sublime The action of a substance changing from the solid state to the gaseous state without going through the liquid state.

subshell A component of a shell in the Schrödinger model, made up of orbitals. Each subshell can be regarded as an energy level that electrons can occupy.

surface area The area of all surfaces of the substance that are exposed to the other reactants. This is proportional to the number of particles available at the surface to react.

surface tension The resistance of a liquid to increasing its surface area.

surroundings The rest of the universe around a particular chemical reaction. The chemical reaction is the system. Energy moves between the system and surroundings in exothermic and endothermic reactions.

symmetrical molecule A molecule in which the polar bonds are evenly (or symmetrically) distributed. The bond dipoles cancel out and do not create an overall molecular dipole.

synthesis The formation of a more complex substance or compound by a chemical reaction combining two or more simpler substances (elements and/or simpler compounds).

synthesis reaction A reaction in which two or more simple substances combine to form a more complex substance. The product is always a compound.

system In chemistry, a chemical reaction. A system operates within its surroundings. Energy can move between the two. Energy is considered to be absorbed or released from the perspective of the system; in other words, energy is absorbed into the system from the surroundings, or energy is released by the system to the surroundings.

systematic error An error that produces a constant bias in measurement. (Systematic errors are eliminated or minimised through calibration of apparatus and careful design of the procedure.)

T

table of standard reduction potentials A list of half-equations, written as reduction reactions, arranged in order so that the strongest oxidising agents are on the bottom left-hand side of the list.

temporary dipole A net dipole formed in a molecule due to temporary fluctuations in the electron density in the molecule.

tensile strength The maximum resistance of a material to a force that is pulling it apart before breaking, measured as the maximum stress the material can withstand without tearing.

tetrahedral shape The shape of a molecule with a central atom surrounded by four other atoms. The bond angle between two outer atoms and the central atom is 109.5°.

thermal decomposition A reaction in which a substance is broken down into at least two other substances by heat.

thermochemical equation A chemical equation that includes the enthalpy change of the reaction, ΔH. For example:
$2H_2(g) + O_2(g) \rightarrow 2H_2O(g)\ \Delta H = -572$ kJ mol^{-1}

thermodynamics The science of heat and energy.

transition element An element (such as iron, copper, nickel or chromium) that is found in the middle block of the periodic table, in groups 3–12.

transition metal An element in groups 3–12 in the periodic table.

transition state An arrangement of atoms in a reaction that occurs when sufficient energy is absorbed for the activation energy to be reached. It represents the stage of maximum potential energy in the reaction. Bond breaking and bond forming are both occurring at this stage, and the arrangement of atoms is unstable.

trend An observed pattern of data in a particular direction.

trend line A line indicating the general trend of data on a graph. Also known as a line of best fit.

triple covalent bond A covalent bond in which six electrons are shared between two nuclei. It is depicted in a valence structure as three lines between the two atoms involved.

U

uncertainty An error associated with measurements made during experimental work.

V

valence electron An electron found in the valence shell; an outermost electron in an atom or ion.

valence shell The highest energy shell (outer shell) of an atom that contains electrons.

valence shell electron pair repulsion (VSEPR) theory A model used to predict the shape of molecules. The basis of VSEPR is that the valence electron pairs surrounding an atom mutually repel each other, and therefore adopt an arrangement that minimises this repulsion, thus determining the molecular shape.

valence structure A representation of the electron arrangement in a molecule in which covalent bonds are represented by lines.

validity The extent to which an experiment or investigation accurately tests the stated hypothesis and purpose.

variable Any factor that can be controlled, changed and measured in an experiment.

velocity A measurement of the rate and direction of motion of an object. It has the symbol v and the unit m s^{-1}.

viscosity A measure of a liquid's resistance to flow. Honey has a high viscosity and petrol has a low viscosity.

volatile Easily evaporated; having a high vapour pressure.

volt The unit of potential difference.

voltage See *potential difference*.

voltmeter An instrument for measuring the electrical potential difference between two points in a circuit.

volume The amount of space that a substance occupies. That of a regular solid can be calculated by multiplying length by width by depth. Otherwise, it can be determined by finding the volume of water displaced by the substance.

volumetric flask A laboratory flask calibrated to contain a precise volume.

Y

yield The amount of product formed in a chemical reaction.

123

% efficiency (calorimetry) The percentage of heat energy that has been transferred from the combustion of a material to the heating of the water.

Index

Attributions

The following abbreviations are used in this list: t = top, b = bottom, l = left, r = right, c = centre.

Cover image: David Scharf/Science Photo Library.
123RF: Elnur Amikishiyev, p. 272l; Shawn Hempel, p. 113; Sebastian Kaulitzki, p. 77. Olivier Le Queinec, p. 2–3; Ivan Mikhaylov, p. 13; mite, p. 444b; ojimorena, p. 313tl; Aleksandr Prokopenko, p. 440–41; saiko3p, p. 353cr; Rui Santos, p. 324br; scanrail, p. 390tl; scyther5, p. 352; stillfx, p. 371; thesupe87, p. 348l; tieury, p. 411; George Tsartsianidis, p. 254cr; Tatyana Vyc, p. 425tr; Maksym Yemelyanov, p. 392tr. **Alamy Stock Photo:** Alchemy, p. 139tr; Caspar Benson/fStop Images GmbH, p. 320t; Nir Ben-Yosef/PhotoStock-Israel, p. 455; bilwissedition/bilwissedition Ltd. & Co. KG, p. 284tl; Classic Image, p. 206; E.R. Degginger, p. 353bl; Susan E. Degginger, p. 169cc; Paul Fearn, p. 213; Ruddy Gold/age fotostock, p. 314tr; Granger, NYC./Granger Historical Picture Archive, pp. 207, 289t; Charles Hood, p. 273br; Peter Horree, p. 317br; Huntstock, Inc, p. 14; Keystone Pictures USA/ZUMAPRESS, p. 507l; Antoine Laurent Lavoisier/World History Archive, p. 204–05; Elke Meitzel/Cultura Creative, p. 299l; Stocksearch, p. 314tl; Metta foto, p. 139r; Christopher Nash, p. 331cr; Lukasz Olek/caia image, p. 71r; Zev Radovan/BibleLandPictures/www.BibleLandPictures.com, p. 71b; Tony Rogers, p. 446; sciencephotos, pp. 315r, 329tr, 373b; studiomode, p. 495; David Taylor, p. 9. Travelscape Images, p. 345t; Kevin Twomey, p. 500–01; Philippe Voisin/Phanie, p. 84t.
Art Gallery of NSW: Aurukun/Cape York/Queensland/Australia Aurukun/West Cape region Mavis Ngallametta (Australia, b.1944) *Untitled* 2010 natural pigments, charcoal and acrylic on canvas, 101 × 107 cm. Art Gallery of New South Wales Gift of the artist. Donated through the Australian Government's Cultural Gifts Program 2015 Photo: Jenni Carter, AGNSW © Mavis Ngallametta, Licensed by Martine Brown, 2017, 156.2015, p. 143r; John Olsen (Australia; England; Spain; Portugal, b.1928) *Five Bells* 1963 oil on hardboard, 264.5 × 274 cm Art Gallery of New South Wales. Purchased with funds provided by the Art Gallery Society of New South Wales 1999 Photo: Mim Stirling, AGNSW © John Olsen. Licensed by Viscopy, 2017, 133.1999, p. 143l.
Australian Saltworks, South Australian Operations: pp. 313bl, 313br.
Biochar.org: Christoph Steiner/UrbanFoodPlus Project/University of Kassel, p. 190 (inset).
CSIRO: scienceimage, p. 189b.
DK Images: Mike Dunning, p. 51; Steve Gorton, p. 169cl; Ian O'Leary, p. 442tl; Tim Ridley, p. 314bl; Howard Shooter, p. 415tr.
Fairfax Photo Sales: Phil Carrick, p. 79b; Orlando Chiodo, p. 419t; Graham Tidy/Canberra Times, p. 118.
Brandon Findlay: © Brandon Findlay, Concordia University, pp. 50bc, 50br.
Fotolia: Leonid Andronov, p. 39; ASP Inc, p. 260; Barabas Attila, p. 229; bluesnote, p. 444tr; bit24, p. 86tr; dabjola, p. 175l; EcoView, p. 88r.efired, p. 52l;KAR, p. 11; Georgios Kollidas, p. 279tr; Jan Matoska, p. 341tr; Mirco251280, p. 140; Monticellllo, p. 442br; Tyler Olson, p. 255; Pabijan, p. 223tr; Nicholas Piccillo, p. 223bc; rekemp, p. 273cr; piotr_roae, p. 150cl; Federico Rostagno, p. 16c; shaiith, p. 272c; sehbaer_nrw, p. 175r; Tatiana Shepeleva, p. 194c; tadamichi, p. 26; Teteline, p. 295; vermontalm, p. 283; WoGi, pp. 42b, 42c, 42t.
Getty Images: andresr/ E+, p. 419b; ATU Images/Photographer's Choice, p. 189t; Ross Barnett/Lonely Planet Images, p. 152; Lowell Georgia/Corbis Documentary, p. 188c; Matt Mawson/Corbis Documentary, p. 19b; Marty Melville/AFP, p. 54l; Paul J. Richards/AFP, p. 262; Universal History Archive/Universal Images Group, p. 282bl.
Richard Hecker: pp. 502l, 502r.
ImageFolk: Alfred Pasieka/Science Photo Library, p. 517t; Steffen & Alexandra Sailer/ardea.com, p. 344r.
Images of Elements: images-of-elements.com, p. 142b.
Ixom Operations Pty Ltd: p. 17.
Lake's Folly: pp. 254tl, 254bl.
Dan Lindsay: © 2009 Porsche Ceramic Composite Brakes on Porsche Carrera S (997)., p. 149br.
Museums Victoria: Photographer: Caroline Harding. A twined conical basket or bathi made of pandanus, Milingimbi, 1930. Collected by Rev. T.T. Webb. Courtesy Museums Victoria (37648) p. 185 (inset).
NASA Images: MSFC/David Higginbotham/Emmett Given, p. 141; NASA/JPL, p. 354; NASA Goddard Space Flight Center. Licensed under Creative Commons Attribution 2.0 Generic (CC BY 2.0), p. 97 (inset).
National Museum of Australia: Jason McCarthy, p. 193bc; Katie Shanahan, p. 193br. **Newspix:** p. 345b. **Pearson Education Ltd:** Martyn F. Chillmaid, p. 336; Trevor Clifford, pp. 73r, 142t; Jules Selmes, p. 51tr. **Sandia National Laboratories:** © Sandia National Laboratories, p. 51cr.
Science Photo Library: pp. 19t, 57bl, 72l, 96br, 97cl, 99, 119, 169cr, 208br, 310–11, 319tr, 321tr, 324tl, 342, 373t, 390cl, 396tl, 418l, 510tl;Ramon Andrade 3Dciencia, p. 191br; George Bernard, p. 344l; J. Bernholc et al/North Carolina State University, p. 190bl; British Antarctic Survey, pp. 272r, 280; Andrew Brookes, National Physical Laboratory, p. 88l; Chemical Heritage Foundation, p. 49t; Martyn F. Chillmaid, pp. 61, 224, , 264tl, 264bc, 340bl, 265, 353br, 356, 373c, 451tr; Carlos Clarivan, pp. 86tl, 87; Trevor Clifford Photography, p. 491; Custom Medical Stock Photo, p. 139bl; Colin Cuthbert, p. 261; Dorling Kindersley, p. 50bl; Equinox Graphics, p. 57tl; Jim Edds, p. 449; Clive Freeman/Biosym Technologies, p. 468t; Dr David Furness, Keele University, p. 493; Adam Hart-Davis, p. 422l; Mikkel Juul Jensen, pp. 55l, 55c, 55r; Russell Kightley, p. 194t; Dennis Kunkel Microscopy, p. 492tl; Laguna Design, p. 468b; Andrew Lambert Photography, pp. 70b, 230, 250b, 264bl, 319b, 349, 353bc, 357, 364br, 364tr, 383, 392l, 416l, 418r, 479r;David Leah, p. 264br; Library of Congress, p. 281cr; Dr P. Marazzi, p. 340tr; Astrid & Hans Frieder Michler, p. 131; B. Murton/Southampton Oceanography Centre, p. 329l; Natural History Museum, London, p. 340br; Alexis Rosenfeld, p. 293b; David Parker, p. 78b; P. Plailly/E.Daynes, p. 425c; John Reader, p. 84b; David Scharf, p. i.Science Picture Co, p. 194b; SPUTNIK, p. 313tc; Sinclair Stammers, pp. 281cc, 281cl; TEK Image, p. 188t; Sheila Terry, pp. 281tr, 331br, 480; Visuals Unlimited, p. 350–51; Dr. Mark J. Winter, p. 136–37.
Science Source - Photo Researchers, Inc: Charles D. Winters, pp. 329br, 358.
Shutterstock: p. 16t; 2728747, p. 231; 56278, p. 314cl; 84177, p. 70 (gold rings); abimages, p. 320b; Africa Studio, pp. 52c, 270–71, 326; Adellyne, p. 46–47; Anette Andersen, p. 433–38; Ardely, p. 492bl; Artbox, p. 49b; Marijus Auruskevicius, p. 149cr; John Austin, p. 145; Eugene Berman, p. 139ct; Eva Blanda, p. 341b; Casimer, p. 240tl; Marcel Clemens, p. 122; CnOPhoto, p. 70 (engine); Comaniciu Dan, p. 216; Concept W, p. 115; Creativa Images, p. 507r; d13, p. 220–21; Daxiao Productions, p. 60l; Christian Delbert, p. 49 (inset); dio5050, p. 183b; Fly_dragonfly, p. 52r; Flying Mouse, p. 344c; focal point, p. 250t; fotohunter, p. 199–201; Laura Gangi Pond, p. 72r; Iculig, p. 6; Idea tank, p. 208bl; irin-k, p. 190br; jamakosy, p. 139cb; Szasz-Fabian Jozsef, p. 51br; JustAnotherPhotographer,

p. 504; kenary820, pp. 268l, 268r; Krunja, p. 422tr; Volodymyr Kyrylyuk, p. 70 (nails); JP Lavoie, p. 150bl; Chris Lenfert, p. 321cr; Rainer Lesniewski, p. 16bc, 16br; Viacheslav Lopatin, p. 139tl; Joana Lopes, p. 29; Nik Merkulov, p. 304–07; mroz, p. 96t; my-summit, p. 277; Nuttapong, p. 223br; operafotografca, p. 71l; Alon Othnay, p. 147; papa1266, p. 252tl; siamionau pavel, p. 223bl; Gunnar Pippel, p. 323; Plati Photography, p. 40; Tatiana Popova, p. 208t; Ian Redding, p. 249; RGtimeline, p. 59; Rojarin, p. 64; Sten Roosvald, p. 285; Albert Russ, p. 355; SJ Travel Photo and Video, p. 509; sknol, p. 57tr; Stable, p. 223cr; Standard Studio, p. 16bl; testing, p. 427; Tooykrub, p. 252bl; The Art Archive/REX, p. 81t; Pamela Uyttendaele, p. 313tr; Jiri Vaclavek, p. 353tr; Peter J. Wilson, p. 522–26; Wire_man, p. 477; xfox01, p. 70 (copper wire); Shougui Yang, p. 409; ymgerman, p. 321bl.

Sydney Water: p. 60r.

Westlab: Courtesy of Westlab, www.westlab.com.au, p. 266bl.

Tocomwall Pty Ltd: Tocomwall Pty Ltd 2016, p. 193bl.

Groups

ATOMIC NUMBER → 12 | RELATIVE ATOMIC MASS ( ) indicates most stable isotope → 24.31
SYMBOL → Mg | BOILING POINT °C → 1090 | MELTING POINT °C → 650 | ELECTRONEGATIVITY → 1.3
ELECTRON STRUCTURE → $[Ne]3s^2$ Magnesium

s BLOCK | d BLOCK | p BLOCK

Periods	1	2	3	4	5	6	7	8	9	10	11	12	13	14	15	16	17	18
1									1 H 1.008 -252.9 -259 2.2 $1s^1$ Hydrogen									2 He 4.003 -268.9 -272 – $1s^2$ Helium
2	3 Li 6.941 1342 180 1.0 $1s^22s^1$ Lithium	4 Be 9.012 2468 1287 1.6 $1s^22s^2$ Beryllium											5 B 10.81 4000 2077 2.0 $1s^22s^22p^1$ Boron	6 C 12.01 4827 3500 2.6 $1s^22s^22p^2$ Carbon	7 N 14.01 -195.8 -210 3.0 $1s^22s^22p^3$ Nitrogen	8 O 16.00 -183 -219 3.4 $1s^22s^22p^4$ Oxygen	9 F 19.00 -188.1 -220 4.0 $1s^22s^22p^5$ Fluorine	10 Ne 20.18 -246 -249 – $1s^22s^22p^6$ Neon
3	11 Na 22.99 882.9 97.8 0.9 $[Ne]3s^1$ Sodium	12 Mg 24.31 1090 650 1.3 $[Ne]3s^2$ Magnesium											13 Al 26.98 2519 660.3 1.6 $[Ne]3s^23p^1$ Aluminium	14 Si 28.09 3265 1414 1.9 $[Ne]3s^23p^2$ Silicon	15 P 30.97 280.5 44.2 2.2 $[Ne]3s^23p^3$ Phosphorus	16 S 32.07 444.6 115.2 2.6 $[Ne]3s^23p^4$ Sulfur	17 Cl 35.45 -34 -102 3.2 $[Ne]3s^23p^5$ Chlorine	18 Ar 39.95 -185.8 -189 – $[Ne]3s^23p^6$ Argon
4	19 K 39.10 758.8 63.4 0.8 $[Ar]4s^1$ Potassium	20 Ca 40.08 1484 842 1.0 $[Ar]4s^2$ Calcium	21 Sc 44.96 2836 1541 1.4 $[Ar]3d^14s^2$ Scandium	22 Ti 47.87 3287 1670 1.5 $[Ar]3d^24s^2$ Titanium	23 V 50.94 3407 1910 1.6 $[Ar]3d^34s^2$ Vanadium	24 Cr 52.00 2671 1907 1.7 $[Ar]3d^54s^1$ Chromium	25 Mn 54.94 2061 1246 1.6 $[Ar]3d^54s^2$ Manganese	26 Fe 55.85 2861 1538 1.8 $[Ar]3d^64s^2$ Iron	27 Co 58.93 2927 1495 1.9 $[Ar]3d^74s^2$ Cobalt	28 Ni 58.69 2913 1455 1.9 $[Ar]3d^84s^2$ Nickel	29 Cu 63.55 2560 1085 1.9 $[Ar]3d^{10}4s^1$ Copper	30 Zn 65.38 907 419.5 1.6 $[Ar]3d^{10}4s^2$ Zinc	31 Ga 69.72 2229 27.8 1.8 $[Ar]3d^{10}4s^24p^1$ Gallium	32 Ge 72.64 2833 938.2 2.0 $[Ar]3d^{10}4s^24p^2$ Germanium	33 As 74.92 613 816.8 2.2 $[Ar]3d^{10}4s^24p^3$ Arsenic	34 Se 78.96 684.8 220.8 2.6 $[Ar]3d^{10}4s^24p^4$ Selenium	35 Br 79.90 58.8 -7.1 3.0 $[Ar]3d^{10}4s^24p^5$ Bromine	36 Kr 83.80 -153.4 -157 – $[Ar]3d^{10}4s^24p^6$ Krypton
5	37 Rb 85.47 687.8 39 0.8 $[Kr]5s^1$ Rubidium	38 Sr 87.61 1377 769 1.0 $[Kr]5s^2$ Strontium	39 Y 88.91 3345 1522 1.2 $[Kr]4d^15s^2$ Yttrium	40 Zr 91.22 4406 1854 1.3 $[Kr]4d^25s^2$ Zirconium	41 Nb 92.91 4741 2477 1.6 $[Kr]4d^45s^1$ Niobium	42 Mo 95.96 4639 2622 2.2 $[Kr]4d^55s^1$ Molybdenum	43 Tc (98) 4262 2157 2.1 $[Kr]4d^55s^2$ Technetium	44 Ru 101.1 4147 2333 2.2 $[Kr]4d^75s^1$ Ruthenium	45 Rh 102.9 3695 1963 2.3 $[Kr]4d^85s^1$ Rhodium	46 Pd 106.4 2963 1555 2.2 $[Kr]4d^{10}5s^0$ Palladium	47 Ag 107.9 2162 961.8 1.9 $[Kr]4d^{10}5s^1$ Silver	48 Cd 112.4 766.8 321.1 1.7 $[Kr]4d^{10}5s^2$ Cadmium	49 In 114.8 2072 156.6 1.8 $[Kr]4d^{10}5s^25p^1$ Indium	50 Sn 118.7 2602 231.9 2.0 $[Kr]4d^{10}5s^25p^2$ Tin	51 Sb 121.8 1587 630.6 2.0 $[Kr]4d^{10}5s^25p^3$ Antimony	52 Te 127.6 987.8 449.5 2.1 $[Kr]4d^{10}5s^25p^4$ Tellurium	53 I 126.9 184.4 113.7 2.7 $[Kr]4d^{10}5s^25p^5$ Iodine	54 Xe 131.3 -108.1 -112 2.6 $[Kr]4d^{10}5s^25p^6$ Xenon
6	55 Cs 132.9 670.8 28.5 0.8 $[Xe]6s^1$ Caesium	56 Ba 137.3 1845 727 0.9 $[Xe]6s^2$ Barium	57 La 138.9 2464 920 1.1 $[Xe]5d^16s^2$ Lanthanum	72 Hf 178.5 4600 2233 1.3 $[Xe]4f^{14}5d^26s^2$ Hafnium	73 Ta 180.9 5455 3017 1.5 $[Xe]4f^{14}5d^36s^2$ Tantalum	74 W 183.9 5555 3414 1.7 $[Xe]4f^{14}5d^46s^2$ Tungsten	75 Re 186.2 5596 3454 1.9 $[Xe]4f^{14}5d^56s^2$ Rhenium	76 Os 190.2 5008 3033 2.2 $[Xe]4f^{14}5d^66s^2$ Osmium	77 Ir 192.2 4428 2446 2.2 $[Xe]4f^{14}5d^76s^2$ Iridium	78 Pt 195.1 3825 1768 2.2 $[Xe]4f^{14}5d^96s^1$ Platinum	79 Au 197.0 2836 1064 2.4 $[Xe]4f^{14}5d^{10}6s^1$ Gold	80 Hg 200.6 356.6 -38.8 1.9 $[Xe]4f^{14}5d^{10}6s^2$ Mercury	81 Tl 204.4 1473 303.8 1.8 $[Xe]4f^{14}5d^{10}6s^26p^1$ Thallium	82 Pb 207.2 1749 327.5 1.8 $[Xe]4f^{14}5d^{10}6s^26p^2$ Lead	83 Bi 209.0 1564 271.4 1.9 $[Xe]4f^{14}5d^{10}6s^26p^3$ Bismuth	84 Po (210) 962 253.8 2.0 $[Xe]4f^{14}5d^{10}6s^26p^4$ Polonium	85 At (210) 366.8 301.8 2.2 $[Xe]4f^{14}5d^{10}6s^26p^5$ Astatine	86 Rn (222) -61.8 -71.2 – $[Xe]4f^{14}5d^{10}6s^26p^6$ Radon
7	87 Fr (223) 676.8 27 0.7 $[Rn]7s^1$ Francium	88 Ra (226) 1140 699.8 0.9 $[Rn]7s^2$ Radium	89 Ac (227) 3200 1050 1.1 $[Rn]6d^17s^2$ Actinium	104 Rf (261) – – – $[Rn]5f^{14}6d^27s^2$ Rutherfordium	105 Db (262) – – – $[Rn]5f^{14}6d^37s^2$ Dubnium	106 Sg (266) – – – $[Rn]5f^{14}6d^47s^2$ Seaborgium	107 Bh (264) – – – $[Rn]5f^{14}6d^57s^2$ Bohrium	108 Hs (267) – – – $[Rn]5f^{14}6d^67s^2$ Hassium	109 Mt (268) – – – $[Rn]5f^{14}6d^77s^2$ Meitnerium	110 Ds (271) – – – $[Rn]5f^{14}6d^87s^2$ Darmstadtium	111 Rg (272) – – – $[Rn]5f^{14}6d^97s^2$ Roentgenium	112 Cn (285) – – – $[Rn]5f^{14}6d^{10}7s^2$ Copernicium	113 Nh (284) – – – $[Rn]5f^{14}6d^{10}7s^27p^1$ Nihonium	114 Fl (289) – – – $[Rn]5f^{14}6d^{10}7s^27p^2$ Flerovium	115 Mc (289) – – – $[Rn]5f^{14}6d^{10}7s^27p^3$ Moscovium	116 Lv (292) – – – $[Rn]5f^{14}6d^{10}7s^27p^4$ Livermorium	117 Ts (294) – – – $[Rn]5f^{14}6d^{10}7s^27p^5$ Tennessine	118 Og (294) – – – $[Rn]5f^{14}6d^{10}7s^27p^6$ Oganesson

f BLOCK

58 Ce 140.1 3443 799 1.1 $[Xe]4f^25d^06s^2$ Cerium	59 Pr 140.9 3520 931 1.1 $[Xe]4f^35d^06s^2$ Praseodymium	60 Nd 144.2 3074 1016 1.1 $[Xe]4f^45d^06s^2$ Neodymium	61 Pm (145) 3000 1042 – $[Xe]4f^55d^06s^2$ Promethium	62 Sm 150.4 1794 1072 1.1 $[Xe]4f^65d^06s^2$ Samarium	63 Eu 152.0 1794 1072 – $[Xe]4f^75d^06s^2$ Europium	64 Gd 157.3 3273 1313 1.2 $[Xe]4f^75d^16s^2$ Gadolinium	65 Tb 158.9 3230 1359 – $[Xe]4f^95d^06s^2$ Terbium	66 Dy 162.5 2567 1412 1.2 $[Xe]4f^{10}5d^06s^2$ Dysprosium	67 Ho 164.9 2700 1472 1.2 $[Xe]4f^{11}5d^06s^2$ Holmium	68 Er 167.3 2868 1529 1.2 $[Xe]4f^{12}5d^06s^2$ Erbium	69 Tm 168.9 1950 1545 1.3 $[Xe]4f^{13}5d^06s^2$ Thulium	70 Yb 173.1 1196 824 – $[Xe]4f^{14}5d^06s^2$ Ytterbium	71 Lu 175.0 3402 1663 1.0 $[Xe]4f^{14}5d^16s^2$ Lutetium
90 Th 232.0 4875 1750 1.3 $[Rn]5f^06d^27s^2$ Thorium	91 Pa 231.0 – 1572 1.5 $[Rn]5f^26d^17s^2$ Protactinium	92 U 238.0 4131 1135 1.7 $[Rn]5f^36d^17s^2$ Uranium	93 Np (237) 3902 644 1.3 $[Rn]5f^46d^17s^2$ Neptunium	94 Pu (244) 3228 640 1.3 $[Rn]5f^66d^07s^2$ Plutonium	95 Am (243) 2011 1176 – $[Rn]5f^76d^07s^2$ Americium	96 Cm (247) 3110 1340 – $[Rn]5f^76d^17s^2$ Curium	97 Bk (247) – 986 – $[Rn]5f^96d^07s^2$ Berkelium	98 Cf (251) – 900 – $[Rn]5f^{10}6d^07s^2$ Californium	99 Es (252) – – – $[Rn]5f^{11}6d^07s^2$ Einsteinium	100 Fm (257) – – – $[Rn]5f^{12}6d^07s^2$ Fermium	101 Md (258) – 827 – $[Rn]5f^{13}6d^07s^2$ Mendelevium	102 No (259) – – – $[Rn]5f^{14}6d^07s^2$ Nobelium	103 Lr (262) – – – $[Rn]5f^{14}6d^17s^2$ Lawrencium